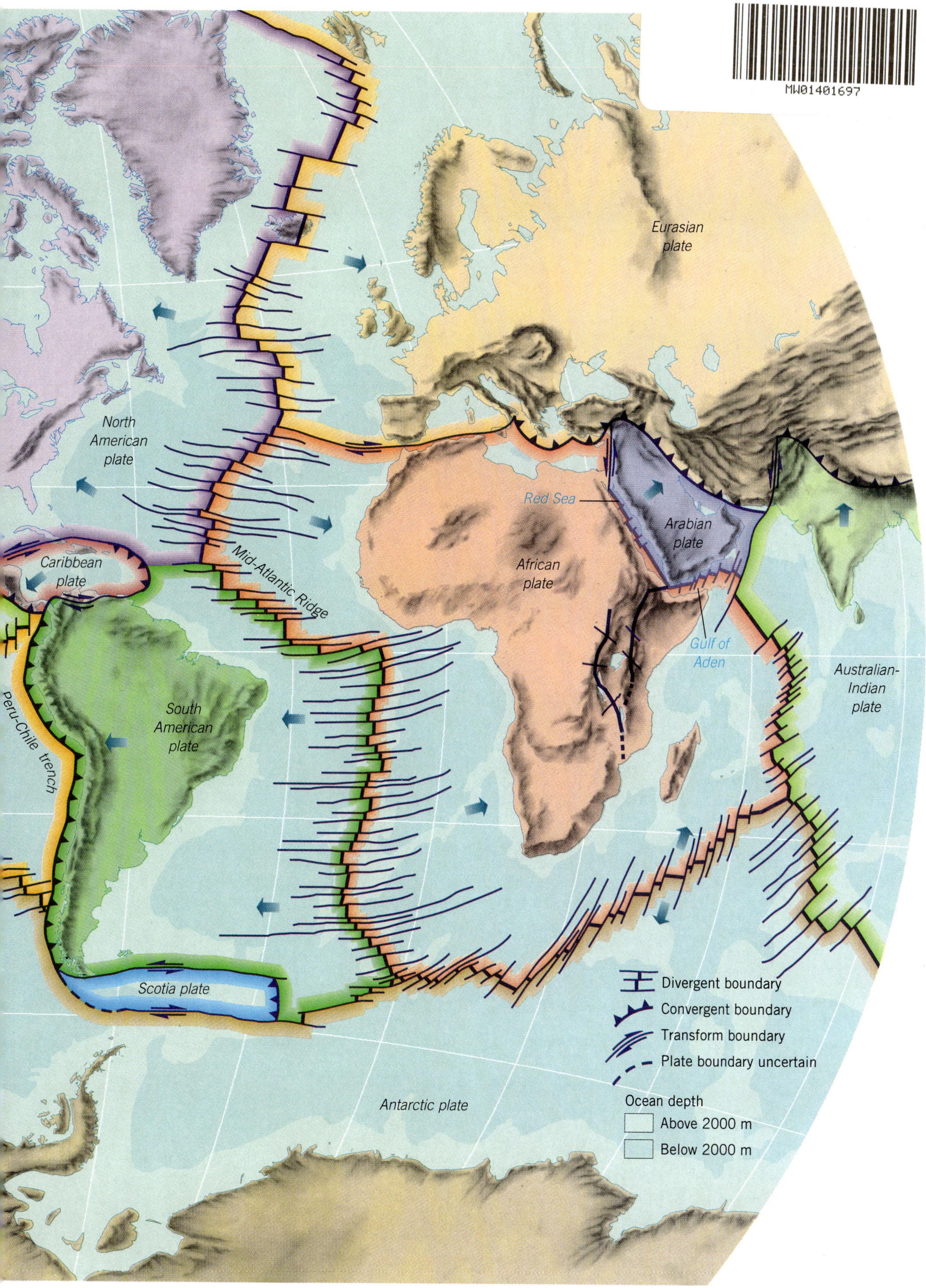
Eurasian plate
North American plate
Red Sea
Arabian plate
Caribbean plate
Mid-Atlantic Ridge
African plate
Gulf of Aden
Australian-Indian plate
South American plate
Peru-Chile trench
Scotia plate
Antarctic plate
Divergent boundary
Convergent boundary
Transform boundary
Plate boundary uncertain
Ocean depth
Above 2000 m
Below 2000 m

ESSENTIALS OF
PHYSICAL

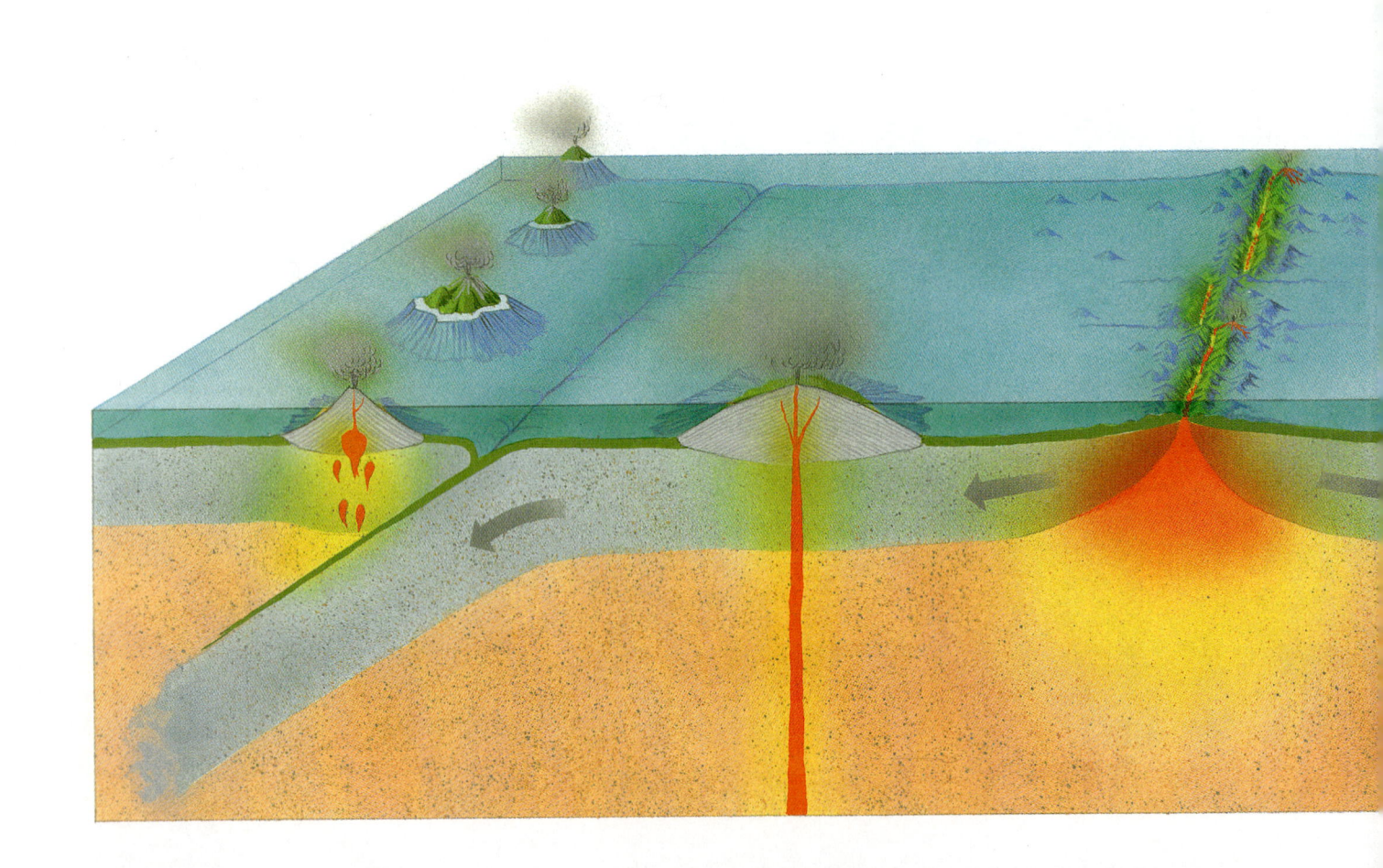

GEOLOGY

ANATOLE DOLGOFF

City University of New York–New York City Technical College

with

MARY FALCON

HOUGHTON MIFFLIN COMPANY
Boston   New York

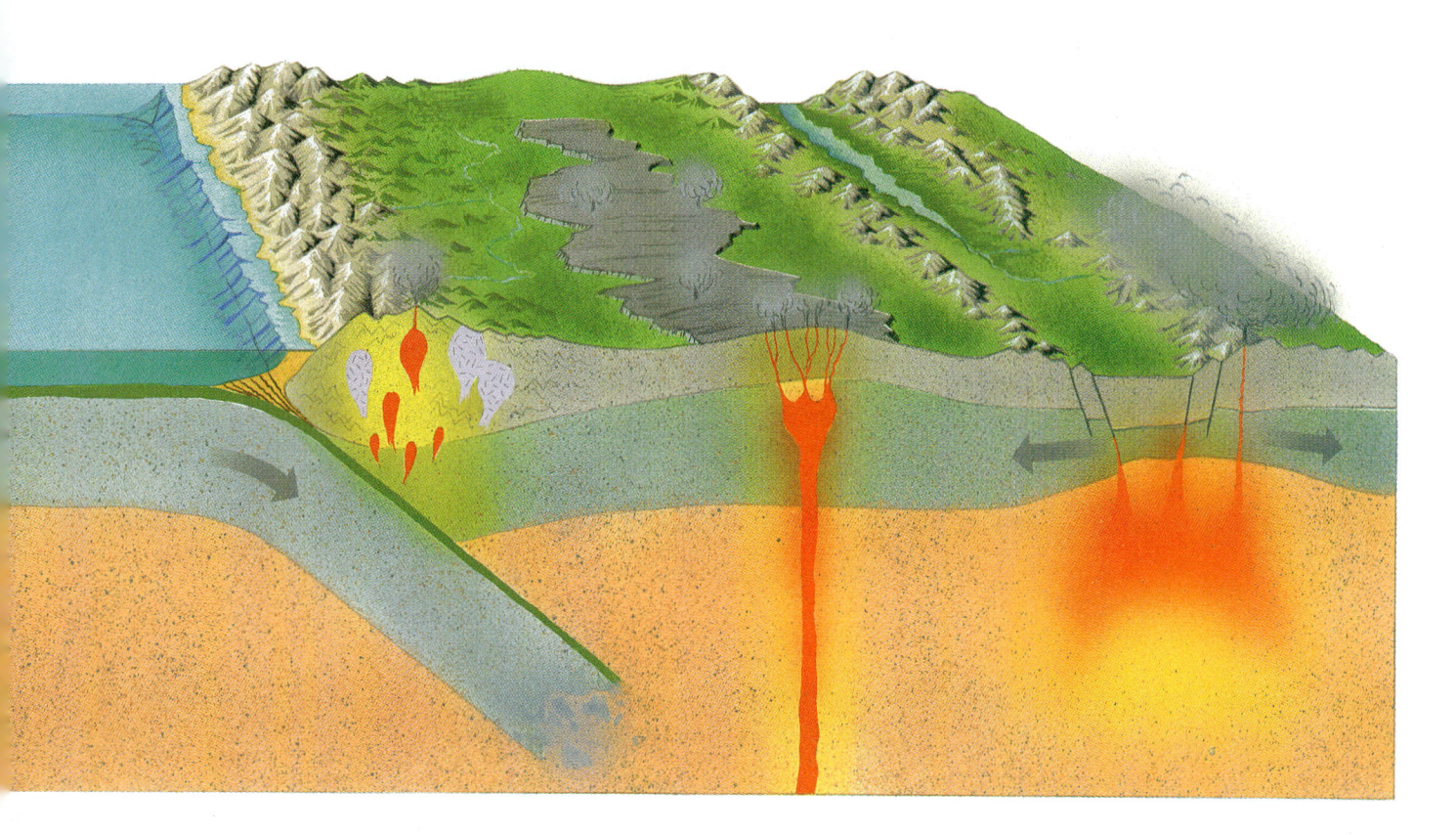

Editor-in-Chief: Kathi Prancan
Project Editor: Elizabeth Gale Napolitano
Senior Production/Design Coordinator: Jill Haber
Manufacturing Coordinator: Sally Culler
Marketing Manager: Karen Natale

Cover design: Stoltze Design
Cover image: Whitewater canoeing in Austria. Jean-Pierre Vollrath/Adventure Photo & Film

Acknowledgments for reprinted material appear beginning on page 447.

Copyright © 1998 by Houghton Mifflin Company. All rights reserved.

No part of this work may be reproduced or transmitted in any form or by any means, electronic or mechanical, including photocopying and recording, or by any information storage or retrieval system without the prior written permission of Houghton Mifflin Company unless such copying is expressly permitted by federal copyright law. Address inquiries to College Permissions, Houghton Mifflin Company, 222 Berkeley Street, Boston, MA 02116-3764.

Printed in the U.S.A.

Library of Congress Catalog Number: 97-72460

ISBN: 0-669-33916-4

123456789-DW-01-00-99-98-97

This book is dedicated to the memory of my parents, Esther and Samuel Dolgoff.

Preface

This textbook is designed for students enrolled in a one-semester geology course. This course typically attracts a variety of students fulfilling a science requirement. My purpose in writing this book has been to present geological principles in a manner that conveys the delight that geologists take in their subject. I have been guided by several major goals:

- *To provide basic information about geology in a way that is inviting for students who have probably taken little interest in the subject before this course.* Many students have grown up in an environment in which the indoor shopping mall has almost completely supplanted the natural world. The sense of connection they should feel with the Earth has largely been stifled, and a good geology textbook must work toward motivating students to reestablish this link. To this end, I have woven the dramatic interactions between geological processes and human activities into the very fabric of the book. I have highlighted such timely issues as successful and failed efforts to design earthquake-resistant buildings; the advantages and disadvantages of flood-control measures on the Mississippi River; the impact of human activity on fragile beaches and hillsides; and the threat of human-induced global warming. I have explored such serious environmental problems as famine and land degradation in the Sahel; the undesirable effects of water diversion in arid climates; and groundwater pollution and overuse. Throughout the book I have discussed geologists' practical contributions toward solving these problems.

  I have not hesitated to stress that many problems geologists, engineers, and regional planners thought they had overcome or were on the verge of overcoming have turned out to be thornier than they originally believed—for example, predicting earthquakes in light of such recent events as the Northridge and Kobe disasters or constructing artificial levee systems strong enough to withstand the likes of the Mississippi floods. Probing these matters emphasizes the challenge of geology and reinforces the concept that science is largely about unfinished business. Indeed, there is much that we leave for the next generation of geologists to accomplish.

- *To maintain the traditional core of the subject while giving proper weight to the advances of recent years.* The list of recent geological advances is virtually endless, and for the author of an introductory textbook, the main problem that arises is determining what to include and what to leave out. The material I have chosen to include strikes a reasonable balance between the theoretical and the practical, between classical ideas and topics now of wide interest. For example, the chapter on earthquakes treats plate tectonics and elastic-rebound theory on the one hand, and engineers' new findings as a result of the Mexico City, Northridge, and Kobe disasters on the other.

- *To involve students actively in the scientific process.* For most of the students taking the one-semester geology course it may be the only science studied in college. For this reason, an introductory textbook must demonstrate how scientists look at problems and attempt to solve them.

  I have set the tone of science as inquiry in Chapter 1, where I discuss Hutton's uniformitarianism not simply as an axiom but as a case study of a basic principle derived through direct observation of the physical world. The emphasis in this book is not so much on the answers to scientific questions but on how geologists go about *seeking* the answers.

  I have maintained the same inquiry-based approach throughout the book. No topic, no matter how familiar, is presented as a finished piece of business. Rather, I have described explanations and theories as working hypotheses, some more firmly established than others.

- *To demonstrate geology's core role in advancing human knowledge.* Finally, I have tried to convey the pivotal knowledge that geology has contributed to human intellectual development. This knowledge includes the concepts of the Earth's vast age and of slow evolutionary change; geologists' explanations of the natural causes of catastrophic events such as earthquakes and volcanoes; and the application of rational methods in evaluating competing ideas.

Organization

I believe that the introductory physical geology course is most solid when it is built upon a foundation of plate tectonics. In this book I have laid the cornerstone of plate tectonics early. Chapter 1 takes readers from the basics of classical geology—uniformitarianism, the age of the Earth, and the rock cycle—to continental drift, the precursor of plate tectonics, to the process of scientific inquiry. In Chapter 2, I have employed an inductive, quasi-historical approach in developing the case for plate tectonics, describing the features of the ocean floor and explaining the significance of the supporting evidence as I go along. The chapter concludes with a preview of the impact of plate tectonics on continental evolution. Thus Chapters 1 and 2 give students the solid foundation and context needed to understand subsequent concepts.

Chapter 3 focuses on earthquakes and the Earth's interior. I then discuss minerals, rocks, volcanism, and rock deformation (Chapters 4–9). Students are now in a position to appreciate the surficial processes (Chapters 10–15) that act upon the continental crust.

The book closes with four chapters on geologic time and Earth history. Chapter 16, Geologic Time, describes the principles of relative and radiometric dating, and how they are used to construct the geologic time scale. Chapter 17, Formation of the Continental Crust, describes the processes involved in mountain building and the assembly of continents as an amalgamation of plates. Recounted, in outline form, are several key events, such as the Appalachian orogeny and the assembly of western North America. These chapters prepare the student for a two-chapter summary of the chronological history of the Earth.

The challenge in writing a brief, two-chapter version of Earth history is to present coherent themes without losing the student in a thicket of terminology and subplots. Furthermore, while it is easier for academics to study the geologic, climatic, and biological history of the Earth separately, I believe it is more valuable to expose students—especially nonscience majors—to an integrative treatment of these factors over time. Chapters 18 and 19 were developed with this goal in mind.

The framework of Chapter 18, Earth History: From the Origin of the Planet Through the Proterozoic Eon, is based on four simple ideas:

1. The Archean and Proterozoic crust, as well as modern plate tectonics, are continuations of the same evolutionary processes that formed the planets.
2. The Earth's modern atmosphere and oceans have developed as consequences of the complex interplay of physical, chemical, and biological processes.
3. The basic cellular structure of organisms evolved during the Archean and Proterozoic eons.
4. The proliferation of life on Earth is a result of biological evolution by genetic alteration and natural selection.

You may notice that I devote more attention to organic evolution and natural selection than is usual for a chapter of this type. My quick tour through the vast store of evidence that proves evolution correct was purposefully designed to counteract the tenacity of creationist beliefs among today's college students. I believe that the concept of evolution is one of the most important components of a liberal arts education, yet most nonscience majors take only one science course. If they don't get it here, then where will they get it?

Chapter 19, Earth History: From the Paleozoic to the Present, is organized around three themes:

1. The breakup of the late Proterozoic supercontinent and its reassembly into Pangaea and the breakup of Pangaea and its reassembly into the modern world.
2. The impact of the organization and reorganization of the Earth's land masses on the Earth's climate.
3. The influence of these climatic and physical changes on the course of biological evolution.

Students like the sequence of topics in this book—which I have used with minor variations in my own classes for many years—because it presents the evidence in context. Moreover, earthquakes, continental drift, and plate tectonics intrigue them, so introducing these topics early in the book affords the opportunity to exploit a keen interest. As I have learned in the classroom, the earlier you capture students' interest, the more likely they will engage actively with the rest of the course.

I am sensitive, however, to the fact that instructors have personal preferences and practical needs that determine the sequence of topics in their courses. I have therefore made the chapters as self-contained as possible, so as to ensure maximum flexibility of use. For example, for those who prefer a traditional approach, Chapter 4, on minerals, can be assigned either immediately before or after Chapter 2, on plate tectonics; and Chapter 3, on earthquakes and the Earth's interior, can be taught after the mineral and rock chapters (Chapters 4–8) without any loss of coherence.

Humans and the Natural Environment

For purposes of this book, an environmental "issue" or "problem" is defined as *any human intervention in the natural environment that may cause actual or potential harm to either humans or the natural environment*. The criterion is broad enough to include the material most of us want to teach, but places limits on what is characterized as "environmental." Thus hurricanes and earthquakes themselves are not environmental issues, but groins and retaining walls that strip beaches of their ability to withstand hurricanes—and dangerous construction and zoning practices in earthquake-prone regions—are.

Topics relating to the environment are bracketed with icons. For those instructors who wish to emphasize the practical application of geology to environmental issues, the icons point to where such topics begin and where they end.

Pedagogy

Some introductory physical geology textbooks divide information into short, easily digestible bits, presumably to make the material accessible to students. Without the context that supports the facts, however, students have no choice but to memorize the material; moreover, educational studies have shown that they retain very little of such information after the final exam. In contrast, the chapters in this book are narratives built around themes and concepts; the details and definitions evolve naturally from the flow of the discussion. For example, in Chapter 7, on sedimentary rocks, the characteristics of coarse detrital sediments emerge logically from a description of their weathering, transport, and depositional history as we travel hypothetically from mountaintop to sea basin. Of course, geology is a very visual science, and I have illustrated and reinforced the major themes of this textbook with a rich and varied program of photographs, diagrams, graphs, and maps.

Three special features in each chapter allow students the freedom to read the material for understanding, without the need for underlining (or highlighting) passages and without the annoyance of breaking their train of thought to look up a word elsewhere.

- **Study Outline.** This detailed outline at the end of each chapter covers all the important concepts and uses every key term. The outline provides a more functional cognitive map than the ordinary paragraph-style summary because it visually conveys the relationships between topics and subtopics. Furthermore, in the paragraph format, many extra words are needed to complete full sentences and provide transitions from one sentence to the next. Because the outline format is economical, more information can be packed into the same number of words.

- **Marginal Definitions.** Each study term is defined in the margin next to the place in the text where it is first discussed.

- **Bold-faced and Italicized Terms.** An important contribution of geologists to the natural sciences has been to name the features and processes we have observed. As a consequence, over the years geologists have coined enough terms to fill an entire dictionary. (I am referring, of course, to the AGI's *Dictionary of Geological Terms*.) As I see it, my job as an introductory textbook author is to use those terms selectively in a way most beneficial to students. Thus I have divided geological terms into three groups:

    1. Study terms. These are words that I believe students should commit to memory and learn to employ as part of their speaking vocabulary. I have tried to make the study terms as accessible as possible by boldfacing them when they are first discussed in the body of the text and by placing their working definitions in the margin directly next to their first discussion. The list of study terms at the end of each chapter tells the student on which page each term is first discussed and defined in the margin, and the study terms and their definitions appear again in the alphabetized glossary at the end of the book.

    2. Terms to remember. These words are important enough that nonmajors ought to get the gist of them by seeing them employed in context. Hopefully, they will recall the general meaning of these words when they come across them again in other contexts, such as magazine articles or TV documentaries. Terms that I feel students ought to remember appear italicized the first time they are discussed, and a number of them are defined in the end-of-book glossary.

    3. Terms that I have found in my own teaching experience to be of little use to undergraduates. Accordingly, in this text written primarily for nonscience majors, I have made every effort to avoid jargon and have painstakingly replaced such terms with common everyday language.

I understand that my choices regarding appropriate vocabulary cannot possibly mirror the preferences of all instructors, but I think textbook authors who include every term so as not to "offend" anyone make the greater mistake in terms of students' needs.

Also, to help students master material and study for exams, the following pedagogical devices are employed in each chapter:

- **Chapter Outline.** Each chapter begins with a preview outline of the headings within the chapter.
- **Study Terms.** This end-of-chapter list of study terms features references to the page where each term is first used and defined.
- **Thought Questions.** A brief set of questions at the end of each chapter tests students' recall *and understanding* of the chapter material.

And in the back of the book:

- **Appendixes.** The appendixes feature a conversion table of metric and English units of measurement, the periodic table of the elements, mineral identification tables, and a basic guide to geologic maps.
- **Glossary.** The glossary provides complete definitions of all boldfaced study terms and useful working definitions of many italicized terms. Each entry is followed by a reference to the text page where the term was first discussed.
- **Selected Bibliography.** This brief supplementary reading list, organized by chapter, includes both classic books and current articles.

A Mixed Majors and Nonmajors Alternative

There may be instructors who are interested in the organization and approach of this book but prefer to use a text designed for science majors as well as nonmajors. To better meet the needs of these instructors, I have prepared an Updated Version of my full-length text *Physical Geology*.

For those instructors who prefer the full-length Updated Version of this book but also want to include more historical geology in their course, the two Earth history chapters may be packaged with the Updated Version. Contact your local Houghton Mifflin sales representative for further information.

Supplements

This textbook is supplemented by a number of useful learning and teaching aids:

- **Student Study Guide.** For students who benefit from pencil-and-paper exercises, drills, and practice tests, the study guide offers an additional avenue of review. Dozens of interactive drills, along with practice multiple-choice exams with answers and self-evaluation charts, allow students to identify areas of weakness and pinpoint the text pages that they should review again.
- **Laboratory Manual for Students.** This manual offers lab studies and activities on topics closely tied to *Physical Geology* as well as *Essentials of Physical Geology*. Included are labs on plate tectonics, mineral identification, rock deformation, streams, groundwater, and glacial landscapes. Other units help to develop and hone students' skills in working with and interpreting geologic and topographic maps. *Physical Geology Interactive* is a CD-ROM version of the lab manual that extends all activities, using web resources for information and data to be used in the exercises.

- **Test Item File and Computerized Testing.** The printed test bank, offering over 1000 multiple-choice questions and approximately 160 additional essay and illustration-based questions, is conveniently organized by chapter. Also featured are 40 questions, covering the material in the first 9 chapters, that are suitable for use in a midterm exam, plus 90 questions, covering the material in the entire 19 chapters, that are designed for use in a final exam. The computerized testing program offers the same questions that appear in the printed test bank but in handy electronic format for IBM-compatible and Macintosh computers.
- **Instructor's Manual.** An invaluable aid for the instructor, this comprehensive manual, written by Michael Scanlin of Elizabethtown College, PA, includes brief chapter summaries, detailed chapter outlines, lecture suggestions, ideas for student activities, teaching tips, and up-to-date information on media resources.
- **Overhead Transparencies.** This set of some 100 color acetates of useful illustrations is grouped by topic for ease of presentation. Most illustrations are taken from the textbook, but some supplementary visuals are included to expand the range of options for the instructor.
- **The Earth Sciences Videodisc Set.** This state-of-the-art videodisc set features full-motion video with narration and numerous animations, as well as more than 3000 still images. All the core course topics are covered, including plate tectonics, minerals, volcanoes, faulting and folding, and the rock cycle, among many others.
- **The Earth Sciences Geology Slide Set.** This generous collection of full-color images features original photographs of rocks, minerals, landforms, and important geologic sites and phenomena from around the world.

Acknowledgments

Writing both versions of this book was a major undertaking, and its completion would have been impossible without the unselfish contributions of numerous individuals. Their interest, support, and advice have sustained me throughout the process. It is only proper that I acknowledge them here, with deep thanks.

I am very grateful to Luigia and Herbert Miller, my oldest and dearest friends, who encouraged me every step of the way and kept me focused on my writing. Isolda, my wife, put up with my long, work-related absences and occasional ill humor with a good grace I did not deserve. My brother, Abraham Dolgoff, an engineering geologist, was a patient and expert consultant. My special thanks go to Kathi Prancan, Editor-in-Chief, Science, and sponsoring editor of this project at Houghton Mifflin, and to Karen Natale, Associate Director of Marketing and marketing manager for this project. Their hard work and enthusiastic support will not be forgotten. Many of the more complex line illustrations in this book were the work of a wonderful artist, Elizabeth Morales, an anthropologist by training, but with a good deal of formal background in geology. The editorial staff calls her work reflective art, but I call it suitable for framing.

There follows a long list of geologists who reviewed the manuscripts for one or both versions of this book and offered detailed suggestions and criticisms, often line by line; most were points well taken. Several of these reviewers deserve special mention. Craig Manning at the University of California, Los Angeles, urged me to trust my judgment concerning the decision to place plate tectonics at the beginning of the book and supplied expert advice on a number of chapters. Wang-Ping Chen at the University of Illinois, Urbana-Champaign, reviewed every manuscript chapter and most of the illustrations for accuracy. No concept or detail was too insignificant to escape his scrutiny, and the book is infinitely better for his careful attention. John Nicholas at the University of

Bridgeport; Pamela Martin at the University of California, Santa Barbara; and John Geissman at the University of New Mexico also devoted many tedious hours to checking the illustrations for accuracy. I thank them and all of the other excellent reviewers for their time and wise counsel:

Reviewers of *Essentials of Physical Geology*

Jeffrey Bauer, *Shawnee State University*
Craig Chesner, *Eastern Illinois University*
Sharon Diekmeyer, *Xavier University*
David Dockstader, *Jefferson Community College*
Peter Harries, *University of South Florida*
Simon Peacock, *Arizona State University*
Donal Ragan, *Arizona State University*
L. Don Ringe, *Central Washington University*
Vernon Scott, *Oklahoma State University*
Laura Serpa, *University of New Orleans*
Anne Sturz, *University of San Diego*
C.J. Yorath, *University of Victoria*

Reviewers of First Edition and Update of *Physical Geology*

Charles Alpers, *United States Geological Survey*
Robert Behling, *West Virginia University*
Mary Lou Bevier, *University of British Columbia*
Marcia Bjornerud, *Miami University*
Thomas Broadhead, *University of Tennessee, Knoxville*
Donald Burt, *Arizona State University*
Karl Chauffe, *St. Louis University*
Chu-Yung Chen, *University of Illinois, Urbana-Champaign*
Kevin Cole, *Grand Valley State University*
Lorence Collins, *California State University, Northridge*
Brian Cooper, *Sam Houston State University*
Spencer Cotkin, *University of Arkansas*
Robert Corbett, *Illinois State University*
Larry Davis, *Washington State University*
Joseph DiPietro, *University of Southern Indiana*
David Dockstader, *Jefferson Community College*
William Dupré, *University of Houston*
Goran A. Ekström, *Harvard University*
Terry Engelder, *Pennsylvania State University*
G. Lang Farmer, *University of Colorado, Boulder*
Michael Gibson, *University of Tennessee, Martin*
Bryce Hand, *Syracuse University*
John Howe, *Bowling Green State University*
Larry Knox, *Tennessee Technological University*
Lawrence G. Kodosky, *Oakland Community College*
David Lea, *University of California, Santa Barbara*
Jonathan Lincoln, *Montclair State University*
Steve Loftouse, *Pace University*
Constantine Manos, *State University of New York, New Paltz*
Sandra McBride, *Queen's University*
Elizabeth McClellan, *Western Kentucky University*
James McClurg, *University of Wyoming*
Eileen McLellan, *University of Maryland, College Park*

James I. Mead, *Northern Arizona University*
David Mogk, *Montana State University*
Anne Pasch, *University of Alaska*
Lincoln Pratson, *University of Colorado*
Fredrick Rich, *Georgia Southern University*
Mary Jo Richardson, *Texas A & M University*
Jeanette Sablock, *Salem State College*
Robert Schoch, *Boston University*
Karl Seifert, *Iowa State University*
Lynn Shelby, *Murray State University*
William Smith, *Western Michigan University*
Donald Spano, *University of Southern Colorado*
Neptune Srimal, *Florida Atlantic University*
George Stephens, *The George Washington University*
Monte Wilson, *Boise State University*

Finally, I wish to express my deepest gratitude to Mary and Ricardo Falcon. For untold hours over the past five years, Ricardo has devoted his computer skills, keen intelligence, and dedication to the project. Mary tirelessly applied her extensive publishing experience, professional know-how, and editorial talent to the development of both the full-length and essentials versions of this book. Every page bears the imprint of her sound judgment. No one could ask for a better editor and adviser.

A. D.

Brief Contents

Preface vii
Chapter 1 Basic Geological Concepts 2
Chapter 2 Plate Tectonics and the Ocean Floor 28
Chapter 3 Earthquakes and the Interior of the Earth 58
Chapter 4 Minerals 88
Chapter 5 Igneous Activity 108
Chapter 6 Volcanism 128
Chapter 7 Sedimentary Rocks 148
Chapter 8 Metamorphic Rocks 174
Chapter 9 Rock Deformation 192
Chapter 10 Weathering, Soils, and Mass Wasting 210
Chapter 11 Streams 240
Chapter 12 Groundwater 260
Chapter 13 Glaciers and Climate 282
Chapter 14 Deserts and Winds 304
Chapter 15 Coasts and Shoreline Processes 324
Chapter 16 Geologic Time 340
Chapter 17 Formation of the Continental Crust 360
Chapter 18 Earth History: From the Origin of the Planet Through the Proterozoic Eon 378
Chapter 19 Earth History: From the Paleozoic Era to the Present 402
Appendix A Conversion Table for Metric and English Units 421
Appendix B Periodic Table of the Elements 423
Appendix C Mineral Identification Table 425
Appendix D Basic Guide to Geologic Maps 429

Glossary 433

Selected Readings 443

Acknowledgments 447

Index 449

Contents

Preface vii

1 Basic Geological Concepts 2

The Huttonian Revolution 4

Establishing Geologic Time 6
  Absolute Time 8

The Earth Machine 8
  External Processes 8
  The Rock Cycle 10

The Earth's Interior 12

Continental Drift 14

Physical Evidence 16
  Structural Evidence 17
    Evidence in the Distribution of Ancient Glaciers 18
    Evidence in the Distribution of Ancient Life-Forms 18

Paleomagnetic Evidence 20

Continental Drift and Plate Tectonics 21

Research and Theory in the Geosciences 22

The Scientific Method 22
  Questions and Tentative Answers 22
  The Search for Evidence 23
  Evaluation by the Scientific Community and Further Research 23

Development of a Theory 24

2 Plate Tectonics and the Ocean Floor 28

An Emerging Picture of the Sea Floor 31

The Sea-Floor Spreading Hypothesis 35
Evidence in Magnetic Reversals 36
Evidence in Sea-Floor Sediment 39
Evidence in Sea-Floor Topography 39

Plate Tectonics 41
Divergent Boundary Processes 46
Convergent Boundary Processes 48
    Ocean-Ocean Convergence 48
    Ocean-Continent Convergence 49
    Continent-Continent Convergence 50
Transform Boundaries 52
The Causes of Plate Motion 52

3 Earthquakes and the Interior of the Earth 58

The Causes of Earthquakes 59

Seismic Waves 61
The Seismograph 62
    Locating the Earthquake 64
Measuring Earthquake Magnitudes and Energy 64

Earthquakes and Plate Tectonics 66
The San Andreas Fault 68
The Los Angeles Faults 68
Intraplate Earthquakes 69

Forecasting Earthquakes 70
Seismic Gaps 71
Recurrence Studies 71
Precursor Studies 72

Minimizing Earthquake Damage 74
Construction Problems 75

Seismic Waves in the Earth's Interior 77
Wave Refraction and Reflection 77
Detection of the Crust, Mantle, and Core 78

Structural and Compositional Divisions of the Interior 79

The Earth's Internal Heat 81
The Geothermal Gradient 81
Mantle Convection 82
Sources of the Earth's Internal Heat 82
The Earth's Core and the Magnetic Field 83

4 Minerals 88

Atoms and Elements 89
Atomic Structure 90

Bonding 92
Ionic Bonds 93
Covalent Bonds 93
Metallic Bonds 94

The Physical Properties of Minerals 94
Crystal Form 94
Cleavage 95
Fracture 95
Hardness 96
Color 96
Streak 96
Luster 97
Specific Gravity 97
Other Properties 98

Common Minerals of the Crust 98
The Classification of Common Minerals 100

The Structure of Silicate Minerals 100
Tetrahedral Linkages 101
Properties Related to Internal Structure 101

Other Mineral Groups 103

The Geologic Origin of Minerals 103

5 Igneous Activity 108

Igneous Rock Bodies 109

The Formation of Rocks from Magma 111
Igneous Rock Textures 111
The Classification of Igneous Rocks 113
- The Peridotite Family 114
- The Basalt-Gabbro Family 114
- The Andesite-Diorite Family 114
- The Granite-Rhyolite Family 115

The Igneous Rock Classification Chart 115
The Sequence of Crystallization of Igneous Rocks 116

The Geologic Settings of Igneous Activity 119
Igneous Activity at Mid-Ocean Ridges 120
Igneous Activity in Subduction Zones 122
- The Ascent of Magma in Subduction Zones 122

Mineral Concentration and Igneous Activity 124
Mineral Concentration at Subduction Zones 124

6 Volcanism 128

The Anatomy of a Volcano 130

Tectonic Settings of Volcanism 131

The Mechanics of a Volcanic Eruption 132

The Materials of Volcanic Eruptions 133
Lavas 133
- Basaltic Lava Flows 134
- Silicic Lava Flows 134

Pyroclastic Materials 135
Volcanic Gases 136

Volcanic Structures and Eruptive Styles 137
Basaltic Volcanoes 137
- Shield Volcanoes 137
- Cinder Cones 139
- Fissure Eruptions and Lava Plateaus 139

Andesitic and Silicic Volcanoes 140
- The Eruption of Mount Saint Helens 140
- Ash Flows 142

Forecasting Eruptions 142

Volcanism and Climate 143

Constructive Aspects of Volcanism 144

7 *Sedimentary Rocks* 148

Weathering, Transport, and Deposition of Sediments 150
Weathering 150
Transport and Deposition 151

Lithification 151

Classification of Sedimentary Rocks 151
Detrital Rocks 152
- Coarse Detrital Rocks 154
- Fine Detrital Rocks 156

Precipitated Sedimentary Rocks 157
- The Limestones 157
- Evaporites 159
- Chert 160
- Coal 160

Sedimentary Structures 160
Fossils and Minor Structures 164

Oceanic Sedimentary Environments 164
The Continental Margins 165
The Deep-Ocean Environment 165
Plate Tectonics and Ocean Sediments 166

Energy Resources in Sedimentary Rocks 167
Coal 167
Petroleum 168
- Oil Traps 169
- Oil Shales and Oil Sands 169
- Plate Tectonics and Petroleum 170
- The Petroleum Supply 170

8 Metamorphic Rocks 174

The Agents of Metamorphism 176
Heat 176
Pressure 177
Chemically Active Fluids 177

Types of Metamorphism 178

Metamorphic Rocks 179
The Minerals of Metamorphic Rocks 179
Textures of Metamorphic Rocks 180
Classification of Metamorphic Rocks 180
- Foliated Rocks 180
- Nonfoliated Rocks 183

The Field Occurrence of Metamorphic Rocks 183
- Contact Metamorphic Rocks 184
- Regional Metamorphic Rocks 184

Metamorphism and Plate Tectonics 188
Metamorphism of the Sea Floor 189
Subduction Zone Metamorphism 189

9 Rock Deformation 192

Stress and Rock Deformation 193
The Response of Rocks to Stress 194

The Map Depiction of Planar Features 195

Folds 196
Plunging Folds 196
Monoclines, Domes, and Basins 199

Fractures 200
Joints 200
Faults 200
- Strike-Slip Faults 202
- Normal Faults 202

Thrust Faults 203

Impact Craters 204
The Stucture of an Impact Crater 205
Frequency of Impact 206

10 Weathering, Soils, and Mass Wasting 210

Weathering, Mass Wasting, and Gradation 211

Weathering Processes 213
Mechanical Weathering 214
- Frost Wedging 214
- Salt Crystallization 214
- Sheeting, Exfoliation, and Spheroidal Weathering 214
- Weathering Effects of Plants and Animals 215

Chemical Weathering 216
- Direct Solution 216
- Hydrolysis 218
- Oxidation 219

Rates of Weathering 219
Differential Weathering 219

Soil Formation 220
The Soil Profile 221
Nutrient Cycling 222
Climate and Soil Types 222
- Soils of Tropical, Humid Climates 222
- Soils of Arid and Semiarid Climates 223
- Soils of Cold to Temperate Climates 223

Mass Wasting 224
Falls 224
Slides 225
- Conditions Favoring Slides 228
- Slumps 228
Flows 230
Creep 232

Predicting and Preventing Mass-Wasting Disasters 233

11 Streams 240

The Energy of Streams 242
Stream Discharge 243
Stream Flow and Channel Erosion 243
Transport and Deposition 244

The Graded Stream 245
Waterfalls 245

Two Contrasting Stream Types   246
  The Braided Stream   246
  The Meandering Stream   247

Mineral Deposits in Streams   248

The Flood Plain   249
The Debate Over Flood Control   250

Deltas   252

Stream Rejuvenation   254

Drainage Patterns   254
Reorganization of Drainage (Stream Piracy)   256

12   Groundwater   260

Accumulation of Groundwater   261
Porosity and Permeability   263
Aquifers   264

Movement of Groundwater   265

Wells   265
Artesian Wells   266
Well Extraction Problems   267
  Land Subsidence   267
  Saltwater Encroachment   268
  Loss of Hydraulic Pressure   269

The Geologic Work of Groundwater   270
Caves   270
Karst Topography   272
Geysers, Hot Springs, and Fumaroles   273

The Quality of Groundwater   275
Groundwater Pollution   276

13   Glaciers and Climate   282

The Formation of Glaciers   284
The Movement of Glacial Ice   285
The Glacial Budget   286

Erosional Features of Glaciers 289
Ice-Sculpted Mountain Topography 289

Depositional Features of Glaciers 290
Moraines 292
Landforms of Stratified Drift 293

Glacial Lakes 295

Glacial and Interglacial Ages 296
Causes of the Ice Ages 296
    Plate Tectonic Effects 296
    Variations in Earth's Orbital Geometry: The Milankovitch Theory 296
    Variations in Atmospheric Carbon Dioxide and Dust 298

Human Induced Global Warming 299
The Impact of Global Warming on Human Society 300

14 Deserts and Winds 304

The Formation of Deserts 306
Winds 306

Weathering and Erosion in Arid Climates 308
Structural Influences on Desert Topography 310
    The Colorado Plateau 310
    Basin-and-Range Topography 311

The Geological Work of Wind in Arid Climates 313
Features of Wind Erosion 313
Features of Wind Deposition 314
    Loess Deposits 314
    Sand Dunes 315

Water Resources in Arid Climates 318

On the Fringe of the Desert: The Sahel 319
The Contribution of Climate 319
The Human Contribution 320

15 Coasts and Shoreline Processes 324

Wind-Driven Waves 325
Headland Erosion and Bay Deposition 327

Beach Formation and Shoreline Processes 328
The Nearshore Circulation Loop 329
Beach Types 329
The Sediment Budget 330
    Tampering with Beaches 331
The Work of Tides 334

Coasts 335
Tectonic Setting 335
Sea-Level Changes 336

16 *Geologic Time* 340

Relative Time 341
Deciphering Local Geologic History 341
    Horizontality and Superposition 342
    Unconformities 342
    Cross-Cutting Relationships 343
Correlation 344
    Statigraphic Correlation and Faunal Succession 344
    The Relative Geologic Time Scale 346
    Correlation Tools: Index Fossils and Overlapping Ranges 346
    The Terminology of the Time Scale 346

Absolute Time 348
Modes of Decay 348
The Decay Principle 349
Radioisotopes Useful in Dating 350
    Uranium- (and Thorium-) to-Lead Dating 351
    Potassium-to-Argon Dating 351
    Rubidium-87-to-Strontium-87 Dating 351
    Carbon-14 Dating 351
Dating the Geologic Time Scale 353
The Age of the Earth 355

17 *Formation of the Continental Crust* 360

Subdivisions of the Continental Crust 361
Cratons 362
Mountain Belts 362
Continental Margins 364
    Passive Margins 365
    Active Margins 365

Mountain Building and Plate Tectonics 366
Subduction and Mountain Building 367
Mountain Building by Accretion 368
  Accreted Terranes of Western North America 369
  The Himalayan Orogeny 370
  The Appalachian Orogeny 372
Growth and Evolution of the Continental Crust 374

18 Earth History: From the Origin of the Planet Through the Proterozoic Eon 378

Origin of the Planets 379

Primitive Planet Earth: The Hadean Eon 381

The Archean Eon 383

The Proterozoic Eon 385
The Precambrian Atmosphere and Oceans 386
The Origin of Life 386
The Late-Proterozoic World 387
Proterozoic Life 388

The Evolution of Life on Earth 389
Evidence of Evolution in Fossils and Comparative Anatomy 389
  Transitional Fossils 390
  Vestigial Organs 390
Evidence of Evolution in Embryology 392
Evidence of Evolution in Genetics 393
The Mechanisms of Evolution 395
Geology and Natural Selection 396
The Status of Evolution 398

19 Earth History: From the Paleozoic Era to the Present 402

The Paleozoic Era 403
Plate Tectonics of the Paleozoic 404
Paleozoic Life 406
  The Vertebrates 409
  Early Amphibians and Reptiles 410
  Causes of Paleozoic Radiations and Extinctions 411

The Mesozoic and Cenozoic Eras 412
Plate Tectonics of the Mesozoic and Cenozoic 412
　　The North American Cordillera 413
Mesozoic and Cenozoic Climate 414
Mesozoic and Cenozoic Life 417

APPENDIX A *Conversion Table for Metric and English Units* 421

APPENDIX B *Periodic Table of the Elements* 423

APPENDIX C *Mineral Identification Table* 425

APPENDIX D *Basic Guide to Geologic Maps* 429

Glossary 433

Selected Readings 443

Acknowledgments 447

Index 449

ESSENTIALS OF PHYSICAL GEOLOGY

Chapter 1

The Huttonian Revolution
   Establishing Geologic Time
      Absolute Time
   The Earth Machine
      External Processes
      The Rock Cycle
      The Earth's Interior

Continental Drift
   Physical Evidence
      Structural Evidence
      Evidence in the Distribution of Ancient Glaciers
      Evidence in the Distribution of Ancient Life-Forms
   Paleomagnetic Evidence

Continental Drift and Plate Tectonics

Research and Theory in the Geosciences
   The Scientific Method
      Questions and Tentative Answers
      The Search for Evidence
      Evaluation by the Scientific Community and Further Research
   Development of a Theory

Basic Geological Concepts

Geology is the science of the Earth, the study of its composition, structure, and history. Above all, it is the study of those processes that shaped the Earth of the past and those that continue to mold the Earth of the present. It is a science concerned with everything from the migration of sand dunes across the desert to the migration of continents across the globe; from the flow of molten rock down the sides of volcanoes to the flow of rivers to the sea. In this age of space exploration, geology has even exceeded its earthly boundaries. Planetary geology has become a thriving branch of the science.

The scope of geological investigations ranges from the atomic to the global, from events that occur in seconds to those that take billions of years to unfold. Much of what **geologists** do involves solving problems that affect millions of people, such as earthquake prediction, estimations of groundwater reserves, or petroleum exploration. Geologists also address questions of theoretical interest just for the joy of discovery, such as whether asteroids, volcanic eruptions, or neither caused the extinction of the dinosaurs.

Geology is a relatively young science, having begun to evolve into something like its modern form in the late eighteenth century. Little was known in those early days of the composition and structure of the Earth's crust. The interior of the Earth was a complete mystery, the vast expanses of the ocean floor a matter of conjecture. If you wanted to visit a region, you had to travel by boat, horse, or foot. Nevertheless, many concepts established during that period remain at the core of the science.

We begin this chapter with a discussion of the concepts of classical geology, which, as the science developed, were applied to an ever-expanding base of knowledge of the rocks and structures of the Earth. In the second half, we discuss a triumph of the application of these concepts, the theory of continental drift. Drift was the forerunner of plate tectonics, the unifying theory of modern geology. The development of plate tectonics is

geology
The study of the materials, processes, products, and history of the Earth.

geologist
One who investigates the materials, processes, products, and history of the Earth. Geologists conduct basic scientific research in order to increase our understanding of the Earth, and they apply their knowledge to improve our lives in many ways.

◂ Much of what geologists know about the composition of the Earth's surface and the processes that shape it comes from the study of rocks in their natural settings.

discussed in full in Chapter 2, and its applications will be evident throughout this text. Of the many threads that connect the pioneering geologists to their modern spiritual descendants, the most important is the widening recognition that the Earth is an ever-changing, dynamic planet.

The Huttonian Revolution

The founding of modern geology is called the Huttonian revolution, in honor of the contributions of the Scottish physician and farmer, James Hutton (1726–1797). His ideas may be expressed in three closely related principles.

1. The processes that presently shape the Earth have acted in much the same manner throughout geologic time, so direct observation of the Earth of today enables us to interpret the past.
2. Observation of the Earth reveals that these processes cause change in very small increments, and yet the sum of the changes we observe is enormous. Thus we conclude that the Earth's age is also enormous—far greater than all of human history.
3. Earth is a dynamic planet whose surface is in a constant state of change and whose materials are ceaselessly cycled and recycled.

Hutton's first principle—which in effect established that the Earth is subject to direct scientific inquiry—is called **uniformitarianism.** It is often summarized by the simple statement, "The present is the key to the past."

Uniformitarianism was an idea not widely accepted in Hutton's day. Some held that the Earth is quite young, having reached its present form through a series of sudden violent events. This position was called **catastrophism.** Its advocates struggled to find a place in science for such biblical events as Noah's Flood, and they sought to justify through science the biblical interpretation that the Earth is only a few thousand years old. In contrast, the uniformitarians pointed to numerous examples of how gradual change could cumulatively account for most of what was known about the Earth. Hence, uniformitarianism became the accepted view of most geologists. Sudden, violent episodes such as landslides, earthquakes, and volcanoes were seen as events that occur within an overall context of gradual change.

Hutton reached his conclusions regarding the Earth's great age and dynamism by applying the principle of uniformitarianism to rocks he observed in the field. In his trips around the British Isles, Hutton observed the everyday effects of rain and wind as they beat down upon rocks and soil. He saw rivulets of water carrying rock fragments, sand, and soil to the streams. He saw muddy waters flowing downstream and, in rainy seasons, sand and gravel moving with the swift currents along the stream bottoms. He thus realized that running water ceaselessly wears down the surface of the land and carries it, particle by particle, from the high mountains down to the oceans, where these particles are deposited as **sediment** on the sea floor.

Hutton also saw rocks in the mountains that were made of cemented sand particles and rock fragments, and he wondered how these sediments had gotten back up the mountains from the sea. Then one day, as he was exploring the low cliffs of Siccar Point along the east coast of Scotland, the answer hit him so suddenly that he was reported to have jumped up and down in delight. Hutton visited Siccar Point by boat in 1788; we will visit it using Figure 1.1.

Hutton observed that the layers, or **strata,** at Siccar Point were **sedimentary rocks,** consisting of fragments derived from the wearing away of older rocks. He also noted that there were two distinct sequences of strata present: a nearly horizontal sequence resting atop one that had been compressed into a series of wavelike folds. So tight were the folds in the lower sequence that the strata appeared to stand upright, or vertically—especially so because the folds were truncated. The boundary between the two sequences is what geologists today call an *unconformity*.

---

uniformitarianism
The principle that is based on the concept that past geological events can be explained by forces occurring today. "The present is the key to the past."

catastrophism
The doctrine that a series of sudden, violent, and short-lived worldwide events are responsible for the state of the Earth's crust and for the variety of life-forms that live on it.

sediment
Solid particles transported by water, wind, or ice and deposited in loose layers on the Earth's surface.

strata
Visually distinguishable layers of sedimentary rock.

sedimentary rock
A layered rock formed from the consolidation of sediment.

Basic Geological Concepts

Figure 1.1 The rock outcrop at Siccar Point, Scotland, from which Hutton drew profound inferences concerning the Earth's age and ability to cycle materials.

Hutton knew that sediments are deposited on the sea floor in horizontal layers, parallel to the Earth's surface, so that the oldest layer is on the bottom with successively younger layers resting on top. This simple principle, called **superposition,** was established long before Hutton. He had seen vertical layers similar to the lower strata at Siccar Point many times before in the folded strata of the highlands of the British Isles. Therefore, it was immediately clear to him how the two sequences of vertical and horizontal rocks at Siccar Point had been formed (Figure 1.2). The vertical sequence had originally been deposited horizontally on the sea floor and was the remnant of a much thicker accumulation of sedimentary rock. Subsequently, the rocks were folded and uplifted by powerful forces from within the Earth to form a mountain range. With uplift came the renewed attack by running water and all the other things that wear away rocks exposed to the Earth's surface. Eventually, the folded rocks were worn down to sea level, and the younger, nearly horizontal strata were spread over the truncated remnants of the once proud mountains. Thus clearly visible at Siccar Point is the story of how each sequence of strata formed, but missing at the unconformity is the entire record of the time it took to wear the mountain range to a nub.

superposition
The principle that in a sequence of sedimentary strata, the oldest layer is located on the bottom and followed in turn by successively younger layers, on up to the top of the sequence.

Figure 1.2 The sequence of events deduced by Hutton from the outcrop at Siccar Point. The surface between the lower and upper strata is erosional—an unconformity that separates rocks of markedly different ages.

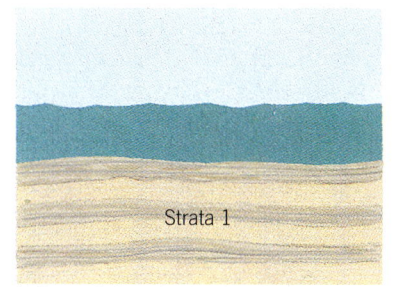

1. Strata 1 deposited on sea floor
2. Strata 1 folded and uplifted above sea level
3. Strata 1 eroded
4. Strata 2 deposited on top of eroded surface
5. Both strata sequences tilted and uplifted above sea level. Strata 2 eroded

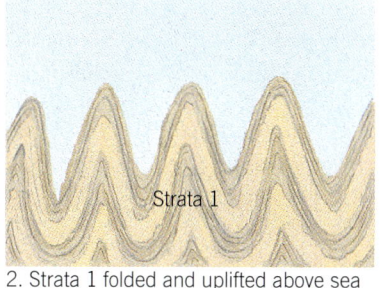

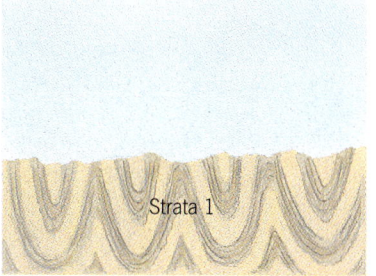

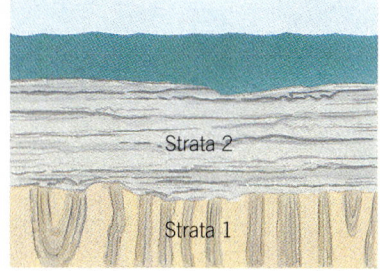

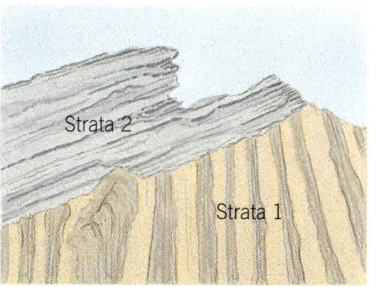

Having observed the slow rate at which streams erode the land, Hutton knew this gap in the record amounted to an enormous time span compared with that of a human life. He also realized that it represented but a short chapter in the history of the Earth, because the Siccar Point rocks had been derived from still more ancient rocks. Furthermore, Hutton saw that the processes that had created the unconformity remained at work at Siccar Point, for the very cliffs he stood upon were being worn low and were supplying sediment to the adjacent beach and ocean floor. He knew this sediment was the raw material of future mountains, because much of the British highlands were composed of identical stuff—uplifted marine strata made of sediment washed in from former landmasses.

For these reasons, Hutton interpreted the events recorded in the strata at Siccar Point as linked to an apparently endless cycle of uplift, erosion, deposition, and renewal—the sum of which constitutes the Earth's past, present, and future.

Establishing Geologic Time

Hutton was able to determine the sequence in which the rocks at Siccar Point formed—that is, their **relative ages**—by tracing the physical relationships among the rocks. However, such methods cannot be applied to rocks that are in separate locations. In order to compare the ages of the rocks at Siccar Point to those in the Grand Canyon or in China, for example, we require the existence of objects that are preserved in strata throughout the world and are the same relative age wherever they are found. Fortunately, such objects do exist. They are called **fossils,** the physical evidence of past plant and animal life preserved in rocks (Figure 1.3).

Wherever geologists have studied thick sequences of strata, they have found that the fossils in them change from bottom to top. Some types of fossils are found only in the lower levels; others at intermediate levels; still others at the highest levels. In other words, organisms exist over a certain range, or time span, and then become extinct. What is more, study of the entire fossil record proves that fossil organisms succeed one another in definite and recognizable order, so that rocks containing similar fossils are similar in age. This is a statement of an important principle, **faunal succession,** and it holds true no matter how widely separated are the rock strata that contain the fossils. When you climb the cliffs of North Dakota and Montana, you discover the bones of dinosaurs in layers below those that contain the bones of ancestral horses. At no place on Earth is it the other way around.

Faunal succession is, of course, a manifestation of organic evolution—the fact that living things have changed through time, giving rise to new species; that the life on Earth today, including human life, is derived from previous life-forms. The biology of evolu-

---

relative age
The age of an object or event expressed relative to the age of other objects or events but not in relation to time units such as years.

fossil
The remains, trace, or imprint of a plant or animal preserved in rock.

faunal succession
The principle that fossil organisms succeed one another in definite and recognizable order, so that rocks containing identical fossils are identical in age.

Figure 1.3 A typical fossil assemblage in ancient limestone strata. We observe that shells of contemporary organisms are incorporated in present-day sediments at the time of deposition. Thus we may confidently assume that the shells, or fossils, of ancient organisms are the same age as the strata within which they are found.

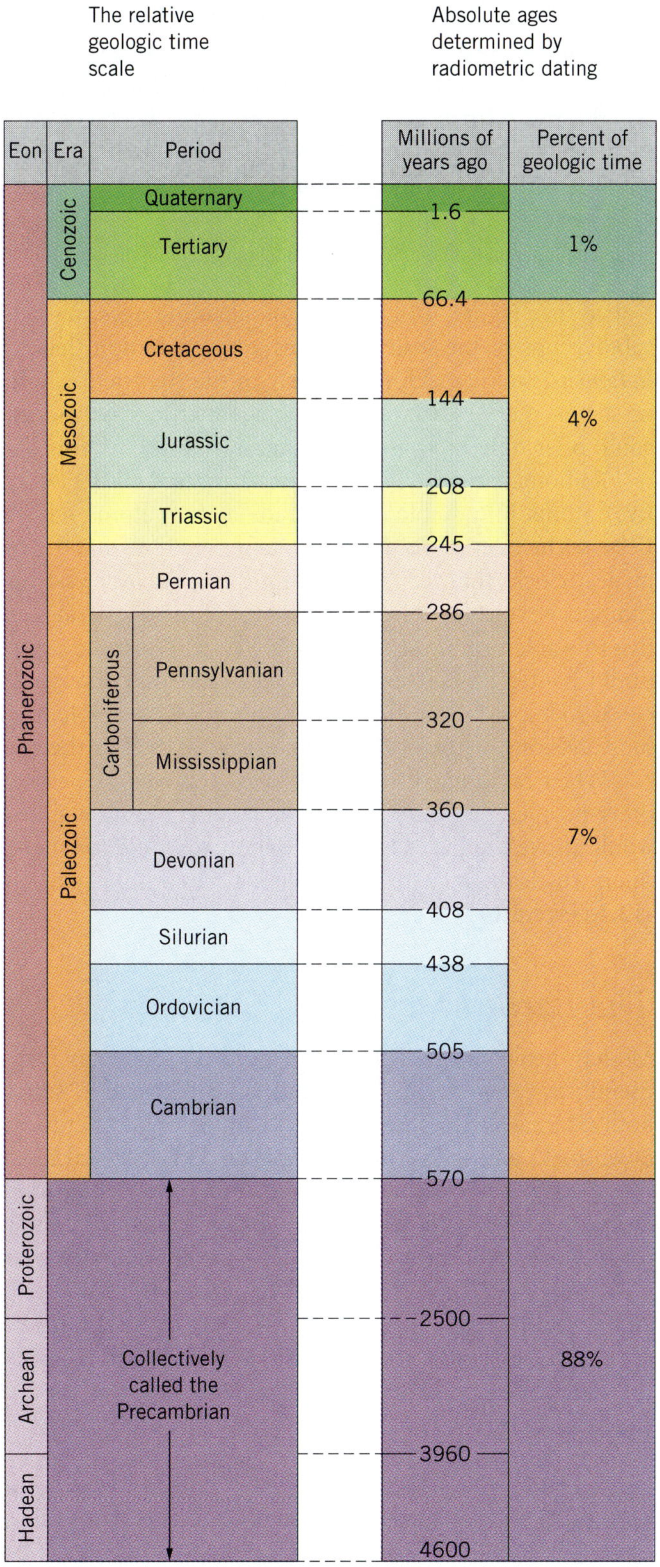

Basic Geological Concepts 7

The relative geologic time scale			Absolute ages determined by radiometric dating	
Eon	Era	Period	Millions of years ago	Percent of geologic time
Phanerozoic	Cenozoic	Quaternary	1.6	1%
		Tertiary	66.4	
	Mesozoic	Cretaceous	144	4%
		Jurassic	208	
		Triassic	245	
	Paleozoic	Permian	286	7%
		Pennsylvanian (Carboniferous)	320	
		Mississippian (Carboniferous)	360	
		Devonian	408	
		Silurian	438	
		Ordovician	505	
		Cambrian	570	
Proterozoic		Collectively called the Precambrian	2500	88%
Archean			3960	
Hadean			4600	

Figure 1.4 The units of the relative geologic time scale are arranged with the oldest divisions at the bottom and successively younger ones on top. The development of radiometric dating techniques enabled geologists to transform the relative time scale into an absolute time scale that shows the age of each unit in years.

tion is a fascinating subject in its own right, but geologists are also interested in faunal succession because it allows them to arrange rocks in chronological order and to *correlate* them (that is, to prove their age equivalence). Once the rocks are arranged and correlated, so too are the events they record.

By comparing and correlating rock strata around the world, geologists were able to construct a chronological ordering of geologic events called the relative **geologic time scale** (Figure 1.4). The scale is *relative* because it denotes the sequence of events and not

geologic time scale
A chronological ordering of geologic events and time units listed in sequence from the oldest on the bottom to the youngest on the top.

the actual time that they occurred. Largely the work of nineteenth-century geologists, it is one of the great achievements of science. As shown in Figure 1.4, the earliest division of the time scale is listed on the bottom, with successively younger ones on top, just the way the ages of rock strata are deduced in the field. We will have much more to say about the time scale in Chapter 16, Geologic Time. For the present, it is sufficient to acquaint yourself with the general vocabulary and subdivisions of the scale.

Absolute Time

Notice that Figure 1.4 also lists the **absolute ages,** in millions of years, of the various subdivisions of the scale. This was a later addition that came with the development of *radiometric-dating* techniques early in the twentieth century. Until then, geologists had no actual way of determining the age of rocks or, indeed, the age of the Earth. They could offer only the informed opinion that these things must be very old.

Radiometric age determinations are based on the decay of unstable (that is, radioactive) atoms into stable (or nonradioactive) atoms within rocks and other materials. Because the rates of decay are constant, we know that the radioactive atoms are, in effect, nuclear clocks that tick away within the rocks over vast time periods. The age of a rock can thus be calculated from knowledge of the decay rates and the proportion of radioactive to stable products present. We will also postpone discussion of radiometric dating until Chapter 16. For the present, let us just note that geologists have dated a wide variety of rocks, and from these dates, they have been able to determine the ages of the various divisions of the geologic time scale, as well as the age of the Earth.

When radiometric dating revealed the true scope of geologic time, geologists, like other people, were quite surprised. Who would have guessed that analysis of meteorites would prove the Earth to be 4.6 billion years old, or that nearly the entire fossil record upon which the relative time scale is based would span only the past 570 million years (12.4 percent of the Earth's age)?

The Earth Machine

Modern geologists find that thinking of the Earth as if it were a gigantic machine is a useful way to understand its complexities. The energy sources that drive the machine are both external and internal. Solar and gravitational energy power those processes that wear down rocks and redistribute fragmental and dissolved materials over the Earth's surface. Internal heat, much of it generated by the slow decay of radioactive atoms, is ultimately responsible for volcanoes, mountains, earthquakes, the creation and destruction of oceanic crust, continental drift, and related phenomena. The Earth machine constitutes a closed system, one whose mass has remained essentially constant since the time the Earth solidified as a planet. It is only energy that enters or leaves the system (aside from meteorites, a few astronauts and space probes, and negligible quantities of cosmic dust).

External Processes

James Hutton knew that each stream is linked to the continuous circulation pattern of the world's water, called the **hydrologic cycle** (Figure 1.5). The cycle begins when ocean water, having absorbed solar heat, evaporates and enters the atmosphere, leaving its dissolved salts behind. Some of the water vapor gets caught up in the wind systems, condenses, and falls on the land as rain or snow. A portion of the water striking the land surface evaporates back to the atmosphere. Some of it is also taken up in the roots of plants and then passed back to the atmosphere. About a third of it, however, either returns directly to the sea as stream runoff, seeps beneath the surface as underground water, or freezes and becomes incorporated in glaciers. Glaciers melt at their edges, however, and most groundwater eventually seeps back to the surface; so in time, both join the general stream runoff back to the sea.

---

absolute age
The age of an object or event in years as determined radiometrically or by other quantifiable means.

Meteorites that have fallen to the Earth's surface from the Solar System are believed to be similar in composition to the matter from which the Earth was formed.

hydrologic cycle
The continuous circulation pattern of the world's water, from ocean to atmosphere to Earth's land surface to ocean again.

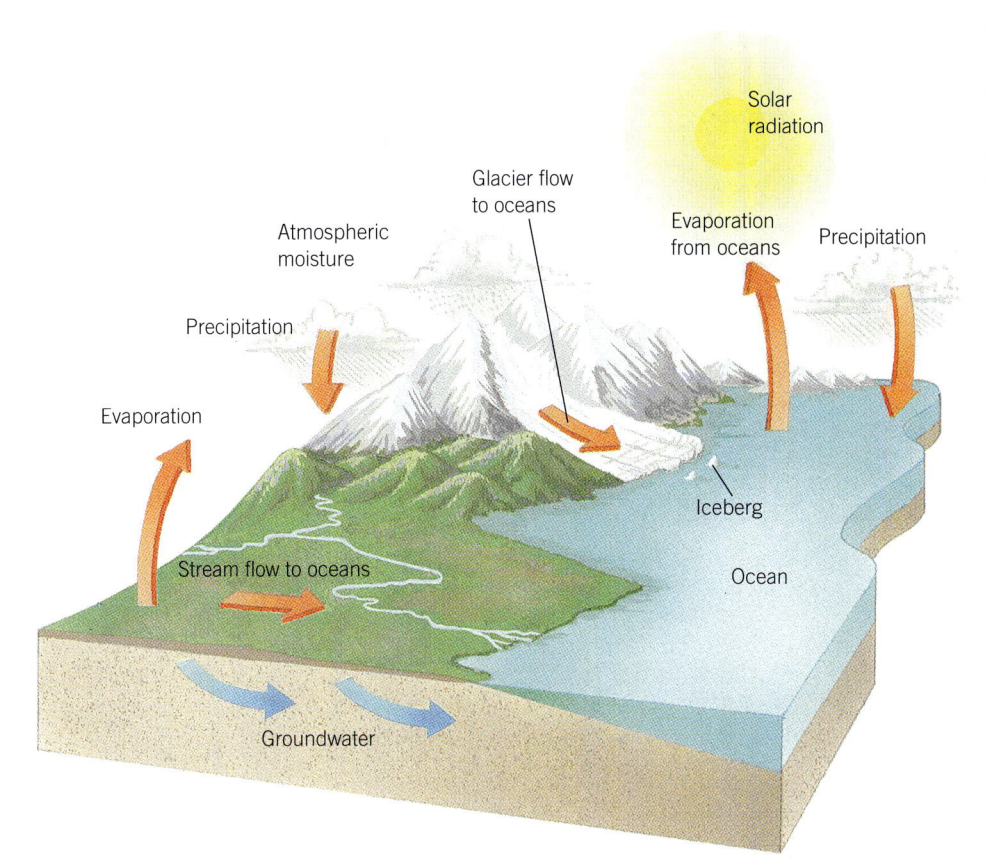

Figure 1.5 Solar energy powers the Earth's hydrologic cycle. The circulation of water among land, atmosphere, and ocean is unique to our planet and is largely responsible for the redistribution of matter over the Earth's surface.

Water plays the key role in a vast system of **erosion** and **deposition** in which the rocks of the continents are broken down and the fragments are transported to the lowlands and margins of the continents. The system begins with **weathering,** the physical and chemical alteration of rocks exposed on the Earth's surface. Though many processes are involved, naturally occurring acids play the key role of converting the hard minerals of the rock to clay and other products. Bacteria, plants, and animals, themselves dependent on water, also aid in the breakdown; and the weathering by-products and organic residues mix to produce soil, the prerequisite of plant and animal life on the land.

Water that falls on the land as rain initially flows downhill in thin sheets. The sheets seek out weaknesses and irregularities in the land surface, and the water soon becomes funneled into stream channels. These channels concentrate the energy of the flowing water as it runs downhill. For this reason, streams are efficient implements for carving valleys. As they do so, hillside slopes are made unstable, and the loose weathering products slide or slump downhill to the streams under the influence of gravity—a process called **mass wasting**. Thus valleys are widened, and the material carved out of the land is eventually brought to the sea. Along the way, the sediments are spread on the flood plains, deltas, marshes, and beaches and on the margins of the continents beneath the sea. Some of the fine clays and silts are carried in suspension far from the coasts and settle to the bottom, forming deep-sea muds.

The geologic work of water on the Earth's surface is both destructive and constructive. The destructive work—erosion—involves weathering, mass wasting, and transport of sediment by streams, and it reduces the elevation of the continents. The constructive work—deposition—redistributes the continental real estate. Glacial ice, wind, ocean currents, and waves also contribute to these constructive and destructive processes. This ceaseless movement of Earth materials downhill, from high places to low, should reduce the surface of the Earth to a featureless sea-level plain. However, such

erosion
The wearing down of rocks or soil, by weathering, mass wasting, running water, wind, and ice.

deposition
The gravitational settling of rock-forming materials out of such natural agents as water, wind, or ice.

weathering
The physical and chemical alteration of rocks exposed to atmospheric influences on the Earth's surface.

mass wasting
The downslope movement of soil and rock material under direct influence of gravity.

Figure 1.6 (a) This geologist is collecting a magma sample of a recent Hawaiian eruption. The magma will eventually cool and solidify to form igneous rock. (b) Four types of igneous rocks and the volcanic glass, obsidian.

magma
Molten (hot-liquid) rock material.

igneous rock
Rock that has cooled from the molten, or magmatic, state.

lithification
The process by which sediment is converted to sedimentary rock through compaction, cementation, or crystallization.

leveling has not occurred on a worldwide basis because forces from within the Earth uplift the land and form mountains and plateaus. Thus internal and external forces work together to disperse matter over the surface of the Earth.

The Rock Cycle

Mountain building requires energy. Hutton believed an abundance of it was available in the form of internal heat. If this heat were sufficient to uplift and deform rocks, he said, it was also capable of melting them deep inside the Earth. The molten masses could then rise and pierce the rocks of the Earth's thin outer layer, the crust. Eventually, some of them would spill onto the surface as lava. Although this Scottish physician and farmer had never seen a live volcano, he pioneered the idea that **magma** is derived from the melting of solid rock and that much of the Earth's surface is composed of **igneous rocks**—rocks that have cooled from the molten state (Figure 1.6).

If the Earth machine continuously *cycles* materials, then somehow the sediment deposited on the continents and on the ocean floor must later be transformed into magma. Hutton strongly suspected this to be the case, but it was left to later geologists to prove that, under certain conditions, sediments are indeed transformed in stages into magma. No one can actually see the process as it occurs, but we can trace the change from sediment to magma in the field. For example, we can see that the major difference between loose sediment and sedimentary rocks is that in rocks, particles are bound more or less tightly together (Figure 1.7). Binding occurs after burial, when the sediment is compacted and the grains are held together either by chemical cements or clay. The process by which sediment is converted to sedimentary rock is **lithification**.

Most sedimentary rocks are simply uplifted and eroded and the particles deposited once again, as at Siccar Point—but not all. We can trace some rock formations from the flanks into the deeply eroded cores of mountain chains and observe the changes in them that occur along the way. We see that the original horizontal layering of the strata becomes increasingly crumpled, or folded, toward the center of the mountain range. Pebbles in the strata, rounded by stream transport and deposition, are now found to be elongated in the deformed strata (Figure 1.8a). Farther into the mountain belt, we see that the original sedimentary features have become fainter, almost ghostlike in appearance, while traces of new minerals and structures have grown to dominance. We can see clearly that the sedimentary rock has been transformed into another rock. But what is the nature of the transformation? Had the rock melted, the sedimentary structures would not have faded; they would have been obliterated. Therefore, we must conclude

Figure 1.7 (a) Unconsolidated sediments typical of a changing coastal environment. The dark layer of bay mud, rich in organic matter, lies between windblown beach sand deposits. (b) The textures, colors, and mineral contents of these sedimentary rocks are largely determined by their erosional and depositional histories.

that the transformation took place while the original materials remained solid. The term for this process is **metamorphism,** and rocks transformed in this manner are called **metamorphic rocks** (Figure 1.8). The agents of metamorphism are heat, pressure, and, in many cases, chemically active fluids that bathe the rock.

Still farther into the mountain belt, we find that the structures of the metamorphic rock break down, and the rock blends with granite or similar igneous rock. Evidently, the heat and pressure generated during mountain building were so great that the metamorphic rock finally began to melt, producing magma. Thus igneous, sedimentary, and metamorphic rocks form a closed loop, the **rock cycle,** reflective of the manner in which internal heat, solar energy, water, and gravity act upon and transform the materials of the Earth's crust (Figure 1.9).

The rock cycle is not a smooth-functioning loop. Sediment is transformed to metamorphic rock only if it is buried deeply and in a region undergoing mountain building or igneous activity. We will soon learn, too, that most magma is not derived from recycled crustal rocks in the manner we have just described, but rather from the melting of solid rock far beneath the crust. In addition, not all materials are cycled at an equal rate,

metamorphism
The structural and mineralogical changes that occur in solid rock through the action of heat, pressure, and chemically active fluids.

metamorphic rock
A rock that has been altered from its original state through metamorphism.

rock cycle
A model that describes the formation, breakdown, and re-formation of a rock as a result of sedimentary, igneous, and metamorphic processes.

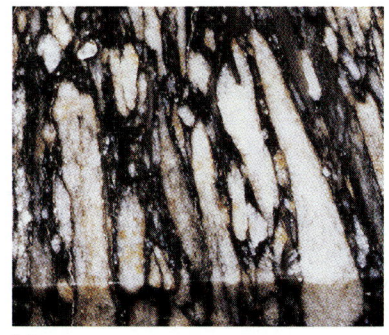

Figure 1.8 (a) Distorted sedimentary structures showing early-stage metamorphism of a sedimentary rock from Death Valley, California. Compression has elongated and aligned the pebbles. (b) This gneiss sample represents a higher degree of metamorphism in which new minerals have grown in size and are segregated into dark and light bands.

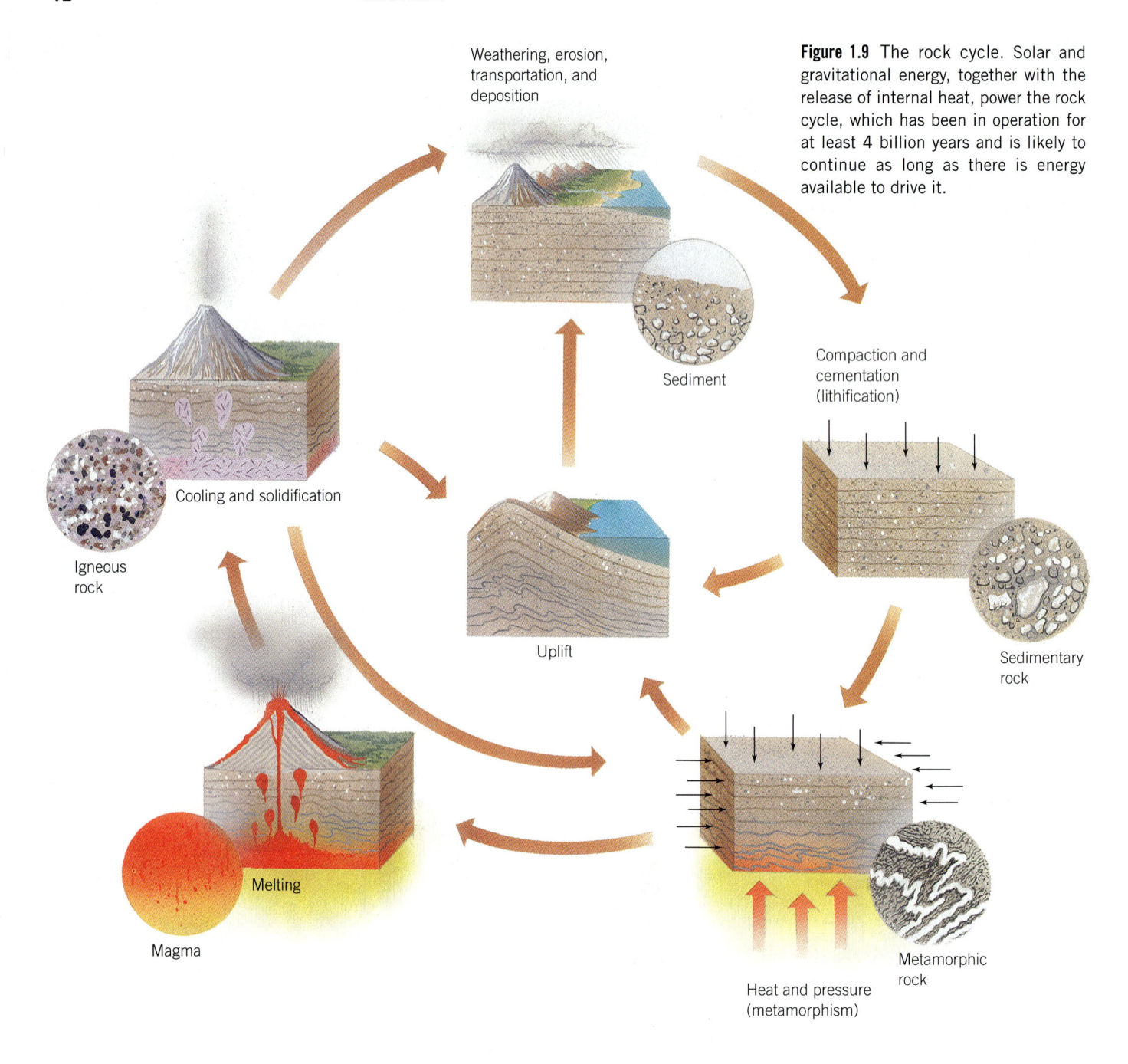

Figure 1.9 The rock cycle. Solar and gravitational energy, together with the release of internal heat, power the rock cycle, which has been in operation for at least 4 billion years and is likely to continue as long as there is energy available to drive it.

because external and internal processes do not act with equal intensity throughout the Earth. Nevertheless, the rock cycle is a powerful concept. Together with the principle of uniformitarianism and the recognition of Earth's great age, it remains a central theme of geology.

The Earth's Interior

Throughout the nineteenth and early twentieth centuries, the concepts of the Huttonian revolution were refined and applied to an expanding base of factual knowledge concerning the Earth. It is a process that continues to this day. One area critical to modern geology is our evolving understanding of the physical and chemical properties of the Earth's interior—the source of volcanism, earthquakes, mountain building, and metamorphism. In fact, Hutton rightly considered the Earth's surface to be largely the expression of its internal activity.

Basic Geological Concepts 13

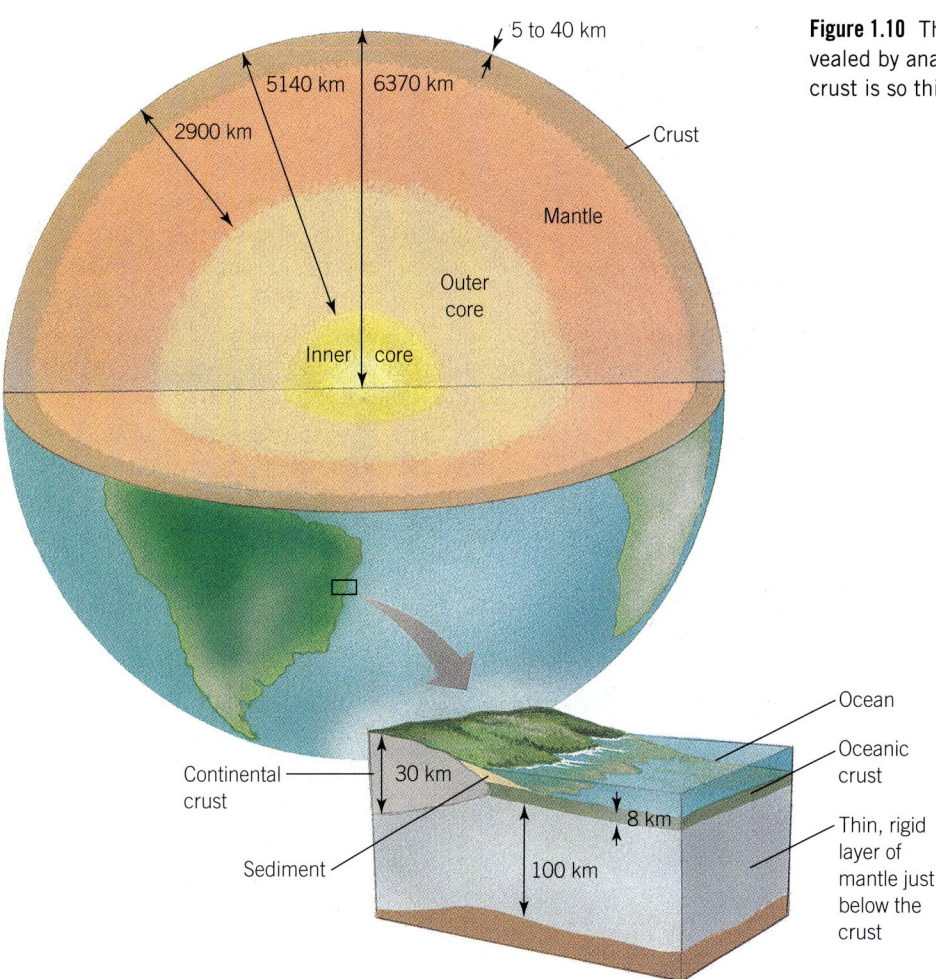

Figure 1.10 The structure of the interior of the Earth as revealed by analysis of seismic waves and other means. The crust is so thin it cannot be drawn to scale.

Of course, the major problem in studying the interior has always been our inability to drill far beneath the surface. (The deepest hole drilled so far is about 15 kilometers.) Geologists have had to infer the properties of the interior by measuring the Earth's gravitational and magnetic fields, by sampling the heat that escapes through the surface, and, most important, by detecting the vibrations set off by earthquakes, or *seismic waves,* that travel through the Earth's interior. By the early twentieth century, geologists had determined that the Earth is largely solid and consists of four major concentric layers: crust, mantle, outer core, and inner core, which differ in thickness, density, composition, and other physical properties (Figure 1.10).

The outermost layer, the **crust,** is the thinnest of the four. It is divided into high-standing *continental crust,* from 20 to 40 kilometers thick, and low-standing *oceanic crust,* from 2 to 10 kilometers thick. Their contrasting thicknesses reflect a fundamental difference in their compositions and densities. The continents are composed largely of coarse-grained, light-colored *granitic* rocks rich in silicon, aluminum, sodium, and potassium—all light **elements.** Oceanic crust consists of dark, fine-textured *basaltic* rocks, which contain some of these same elements, but in lesser proportions, and, in addition, are composed of a large percentage of the denser elements iron and magnesium.

Both continental and oceanic crust rest on the denser **mantle.** Extending from the base of the crust to the Earth's outer core at depths of 2900 kilometers, this vast layer comprises most of the Earth's volume and mass. The mantle is composed of silicate rock rich in iron and magnesium. No doubt the mantle is the ultimate source of all crustal materials. What is more, like an iceberg in water, the lighter crust floats on the denser mantle—a state of balance called **isostasy** (Figure 1.11). Because continental crust is thicker and lighter than oceanic crust, it sinks deeper into the mantle at the same time that it rises higher above it.

crust
The outermost zone, or layer, of the Earth, consisting of continental and oceanic components of distinctly different density and composition.

element
A substance that cannot be broken down into other substances by ordinary chemical or physical means. Water, for example, is not an element because it can be broken down into two elements, oxygen and hydrogen.

mantle
The zone of the Earth's interior between the crust and the outer core.

isostasy
The condition of balance, or equilibrium, in which the crust floats on the mantle.

Figure 1.11 Like wood floating in water, the Earth's crust is in a state of flotational balance, or isostasy, with the mantle.

outer core
The molten outermost zone of the Earth's core, between the mantle and the inner core.

inner core
The solid spherical center of the Earth.

continental drift
A hypothesis suggesting that the continents move over the Earth's surface.

Pangaea
The name of the supercontinent that existed between 200 and 300 million years ago, which has since fragmented into the present continents.

Thus the mantle exhibits dual properties. In response to short-term stresses, such as the vibrations triggered by earthquakes, it behaves more rigidly than cold steel. However, in response to long-term stresses, such as the weight of the continents, it yields like a viscous fluid.

At the boundary between the mantle and the **outer core** seismic waves capable of traveling only through solids die out, indicating that the Earth is molten at that depth. The outer core is composed mostly of iron mixed with lighter elements, possibly carbon, sulfur, or silicon. Later in this chapter, we will discuss an important aspect of the Earth—its magnetic field. Geologists believe that motions within the molten outer core generate this field. The outer core reaches to a depth of 5140 kilometers. From the behavior of seismic waves and the estimated pressures and temperatures at that depth, seismologists infer that the **inner core** is a solid metallic ball about 2460 kilometers in diameter. The inner core is probably composed of iron alloyed with other heavy metals, such as nickel.

Such was the approximate state of knowledge concerning the interior up until the middle of this century. It was essentially a simple picture of four cleanly defined, independent layers. Since then, our knowledge of the interior has grown more sophisticated, spurred by a more sophisticated technology. For example, the mantle has turned out to be far more complex than previously imagined and, in fact, consists of at least three subdivisions exhibiting varying degrees of fluidity and rigidity. The thin, uppermost layer of the mantle is very rigid and is welded to the rigid crust as a structural unit. This rigid zone of crust and upper mantle, known today as the *lithosphere,* rests on the *asthenosphere,* a warm plastic zone that stays close to its melting point. The properties of the lower mantle are still a matter of debate. Most geologists agree that although it is extremely rigid, over millions of years, it is capable of slowly flowing and deforming.

Continental Drift

It is ironic that for nearly two centuries, mainstream geologists who championed Hutton's concept of a restless, mobile Earth had, in fact, grossly underestimated the restlessness and mobility of our planet. They thought that it was fundamental to the idea of uniformitarianism that the continents and ocean basins be permanently frozen in place.

The rival hypothesis to permanence was **continental drift,** which envisioned a far more restless, mobile Earth. Its central concept was that the continents have not been locked permanently in their present positions on the Earth's surface but have moved relative to one another, and to the ocean floor, in the past. The twentieth-century scientist deservedly given the most credit for marshaling the facts that refute the idea of permanently fixed continents and ocean basins is the German astronomer, meteorologist, and geophysicist Alfred Wegener (1880–1930). He presented his case for continental drift in *The Origin of the Continents and Oceans,* first published in 1915. Wegener theorized that all the world's continental crust had existed as a single landmass, which he called **Pangaea** (whole Earth), and that the continents we see today are the result of its fragmentation. Figure 1.12a shows Pangaea according to the interpretation of Wegener's colleague Alexander du Toit. He divided Pangaea into two subcontinents: *Laurasia,* consisting of North America, Greenland, and Eurasia; and *Gondwanaland,* consisting of South America, Africa, India, Australia, and Antarctica. Oceans covered the rest of the world, and a triangular sea, the Tethys, partially divided the two subcontinents. Figures 1.12b–c show a later interpretation of how the supercontinent Pangaea

Basic Geological Concepts

(a) 200 million years ago

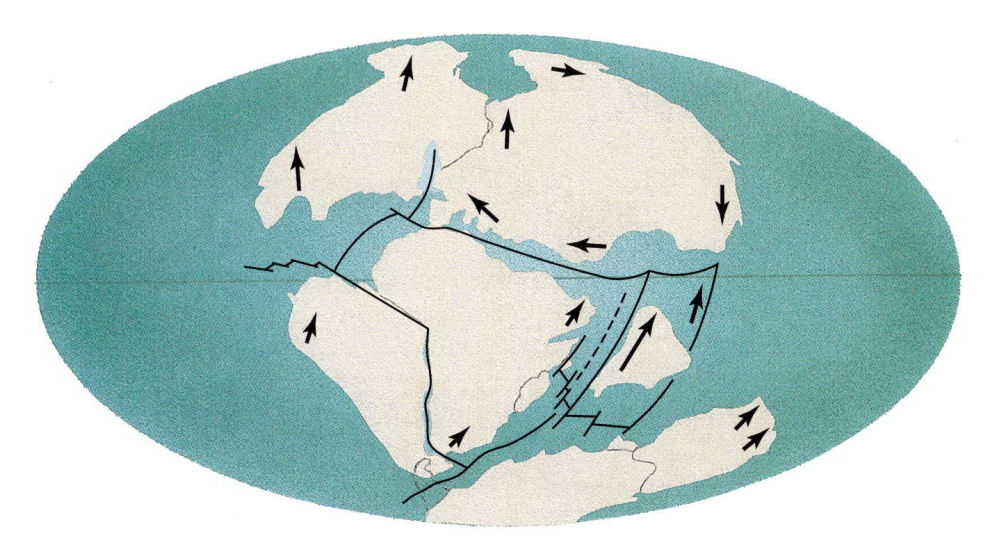

(b) 135 million years ago

(c) Present

Figure 1.12 Stages in the fragmentation of Pangaea leading to the modern configuration of the continents. Note the break-up of Gondwanaland and the collision of Africa and India with Eurasia, closing out the ancient Tethys Sea. Note also that the Atlantic Ocean did not form all at once. The equatorial and southern segments opened first, followed by the northern Atlantic. (Adapted from Robert S. Dietz and John C. Holden, *Journal of Geophysical Research* 75: pp. 943–956.)

divided and the continental blocks that we know today drifted to their present positions.

The key weakness in the hypothesis of continental drift was the apparent lack of an adequate mechanism by which the continents can move across the globe in the manner that Wegener had envisioned. This led some early advocates to search for catastrophic or single-event explanations for drift. Wegener, however, viewed drift as an ongoing, continuous process, explainable according to sound uniformitarian principles. He suggested that as a result of both the Earth's rotation about its axis and the tidal attraction between the Moon and the Earth, the continents are subjected to forces that impel them westward and toward the equator. Like ships plowing through water, the continents push their way through the yielding oceanic crust.

Opponents were quick to point out several flaws in Wegener's proposed mechanism for drift. First, ships are obviously more rigid than the water they plow through; in contrast, ocean floor rock is quite rigid. In fact, experimental studies and the analysis of seismic waves reveal oceanic rock to be more rigid than continental rock. How, then, could the continents push their way through oceanic crust without being torn apart or completely distorted in the process? Second, the Earth's rotation and tidal forces are about a million times too small to accomplish the task of propelling the continents.

Another proposed drift mechanism was internal heat convection that drives the continents from below. However, prominent physicists at the time mistakenly ruled out that possibility. With external forces too small and internal forces considered nonexistent, the hypothesis of continental drift seemed to be impaled on a self-generated contradiction.

Still, the concept of continental drift would not go away. The crux of the debate can be summed up in the following analogy: If the claim is made that there is an elephant in the next room, the only action that will positively prove or disprove the assertion is to check. If the elephant turns out to be there after all, you cannot explain it away by saying that you do not know how it got there. In science, facts are the final arbiter of theory. Drift advocates attempted to provide incontrovertible evidence that the "elephant" was indeed in the next room—that the continents had indeed moved. For drift advocates, precisely *how* the continents had moved was an important problem to be addressed *after* the fact that they had moved was firmly established. In the following sections, we will examine some of the evidence for drift.

Physical Evidence

Consider the matching coastlines of South America and Africa in Figure 1.13. Their complementary shapes were noted as far back as 1620 by the natural philosopher Francis Bacon, using the crude maps of a largely unexplored Earth. Every curious schoolchild since then has noticed that the coastlines of the continents display a jigsaw puzzle fit.

Modern geologists do not consider the coastlines the true edges of the continents, because they know that even slight sea-level fluctuations can cause large-scale changes in the configuration of the coast, especially in low areas. They take the true edge of the continents to be considerably deeper, about halfway down the continental margin. Figure 1.13 shows a computer-generated jigsaw puzzle fit of the continents at the 900-meter depth. Notice that the fit is far more exact than is implied from the shape of the coastlines. The fit holds not only for South America and Africa but also for North America and Eurasia. Putting these landmasses together produces a world that closely matches the Pangaea supercontinent of Figure 1.12.

Nevertheless, drift supporters knew that it was not enough to show that the continents *could* fit together—it had to be demonstrated that they had indeed been together. The way to do this was to establish that various patterns and events are explainable only when the continents are assembled in Wegener's jigsaw puzzle configuration. After the jigsaw puzzle is assembled, the pieces must make a coherent picture.

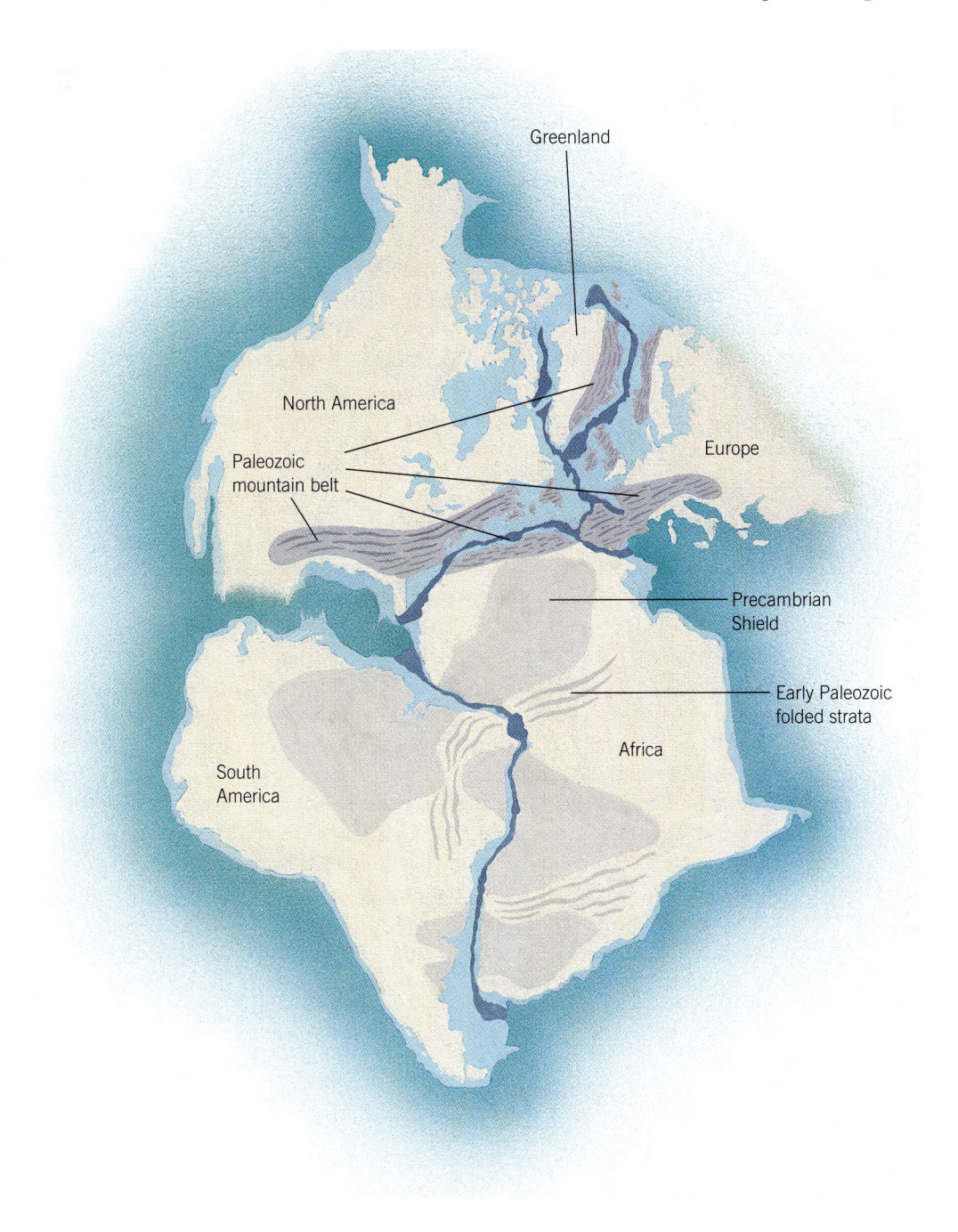

Figure 1.13 Strong evidence in support of continental drift includes the jigsaw puzzle fit of the continents (about halfway down the continental margins). Note the positions of Paleozoic mountain belts of the Northern Hemisphere and the complex structural trends of the Southern Hemisphere. All of these geological patterns match on either side of the junction. (Adapted from E. C. Bullard, J. E. Everett, and A. G. Smith, "The Fit of Continents Around the Atlantic," in P. M. S. Blacket, E. C. Bullard, and S. K. Runcorn, eds., *A Symposium on Continental Drift,* London: Royal Society of London, 1965, vol. 1088, p. 41.)

Structural Evidence

If a continent divides, rock structures located where the division occurred should be interrupted, like a picture torn in two. Unless the rocks have been worn away or buried beneath younger deposits of the newly formed continental margins, the structures should align again when the pieces are reassembled properly.

And they do! The most remarkable case is between Africa and South America. In Ghana, Nigeria, and Guinea, metamorphic rock formations of the Sahara shield, formed some 2 billion years ago, are overlain by younger, less deformed ones, 600 million years of age. The unconformity between the two formations can be traced right out to the Atlantic Ocean. If South America is nestled against Africa, in accordance with the computer fit, the same rocks, bearing the same relationship to one another, continue through the coastal city of São Luis, Brazil, and wind the same snakelike path that they do in Africa.

Less immediately striking, but no less convincing, is the correspondence of the Appalachian, European, and African mountain belts. The Appalachians had an extremely complicated history during the Paleozoic era, 245 to 570 million years ago. Four distinct periods of mountain building can be deciphered from their intricate structures. Similar structures of the same age are found in the British Isles, Scandinavia, southern Europe, and Africa. Reassembled as Pangaea, they form a continuous chain of identical rocks and structures with the Appalachians (see Figure 1.13).

Evidence in the Distribution of Ancient Glaciers

What was an enormous glacier doing astride the African equator about 300 million years ago? This was no mere high-altitude mountain glacier but a continental ice sheet similar in extent to the one atop present-day Antarctica. Markings made by the ice as it moved over underlying rock and distinctive deposits left upon melting leave no doubt that there was indeed such a glacier centered in Africa. On the other hand, North America and Europe show no evidence of glaciation during that time. Even in northern Canada, the climate apparently was warm, as recorded by coal deposits, rocks lithified from desert sands, and reptile fossils.

Field investigations leave little doubt that glacial ice also moved over parts of South America, India, and Australia at the same time that the African ice sheet was advancing. If these continents were in their present locations at that time, where did the ice come from? Tracing the direction of ice movement on each continent places the source of the ice in each case in the deep ocean, an impossibility (Figure 1.14a). (As we will see in Chapter 13, Glaciers and Climate, glaciers require the support of solid land beneath them in order to grow and spread.) However, if one assumes the jigsaw puzzle fit of Figure 1.14b, the continents cluster around Africa and Antarctica at the South Pole, and the spread of glaciers from there is now more easily understood.

Evidence in the Distribution of Ancient Life-Forms

The facts of the distribution of ancient life as seen in fossils are equally hard to ignore. For example, present-day shallow-water corals attach to the sea bottom and often grow in reef-forming colonies. The reefs occur in the warm waters from 30 degrees north to 30 degrees south of the equator. Yet extensive fossil deposits of extinct corals and associated organisms are found as far north as the Arctic Circle, and they seem to have lived in shallow seas that cut sharply across today's climatic zones. When the continents are reassembled as Pangaea, the ancient reefs are restored to latitudes that match the distribution of modern coral reefs.

Consider, too, the curious case of *Lystrosaurus,* a prime example of the puzzling distribution of the fossil record of certain land organisms. *Lystrosaurus* was a reptile, about 1.5 meters long, whose habitat was the swamps and river deltas of the Triassic period, some 240 million years ago (Figure 1.15). *Lystrosaurus* bones are found in India and South Africa. They have also been hacked out of rocks of the same age in Antarctica. How could *Lystrosaurus* have made it to these far-flung locations if the continents were locked permanently in their present positions? Did it swim? Its skeleton indicates that it was a land animal incapable of swimming thousands of kilometers. Assuming that it was a cold-blooded reptile, unable to regulate its body temperature the way mammals can, how could it have survived Antarctica's climate, even if that climate was warmer than today's?

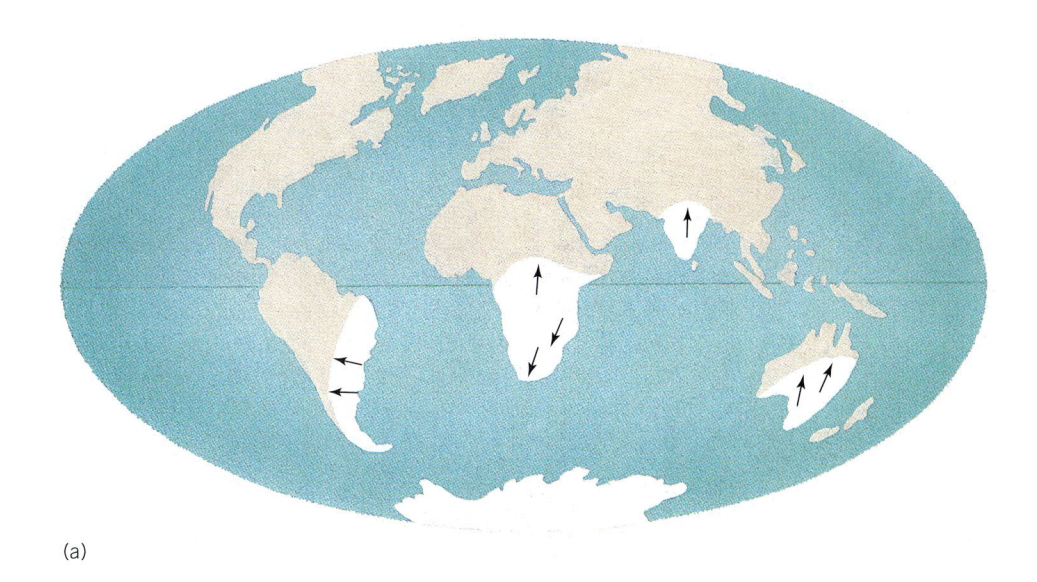

Figure 1.14 (a) The distribution of ancient glacial features suggests that equatorial Africa was covered by a continental ice sheet 300 million years ago. Also, glacial features of the same age located on the other southern continents appear to have sources in the deep ocean. (b) When the continents are reassembled as Pangaea, these same patterns point to a huge southern hemisphere glacier centered in Africa, which was close to the South Pole at the time. (Adapted from C. K. Seyfert and L. A. Sirkin, *Earth History and Plate Tectonics,* New York: Harper and Row, 1973, Fig. 7.6).

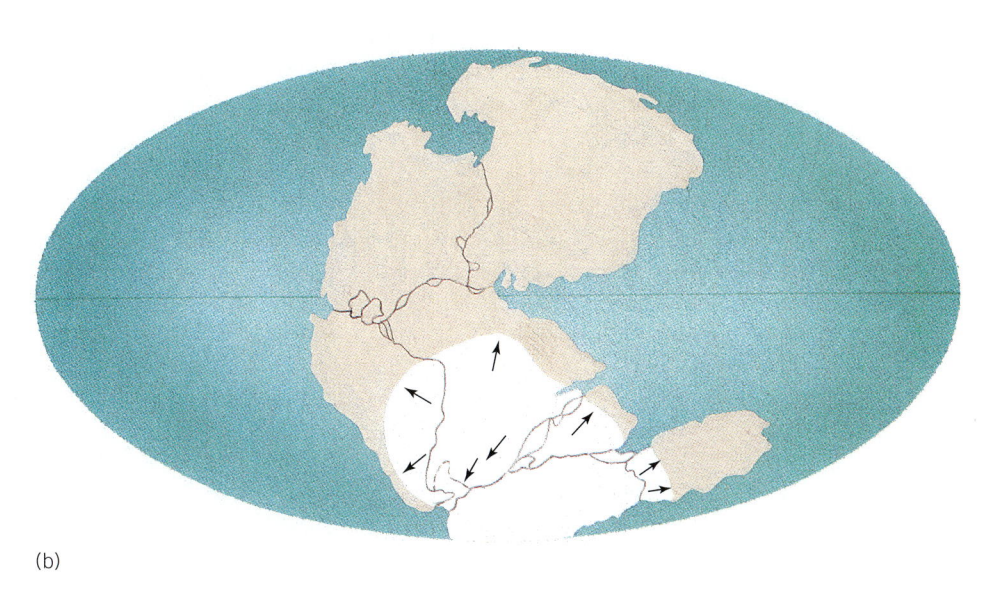

A similar paradox is presented by the distribution of *Glossopteris* (Figure 1.16), an ancient seed fern found in all continents of the Southern Hemisphere but only in India in the Northern Hemisphere. Drift opponents suggested that in the past, there were land bridges, which were similar to today's Isthmus of Panama or to island chains like the Aleutians, that would have provided an avenue for the spread of life into these far-flung continents. Yet surveys of the ocean bottom around Antarctica and other southern continents reveal no evidence of such features.

Lystrosaurus and *Glossopteris* are not oddball fossils. They are but two of the many groups of animals and plants that occupied vast regions of the present Southern Hemisphere and India. When the continents are reconstructed to resemble Wegener's Pangaea, these flora and fauna make sense as occupants of a broad habitat, probably a vast swampy plain built of the sediment washed in by streams issuing from mountain chains to the west and south.

Figure 1.15 Because *Lystrosaurus* could not have swum across oceans, it is difficult to imagine how it migrated to all the southern continents. However, if the continents were once a single unit, this large land reptile could have migrated across the huge landmass. (Illustration by Juan Barbera, *New York Times,* July 19, 1988.)

Figure 1.16 *Glossopteris* is a fossil seed fern widely distributed throughout the Gondwanaland continents.

paleomagnetism
The study of the Earth's past magnetism as recorded in rocks at the time of their formation.

This broad expanse of the Earth's surface remained a coherent geologic setting for a far longer span of time than that represented by the 100-million-year life span of *Lystrosaurus*. This setting is preserved in a sequence of strata, the *Gondwana succession,* named for the region in India where they were first described in detail. The Gondwana strata are a record of the environmental changes that occurred over hundreds of millions of years. From coal deposits that indicate lush swamps, to rocks made of windblown sands that tell of deserts, to coarse debris that tells of melting glaciers, this record is mirrored in striking detail on all the southern continents (South America, Africa, Antarctica, and Australia).

Paleomagnetic Evidence

Though impressive, much of the evidence of continental drift presented so far is circumstantial and subject to differences of interpretation. It is the study of the Earth's past magnetism, or **paleomagnetism,** that compellingly proves continental drift. It also proves that continental drift has been an ongoing phenomenon that began long before the breakup of Pangaea. To understand this evidence, we must first discuss briefly the Earth's magnetism and how it is recorded in the rock record.

At any given location, a compass needle will rotate horizontally and point in the direction of the Earth's magnetic North and South poles, which are close to, but not the same as, the geographic poles. This is how a compass is generally used. A compass needle designed to rotate vertically forms an angle with the horizontal plane (or the horizon) called its *magnetic inclination*. Figure 1.17 shows that the magnetic inclination

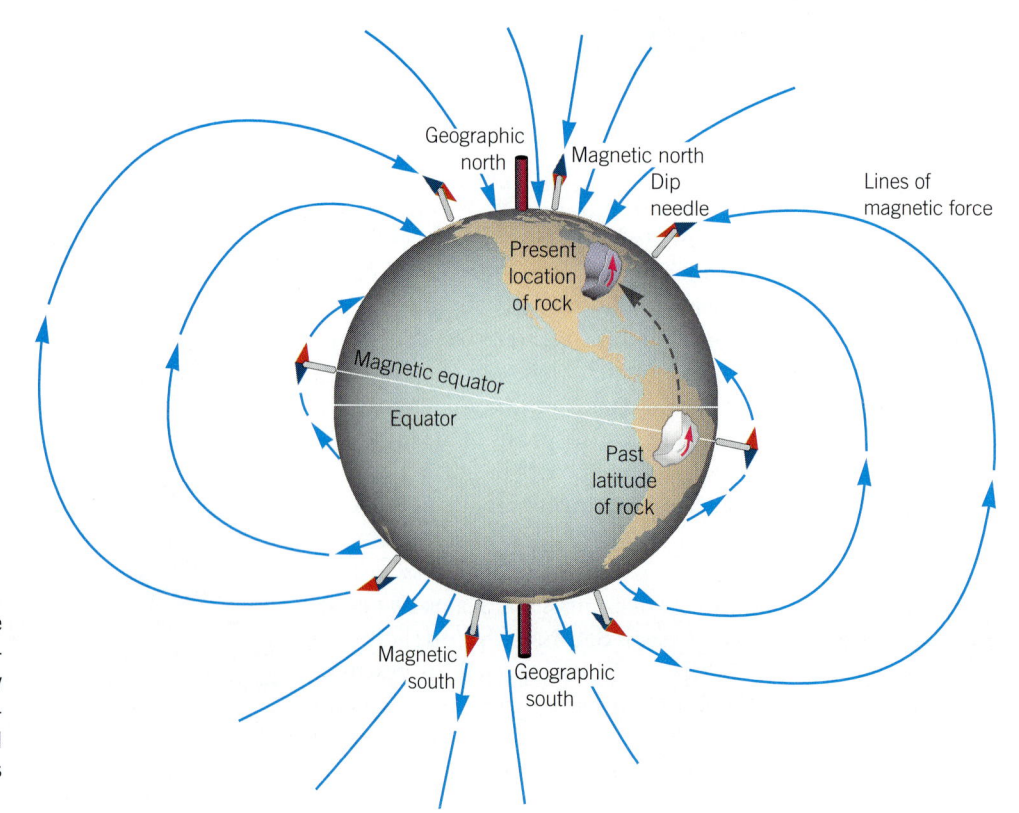

Figure 1.17 Lines of force delineate the Earth's magnetic field and show the inclination of a compass needle at any given latitude. The horizontal inclination of the rock sample presently found in North America indicates that it was formed at the equator.

Figure 1.18 Paleomagnetic evidence strongly supports continental drift. When the polar-wandering curves of Europe and North America are aligned so that they match, the Atlantic Ocean is closed out and the continents fit together in a manner that closely resembles Pangaea.

varies between perfectly horizontal (zero) at the magnetic equator and perfectly vertical (pointing straight into the ground) at the magnetic North Pole. Note again that the magnetic poles and equator are also close to their geographical counterparts, so that the inclination of the compass needle closely measures latitude. Now, how can we use this information to measure the past positions of the continents?

The magnetite crystals of a lava or of a sedimentary rock are in effect tiny compass needles that preserve the inclination of the field at the latitude where the rock formed. For example, the magnetic inclination of the rock shown in Figure 1.17 is horizontal. This orientation indicates that the landmass in which the rock is embedded once straddled the equator, as South America does today.

Because the rock's present location is North America, we may conclude that the landmass within which it is embedded has been moved from the equator to its present position. By examining the changes of inclination in a sequence of strata in a given location, such as from the bottom to the top of the Grand Canyon, we can determine the path of North America through time with respect to magnetic north. These changes, when plotted on a present-day globe, define an *apparent polar-wandering curve* (Figure 1.18a).

Things really get interesting when we plot the European and North American polar-wandering curves on the same globe (Figure 1.18b). The curves of the two continents, Europe and North America, are nearly identical but are separated by about 40 degrees longitude. But wait! How can there be *two* curves if there is only *one* North Pole at any given time, whether the rocks are located in Europe, North America, or Tasmania?

The answer is clear if we consider that 40 degrees is the width of the Atlantic Ocean. When the curves are fitted together, the Atlantic Ocean closes and the margins of Europe and North America fit together as well (Figure 1.18c). This demonstrates that it was the opening of the Atlantic Ocean and the resulting drift of the continents that created the two curves. In fact, we can rearrange all the continents using paleomagnetic evidence. The resulting configuration confirms the jigsaw puzzle fit of the continental margins, the matching geologic structures, and the climatic patterns—in sum, a world closely resembling Wegener's Pangaea.

The paleomagnetic data indicate further that the clustering of the continental blocks into the single landmass, Pangaea, was a relatively short-term episode in the long-running history of continental drift. Continental crust, in one configuration or another, drifted over the globe prior to the breakup of Pangaea and probably has done so throughout geologic time. For example, the paleomagnetic evidence shows that 400 million years ago, ancestral North America straddled the equator. Now the northward distribution of extinct corals mentioned earlier becomes less exotic; the corals had not spread north to the Arctic Circle but had once occupied roughly the same latitude zones in which present-day corals are living.

The paleomagnetic confirmation of continental drift came too late to comfort Alfred Wegener, who was routinely ridiculed for his beliefs; he died in 1930 while on an expedition to the Greenland ice cap in search of evidence to support his ideas.

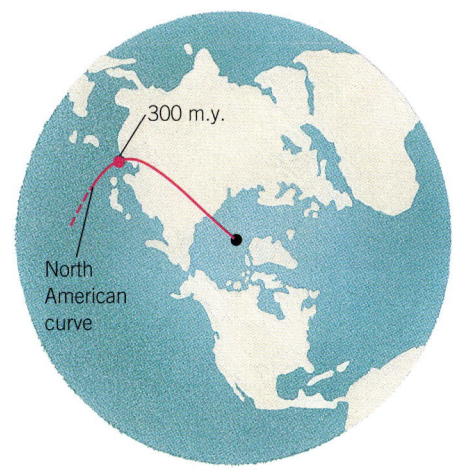

(a) North American polar-wandering curve

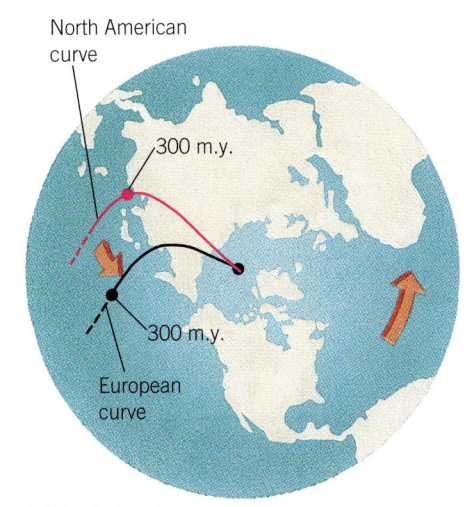

(b) North American and European curves

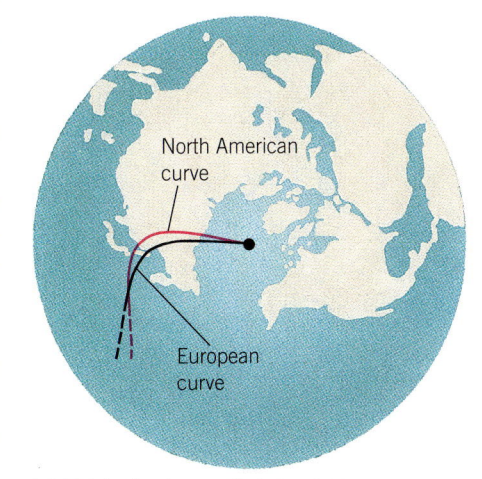

(c) Match of polar-wandering curves

Continental Drift and Plate Tectonics

Continental drift is now viewed as one aspect of a more comprehensive process, *plate tectonics*. Instead of being thought of as rafts plowing *through* crust, the continents are now considered "cargo" embedded in larger rafts, called *lithospheric plates.* These plates consist of oceanic as well as continental crust and include a thin part of the upper mantle down to the soft, ductile zone mentioned previously, the asthenosphere. The plates move as coherent slabs over the asthenosphere; and as they wander, so do the embedded continents. In their nomadic wanderings over the Earth, the continents have split and reassembled in different combinations many times in the geologic past.

The evidence for plate tectonics was gathered in the decades following World War II, when geologists put to sea with a technology Wegener would probably have considered science fiction. This technology enabled them to detect an ocean floor topography as varied as that on land, to study the record of the soft muds that lay on the hard-rock oceanic crust, to probe the secrets of that crust, and to get a clearer picture of the Earth beneath the crust itself. In the next chapter, we will describe plate tectonics fully, beginning with the evidence revealed by studies of the ocean floor. In later chapters, we will explore the close connection between plate tectonics, mountain building, earthquakes, and volcanic activity. In Chapter 17, Formation of the Continental Crust, we will discuss the influence of plate tectonics on continental evolution.

Research and Theory in the Geosciences

Science is the knowledge of the natural or material world gained through systematic observation and experimentation. A key word in this definition is *observation*. Science differs from other human endeavors in its insistence that observations about the world take precedence over assertions that are based on how the world should be or how we would like it to be. Another key word in the definition is *systematic;* it means "having a plan or method." Within the story of any geological inquiry, we can discern the various components of the systematic process of science, or the **scientific method.**

The Scientific Method

The scientific method can be narrowly defined as a series of steps that begins with a question or problem to be solved and ends with an answer or solution (Figure 1.19). The solution is based on a solid body of meticulously collected evidence that has withstood rigorous testing and evaluation by many scientists over time. In broader terms, however, the scientific method is less a procedure than a way of thinking. Its purpose is to ensure objectivity and the dogged determination to get at the truth—no matter what it is! What cannot be described in textbook fashion is that scientific progress often leaps ahead on bursts of insight and key discoveries that change perspectives overnight.

Questions and Tentative Answers

All scientific inquiries begin with a question or series of questions about the natural world. The subject of inquiry may never have been addressed previously, or it may have been addressed by others but not satisfactorily answered. For this reason, researchers usually pay close attention to what other scientists have written on the subject before proceeding with their own investigation.

Like a detective who uses the evidence gathered at the scene of a crime to identify a suspect, geologists use evidence gathered from the field to develop a **hypothesis** or **model**—that is, a tentative answer to the question. Exactly when the geologists feel ready to make an educated guess about the answer to the question depends upon the extent of previous knowledge about the phenomenon being researched; the speed with which they are able to obtain evidence; and, perhaps most important, the knowledge, experience, and intuition of the researchers themselves.

Agatha Christie's detective hero Hercule Poirot proclaimed, "Until the real culprit is found, everyone is a suspect!" Geologists may approach the solution to a problem in

---

scientific method
A process of investigation in which a problem is identified, data are collected and analyzed, and a hypothesis is formulated and tested.

hypothesis
A tentative explanation of a phenomenon or process that is tested for validity by repeated observation or experimentation

model
A hypothesis expressed as a visual or statistical simulation, or as a description by analogy of phenomena or processes that are difficult to observe and describe directly.

a similar manner using the method of **multiple working hypotheses.** The advocates of a new model may have gathered an impressive array of evidence for their hypothesis, but there may also be substantial evidence in support of another hypothesis. Therefore, other researchers pursuing an answer to the same question will use both models as their multiple working hypotheses, gathering data and following all leads as impartially as possible, ruling out neither explanation until the evidence unequivocally contradicts it.

The Search for Evidence

The systematic collection of **data**—that is, information about the subject of inquiry—is a crucial step in attempting to prove or disprove a hypothesis. The data may be in the form of physical evidence obtained directly from a location in question, or the evidence may be measurements obtained from instruments. Geological researchers today may also access data from a host of other sources, such as the satellite images, sea-floor maps, and other types of information routinely collected by government agencies like the United States Geological Survey and research institutions such as the Lamont-Doherty Earth Observatory.

Having collected the evidence, geologists next need to organize and analyze it. The primary goal of analysis is to highlight relevant similarities and relationships among the data. Only then can geologists reach conclusions regarding what they have found. This stage may well be the most creative and challenging aspect of the endeavor.

Figure 1.19 Flowchart showing the major steps involved in the scientific method.

Evaluation by the Scientific Community and Further Research

When they are satisfied that the evidence supports their hypothesis and that their findings are significant enough to share with the scientific community, the researchers prepare a report on their findings, commonly called a "paper," and submit it for publication to a scientific journal. The paper is then sent out for peer review to experts on the subject, who transmit their confidential opinions to the journal editors. Their role is to determine whether the paper is of a quality worthy of wider circulation, not to determine whether the hypothesis is correct or to argue why they do or do not agree with it. Upon publication, the paper is then informally evaluated by the geological profession at large, especially by those working on the same or similar problems. The authors may also be invited to present their paper at academic meetings of professional associations (Figure 1.20). The largest of these include the Geological Society of America and the American Geophysical Union, and there are myriad specialized groups affiliated with these organizations.

Now that the researchers' work has been presented to the scientific community for evaluation, the next step is taken by other researchers who have been intrigued by the paper and begin their own investigations. Indeed, hypotheses are the lifeblood of geology, and of science in general, because they stimulate debate and further research. A new hypothesis may have little impact; or it may be debated and tested immediately, acting as a springboard to fresh investigations—the ultimate compliment, as many geologists would argue.

It also is entirely possible that the hypothesis will be ignored or even scorned at first, only to be revived decades later when new data and fresh insights prove its worth.

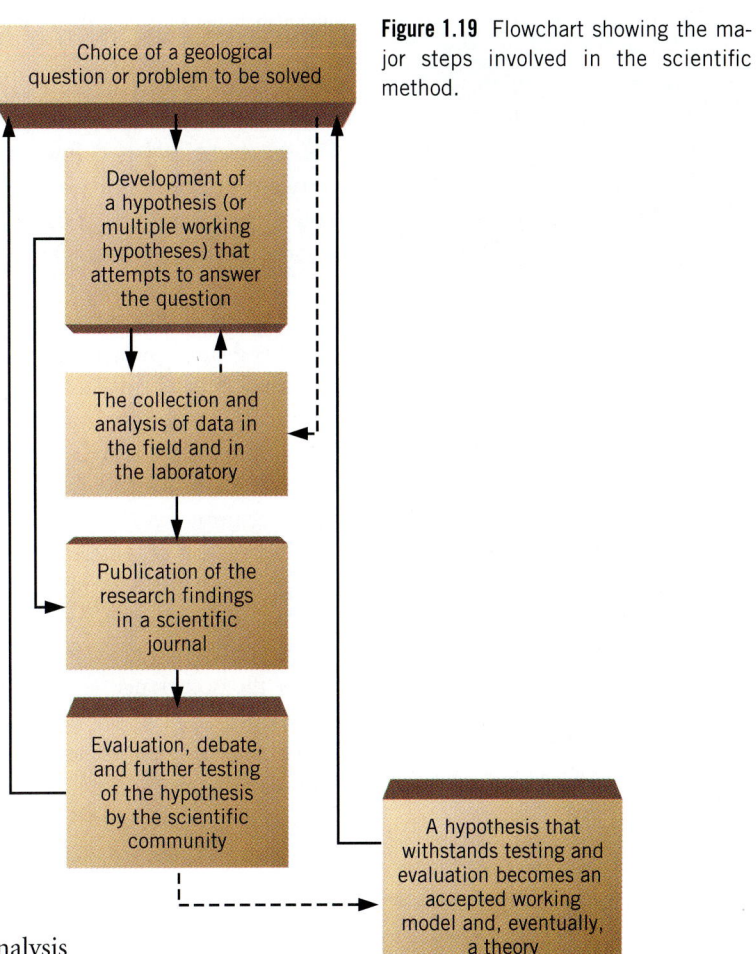

multiple working hypotheses
An approach to geological research in which several possible explanations of a phenomenon are developed and evaluated simultaneously and impartially.

data
Items of factual or statistical information.

Development of a Theory

All hypotheses contain the germ of a theory. Hypotheses become theories when, over time, they meet the challenges presented by newly discovered data and competing hypotheses. Scientists define a **theory** as a set of propositions that is widely accepted by the scientific community as the explanation of a broad group of natural phenomena. A good theory is a work of creative imagination, similar in some respects to an oil painting or sculpture. Painters and sculptors, however, are free to bend their materials to their will. They are governed only by the restrictions they choose to place on the forms they create, whereas the restrictions placed on scientists are far more stringent. A scientific theory is grounded in physical reality, and it is valid only insofar as it is able to explain and predict that reality as accurately as possible. A scientific theory must proceed from cause to effect in a logical manner and account for *all* the relevant facts, not just those that support the theory. Also, it must be consistent with the laws of nature—that is, observations proven true time and again. A theory should not violate the law of gravity, for example.

The theory of *plate tectonics* is the backbone of modern geology. It plays a role analogous to the theory of evolution in biology in that it provides a unifying set of explanations for what had previously been considered unrelated geological phenomena. Plate tectonics explains such things as the origin of oceans and mountain belts, the drift of continents, the occurrence of earthquakes and volcanoes, and much more. Beginning with the next chapter, we will see how plate tectonics amply fulfills the criteria of a good theory.

Finally, keep in mind that the establishment of a theory, no matter how widely accepted, does not place it beyond debate. In science, nothing is beyond debate, although the validity of an argument against an established theory will be judged by the strength of the supporting evidence. The revision of established theories is considered to be part of the scientific process.

Figure 1.20 The frank exchange of ideas is an essential element of geological research. These geologists are participating in a seminar at a conference on oil and energy resources.

theory
A widely accepted explanation for a group of known facts. A theory is a hypothesis that has been elevated to a high level of confidence by repeated confirmation through testing and experimentation.

STUDY OUTLINE

Geology is the study of the composition, structure, and history of the Earth and the processes that shape it. **Geologists** conduct geological research studies and apply their knowledge to benefit human society.

THE HUTTONIAN REVOLUTION. **Uniformitarianism,** the basic principle proposed by James Hutton, says that we can interpret the past through observation of the Earth today because geologic processes have remained the same throughout geologic time.

   A. Hutton saw the Earth's surface in a constant state of change, its materials ceaselessly cycled and recycled. He observed that the **strata** at Siccar Point were **sedimentary,** consisting of fragments of older rocks. He knew the simple principle of **superposition,** which states that sediments are deposited in horizontal layers, with the oldest on the bottom and successively younger layers on top. Thus he realized that enormous geologic changes occur in very small increments, and he concluded that the Earth's age is also enormous.

   B. The opposing view, **catastrophism,** held that the Earth is quite young, having reached its present form through a series of sudden violent events.

   C. Establishing geologic time
      1. The principle of **faunal succession** holds that, because **fossils** succeed one another in order, rocks containing similar fossils are similar in age. This principle allowed geologists to create the **geologic time scale,** by which the **relative ages** of rocks can be measured.
      2. Later development of *radiometric dating* enabled geologists to determine the absolute ages of rocks. The age of the Earth is 4.6 billion years.

D. The Earth machine
   1. In the **hydrologic cycle,** seawater evaporates and enters the atmosphere; it condenses and falls to the land, where it eventually flows back to the sea. Water is the key agent of **erosion** and **deposition.** It carries sediment from high to low places and in this manner levels the land surface. The leveling process begins with **weathering,** the physical and chemical alteration of rocks. As running water forms streams and carves valleys, loose weathering products slump downhill toward the streams in **mass wasting.** As external processes work to level the land, internal forces work to uplift it.
   2. In the **rock cycle,** sediment is cemented into sedimentary rock (**lithification**), which is buried beneath other layers; heat and pressure transform it into **metamorphic rock,** which is melted and turned into magma. **Igneous rock** results when **magma** is cooled. The igneous rock is weathered and converted to sediment, and the rock cycle begins again.

E. The Earth's interior.
   1. The Earth is largely solid and consists of four major concentric zones: **crust, mantle, outer core,** and **inner core,** which differ in thickness, density, composition, and other physical properties.
   2. The crust floats on the mantle in a state of balance called **isostasy.** Because continental crust is thicker and is made of lighter **elements** than oceanic crust, the continents sink deeper but also rise higher than the ocean basins.

II. **CONTINENTAL DRIFT.** Early in this century, Alfred Wegener proposed that all the world's continental crust had existed as a single landmass, called **Pangaea,** and that the continents we see today are the result of its fragmentation.
   A. Physical evidence. The jigsaw puzzle fit of the continental margins suggests that the continents were a single landmass that was later torn apart.
      1. When the continents are recombined as Pangaea, the trends of rock structures make a continuous line, as though they had been torn in two.
      2. Ancient sedimentary rocks currently located near the equator contain evidence of past glaciation. However, if the continents are moved into their former positions as Pangaea, the contradiction is cleared up.
      3. The locations of many animal and plant fossils also make sense only if the continents were a single landmass in the past.
      4. Sequences of strata found on all the southern continents display remarkable similarities of rock types and fossils.

   B. Paleomagnetic evidence. Above all other evidence, it was **paleomagnetism,** the study of Earth's past magnetism, that proved continental drift.
      1. An unconstrained compass needle forms an angle with the horizontal plane of the Earth, called its *magnetic inclination.* Magnetized crystals within sedimentary rocks are "fossil magnets" that point to the position of the magnetic poles of the past.
      2. Geologists used the magnetic inclination of ancient rocks to trace the *apparent polar-wandering curves* for each continent. Because the magnetic poles remain essentially stationary, the curves strongly suggest that the continents moved. The polar-wandering curves are aligned only when the continents are in their Pangaea positions.

III. **CONTINENTAL DRIFT AND PLATE TECTONICS.** Continental drift is one aspect of *plate tectonics.* The continents are imbedded in *lithospheric plates,* which move as coherent slabs over the *asthenosphere.* The continents have split and reassembled in various combinations many times in the geologic past.

IV. **RESEARCH AND THEORY IN THE GEOSCIENCES**
   A. Geologists employ the **scientific method,** a systematic approach to gaining knowledge of the natural world.
      1. Geologists begin an investigation by asking a question about a geological phenomenon.
      2. Development of a **hypothesis** or **model,** a tentative answer to the question, may be the next step in the investigation, or it may come later during the collection of data. Sometimes, in geological research, it is best to employ **multiple working hypotheses,** ruling out no proposed explanation until the evidence unequivocally contradicts it.
      3. The evidence, or **data,** may be physical samples, instrument measurements, or statistics.
      4. When they are ready to share what they have found with the scientific community, the researchers prepare a paper describing their hypothesis and supporting evidence, and submit it to a scientific journal for publication.
      5. The hypothesis is subjected to evaluation, debate, and further testing by the scientific community. This assessment leads to new research, and the process begins again.
      6. A hypothesis becomes a **theory** when it has been elevated to a higher level of confidence through extensive testing and debate. In science, however, no theory is ever elevated beyond debate.

STUDY TERMS

absolute age (p. 7)
catastrophism (p. 4)
continental drift (p. 14)
crust (p. 13)
data (p. 23)
deposition (p. 9)
element (p. 13)
erosion (p. 9)
faunal succession (p. 6)
fossil (p. 6)
geologic time scale (p. 7)
geologist (p. 3)
geology (p. 3)
hydrologic cycle (p. 8)
hypothesis (p. 22)
igneous rock (p. 10)
inner core (p. 14)
isostasy (p. 13)
lithification (p. 10)
magma (p. 10)

mantle (p. 13)
mass wasting (p. 9)
metamorphic rock (p. 11)
metamorphism (p. 11)
model (p. 22)
multiple working hypotheses (p. 23)
outer core (p. 14)
paleomagnetism (p. 20)
Pangaea (p. 14)
relative age (p. 6)
rock cycle (p. 11)
scientific method (p. 22)
sediment (p. 4)
sedimentary rock (p. 4)
strata (p. 4)
superposition (p. 5)
theory (p. 24)
uniformitarianism (p. 4)
weathering (p. 9)

THOUGHT QUESTIONS

1. How might the concept of uniformitarianism, if taken to an extreme, inhibit the advance of geological knowledge? Give two examples cited in this chapter.
2. Why does an unconformity represent a gap in the geologic record?
3. Explain the relationship between the principles of superposition and faunal succession.

4. Explain the relationship between gravity and the hydrologic cycle.
5. What conditions must exist for a planet to exhibit a rock cycle?
6. Can you think of materials other than the Earth's mantle that exhibit the dual properties of rigidity and plasticity?
7. Why is the representation of the paths of the continents over the globe through geologic time referred to as an *apparent* polar-wandering curve?
8. Were Wegener's opponents merely short-sighted and intolerant, or did they have sound reasons for disagreeing with his hypothesis of continental drift? Explain.
9. A geologist is often compared to a detective visiting the scene of an event millions of years after it occurred. In what ways is this comparison reasonable? In what ways is it not? How does a scientist's work compare to an artist's, a minister's, or a politician's work?
10. Explain the difference between a hypothesis and a theory.

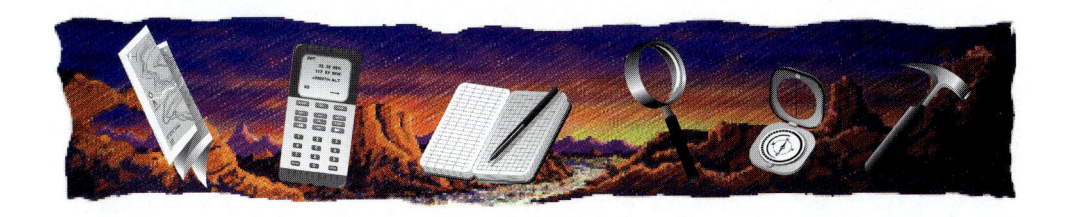

Chapter 2

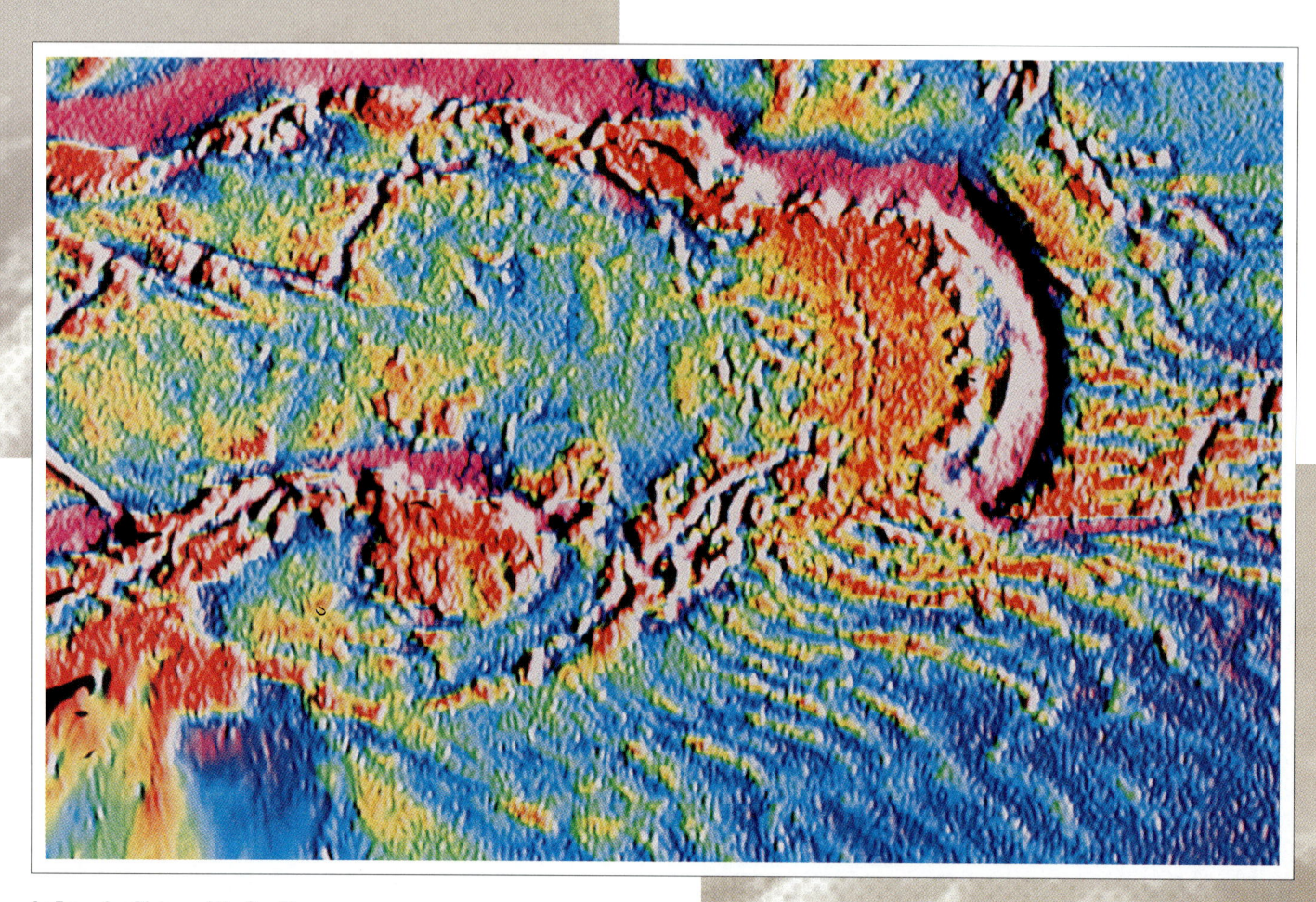

An Emerging Picture of the Sea Floor

The Sea-Floor Spreading Hypothesis
   Evidence in Magnetic Reversals
   Evidence in Sea-Floor Sediment
   Evidence in Sea-Floor Topography

Plate Tectonics
   Divergent Boundary Processes
   Convergent Boundary Processes
     Ocean-Ocean Convergence
     Ocean-Continent Convergence
     Continent-Continent Convergence
   Transform Boundaries
   The Causes of Plate Motion

Plate Tectonics and the Ocean Floor

Save for a few scattered expeditions employing crude equipment, the ocean floor was uncharted territory through the first half of the twentieth century. World War II changed all that. By the early 1950s, the United States government, oil companies, universities, and foundations were pouring millions of dollars into marine geology. It was the beginning of the age of "big science" and team research, modeled in some ways after wartime projects. It was also an age of great technological advances, as instruments designed for the war effort were adapted to probe the sea floor and new ones were invented. The ocean floor was the new frontier of the geosciences, offering many opportunities for creative people who were willing to work hard.

Bruce Heezen and Marie Tharp of the Lamont Geological Observatory* were among the many scientists seeking to forge careers in marine geology during the 1950s and 1960s. Although both were in their twenties, they were entrusted with the important task of making sense of an immense amount of data collected by Lamont research ships in the North Atlantic. Tharp had the difficult assignment of turning thousands of depth-sounding numbers into a series of highly detailed profiles, or cross sections, of the North Atlantic floor. As shown in Figure 2.1, the submerged mountainous region towering above the general level of the ocean floor in the center of each profile is the Mid-Atlantic Ridge. From the form of the mountains and from samples dredged from the ocean floor, it was apparent that the Mid-Atlantic Ridge consists of a series of submarine volcanoes, many of them active. Tharp noted, "I was struck by the fact that

---

*Now called the Lamont-Doherty Earth Observatory.

◀ Topographic image of the South Atlantic Ocean floor, based on data gathered by the U.S. Navy's Geosat satellite. To the right of the southern tip of South America (black, top-left) and the northern tip of Antarctica (black, bottom-left) lies the Scotia plate. Just right of the plate is the curved South Sandwich trench. At the upper right of the image is the southern end of the Mid-Atlantic Ridge, an undersea mountain chain that extends down the entire length of the Atlantic Ocean. The theory of plate tectonics offers a compelling explanation of these and other large-scale features of the Earth's crust.

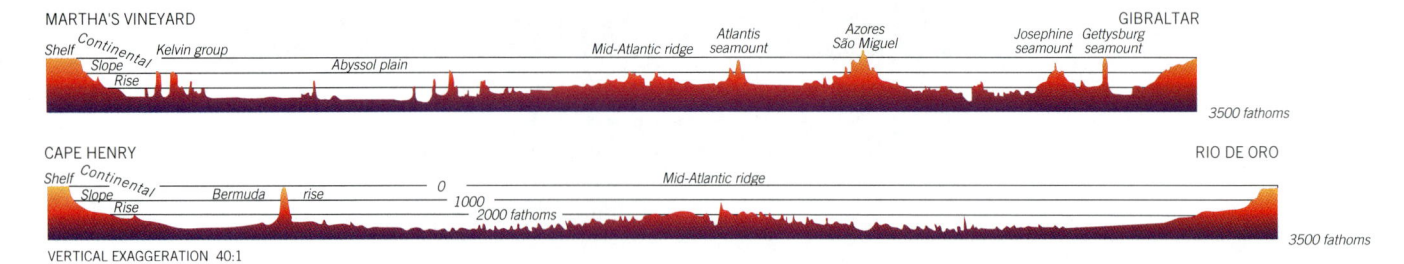

Figure 2.1 These profiles of the Mid-Atlantic Ridge were constructed by Bruce Heezen and Marie Tharp from data collected by Lamont Geological Observatory expeditions. Near the center of each profile is a prominent cleft, which Tharp identified as a central rift valley. (Marie Tharp)

fault
A fracture in bedrock along which rocks on one side have moved relative to the other side.

mid-ocean ridge
A seismically and volcanically active mountain range, marked by a central rift valley, that extends continuously through the major ocean basins.

the only consistent match-up when I compared the profiles was a V-shaped indentation located in the center of each profile. The individual mountains didn't match up, but the cleft did."* Tharp believed she had discovered a *central rift valley* in the Mid-Atlantic Ridge.

Tharp's belief was based on the cleft's form, identical to regions of continents where a central valley has dropped down between huge parallel **faults,** or fractures in the crust. The valleys are tensional features, formed by stretching or extension of the crust. Many continental rifts are still active and marked by numerous shallow earthquakes that are triggered when the crustal blocks, hung up along the faults by friction, suddenly slip. So Heezen proposed that they plot the locations of several thousand Atlantic earthquakes to see whether they matched the central rift valley of the ridge. It was a perfect fit.

The next step was to refer to the earthquake records and plot the locations of earthquakes in the world's oceans. The researchers found that a line of earthquakes, continuous with the Mid-Atlantic Ridge, extended around the globe. What is more, the earthquake belt could be traced into the continental East African Rift Valley, whose profile was identical to the central rift valley of the Atlantic (Figure 2.2). Heezen and Tharp had discovered "a huge tension crack, caused by a splitting apart of the Earth's crust." Actually, as Tharp pointed out, it is a branching network of cracks splitting all the ocean basins of the world and, as Figure 2.3 shows, parts of the continents as well. Stretching more than 64,000 kilometers, the **mid-ocean ridge** is the longest continuous feature of the Earth's crust.

Heezen was somewhat ambivalent about their discovery in the beginning. According to Tharp, "He groaned and said, 'It cannot be. It looks too much like continental drift.' He and almost everyone else at Lamont, and in the United States, thought continental drift was impossible. North American geologists considered it to be almost a form of scientific heresy, and to suggest that someone believed in it was comparable to saying there must be something wrong with him or her."[†]

Heezen knew that the discovery of the mid-ocean ridge system was a serious blow to the theory of permanently fixed continents and ocean basins. He saw the striking symmetry between the Mid-Atlantic Ridge and the margins of the continents on either side; and he realized that this huge tension crack was proof that the crust was being pulled apart in the same directions predicted by the theory of continental drift.

In this chapter, we will see that Heezen and Tharp's discovery was one of a series that spurred a revolution in geology not seen since James Hutton's day. That revolution is plate tectonics, the unifying theory of the Earth's crust. Plate tectonics provides the

---

*Marie Tharp, "Discovery of the Mid-Ocean Rift System," in *Yearbook* (New York: Lamont-Doherty Geological Observatory of Columbia University, 1989).

[†]*Ibid.*

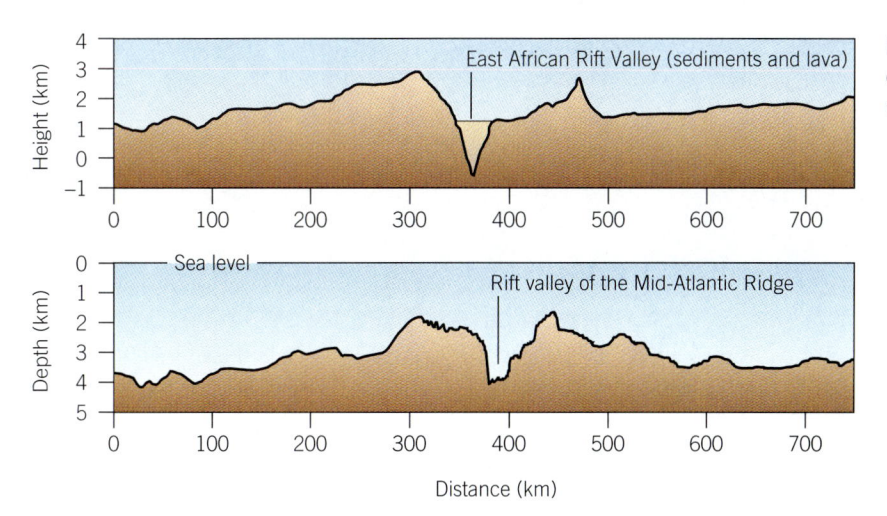

Figure 2.2 The similarity in the profiles of the East African and Mid-Atlantic rift valleys is striking.

mechanism that validates Wegener's hypothesis of continental drift; it explains the origins of the ocean basins, continents, and mountains. Its impact is felt in all branches of the geosciences. From forecasting violent volcanoes and earthquakes to searching for oil and mineral deposits, the theory of plate tectonics is also a practical theory.

The story of the plate tectonics revolution begins on the ocean floor, and Marie Tharp's detailed depiction of ocean floor topography is a useful guide to the following discussion (see Figure 2.3).

An Emerging Picture of the Sea Floor

Once the mid-ocean ridge was discovered, marine geologists proceeded to map its topography in detail. They found that many seemingly isolated volcanic islands, such as Iceland, are located on the ridge system. Also, the mid-ocean ridge is crossed by narrow canyons dividing it into segments that do not connect; in other words, the segments are offset (see Figure 2.3). The narrowness of the canyons, plus the occurrence of numerous earthquakes between the offset rift valley segments, indicate that the canyons are in fact long cracks, or **fracture zones**. A number of them have been traced hundreds, even thousands, of kilometers until they pass under ocean sediments. Fracture zones, although strikingly displayed on the ocean floor, are not exclusively marine features; the San Andreas fault of California is a prominent fracture zone in the continental crust.

Marine geologists also employed a variety of techniques to determine what lies beneath the mid-ocean ridge. Heat probes across the ridge recorded abnormally high quantities of heat escaping from the central rift valley. Tests using sensitive pendulums and weights attached to springs also recorded relatively weak gravitational attraction across the rift valley, pointing to lighter-than-normal materials beneath the surface. Natural and artificially induced seismic waves slowed beneath the rift, indicating that they had passed through softer, lighter, and presumably warmer substances. The combined data made a strong case for the existence of a thick wedge of semimolten rock beneath the ridge crest, the likely source of the erupting lava. This inference was later confirmed by geologists who descended in submersibles through the dark, deep water to the rift valleys of the North Atlantic and the East Pacific Rise. There they observed, firsthand, lava oozing from cracks in the ocean floor and jets of hot fluids issuing from chimney-like vents (Figure 2.4).

A striking, counterintuitive fact is that the shallowest features of the ocean basins—the mid-ocean ridges—are located toward the basin centers, whereas the deepest features are found along the basin margins. Long known from the laying of transoceanic

fracture zone
A region of closely spaced cracks or faults in rocks. In the context of this chapter, a prominent crack or fault that runs perpendicular to, and offsets, the mid-ocean ridge.

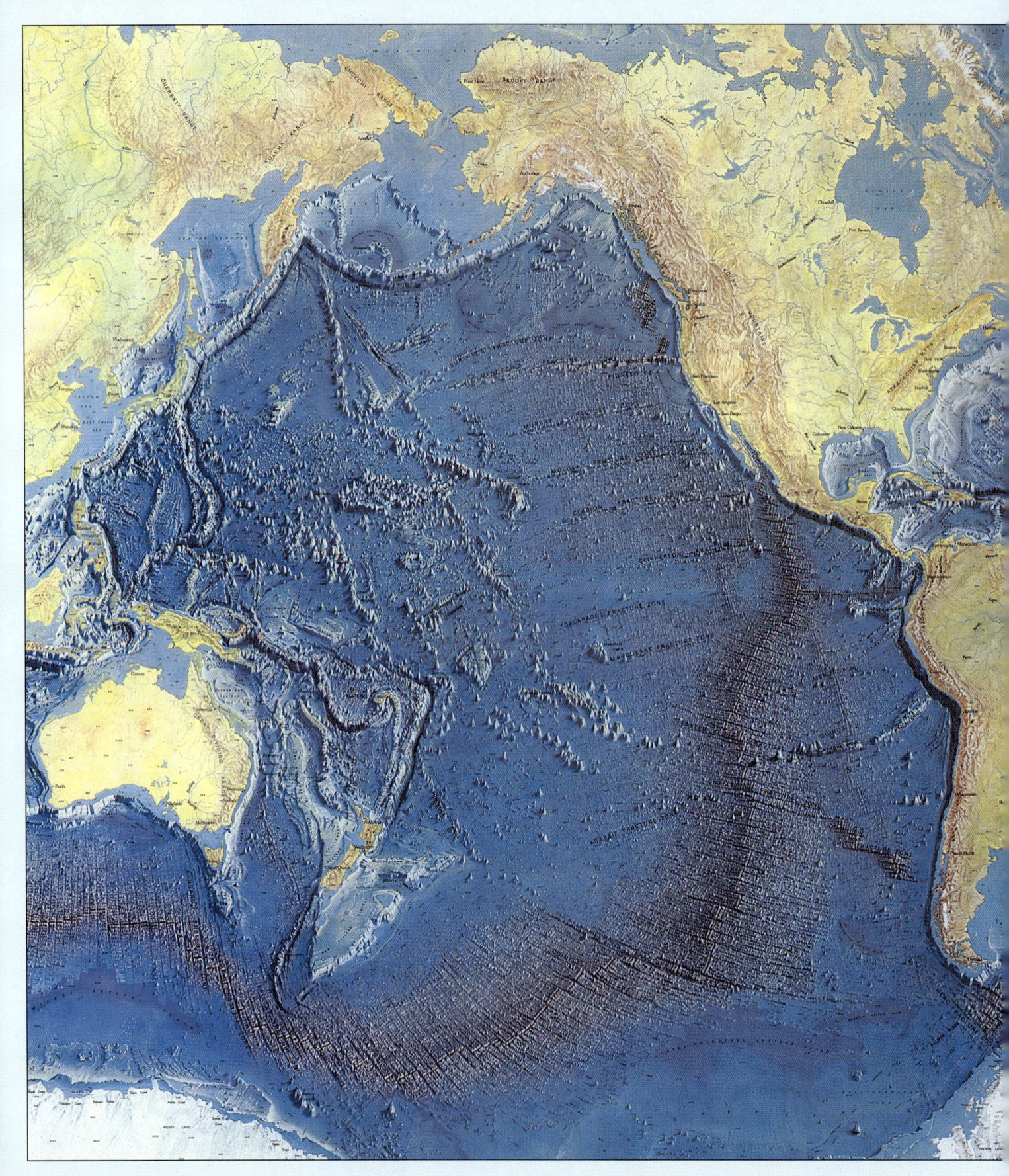

Figure 2.3 The main features of the ocean floor are shown in this model, which was created from Marie Tharp and Bruce Heezen's research. The mid-ocean ridge system divides the world's ocean basins and intersects the continents (note, for example, the Red Sea and the Gulf of California). Deep-sea trenches rim the Pacific Ocean. (Marie Tharp)

WORLD OCEAN FLOOR
BY BRUCE C. HEEZEN AND MARIE THARP

Based on Research and Exploration Initiated and Supported by the

UNITED STATES NAVY · OFFICE OF NAVAL RESEARCH

1977

Figure 2.4 Jets of hot fluid escaping from submarine vents along the East Pacific Rise. These features, plus lava erupting onto the sea floor, are evidence that molten rock is present beneath the mid-ocean rift valleys.

deep-sea trench
A deep, narrow, elongated depression in the sea floor.

island arc
A curved chain of islands with its convex side to the ocean, emerging from the deep-sea floor. Generally located close to a continent.

telephone cables, the **deep-sea trenches** are a series of narrow, curving depressions that together form the lowest portions of the Earth's surface (see Figure 2.3). The Marianas trench, for example, is at a greater depth beneath sea level than the Himalayas are above it. The trenches mainly encircle the Pacific and parts of the Indian Ocean, a notable Atlantic exception being the Puerto Rico trough.

The great archipelagos, or volcanic **island arcs,** of the world—the Aleutians, Japan, the Philippines, Indonesia, and the West Indies—are closely associated with the trenches. The island chains are convex toward the ocean, and the trenches run parallel to them along their seaward side (see Figure 2.3). Heat escapes from the trenches at a lower rate than it does from the mid-ocean ridge and the ocean floor in general—evidence that the oceanic crust beneath the trenches is cold and, therefore, dense.

In the eastern Pacific, deep-sea trenches border continental landmasses rather than island arcs. There too, however, the trenches are closely associated with volcanic activity. The Peru-Chile trench is virtually offshore of the west coast of South America, and not surprisingly, it is close by a volcanic chain, the western Andes Mountains (see Figure 2.3). Central America, itself largely of volcanic origin, is bordered by a similar trench. Farther north, the mid-American trench is directly off the west coast of Mexico. This pattern, interrupted in southern California, is resumed again off the Oregon-Washington coast, seaward of the volcanic Cascade mountain range.

The volcanoes that rim the Pacific are active, flickering on and off like fireflies in the dark. Indeed, the Pacific Rim is often referred to as the *Ring of Fire.* Most of the world's volcanoes are confined to this ring, a few other regions of active mountain building, and the mid-ocean ridge system. Except for isolated hot spots such as the volcanic Hawaiian Islands, the vast areas in between are calm by comparison.

Over 99 percent of the world's earthquakes follow essentially the same narrow bands as the volcanoes (Figure 2.5). Earthquakes are triggered by the sudden slippage of rock along faults. At the central rift valley of the mid-ocean ridge, they occur in the uppermost crust. Those occurring landward of the deep-sea trenches are of a different nature, however, as Kiyoo Wadati, a Japanese seismologist, found in 1933. Wadati traced a narrow, slanting zone of earthquake activity beneath the Kurile trench north of Japan to depths of 720 kilometers into the mantle (Figure 2.6). Twenty years later, surveys by American seismologist Hugo Benioff confirmed Wadati's discovery, which is now known as the *Wadati-Benioff zone.* It is only beneath the world's trenches that such a pattern of earthquakes is found.

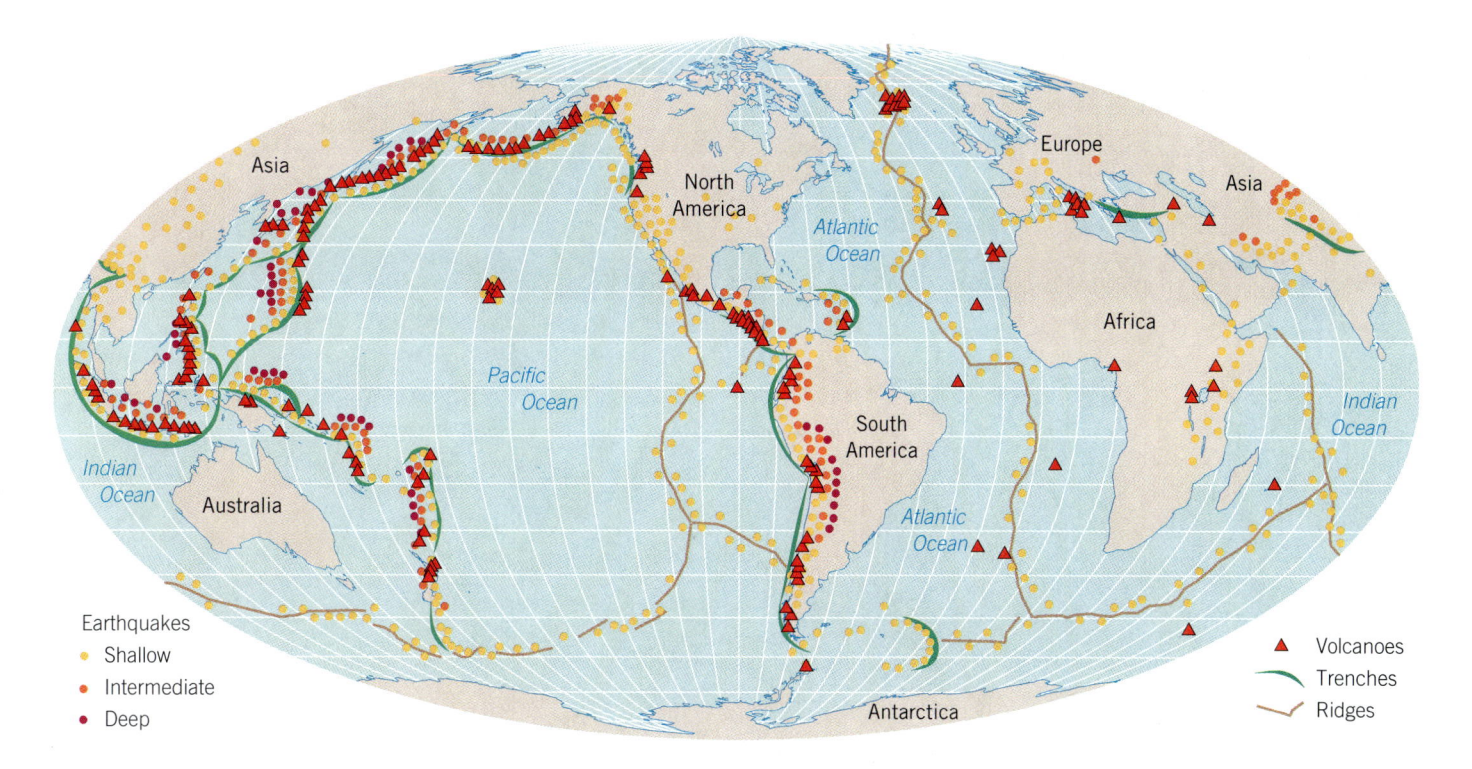

Plate Tectonics and the Ocean Floor   35

Earthquakes
- Shallow
- Intermediate
- Deep

▲ Volcanoes
⌒ Trenches
⌒ Ridges

Figure 2.5 Most earthquakes occur along deep-sea trenches, mid-ocean ridges, and regions of active mountain building. The mid-ocean ridges are sites of shallow-focus earthquakes exclusively. Intermediate- and deep-focus earthquakes are confined to zones beneath and landward of the trenches. Also, note the close association between the location of active volcanoes and earthquakes in the vicinity of the trenches.

The Sea-Floor Spreading Hypothesis

In the early 1960s, Harry Hess of Princeton University offered an elegant explanation for the apparent symmetry between mid-ocean ridges and deep-sea trenches—that is, hot magma rising at the ridge crests and cold oceanic crust sinking at the trenches. He proposed that, like a wound refusing to heal, oceanic crust separates along the rift valley in the mid-ocean ridge, allowing magma to bleed to the surface and form fresh oceanic crust. The crust then spreads laterally, cooling and contracting as it moves toward the trenches. There it descends back into the mantle and triggers the deep line of earthquakes beneath the trenches. Thus the ocean floor is in a state of slow motion—a vast conveyor belt. American geologist Robert Dietz, an early advocate and contributor to this concept, named the process **sea-floor spreading** (Figure 2.7).

Hess's idea was exciting, but it also seemed unreasonable to geologists schooled in the idea that the continents are locked permanently in place. After all, the mid-ocean ridge and the deep-sea trenches are typically separated by thousands of kilometers; in

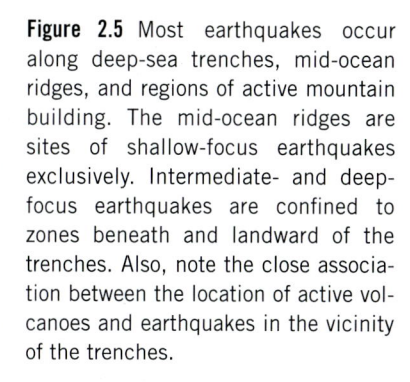

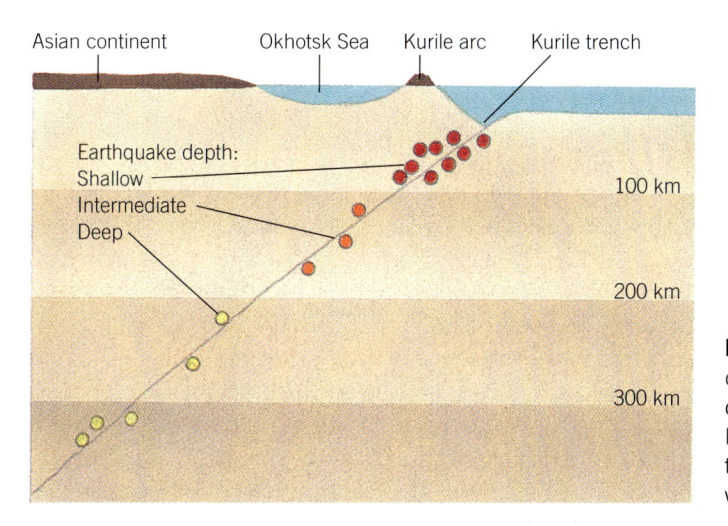

sea-floor spreading
The theory that new oceanic crust is created at the mid-ocean ridges, spreads laterally, and descends back into the mantle at the deep-sea trenches.

Figure 2.6 Shallow-, intermediate-, and deep-focus earthquakes occur on a deep-slanting plane in the Wadati-Benioff zone beneath the Kurile trench. The Kurile island arc is of volcanic origin.

Figure 2.7 A representation of the sea-floor spreading hypothesis. Newly created oceanic crust spreads from mid-ocean rift valleys as fresh magma wells up to replace it. Old oceanic crust descends into the mantle beneath deep-sea trenches. There it triggers deep-focus earthquakes and generates magma that feeds bordering volcanic arcs.

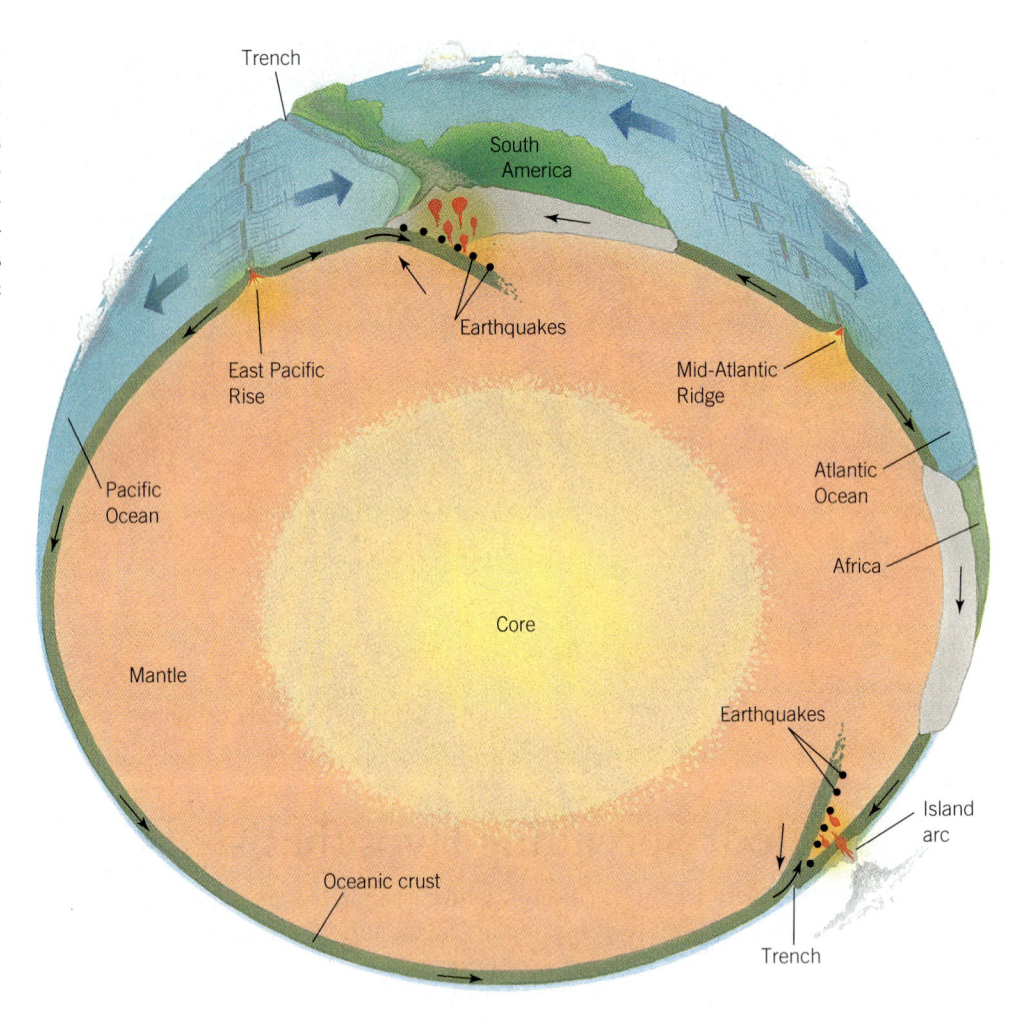

many parts of the world, entire continents intervene. More evidence was needed of the ocean floor between these features and of the mantle beneath the crust in order to test the spreading hypothesis.

What constitutes a valid test of this hypothesis? One approach is to reason as follows: If the sea floor is created at the mid-ocean ridges and spreads toward the trenches, then we should expect the age of the sea floor to increase steadily from ridge crest to deep-sea trench. On the other hand, if the ocean basins are locked permanently in place, then we should expect the sea floor to be the same age throughout. Because the sea floor presumably formed early in the history of the Earth, if it is locked in place, it should be billions of years old.

Evidence in Magnetic Reversals

Clearly, in order to test sea-floor spreading, geologists had to find a way to date the age of the ocean floor—no easy task considering its vast extent and depth. However, there was a solution to the problem, and it relates to a seemingly strange bit of evidence that British geologist Ronald Mason uncovered in the mid-1950s, when he had a magnetometer towed from a naval vessel off the California-Oregon coast. The magnetometer detected parallel bands of strong and weak magnetism in the basaltic ocean floor that run parallel to the axis of a mid-ocean ridge segment, the Juan de Fuca Ridge (Figure 2.8). The bands extend very far on both sides of the ridge until they are lost under sediment.

As we discussed in the previous chapter, lava preserves the direction of the Earth's magnetism at the time of eruption because tiny magnetic crystals act like compass nee-

Plate Tectonics and the Ocean Floor   37

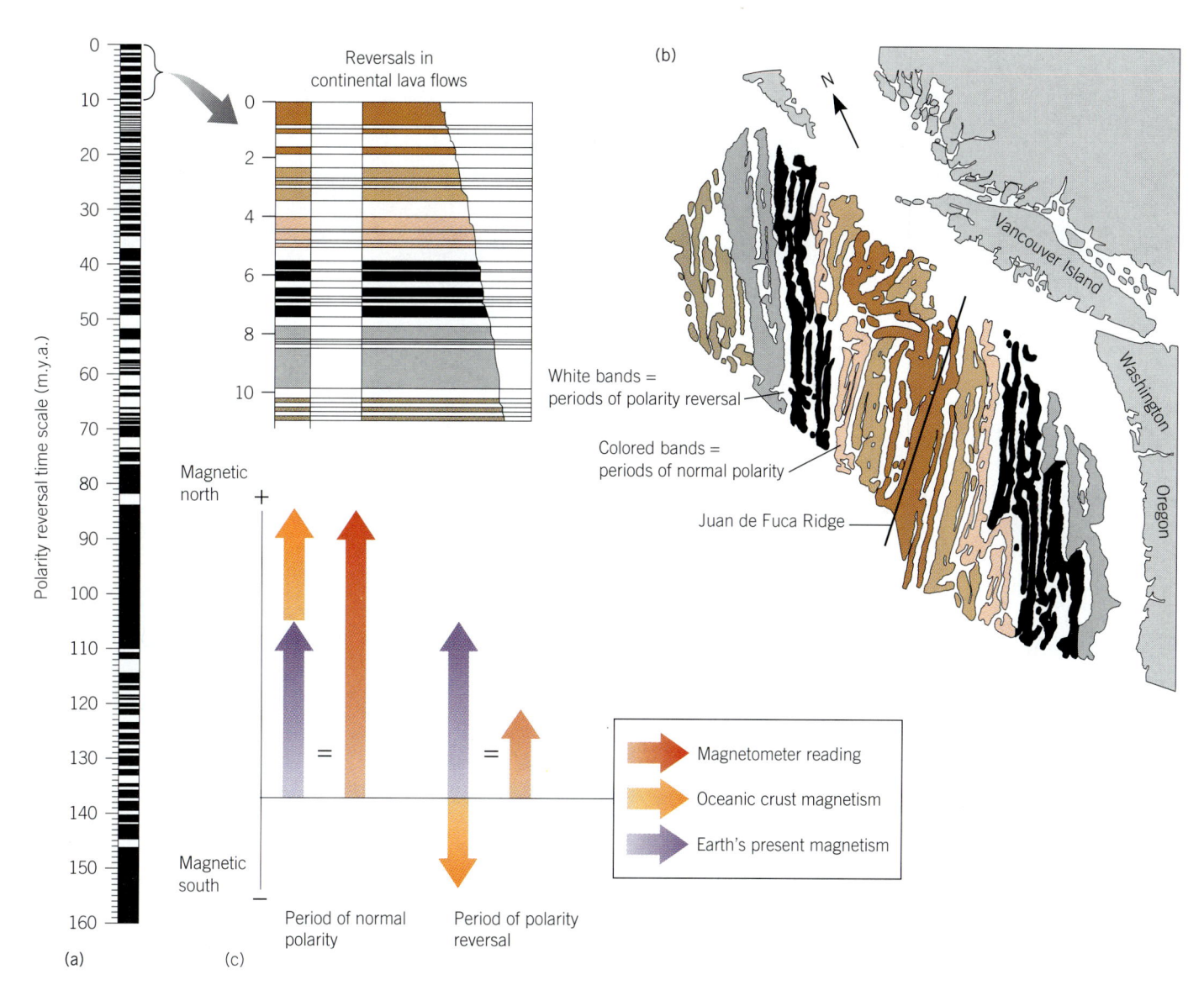

Figure 2.8 Magnetic polarity reversals are used to determine the age of oceanic crust. (a) Scientists detected a pattern of magnetic reversals in successive lava flows on land and dated them by radiometric methods, thus establishing the polarity reversal time scale. (b) In the Juan de Fuca Ridge, an identical pattern occurs on both sides of the central rift valleys of the mid-ocean ridges, with the youngest reversals closest to the ridge crests. (c) Determination of a polarity reversal. A magnetometer records a stronger-than-average reading where the magnetism of the oceanic crust adds to the polarity of Earth's magnetic field. A weaker-than-average reading occurs where sea-floor magnetism is reversed—that is, where it is opposite to the polarity of the Earth's field. (Adapted from H. W. Menard, *Marine Geology of the Pacific*, New York: McGraw-Hill, 1964.)

dles and are frozen in place when the basaltic lava solidifies. The magnetism that Mason measured is the sum of the Earth's present magnetism and the magnetism frozen into the oceanic crust at the time the lava erupted (see Figure 2.8). When the magnetic field of each is pointing in the same direction, the total magnetic field is stronger than normal; when the magnetic field of the oceanic crust is opposite to the polarity of the Earth's magnetic field, the total intensity is weaker than normal. In other words, the alternating bands of sea-floor magnetism are the record of past reversals of the Earth's magnetic field. If the sea-floor spreading hypothesis is correct, then the ages of the magnetic bands should increase steadily away from the mid-ocean ridge. However, to prove this hypothesis, geologists had to establish the ages of the **magnetic reversals.**

magnetic reversal
A switch in the direction of the Earth's magnetic field such that a compass needle that today points north would, at the time of reversal, have pointed south.

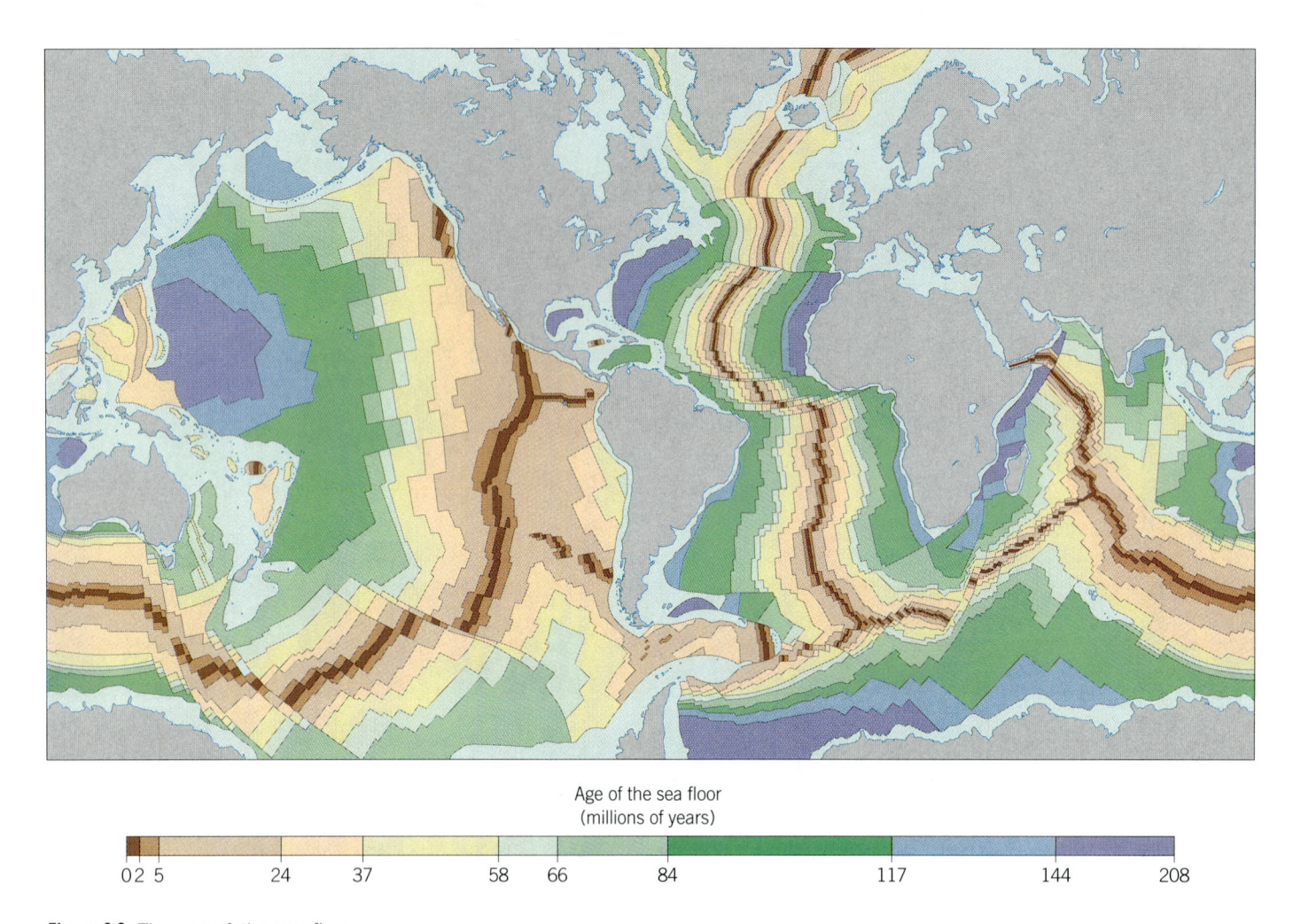

Figure 2.9 The age of the sea floor as determined primarily by polarity reversal data. Notice the similarity of the pattern in all the ocean basins. The Pacific basin is spreading at a faster rate than the Atlantic, as indicated by the broader pattern. (Adapted from R. L. Larson and W. C. Pitman, *The Bedrock Geology of the World,* New York: W. H. Freeman, 1985.)

When they made their discovery, Fred Vine was a young graduate student at Cambridge University and Drummond Matthews was his advisor.

With this purpose in mind, Alan Cox and his colleagues at the United States Geological Survey set out to measure the magnetism in lava layers of volcanoes along the Hawaiian coast. The volcanoes are constructed of thick layers of lava that record almost continuous eruptions going back millions of years; landslides have left a thick cross section of these layers exposed in the cliffs facing the ocean. The geologists climbed the cliffs and carefully sampled each layer, bottom to top. Back in the laboratory, the age of each sample was determined from the decay rates of radioactive atoms present, and its magnetization was identified by using a magnetometer. Combining the data, researchers were able to date the age of the magnetic reversals in the layers reaching back 4.5 million years. Later, other geologists would extend their data several hundred million years into the past by measuring the magnetism of older lavas in all parts of the world. The results are displayed as the *polarity reversal time scale* of Figure 2.8. Notice that the upper 10-million-year part of the scale matches perfectly the reversal pattern of the Juan de Fuca Ridge.

Now, refer to Figure 2.9 and notice that the irregular, distinctive pattern of the polarity reversal time scale is reproduced in uncanny detail by the magnetic bands on either side of the worldwide mid-ocean ridge. The match is perfect; the most recent reversals occur at the mid-ocean ridge, and progressively older ones are found progressively farther away toward the flanks of the ocean basins. Reversals of short duration are matched by relatively narrow bands, and those of long duration by broad bands.

In the early 1960s, British geologists Fred Vine and Drummond Matthews realized that there is only one way to explain this correspondence. As lava bleeds out of the rift valley of the mid-ocean ridge, it acquires the polarity of the Earth's magnetic field at the time of eruption (Figure 2.10). Upon solidification, the lava forms oceanic crust, which is moved aside as younger lava wells up from beneath the rift valley. In this manner, subsequent magnetic reversals are then recorded in fresh lava. Therefore, each segment of the oceanic crust was originally formed at the rift valley and has moved gradually away from the mid-ocean ridge, bearing the magnetic polarity of its time of solidification.

Referring again to Figure 2.9, we see from the magnetic reversal pattern that the age of the Pacific Ocean floor increases steadily from the mid-ocean ridge to the deep-sea trenches on either side of the ocean basin, where the pattern ends abruptly. This evidence lends strong support to the idea that the spreading ocean floor descends beneath the trenches. Thus geologists came to visualize the oceanic crust spreading from the ridge crests as a very wide, very slow-motion conveyor belt—or better yet, because polarity reversals are imprinted on it, as a gigantic tape recording.

We can calculate the velocity of sea-floor spreading by measuring the distance from the ridge crest to a given magnetic band and then dividing it by the age of that band as determined by its position in the polarity reversal time scale. Notice in Figure 2.9 that spreading rates vary considerably. The Atlantic Ocean floor is spreading from the Mid-Atlantic Ridge about as fast as your fingernails grow. The Pacific Ocean floor is spreading from the East Pacific Rise at up to ten times that speed. The more rapid Pacific spreading rate is reflected in the wider magnetic bands across the Pacific floor.

Evidence in Sea-Floor Sediment

Studies of sea-floor sediments provide an independent check of the sea-floor spreading hypothesis. To collect these samples, a consortium of research institutions launched the Joint Oceanographic Institution's Deep Earth Sampling project (JOIDES). At the heart of the project was the *Glomar Challenger,* a vessel designed specifically to recover cores of sediment and basaltic crust from the deep ocean floor. Its six independent engines, coordinated by computer, kept the ship on target over kilometers of ocean while the crew drilled through thousands of meters of marine mud, sand, gravel, and rock. The geologists, rotating 24-hour shifts aboard the *Glomar Challenger,* sampled the world's ocean floors and discovered a remarkably consistent pattern.

Toward the margins of the continents, the oceanic crust is covered by a wedge of sediment thick enough to bury the irregularities of the ocean floor and produce a flat, nearly featureless **abyssal plain.** However, the wedge gradually tapers off toward the mid-ocean ridge, where the rocks of the crust are bare—save for thin patches of recent plankton shells that have settled out of the ocean above the ridge. Furthermore, as we approach the mid-ocean ridge, progressively younger sediment is found in contact with basaltic crust of the same age. Both the thinning of the sediment and the age pattern are to be expected if the sea floor is spreading. As new crust forms along the mid-ocean ridge, older crust and sediment are moved aside, and the younger sediment blankets not only the older sea floor but also the newly exposed crust. One of the triumphs of the *Glomar* expeditions was the discovery that the ages of the oldest parts of the Atlantic Ocean basin—that is, its flanks—match the time of fragmentation of Wegener's Pangaea.

Another significant point concerns the sediment found on the floor of the western Pacific. At 200 million years of age, it is the oldest known sediment of the ocean and rests upon the oldest known oceanic crust, also 200 million years old—which represents a mere 5 percent of the Earth's 4.6-billion-year age. Thus, surprisingly, the present ocean basins—which compose 60 percent of the Earth's surface—have been in existence for less than 5 percent of geologic time. The ocean basins are young features that have been cycled and recycled many times in the past.

Evidence in Sea-Floor Topography

Sea-floor spreading offers a plausible explanation for many formerly puzzling aspects of ocean topography, including the fact that depths increase with distance from the mid-ocean ridge crests. The mid-ocean ridge stands high above the general level of the ocean

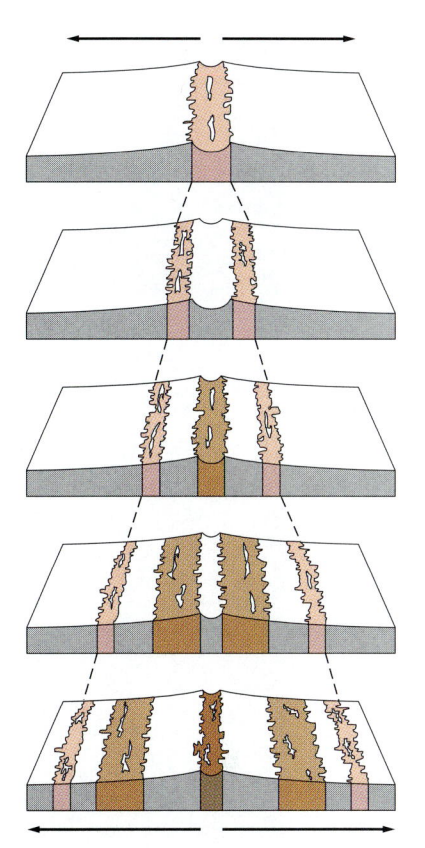

Figure 2.10 The explanation of the polarity reversal pattern supports the theory of sea-floor spreading. As oceanic crust spreads away from the ridge crest, younger oceanic crust is created by the upwelling of fresh magma from beneath the rift valley. Thus the age of the sea floor increases steadily away from the ridge crest.

abyssal plain
A flat, level area of the ocean floor that begins at the foot of the continental rise.

Figure 2.11 The Kagangel Atoll, Belau, is a coral reef surrounding a lagoon. The reef is anchored to a submerged volcanic island.

floor because it is warmer, which makes it less dense and more buoyant. As oceanic crust spreads from the central rift valley, it cools, grows denser, and sinks lower into the mantle. For this reason, the mountainous topography of the mid-ocean ridge does not extend across the entire ocean floor. However, like a hot skillet just removed from the stove, cooling is most rapid in the portion of the crust that has most recently moved away from the heat source below the rift valley. The cooling rate then levels off with time—that is, with distance from the ridge crest. These differences in cooling rates are reflected in the profile of the ocean floor: steep at the mid-ocean ridge and gentle at the margins of the basins. In general, the depth of each location on the ocean floor is determined by its age. Differences in cooling rates are also responsible for the subtle contrast between the topography of the Atlantic and Pacific floors across their respective mid-ocean ridges. The slow-spreading Atlantic basin has had more time to deepen; hence, the profile of the Mid-Atlantic Ridge is generally steep and rugged. On the other hand, the Pacific crust, spreading more rapidly from the East Pacific Rise, has had less time to cool; its broader, gentler profile is the reason it is called a "rise."

The ocean floor is pockmarked with thousands of volcanoes, and their distribution and depth confirm the effects of sea-floor spreading on sea-floor topography. Most are **seamounts**—recognized by their cone-shaped cross sections—that grew to their maximum height without ever reaching the surface. Other undersea volcanoes were once islands, as shown by cores containing ancient coral reefs, beach gravel, and shallow-water fossils. Wave erosion has planed the tops of these **guyots** so thoroughly that their older name *tablemount* is descriptive. Many are currently thousands of meters below sea level; their submergence offers strong evidence that the sea floor sinks as it spreads.

Charles Darwin made some shrewd observations concerning the origin of the Pacific Islands when he visited them while employed as a naturalist aboard the *HMS Beagle* in the 1830s (Figure 2.11). He noted that a newly emergent volcano will acquire a *fringing*

seamount
A sea-floor volcano that has never risen above sea level.

guyot
A sea-floor volcano that at one time rose above sea level, where its top was planed by wave erosion.

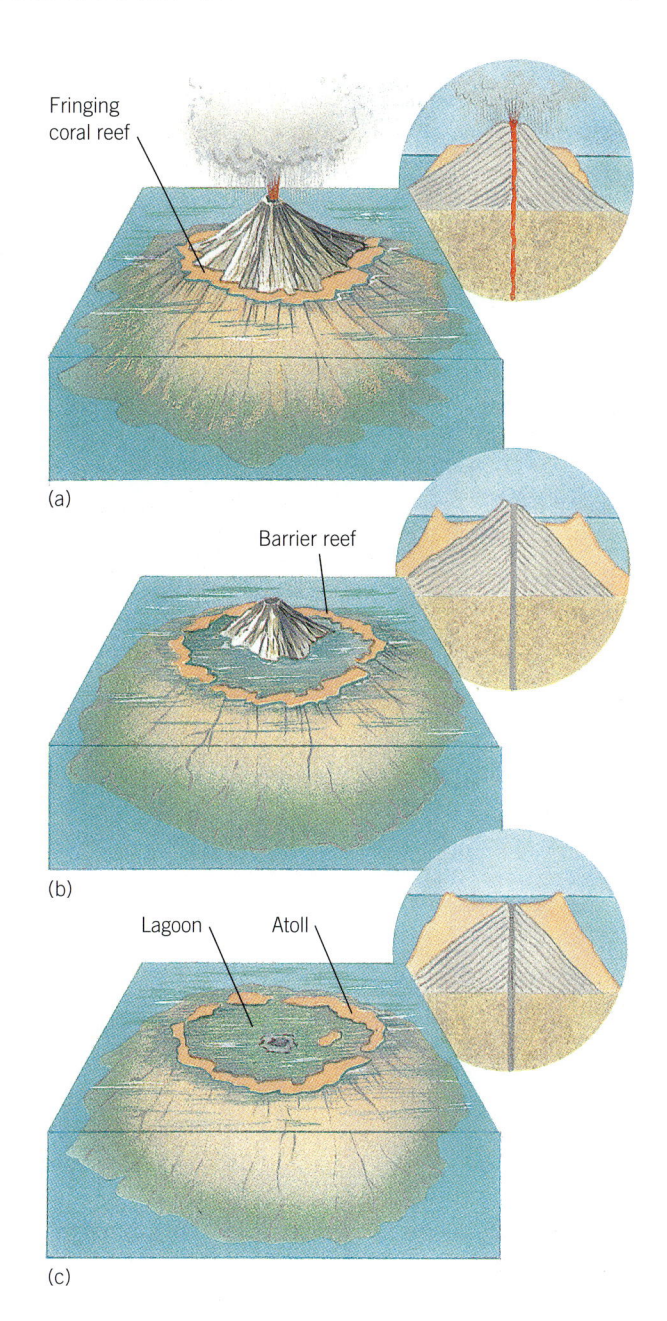

Figure 2.12 Darwin's subsidence theory of Pacific island coral reefs. As a volcanic island sinks, the attached coral continues to grow upward toward the sunlight at sea level, and, in the process, evolves from a fringing reef to a barrier reef to an atoll.

reef that grows directly offshore. But as the volcanic island sinks, the reef—needing sunlight—continues to grow up to sea level. In time, it evolves into a barrier reef, with a lagoon between the reef and the island. The last stage is a roughly circular-shaped *atoll* encircling the submerged volcano, which twentieth-century surveys would show to be a guyot (Figure 2.12).

In the 1880s, American geologist James Dana surveyed another group of Pacific islands, the Hawaiian chain. There he observed that active or recent volcanism was confined to the islands of Hawaii and Maui. Also, the more northwestward the islands that he visited, the more eroded their volcanoes were. He therefore concluded that the ages of the islands increase from southeast to northwest; radiometric dating of the island lavas in the mid-twentieth century bore out his conclusion. At about the same time, oceanographic surveys discovered that the Hawaiian Islands are the most visible of a 7000-kilometer-long line of seafloor volcanoes. The line extends from the active Loihi Seamount off the southeast coast of Hawaii and northwest to Midway Island; then it bends northward as the Emperor Seamount chain toward the Aleutian trench (Figure 2.13). The ages of the volcanoes increase steadily from Hawaii to the trench. The oldest of them last erupted 65 million years ago.

Significantly, of this long chain, only the southeasternmost Hawaiian volcanoes are active. They are situated over a magma reservoir 150 kilometers within the mantle. If the sea-floor spreading hypothesis is correct, then the oceanic crust now at the Aleutian trench was created at a mid-ocean ridge far to the southeast and passed over the spot now occupied by Hawaii. For this reason, Canadian geologist J. Tuzo Wilson proposed in 1963 that the volcanoes of the Hawaiian-Emperor Seamount chain were punched out as the Pacific oceanic crust passed over the same magma source that spews out the lava currently building Hawaii. Wilson called this long-active region of volcanic activity a **hot spot**. Figure 2.14 shows that many hot spots have since been identified; they are distributed throughout the continents and ocean basins, although they are heavily concentrated in Africa.

J. Tuzo Wilson of the University of Toronto was an especially insightful geoscientist who contributed significantly to many aspects of the theories of sea-floor spreading and plate tectonics.

hot spot
An area of volcanic activity produced by a plume of magma rising from the mantle.

Plate Tectonics

By the early 1960s, geologists working primarily on land had assembled overwhelming evidence of the essential correctness of Alfred Wegener's theory that the continents had for a time existed as a single landmass and have since fragmented and drifted to their present positions. At about the same time, marine geologists had assembled overwhelming evidence of sea-floor spreading. What is more, it was obvious from the timing and geometry of drift and spreading that the two processes are related. The pursuit of an explanation for this relationship led to a new synthesis, **plate tectonics,** which, as we will see, goes far beyond merely combining drift and spreading to offer a comprehensive theory of Earth's crustal structure.

plate tectonics
The theory that proposes that the lithosphere is divided into plates that interact with one another at their boundaries, producing tectonic activity.

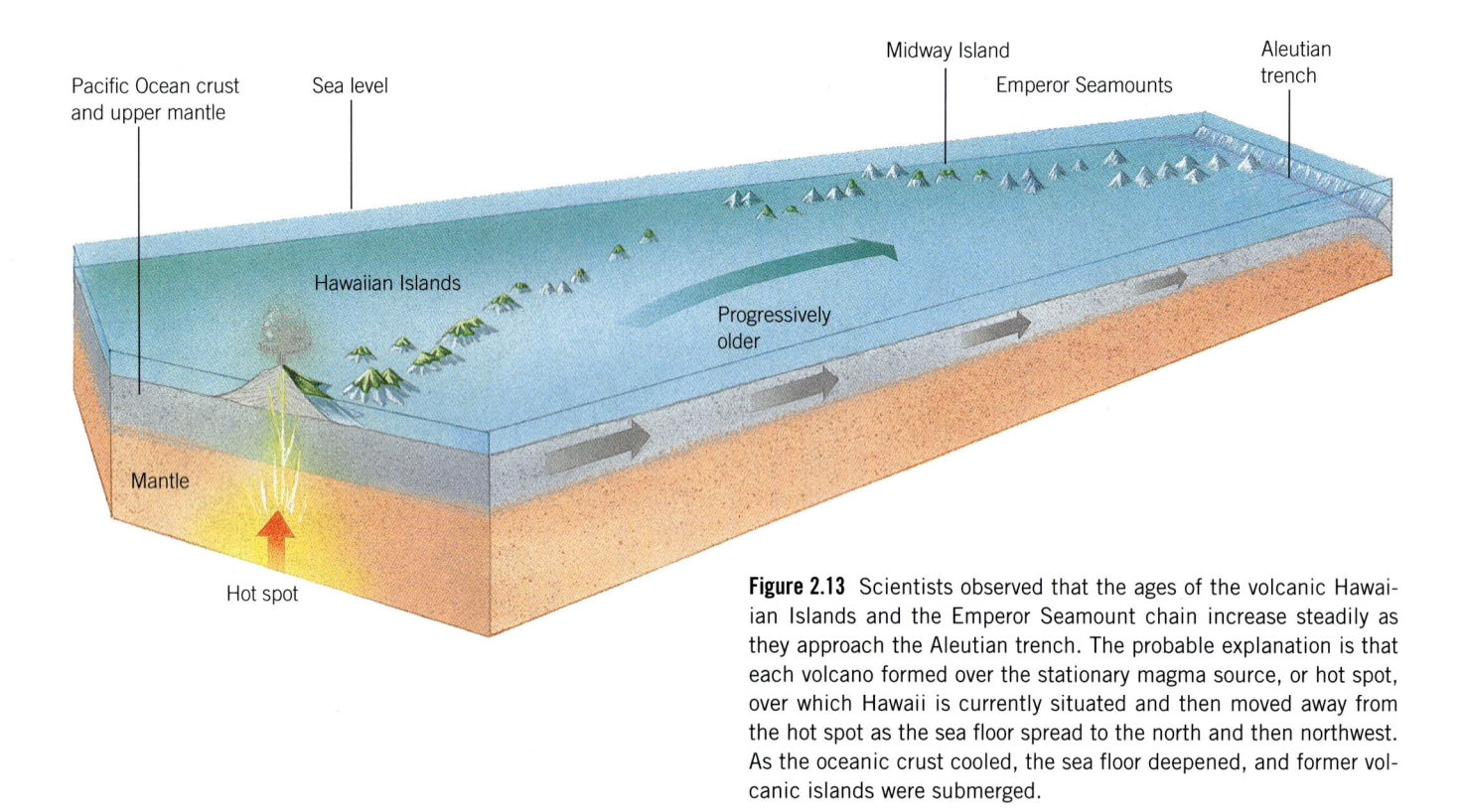

Figure 2.13 Scientists observed that the ages of the volcanic Hawaiian Islands and the Emperor Seamount chain increase steadily as they approach the Aleutian trench. The probable explanation is that each volcano formed over the stationary magma source, or hot spot, over which Hawaii is currently situated and then moved away from the hot spot as the sea floor spread to the north and then northwest. As the oceanic crust cooled, the sea floor deepened, and former volcanic islands were submerged.

Confirmation of sea-floor spreading established that new oceanic crust is created at the mid-ocean ridge by magma erupting from beneath the central rift valley floor, and that the new crust is then moved as if it were on a conveyor belt to the deep-sea trenches, where it descends back into the mantle. Moreover, these same features—trenches and mid-ocean ridge—coincide with the major earthquake or **seismic belts;** that is, the earthquakes trace a network of stress lines around the globe. Furthermore, geologists realized that the crust within the seismic belts must behave rigidly; if the sea floor behaved like soft clay, then the parallel bands of alternating magnetism would deform beyond recognition on their journey from ridge crest to trench.

In 1968, a handful of geologists, independently of one another, proposed the theory of plate tectonics: that the regions encircled by seismic belts are rigid slabs, or **plates,** and that the earthquakes are generated when these plates grind against one another as they move. According to the theory, the seismic belts—that is, the blizzard of dots confined to the narrow strips shown in Figure 2.5—delineate plate boundaries. These boundaries segment the Earth's surface into a mosaic of about seven or eight large plates and many smaller ones (Figure 2.15). Some, such as the Nazca and Pacific plates, contain oceanic crust exclusively. Most of the others include continental as well as oceanic crust; as these plates spread away from the mid-ocean ridge and toward the trenches, the continents move along with them. So the continents really do drift—but passively, embedded within the moving plates, and not like ships plowing through oceanic crust, as Alfred Wegener had envisioned.

How thick are the plates? The answer to this question must be found indirectly through analysis of seismic waves, for we lack the technology to drill deep into the Earth. The velocity of a seismic wave is determined by the properties of the substance through which it travels; the more rigid the substance, the higher is its velocity. Figure 2.16 (on page 46) shows that wave velocities increase to a depth of about 100 kilometers, confirming that this outer zone of the Earth is quite rigid. This depth marks the base of the **lithosphere,** which is composed of the crust and a thin layer of the upper mantle. Its thickness amounts to a sixty-fourth of the radius of the Earth, so on the scale of Earth's dimensions, it is a mere eggshell. The lithosphere is divided into the mosaic of litho-

---

seismic belt
A long, narrow zone of earthquake activity associated with lithospheric plate boundaries.

plate
A rigid segment of the lithosphere that moves as a unit over the asthenosphere.

The scientists credited with establishing the plate tectonics model include Dan McKenzie of Cambridge University and Robert Parker of the Scripps Institute of Oceanography; Jason Morgan of Princeton University; Xavier LePichon of Lamont-Doherty Geological Observatory; and Don Isacks, Jack Oliver, and Lynn Sykes, also of Lamont. At the time, they were all in their twenties or early thirties and were either graduate students or recent Ph.D.s.

lithosphere
The rigid outermost layer of the Earth, which includes the crust and a sliver of the upper mantle.

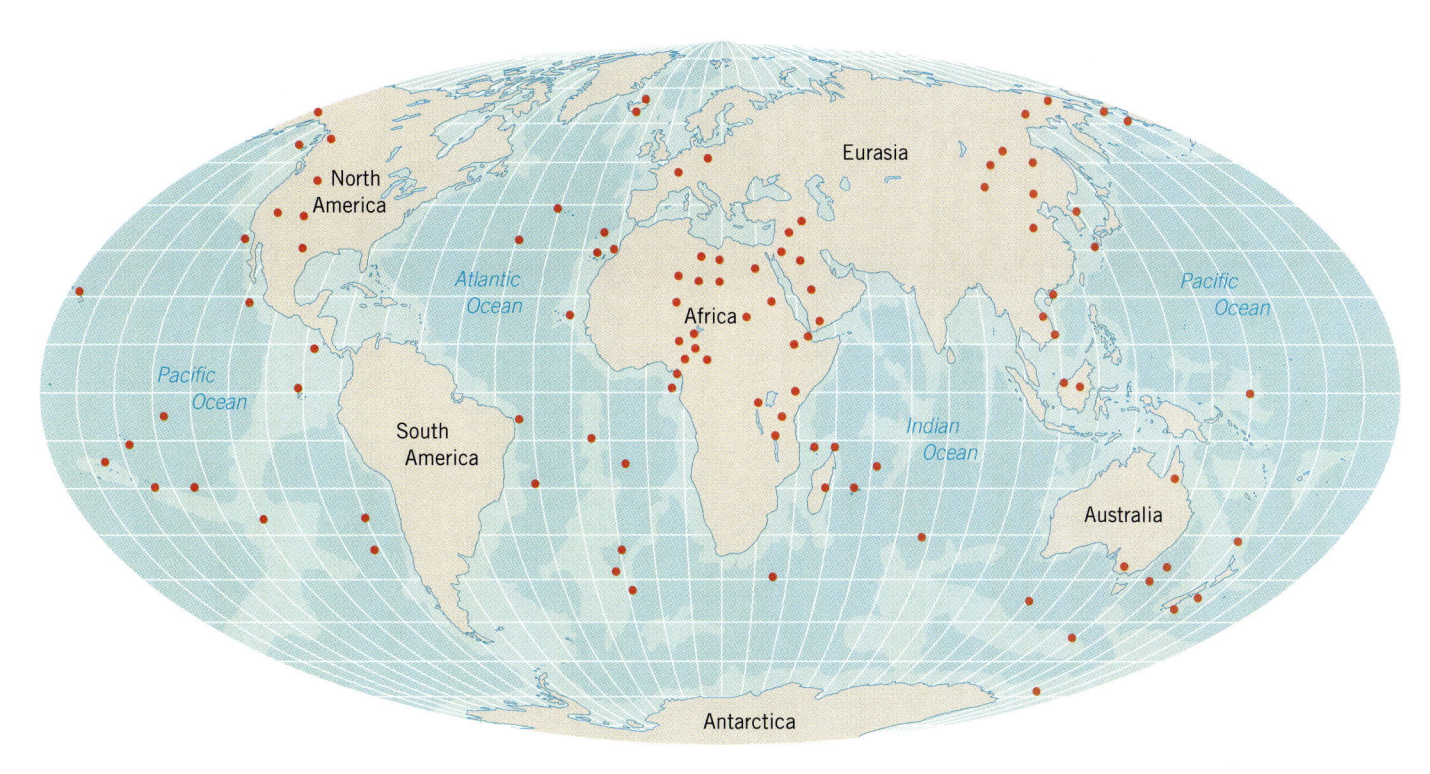

Figure 2.14 The distribution of some prominent hot spots.

spheric plates that constitute Earth's outer surface; these plates are what geologists refer to in discussing plate tectonics. Seismic-wave velocities decrease markedly beneath the lithosphere and do not recover until they reach 250 kilometers, which suggests a zone in the mantle that is close to melting and has lost some of its rigidity. The plates glide over, or are carried along on, this soft yielding zone in the mantle, the **asthenosphere.**

Up to this point, we have said a great deal about the *plate* in plate tectonics but have neglected the latter term. *Tectonics* is derived from the same Greek root as architecture, *tekton*, meaning "to construct." It refers to those processes responsible for the origin of the large-scale structural elements of the crust. Thus plate tectonics is a theory stating that the Earth's surface is divided into lithospheric plates and that the interactions occurring at plate boundaries are responsible for the production and destruction of oceanic crust; the rifting, drifting, and collision of continents; the formation of mountains, volcanoes, and earthquakes; and more.

Because the plates are rigid slabs packed tightly together on the surface of a sphere, there are only three ways they can interact (Figure 2.17 on page 46): they may diverge, converge, or slide by one another. The narrow zones where these interactions occur constitute the three principal types of plate boundaries, each displaying characteristic landform features, earthquake patterns, and igneous activity. **Divergent boundaries,** where the plates separate and grow, are characterized by mid-ocean ridge rift valleys and their continental extensions. **Convergent boundaries,** where the plates collide and cycle lithosphere back to the mantle, are characterized by deep-sea trenches, volcanic arcs, and fold-mountain belts. **Transform boundaries,** where the plates slide by one another, are characterized by narrow fracture zones that connect offset segments of the mid-ocean ridge or deep-sea trenches.

Figure 2.15 (on pages 44–45) According to the theory of plate tectonics, the Earth's surface is divided into a mosaic of lithospheric plates whose boundaries are delineated by narrow seismic belts. Some of the plates consist solely of oceanic crust; others include continental crust as well. The arrows show that the plates move away from the ridge crests and toward the trenches. (Adapted from "This Dynamic Planet: World Map of Volcanoes, Earthquakes, and Plate Tectonics," by Tom Simkin and Robert I. Tilling of the Smithsonian Institution and James N. Taggart, William J. Jones, and Henry Spall of the U.S. Geological Survey, 1989.)

asthenosphere
A semimolten zone of the mantle just below the lithosphere.

divergent boundary
The border between two plates that are moving away from one another as new crust is formed.

convergent boundary
The border between two plates that are colliding with one another. Crustal material is subducted back into the mantle at these boundaries.

transform boundary
The border between two plates that are sliding by one another horizontally without either creating or destroying oceanic crust. It connects offset ridge segments, offset trenches, or ridge segments to trenches.

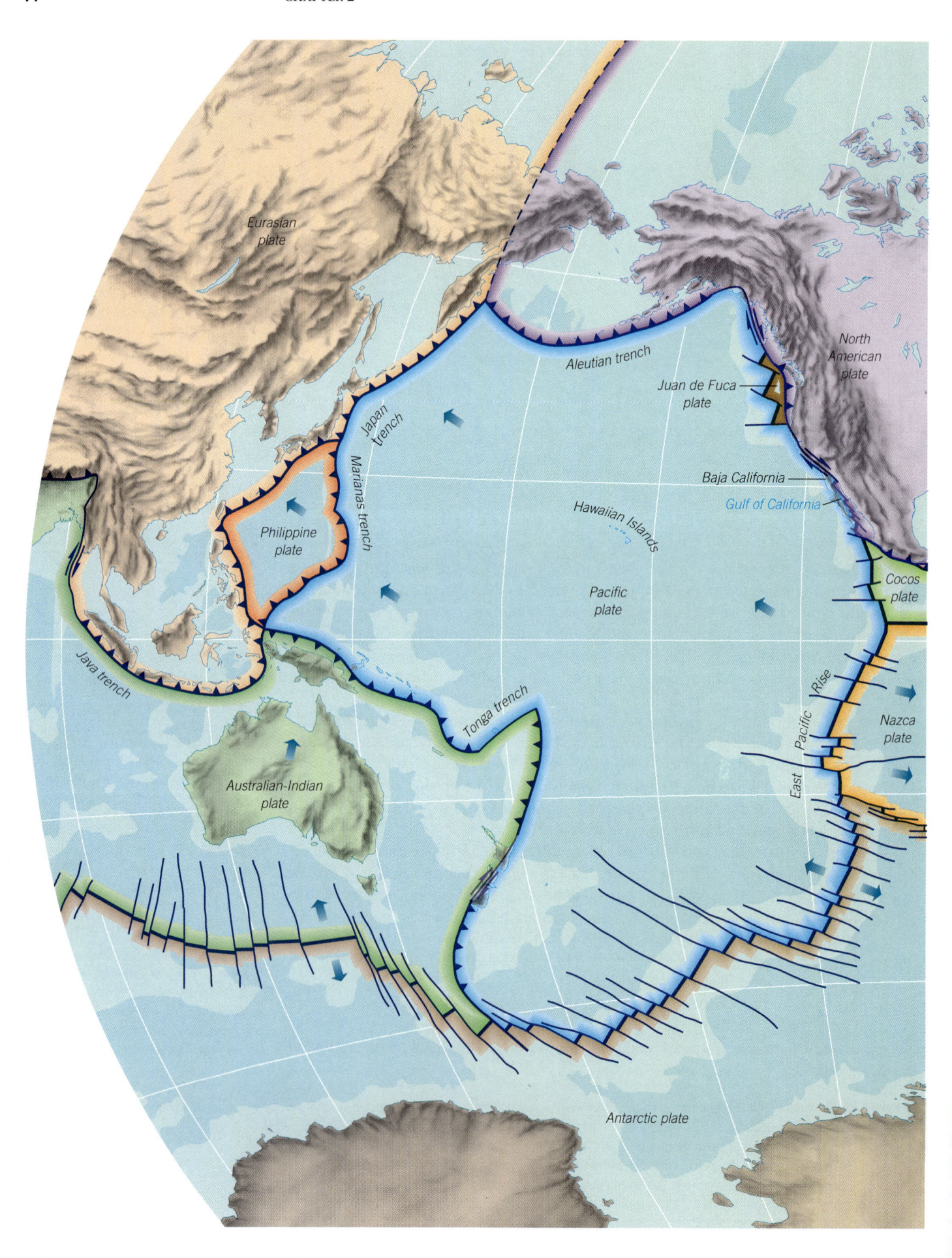

CHAPTER 2

Plate Tectonics and the Ocean Floor 45

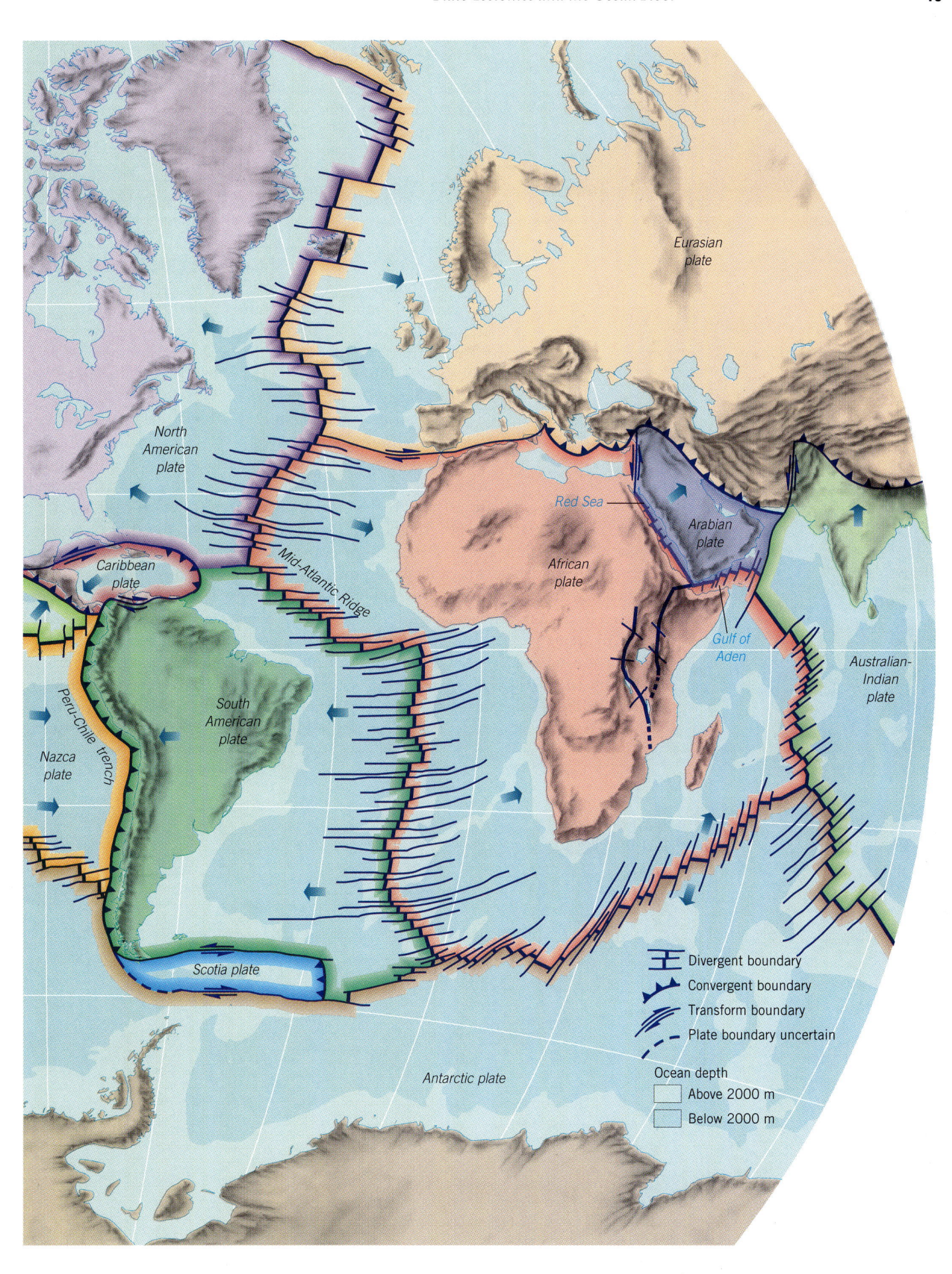

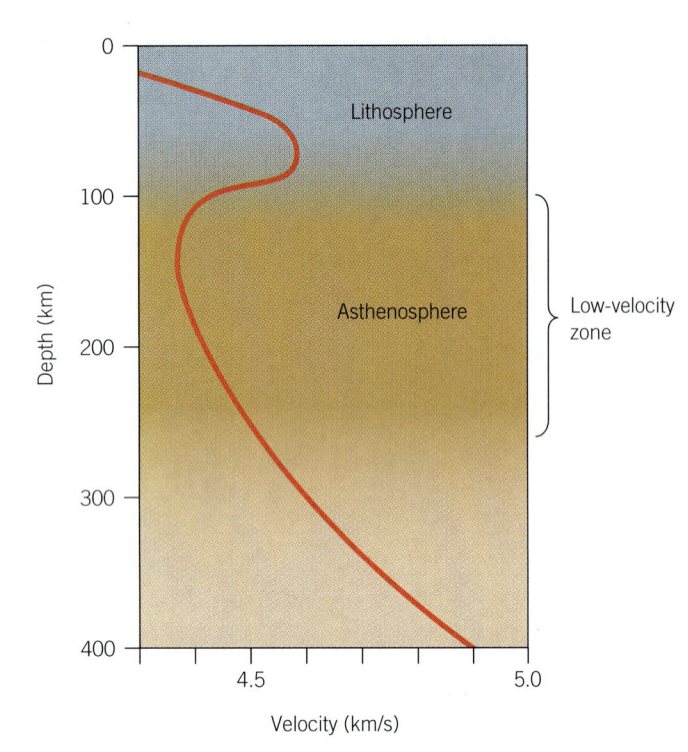

Figure 2.16 The abrupt change in seismic velocities at depths of about 100 kilometers marks the boundary between the lithosphere and the asthenosphere, over which the plates slide.

Divergent Boundary Processes

The narrow band of earthquakes that trace the elevated central rift valley of the mid-ocean ridge and its continental extensions defines the divergent boundaries of the world. The earthquakes there are shallow, and analysis of the associated fault motions that trigger them indicates that the plates are under tension along the boundary. This tension is responsible for the rift valley, which has dropped downward along steep faults as the plates on either side are pulled apart. The earthquakes are triggered when the rocks, restrained on either side of the fault plane by friction, suddenly break free.

Divergent boundaries, the sites of simultaneous plate separation and growth, are also referred to as *spreading centers.* As the plates move apart, magma that wells up from the partially molten asthenosphere beneath the central rift valley is accreted to the trailing edges of the plates. In this way, the plates widen in parallel strips at the rate of 2 to 10 centimeters per year as they diverge from the ridge crest.

If new oceanic crust is being created at divergent boundaries, then the continents on either side have been separating for as long as these basins have been in existence. Thus, for example, North America, South America, Europe, and Africa were joined prior to the opening of the Atlantic Ocean. Hence, the hypothesis of continental drift, so long the subject of intense debate, fits easily within the framework of plate tectonics.

It follows, therefore, that the divergent boundary that today is the Mid-Atlantic Ridge began as a fracture in continental crust. Indeed, we can observe what may be a very early stage of the same process at the Afar Triangle, where northeast Africa and the Arabian Peninsula fit hand-in-glove. Notice that three rift zones meet there and form a *triple junction* (Figure 2.18). Two of them, the Red Sea and Gulf of Aden, are continuous with the volcanic Carlsberg Ridge that splits the Indian Ocean; the third, the East African Rift Valley, intersects continental crust. The triple junction is in fact located over a hot spot, a point

(a) Divergent boundary   (b) Transform boundary   (c) Convergent boundary

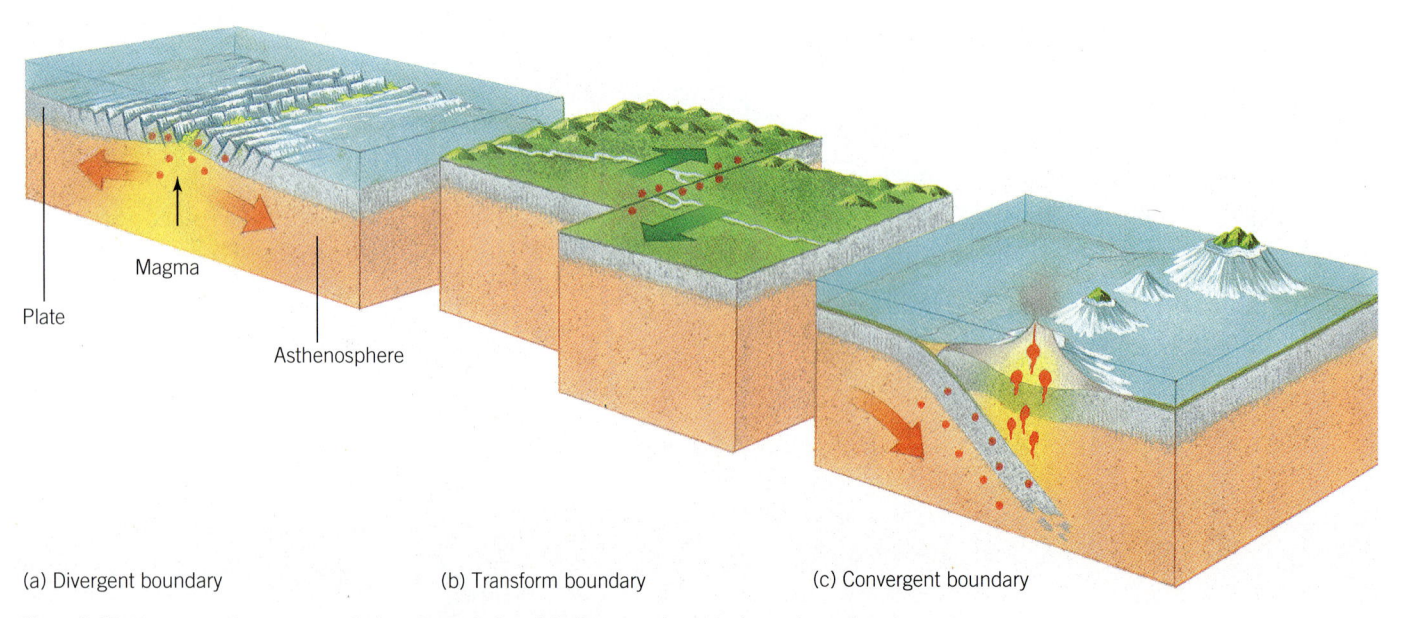

Figure 2.17 There are three types of plate boundaries: (a) divergent boundaries, where the plates are created and spread away from a mid-ocean ridge; (b) transform boundaries, where the plates slide by one another horizontally; and (c) convergent boundaries, where the plates collide, causing one to descend into the mantle beneath the deep-sea trench.

Plate Tectonics and the Ocean Floor   47

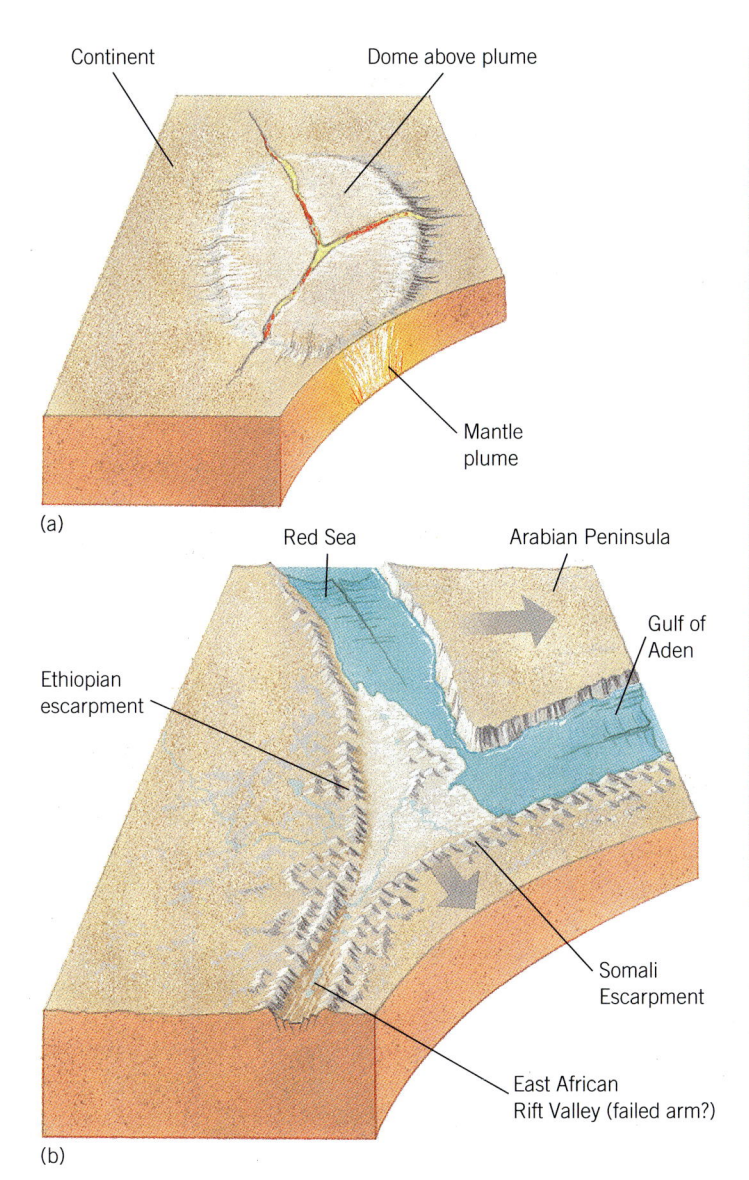

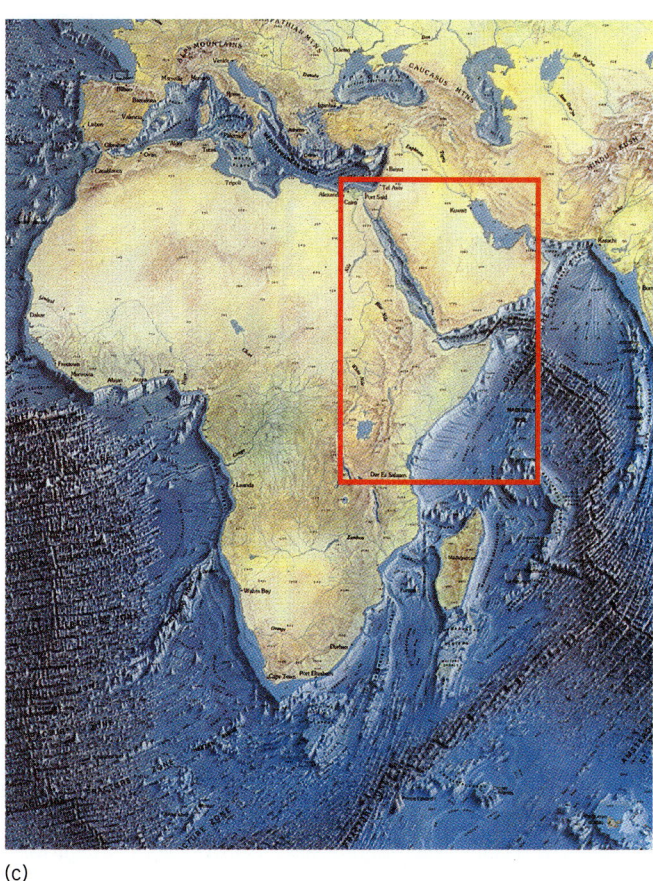

Figure 2.18 Hypothesis of how a divergent plate boundary develops. (a) A rising mantle plume beneath the continental crust forms a dome with three branching fractures (a triple junction). (b) The fractures spread apart, forming narrow rift valleys that widen and deepen far enough for magma to well up from the mantle and form oceanic crust. The third rift does not completely separate the continental crust; instead, it remains a failed arm. (c) This detail of Marie Tharp's model shows the triple-junction relationships among the Red Sea, the Gulf of Aden, and the East African Rift Valley. Notice the connection of the triple junction to the Carlsberg Ridge, the segment of the mid-ocean ridge directly to the east. (Marie Tharp)

of intense volcanic activity formed by a plume of magma and hot rock rising from the mantle. Earlier, we described how these plumes of magma punch out volcanoes on the thin, moving oceanic crust; but they may act differently on thicker, stationary continental crust.

Geologists believe that about 15 million years ago, a plume that rose beneath what is now the Afar Triangle heated and stretched the overlying continental crust until it fractured in three directions. The continents separated cleanly along two of the fractures, and basaltic magma welled up from the mantle between the continental blocks, forming denser oceanic crust. Eventually, the Indian Ocean flooded these canyons to create what are now the Red Sea and Gulf of Aden. Today, these narrow seaways are volcanically active spreading centers, and they continue to widen with the passage of time.

Volcanic and seismic activity are less frequent along the third branch of the Afar Triangle, the East African Rift Valley, which is still underlain by continental crust. Evidently, the East African rift is not as fully developed as the Red Sea and Gulf of Aden rifts. It may well remain as a deep, linear trough or *failed arm* in the African continental crust.

The Atlantic Ocean of today is in a mature stage of development. It is characterized by a wide ocean basin between separated continents and by **passive continental margins** that are nearly free of earthquakes and igneous activity because they are far

passive continental margin
A margin between an ocean and a continent that does not include a plate boundary. It is also free of volcanic or seismic activity.

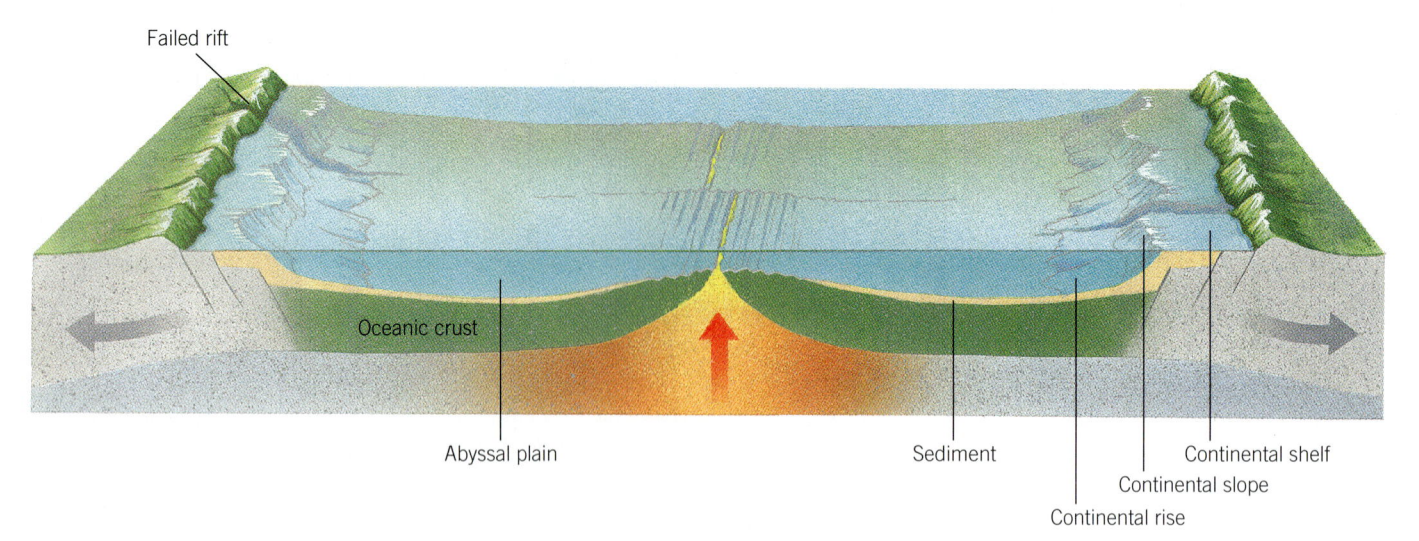

Figure 2.19 Late-stage development of an ocean basin from a triple junction. The ocean basin deepens and widens and elaborate passive margins are built of thick sediment wedges deposited on the faulted oceanic crust. Rivers flowing to the sea may follow the route of ancient failed arms.

continental shelf
A very gently sloping surface that extends from the shoreline to the continental slope.

continental slope
A more steeply sloping surface that extends from the continental shelf to the continental rise or oceanic trench.

continental rise
That part of the continental margin that extends gently downward from the continental slope to the abyssal plain.

subduction
The downward plunging of one lithospheric plate under another.

from plate boundaries (Figure 2.19). The old oceanic crust near the continental margins has cooled and deepened, and the adjacent continental crust has itself subsided well below sea level. The continent-ocean boundary, now stable for many millions of years, has accumulated thousands of meters of land-derived sediment, which thins toward the mid-ocean ridge because of sea-floor spreading.

The wide passive continental margin, built of this sediment brought down from the continents, includes an extensive, shallow **continental shelf** and a broad **continental slope** that leads down to the deep water (see Figure 2.19). An apronlike **continental rise** fringes the base of the margin and trails off into the flat abyssal plains that smother the irregularities of the basaltic oceanic crust. These margins, located on the trailing edges of the separating continents, retain vestiges of the rift valleys, lavas, and continental sediments associated with the ancient separation of the continents.

You may have noticed that the discussion of plate divergence has been confined mainly to the Atlantic Ocean, which opened up as Pangaea was rifted apart. However, not all plates are created by the separation of continents. For example, those spreading from the East Pacific Rise may have formed independently of continental crust.

Convergent Boundary Processes

The narrow bands of earthquakes that occur along deep-sea trenches delineate most of the world's convergent boundaries. These boundaries are the sites where two plates, spreading toward one another from the mid-ocean ridges, collide. Where they meet, the denser, colder plate bends and descends into the mantle in a process called **subduction,** meaning "to pass under." The descent of the subducting plate is traced by the curving Wadati-Benioff zone of earthquakes deep into the mantle. *Subduction zones* are areas where the plates that are created at the mid-ocean ridge are eventually destroyed or, more precisely, recycled into the mantle.

The most numerous and varied examples of convergence are located along the margins of the Pacific and Indian oceans. There we are able to distinguish three convergence modes based on the nature of the lithosphere involved: *ocean-ocean*, *ocean-continent*, and *continent-continent*.

Ocean-Ocean Convergence

Ocean-ocean convergence produces island arc and deep-sea trench systems, of which there are about twenty, mainly in the western Pacific. No two are identical, but the Japan arc displays features and processes common to most of them (Figure 2.20).

Plate Tectonics and the Ocean Floor

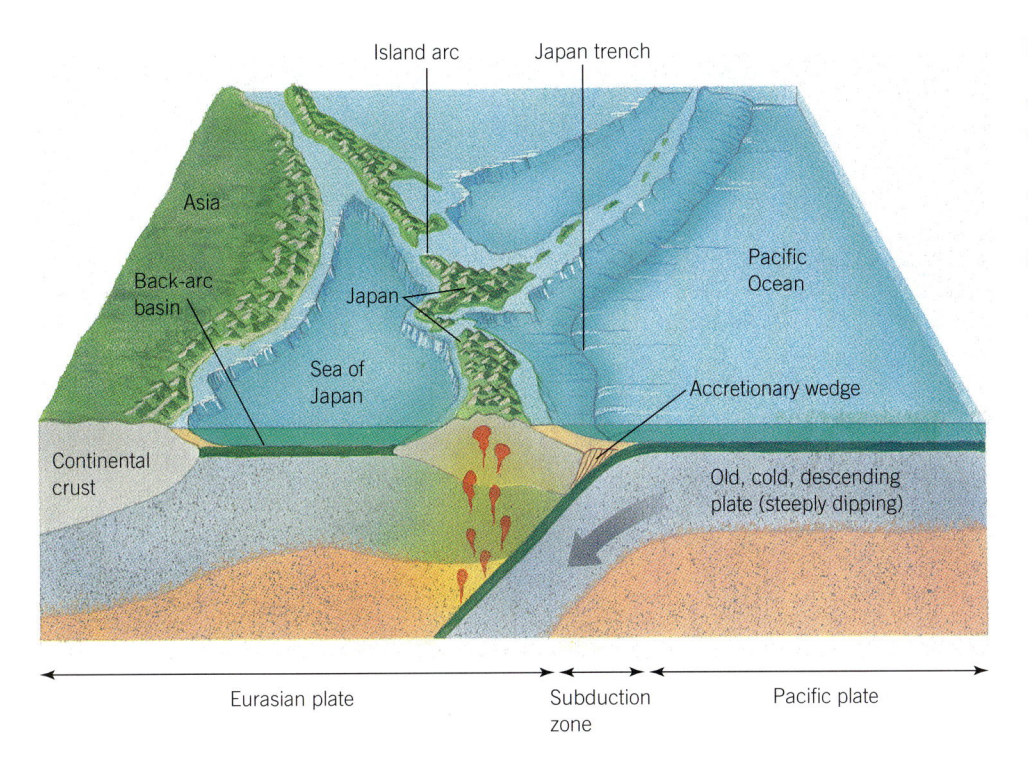

Figure 2.20 Ocean-ocean convergence. Features of this type of plate interaction include: (1) a deep-sea trench where the denser plate is subducted; (2) an accretionary sediment wedge scraped onto the adjacent plate; (3) a volcanic island arc fed by the magma generated by plate subduction; and (4) a back-arc basin between the island arc and the continental mainland.

We can see that as the cold and dense Pacific plate plunges beneath the Eurasian plate, it forms a deep-sea trench. Scraped off the descending plate is much of the sediment deposited on the sea floor during its long journey from mid-ocean ridge to subduction zone. This sediment is then plastered to the edge of the bordering plate as an **accretionary wedge,** which forms a low ridge that rims the trench.

At about 100 kilometers beneath the surface, the subducting plate generates magma that rises through the denser mantle to the surface and fuels the island arc volcanoes just landward of the trench. Precisely how the magma is generated remains unknown. One hypothesis is that the descending plate absorbs enough heat from the warm mantle to partially melt. Another hypothesis suggests that the subducting plate triggers melting in the wedge of upper mantle above it because of water rising from the plate—water that was trapped within the subducting plate and was expelled because of the increasing pressure. The water in turn lowers the melting point of the hot mantle, generating the magma.

Ocean-Continent Convergence

In contrast to the aged crust of the western Pacific, the oceanic crust of the eastern Pacific is quite young, as shown by the sea-floor age patterns on the map in Figure 2.9. The reason for the age difference is also shown on the map. The North and South American plates, spreading westward as a consequence of the breakup of Pangaea and the opening of the Atlantic, are in the process of overriding the plates that are spreading eastward from the Pacific mid-ocean ridge. All that remains of a presumably huge expanse of plate that lay west of North America is the tiny Juan de Fuca plate. The Cocos and Nazca plates to the south are much larger than the Juan de Fuca plate but are small in comparison to the huge Pacific plate that is spreading from the East Pacific Rise.

Subduction zones that border nearly the entire length of South, Central, and North America mark the boundaries where plates spreading from the east and from the west meet and clash. This vast stretch of the Earth's surface is undergoing ocean-continent convergence wherein a subducting plate descends beneath a deep-sea trench directly off the shore of continental crust (Figure 2.21). Consequently, the magma generated by the

accretionary wedge
A large mass of sediment that accumulates in a deep-sea trench and is scraped onto the neighboring plate by the subducting slab.

Figure 2.21 Ocean-continent convergence. Buoyant oceanic crust subducts into the mantle at a relatively low angle. Volcanoes erupt on the bordering continental crust and build mountain chains such as the Andes and the Cascades.

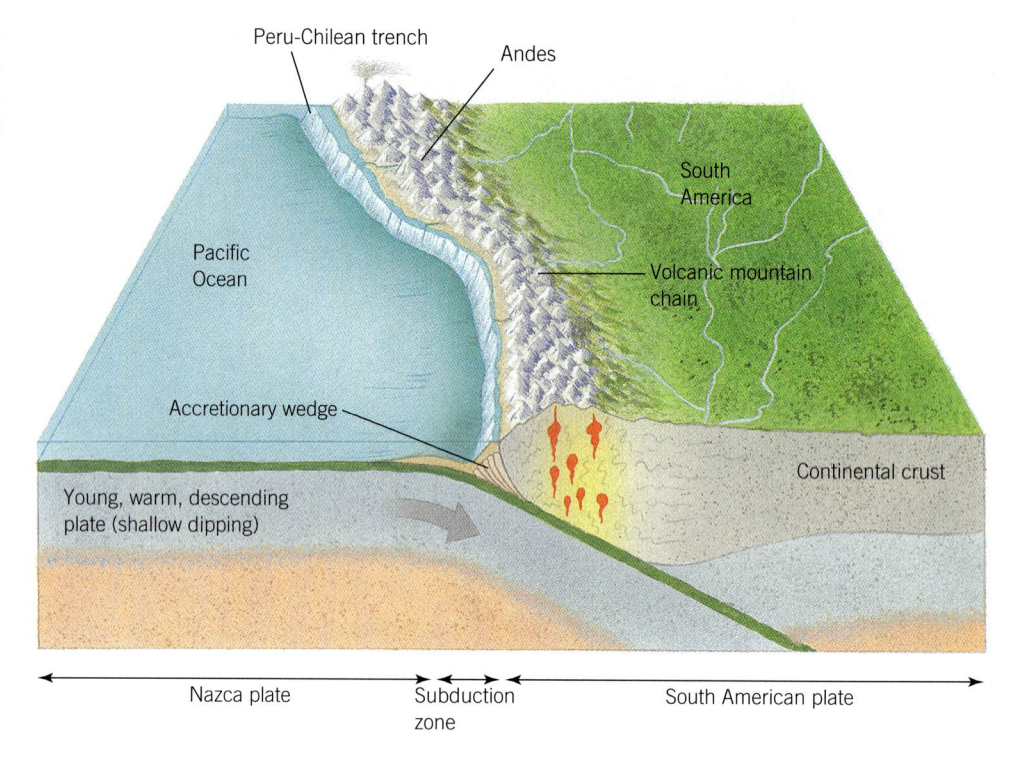

subducting plate erupts on land to build the great volcanic chains that parallel the west coasts of these continents—the Andes and the Cascades, for example.

Ocean-continent convergence is responsible for the rugged coast of western South America and the Andes. In addition, the region is shaken by frequent and powerful earthquakes set in motion by the subducting Nazca plate. The continental margin is narrow and steep; sediments brought to the coast are washed quickly into the Peru-Chile trench directly offshore. Western South America is an **active continental margin,** because it is on the leading edge of a continental plate astride a subduction zone. Contrast it with the South American Atlantic coast. Far from plate boundaries, the eastern coast is nearly earthquake-free, and its only rugged topography consists of the resistant remnants of very ancient mountains that predate Pangaea. Its passive margin consists of a wide continental shelf and slope built of sediment deposited on a stable crust (see also Figure 2.19). Refer again to Figure 2.15, and you will see that the margins of North America also have features and tectonic settings that are broadly similar to what we have just described: a rugged, active western margin at the intersection of plate boundaries, and a broad, passive eastern margin far from plate boundaries.

active continental margin
A boundary between an ocean and a continent marked by plate interaction and thus by frequent volcanic or seismic activity.

Continent-Continent Convergence

If rifting separates continental blocks, subduction brings them together, and the resulting collision builds mountains. The Himalayas are the classic example of continent-continent convergence. They were formed as a result of the breakup of Pangaea when India separated from what is now Antarctica and rode northward on a spreading plate toward the stationary Eurasian plate (Figure 2.22). The leading edge of the Australian-Indian plate was subducted beneath Asia in ocean-continent convergence until the intervening ocean basin, the *Tethys,* closed up. Too light to sink, the Indian landmass had no alternative but to collide with Asia. Trapped between the closing landmasses, as in a vise, was all the sediment that had accumulated in the Tethys Sea before and during India's long journey and that had been scraped off the descending plate. The Himalayas are constructed of this highly deformed sedimentary wedge. They rest on the broad, thickened foundation of continental crust formed by the collision, as does the high Tibetan plateau directly to the north. The thickened foundation inhibits further subduction.

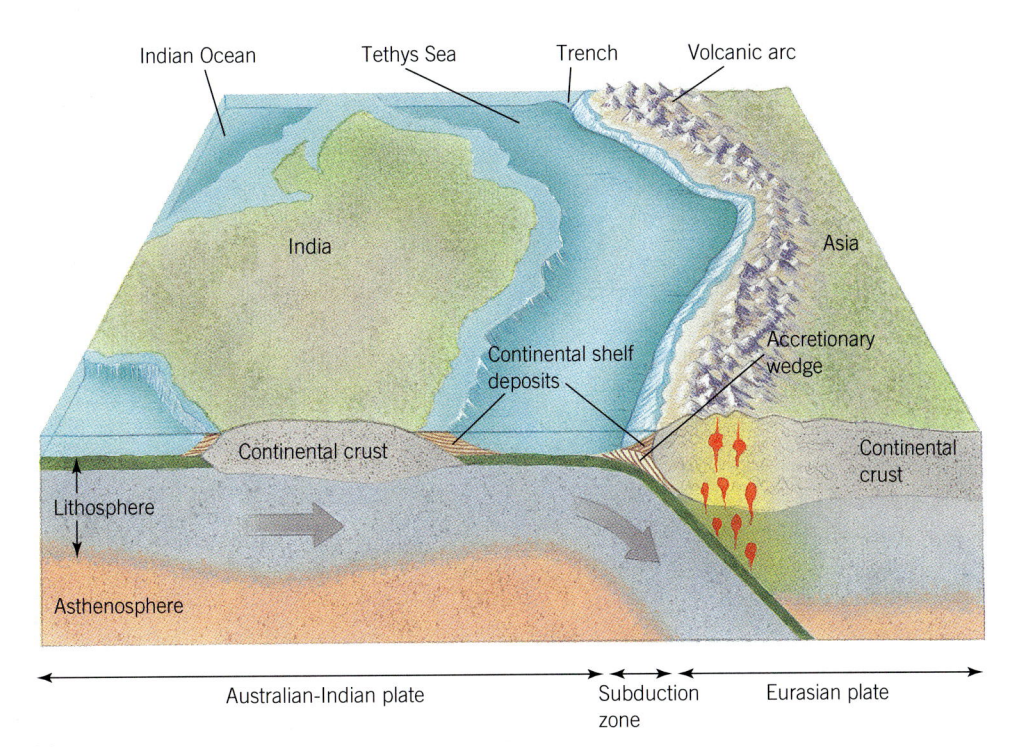

Figure 2.22 Continent-continent convergence. The Himalayas were created by this type of plate interaction. (a) Sediment accumulated in the now-vanished Tethys Sea as the Australian-Indian plate subducted beneath the Eurasian plate. (b) As the Indian and Eurasian landmasses collided, the Himalayas developed from the uplifted and deformed Tethys sediment.

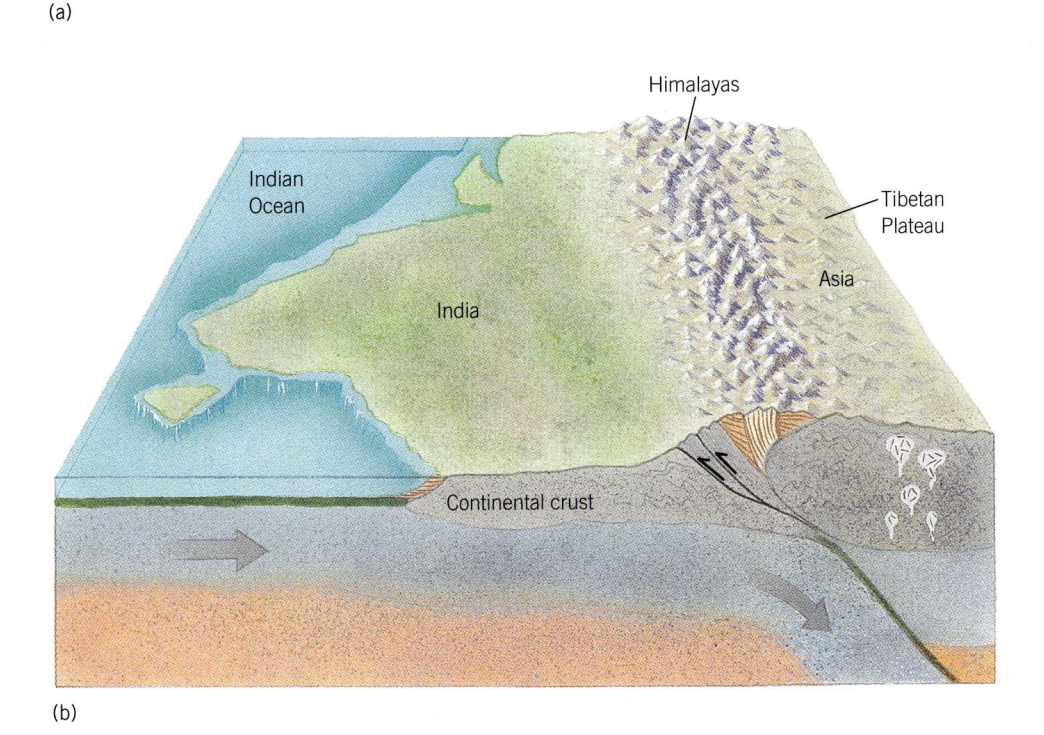

The Tethys Sea no longer exists. The evidence of its former existence is, however, preserved in the sediment and slivers of oceanic crust that compose the Himalayas. Indeed, the places to search for any former oceans, and all the history of subduction and collision they represent, are in the mountain belts of continents. As you have learned, the ocean basins are recycled every 200 million years or so, while continental crust, which is too light to sink into the mantle, remains on the Earth's surface to absorb the impact of this recycling.

The continents are a collage of fragments thrown together by collision, welded by metamorphism and igneous activity, torn apart by rifting, and reshuffled by huge and prolonged horizontal displacements. Their origin and history are influenced by plate tectonics as much as the ocean basins are. Just as processes associated with plate divergence play the key role in creating oceanic crust, those associated with plate convergence are mainly responsible for the evolution of continental crust. It is at convergent margins that new continental crust is generated. In Chapter 17, we will explore fully the role of plate tectonics in building the continents.

Transform Boundaries

transform fault
Type of fault where the plates slide by one another horizontally without either creating or destroying oceanic crust.

Refer again to Figure 2.15; notice that plates not only separate and collide but also slide by one another horizontally. The transform boundaries along which sliding occurs are named after the kinds of faults that create them, **transform faults,** first described by J. Tuzo Wilson. Inspection will reveal the reason for the name: they connect either offset mid-ocean ridge segments, offset deep-sea trench segments, or ridges to trenches (Figure 2.23). In each case, the relative motion between plates is altered or transformed along the connection. For example, divergence at a mid-ocean ridge spreading center is converted to convergence at a trench subduction zone.

Transform faults may displace continental as well as oceanic crust. The dangerous San Andreas fault of California is a transform fault boundary connecting two offset mid-ocean ridge segments: the East Pacific Rise in the Gulf of California and the Juan de Fuca Ridge off the northwest Pacific coast. Between these segments, southern California is moving northward relative to the rest of North America, and earthquakes result. (More on this in the next chapter.)

The Causes of Plate Motion

One major unresolved problem posed by the theory of plate tectonics is that no one has been able to pin down precisely what causes the plates to move. Most theories emphasize the role of mantle *convection currents,* in which warm rock rises from the interior, surrenders its heat as it spreads, and sinks back into the mantle where it is reheated, much like spaghetti in a pot of boiling water. In a simple model, mid-ocean ridge spreading centers overlie the rising portions of the convection current, and subduction zones overlie the cooler sinking portions (Figure 2.24). The plates created by the upwelling magma at the spreading centers are dragged along by the spreading currents toward the subduction zones.

While acknowledging the role of some sort of convection, other theories give greater emphasis to the way in which the plates move with respect to the soft, hot asthenosphere. Figure 2.25 illustrates three proposed mechanisms. *Ridge-push* postulates that the plates are moved aside and toward the trenches by the injection of lava at the mid-ocean ridge. *Gravity sliding* emphasizes the difference in elevation between ridge crest and trench, and suggests that the plates simply slide downhill over a sloping astenosphere, thus keeping a central rift valley open and allowing magma to rise. *Slab-pull* suggests that the hot plate is pulled from the ridge crest by the cold, dense part of the slab that descends beneath the trenches. It is rather like soaking the edge of a towel and letting it hang over the edge of a table; the towel will slide off the table because of the weight of the wet end.

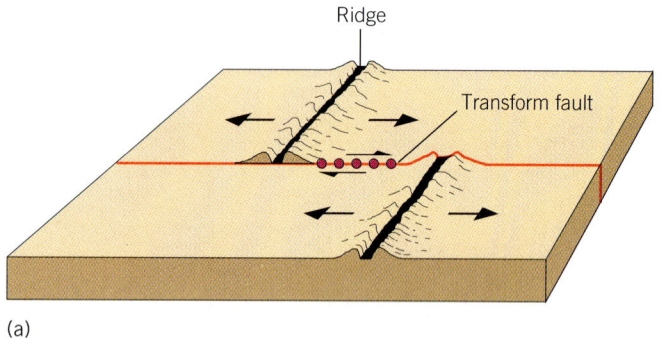

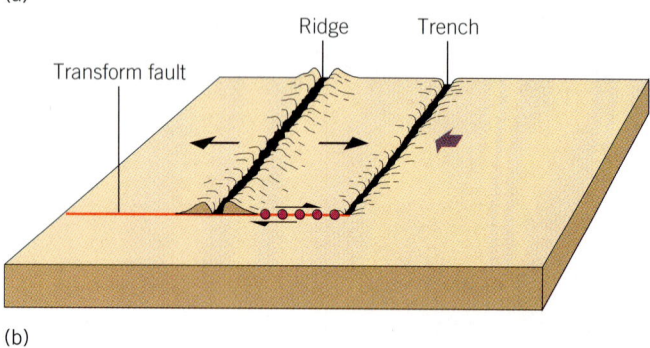

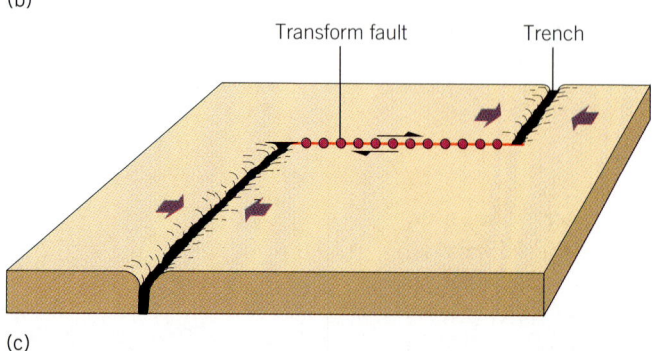

Figure 2.23 A transform fault may connect (a) two offset ridge segments, (b) a ridge segment to a trench segment, or (c) two offset trench segments. Notice that earthquakes (indicated by dots) occur only within the offset segments of the fault.

Plate Tectonics and the Ocean Floor 53

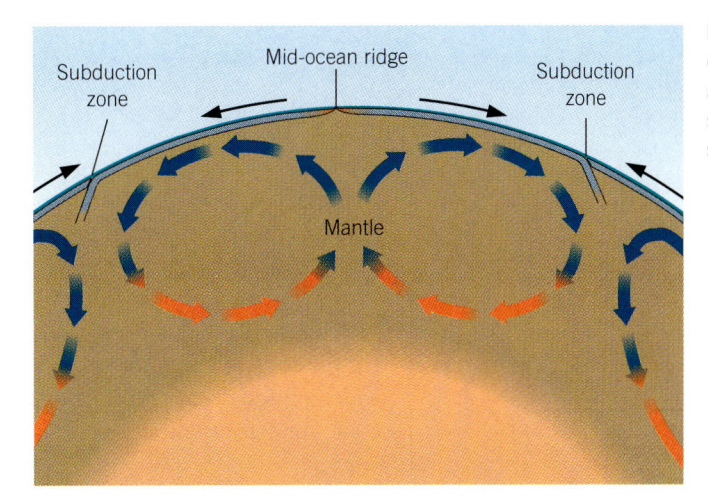

Figure 2.24 A simple model of a mantle convection current: the plates separate above the rising limb of the current and sink into the mantle above the descending limb.

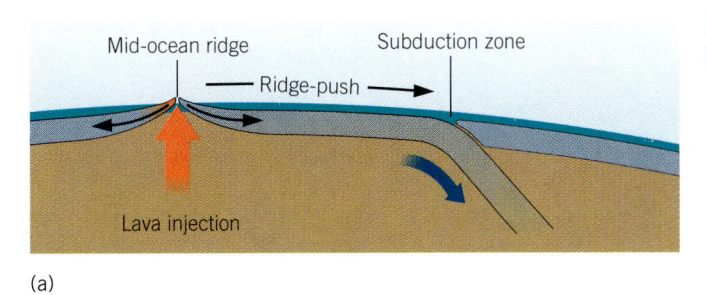

Figure 2.25 Three hypothesized mechanisms of plate motion.

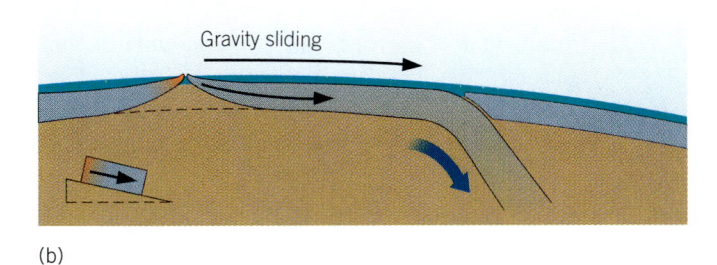

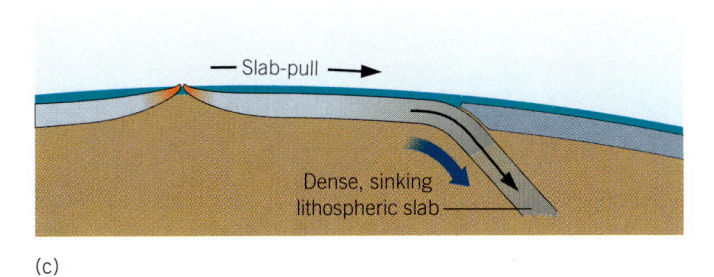

Disguised in these models is the age-old chicken-and-egg problem. Does eruption of magma initiate plate motion, as in the ridge-push model, or does plate motion initiate eruption of magma, as in the other models? Each of these proposed mechanisms has strong points and drawbacks. Indeed, it may be that each contributes to plate motion.

There, painted with a wide brush, is the theory of plate tectonics. Many of the features and processes that it explains are summarized in Table 2.1, and we will be adding to and refining that list in subsequent chapters. Nevertheless, you should not think that all geologists agree with every aspect or interpretation of the theory. There is no such thing as a closed subject in science—and certainly not in geology, where so much theory must, of necessity, rest on fragmentary information.

Table 2.1 Characteristics of Plate Boundaries

Boundary Type	Lithosphere Involved	Relative Motion	Predominant Process	Features	Examples
Divergent	Continent	Away from one another	Continental rifting	Narrow rift valleys, some traceable to triple junctions; shallow earthquakes; lava flows; volcanoes	Afar Triangle, Red Sea, East African Rift Valley; Gulf of California
Divergent	Ocean	Away from one another	Sea-floor spreading	Mid-ocean ridges and rises; central rift valleys; shallow earthquakes; widening and deepening oceanic crust; submarine volcanic peaks	Atlantic Ocean
Convergent	Ocean-ocean	Toward one another	Subduction	Deep-sea trenches, island arcs, back-arc basins; active volcanism; huge quantities of magma added to crust. Earthquakes trace descending plates deep into mantle	Marianas, Aleutians, Japan, Philippines
Convergent	Ocean-continent	Toward one another	Subduction	Deep-sea trenches bordering on continents; volcanoes erupting on land, forming coastal chains; numerous powerful earthquakes triggered by downward action of subducting oceanic lithosphere	Western South America, Washington, Oregon coast
Convergent	Continent-continent	Toward one another	Continental collision	High continental mountain chains formed of sediment deposited in closed-out ocean basins. Continental crust folded and thickened	Himalayas, Alps
Transform	Ocean or continent	Past one another	Horizontal slippage	Long, narrow fracture zones that run perpendicular to mid-ocean ridges and/or deep-sea trenches; powerful shallow earthquakes between offset ridge segments	San Andreas fault; Alpine fault, New Zealand; Mendocino fracture zone

STUDY OUTLINE

Discovered in the 1950s, the high-standing **mid-ocean ridge** is a string of active submarine volcanoes cut down the middle by a central rift valley. Similar tensional features caused by **faults** had been found previously on the continents. The mid-ocean ridge traces a seismic (earthquake) belt that extends through the ocean basins of the world.

I. **SEA-FLOOR TOPOGRAPHY.** The mid-ocean ridge is the shallowest feature of the ocean basins. Segments of the rift valley are offset and marked by **fracture zones.** Tests and observations found high heat flow beneath the ridge crest and seismic activity confined to the fracture zones between offset ridge segments.

  A. **Deep-sea trenches,** found around the margins of the Pacific Ocean, are the deepest features of the ocean basins. They are characterized by seismic activity and low heat flow. Volcanically active continental mountain chains, or **island arcs,** are located parallel to the trenches.

II. **THE SEA-FLOOR SPREADING HYPOTHESIS.** In the early 1960s, geologists proposed that, in a process called **sea-floor spreading,** oceanic crust separates along the rift valley in the mid-ocean ridge, allowing magma to bleed to the surface and form fresh oceanic crust. The crust then spreads laterally, cooling and contracting as it moves toward the trenches. There it descends back into the mantle and triggers the deep line of earthquakes beneath the trenches.

   A. Evidence in magnetic reversals
      1. Parallel bands of strong and weak magnetism in the ocean floor run parallel to the axis of the mid-ocean ridge. Identical alternating bands of magnetism, found in continental lava layers, record **magnetic reversals**—periods in the Earth's history when a compass needle that would today point north pointed south, alternating with periods when the needle pointed north again.
      2. The most recent reversals occur at the mid-ocean ridge, and progressively older ones are found toward the flanks of the ocean basins, evidence that each segment of the oceanic crust was originally formed at the rift valley and has been moved gradually away from the ridge.

   B. Evidence in sea-floor sediment. Further evidence is found in the flat **abyssal plain** along the continental margins, a wedge of sediment that becomes thinner and younger toward the ridge crest. The oldest known sediment and oceanic crust, found in the western Pacific, is 200 million years old.

   C. Evidence in sea-floor topography
      1. Differences in cooling rates are reflected in the topography of the ocean floor: steep at the ridge crest and gentle at the margins of the basins. In general, each portion of the ocean floor is at a depth determined by its age.
      2. **Seamounts** and **guyots** are submerged islands—strong evidence that the sea floor sinks as it spreads. Scientists proposed that the volcanoes of the Hawaiian-Emperor Seamount chain were punched out as the Pacific oceanic crust passed over the same magma source that is currently beneath Hawaii. Many such **hot spots** are distributed throughout the continents and ocean basins.

III. **PLATE TECTONICS.** Proposed in the mid-1960s, **plate tectonics** went beyond continental drift and sea-floor spreading to offer a comprehensive theory of the Earth's crustal structure. The **lithosphere,** consisting of oceanic and continental crust and upper mantle, is divided into rigid slabs, or **plates.** Plate boundaries occur along **seismic belts** and are marked by earthquakes generated when the plates grind against one another as they move over the **asthenosphere.** There are only three ways in which the plates can interact: they may diverge, converge, or slide by one another.

   A. **Divergent boundaries** are the sites of simultaneous plate separation and growth. As the plates move apart, magma that wells up through the central rift valley is accreted to the trailing edge of the plates.
      1. The Mid-Atlantic Ridge is a divergent boundary that began as a fracture in the continental crust of Pangaea. In a similar rifting process happening today in the Afar Triangle, a hot spot stretched and fractured the crust, forming a triple junction. The Mid-Atlantic Ridge was probably formed as a series of connecting triple junctions.
      2. The Atlantic Ocean of today is in a mature stage of development. It is characterized by a wide ocean basin between separated continents and by **passive continental margins**—divided into shallow **continental shelf,** a **continental slope,** and an apronlike **continental rise**—that are nearly free of earthquakes and igneous activity because they are far from plate boundaries.

B. Where plates meet at **convergent boundaries**, the denser, colder plate bends and descends into the mantle in a process called **subduction**. There are three types of convergence.
  1. Ocean-ocean convergence. Where one plate plunges beneath another, it forms a deep-sea trench. Sea-floor sediment is scraped off the descending plate and plastered to the edge of the bordering plate as an **accretionary wedge.** Magma generated by the subducting plate rises to the surface and fuels the island arc volcanoes just landward of the trench.
  2. Ocean-continent convergence. Where a plate descends beneath a deep-sea trench directly off the shore of continental crust, the magma generated by the subducting plate erupts on land to build the great volcanic mountain chains that parallel the coasts. Characterized by frequent and powerful earthquakes, the **active continental margin** is narrow and steep.
  3. Continent-continent convergence. When continental crust on both sides of a subduction zone converges, ocean sediment trapped between is deformed and pushed up into a mountain belt as the continents collide.

C. **Transform boundaries** occur where two plates slide by one another horizontally. As new crust spreads away from the central rift valleys, the plates move in opposite directions in the narrow section of the fracture zone between the offset ridge crests, called a **transform fault.** The motion triggers the earthquakes that are confined only to that zone. Beyond the ridge crests, the crust on either side of the fracture moves in the same direction at the same speed.

D. Most theories of plate motion emphasize the role of mantle *convection currents,* in which warm rock rises from the interior, surrenders its heat as it spreads, and sinks back into the mantle, where it is reheated. Three proposed mechanisms of plate motion are ridge-push, gravity sliding, and slab-pull.

STUDY TERMS

abyssal plain (p. 39)
accretionary wedge (p. 49)
active continental margin (p. 50)
asthenosphere (p. 43)
continental rise (p. 48)
continental shelf (p. 48)
continental slope (p. 48)
convergent boundary (p. 43)
deep-sea trench (p. 34)
divergent boundary (p. 43)
fault (p. 30)
fracture zone (p. 31)
guyot (p. 40)
hot spot (p. 41)

island arc (p. 34)
lithosphere (p. 42)
magnetic reversal (p. 37)
mid-ocean ridge (p. 30)
passive continental margin (p. 47)
plate (p. 42)
plate tectonics (p. 41)
sea-floor spreading (p. 35)
seamount (p. 40)
seismic belt (p. 42)
subduction (p. 48)
transform boundary (p. 43)
transform fault (p. 52)

THOUGHT QUESTIONS

1. How did the discovery of magnetic reversals confirm the sea-floor spreading hypothesis?
2. Use the sea-floor spreading concept to explain why the shallowest part of an ocean basin is toward the middle and the deepest part of the basin is along its margins.
3. In what ways does the worldwide pattern of earthquakes support the theory of plate tectonics?

4. What is the importance to plate tectonics theory of the decrease in seismic wave velocities at depths of about 100 kilometers?
5. In Chapter 1, we learned that 200 million years ago, the continents existed as a single landmass called Pangaea. Use the concepts of plate tectonics to explain how the continents moved to their present locations on the Earth.
6. Compare and contrast the features of active and passive continental margins. How does plate tectonics account for the differences?
7. Assuming that the processes initiated by the hot spot at the Afar Triangle continue, sketch and describe the geologic future of the Red Sea and the Gulf of Aden.
8. As its name implies, the mid-ocean ridge runs down the center of the Atlantic Ocean. Explain why the ridge does not run down the center of the Pacific Ocean but is located along its eastern margin.

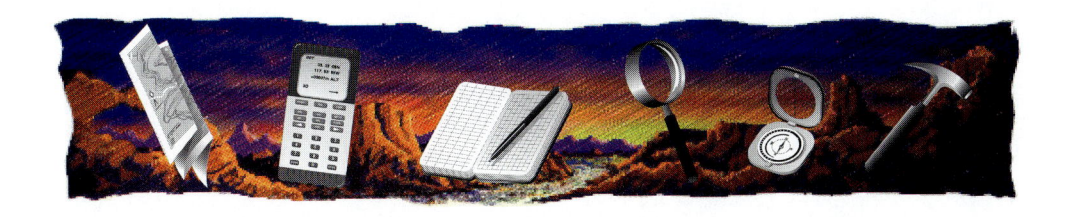

Chapter 3

The Causes of Earthquakes

Seismic Waves
   The Seismograph
      Locating the Earthquake
   Measuring Earthquake Magnitudes and Energy

Earthquakes and Plate Tectonics
   The San Andreas Fault
   The Los Angeles Faults
   Intraplate Earthquakes

Forecasting Earthquakes
   Seismic Gaps
   Recurrence Studies
   Precursor Studies
   Monitoring Tsunamis

Minimizing Earthquake Damage
   Construction Problems

Seismic Waves in the Earth's Interior
   Wave Refraction and Reflection
   Detection of the Crust, Mantle, and Core

Structural and Compositional Divisions of the Interior

The Earth's Internal Heat
   The Geothermal Gradient
   Mantle Convection
   Sources of the Earth's Internal Heat
   The Earth's Core and the Magnetic Field

Earthquakes and the Interior of the Earth

Earthquakes are shock waves, vibrations. Most are triggered by the sudden slippage of rock along fault planes in the crust and in the subduction zones of the upper mantle. Human history is replete with examples of catastrophic earthquakes (Table 3.1). In this chapter, we examine the causes of earthquakes, the close relationship between earthquakes and plate tectonics, the detection and analysis of earthquakes through instruments, and, finally, the efforts of geologists to predict them and minimize their impact. We also examine the ways in which geologists (seismologists) gather the information provided by earthquakes to render an ever-more sophisticated picture of the Earth's interior.

earthquake
Vibrations within the Earth set in motion by the sudden release of accumulated strain energy.

The Causes of Earthquakes

In 1884, geologist G. K. Gilbert was engaged in constructing a geological map of a part of the rugged Basin and Range province of Nevada and Utah. Following the occurrence of a swarm of earthquakes in the vicinity of Salt Lake City, he observed the appearance of freshly exposed steep cliffs, or *scarps,* at the base of the mountains nearby (Figure 3.1). The fresh scarps were proof that fault movement was uplifting the mountains, and Gilbert strongly suspected that the earthquakes had been generated by the fault motion. He concluded:

> The upthrust [of the mountains] produces a strain in the crust, involving a certain amount of distortion, and this strain increases until it is sufficient to overcome the starting friction along the fractured surface. Suddenly, and

◀ This photograph was taken in the wake of the earthquake in Kobe, Japan, on January 16, 1995.

Table 3.1 Selected Historic Earthquakes

Year	Location	Magnitude	Number of Deaths
365	Crete, Knossos		50,000
1201	Egypt, Syria		1,000,000
1755	Portugal, Lisbon		70,000
1811	Missouri, New Madrid		Several
1886	South Carolina, Charleston		60
1906	California, San Francisco	8.2	700
1906	Ecuador	8.9	1,000
1920	China, Kansu	8.5	180,000
1923	Japan, Kwanto	8.2	143,000
1939	Turkey, Erzincan	8.0	23,000
1960	Southern Chile	8.5	5,700
1964	Alaska	8.6	131
1970	Peru	7.8	66,000
1971	California, San Fernando	6.5	65
1975	China, Haicheng	7.4	300
1976	China, Tangshan	7.6	650,000
1985	Mexico, Mexico City	8.1	5,600
1988	Armenia	6.9	25,000
1989	California, Loma Prieta	7.1	62
1990	Iran	7.3	50,000
1992	California, Landers	7.5	1
1994	California, Northridge	6.7	55
1995	Japan, Kobe	6.9	5,160

Source: U.S. National Oceanic and Atmospheric Administration.

Figure 3.1 A freshly exposed fault scarp at the base of Borah Peak, Idaho. G. K. Gilbert hypothesized that earthquakes are triggered by the release of strain energy that accompanies sudden movement along faults. Fault movement creates scarps such as this one.

almost instantaneously, there is an amount of motion sufficient to relieve the strain, and this is followed by a long period of quiet, during which the strain is gradually reimposed.*

Gilbert's insight gained important quantitative support from geologist H. Fielding Reid's classic investigation into the cause of the catastrophic San Francisco earthquake of 1906. The earthquake occurred near Point Reyes, a few kilometers north of the city along a segment of the San Andreas fault. Reid and his colleagues mapped such disrupted features as stream courses, fence posts, roads, and railroad tracks (Figure 3.2). Then they compared the offset position of these features with their position before the earthquake, as located by surveys taken prior to 1906. They also measured the orientation of ground breaks in the fault zone. Their observations enabled Reid to determine that, in the San Francisco region at the time of the earthquake, the west side of the San Andreas fault moved as much as 6.5 meters north with respect to the east side. This movement represented the sudden release of huge energy, enough to trigger a great earthquake.

Figure 3.2 This classic photograph of a fence offset was taken by G. K. Gilbert following the 1906 San Francisco earthquake.

Exert an external force on a body and you are subjecting it to *stress*. The deformation (change of shape and/or volume) that the body undergoes in response to the stress is called *strain*; it is a way of storing the energy that the force has transmitted to the body. The Reid-Gilbert explanation of the cause of earthquakes, now known as the **elastic-rebound theory,** is based on this essentially simple relationship (Figure 3.3). The crux of the theory is that earthquakes result from the sudden release of strain energy that occurs when rocks rupture and lurch past one another along fault planes.

Most of the huge amount of energy released during an earthquake is expended in pulverizing rock and generating heat. Only a relatively minor portion of the energy causes vibrations, or **seismic waves.** The waves fan out in all directions from the earthquake **focus,** which is the point of initial rupture beneath the surface of the Earth. The vertical projection of the focus onto the surface of the Earth is the earthquake **epicenter** (Figure 3.4). For example, the focus of the 1906 San Francisco earthquake was several kilometers below the Point Reyes epicenter.

elastic-rebound theory
The concept that earthquakes are generated by the sudden slippage of rocks on either side of a fault plane. In the process, the rocks release gradually accumulated strain energy and are returned to an unstrained condition.

seismic wave
A wave generated within the Earth by sudden fault slippage or an explosion.

focus
The initial point of rupture within the Earth from which seismic waves are generated.

epicenter
The point on the Earth's surface that is directly above the focus of an earthquake.

Seismic Waves

An earthquake releases two classes of seismic waves: **body waves** that travel through the interior and **surface waves** that travel along the surface. Body waves are in turn classified into two types: the faster-traveling **primary (P) waves** and the slower **secondary (S) waves.** P waves alternately compress and expand the rocks they pass through by shortening and lengthening atomic bonds. As shown in Figure 3.5(a), the rock vibrates back and forth along the wave in an accordionlike motion. P waves are virtually identical to common sound waves; the major difference is that they generally have vibration rates (or frequencies) lower than the human ear can detect. An S wave causes the rock to vibrate at right angles to the direction of the wave path, much like what happens when you shake the free end of a rope tied to a pole (Figure 3.5b). Because the rock particles slide by one another as they vibrate, S waves are also called *shear* waves.

The velocities of seismic waves depend on the elastic properties and density of the material through which they pass. Thus we can learn a great deal about the composition

body wave
A seismic wave that travels through the Earth's interior.

surface wave
A seismic wave that travels along the Earth's surface and affects the interior to depths that are dependent upon its wavelength.

primary (P) wave
A seismic wave that propagates through the Earth as a series of compressions and expansions.

secondary (S) wave
A seismic wave that causes the components of a rock to vibrate perpendicularly to the direction of wave propagation. Also called a shear wave.

---

*G. K. Gilbert, "A Theory of Earthquakes of the Great Basin with a Practical Application," *American Journal of Science,* 1884.

and physical state of the Earth's interior by studying these waves. For example, P waves travel more rapidly through rigid, relatively dense crustal rocks than through softer rocks or liquids. S waves cannot travel through liquids at all, because liquids lack the internal strength to support shearing motions. The fact that S waves do not penetrate beyond 2900 kilometers is strong evidence that the Earth is molten at that depth.

Surface waves are a combination of two complex vibrations. One sets the Earth's surface swaying in a horizontal, snakelike motion at right angles to the wave path. These vibrations are similar to S waves and, like them, cannot travel through a lake or ocean—although they can shake the basins holding these water bodies. The other vibration causes the surface to rise and fall like ocean waves. During a powerful earthquake, the ground rises and falls at the same time it sways from side to side, which is one reason for damage to foundations, sewers, and pipelines.

The Seismograph

The waves generated during an earthquake are detected by an instrument called a **seismograph** (Figure 3.6). Modern versions are sensitive enough to record the Earth's every hiccup, and they never sleep. Several large-scale seismograph networks cover the globe. Each consists of hundreds of stations that constantly monitor for earthquakes and routinely exchange vast amounts of data.

A simple seismograph can be constructed by suspending a large mass (that is, a heavy weight) by thin springs from a rigid frame that is attached firmly to a platform embedded in bedrock. Also attached to the platform is a drum that rotates at a fixed rate. When the seismic waves of an earthquake arrive, they shake the platform, frame, and drum, but the suspended weight remains nearly stationary because of its inertia. A sensor—in simplest form, a pen attached to the weight—records on the rotating drum the difference in motion between the vibrating platform and the stationary weight. The resulting wave pattern, the record of the earthquake, is a *seismogram*. At least three seismographs are necessary to fully record the ground motion of an earthquake: two detect horizontal north-south and east-west vibrations, and the third detects vibrations in the vertical plane. Modern seismographs use microelectronics to record and transmit ground-motion data directly to digital computers.

Seismograms yield valuable information. Reading them properly enables us to locate the epicenter and focal depth of an earthquake, gauge its magnitude and energy, and determine the fault motions responsible for it in the first place. Seismograms are also our main sources of information concerning the forces that drive the Earth's lithospheric

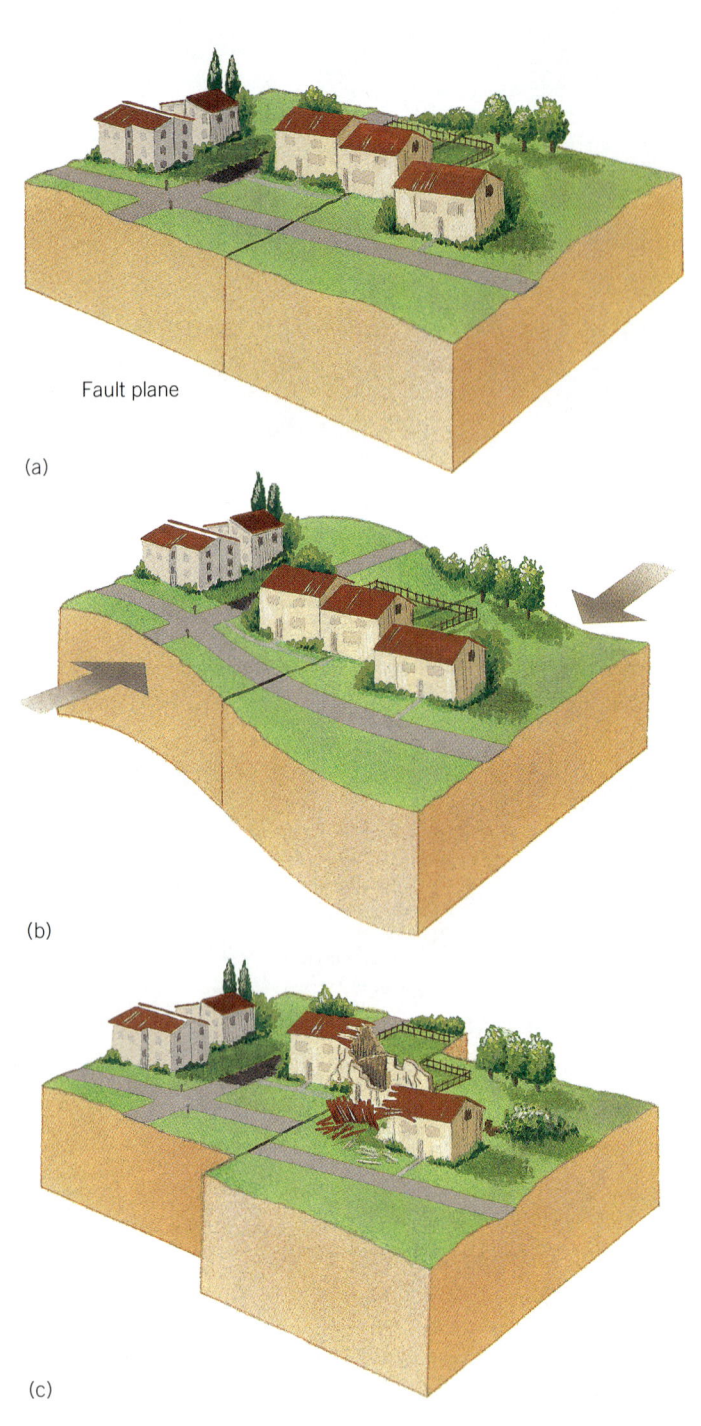

Figure 3.3 Elastic rebound. (a) Rocks along a fault plane in unstrained condition. (b) Strain builds as stress is applied and the blocks on either side of the fault plane are held together by friction. (c) The blocks suddenly lurch past one another to unstrained positions, releasing strain energy in the form of an earthquake.

seismograph
A device that detects and records seismic waves.

Earthquakes 63

Figure 3.4 The focus of an earthquake is the point of initial rupture, where seismic waves originate. The epicenter is the surface projection of the focus.

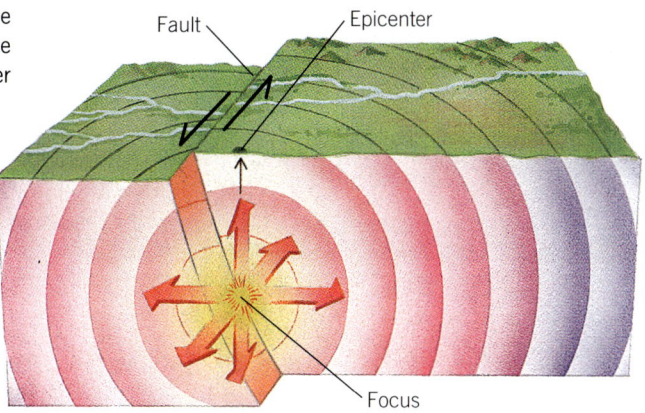

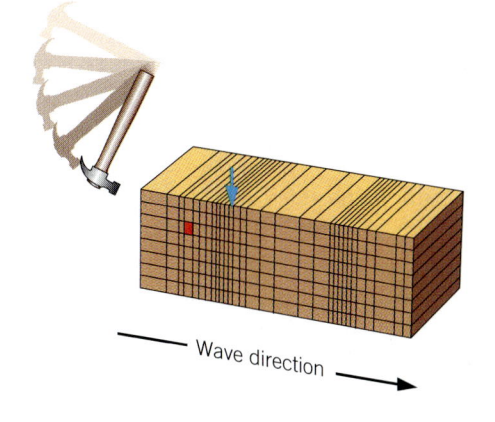

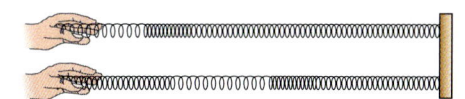

(a) P wave

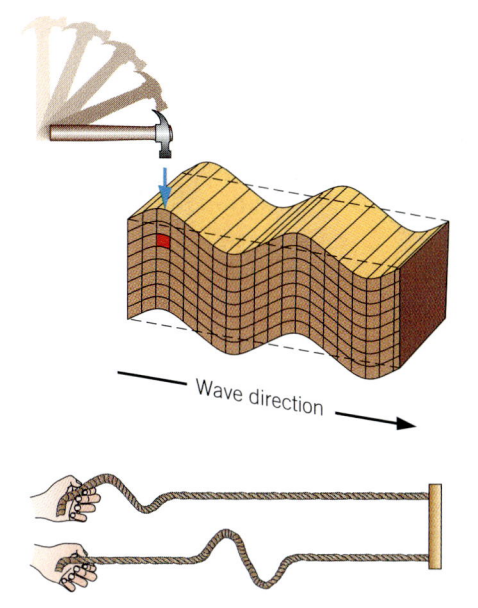

(b) S wave

Figure 3.5 The principal seismic body waves. (a) P waves generate a series of alternating compressions and expansions in the direction of wave propagation. (b) S waves generate a series of vibrations perpendicular to the direction of wave propagation.

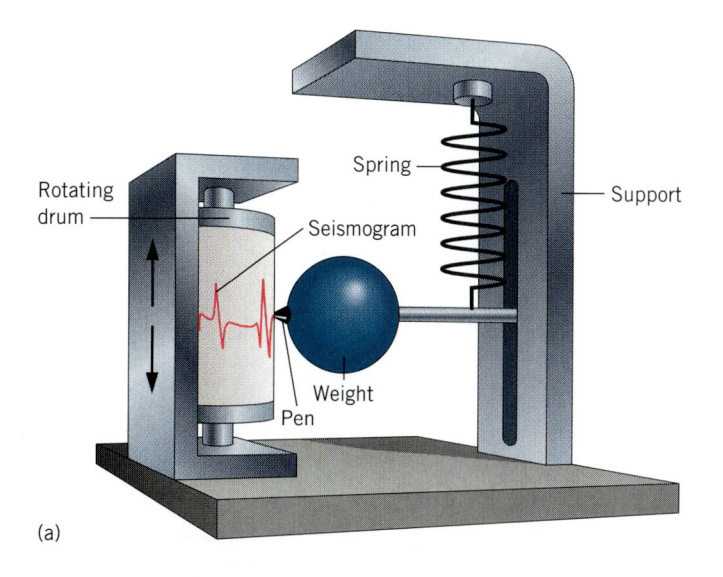

Figure 3.6 (a) A simple model of a seismograph designed to detect vertical ground motion. (b) A modern station employs a variety of instruments to detect and analyze seismic waves. The map shows the region that was subjected to the most intense ground shaking during the earthquake in Loma Prieta, California, in 1989.

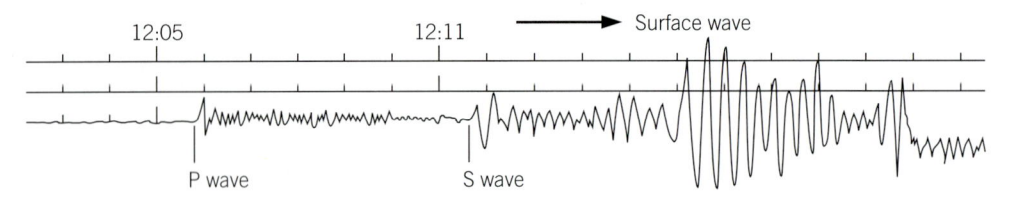

Figure 3.7 A typical seismogram, which records the time of arrival of P, S, and surface waves.

plates, and they are essential in determining the structure and composition of the interior. This use of seismograms will be discussed in the next chapter.

Locating the Earthquake

Figure 3.7 is a typical seismogram of a distant earthquake. Note first the series of 1-minute markers across the top of the seismogram that enable us to precisely date arrival times of waves at the station. Reading the trace from left to right, we first encounter a line that is nearly straight. The slight wiggles are the equivalent of background radio static. These *microseisms* have many causes: local traffic, landslides, waves breaking on a nearby shore. They prove that, like the ocean, the solid Earth is never at rest.

The abrupt change at 12:06 in the nearly straight-line pattern of Figure 3.7 marks the arrival of the P waves spreading from an earthquake. To the right (which, on the seismogram, means a later time), we see other changes in the pattern that record in sequence the arrival of the S waves and surface waves. The spacing between P and S wave arrivals is caused by velocity differences; because P waves outrun the S waves in their dash through the interior, they reach the station first. The last to arrive are the surface waves, because they are the slowest and travel the longer route of the Earth's circumference.

Since all seismic waves originate at the same point and then spread from that point at different speeds, the farther they travel, the greater the interval is between their arrival times. In Figure 3.8, we see that an interval of 6 minutes between P and S arrivals corresponds to an epicenter that is 4100 kilometers from the station. This information enables us to determine the distance to the epicenter but not its location, because the waves could have come from any direction. That is, as far as we know, the epicenter may lie anywhere on a circle of a 4100-kilometer radius. To pinpoint the epicenter, we need to pool information with at least two other stations that have followed our methods. Then we can construct three circles on a globe; their point of intersection marks the epicenter (Figure 3.9).

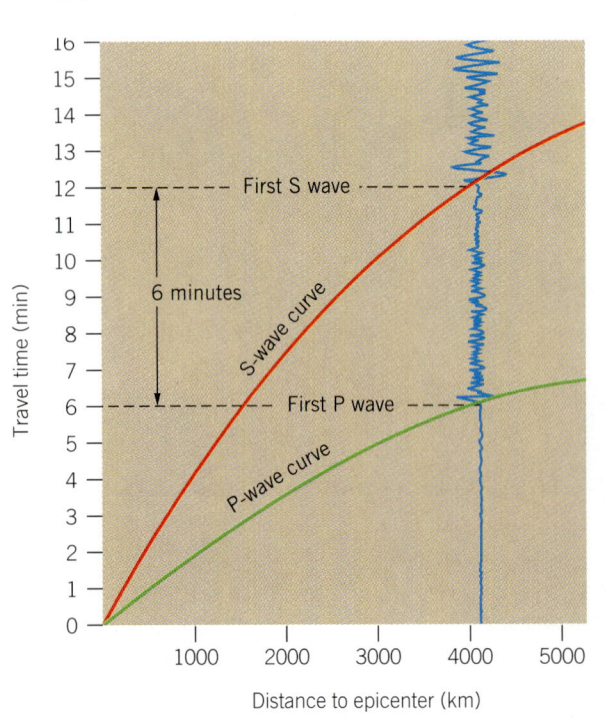

Pinpointing the focus of an earthquake is a trickier problem than finding the epicenter. However, the principle used is the same: the time gap between wave arrivals depends upon the distance traveled. One method compares the time interval between the P waves that take the direct route to the station and those that reach the station after having first reflected off the Earth's surface. The greater the difference in arrival times, the greater the focal depth of the earthquake.

Measuring Earthquake Magnitudes and Energy

The more energy you impart to a rope you shake or a tuning fork you strike, the greater will be the *amplitude* of their vibrations—that is, the greater their displacement from the undisturbed position. The same is true of seismic wave amplitudes: the greater the energy

Figure 3.8 The interval between the arrival times of P and S waves depends upon the distance from the epicenter to the station. Therefore, a specific interval is associated with a specific distance. For example, the 6-minute interval shown in this graph corresponds to a distance of 4100 kilometers.

released during an earthquake, the larger the amplitudes recorded on the seismogram. Therefore, wave amplitude can serve as an objective measure of an earthquake's size or **magnitude**. In 1935, California Institute of Technology seismologist Charles Richter developed a practical application of this concept, which has since become known as the **Richter magnitude scale.**

A Richter magnitude is a measurement of the amplitude of the highest wave recorded on a standard seismograph divided by its *period*, which is the time required for that wave to complete the vibration. Because wave amplitudes diminish with distance from the epicenter, Richter added a factor that corrects the measurement for distancing effects and adjusts it to the amplitude the seismograph would record if it were located 100 kilometers from the epicenter. Thus an earthquake should register the same magnitude at close and distant stations.

The Richter scale is logarithmic, which means that adjacent numbers on the scale signify wave amplitudes that differ by a factor of 10. For example, the amplitude of a magnitude 5 earthquake is 10 times that of a magnitude 4 earthquake. Earthquakes of magnitude 8 and above, which occur on average once every few years, are classified as *great earthquakes* (Table 3.2). Though the scale has no upper limit, no earthquake greater than 8.9 has ever been recorded. The magnitude of the 1906 San Francisco earthquake is in debate owing to the use of unstandardized instruments at that time, but it is estimated to have been 8.2. The 1964 Anchorage earthquake registered 8.6, and the 1985 Mexico City earthquake 8.1.

Keep in mind that the difference in effect between two magnitude points on the low end of the scale is not the same as the difference in effect between two points on the high end. For example, a magnitude 2 quake, although 10 times the amplitude of a magnitude 1, is still a very mild quiver, undetectable by all but the most sensitive instruments. By contrast, a magnitude 8 earthquake is an enormous jolt 10 times the amplitude of a very powerful magnitude 7 earthquake.

Researchers have established that energy increases roughly 30 times for each magnitude number on the Richter scale. Thus a magnitude 7 earthquake releases 900 (30 × 30) times the energy of a magnitude 5 earthquake. Also, two earthquakes that register closely on the Richter scale nevertheless differ significantly in their energy release. For example, the magnitude 8.2 San Francisco earthquake of 1906 released three times the energy of the magnitude 8.1 Mexico City earthquake of 1985. Because of the magnitude-energy relationship, the energy released during a great earthquake far exceeds the combined energy of the hundreds of thousands of small earthquakes that occur each year.

In recent years, seismologists have been using a magnitude scale that differs slightly from the Richter scale because it is based on more refined instrumentation and greater knowledge of seismic waves. Neither the old scale nor the new, however, is a measure of the destructiveness of an earthquake. Certainly, the greater the magnitude of an earthquake, the greater is its potential for destruction. The extent of the damage, however, depends upon other factors, too, such as the depth of the earthquake,

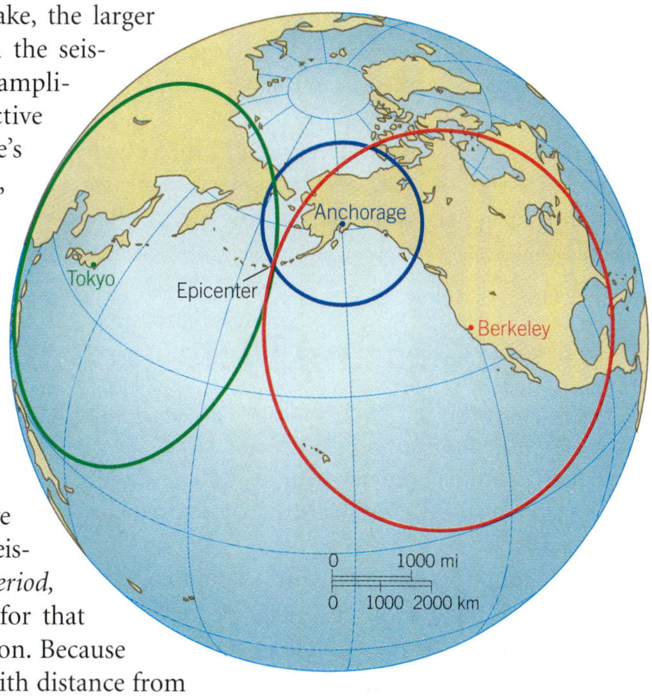

Figure 3.9 The intersection of circles drawn from three recording stations locates the epicenter. The radius of each circle is the distance from the epicenter to the station.

magnitude
A measure of earthquake strength as interpreted from the maximum wave amplitude recorded by a seismograph.

Richter magnitude scale
A logarithmic scale of earthquake magnitudes based on the maximum-amplitude wave recorded by a standard seismograph and corrected for distance of the seismograph from the epicenter.

Table 3.2 Frequency of Earthquakes Worldwide

Magnitude	Average Number per Year
Over 8.0	1 to 2
7.0 to 7.9	20
6.0 to 6.9	120
5.0 to 5.9	800
4.0 to 4.9	6,200
3.0 to 3.9	49,000
2.0 to 2.9	300,000

Source: Adapted from *Earthquake Information Bulletin.*

Mercalli intensity scale
A 12-point scale that measures earthquake severity in terms of the damage inflicted.

the types of rocks and soil through which the waves travel, the proximity of the epicenter to population centers, and the structures and utilities affected. The modified **Mercalli intensity scale** measures the severity of an earthquake in terms of the level of destruction and panic it causes. Values in the form of Roman numerals I–XII are assigned to objective measures, such as the extent of damage to buildings constructed of brick, mortar, or steel, and to subjective measures, such as the degree of social disruption (hardly felt, awakened during sleep, general panic) caused by the earthquake (Table 3.3).

Earthquakes and Plate Tectonics

In the relationship between earthquakes and plate tectonics, we have a near-perfect fit of observation and theory. On the one hand, the global distribution of earthquakes and their fault motions confirm plate tectonics—indeed, they underpin the theory. On the other hand, plate tectonics offers a potent explanation of the stresses that cause most earthquakes.

To better understand the connection between earthquakes and plate tectonics, consider this fact: the epicenters of over 99 percent of the earthquakes that occur each year are confined to narrow seismic belts. Figure 3.10 shows that these belts are located around the Pacific Rim, the Indonesian arc, the Alpine-Himalayan mountain belt, and

Table 3.3 The Modified Mercalli Intensity Scale (Abridged)

Intensity	Effects
I	Not felt.
II	Felt by persons at rest.
III	Felt indoors; hanging objects swing; vibration like passing of light trucks.
IV	Vibration like passing of heavy trucks; windows, dishes rattle.
V	Felt outdoors; sleepers wakened; liquids disturbed, some spilled; doors swing open, closed.
VI	Felt by all; many are frightened and run outdoors; windows, dishes, glassware broken; weak plaster and masonry cracked; furniture moved or overturned.
VII	Difficult to stand; hanging objects quiver; fall of plaster, loose bricks, stones, tiles, and architectural ornaments; waves on ponds.
VIII	Steering of cars affected; fall of stucco and some masonry walls; twisting, fall of chimneys, factory stacks, monuments, elevated tanks; branches broken from trees.
IX	General panic; frame structures, if not bolted, shifted off foundations; underground pipes broken; conspicuous cracks in ground; in alluviated areas, sand and mud ejected; earthquake fountains, sand craters.
X	Most masonry and frame structures destroyed; some well-built wooden structures and buildings destroyed; serious damage to dams and embankments; large landslides.
XI	Rails bent greatly; underground pipelines completely out of service.
XII	Damage nearly total; lines of sight and level distorted; objects thrown into the air.

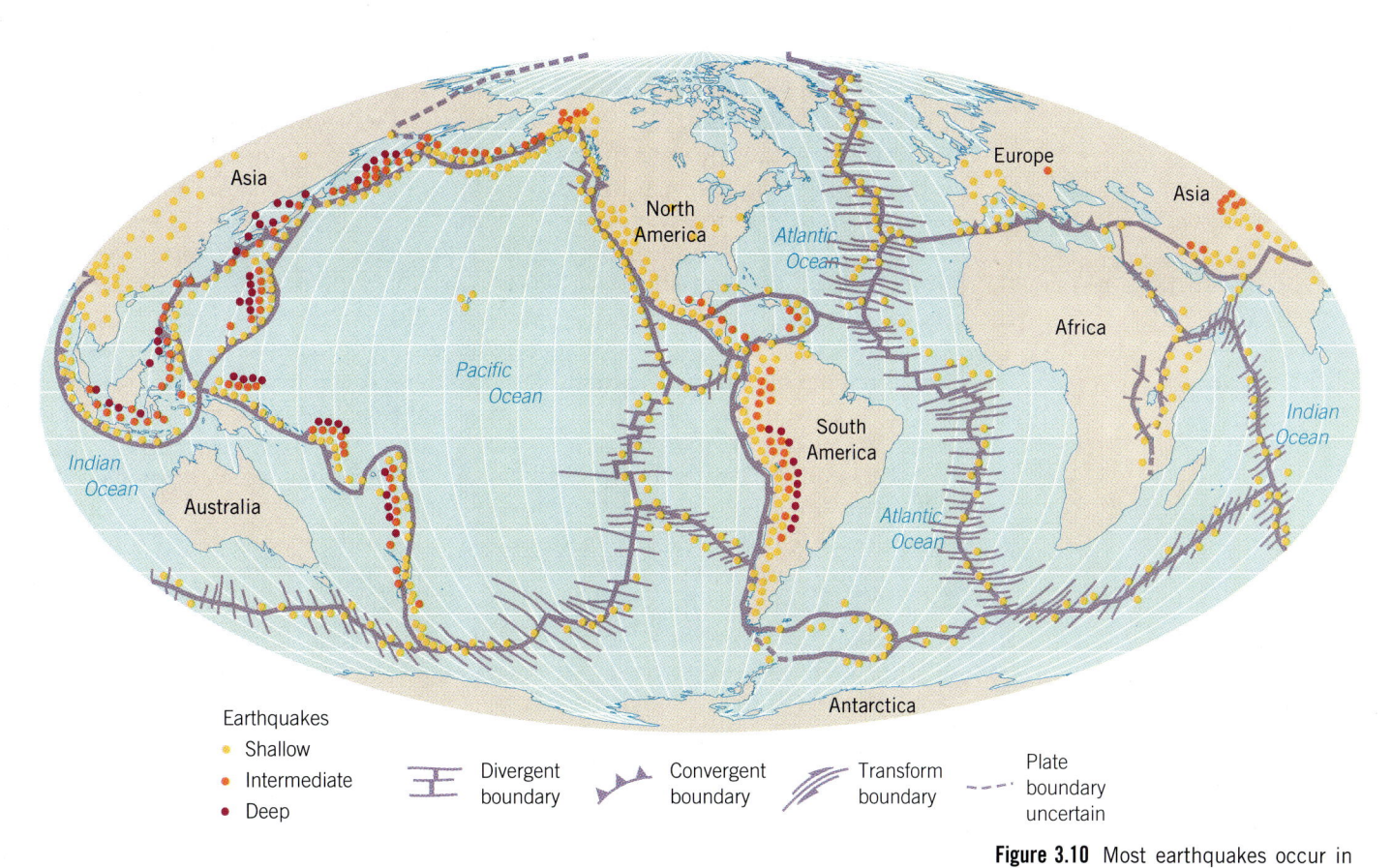

Earthquakes 67

Figure 3.10 Most earthquakes occur in narrow belts that define the plate boundaries. Intermediate- and deep-focus earthquakes follow the path of descending plates within subduction zones, while rift valleys and fracture zones are the sites of shallow earthquakes.

the Caribbean Sea. Another thinner seismic belt traces the Earth-girdling mid-ocean ridge system. There is a striking contrast in activity between these seismic belts and the vast intervening regions. According to plate tectonics theory, the seismic belts mark the boundaries of the Earth's lithospheric plates. Only a few great earthquakes occur far from plate boundaries, and many of them may also have connections to plate movements, as we shall soon see.

The distribution of earthquake focal depths proves as informative as the distribution of epicenters. Those within the narrow mid-ocean ridge seismic belts are shallow; they occur high up on the ridge crest in the rigid basaltic crust of the central rift valley. This observation fits well with plate tectonics theory, which indicates that the rift valleys are the sites of plate separation.

More striking is the pattern of focal depths along subduction zones. There are shallow earthquakes near the trenches, but there are also intermediate and deep earthquakes to depths of 700 kilometers. In fact, virtually all the Earth's deep-focus earthquakes are associated with subduction zones. The patterns of deep-earthquake focal points form continuous lines within the oceanic crust that trace the paths of subducting plates as they descend into the mantle.

Not only are the locations and depths of earthquakes consistent with plate tectonics theory, but also the directions of fault movements suggested by P-wave patterns match the motions at plate boundaries predicted by the theory. P waves are simply a series of compressions and expansions in bedrock, with compressions recorded as upticks (∧) and expansions recorded as downticks (∨) on the seismogram. Studies of P waves confirm that earthquakes are generated by compression at convergent boundaries, by tension at divergent boundaries, and by horizontal shear at transform boundaries. Earthquakes at transform boundaries also satisfy the theory's requirement that they occur between offset mid-ocean ridge segments.

foreshocks
A series of lower-magnitude earthquakes that may directly precede a higher-magnitude earthquake.

aftershocks
A series of lower-magnitude earthquakes that may directly follow a higher-magnitude earthquake.

Figure 3.11 The San Andreas is a transform fault that formed when the North American plate intersected the East Pacific Rise.

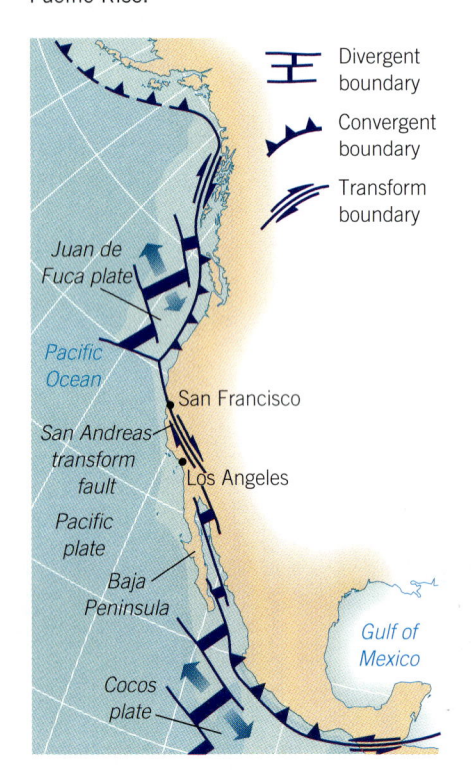

Thus there is compelling evidence that plate boundary interactions—divergence, convergence, and transform—are responsible for over 99 percent of earthquakes. With the exception of deep-focus earthquakes of subduction zones, where the cold lithosphere is jammed far into the mantle, the vast majority of earthquakes occur within 30 to 40 kilometers of the surface—that is, within the upper parts of the plates. There, the rocks are rigid enough to store strain energy and brittle enough to rupture suddenly and generate an earthquake. Toward the base of the plates, however, higher temperatures and pressures soften the rocks somewhat, which causes them to flow plastically, rather than rupture, in response to stress.

The difference in response to stress between the upper and lower parts of the plate may be responsible for most earthquakes. According to one hypothesis, as two plates move past one another, the warm, soft lower layers of the plates move freely and smoothly, dragging the rigid upper layers of the plates with them. The progress of the upper layers is halted wherever they get hung up at an *asperity*, a rough point along the fault plane. Strain builds as the lower layers continue to move and the upper layers remain locked. Finally, the shear stress overcomes the hang-up, and the rocks lurch to unstrained positions, releasing energy in the form of seismic waves—the elastic-rebound effect. The greater the accumulation of strain at the asperity, the more powerful the earthquake will be. **Foreshocks** usually precede the main event as the rocks in the vicinity of the asperity begin to crack. **Aftershocks** follow as the rocks along the fault readjust and strain is transferred to the next asperity. Then there is a period of quiet while strain accumulates again.

The San Andreas Fault

The San Andreas fault is a transform boundary connecting two offset segments of the mid-ocean ridge, one in the Gulf of California and the other off the California-Oregon coast (Figure 3.11).

A slice of California that includes Los Angeles and the Baja Peninsula is attached to the Pacific plate and is gliding northwestward relative to the North American plate along the San Andreas fault. The overall rate of movement is about 5 centimeters per year. Judging from the location of identical rock formations that are now separated by the fault, the total displacement of the west side of the San Andreas fault relative to the east side has amounted to at least 350 kilometers since the fault has been in existence (Figure 3.12). Some segments of the fault appear to be moving smoothly; they are peaceful and earthquake-free. Great earthquakes occur along fault segments that remain locked and release their strain energy sporadically.

The Los Angeles Faults

California's problems are not confined to the San Andreas fault; the state is sliced by active faults of all kinds. The region is subject to complicated stresses generated as the San Andreas fault grinds northward and shears, compresses, and extends the crust at many locations along the way (Figure 3.13). Southern California in general and the Los Angeles area in particular are especially vulnerable.

Notice in Figure 3.13 that the San Andreas fault makes a sharp westward bend about 100 kilometers northeast of Los Angeles. As a result, the Pacific plate, whose relative motion is to the northwest, abuts obliquely against the North American plate. The strain energy thus generated is transferred to the Los Angeles region, which is being compressed at the average rate of 1 centimeter per year. The crust has responded by buckling and fracturing. A 9400-square-kilometer network of faults underlies Los Angeles and the San Fernando Valley. Both are deep basins that hold thick accumulations of sedimentary rock, and many of the faults are in these rocks.

Earthquakes 69

Figure 3.12 Offset stream beds show the direction of movement along the San Andreas fault. The San Andreas is a right lateral fault, meaning that when you face the fault, the movement is to the right.

The magnitude 6.7 Northridge earthquake of January 17, 1994, was typical of those that plague this region. Its focus was located about 14 kilometers deep along a gently dipping *thrust fault*. In a fault of this type, the block of the crust overlying the fault plane slides upward toward the Earth's surface with respect to the block beneath the plane. The fault occurs in response to compression, and the result shortens the crust. Strain energy may remain locked within a Los Angeles or San Fernando Valley thrust fault for a decade or a century, until the fault suddenly ruptures and moves a meter or two, setting off an earthquake. The Northridge event and the aftershocks that followed knocked down freeway overpasses, destroyed or severely damaged hundreds of buildings (including a large shopping mall), ignited numerous fires, burst water mains and gas lines, necessitated the evacuation of hospitals, and rendered unsafe the enormous Los Angeles Coliseum. In all, 55 people were killed and many billions of dollars in damage occurred.

Intraplate Earthquakes

Fewer than 1 percent of all earthquakes occur away from plate boundaries, but some of the more powerful of these

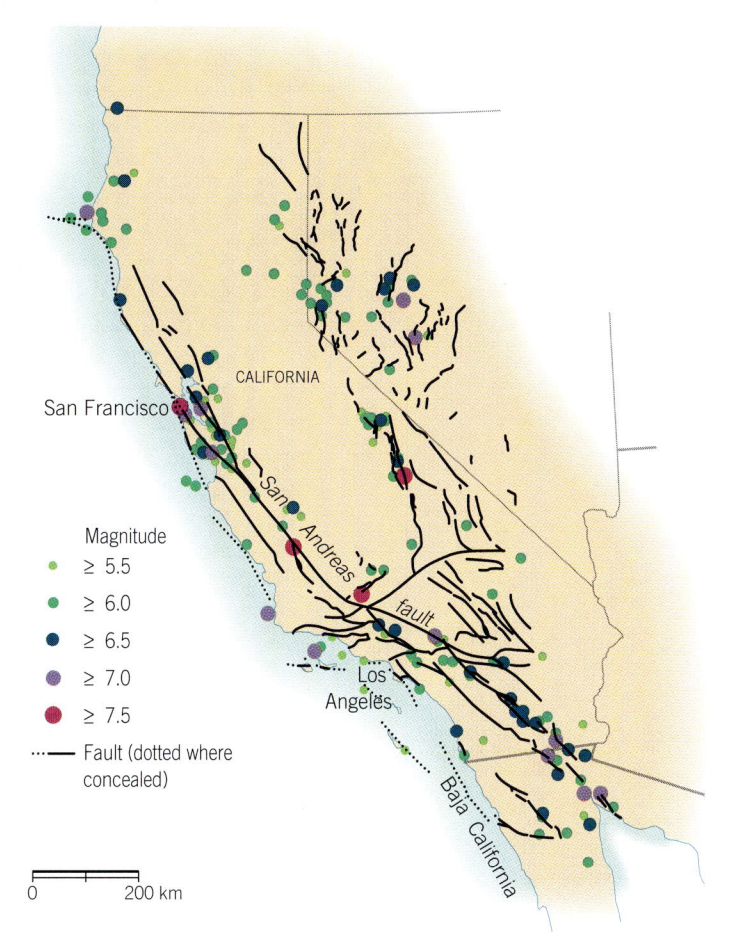

Figure 3.13 Some branches of California's complex network of faults. The stress that drives the Los Angeles faults is caused by the sharp bend in the San Andreas fault about 100 kilometers northeast of the city.

Figure 3.14 This seismic risk map of the United States shows that there are regions far away from plate boundaries where we can expect significant earthquake activity. Note the high degree of risk attributed to the New Madrid, Missouri, region. (Courtesy of National Oceanic and Atmospheric Administration.)

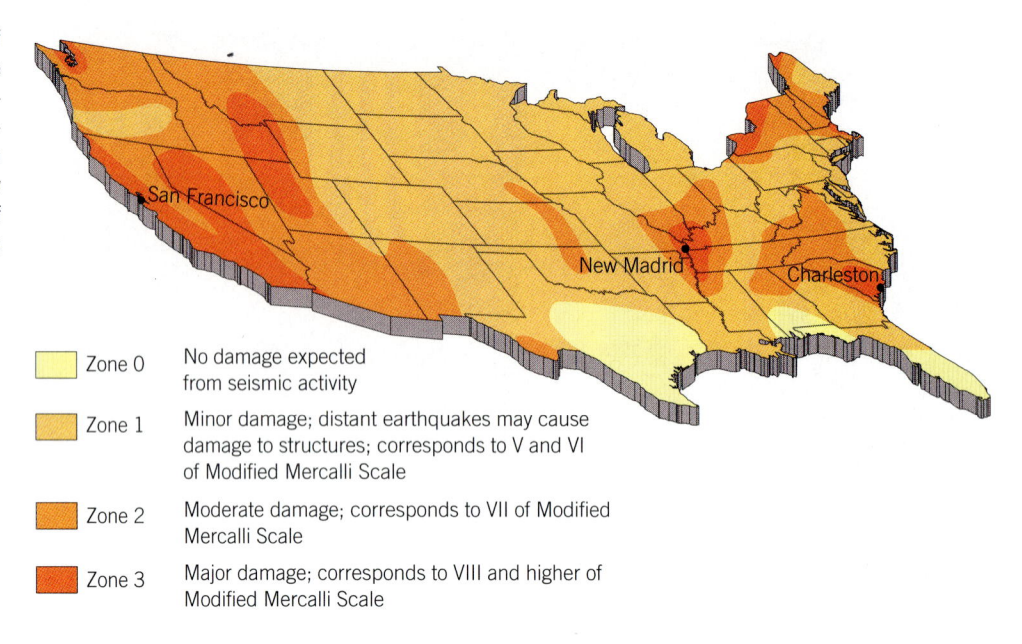

Zone 0 — No damage expected from seismic activity

Zone 1 — Minor damage; distant earthquakes may cause damage to structures; corresponds to V and VI of Modified Mercalli Scale

Zone 2 — Moderate damage; corresponds to VII of Modified Mercalli Scale

Zone 3 — Major damage; corresponds to VIII and higher of Modified Mercalli Scale

intraplate earthquake
An earthquake whose epicenter is far from a lithospheric plate boundary.

intraplate earthquakes have caused rampant destruction to populated regions. The magnitude 7.6 earthquake that hit the heavily populated industrial city of Tangshan, China, in 1976 killed at least 250,000 people.* In fact, China, far from plate boundaries, has a long history of devastating earthquakes.

Closer to home, the most powerful series of earthquakes in the history of the United States took place not in San Francisco or Anchorage but in New Madrid, Missouri, in 1811 and 1812 (magnitudes were later estimated to have been as high as 8.5). The shifting crust rerouted the Mississippi River, and significant damage was recorded as far away as Cincinnati, Ohio. From time to time, a number of moderate-sized earthquakes have shaken other supposedly stable regions of the interior of North America. The 1886 Charleston, South Carolina, earthquake not only wrecked the city but also sent out vibrations powerful enough to rock Boston and Chicago. No location in the United States—or the world, for that matter—should be considered risk-free (Figure 3.14).

We do not fully understand the mechanics of intraplate earthquakes. One plausible theory relates them to the reactivation of ancient faults, many deeply buried in the continental crust, that formed during episodes of rifting and subduction that predate present plate boundary interactions. As the present plates slide slowly over the asthenosphere, stresses are transferred to the old faults. Like the weak links in a chain, these faults then release their strain energy as earthquakes.

Forecasting Earthquakes

In 1975, a magnitude 7.4 earthquake that struck Haicheng, China, set off 90 seconds of intense ground shaking that reduced 90 percent of the city's buildings to rubble. With 3 million people living near the epicenter, this earthquake could have been another mind-numbing tragedy, but it was not. Chinese seismologists were aware that Haicheng was situated astride a major fault zone. They had also noted that over a period of years, the earthquake epicenters along the fault were drawing ever closer to the city and that the earthquakes were increasing in magnitude and frequency. Analysis of this trend enabled them to predict the earthquake with enough lead time to allow for evacuation of the city and surrounding villages, thus saving tens of thousands of lives.

---

*The official death toll was 250,000 people, but other estimates have placed it at 650,000.

Haicheng 1975 marked the first fully documented, successful earthquake prediction. Unfortunately, the Chinese were unable to prepare for the magnitude 7.6 earthquake that struck Tangshan without warning one year later. Because Haicheng is closer to Tangshan than Los Angeles is to San Francisco, no two events so closely matched in time, space, and magnitude better illustrate the potential benefit of earthquake prediction research.

Given the imperfect state of our knowledge regarding earthquakes, many geologists prefer to distinguish between a *prediction* and a *forecast*. A prediction specifies the time, location, and magnitude of an event within narrow limits of uncertainty, whereas a forecast projects trends over a broad time frame. Each has its purpose. Forecasts are used for long-range planning: legislating zoning and building codes, setting up evacuation routes and emergency procedures. Predictions are used for immediate preparations: evacuating people, turning off gas lines, stopping trains, closing off bridges. Unfortunately, methods of prediction have so far proved to be largely unreliable, although we have made progress in forecasting earthquakes. In the following section, we describe some of the strategies geologists employ to forecast and predict earthquakes.

Seismic Gaps

Close examination of earthquake records for the seismic belt rimming the Pacific reveals that segments that have recently experienced a great earthquake alternate with those that have long been quiescent—so-called **seismic gaps.** An interesting hypothesis uses this observation to forecast great earthquakes. As an earthquake relieves the strain at an asperity, the strain shifts to neighboring asperities that are currently quiescent—the seismic gaps. According to the seismic gap hypothesis, quiet segments of the fault bordered by the epicenters of recent great earthquakes are the places to look for the next great earthquake.

The gap hypothesis has had some successes; for example, the magnitude 7.1 Loma Prieta earthquake of 1989, which devastated downtown Santa Cruz and rocked San Francisco 95 kilometers to the north, occurred within a segment of the San Andreas fault that USGS geologists had identified as a seismic gap. However, in other large earthquakes, the epicenters did *not* fall within seismic gaps; that is, they occurred in places not predicted by the model. Some geologists thus conclude that the model is either mistaken or in need of refinement.

seismic gap
A seismically inactive segment of a fault within which strain is accumulating; these gaps are bracketed by the epicenters of relatively recent great earthquakes.

Recurrence Studies

One of the problems in testing forecasting tools such as seismic gaps is the insufficiency of historical records. They do not extend back far enough in time to establish reliably how often major earthquakes have occurred in a given locale—that is, to establish their *recurrence intervals.*

Paleoseismology, as applied by geologist Kerry Sieh and his colleagues at the California Institute of Technology, is probably the most accurate method of extending the record back in time; their work in southern California has added enormously to our knowledge of the history of the San Andreas fault. They have found that Pallet Creek, which crosses the fault about 55 kilometers north of Los Angeles, has been repeatedly disrupted by sudden movements along the fault. Each episode accomplished two things: it cut the layers of riverbed sediments previously deposited, and it temporarily dammed the river, allowing a layer of organically rich peat to be deposited directly on top of the faulted sediment. Sieh obtained the age of each peat layer by the carbon-14 method, and, using these dates, he was able to closely date each episode of fault movement.

Sieh has been able to discern 12 major earthquakes during the past 1400 years in the Pallet Creek sediment. Recurrence intervals vary between 50 and 300 years, the average being 140 to 150 years. In fact, it has been almost 140 years since the last great earthquake struck the region in 1857, an interval uncomfortably close to the average determined by

See Chapter 16, Geologic Time, for a discussion of the carbon-14 and other radiometric dating methods.

Figure 3.15 Probabilities of major earthquakes occurring along various segments of the San Andreas and Hayward faults over the next 30 years. There is a 90 percent probability of an earthquake of magnitude 6 or greater occurring at the tiny village of Parkfield, California, population 34. In the late 1980s, seismologists made the firm prediction that this earthquake would occur before 1993. That the event still had not occurred by late 1997 illustrates the uncertainties of earthquake prediction.

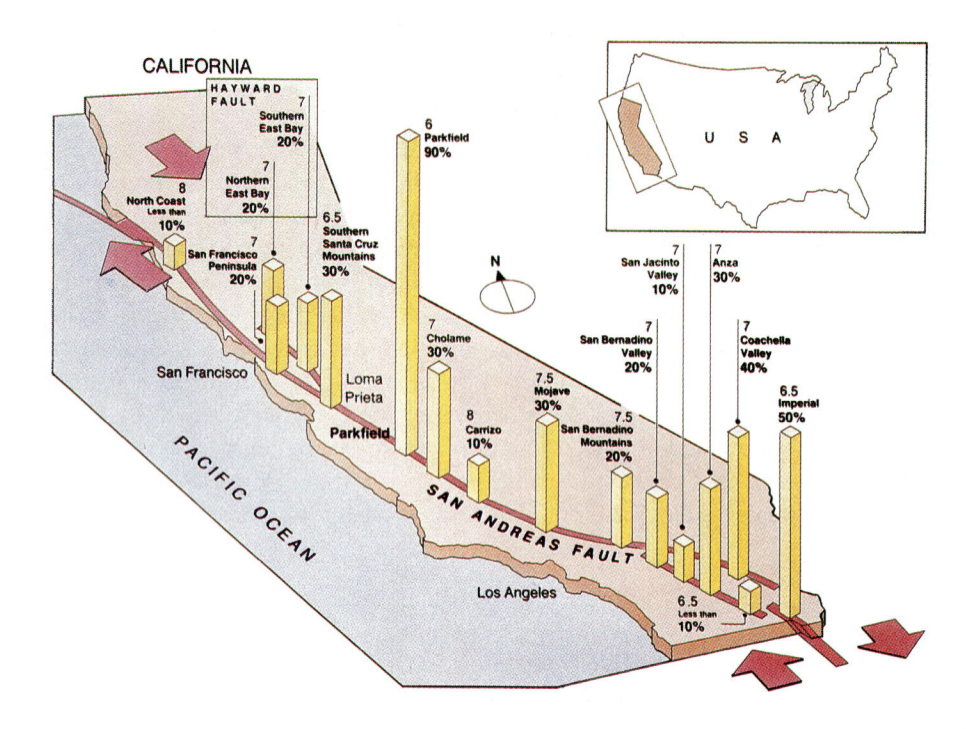

Sieh. For each year that an earthquake does not occur, the chance that one will occur increases by about 2 to 5 percent, and each year, there is a greater probability that it will carry a more wicked punch.

The recurrence studies suggest that Los Angeles is at greater risk than previously thought. San Francisco, on the other hand, may have more breathing room (Figure 3.15). The recurrence interval for the northern segment of the San Andreas fault is about 150–300 years, and the last great earthquake struck in 1906, not quite a century ago. However, we must remember that recurrence studies offer only one line of evidence. According to seismic gap evidence, the rocks of the San Francisco region remain locked, and estimates of the strain accumulating in the region are similar to 1906 levels. The 1989 Loma Prieta earthquake that rocked San Francisco was not the "big one" forecast for that great city; its epicenter was 95 kilometers to the south. More important, the "big one" is expected to release 50 times the energy and to be accompanied by four times the fault displacement of the Loma Prieta rupture.

Precursor Studies

Subtle changes that warn us that the rocks are close to failure and an earthquake is imminent may take place in the vicinity of an active fault. These changes, deviations from normal or expected patterns, are called *precursors.* Some of the more promising precursors are accelerated land uplift, increased emissions of radon gas from wells, a precipitous rise in groundwater levels, and fluctuations in the frequency of small earthquakes and in the P-wave velocity of the waves generated. All of these changes are related to the response of the rocks to accumulation of strain energy. Ideally, changes would build to a crescendo just before the earthquake so that the geologist tracking the trends could issue warnings to the appropriate authorities.

Though all of these precursors have been detected prior to one earthquake or another, not all of them have been observed prior to the *same* earthquake, nor do they always follow the same patterns. In some instances, precursors have pointed to earthquakes that did not, in fact, occur; by the same token, earthquakes have struck without

Figure 3.16 Upon approaching shallow water, a tsunami is slowed by friction, its length is shortened, and it rises to great height before it breaks against the shore.

Speed: 60 km/hr 300 km/hr 800 km/hr
Depth: 20 m 800 m 5000 m

Displacement

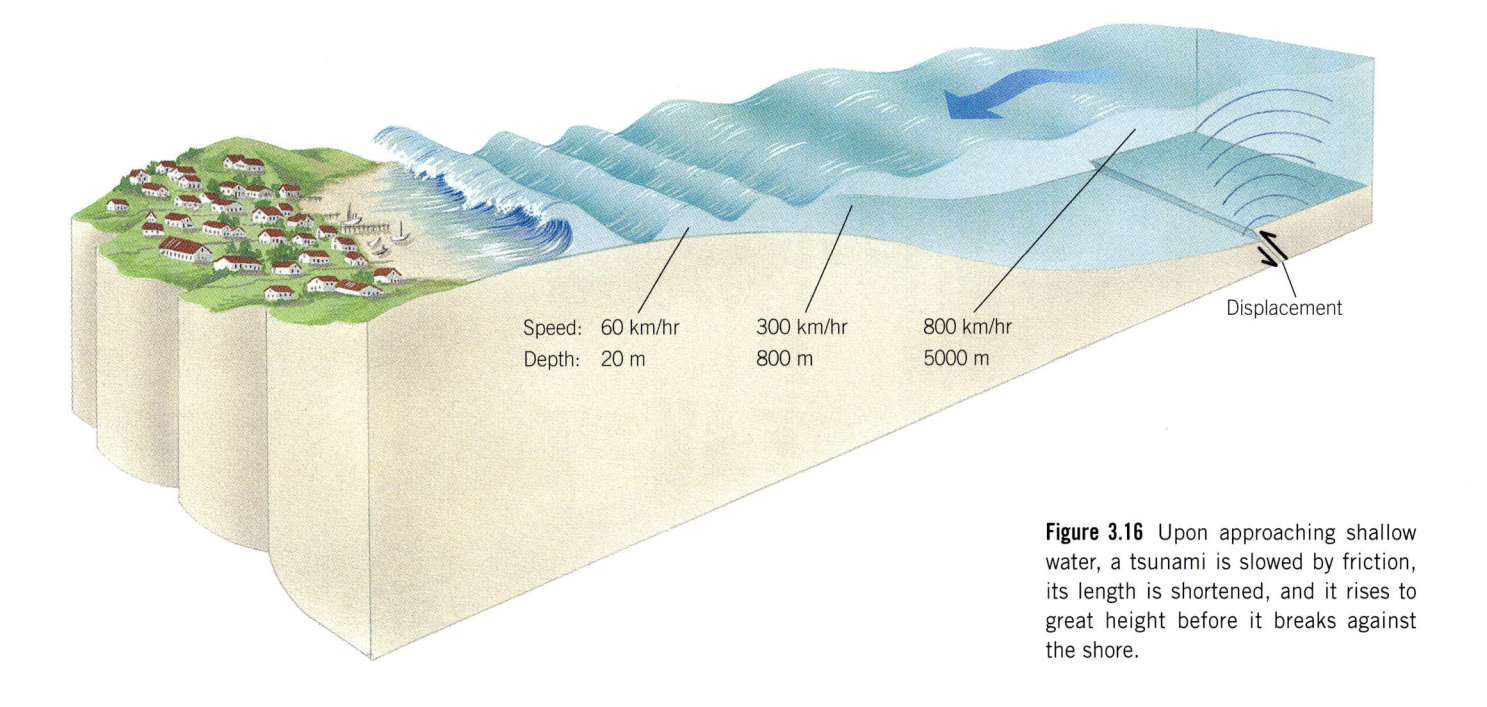

these warning signs. For these reasons, geologists have yet to determine if any precursors are reliable enough to form the basis of a prediction or a forecast. At our present level of knowledge, predicting earthquakes is as much an art as it is a science.

Recently, however, there has been some success in identifying those submarine earthquakes that are likely to trigger a devastating **tsunami.** Tsunamis are giant ocean waves generated by submarine earthquakes. The vibrations set off by a fault in the oceanic crust are transferred to the water surface and spread across the ocean in a series of low waves (Figure 3.16). Most tsunamis originate above the subduction zones that rim the Pacific basin. There, cold and brittle oceanic crust fractures as it descends into the mantle at the deep-sea trenches.

With amplitudes of less than a meter and wave-lengths of many kilometers, a tsunami is difficult to detect in the open ocean. Nevertheless, it travels at jet plane speeds (800 to 1000 kilometers per hour). Then when it reaches the shallow depths of the coast, friction slows the wave down and, at the same time, causes it to roll up into a wall of water as high as 30 meters. These effects are intensified in the confined spaces of bays where most harbors are located—thus the name *tsunami,* Japanese for "harbor wave." The wall of water descending upon a harbor is preceded by an emptying of the harbor, and many deaths are attributed to the curiosity of people who run out to examine the exposed harbor bottom.

The Japanese coastline is especially vulnerable; history records at least 15 major disasters there, some killing hundreds of thousands of people. For these reasons, submarine earthquakes are monitored closely, and an early-alert system warns of the possibility of a tsunami striking coastal regions. However, this system is not the complete solution, for a large-magnitude earthquake may not trigger a tsunami, whereas a relatively minor one may send huge waves crashing into an unsuspecting harbor.

Recently, though, geophysicists have developed a promising method of identifying those earthquakes likely to lead to a tsunami. They have found a correlation between the period (vibration rates) of tsunamis and certain very-long-period seismic waves. Earthquakes that have an appreciable amount of their energy in this long-period range are likely to trigger a sizable tsunami. Because seismic waves travel at speeds on the order of kilometers per second, rather than kilometers per hour, they outrun the tsunamis to recording stations. Their arrival may provide seismologists with sufficient lead time to issue warning bulletins.

tsunami
A large, often deadly sea wave set in motion by an undersea earthquake.

Figure 3.17 Rescue workers toil atop a poorly constructed building in Spitak, Armenia, that was pancaked by the magnitude 6.9 earthquake on December 7, 1988.

Minimizing Earthquake Damage

Why is it that in 1988, a magnitude 6.9 earthquake killed 25,000 people but in 1992 a 7.5 earthquake—six times stronger—killed only one unfortunate individual? The answer is that the 6.9 earthquake struck a heavily populated region of Armenia and caused thousands of old and poorly constructed buildings to collapse on their inhabitants (Figure 3.17). The 7.5 quake was centered in Landers, California, in the sparsely populated Mojave Desert region. Buildings there are few and far between, and the newer ones have been constructed according to California's earthquake building codes. These two earthquakes are testimony to the adage that earthquakes don't kill people, buildings do. The major problems of earthquakes are the result of civilization—namely, the types of structures built and the patterns of land use brought on by population growth, custom, and urban-suburban development.

After an earthquake, surveys are taken and the various Mercalli intensities are mapped so that patterns of damage may be established. (For more on the Mercalli intensity scale, see Table 3.3.) We have learned much from these surveys. For example, we can observe the connection between the type of subsurface in the region and the levels of earthquake damage that occurred. In the 1989 Loma Prieta earthquake, the Marina District of San Francisco was the site of severe residential damage. Figure 3.18(a) shows that in the 1906 San Francisco earthquake, most of the direct damage also occurred in the Marina District. Figure 3.18(b) explains why. Homes in the Marina District were built on landfill made of unconsolidated sediment dredged from San Francisco Bay. Saturated sediments have a tendency to lose contact when shaken, a process called **liquefaction.** This is what transformed the solid ground of the Marina District into loose mud incapable of supporting loads; thus foundations buckled and buildings pitched into the street.

Plans to minimize the damage of earthquakes should include surveys of the secondary effects of earthquakes as well as the more obvious damage. Investigators must make careful note of the proximity of population centers to potential landslide hazards and unsafe dams; of the possibility of fires and the disruption of communications and utilities; and of the possibility that damage to water mains and sewage pipelines can lead to contamination and disease.

liquefaction
The transformation of saturated sediment or soil to liquid when ground shaking causes the particles to lose contact.

Construction Problems

Engineers should follow two important design principles to make a building earthquake-resistant. The first is to ensure that all beams and columns supporting the structure are firmly connected to one another and to the ground. The second principle is to make sure that the columns or beams that support the building's walls and floors transmit load directly to the ground. This design will minimize the effects of the strong shearing stresses that develop when the building is shaken from side to side.

Buildings, dams, and bridges are basically inverted pendulums, with their lower ends anchored to the ground and their upper ends free to vibrate. Tall structures are likely to magnify ground shaking, creating a tough problem for engineers working in downtown areas of modern cities. Conditions on the ground may not match those on, say, the thirtieth floor of an improperly designed building, which may sway like wheat in the wind. The engineer must design the building to flex during an earthquake and in this way dissipate energy within the steel, concrete, brick, and mortar of the building. Of course, the building must not flex so much as to rip the floors and roof from the support columns or to shear off the support columns at their base. As you may guess, designing a skyscraper to withstand a great earthquake can be expensive; the cost of damage prevention may increase tenfold for every order of earthquake magnitude required.

Faulty design and defects in construction are the reasons most buildings collapse during an earthquake and the reasons most people are killed. For example, during the 1985 Mexico City earthquake, traditional adobe houses crumbled; modern buildings with odd shapes were twisted apart. Adjacent short and tall buildings, vibrating at different rates, bumped against one another, which caused the short buildings to collapse. On the other hand, certain earthquake-resistant buildings survived; those built on specially designed rollers were effectively decoupled from ground vibrations and survived with little or no damage.

Earthquakes are not new to Mexico City, and many of its buildings were constructed according to strict codes. California's engineers thus took careful notes of the earthquake's effects. Attracting special attention was the poor performance of modern buildings with large open spaces on the lower levels: corner buildings with high lobbies, hotels and office buildings with parking garages, and the like. Intense ground shaking caused the columns in these open areas to fail, because the upper floors of the buildings—packed with equipment, fixtures, and furniture—were less flexible and increased stress on the lower, open portions. Unfortunately, older buildings in California were not constructed with the benefit of this

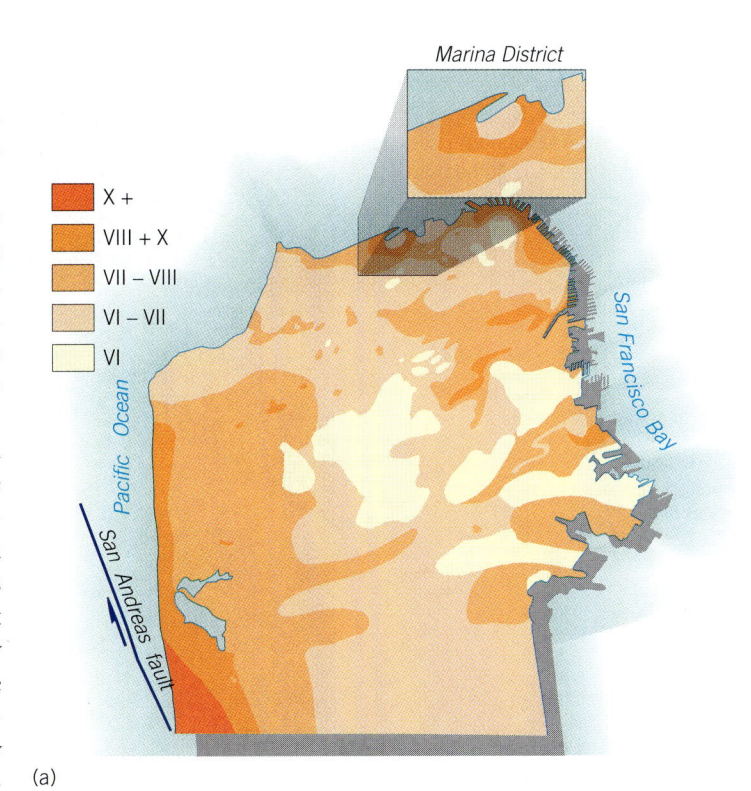

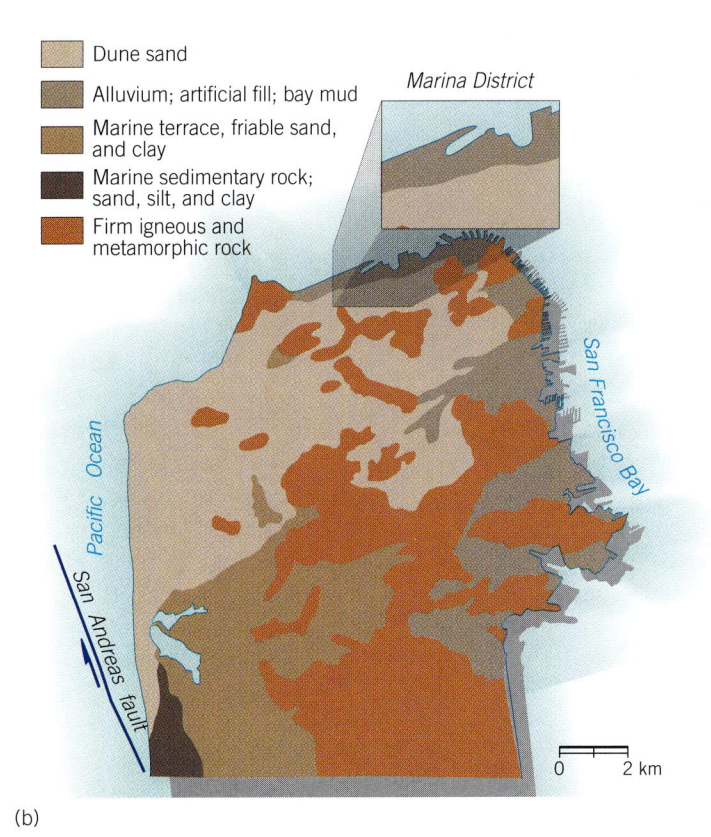

Figure 3.18 The relationship between (a) earthquake intensity and (b) subsurface conditions. Houses built on a landfill made of unconsolidated sediment dredged from San Francisco Bay are especially vulnerable, while those built on firm igneous and metamorphic bedrock are better protected. (Adapted from *Earthquakes: A Primer* by Bolt © 1978 by W. H. Freeman and Company. Used with permission.)

Figure 3.19 Failure of freeway overpasses constructed prior to 1971 caused by the 1994 Northridge earthquake. Those overpasses built or retrofitted to more modern standards held up better.

information. In the 1994 Northridge earthquake, the worst single disaster involved a three-story apartment complex; built over parking garages in the 1970s, it collapsed and killed 15 of its residents. Most of the freeway overpasses that collapsed were built according to pre-1971 standards and were scheduled for upgrading. Those that had been reinforced according to more recent standards did not collapse (Figure 3.19).

Buildings have different vibration periods: short buildings vibrate more rapidly than tall ones. In designing an earthquake-resistant building, engineers must minimize *resonance effects*—conditions that result when the building and ground vibrate at the same rate. Because of resonance, the building will sway with ever greater amplitude, a nightmarish effect. Without proper design precautions, a resonance condition can turn a low-magnitude earthquake into a major disaster.

No amount of planning can prevent catastrophe when a strong earthquake causes ground shaking beneath a densely populated city. Nevertheless, seismologists and engineers were unpleasantly surprised at the scope of the damage sustained during the magnitude 6.9 earthquake that struck Kobe, Japan, in January 1995, one day before the first anniversary of the Northridge disaster. The ground shaking, which lasted only 20 seconds, was especially intense because the shallow focal depth of the earthquake—only 10 kilometers—caused the energy released to be coupled more directly to surface structures such as buildings and elevated roadways. Although many were built according to the world's strictest design codes, over 50,000 structures collapsed, pitched over, or were damaged beyond repair. Over 5000 people were killed and 300,000 were left homeless. The total cost of the damage has been estimated at over $100 billion.

The lesson—and challenge—of Kobe is in how much we have yet to learn about earthquakes. Japanese scientists have been using signals from satellites and distant stars to measure minute shifts in the Earth's crust; they have drilled wells 2 kilometers deep and packed them with equipment to track movements at depth; and they routinely monitor oceanic faults in order to chart the grinding motion of the Philippine, Pacific, and Eurasian plates, the ultimate cause of the earthquakes that plague Japan. Yet the Kobe earthquake went unpredicted. Engineers also have a long way to go in planning for safety. Not only did many supposedly earthquake-resistant structures fail in Kobe, but in the months following the Northridge earthquake, it was discovered that the joints binding many earthquake-resistant buildings in Los Angeles may have sustained dangerous stress fractures.

The velocities of P and S waves are dependent upon the properties of the materials through which they travel. They are calculated using the following formulas:

$$V_P = \sqrt{\frac{C + 4/3R}{\rho}}$$

$$V_S = \sqrt{\frac{R}{\rho}}$$

V_P = P wave velocity

V_S = S wave velocity

C is a measurement of the material's incompressibility, or resistance to volume reduction.

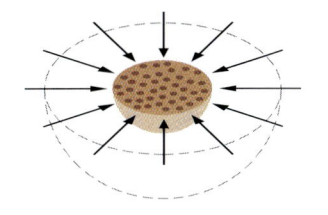

R is a measurement of the material's rigidity, or resistance to shear.

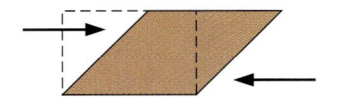

ρ is the density of the material.

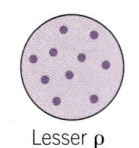

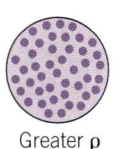

Lesser ρ Greater ρ

Figure 3.20 Important information about the properties of P and S waves. Notice that the velocity of an S wave depends upon the rigidity (R) of the substance through which it travels—that is, the resistance of the substance to shear. S waves can pass only through solids, because liquids offer no resistance to shear. The velocity of a P wave depends upon the incompressibility (C) of the substance as well as its rigidity. Thus a P wave is capable of passing through both liquids and solids.

Seismic Waves in the Earth's Interior

We learned earlier that an earthquake releases several types of seismic waves. Body (P and S) waves travel through the interior of the Earth, and surface waves travel near the circumference of the Earth, dying out with depth. We also learned that the velocities of these waves depend upon the elastic properties and densities of the materials through which they travel (Figure 3.20). Therefore, an abrupt change in the velocity of a body wave at a given depth signifies that the wave has intersected a boundary between rocks of different densities and/or elastic properties.

Wave Refraction and Reflection

A phenomenon common to all waves—light, seismic, sound, and water—is that they will bend, or *refract,* upon entering a medium in which their velocities change (Figure 3.21). If a wave slows down upon entering the new medium, its path will bend away from the boundary surface; if it speeds up, its path will bend toward the surface. No refraction occurs when a wave enters a new medium at right angles to the boundary; although the velocity does change upon entering, there is no deviation in wave path.

A wave may also exhibit varying degrees of *reflection* upon striking a boundary. Figure 3.22 illustrates that reflection may be total or partial, depending upon the nature of the material at the boundary and the angle of the incoming wave. In partial reflection, a portion of the incoming wave energy is also refracted at the boundary, so that a single wave may give birth to two waves whose paths differ from the original.

Figure 3.21 The refraction of a light beam as it passes through a glass of water.

Figure 3.22 The energy of a wave striking the boundary of another medium may generate both reflection and refraction effects.

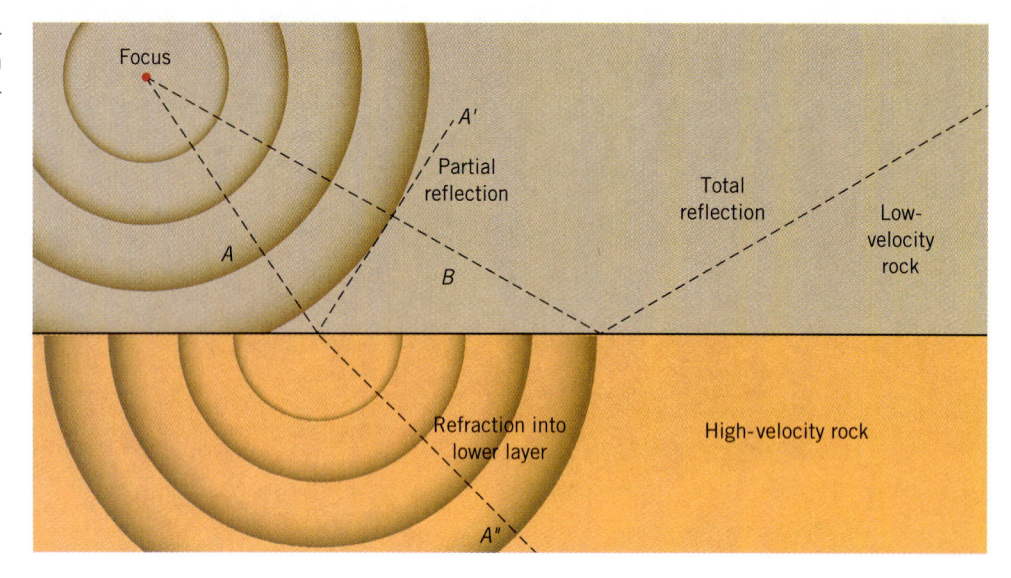

Of course, the Earth is not transparent—we cannot see the boundaries between layers, nor can we see the abrupt changes in wave velocity and path. But we can infer the existence of these features by examining seismograms from stations around the globe. The basic idea is that refraction and reflection alter the paths of the waves and their travel times. For this reason, the waves spreading from an earthquake focus may fail to arrive at certain stations and show up at others earlier or later than expected. By observing which stations were skipped, and by recording the precise arrival times of the waves at the other stations, researchers can determine the paths of waves, the depth of the boundaries, and the velocities of the waves as they travel through the Earth.

Detection of the Crust, Mantle, and Core

It follows from the previous discussion that if the Earth were of uniform composition, seismic waves would travel through the interior with constant velocity. That is not the case, however. Seismic waves speed up as they travel through the interior, indicating gradual changes in density and elastic properties with depth. The reconstruction of their paths, using data from many stations, is shown in Figure 3.23. Notice the thick region between about 20 and 2900 kilometers in depth, where the waves curve smoothly and surface 105° from the epicenter. This enormous concentric layer of the Earth is called the **mantle.** It comprises about 80 percent of the Earth's volume and 67 percent of the Earth's mass. The mantle is bounded by the outer core below and the crust above.

Stations located between 105° and 142° of an earthquake epicenter receive no direct P and S waves. The blanked-out region is called the **shadow zone** (see Figure 3.23). Direct P waves are received beyond the shadow zone, but they are much delayed. Direct S waves, on the other hand, are simply not recorded beyond 105°. As the term *shadow zone* implies, these drastic changes are explainable only by the existence of a major obstacle at great depth that slows down and refracts P waves and completely blocks the S waves. This obstacle is the Earth's 2240-kilometer-thick **outer core.**

The fact that P waves travel through the outer core, whereas S waves do not, leads geologists to believe that the outer core must be molten. Liquids lack the internal strength (or rigidity) to support the shearing motions by which S waves are propagated;

mantle
The layer of the Earth located between the base of the crust and outer core whose depth extends from about 20 to 2900 kilometers.

shadow zone
A region between 105° and 142° from the epicenter of an earthquake where no direct P and S waves reach the Earth's surface, owing to the properties of the outer core.

outer core
The upper layer of the Earth's core whose depth extends from about 2900 to 5140 kilometers. It is presumed to be molten.

Earthquakes

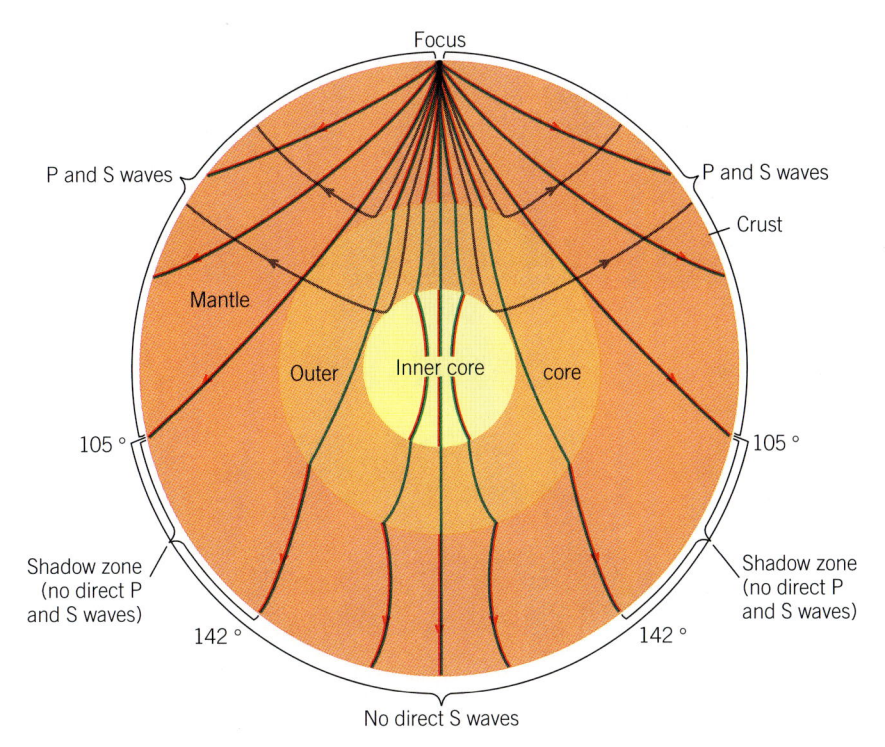

Figure 3.23 Each of the Earth's layers is characterized by distinctive seismic wave patterns. The region of smoothly curving ray paths (up to 105° from the focus) is the mantle. The presence of a shadow zone (from 105° to 142°) and P-wave reflections are evidence of a liquid outer core. P waves that arrive earlier than expected at stations located 180° from the focus are evidence of the existence of a solid inner core.

that these waves are not detected beyond 105° of the epicenter means the interior is molten at outer-core depths. At the same time, a molten outer core explains why P-wave velocity decreases: as shown in the equations of Figure 3.20, one of the factors that adds to P-wave velocity—rigidity, or the resistance to shear—is lost upon entering a liquid. However, P waves that travel through the center of the Earth arrive at the opposite side earlier than would be expected if the core were completely molten. This observation, along with more complex reflection and refraction patterns, implies that the solid **inner core,** 2460 kilometers in diameter, rests like the hard pit in a soft peach at the very center of the Earth.

Because P waves travel faster through the mantle than through the **crust,** studies of P-wave travel times were used to determine the boundary between the *upper mantle* and the crust. The boundary is referred to as the **Mohorovičić discontinuity,** or **Moho,** in honor of its discoverer, the Croatian scientist Andrija Mohorovičić (1857–1936). The depth to the Moho is quite variable. It ranges from 2 to 10 kilometers beneath the ocean floor to an average depth of about 35 kilometers below the continents. It may reach depths of 70 kilometers under certain mountain ranges.

inner core
The solid spherical center of the Earth about 2460 kilometers in diameter.

crust
The outermost layer of the Earth, about 2 to 50 kilometers thick, representing less than 0.1 percent of the Earth's total volume.

Mohorovičić discontinuity (Moho)
The surface that defines the boundary between the Earth's crust and mantle.

Structural and Compositional Divisions of the Interior

Recall from Chapter 2 that the crust constitutes the upper part of a much thicker zone, the **lithosphere.** It is this zone that is divided into the rigid plates whose interactions are so closely connected with the occurrence of earthquakes, volcanoes, mountain building, and continental drift. These plates are able to slide across the globe because they rest on a relatively soft zone within the mantle—the asthenosphere. So it is fair to ask, What is the evidence of the existence of the lithosphere and the asthenosphere?

The main evidence rests on a general similarity of seismic wave velocities within each layer that contrasts markedly with the seismic wave velocities between them. We have just seen that there is a jump in velocities at the Moho, but the behavior of these

lithosphere
A rigid zone of the Earth that includes the crust and a sliver of upper mantle and that rests directly on the asthenosphere.

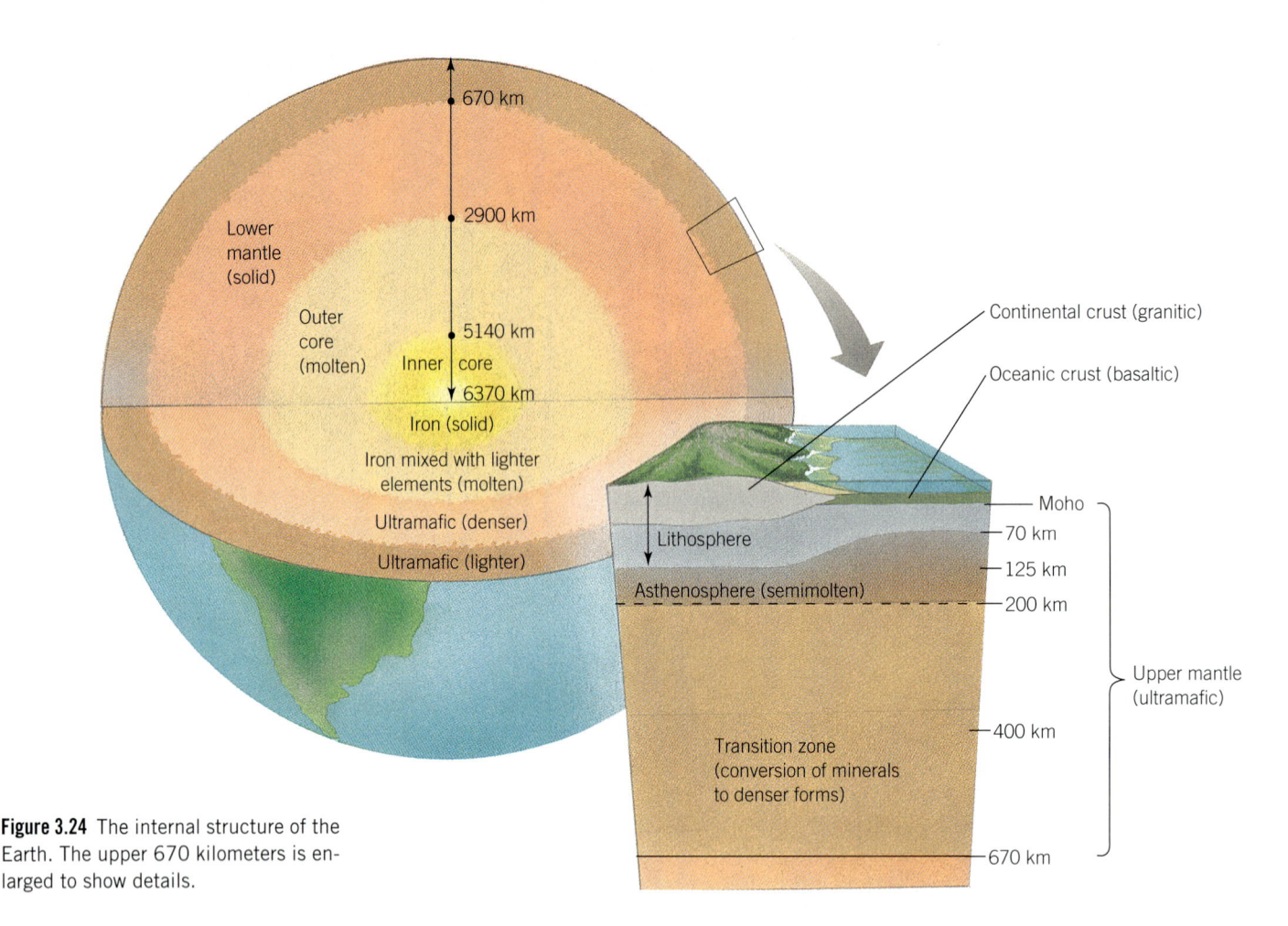

Figure 3.24 The internal structure of the Earth. The upper 670 kilometers is enlarged to show details.

Note about mafic and ultramafic rocks: Mafic refers to rocks with a composition rich in iron and magnesium silicate. Ultramafic rock is so named because it is higher in iron and magnesium content than mafic rocks.

waves nevertheless indicates that they pass through rigid rock from the Earth's surface to a depth of about 100 kilometers (Figure 3.24). We have, then, a relatively thin crust and a part of the upper mantle that display a more or less coherent set of physical properties—namely, rigidity and the ability to move together in large slablike units. Thus the lithosphere is largely defined by similarities in physical properties and contains subunits of different chemical properties: a granitic continental crust, a basaltic oceanic crust, and a thin slice of uppermost mantle that underlies both.

We observed in Chapter 2 that the magma that erupts from beneath the central rift valleys of the mid-ocean ridges and builds the basaltic oceanic crust is derived from the upper mantle. This magma is *mafic* in composition—meaning that in addition to containing lighter elements such as silicon, oxygen, and aluminum, it is fairly rich in the denser elements iron and magnesium. What might be the composition of the upper-mantle rock that gives rise to this magma?

One possibility is that the upper mantle is basalt that melts to yield mafic magma of the same composition. But when geologists experimentally subject basalt to upper-mantle pressures and pass seismic waves through it, they find that the wave speeds do not match the speeds observed in the upper mantle. However, if they conduct the same kinds of experiments on an *ultramafic* rock, one richer in iron and magnesium than basalt, the seismic waves in the sample match perfectly those observed in the upper mantle. Furthermore, geologists find that when they heat the ultramafic rock gradually, it melts in stages, yielding a mafic magma first before it converts completely to liquid. They therefore conclude that the upper mantle is composed of ultramafic rock that partially melts to form the mafic magma of the oceanic crust (see Figure 3.24).

If the upper mantle is ultramafic, then what accounts for the decrease in seismic wave velocities between approximately 100 and 200 kilometers, the zone within the upper mantle known as the **asthenosphere**? Experimenters have shown that if they subject ultramafic rock to the pressures at that depth and raise its temperature to the point where a tiny fraction begins to melt, seismic wave velocities will match those present in the asthenosphere. The high temperatures lessen the rigidity of the rock, slowing the passage of P and S waves. The lithospheric plates rest directly on this relatively soft layer; without it, the plates would be welded to the lower mantle, and there would be no such thing as plate motion (see Figure 3.24). Located close to the surface beneath the mid-ocean ridges, the asthenosphere is also undoubtedly the source of the magma that builds the plates.

Experiments show that the steplike velocity increases of the **transition zone** (between depths of 400 and 700 kilometers) are caused by the conversion of minerals in ultramafic rock into denser forms. The transformations also have the effect of increasing the rigidity and compressibility of the mantle, which accounts for the velocity changes. The gradual increase in wave velocities with depth beneath the transition zone suggests no sharp compositional changes in the **lower mantle** down to the outer-core boundary.

P waves decline sharply and S waves die altogether upon encountering the molten outer core; within the solid inner core, the P waves speed up and the S waves begin again. The outer and inner cores are dense and therefore are probably composed mainly of dense metallic elements. The presence of the Earth's magnetic field is evidence that the metal is probably iron; however, shock wave experiments on possible core materials show that pure iron is a bit too dense for the outer core. A mixture of iron and a lighter element more closely fits the data.

asthenosphere
A semimolten zone of the mantle just below the lithosphere whose depth extends from about 100 to 200 kilometers and is characterized by diminished seismic wave velocities and rocks that are close to melting.

transition zone
The upper-mantle region, between 400 and 700 kilometers in depth, characterized by a series of steplike increases in seismic wave velocities and the conversion of minerals to denser forms.

lower mantle
The part of the mantle that extends from 670 to 2900 kilometers in depth.

The Earth's Internal Heat

A major goal of geoscientists is to fully understand what drives the Earth's lithospheric plates with all of their related tectonic and volcanic activity. Another goal is to learn the secret of the Earth's magnetic field and the possible connections between the forces that generate the field and those driving the plates. To gain insight into these problems, researchers must be able to trace the flow of energy in the Earth—specifically, the transfer of heat from high- to low-temperature regions.

geothermal gradient
The rate of increase of the temperature in the Earth with depth. The gradient averages 25 °C per kilometer of depth in the crust.

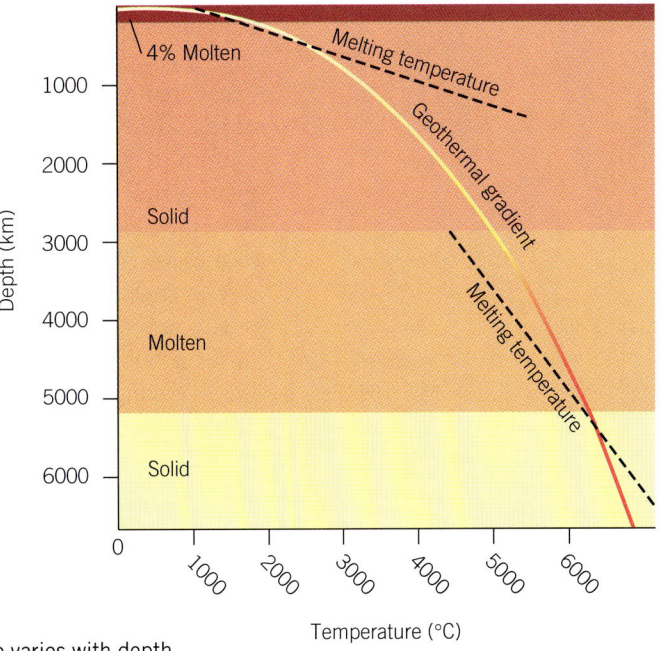

The Geothermal Gradient

The first step in tracing the flow of energy in the Earth is to determine how the temperature changes with depth—that is, the **geothermal gradient** (Figure 3.25).

Measurements taken in mine shafts and boreholes reveal that temperatures near the Earth's surface increase between 20 °C and 30 °C for every kilometer of depth. Geothermal

Figure 3.25 An estimate of how temperature varies with depth in the interior. Notice that the rate of increase is rapid in the crust, but declines in the upper mantle.

- Asthenosphere
- Mantle
- Outer core
- Inner core

gradients are generally higher in tectonically active regions, such as rift zones, than in currently inactive regions far from plate boundaries. In any event, these rates of temperature increase cannot be sustained for more than a few kilometers. If we take the average rate of increase to be 25 °C per kilometer, we would project a temperature of more than 150,000 °C at the Earth's center! An Earth that hot could not remain solid. Indeed, it would probably be a swirling mass of ionized gas. Thus we are sure that the geothermal gradient declines drastically in the upper mantle.

Even at lower rates of temperature increase, deep crustal and mantle rocks become blast-furnace hot within 10 to 20 kilometers deep; if subjected to the same temperatures at the surface, they would undoubtedly melt. What keeps them solid is the fact that they are under enormous pressure, and pressure raises the melting point of the rocks. Therefore, it is highly significant that the mantle begins to melt at depths between 100 and 200 kilometers. This zone is, of course, the asthenosphere, and experiments place its temperature at about 1150 °C.

The mineral transformations of the transition zone, between about 400 and 670 kilometers, mark the next great change in the mantle. Experiments show that for the reactions to occur at the depths indicated, temperatures must be in the neighborhood of 1500 °C at the top of the transition zone and about 2000 °C at the bottom of the zone. Temperatures increase steadily toward the base of the mantle.

Estimates of the temperature at the outer-core–mantle boundary vary. Only in the past few years have techniques been developed to reliably duplicate pressure at this depth, and there is debate over the molten outer core's exact composition. We do know that temperatures must be low enough to keep the lower mantle solid and hot enough to keep the iron outer core molten. Recent experimental work places the temperature of the mantle–outer-core boundary at about 4500 °C and of the Earth's center at greater than 6600 °C. These temperatures are much higher than previously believed. If correct, the temperature at the center of the Earth is similar to the temperature at the surface of the sun. Whatever the generally agreed-upon estimate turns out to be, it is clear that the Earth's core generates more than enough heat to power the Earth's internal engine.

Mantle Convection

The pattern just noted strongly suggests that the mantle is heated from below by the core like a pot of thick tar sitting over a fire. So let us consider how heat circulates in such a pot of tar. The tar at the bottom of the pot absorbs heat, expands, becomes less dense than its surroundings, and rises buoyantly through the colder tar to the surface of the pot. There, it spreads, surrenders heat to the air, becomes dense, and sinks to the bottom of the pot, where it is reheated. This type of heat transfer, involving the movement of warmer materials to colder areas, is called *convection*.

One school of thought does indeed consider the mantle to act in a manner similar to the slowly churning tar pot described. In this view, which is called *whole-mantle convection*, hot and soft rocks at the base of the mantle—having received heat from the outer core—rise as **mantle plumes** and make their way to the surface, where they erupt to form hot spots and lava plateaus (Figure 3.26). Other plumes forge long cracks in the overlying crust (the mid-ocean ridges) and supply the magma that becomes the Earth's lithospheric plates. The plates cool as they spread and are subducted back into the mantle. Once in the mantle, being dense, they sink toward the core.

mantle plume
A pipe-shaped mass of heat-softened light rock that rises from the mantle toward the crust.

Sources of the Earth's Internal Heat

Identifying the sources of the internal energy that powers the Earth machine remains an unsolved problem in the geosciences. Until they *are* identified, we will never fully understand the way our planet—the most geologically active in the Solar System—works.

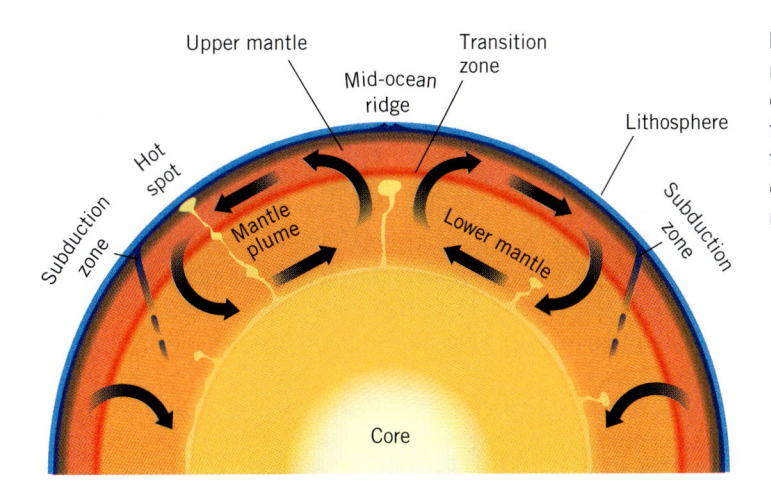

Figure 3.26 Mantle plumes play a key role in some models of whole-mantle convection. The plumes may rise to the surface as isolated hot spots or they may combine to initiate the crustal rifting that results in the formation of lithospheric plates.

The heat released from the disintegration, or decay, of *radioactive isotopes* is an undoubted source of internal energy. These elements have been observed in the crust and in lavas that have erupted from the mantle. Their heat production is imperceptible over the course of a human life time, but over millions of years, enough of it accumulates to fuel all the processes associated with plate tectonics.

The concentration of radioisotopes apparently diminishes with depth, and they are scarce in the lower mantle. Because they do not combine with iron, they are probably not found in the core. Therefore a question remains about whether the radioisotopes are in the proper location to supply the heat for mantle convection.

Other explanations of the Earth's internal heat are more conjectural. One suggests that the outer core is slowly solidifying—a process that releases heat. Another hypothesis relies on the recent discovery that the inner core rotates more rapidly than the rest of the Earth. In this model, the differential motion generates frictional effects that are dissipated as heat.

The Earth's Core and the Magnetic Field

Heat-driven outer-core motions probably play a key role in generating the Earth's magnetic field. Although the precise mechanism eludes them, most geophysicists believe that the outer core operates as a *self-sustaining dynamo* (Figure 3.27). Reasoning by analogy with motors and generators, geophysicists envision an electric current in the churning core that produces an intense magnetic field. This field, in turn, induces a current that feeds back into the core to reinforce the field. Meanwhile, the Earth's rotation keeps the magnetic field aligned in a roughly north-south direction. Each cycle of the "dynamo" dissipates a small amount of energy. If energy were not constantly supplied to churn the fluid core, the dynamo would eventually grind to a halt and the magnetic field would peter out. However, it is unlikely that the churning motions of the core are constant; they probably fluctuate in time and space because of friction or other effects. If they do, the polarity of the magnetic field will fluctuate and may even reverse.

A recent series of remarkable studies, based mainly on extremely detailed analysis of earthquake seismograms, suggests an outer core–inner core linkage in the generation of the magnetic field. One study found that seismic waves directed along the Earth's north-south axis travel 3 to 4 percent faster through the inner core than those that cross the inner core along the equatorial plane. This finding is intriguing because the inner core is composed of iron, and iron forms long, six-sided crystals that transmit waves most rapidly in the direction parallel to their long dimension. One possible conclusion is that the inner core consists of a single gigantic crystal of iron, the

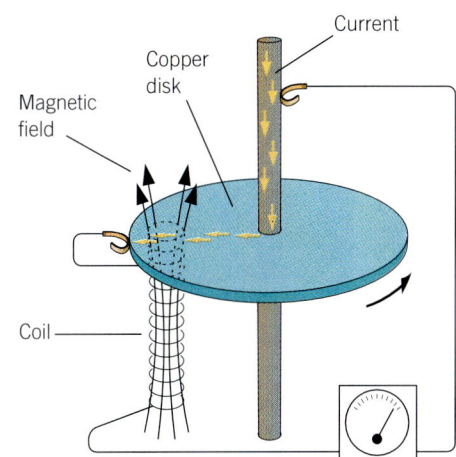

Figure 3.27 The self-sustaining dynamo. A current in a coil produces a magnetic field; a rotating disk cuts the field and induces a current that is fed back into the coil, thus reinforcing the field. A constant input of energy is required to keep the disk rotating.

size of the Moon, whose long axis points north-south! Or perhaps the inner core is composed of billions of crystals all lined up in the same direction, similar to the grain in a block of wood. The crystal (or crystals) could have formed over billions of years by the slow solidification of the outer core, and the heat released in the process may serve as the source of energy for outer-core convection.

A second study focusing on the inner core found that its long axis is actually tilted about 10° to the Earth's rotational axis. This finding led to a third investigation, in which researchers used earthquake records going back many years to find that each day the axis of the inner core shifts ever so slightly eastward. The only logical explanation for this phenomenon is that the inner core spins slightly faster than the Earth as a whole.

The implications of these discoveries are enormous, as they provide our first real insight into the workings and structure of the inner core. Already, mathematical models have been developed that link the motion of the inner core to the generation of the magnetic field. These models propose that the inner core supplies the outer core with the energy necessary to power the self-sustaining dynamo, and that the rotation of the inner core stabilizes the magnetic field.

STUDY OUTLINE

Earthquakes are vibrations triggered by the sudden slippage of rock along fault planes.

I. THE CAUSES OF EARTHQUAKES

  A. The Reid-Gilbert **elastic-rebound theory** states that earthquakes result from the sudden release of strain energy that occurs when rocks rupture and lurch past one another.

  B. Earthquake vibrations, or **seismic waves,** fan out in all directions from the earthquake **focus,** the point of initial rupture. The vertical projection of the focus onto the surface of the Earth is the earthquake **epicenter.**

II. SEISMIC WAVES

  A. There are two basic types of seismic waves.
   1. **Body waves** travel through the Earth's interior. The faster-traveling **primary (P) waves** alternately compress and expand the rocks they pass through. The slower **secondary (S) waves** cause the rock to vibrate at right angles to the wave path.
   2. **Surface waves** travel mainly in the crust. They set the ground swaying horizontally while also causing it to rise and fall like ocean waves.

  B. Seismic waves are detected by **seismographs.** The resulting wave pattern is a *seismogram*. Because seismic waves travel at different speeds, the time interval between the arrivals of P and S waves at a station is used to calculate the distance of that station from the epicenter.

  C. The **Richter magnitude scale** uses wave amplitude as a measure of an earthquake's size. The scale is logarithmic; adjacent numbers differ by a factor of 10. The energy released by an earthquake increases 30 times for each **magnitude** number.

  D. The modified **Mercalli intensity scale** measures the severity of an earthquake in terms of the level of destruction and panic it causes.

III. EARTHQUAKES AND PLATE TECTONICS

  A. Plate boundaries are defined by seismic belts, the sites of 99 percent of all earthquakes.

1. Shallow focal depths at the mid-ocean ridge and deep-focus earthquakes at the trenches are predicted by plate tectonics theory. Studies of P waves confirm the fault movements predicted by the theory.
2. As plates move by one another, the soft lower layers move smoothly, but the upper layers are subjected to the elastic-rebound effect. **Foreshocks** may occur as the rocks begin to crack. **Aftershocks** may follow as the rocks readjust.

B. The San Andreas fault is a transform boundary between the North American and Pacific plates. A bend in the San Andreas fault is transferring stress to the Los Angeles region.

C. **Intraplate earthquakes** occur away from plate boundaries; nevertheless, they can be powerful and devastating to populated areas.

IV. **FORECASTING EARTHQUAKES.** Forecasting is used for long-range planning; prediction is used for immediate preparations.

A. **Seismic gap** theory forecasts that great earthquakes will occur in the quiet segments of the seismic belts bordered by the epicenters of recent great earthquakes.

B. Studies of past episodes of fault movement determine the earthquake *recurrence interval* at a given locale along a fault.

C. *Precursors* are deviations from normal patterns in the vicinity of an active fault that warn of an imminent earthquake.

D. **Tsunamis,** giant ocean waves generated by submarine earthquakes, may be predicted based on their relationship to certain long-period seismic waves.

V. **MINIMIZING EARTHQUAKE DAMAGE.** The extent of death and destruction in a region caused by an earthquake depends upon the concentration of the population and the location and types of structures built there.

A. Intensity surveys have shown the relationship between the type of subsurface and the levels of earthquake damage; for example, vibrations can cause **liquefaction:** the conversion of loose sediments from solid ground to mud.

B. Faulty design and defects in construction are the reasons most buildings collapse during an earthquake.

VI. **SEISMIC WAVES IN THE INTERIOR.** An abrupt change in the velocity of a body wave at a given depth signifies that the wave has intersected a boundary between rocks of different densities and/or elastic properties.

A. As seismic waves intersect boundaries, **refraction** or **reflection** changes their arrival times at seismic stations. By comparing actual and expected arrival times, researchers can trace the paths of the waves.

B. Changes in seismic wave velocities indicate that the Earth's interior is divided into the following boundaries and zones.
1. The boundary between the low-velocity rocks of the **crust** and the high-velocity rocks of the *upper mantle* is called the **Mohorovičić discontinuity, or Moho.**
2. Comprising 80 percent of the Earth's volume, the **mantle** is the largest layer.
3. The **shadow zone,** through which no direct P or S waves travel, is located between 105° and 142° of an earthquake epicenter. Its existence is proof of a molten **outer core.**
4. The early arrival of P waves that travel through the center of the Earth implies that the **inner core** is solid.

VII. STRUCTURAL AND COMPOSITIONAL DIVISIONS OF THE INTERIOR

A. The **lithosphere** is a zone that contains subunits of different chemical properties—a granitic continental crust, a basaltic oceanic crust, and a thin slice of ultramafic upper mantle. The subunits of the lithosphere display a coherent set of physical properties—namely, rigidity and the ability to move together in large slablike units.

B. The rigid upper mantle is composed of ultramafic rock that partially melts to form the mafic magma of the oceanic crust. The **asthenosphere** is a zone of softer, warmer rock within the upper mantle over which the lithospheric plates slide.

C. The **transition zone,** an upper mantle region between 400 and 700 kilometers in depth, is characterized by a series of steplike increases in seismic wave velocities and the conversion of minerals to denser forms.

D. The outer and inner cores are believed to be composed of iron plus some lighter elements yet to be determined.

VIII. THE EARTH'S INTERNAL HEAT

A. The temperatures of the interior generally increase with depth, but the rate of increase, or **geothermal gradient,** differs in the various layers and zones. The gradient averages about 25 °C per kilometer in the crust and declines steeply in the mantle.

B. **Convection** is the means of heat transfer in materials that are heated from below. In the *whole-mantle convection* model, the entire mantle convects by way of hot, rising mantle rocks, or **mantle plumes.**

C. Heat-driven core motions may generate the Earth's magnetic field. The intensity of mantle and core convection may control the Earth's magnetic-field reversals.

D. The source of the Earth's internal heat remains an unsolved problem in the geosciences. One proposed source is the decay of radioactive isotopes.

STUDY TERMS

aftershock (p. 68)
asthenosphere (p. 81)
body wave (p. 61)
crust (p. 79)
earthquake (p. 59)
elastic-rebound theory (p. 61)
epicenter (p. 61)
focus (p. 61)
foreshock (p. 68)
geothermal gradient (p. 81)
inner core (p. 79)
intraplate earthquake (p. 70)
liquefaction (p. 74)
lithosphere (p. 79)
lower mantle (p. 81)
magnitude (p. 65)

mantle (p. 78)
mantle plume (p. 82)
Mercalli intensity scale (p. 66)
Mohorovičić discontinuity (Moho) (p. 79)
outer core (p. 78)
primary (P) wave (p. 61)
Richter magnitude scale (p. 65)
secondary (S) wave (p. 61)
seismic gap (p. 71)
seismic wave (p. 61)
seismograph (p. 62)
shadow zone (p. 78)
surface wave (p. 61)
transition zone (p. 81)
tsunami (p. 73)

THOUGHT QUESTIONS

1. Using the elastic-rebound theory, describe what happens to the rocks at the focus just before, during, and just after an earthquake.
2. If you were a field geologist, what evidence would you search for to indicate recent fault movement?
3. What is the difference between earthquake magnitude and intensity?
4. Why do most earthquakes occur at plate boundaries?
5. Why are most earthquakes confined to the upper 20 or so kilometers of the crust?
6. What is the seismic risk of your present location? How safe or unsafe is the building in which you are currently living? Upon what factors would your answer depend?
7. Sound cannot be transmitted by S waves for the same reason that S waves do not travel through the outer core. Explain.
8. Explain how the properties of the outer core cause the shadow zone.
9. What is the relationship between mantle convection and plate tectonics?

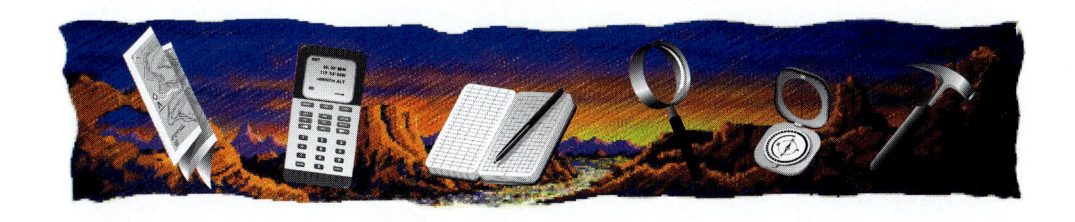

Chapter 4

Atoms and Elements
Atomic Structure

Bonding
Ionic Bonds
Covalent Bonds
Metallic Bonds

The Physical Properties of Minerals
Crystal Form
Cleavage
Fracture
Hardness
Color
Streak
Luster
Specific Gravity
Other Properties

Common Minerals of the Crust
The Classification of Common Minerals

The Structure of Silicate Minerals
Tetrahedral Linkages
Properties Related to Internal Structure

Other Mineral Groups

The Geologic Origin of Minerals

Minerals

The crust of the Earth is composed of igneous, metamorphic, and sedimentary rocks. Each rock is composed of smaller units distinguishable by their colors and shapes. Some are visible to the naked eye; others require a microscope. These constituents are called minerals. Thus a rock is simply an aggregate of one or more types of minerals.

Minerals are composed of tiny particles called **atoms,** which are arranged in regular patterns. A **mineral** is defined as a naturally occurring, inorganic solid, having a definite chemical composition and orderly internal atomic arrangement. All atoms are composed of the same basic subatomic building blocks: **protons, neutrons,** and **electrons.** The various elements that comprise minerals—such as carbon, oxygen, hydrogen, sodium, silicon, and uranium—are merely substances composed of atoms that differ from one another in the number and arrangement of these building blocks. We will see throughout this chapter that the differences between things often comes down to their organization.

Organization, or how things fit together, is the central theme of this chapter. It is repeated, with variations, in our discussion of atoms and elements and of the composition, symmetry, internal structures, and various other properties of minerals.

Atoms and Elements

An **element** is a substance that cannot be further subdivided by ordinary chemical or physical means. There are 92 elements that occur naturally on the Earth, and 17 others have been manufactured in laboratories. Separately or in combination, elements form

atom
The smallest unit of an element that still retains the properties of the element; a dense, positively charged nucleus surrounded by negatively charged electrons.

mineral
A naturally occurring, inorganic solid with a definite chemical composition and orderly internal atomic arrangement.

proton
A positively charged particle in an atomic nucleus.

neutron
A particle in an atomic nucleus with a mass virtually equal to the proton's but with no electric charge.

electron
A tiny particle with a negative charge—equal to a proton's positive charge—that orbits the nucleus of the atom.

element
A substance made up entirely of atoms of the same atomic number that cannot be decomposed into a simpler substance by ordinary chemical or physical means.

◀ This display illustrates the symmetry and color of some gem quality minerals. Included are tourmaline, aquamarine, morganite, heliodor, topaz, kunzite, spodumene, and citrine.

minerals, rocks, water, air, life—in short, the Earth as we know it. The discovery of the various elements and the identification of their physical and chemical properties are the result of centuries of labor by many scientists.

We know that elements are composed of atoms, but what holds atoms together? The answer lies in the concept of electric charge. Electric charge comes in equal units that are either positive or negative. Each proton carries a positive charge unit; each electron a negative charge unit. The equality between protons and electrons does not extend to their mass, however. Protons are 1800 times more massive than electrons.

Most of the mass of an atom is confined to a dense knot of matter called the **nucleus.** Within the nucleus are the positively charged protons and neutrons, particles of nearly equal mass but of zero charge. Neutrons play the crucial role of overcoming the repulsive forces between protons, thus binding the nucleus into a tight structural unit. All atoms of a given element contain the same number of protons, called the **atomic number** of the element; in the most fundamental sense, this number serves to distinguish one element from another. For example, oxygen (atomic number 8) contains 8 protons; nitrogen (atomic number 7) and fluorine (atomic number 9) have 7 and 9 protons, respectively.

The sum of the protons and neutrons in the nucleus constitutes the **atomic mass** of the atom. (Electron mass is so slight that it is generally ignored.) Most elements occur in a variety of atomic masses, for although the number of protons in the atoms of a given element is constant, the number of neutrons may vary. The varieties of a given element that differ in atomic mass are called the **isotopes** of that element. For example, the three isotopes of carbon (atomic number 6) have atomic masses of 12, 13, and 14 and are referred to as carbon-12, carbon-13, and carbon-14, respectively (Figure 4.1). (Appendix B displays the atomic masses of the elements in terms of the weighted contribution of each isotope's atomic mass.)

The isotopes of an element, although nearly identical in their ability to combine with other elements, may vary in geologically important respects. Certain isotopes are subject to **radioactive decay;** their nuclei are unstable, and they change into stable isotopes of other elements at fixed, measurable rates by emitting subatomic particles. For example, the *half-life* of uranium isotope U-238 is 4.5 billion years—that is, half of a quantity of uranium isotope U-238 decays to the lead isotope Pb-206 in 4.5 billion years. They are therefore extremely valuable "nuclear clocks" that allow us to date the age of rocks (which we will learn more about in Chapter 16, Geologic Time).

Atomic Structure

In the neutral state of the atom, the number of protons and electrons is equal; in other words, positive and negative charges are balanced. Electrons orbit at specific distances from the dense, positively charged nucleus; the farther away the orbits are from the nucleus, the greater the energy of the electrons within them (Figure 4.2). The paths of electrons are three-dimensional, so orbits are also called **electron shells.**

Electrons fill the shells of elements of increasing atomic number in predictable sequence. In general, shells closest to the nucleus are filled first; then successively higher shells are filled. The closest shell holds a maximum of two electrons, and the outer shell can hold no more than eight.

The regularity of atomic structure serves as a basis for organizing the elements and displaying them on a chart called the *periodic table of the elements* (Appendix B). Figure 4.3 is a simplified version of the table, showing the structure of elements 1 through 20. The table is arranged so that all members of a vertical group (or family) contain the same number of electrons in their outer shell. For example, all the elements in Group I—lithium, sodium, potassium, and so on—have a single electron in their last orbit. The table is also arranged in horizontal rows, which specify the number of shells present. Each element in row 2, for example, has two electron shells surrounding its nucleus.

---

nucleus
The dense center of an atom composed of protons and neutrons. Nearly all of the mass of an atom is concentrated in the nucleus.

atomic number
The number of protons in the nucleus of an atom.

atomic mass
The average mass of the atoms of an element; to a close approximation, the number of protons plus the number of neutrons in an atom of the element.

isotope
A variety of an element that differs from other varieties of the same element in the number of neutrons it contains and, thus, in its atomic mass.

radioactive decay
The spontaneous release of subatomic particles and energy from the nucleus of an atom.

electron shell
A region surrounding the nucleus occupied by electrons having approximately the same energy.

There is a striking correlation between the chemical and physical properties of the elements and their placement on the periodic table, as shown in Figure 4.3. Compare, for example, the chemical activity of neon (atomic number 10), fluorine (atomic number 9), and sodium (atomic number 11). "Nonactivity" is a good description of the activity of neon. It belongs to the family of *inert* elements, so called because they do not ordinarily participate in chemical reactions. Sodium and fluorine, on the other hand, are extremely active and participate in many reactions with other elements.

The major difference between the structure of neon and the structures of fluorine and sodium is that the outer shell of neon is filled. Fluorine is one electron short of being filled, and sodium is one electron in excess of it. Evidently, the chemically active elements are those with unfilled outer shells. When fluorine gains an electron or sodium loses one, the result in either case is an atom whose shells are filled. In fact, virtually every chemical reaction involves the loss, gain, or sharing of electrons in order to fill shells. Because all members of the same group on the periodic table have the same electron requirements for a filled outer shell, they will react similarly. For this reason, the chemical properties of members of the same group are closely related.

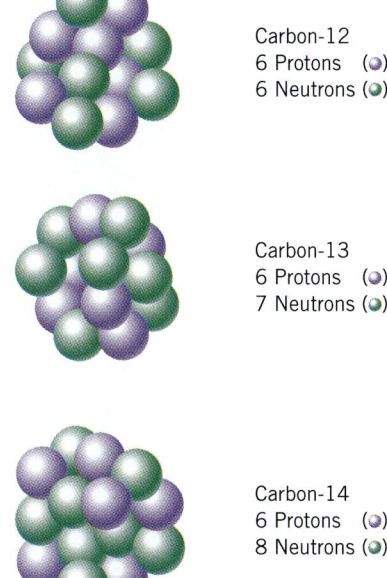

The tendency of atoms to fill their shells is not based on some mysterious property of atoms too difficult to understand. It is simply that the structures of atoms are more stable when the electron shells are filled; less total energy is required to hold electrons to the nucleus. Elements in Groups I through III shed their extra electrons, since this is the most direct way for them to fill their outer shells. Elements having this ability are called *metals;* they donate electrons. Elements of Groups V through VII, especially those of low atomic number, fill their outer shells by accepting electrons; they are called *nonmetals.* Elements of Group IV occupy a midpoint between metallic and nonmetallic elements. Carbon and silicon, for example, have four electrons to either donate or share in order to fill shells, which gives them great chemical flexibility. They combine in a variety of ways with a variety of elements. It is no coincidence that carbon is the key

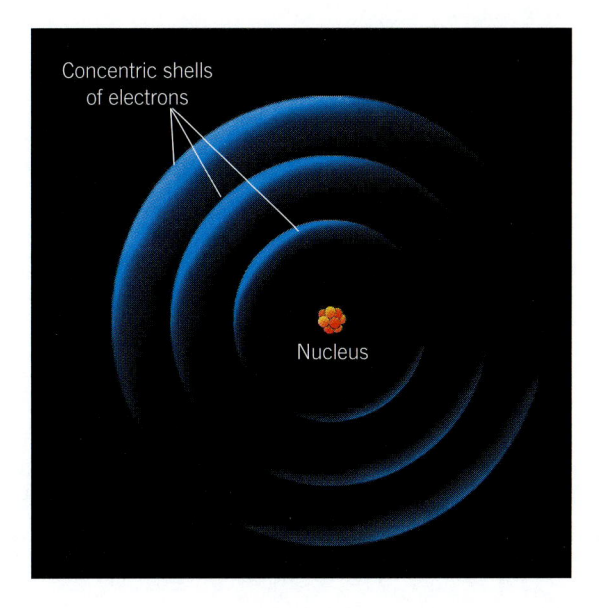

Figure 4.1 Three isotopes of carbon. Isotopes differ from one another in their atomic masses; each contains a different number of neutrons than other isotopes of the same element.

Carbon-12
6 Protons
6 Neutrons

Carbon-13
6 Protons
7 Neutrons

Carbon-14
6 Protons
8 Neutrons

Figure 4.2 In this simplified model of an atom, electrons are confined to concentric shells surrounding a dense nucleus of protons and neutrons.

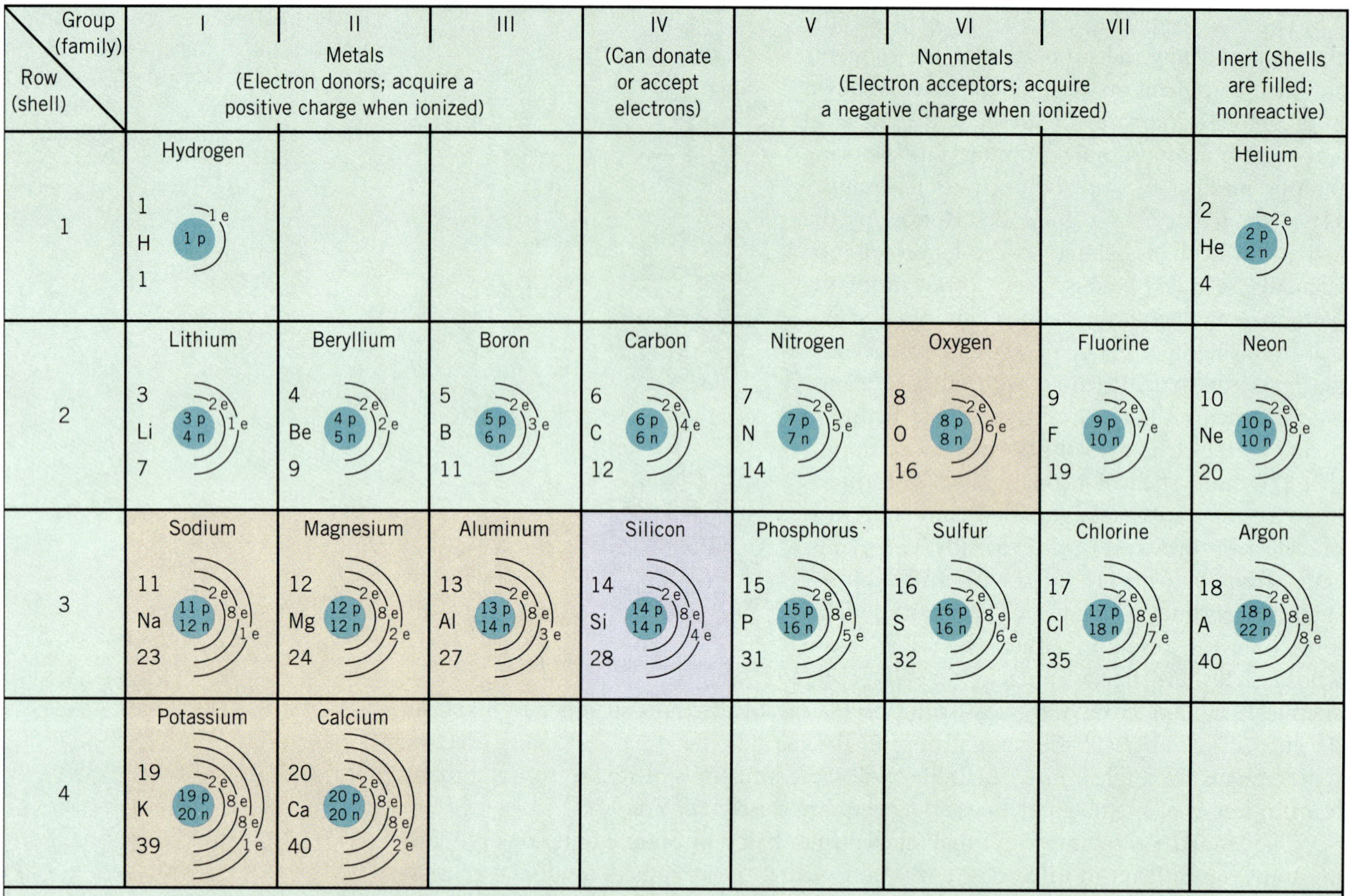

92 CHAPTER 4

Group (family) Row (shell)	I	II	III	IV	V	VI	VII	
	Metals (Electron donors; acquire a positive charge when ionized)			(Can donate or accept electrons)	Nonmetals (Electron acceptors; acquire a negative charge when ionized)			Inert (Shells are filled; nonreactive)
1	Hydrogen 1 H 1 (1p, 1e)							Helium 2 He 4 (2p, 2n, 2e)
2	Lithium 3 Li 7 (3p, 4n, 2e, 1e)	Beryllium 4 Be 9 (4p, 5n, 2e, 2e)	Boron 5 B 11 (5p, 6n, 2e, 3e)	Carbon 6 C 12 (6p, 6n, 2e, 4e)	Nitrogen 7 N 14 (7p, 7n, 2e, 5e)	Oxygen 8 O 16 (8p, 8n, 2e, 6e)	Fluorine 9 F 19 (9p, 10n, 2e, 7e)	Neon 10 Ne 20 (10p, 10n, 2e, 8e)
3	Sodium 11 Na 23 (11p, 12n, 2e, 8e, 1e)	Magnesium 12 Mg 24 (12p, 12n, 2e, 8e, 2e)	Aluminum 13 Al 27 (13p, 14n, 2e, 8e, 3e)	Silicon 14 Si 28 (14p, 14n, 2e, 8e, 4e)	Phosphorus 15 P 31 (15p, 16n, 2e, 8e, 5e)	Sulfur 16 S 32 (16p, 16n, 2e, 8e, 6e)	Chlorine 17 Cl 35 (17p, 18n, 2e, 8e, 7e)	Argon 18 A 40 (18p, 22n, 2e, 8e, 8e)
4	Potassium 19 K 39 (19p, 20n, 2e, 8e, 8e, 1e)	Calcium 20 Ca 40 (20p, 20n, 2e, 8e, 8e, 2e)						

1. Members of groups (or families) contain the same number of electrons in last shell.
2. Each row contains the same number of shells.
3. The number above the element symbol is the element's atomic number; the number below is its atomic mass.
4. The eight most abundant elements of the crust (over 99% by weight) are oxygen, silicon, aluminum, iron, calcium, sodium, potassium, and magnesium. Iron is a metal and an electron donor. It is not on the simplified chart because it has a more complex structure. See complete table in Appendix B.
5. Electron donors (metals) are in Groups I–III; electron acceptors (nonmetals) are in Groups V–VII.
6. Carbon and silicon combine easily with metals and nonmetals; they can be thought of as positive in calculating formulas.

Figure 4.3 The arrangement of the first 20 elements in this chart illustrates the principles behind the complete periodic table of the elements in Appendix B.

structural element of the atomic architecture of living things, and silicon, together with oxygen, is the key structural element of the Earth's crust. You may notice that helium, neon, and argon are classified as inert. The electron shells in this group are filled, making these elements inactive.

chemical bond
The forces exerted between atoms that hold them together.

compound
A substance formed by the chemical combination of two or more elements in definite proportions and commonly having properties different from those of its constituent elements.

Bonding

When atoms combine, they form **chemical bonds.** The bonding may involve the same elements, for example, H_2, O_2, or N_2. Or it may involve several elements whose atoms combine to form **compounds.** For example, carbon dioxide, CO_2, is a compound consisting of a single carbon atom attached to two oxygen atoms. The bonding involves either transferring or sharing of electrons until all atoms have filled outer shells.

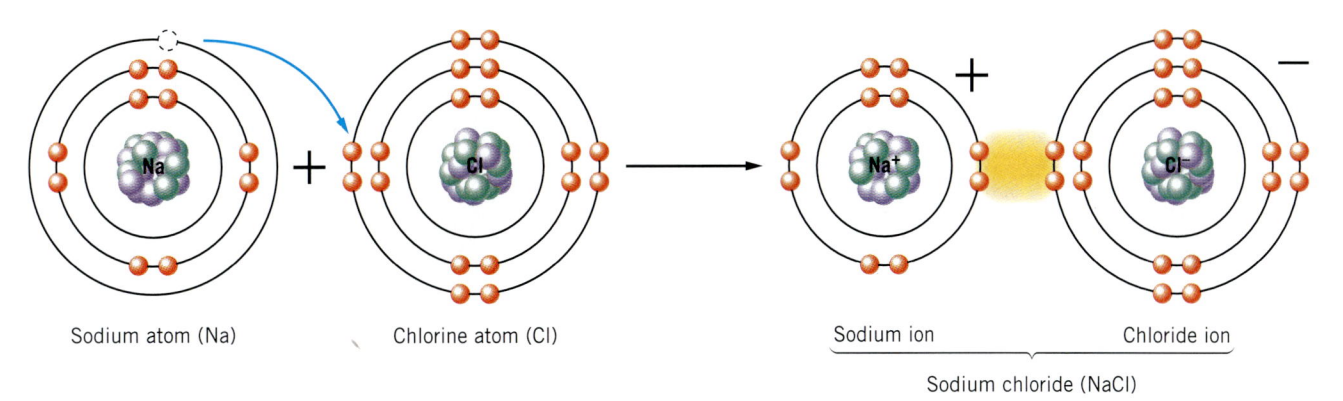

Figure 4.4 The ionic bond involves the transfer of electrons. The opposite charge of the resulting ions holds the ions together.

Ionic Bonds

Ionic bonds are accomplished through electron transfer. For example, in Figure 4.4, the single electron in the third shell of sodium has jumped over to the third shell of chlorine. This transfer leaves sodium with a net positive charge of 1 (written Na^+) and chlorine with a net charge of -1 (written Cl^-). The charged atoms are referred to as **ions**. Because opposite charges attract, sodium and chlorine ions bond, forming sodium chloride (NaCl), better known as table salt and less well known as the mineral halite.

In the process of ionization, atomic sizes change (Figure 4.5). Positive ions like sodium, having lost electrons, become smaller in radius than their neutral atoms. Negative ions like chloride, having gained electrons, generally increase in radius. The relative size of positive and negative ions influences the internal geometry of minerals, and this geometry in turn determines many mineral properties.

ionic bond
A chemical bond that holds two oppositely charged ions together through electron transfer.

ion
An atom with either a positive or a negative charge caused by the loss or gain of electrons.

Atom		Ion	
Si 1.17		Si^{4+} 0.39	
Fe 1.17		Fe^{3+} 0.64	
Al 1.43		Al^{3+} 0.50	
Mg 1.60		Mg^{2+} 0.65	
Na 1.86		Na^+ 0.95	
Ca 1.97		Ca^{2+} 0.99	
K 2.31		K^+ 1.33	
O 0.66		O^{2-} 1.40	
Cl 0.99		Cl^- 1.81	

Figure 4.5 The radius of an atom changes when it is ionized. A positive ion has a smaller radius than the atom in a neutral state; a negative ion has a larger radius than the neutral atom.

Covalent Bonds

Covalent bonds are formed by the equal sharing of electrons in the outer orbit of each element. In Figure 4.6, for example, the shells of four hydrogen atoms, each with one electron to share, combine with the outer shell of carbon, which has room for four electrons. Electrons orbit freely in all the outer shells. The result is the common gas methane (CH_4). Carbon dioxide (CO_2) and water (H_2O) are combinations of different elements held together by covalent bonds. Through covalent bonding, atoms of the same element can combine: oxygen (O_2), hydrogen (H_2), and nitrogen (N_2) are typical examples. These covalently bonded atoms tend to collect in the atmosphere as gases, but not all covalently bonded substances are gases. Diamond is an

covalent bond
A type of chemical bond in which electrons are shared by atoms.

Figure 4.6 Methane is a typical covalently bonded molecule in which shells are filled by electron sharing rather than by electron transfer.

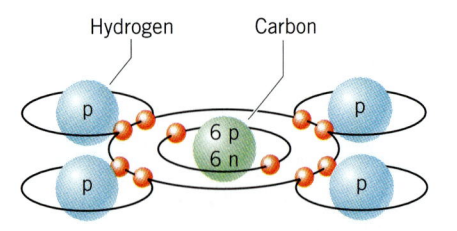

outstanding example of a covalently bonded mineral; its bonds are responsible for the perfectly symmetrical arrangement of its carbon atoms, and therefore, its great hardness and brilliance.

Metallic Bonds

Although metallic elements on the Earth's surface form ionic bonds with nonmetals, especially oxygen, they are also, on occasion, found in the pure or native state. In these cases, the metals form **metallic bonds.** In a manner analogous to the covalent bonding of nonmetals, the metallic elements combine by sharing electrons. However, metals have far more vacant spaces to fill in their shells than nonmetals have. With many electron vacancies to fill, and with each atom capable of sharing but a few electrons, many atoms are involved in the metallic bond. Atoms are thus packed so tightly together that their outer shells tend to merge. Outer electrons no longer belong to a particular shell but migrate freely from atom to atom within the mass. Positively charged nuclei of these metals exist in a "cloud" of loosely held electrons.

Metals in the native state have many familiar properties: they are good conductors of heat and electricity; they also tend to be dense, malleable, and opaque. All these properties stem from the metallic bond.

metallic bond
A chemical bond created in electron-donating elements through the merging of electron shells.

The Physical Properties of Minerals

Most minerals can be identified by inspection of their physical properties, which include crystal form, cleavage, fracture, hardness, color, streak, luster, and specific gravity.

Crystal Form

Long before the development of modern techniques for analyzing the atomic structure of minerals, scientists knew that if minerals were allowed to grow without interference (that is, in an open space), they would develop into **crystals,** objects whose regularity of form, or symmetry, is a manifestation of an orderly arrangement of atoms. The term *crystal* comes from the ancient Greek *kristolis,* meaning "clear ice." The Greeks were referring to transparent six-sided quartz crystals, which lined rock cavities and veins. The quartz reminded them of ice prisms hanging from branches, rock ledges, and cave entrances. Let us look at some of the properties of crystals.

Crystals are bounded by flat, planed surfaces called *crystal faces,* which are made of closely spaced atoms. If we take a number of quartz crystals of differing form—some short and stubby, others long and needlelike, still others displaying prominent pyramids—and measure the angle between neighboring faces in each crystal, we discover an interesting relationship: the angle between adjacent faces in all the specimens is the same, about 120 degrees (Figure 4.7). The *constancy of interfacial angles* is true for a given mineral regardless of the different appearance of particular specimens.

crystal
A solid element or compound whose atoms display a definite, orderly atomic arrangement repeated throughout the solid.

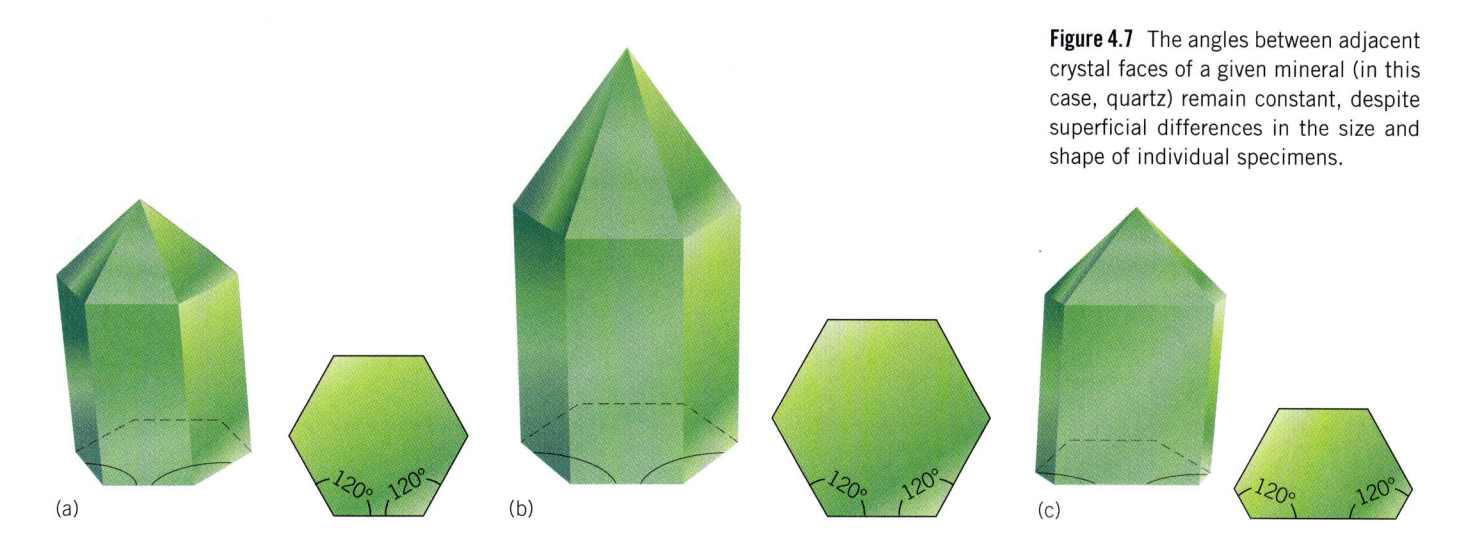

Figure 4.7 The angles between adjacent crystal faces of a given mineral (in this case, quartz) remain constant, despite superficial differences in the size and shape of individual specimens.

Most minerals crystallize when ions migrate through a fluid and bond with one another in orderly atomic arrangements. Under ideal conditions, a mineral will grow into a perfect crystal form. However, the outward symmetry of crystals is usually destroyed by interaction with competing crystals during growth, by circulating fluids after formation, or by mechanical damage from tectonic forces.

Cleavage

Cleavage is the tendency of some minerals to break along definite parallel planes; it develops in the directions of the weakest bonds. Strong bonds occur in planes that are parallel to the crystal face, where atoms are closely spaced; the bonds that attach the planes together are weak. Thus minerals displaying cleavage break either in planes parallel to a crystal face or in planes that bear an angular relationship to it. Because cleavage directions are controlled by the basic atomic arrangement and internal symmetry of the mineral, a given mineral will always display the same cleavage.

Cleavage is often the key trait leading to the identification of a mineral. Mica, for example, has perfect cleavage in one direction, which allows it to be peeled in thin onionlike layers, whereas galena shatters into tiny cubes.

cleavage
The tendency of a mineral to break along parallel planes of weak bonding.

Fracture

As usually applied, the term **fracture** refers to the irregular broken surfaces of minerals that do not display cleavage when they break. Some minerals display smooth fractures; some exhibit rough, splintery, or jagged fractures. Other minerals, such as quartz and malachite, exhibit *conchoidal fracture,* the tendency to break along smoothly curved surfaces (Figure 4.8). Fractures usually develop where average bond strengths are approximately equal in all directions.

fracture
The manner in which minerals break other than along planes of cleavage.

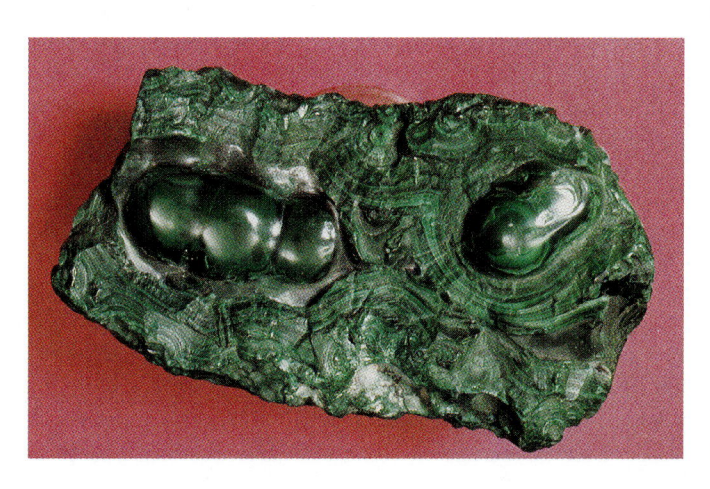

Figure 4.8 Typical conchoidal fracture in malachite.

Table 4.1 Mohs Hardness Scale

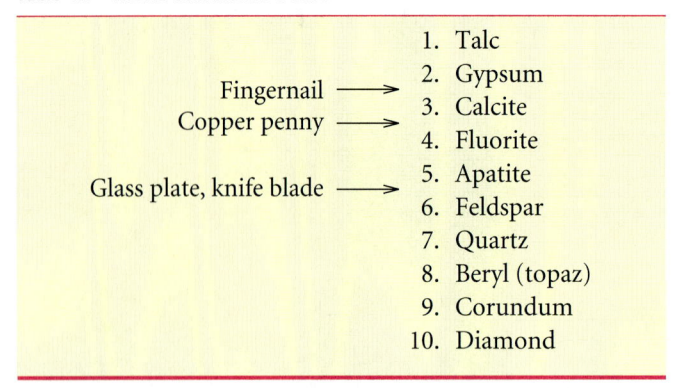

	1. Talc
	2. Gypsum
Fingernail →	3. Calcite
Copper penny →	4. Fluorite
	5. Apatite
Glass plate, knife blade →	6. Feldspar
	7. Quartz
	8. Beryl (topaz)
	9. Corundum
	10. Diamond

Hardness

hardness
The resistance of the surface of a mineral to scratching.

Hardness is related to the strength of chemical bonding and is defined as resistance to scratching. The common way of measuring hardness is by means of the Mohs hardness scale, which consists of ten minerals arranged in sequence from softest to hardest (Table 4.1). Any mineral on the scale will scratch a mineral with a lower number; minerals with the same hardness will scratch one another. Thus the hardness of any mineral may be determined by testing it against one of the minerals on the scale. A convenient reference is the hardness of a fingernail, about 2.5, which means that it will scratch gypsum and talc but not calcite. A copper penny, with a hardness of 3.5, will scratch calcite but not fluorite; a good steel knife blade has a hardness of 5.5 and will scratch apatite but not feldspar. Also, keep in mind that the Mohs scale is only relative, not quantitative. Diamond, for example, is probably thousands of times harder than, say, gypsum and is far harder than its neighbor, corundum.

Color

Although color is an obvious physical trait, it is not necessarily a reliable one. Often, many minerals display the same color; conversely the same mineral may occur in a wide variety of colors (Figure 4.9). For example, quartz (SiO_2) is usually clear, but you can also find varieties that are white (milky), black (smoky), green, pink, or purple (amethyst). However, calcite ($CaCO_3$) is also found in many of these colors. Color variety is usually caused by trace elements that amount to no more than a minute percentage of the atoms present, just as a drop of dye can color a pot of water. Because weathering almost invariably alters color, the true color of a specimen can best be observed on a freshly broken surface.

Some minerals almost invariably display the same color, determined by the presence of distinctive ions in their crystal structure. For example, malachite [$Cu_2CO_3(OH)_2$] is always rich green due to the presence of copper (see Figure 4.8). It is closely associated with azurite [$Cu_3(CO_3)_2(OH)_2$], which owes its distinctive blue color to the oxidation of copper ions (Figure 4.10). Still, color alone is not usually sufficient grounds for sure identification.

Streak

streak
The color of the powder of a mineral when scratched on a porcelain plate.

If you grind a mineral against a hard surface, it will powder; **streak** is the color of the powder. It is often a more consistent property than color, for powdering eliminates differences caused by the effects of minor impurities. The streak is usually observed by rubbing the mineral on an unglazed porcelain *streak plate*. A mineral harder than porcelain (between 6 and 7) will not streak. In some cases, a mineral's streak may not agree with its color in a hand specimen. For example, brass yellow pyrite (fool's gold) streaks black.

Figure 4.9 A collection of quartz specimens. Quartz, like many minerals, comes in a wide variety of colors.

Figure 4.10 Azurite is one of the few minerals that come in only one characteristic color.

Luster

Luster is determined by the way the surface of a mineral reflects light. It is a subtle property dependent upon the ratio of reflected to absorbed light, the depth of light penetration, surface impurities and irregularities that scatter light, and other traits. Yet luster is often a more reliable identifying property of a mineral than color. Probably the most useful determination to make is whether the mineral has metallic or nonmetallic luster (Figure 4.11). Metallic luster is often displayed by metallic and metal sulfide minerals, such as pyrite (FeS_2) and galena (PbS), which gleam like stainless steel. Types of nonmetallic luster include pearly (some varieties of talc or opal), vitreous (glassy), resinous (like resin or pitch), silky, dull, or earthy (clay). Some minerals—for example, hematite (Fe_2O_3)—occur in both metallic and nonmetallic guises.

luster
The quality and intensity of light reflected from the surface of a mineral.

Specific Gravity

The **specific gravity** of a mineral is defined as the ratio of its weight to the weight of an equal volume of water. Because the density of water is one gram per cubic centimeter, specific gravity and density are numerically equivalent terms. For example, a

specific gravity
The density of a substance compared with the density of water.

(a)

(b)

Figure 4.11 Luster is the appearance of a mineral in reflected light. (a) Pyrite displays metallic luster. (b) Limonite displays a nonmetallic (earthy) luster.

mineral that weighs two-and-a-half times as much as an equal volume of water has a specific gravity (or density) of 2.5.

As you might expect, the specific gravity of a mineral is controlled by the atomic mass of the elements that compose it and by the degree of atomic packing—that is, how closely spaced the atoms are. The specific gravity of halite, which is composed of light, widely spaced atoms, is 2.1. By contrast, galena, a densely packed lead sulfide, has a specific gravity of 7.6. In general, native metals and metal sulfides tend to have the highest specific gravities.

Other Properties

Minerals frequently exhibit a host of subtle properties that the geologist, like a good detective, learns to observe and interpret. The ability of a substance to transmit light is one of them. Minerals may be described as *transparent* if you can see through them; as *translucent* if they transmit light, but not well enough to permit seeing through them; and *opaque* if they do not transmit light at all. Clear calcite exhibits *double refraction*, which means that light, upon entering the crystal, divides and vibrates along separate planes; the result is two images (Figure 4.12). Try this experiment. Draw an X on a piece of paper. Place clear calcite over it. You will see two X's. Rotate the calcite and one image will rotate with it. There are other physical properties of minerals as well. Halite, being salt, tastes salty. Magnetite, being magnetic, attracts iron. Talc has a smooth, soapy feel, as talcum powder does.

Common Minerals of the Crust

In examining the abundance of elements in the crust, we are confronted with a striking fact: a mere 8 elements comprise 98.5 percent of the total mass of the crust. They are (in order of abundance by weight): oxygen, silicon, aluminum, iron, calcium, sodium, potassium, and magnesium. The other 84 or so naturally occurring elements amount to a mere 1.5 percent of the crust.

Table 4.2 illustrates that the abundance of oxygen in the crust is overwhelming. The lightest of the eight elements listed, it nevertheless accounts for nearly 46.6 percent of the weight of the crust. The largest of the common elements, it occupies 94

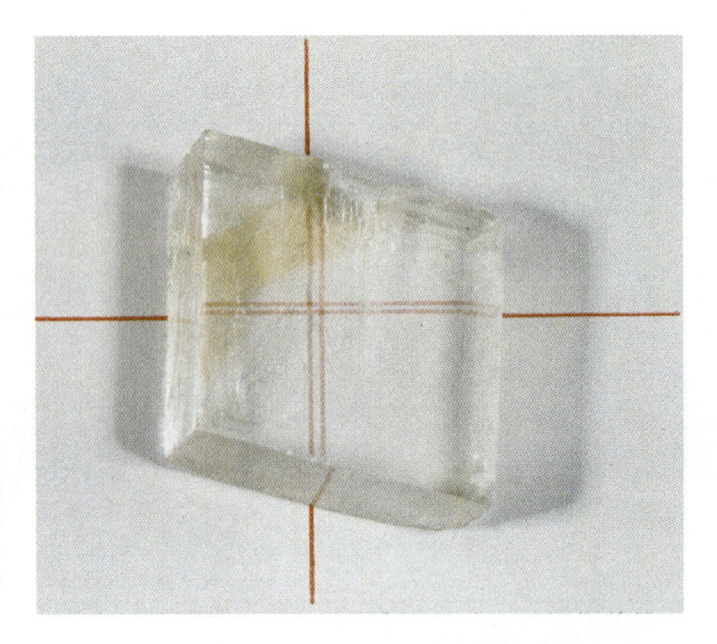

Figure 4.12 Polarization through double refraction in calcite. Light entering the crystal is bent (or refracted) and made to vibrate on separate planes; thus we receive two images of the crossing lines.

Table 4.2 Abundances of the Common Elements

Element	Symbol	By Weight	By Volume	By Atoms
Oxygen	O	46.6%	94.0%	62.6%
Silicon	Si	27.7	0.9	21.2
Aluminum	Al	8.1	0.5	6.5
Iron	Fe	5.0	0.5	1.9
Calcium	Ca	3.6	1.2	1.9
Sodium	Na	2.8	1.1	2.6
Potassium	K	2.6	1.4	1.4
Magnesium	Mg	2.1	0.3	1.8
All others		1.5	0.1	0.1

Percentage of Crust

percent of the volume of the crust. Of all the atoms of the crust, 62.6 percent—nearly two-thirds—are oxygen. We often think of oxygen as a gas, as part of the air we breathe, or perhaps as a component of water; there is more oxygen by far in dense crustal rocks, however, than in the atmosphere and oceans combined. The oxygen ion, with a charge of -2, is the only negative ion among the eight elements listed in Table 4.2. A glance at the periodic table shows that the other seven elements occur as positive ions. Not surprisingly, then, the vast majority of minerals of the Earth's crust are compounds involving oxygen, either as oxides or as *radicals*, clusters of atoms that act as single ions.

Refer again to Table 4.2, and note that silicon is the second most abundant element of the crust and by far outstrips the remaining six. It accounts for 27.7 percent of the weight and 21.2 percent of all atoms of the crust. Therefore, it should also be no surprise that the majority of crustal minerals are **silicates,** compounds involving silicon and oxygen in the form of the silicate radical, $(SiO_4)^{-4}$.

Aluminum is the third most abundant crustal element. As you might expect, the aluminum silicates constitute the most abundant crustal minerals. In fact, the vast bulk of the crust is composed of

1. the *feldspars:* aluminum silicates combined with varying proportions of sodium, calcium, and potassium ions;
2. the *ferromagnesian* minerals: silicate compounds of iron (Fe) and magnesium (Mg);
3. *quartz:* composed of silicon dioxide (SiO_2), which is technically classified as an oxide but obviously has much in common with the silicates.

silicate
A compound containing silicon and oxygen ions arranged as negatively charged ions.

Continental crust is overwhelmingly composed of feldspar and quartz, mostly included in the rock granite. Dense ferromagnesian minerals are relatively minor in the crust of the continents. Oceanic crust, in contrast, is composed overwhelmingly of feldspar and ferromagnesian minerals, mostly included in basalt and similar rocks. Little or no quartz is present. This fundamental difference in composition is the reason that continental crust is of lower density than oceanic crust (specific gravity of 2.7 versus 3.0).

Lighter continental crust tends to rise and denser oceanic crust to sink deeper into the mantle. Thus the major topographic feature of our planet—the difference in elevation between the continents and the ocean basins—is the direct result of the density difference between continental and oceanic rocks. As we learned in Chapter 3, the ultramafic mantle is denser than both continental and oceanic crust. The term *ultramafic* refers to a group of rocks composed almost totally of the ferromagnesian minerals olivine and pyroxene. The most common ultramafic rock is peridotite.

Table 4.3 Simplified Mineral Classification Chart

Chemical Group	Typical Minerals	Formula
Native elements		
Metals	Gold	Au
	Silver	Ag
	Copper	Cu
Nonmetals	Carbon (diamond, graphite)	C
	Sulfur	S
Oxides	Corundum (ruby, sapphire)	Al_2O_3
	Hematite	Fe_2O_3
	Uraninite	UO_2
	Ice	H_2O
Hydroxides	Limonite	$Fe_2O_3 \cdot nH_2O$
	Brucite	$Mg(OH)_2$
Sulfides	Pyrite	FeS_2
	Galena	PbS
	Chalcopyrite	$CuFeS_2$
Halides	Halite	$NaCl$
	Fluorite	CaF_2
Silicates	Feldspar	$Na,Ca,AlSi_3O_8$
	Olivine	$(Fe,Mg)_2SiO_4$
	Mica	$KAl_2(Si_3Al)O_{10}(OH_2)$
Carbonates	Calcite	$CaCO_3$
	Dolomite	$Ca,Mg(CO_3)$
	Siderite	$FeCO_3$
Sulfates	Gypsum	$CaSO_4 \cdot 2H_2O$
	Barite	$BaSO_4$
Phosphates	Apatite	$Ca_5(PO_4)_3(OH,F,Cl)$

The Classification of Common Minerals

There are other minerals in the crust besides the silicates, as shown in Table 4.3. The major mineral groups are typically named after the negative ion or radical present. The most common simple compounds are those in which metals combine directly with oxygen (the oxides and hydroxides); sulfur (the sulfides); and fluorine, chlorine, and iodine (the halides). The major radicals of the more complex minerals include the silicates, the carbonates, and the sulfates. Because the minerals within each group have similar chemical compositions, they often have similar physical properties and atomic structure. As a rule, they are identified through similar chemical tests.

The Structure of Silicate Minerals

There are two fundamental rules of mineral construction. The first is that ions must pack together so tightly that they touch. This packing arrangement is determined by the relative sizes of the positive and negative ions of the mineral. The second rule is that the mineral as a whole must be electrically neutral; the charges of the various ions must balance.

Let us first consider the packing arrangement of the silicates. X rays reveal that the positive silicon ion is quite tiny in comparison to the negative oxygen ion. In fact, only four oxygen ions can fit around the silicon ion. The structural arrangement that results can be regarded as a four-sided figure (a tetrahedron), with oxygen at the ends and silicon enclosed in the center (Figure 4.13). This simple unit, the **silicon-oxygen tetrahedron,** forms the skeleton of the silicates and of virtually the entire crust and upper mantle.

A single silicon-oxygen tetrahedron is incapable of satisfying the second rule of mineral construction—that the mineral as a whole must be electrically neutral. The silicon ion with a charge of +4 cannot balance the combined charge of the four oxygen ions, which is −8. To neutralize the excess negative charge, the tetrahedron forms linkages—either with other tetrahedra or with positively charged ions, such as metals. In this manner, the various silicate minerals are constructed.

(a) Tetrahedron

(b) Silicon-oxygen tetrahedron

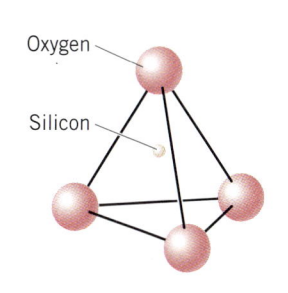

(c) Silicon-oxygen tetrahedron expanded

Figure 4.13 Four oxygen ions surrounding a tiny silicon ion is the fundamental structural unit of silicate minerals. The shape of the unit suggests a pyramid, or tetrahedron. The net charge of the unit is minus four (−4).

silicon-oxygen tetrahedron
An arrangement in which four oxygen ions surround one silicon ion, forming a four-sided structure of negative charge.

Tetrahedral Linkages

The structural arrangements of the common silicate minerals are illustrated in Figure 4.14. In Figure 4.14a, isolated tetrahedra are linked at their edges to positively charged metal ions. This is the structure of the *olivine* group, in which the metal ions are commonly iron (Fe^{2+}) and/or magnesium (Mg^{2+}).

Figure 4.14b shows two oxygen ions shared by adjacent silicon ions. As a result, tetrahedra link to form long single chains. The chains are attached by means of intervening metal ions, iron (Fe^{2+}), magnesium (Mg^{2+}), or calcium (Ca^{2+}). This is the structure of the *pyroxene* group.

In Figure 4.14c, silicon ions that share three oxygen ions alternate with silicate ions that share two oxygen ions. The resulting structures are long double chains, the basic architecture of the *amphibole* group. Double chains are attached through bonds with metal ions.

In the *sheet silicates* (Figure 4.14d), tetrahedra are linked by means of three oxygen ions, and the linkages occur in the same plane. The fourth oxygen ion, pointing either up or down, attaches to metal ions such as aluminum, potassium, iron, magnesium, and calcium, and to hydroxide (OH^-). The result is a series of stacked planes of alternating tetrahedra and intervening ions—an atomic sandwich. The micas, clays, chlorite, and talc are some of the important sheet silicates.

Figure 4.14e shows that quartz (SiO_2) is composed solely of oxygen atoms linked to silicon atoms. Each oxygen is shared by adjacent tetrahedra. The resulting three-dimensional linkage is called a *framework structure.*

Properties Related to Internal Structure

We can readily trace many of the physical and chemical properties of the silicates directly to their atomic structures. For example, cleavage in the silicates is most pronounced in minerals whose internal structures display built-in planes of weakness. These planes are occupied by ions that link the major architectural elements of the mineral: the tetrahedral

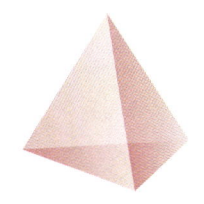

Figure 4.14 The structures of common rock-forming silicate minerals. (a) Isolated tetrahedra (olivine): the tetrahedra are separated in all directions by metal ions and their orientations are reversed in alternating rows. (b) Single chain (pyroxene): two oxygen ions of adjacent tetrahedra link to form a chain. (c) Double chains (amphibole): oxygen ions of adjacent tetrahedra link alternately in threes and twos to form a double chain. (d) Sheets (mica, clay, talc, and graphite): three oxygen ions align on a plane. The fourth points out of the plane. (e) Framework (quartz, feldspar): Tetrahedra are arranged in complex three-dimensional patterns.

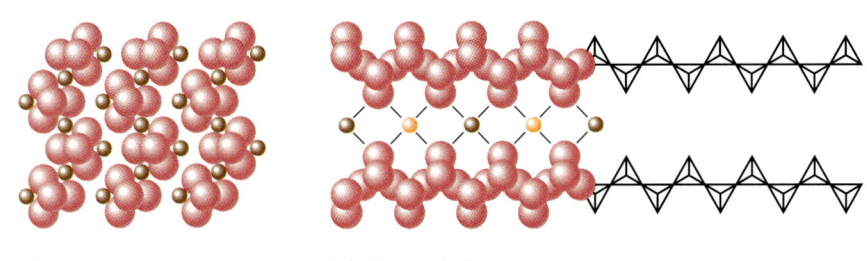

(a) Isolated tetrahedra

(b) Single chain

(c) Double chain

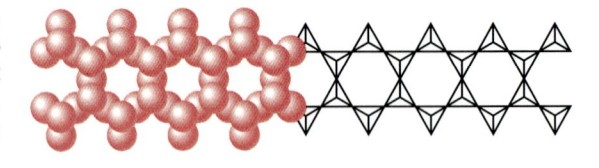

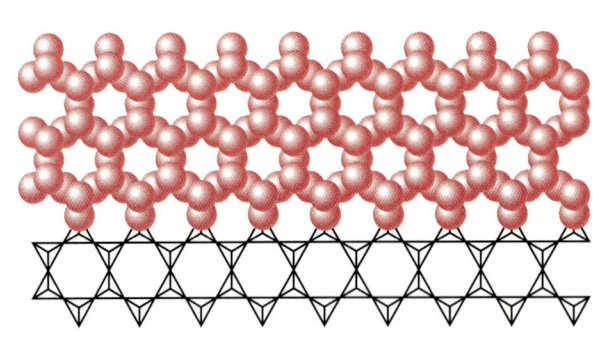

(d) Sheet

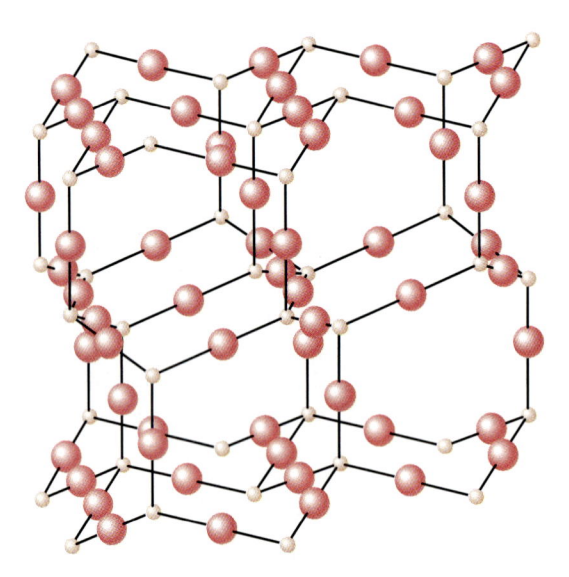

(e) Framework

○ O   ○ Si   ○ Fe   ○ Mg

chains, sheets, and certain framework structures. For this reason, pyroxene (a single chain), amphibole (a double chain), and mica (a sheet) all display prominent cleavage. The cleavage planes of a mineral are parallel, because minerals are built of repetitions of the same pattern. Lack of cleavage is also traceable to the internal structure of minerals. Quartz and olivine fracture conchoidally, like thick glass, rather than along cleavage planes because neither mineral has a pronounced plane of weakness; the bonds are of approximately equal strength in all directions.

Scratching a mineral separates ions—that is, it breaks chemical bonds. But because silicate tetrahedra form especially strong bonds, most silicate minerals have hardness toward the upper end of Mohs hardness scale. However, not all silicates are hard; talc, kaolinite, chlorite, and mica can all be scratched with the fingernail. These minerals have sheet structure in common, and it is the weak bonds linking the sheets that makes the minerals soft. The weak bond is also the reason that mica peels in thin sheets.

The chains and sheets of silicate minerals also allow storage of water in the form of hydroxide $(OH)^-$ radicals, which can fit between these structures. Thus some silicate minerals hold surprisingly large volumes of water; in fact, there may be more water stored in the silicate minerals of the crust and upper mantle than in the oceans. Clay minerals are especially efficient water sponges, which explains why they swell and become slippery when wet.

Many silicate minerals display variable composition, a condition that can be traced to the internal atomic structure of the silicates. Recall that metal ions serve to hold individual tetrahedra together or to link large architectural units such as chains and sheets. The requirements for the job are proper charge, to balance the negative charge of the tetrahedra, and proper size, to fit between the chains, sheets, and tetrahedra without distorting crystal shape. Metals having these qualifications can freely substitute for one another. The result is a group of chemically similar minerals in which certain ions substitute for one another in the chemical formula and, of course, within the structure of the mineral.

The olivine group is a good example. Magnesium (Mg^{2+}) and iron (Fe^{2+}) are of equal charge and are nearly the same size, and they easily substitute for one another in the structure of olivine. The composition of a given olivine mineral can therefore vary in any proportion between 100 percent fosterite (Mg_2SiO_4) and 100 percent fayalite (Fe_2SiO_4).

Iron, however, is heavier than magnesium and absorbs light differently, which makes the density and color of the minerals within the olivine group change with composition. Magnesium-rich varieties are relatively light and are colored pale olive green. Iron-rich olivine is black and dense. Here, again, we see the relationship of internal structure to the physical properties of minerals: both color and density are determined by the ability of the structure to accept ions of a given charge and size.

Compositional variability also characterizes the crust's most abundant mineral groups, the feldspars. These aluminosilicates, the chief minerals of granite and basalt, compose more than half the volume of the crust. A group of feldspars in which calcium and sodium ions substitute for one another is called the *plagioclase* group. Composition within this group varies between 100 percent calcium feldspar and 100 percent sodium feldspar. Any given mineral in the series has a composition somewhere between these extremes.

Other Mineral Groups

We have concentrated on silicate minerals in this chapter because they compose the vast bulk of the crust and mantle; however, this emphasis should not imply that other groups are unimportant. The same principles that govern the atomic structure of the silicates apply equally to them. Charge must be balanced and distributed as evenly as possible

Figure 4.15 Common structures of carbonates, sulfides, oxides, and halides.

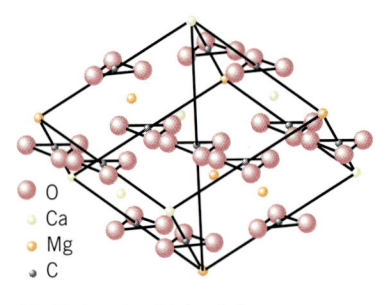

(a) Carbonates (dolomite)
- O
- Ca
- Mg
- C

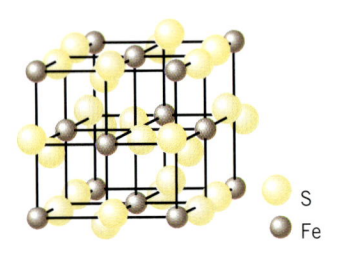

(b) Sulfides (pyrite)
- S
- Fe

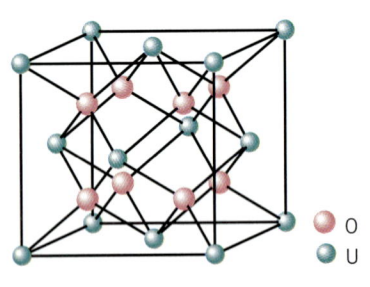

(c) Oxides (uraninite)
- O
- U

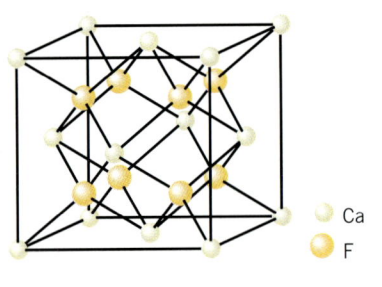

(d) Halides (fluorite)
- Ca
- F

throughout the mineral. Atomic packing arrangements must conform to the relative sizes of the ions involved. In all minerals, properties such as crystal symmetry, cleavage, fracture, hardness, and color depend upon internal atomic arrangement and chemical composition. Ionic substitution is found in many minerals and is not an exclusive property of the silicates. Figure 4.15 illustrates the structures of some other important groups: the carbonates, the sulfides, the oxides, and the halides.

Comparison of the silicates to the carbonates illustrates the importance of relative ionic size in determining the structure of minerals. A glance at the periodic table reveals that although carbon and silicon belong to the same group (IV) and have a charge of +4, carbon is a much smaller element. Because of this difference in size, only three oxygen atoms can fit around the carbon ion, whereas four can fit around the silicon ion. The silicon-oxygen tetrahedron, with its charge of −4, is able to form a large number of linkages, which accounts for the great variability of the silicate structures. The possibilities for the carbonates are more limited, however. Three oxygen atoms form a triangular unit with carbon at the center, and the overall charge of the unit is −2. This leads to the simple alternation of the negative carbonate unit with positive metal ions, such as calcium, magnesium, or iron, which results in a rhombic shape. Figure 4.15a illustrates the structure of dolomite, a carbonate mineral closely related to calcite. The formula for calcite is $CaCO_3$ and for dolomite is $CaMg(CO_3)_2$.

The Geologic Origin of Minerals

For all their myriad complexities and details, there are but four common ways that minerals form in the Earth.

1. Minerals crystallize from solutions of molten silicate rock (magma).
2. Preexisting minerals are altered by the atmosphere and surface water solutions.
3. Minerals precipitate from seawater, groundwater, or surface water solutions.
4. Preexisting minerals are transformed into new minerals while in the solid state.

The first of these origins leads to igneous rocks and includes the vast bulk of Earth materials—oceanic crust and much of the continents. The second involves weathering—for example, alteration by rainwater of feldspar to clay. The third also involves mineral formation by everyday processes that occur at or near the surface of Earth. Some examples are the precipitation of salt and gypsum resulting from evaporation of seawater and lakes and the extraction of calcium carbonate and silica by marine organisms. These minerals are considered sedimentary in origin. The fourth origin involves metamorphism, the application of intense heat and/or pressure to preexisting minerals; the new minerals form as an adjustment to these conditions.

The four origins appear to be distinct from one another; in the real world, however, minerals form as a result of processes that are often intimately related and difficult to define. Some minerals have multiple origins. To fully explore the origin of minerals is to examine igneous, sedimentary, and metamorphic processes. These important subjects are pursued in succeeding chapters.

STUDY OUTLINE

I. ATOMS AND ELEMENTS
   A. A **mineral** is a naturally occurring, inorganic solid, having a definite chemical composition and orderly internal atomic arrangement.
      1. A rock is an aggregate of minerals.
      2. The **elements** that comprise minerals are substances composed of **atoms** that differ from one another in the number and arrangement of their **protons, neutrons,** and **electrons.**

B. An element cannot be broken down or subdivided by ordinary chemical or physical means.

C. Atoms are held together by equal units of positive charge (carried by the protons) and negative charge (carried by the electrons).

D. Most of the mass of an atom is in the **nucleus,** which contains the protons and neutrons—particles of nearly equal mass. Neutrons have zero charge and bind the protons tightly together.
   1. All atoms of a given element have the same number of protons, or **atomic number.**
   2. The number of protons and neutrons is the **atomic mass** of the atom.
   3. Most elements occur in a variety of atomic masses, called the **isotopes** of that element.
   4. Certain isotopes are subject to **radioactive decay.** Their nuclei are unstable, and they decay to stable isotopes of other elements.

E. In the neutral state of the atom, the number of protons and electrons is equal. Regularity of atomic structure is the basis for the periodic table of the elements.
   1. Electrons orbit at specific distances from the nucleus. Their paths are three-dimensional, so orbits are also called **electron shells.**
   2. The first shell, which holds only two electrons and is filled first, is followed by successively higher shells; the outermost shell holds a maximum of eight.

II. **BONDING.** Elements whose outer shells are not filled are active and participate in reactions with other elements.
   A. **Chemical bonds** involve the loss, gain, or sharing of electrons in order to fill shells.
      1. **Ionic bonds** are formed by electron transfer. Charged atoms are called **ions.**
      2. Elements form **covalent bonds** by sharing electrons in their outer orbits.
      3. **Metallic bonds** are like covalent bonds but involve many atoms packed tightly together.
   B. Bonding of different elements results in **compounds.**

III. **THE PHYSICAL PROPERTIES OF MINERALS**
   A. Unrestricted, minerals grow into symmetrical **crystals.** The planed crystal faces are made of closely spaced atoms.
   B. Some minerals display **cleavage,** the tendency to break along parallel planes.
   C. Many minerals **fracture** to form smooth, jagged, or splintery surfaces.
   D. **Hardness,** or resistance to scratching, is measured by the Mohs scale.
   E. Some minerals can be identified by their distinctive color.
   F. Some minerals can be identified by their **streak,** the color of the powder formed when they are rubbed against a porcelain plate.
   G. **Luster** refers to the appearance of a mineral in reflected light.
   H. **Specific gravity** is a mineral's weight relative to an equal volume of water.
   I. Other mineral properties are transparency, opacity, and double refraction.

IV. **COMMON MINERALS OF THE CRUST**
   A. Eight elements comprise 98.5 percent of the total mass of the crust: oxygen, silicon, aluminum, iron, calcium, sodium, pot assium, and magnesium.

B. Most crustal minerals are **silicates,** compounds involving silicon and oxygen, the two most abundant elements of the crust.

C. Aluminum is the third most abundant element. The most abundant crustal minerals are the feldspars, ferromagnesian minerals, and quartz. Minerals of the ultramafic mantle are primarily ferromagnesian.

V. **THE STRUCTURE OF SILICATE MINERALS.** The basic structure of silicate minerals is the tetrahedron: four oxygen ions around one silicon ion.
   A. Tetrahedral linkages involve **silicon-oxygen tetrahedra** and intervening positive metal ions.
      1. Isolated tetrahedra are linked by positively charged metal ions as in the olivine group.
      2. Tetrahedra link to form single chains as in the pyroxene group.
      3. Tetrahedra of the amphibole group form double chains.
      4. Tetrahedra are linked to form sheet silicates, as in the micas.
      5. Tetrahedra that are linked to one another exclusively form framework structures, as in quartz.
   B. Many chemical and physical properties of minerals are related to their internal structure. For example, the cleavage planes of micas and clays occur because of weak bonds between sheets.
   C. Many silicate minerals display variable composition, because certain metal ions that serve as linkages between tetrahedra may substitute for one another.

VI. **OTHER MINERAL GROUPS.** Other important mineral groups are the carbonates, the sulfides, and the oxides. Because only three oxygen units can fit around a carbon unit, the structures of carbonate linkages are less varied than those of the silicates.

VII. **THE GEOLOGIC ORIGIN OF MINERALS.** Minerals form in the Earth in four ways. Minerals crystallize from magma; minerals are altered by weathering; minerals are precipitated from seawater; or preexisting minerals are transformed into other minerals while in the solid state through metamorphism.

STUDY TERMS

atom (p. 89)	ionic bond (p. 93)
atomic mass (p. 90)	isotope (p. 90)
atomic number (p. 90)	luster (p. 97)
chemical bond (p. 92)	metallic bond (p. 94)
cleavage (p. 95)	mineral (p. 89)
compound (p. 92)	neutron (p. 89)
covalent bond (p. 93)	nucleus (p. 90)
crystal (p. 94)	proton (p. 89)
electron (p. 89)	radioactive decay (p. 90)
electron shell (p. 90)	silicate (p. 99)
element (p. 89)	silicon-oxygen tetrahedron (p. 101)
fracture (p. 95)	specific gravity (p. 97)
hardness (p. 96)	streak (p. 96)
ion (p. 93)	

Thought Questions

1. Why are coal, glass, and manufactured diamonds not considered minerals?
2. Distinguish between the following:
   a. neutron—proton
   b. atomic number—atomic mass
   c. stable isotope—radioactive isotope
   d. electron shell—nucleus
   e. metal—nonmetal
   f. ionic bond—covalent bond
3. Explain the organization of the periodic table.
4. Distinguish between the following:
   a. fracture—cleavage
   b. color—streak
   c. luster—translucency
5. Why do some minerals display cleavage and others do not?
6. What is the relationship between the distribution of elements in the crust and the difference in elevation between the continents and ocean basins?
7. Why do the carbonates lack the structural complexity of the silicates?

Chapter 5

Igneous Rock Bodies

The Formation of Rocks from Magma
   Igneous Rock Textures
   The Classification of Igneous Rocks
      The Peridotite Family
      The Basalt-Gabbro Family
      The Andesite-Diorite Family
      The Granite-Rhyolite Family
   The Igneous Rock Classification Chart
   The Sequence of Crystallization of Igneous Rocks

The Geologic Settings of Igneous Activity
   Igneous Activity at Mid-Ocean Ridges
   Igneous Activity in Subduction Zones
      The Ascent of Magma in Subduction Zones

Mineral Concentration and Igneous Activity
   Mineral Concentration at Subduction Zones

Igneous Activity

The Earth's crust is overwhelmingly igneous in origin, as a glance at a world map will confirm. There we see that most of the crust is oceanic, formed by the upwelling of magma from beneath the mid-ocean ridges. The oceanic crust is also very young. As we have seen in our discussion of sea-floor spreading, the entire oceanic crust has been recycled back into the mantle within less than 5 percent of geologic time. So from the dual perspectives of vigor and volume, it is fair to say that igneous activity is the Earth's most important geologic process.

The influence of igneous activity is not confined to the oceans. Too light to be subducted, the continental crust has grown through geologic time by the addition of igneous rocks. As the oceanic lithosphere descends beneath subduction zones, it generates fresh magma, which rises and, upon cooling, intrudes the continents and builds volcanic chains. Where erosion has stripped away the thin sedimentary veneer of the continents, the igneous rocks that formed beneath the surface are exposed. Some are billions of years old, and it becomes clear that deep-seated igneous rocks are the major architectural units of the continental crust.

Igneous Rock Bodies

From field studies and seismic evidence, as well as from the controlled melting of igneous rocks under laboratory conditions, geologists have learned that most **magmas** are generated through the partial melting of the upper mantle. The magmas then rise and spread from their source regions and, through a variety of means, intrude the crust (Figure 5.1). Most of them crystallize beneath the surface, forming intrusive rock

magma
Molten (hot-liquid) rock material, generated within the Earth, that forms igneous rocks when solidified.

◀ Photomicrograph of gabbro, a coarse-grained igneous rock, taken under polarized light. The long, rectangular crystals are plagioclase feldspar.

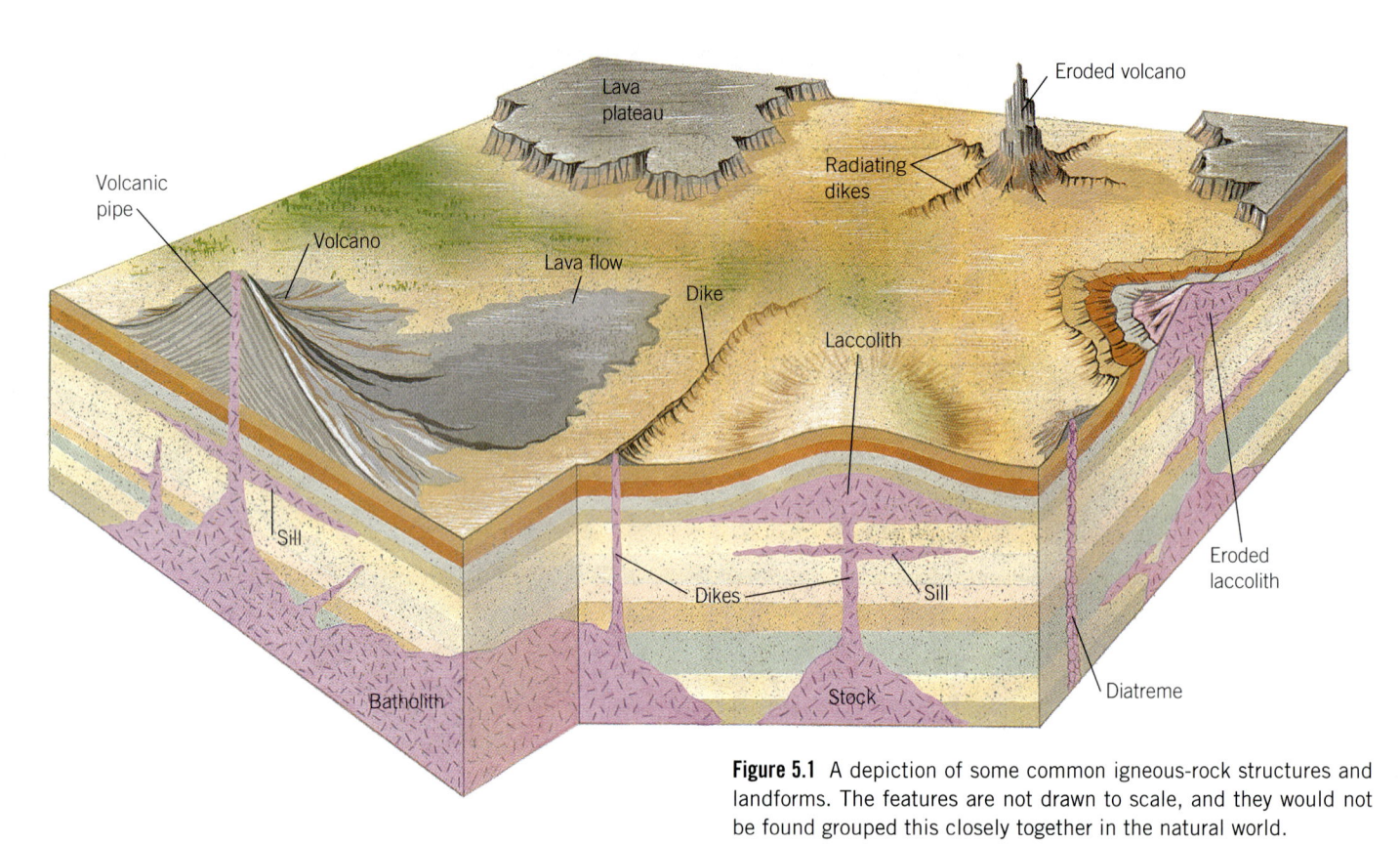

Figure 5.1 A depiction of some common igneous-rock structures and landforms. The features are not drawn to scale, and they would not be found grouped this closely together in the natural world.

bodies or **plutons** (after Pluto, the Greek god of the underworld). Processes connected to the formation of igneous rocks at depth are called plutonism.

Large volumes of magma may also rise through the crust and escape onto the surface—either relatively peacefully, as *lava,* or explosively, as fragments sent into the air. If the magma rises to the surface through a more or less localized vent, the lava and fragments will pile up around it. Repeated eruptions of this sort will build a large conical volcano. If, on the other hand, the lava pours onto the surface through a long fracture or fissure, it will flood the surrounding region. Many such episodes build a thick *lava plateau.* All these expressions of extrusive igneous activity go under the name of volcanism (after Vulcan, the Roman god of fire). Volcanism is the subject of Chapter 6.

Plutons are generally classified by their relation to the layering of the host or country rocks that they intrude. Plutons that are **concordant** form parallel to the layering of the host rock, whereas those that are **discordant** cut across the layering. Concordant plutons come in two main varieties: sills and laccoliths. **Sills** are flat, tabular bodies intruded parallel to the host rock layering (Figure 5.2). They can range from less than a meter to hundreds of meters thick; some extend for hundreds of kilometers, and others disappear in a few short steps. **Laccoliths** are mushroom-shaped bodies. Fed from a lower vent, magma rises and domes the overlying layers while it spreads laterally.

Discordant plutons cut across the layering of host rock; they include dikes, pipes, batholiths, and stocks. **Dikes** are tabular bodies (Figure 5.3). They may occur in a variety of patterns, some forming deep, irregularly shaped intrusions. *Pipes* are cylindrical bodies, intrusions that filled and solidified in the vents that once fed laccoliths, sills, or volcanoes high in the crust. *Radiating dikes* may spread from these pipes like spokes from a bicycle wheel, and still others may occur as concentric *ring dikes. Diatremes* are peculiar cylindrical, pipelike intrusions. Many of them contain diamonds—strong evidence that they are derived from the mantle, hundreds of kilometers below the crust.

Batholiths (*bathos,* meaning "bottom" in Greek; *lith,* meaning "rock") are enormous, complex rock bodies composed mostly of granite: they cover at least a hundred

---

pluton
An intrusive rock body.

concordant
Pertaining to igneous rock bodies that intruded and solidified parallel to the layers of country rock.

discordant
Pertaining to igneous rock bodies that cut across the layers of country rock.

sill
A tabular, concordant igneous intrusion.

laccolith
A mushroom-shaped, concordant igneous intrusion that has domed the overlying crustal rocks.

dike
A tabular, discordant igneous intrusion.

batholith
An igneous intrusion with a large mass, a surface area greater than 100 km², and no known floor.

Figure 5.2 A sill intruded parallel to the layering of the country rock.

Figure 5.3 A dike cutting across the layering of a shale formation in Grand Canyon National Park, Arizona.

square kilometers and have no visible floor. Many show signs that they were formed over millions of years by the repeated injections of relatively small magma bodies. Batholiths are major architectural elements of the continents. Some batholiths, such as the Sierra Nevada batholith in California, cover the greater part of entire states (Figure 5.4). Smaller bodies of similar origin are called *stocks*.

We will discuss the emplacement of igneous rock bodies in the course of this chapter.

The Formation of Rocks from Magma

A typical magma is a complex mixture consisting of molten silicate rock material, dissolved gases, solid minerals that have crystallized from the melt, and chunks of country rock incorporated within the liquid mass. The question is, How do solid rocks form from this mixture? To arrive at an answer, geologists begin with the fact that all rocks, including those that are igneous in origin, are essentially aggregates of minerals. By studying the texture of the aggregate—that is, the size, shape, and arrangement of the minerals—they can extract a great deal of information about the physical conditions that led to the formation of the rock.

Igneous Rock Textures

The texture of an igneous rock develops as minerals crystallize from a cooling magma. There are a few simple experiments, employing water and dissolved ions, that illustrate the origin of some important textural relationships—notably, how large and small crystals form. Take a beaker of hot water and dissolve blue powdery copper sulfate in it until the water cannot hold any more. Let the solution cool to room temperature. At this point, the solution is *supersaturated*, holding more copper sulfate than is normal for that temperature. Next, add a few tiny copper sulfate crystals, or seeds, that sink to the bottom. Let the solution stand for a few days and observe the results. You will find that the seed has grown into a large, symmetrical copper sulfate crystal and that the blue color of the water has lightened noticeably. Evidently, the copper and sulfate ions in solution were attracted to the seed and crystallized. We may conclude, therefore, that slow, undisturbed cooling of a solution that contains a few seeds tends to favor the precipitation of large, well-formed crystals.

Now, repeat the experiment, but this time add a number of seeds. You will find that a number of tiny crystals have precipitated. Next, try the experiment without adding

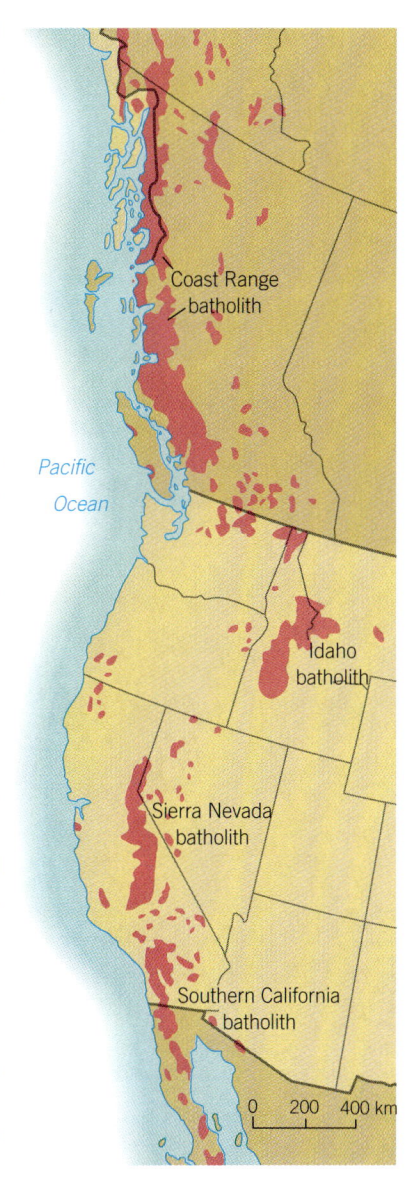

Figure 5.4 Batholiths cover wide areas of western North America.

Figure 5.5 Hand sample of a typical gabbro, showing phaneritic (coarse-grained) texture. The dark minerals are amphibole (hornblende), the lighter ones are plagioclase.

Figure 5.6 A typical basalt, displaying aphanitic texture and dark color. Compare with gabbro (see Figure 5.5), its coarse-grained equivalent.

phaneritic
An igneous-rock texture in which the mineral components are visible to the unaided eye.

aphanitic
An igneous-rock texture in which the mineral components are too small to be identified by the unaided eye.

glassy
The texture of an igneous rock having a high content of glass.

intrusive
An igneous rock derived from magma that solidified within the mantle or crust.

extrusive
Igneous rock formed from lava flows or pyroclastic materials that were spewed to the Earth's surface.

porphyritic
An aphanitic igneous-rock texture in which large crystals are embedded within an aphanitic or glassy matrix.

the seeds. Instead, cool the copper sulfate solution very rapidly and shake it at the same time. You will find again that tiny crystals have precipitated. Thus we may conclude that small crystals precipitate when there are many seeds present and that these seeds may form from a liquid that is agitated as it cools rapidly.

Although a magma is far more complicated than our simple copper sulfate solution, minerals will crystallize from it in an analogous manner. The major difference is that magma is a silicate solution capable of holding a far greater concentration of ions than water. Crystal growth within a magma begins with the precipitation of seeds. These seeds contain the basic structural plans of the minerals, and as ions migrate to them, they grow in size. Eventually, their surfaces interfere with those of other growing crystals. At this point, the crystals interlock and growth ceases. For this reason, igneous rocks typically display a mosaic texture of interlocking mineral grains.

In our experiments, we also saw that cooling rates control crystal growth. In a slowly cooling magma, relatively few seeds form; the minerals that grow around them are large. If the minerals grow large enough to be seen with the naked eye, we say the rock has a coarse or **phaneritic** texture (Figure 5.5). In a rapidly cooling magma, by contrast, numerous seeds compete for ions, and many tiny minerals crystallize; this gives the rock an **aphanitic** texture, meaning that the minerals are visible only under a microscope (Figure 5.6). Ultimately, if the magma cools too rapidly, ions are robbed of the energy to migrate to the seeds, and the magma solidifies without crystallizing. Glass, a smooth and hard substance, forms instead. It has the amorphous internal structure of a liquid, rather than the organized structure of minerals. Thus rapid quenching, or supercooling, produces a **glassy**-textured rock (Figure 5.7).

In the field, phaneritic textures are characteristic of **intrusive** rocks—those that cool beneath the surface. Large mineral crystals form because magmas are well insulated at depth and cool slowly. By contrast, aphanitic and glassy textures are characteristic of **extrusive** rocks—those that cool on the surface. Because the temperature contrast between air and lava is very high and heat loss is rapid, only small crystals form.

The link between cooling and depth of burial also provides an explanation for the origin of most **porphyritic** textures (Figure 5.8). Porphyritic rocks display a mixture of coarse-grained phaneritic and fine-grained aphanitic or glassy textures. In other words, rocks having this interesting texture contain minerals of two distinct sizes: large minerals embedded in a matrix of small minerals. As such, porphyritic rocks represent two generations of cooling. Picture a magma cooling slowly in a large chamber thousands of meters below a volcano. Large minerals characteristic of that depth crystallize in the

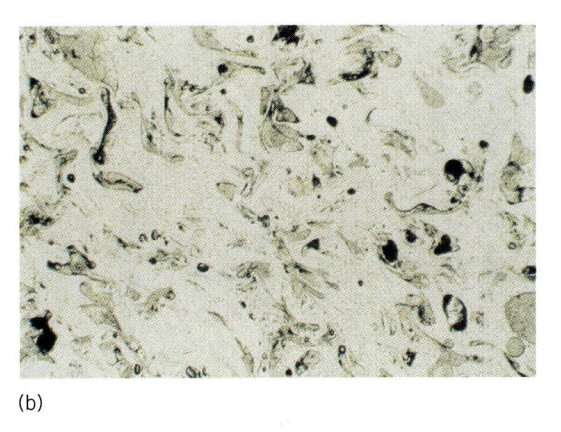

Figure 5.7 (a) A hand sample of the volcanic glass pumice. (b) This photomicrograph of pumice shows the vesicles and glass fibers that make it useful as an abrasive.

melt. Next, pent-up gases within the chamber burst through the plug that caps the volcanic vent. The magma then erupts onto the surface as lava, carrying the coarse crystals with it. As the lava cools rapidly, the aphanitic minerals that crystallize from it form the matrix of the porphyritic rock.

Up to this point, we have discussed the textures of rocks that cool either slowly or rapidly from a coherent body of magma. But that is not always the case. As we have mentioned, many volcanoes pass through an explosive phase, triggered by the escape of pent-up gases. The lava is literally torn to bits in the process, and clots ranging in size from minute droplets to small cars go flying out of the vent. On the way down, they cool; when the fragments land in a heap, they are cemented together. The rock that results has a **pyroclastic** texture (Figure 5.9). We will discuss pyroclastic rocks further in Chapter 6, Volcanism.

pyroclastic
Pertaining to rock material formed by an explosive ejection from a volcanic vent.

The Classification of Igneous Rocks

If the texture of an igneous rock points to the physical conditions at the time that the magma crystallized, then the mineral content of the rock suggests the origin and chemical evolution of the magma. Fortunately, of the more than 3000 minerals that exist, only 7 compose over 95 percent of all igneous rocks. They are the ferromagnesian minerals olivine, pyroxene, amphibole, and biotite; the aluminum silicate feldspars (both the plagioclase and potassium varieties); and quartz. To simplify matters still further, no one rock contains all these minerals; no more than 3 or 4 are commonly present, and they form characteristic groupings. This information forms the basis for classifying igneous rocks into *rock families,* which are named after their most common rock members.

Figure 5.8 A typical hand sample of rhyolite displaying porphyritic texture.

Figure 5.9 Pyroclastic texture is exhibited by this volcanic mudflow breccia, which consists of angular pyroclastic fragments of various sizes cemented in a volcanic mud matrix (from the 1963 eruption of Mount Agung, in Indonesia).

Figure 5.10 Peridotite, an ultramafic rock. This specimen consists almost entirely of the ferromagnesium mineral olivine.

Figure 5.11 Hand sample of serpentinite. Peridotite is often altered by hot-water solutions to serpentinite, a metamorphic rock.

The Peridotite Family

Peridotite is a dark, coarse-grained intrusive rock composed mainly of olivine, with lesser amounts of pyroxene and containing little or no plagioclase (Figure 5.10). It is believed to form the bulk of the upper mantle. As such, it is either directly or indirectly the source rock of the magmas from which most other igneous rocks are derived. This important igneous rock is relatively rare in the crust, however. It sometimes occurs as inclusions in other igneous rocks of the oceanic crust or as layers in deep fracture zones of the oceanic crust. It is also preserved in slivers of oceanic crust that have been sheared off descending plates and thrown up onto the continents. Peridotite may often be altered to the soft, greenish rock *serpentinite* by hot-water solutions that penetrate the crust at mid-ocean ridges and other fracture zones (Figure 5.11). Because of its very high content of magnesium and iron, peridotite is often referred to as an *ultramafic* rock.

The Basalt-Gabbro Family

basalt
A dark, fine-grained, mafic, extrusive igneous rock composed largely of plagioclase feldspar, pyroxene, and olivine.

Basalt is a fine-textured, dark brown to black extrusive rock, composed primarily of calcium-rich plagioclase and pyroxene and, commonly, lesser amounts of olivine (see Figure 5.6). *Gabbro* is its coarse-textured, deep intrusive equivalent (see Figure 5.5). *Diabase*, whose texture lies between that of basalt and gabbro, often forms in intrusions that occur close to the surface. The ferromagnesian minerals remain important components of this family, though they are less plentiful than in the peridotite family. Therefore, basalt and gabbro are also known as *mafic* rocks.

Mafic rocks compose the entire oceanic crust—basalt forming the upper layers and gabbro forming the thicker internal zone upon which the basalt rests. The oceanic volcanoes of the Hawaiian Island–Emperor Seamount chain, the entire island of Iceland, and great oceanic plateaus of the sea floor are all composed of basalt. Vast outpourings of basaltic lava on the continents have built the Deccan Plateau of India, the Columbia Plateau of Washington, and similar features the world over.

The Andesite-Diorite Family

Andesite is a gray, fine-grained volcanic rock consisting of plagioclase and a ferromagnesian mineral such as hornblende (amphibole) or biotite (mica). The plagioclase contains about equal amounts of calcium and sodium ions. *Diorite* is the coarse-grained equivalent (Figure 5.12).

The term *andesite* was coined in 1826 by pioneering field geologists for the rocks and lavas characteristic of the Andes Mountains of South America. About one hundred forty years later, geologists learned that the Andes range borders an active subduction zone, and indeed, rocks of the andesite family are typical of the volcanic island arcs and continental chains that border these zones.

Figure 5.12 A hand sample of diorite, which is commonly lighter-colored than gabbro because it contains a higher proportion of sodium-rich plagioclase.

The Granite-Rhyolite Family

Granite is a light-colored, coarse-grained intrusive rock consisting primarily of quartz, potassium feldspar, and/or sodium plagioclase (Figure 5.13). Ferromagnesian minerals such as hornblende or biotite are either absent or present only in minor quantities. However, this simple description does not do justice to the extremely varied textural and mineralogical complexity of granite. Members of the granite family run from the extremely coarse-grained *pegmatite* to the sugary-textured *aplite*.

Granite and its slightly more mafic variety *granodiorite* are the most common igneous rocks of the continental crust. Indeed, to a rough approximation, granitic rocks *are* the continental crust. It is also significant that granite is found only on the continents, whereas the oceanic crust consists entirely of basaltic rocks. Until recently, this division was never adequately explained. One of the triumphs of the theory of plate tectonics is that it provides this explanation (see Chapters 2 and 17).

Rhyolite is the extrusive equivalent of granite and is also generally confined to the continental crust (see Figure 5.8). Most of the violent volcanoes that border subduction zones involve either andesite, *dacite* (the extrusive equivalent of granodiorite), or rhyolite lava. Because of their extremely high silica content, members of the granite-rhyolite family are also referred to as *silicic* rocks.

Figure 5.13 A typical granite displaying coarse-grained phaneritic texture. The rusty pink color is the consequence of an abundance of potassium feldspar. The colorless minerals are quartz.

granite
A coarse-grained, intrusive igneous rock containing quartz and feldspar (primarily potassium feldspar).

The Igneous Rock Classification Chart

Texture and mineral content form the basis for identifying and classifying igneous rocks, as shown in Figure 5.14. Using the chart, we see that granite is a coarse-grained (phaneritic) rock containing feldspar and abundant quartz—a mineral that in some rocks assumes the look of dull glass and in others looks like glass sparkling in the sun. Basalt, on the other hand, is a fine-grained (aphanitic) rock consisting of calcic plagioclase, pyroxene, and, commonly, olivine.

The classification chart shows that the igneous rocks vary systematically in mineral content, chemical composition, density, and temperature of crystallization. Note first the gradational nature of the mineral content. As we proceed from the peridotite end toward the granitic end of the chart (from right to left), olivine and pyroxene phase out and are replaced in sequence by hornblende and biotite. The ferromagnesians decline in abundance, and the percentages of feldspar and quartz increase. At the same time, the plagioclase feldspars change in composition from calcium-rich to sodium-rich varieties.

The changes in mineral content across the chart are broadly reflective of the chemical trends encountered as one rock type blends into another. Most significant is the increasing silica content, grading from about 45 percent at the peridotite end to up to 70 percent at the granitic end. Also notice that iron, magnesium, and calcium are

Figure 5.14 Mineral content and texture form the basis for classifying igneous rocks, as shown in this chart. The chart also illustrates systematic changes in the mineralogy, chemical composition, density, and crystallization temperature of the igneous rocks.

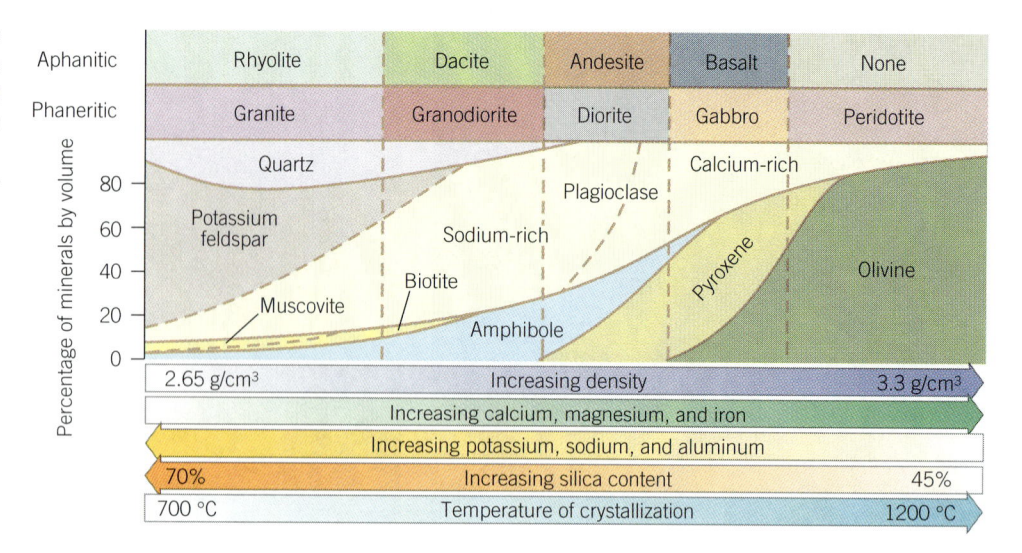

important at the peridotite end, decline markedly, and are replaced by sodium and potassium toward the granitic end of the chart. These changes in mineralogy and chemistry are responsible for the marked decrease in density toward the granitic end of the chart; peridotite and basalt are heavy rocks, and granite and rhyolite are light rocks. Finally, note that the minerals at the peridotite end crystallize at about 1200 °C, and those of granitic rocks begin to crystallize at about 700 °C.

The Sequence of Crystallization of Igneous Rocks

When viewed under a microscope, igneous rock textures bear out the crystallization sequence suggested by the classification chart. The figures that follow show thin sections of some of the minerals of igneous rocks viewed under polarized light. They preserve the evidence of reactions, frozen in time, in which olivine converted to pyroxene, pyroxene to amphibole, and so on. The evidence is preserved because, in each case, the magma crystallized before the reaction proceeded to completion. Figure 5.15(a), for example, shows olivine cores encircled by pyroxene *reaction rims*. A feature of this type can occur only if the olivine first crystallized and reacted with the magma at a lower temperature, at which time its outer rim was converted to pyroxene. The same kind of reaction holds for the other ferromagnesian minerals of igneous rocks (Figure 5.15b). These reactions constitute a *discontinuous series;* they occur in steps, at specific temperatures, rather than continuously.

Figure 5.15 Photomicrographs under polarized light of discontinuous reactions. (a) The brightly colored minerals are olivine. Each is encircled by a pyroxene reaction rim of a different color. (b) A pyroxene core is surrounded by an amphibole reaction rim. The grayish, rectangular minerals in both (a) and (b) are plagioclase.

(a)

(b)

Igneous Activity

Figure 5.16 Photomicrograph under polarized light of zoning in plagioclase feldspar. The layers or zones surrounding the calcium-rich core become progressively enriched in sodium and depleted in calcium toward the edge of the mineral. The reaction occurs continuously over the entire range of falling temperatures within the magma.

Figure 5.17 Photomicrograph under polarized light of granite with a complex mineral assemblage. The rectangular minerals with light and dark bands are plagioclase; the mottled gray ones are potassium feldspar; the bright-colored ones are biotite. The clear minerals showing shades of gray are quartz, the last to crystallize from the magma.

The plagioclase feldspars exist side by side with the ferromagnesian minerals in igneous rocks, and they also exhibit reactions frozen in time. Figure 5.16 shows a crystal of plagioclase displaying a series of rings or *zones* wrapped around a central, calcium-rich core. The succeeding zones mark a series of gradations in which the calcium content decreases and the sodium content increases. Clearly, the first plagioclase to crystallize was the calcium-rich core; as temperatures fell, it continued to react with the magma, exchanging calcium for sodium. Since there is no interruption of the reaction and there are no drastic conversions, we say that the plagioclase feldpars constitute a *continuous reaction series.*

Lastly, see the granite thin section of Figure 5.17. Quartz seems to exist as a formless mass between crystals of feldspar. This thin section shows that the last remaining liquid of the magma was pure silica, and it crystallized as quartz in the remaining spaces.

Figure 5.18 summarizes the sequence of mineral crystallizations expected from a typical basaltic magma. It is known as **Bowen's reaction series,** in honor of the Scottish-American scientist Norman Levi Bowen, whose *Evolution of the Igneous Rocks,* published in 1925, is a classic in the field of experimental geology. Note its similarity to the classification chart of Figure 5.14 and to the mineral sequences preserved in the frozen reactions we just described.

Bowen prepared glass samples whose compositions resembled simplified basaltic magmas, and subjected them to high pressures and temperatures that simulated conditions of the shallow crust. Then he cooled his samples rapidly to freeze the reactions and see what minerals, if any, had crystallized from the artificial magma. He found that olivine crystallized at about 1200 °C; at about 1050 °C it became unstable, reacted with the magma, and converted to pyroxene. Indeed, by using glass magmas of the appropriate compositions and by adjusting the temperature-pressure relationships, he was able to reproduce all the relationships shown in Figure 5.18.

In Bowen's experiments, the minerals that crystallized maintained contact with the artificial magmas in a state of equilibrium. For this reason, the minerals and magma were able to freely react and exchange ions. He found that as the temperatures were lowered, more minerals formed and less liquid remained. The last drop of liquid had a composition very high in silica, and the minerals consisted of a mixture of pyroxene and plagioclase. When all the liquid was used up, the minerals that had formed, taken together, had the same composition as the original magma. The basaltic magma and the minerals that crystallized from it thus went through a series of interactions that resulted in a basaltic rock.

Bowen's reaction series
A proposed sequence of mineral crystallization from basaltic magma, based on experimental evidence.

Figure 5.18 Bowen's reaction series. This chart shows the experimentally determined sequence in which minerals crystallize from a cooling basaltic magma. The ferromagnesium silicates form a discontinuous reaction series and the plagioclase feldspars a continuous series. Potassium feldspar, muscovite (colorless mica), and quartz crystallize at the lowest temperatures.

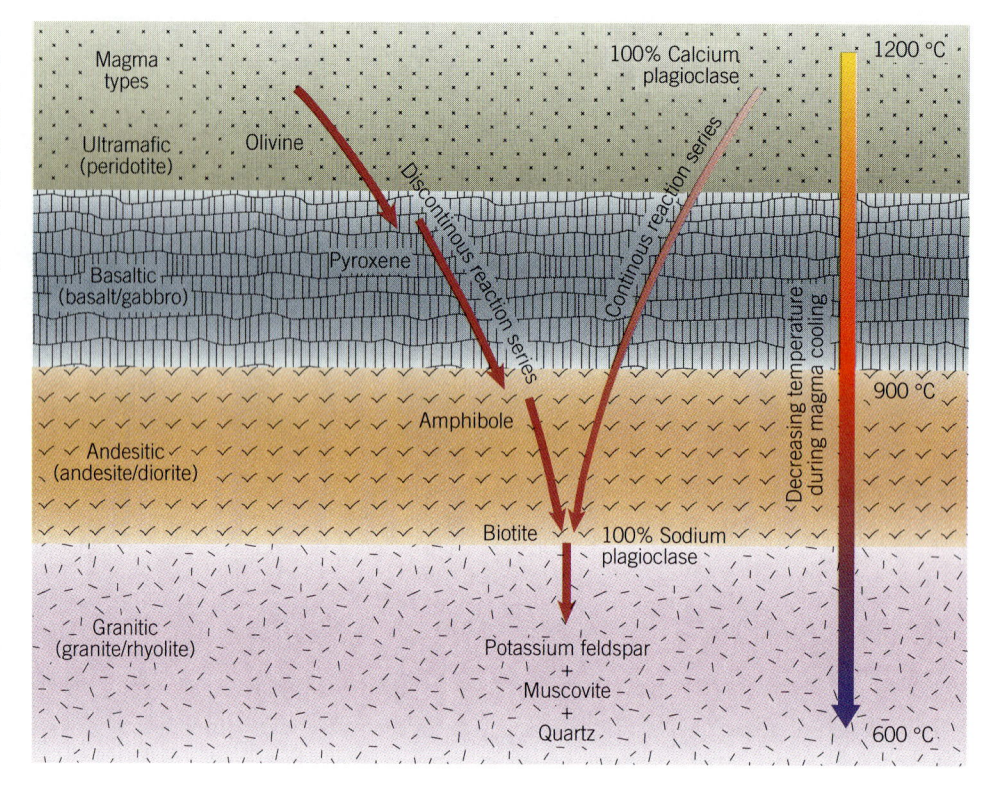

Bowen then considered what might happen if an early-formed mineral, such as olivine, is separated from the basaltic liquid. This mineral extracts a high percentage of iron and magnesium from the liquid, thereby shifting the liquid's composition toward the silica-rich end of the magmatic spectrum. A magma of this composition can remain liquid at lower temperatures than a basaltic magma, and more silica-rich minerals are able to crystallize from it. Therefore, the separation of early-formed ferromagnesian minerals and calcium-rich feldspar from a basaltic magma may prolong the life of the melt and produce a wide range of rocks. This process—in which early-formed crystals are prevented from reacting with the melt while changing the melt's composition through extraction—is called *fractionation*.

Fractionation takes several forms in natural settings. *Gravitational settling* is effective in the fractionation of basaltic magma. It occurs as olivine and pyroxene, which are among the first minerals to crystallize, sink to the bottom of the magma chamber, out of contact with the liquid. These minerals collect to form peridotite. As in Bowen's experiments, their separation from the magma extracts a high proportion of iron and magnesium. Thus the remaining magma, enriched in lighter elements, will form basaltic and, in extreme cases, dioritic rocks.

Although gravitational settling works well in extremely fluid magmas, geologists have come to question its effectiveness as a fractionation mechanism in most magmas. They doubt that the density differences between the early-formed minerals and the magma are great enough to overcome the magma viscosity, or resistance to flow. The situation is analogous to nuts mixed in a jar of honey; the nuts may be denser than the honey, but the honey will not let them sink. Geologists thus believe that other types of fractionation mechanisms may be more common. One such mechanism is *flow segregation*. As the magma rises in the crust, the early-formed minerals coat the walls of the enclosing country rock and are separated from the magma.

The late-stage silicate components of the magma are rich in water vapor and other volatiles. Superheated under high pressure, this hot mixture mobilizes and bursts through the soft, freshly formed igneous rock and out into the enclosing country rocks of the crust. There, it may blast out cavities and crystallize. Ions are able to migrate so rapidly in this environment that they form the huge crystals of coarse-textured pegmatite (Figure 5.19).

Figure 5.19 A coarse-grained pegmatite dike cutting across country rock. Some pegmatites contain rare gems and elements that were retained in the late-stage fluids from which the igneous rock crystallized.

These rocks are an exception to the general rule that coarse textures are associated with slow cooling.

The Geologic Settings of Igneous Activity

The global occurrence of igneous activity is confined primarily to plate boundaries or to intraplate regions above mantle hot spots (Figure 5.20). Each setting is characterized by magmas of distinctive composition and by distinctive styles of intrusive and extrusive activity. At divergent boundaries, basaltic magma is generated as the plates separate. The magma is derived from partial melting of the upper mantle, and it erupts from beneath the mid-ocean ridges to build basaltic crust. All this crust is recycled within about 200 million years as the oceanic lithosphere descends into the mantle at convergent boundaries.

Magmas display a more varied composition at convergent plate boundaries and are commonly richer in silica than the mafic magmas of divergent boundaries. Those plates that descend beneath the oceanic lithosphere generate magmas that are generally of andesitic composition. The magmas rise and build island arcs. Those that descend beneath continental crust generate both andesitic and granitic (silicic) magmas, which intrude the base of the continent. Most volcanoes that border subduction zones are explosive, because their highly viscous silica magmas prevent gases from escaping. When pressures reach a critical point, the volcanoes erupt violently.

Mantle hot spots occur beneath both oceanic and continental crust. Those that occur beneath oceanic crust generate basaltic magmas and give rise to huge volcanoes. The Hawaiian Islands were formed over such a hot spot. Constructed layer by layer from the ocean floor by relatively peaceful liquid lava eruptions, the islands consist of some of the world's largest volcanoes, and they would be far larger if not for the fact that the Pacific plate upon which they rest is carrying them away from the hot spot. Those mantle hot spots that occur beneath continental crust have a more elaborate sequence of activity spanning millions of years. Early on, the hot spot melted the crust above it, which generated highly silicic and viscous magma that erupted explosively in huge ash flows. Later, the upper mantle was tapped directly, and basalt magma erupted.

Long fracture zones in the oceanic and continental crust constitute another major setting of igneous activity. These fractures extend to the mantle, where they tap voluminous sources of highly fluid basaltic magmas. Extensive lava plateaus are built of layer upon layer of basalt that erupts over millions of years. These fractures are associated with crustal rifting during the early stages of plate divergence.

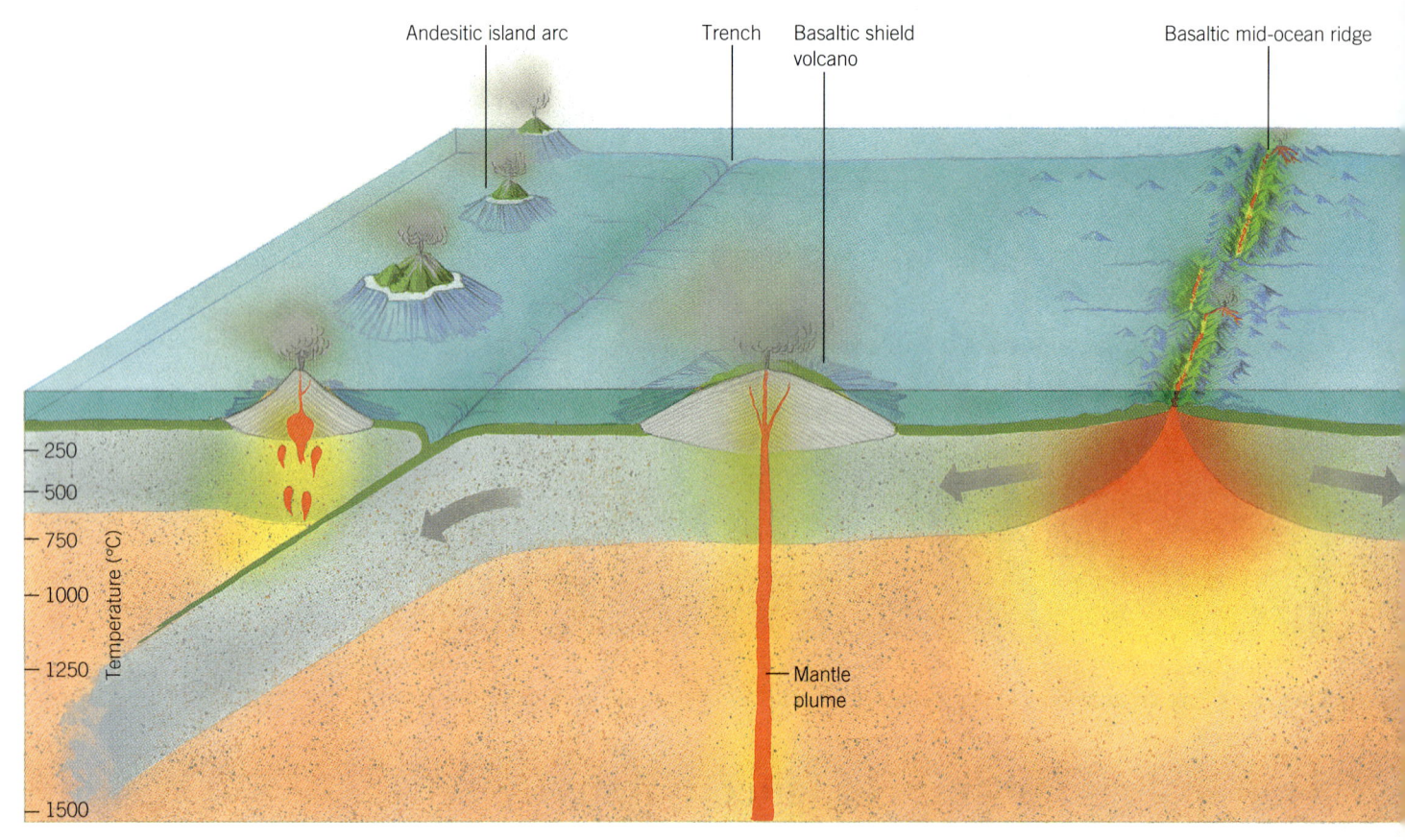

Figure 5.20 The structural settings of the various forms of igneous activity. Most of the activity occurs at divergent and convergent plate boundaries. Where they border island arcs, subduction zones generate andesitic magma; where they border continents, andesitic or granitic magma. Depending on the depth at which melting occurs, continental hot spots and rifts generate rhyolitic and granitic magmas (from crustal melting) or basaltic magma (from the mantle). Deep-seated mantle plumes feed oceanic hot spots.

Igneous Activity at Mid-Ocean Ridges

At the same time that Bruce Heezen and Marie Tharpe were mapping the global extent of the mid-ocean ridge system, other geologists were discovering that the oceanic crust is created by igneous processes within and beneath the ridge system. In the succeeding decades, geologists have studied these processes using seismic probes, deep-sea drilling, and submersible craft that descended to the rift valleys. In some locations, geologists have observed lava pouring out of cracks in the rift valley floor, and huge hydrothermal vents—called chimneys or black smokers—spewing mineral-rich hot water many meters high. Magma lying at shallow depths beneath these locations provides the heat to drive these submarine hot springs. These on-site investigations are supplemented by inspection of slices of oceanic crust that have been uplifted and thrown onto the continents at subduction zone boundaries. The major result of these studies is the determination that the oceanic crust is basically a four-layered structure. Magma rises beneath the rift valley and crystallizes to form three of these layers; sediments form the fourth (top) layer (Figure 5.21).

Geologists have a reasonably good idea how this four-layer oceanic crust is constructed. First, basaltic magma, less dense than the surrounding mantle rock, rises beneath the ridge crest and accumulates in a space, or magma chamber, it has made for itself. There, it begins to crystallize in stages. Slow cooling forms the lower, massive gabbro layer. However, in the cooling stage, ultramafic olivine and pyroxene crystallize first and sink to the bottom of the magma chamber. These minerals form the peridotite of the uppermost mantle. While all this is happening, the crust continues to fracture and spread. The remaining magma forces its way through the vertical fractures in the gabbro and forms the sheeted basalt dikes. These dikes act as conduits for magma spilling onto the cold ocean floor, where it congeals rapidly to form the pillow basalts (Figure 5.22). As the solidified oceanic crust spreads away from the ridge, marine sediments gradually accumulate on the pillow basalts.

Igneous Activity 121

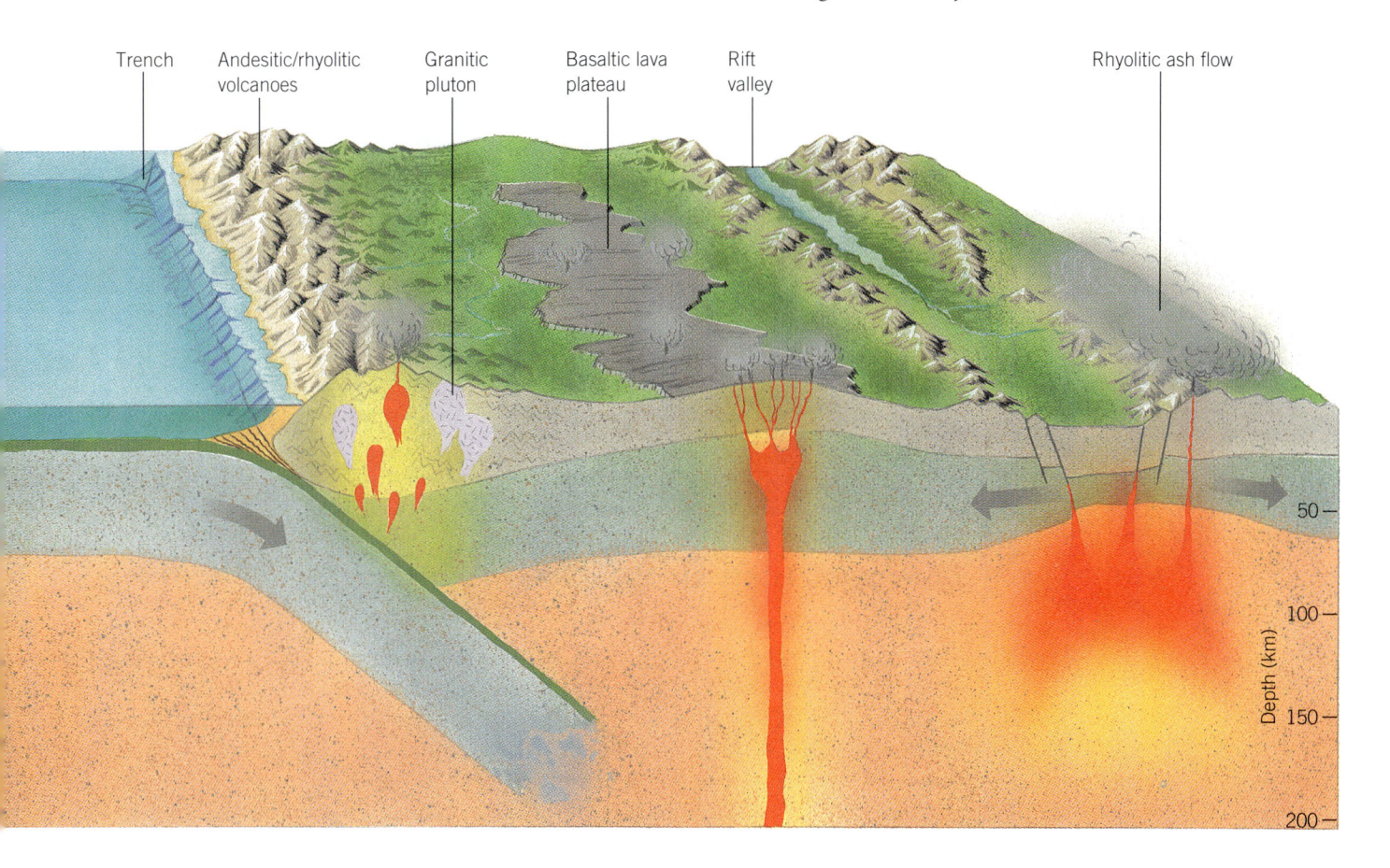

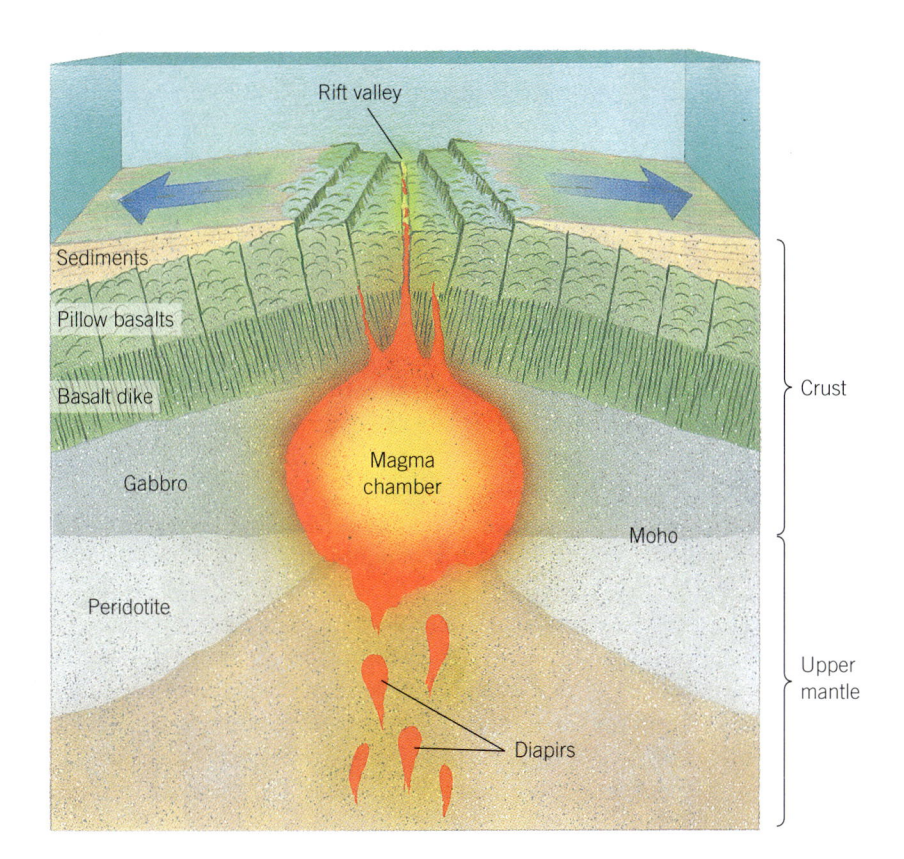

Figure 5.21 The layered structure of the oceanic crust and its relation to magmas beneath the rift valley of the mid-ocean ridge. The magma chamber is fed by partial melting of the asthenosphere. Olivine and pyroxene settle to the bottom of the chamber and form the peridotite of the upper mantle. Coarse-grained gabbro crystallizes from the main mass of the magma within the chamber. The basaltic dikes form as the crust separates and magma escapes the chamber through fractures. Basaltic magmas erupt through the dikes to the surface, cool rapidly when they come in contact with sea-water, and then congeal as pillow lavas.

Figure 5.22 Recent lava flows on the crest of the East Pacific Rise, showing rounded pillow structures.

Igneous Activity in Subduction Zones

Subduction zone magmas are far richer in silica and other lighter elements than the mafic magmas of the central rift valley. Because subduction zone magmas form at lower temperatures than mafic magmas, it is tempting to imagine a simple system in which the lithospheric plates carry the basaltic rocks of the oceanic crust to the subduction zones, where they descend, partially melt at low temperature, and generate higher-silica magmas. This sequence may indeed happen to a small degree, but most geologists who have studied the process doubt that the cold subducting plates can absorb enough heat from their surroundings to melt appreciably; consequently, they seek other causes.

One explanation of the origin of subduction zone magmas involves water. As we have seen, it lowers the melting point of rocks at depth. The descending plates probably carry a great deal of water, some of it held by the sediments riding on the plates and some of it held by the minerals within the basalt. At depths beginning at about 75 kilometers, the pressures are so great that water is expelled from the plates to the hot, overlying mantle rocks. The melting points of the rocks are thus lowered enough to initiate partial melting, and given the lower temperatures that prevail in subduction zones, high-silica magmas are generated. The very-high-silica magmas that form the continental granites may themselves be generated by remelting of the granitic crust at the base of the continents.

The Ascent of Magma in Subduction Zones

How does magma rise in subduction zones and invade the crust? To answer this question, we must begin with the melting of mantle and lower crustal rock just above the subducting plate. The melting of source rock probably proceeds in stages. Early in the process, the rock is largely solid, with melting confined to the matrix between the grains. Eventually, a point is reached at which the rock loses its integrity as a solid mass, even though much solid material is present in the melt. We may envision the magma at this point as having the consistency of slush (partially melted snow) within which the liquid and solid constituents move together as a unit.

For several reasons, this magma is less dense than an equal volume of surrounding rocks and the difference in density between the magma and the enclosing rock creates an unstable condition. In experiments simulating these circumstances, a layer of magma tends to bunch into a mushroomlike form, which buds off and rises buoyantly (Figure 5.23). The rising bud—which may reach tens of kilometers in diameter—is called a *diapir*. When it comes to rest in the crust, it crystallizes into a granitic or granodioritic batholith (Figure 5.24).

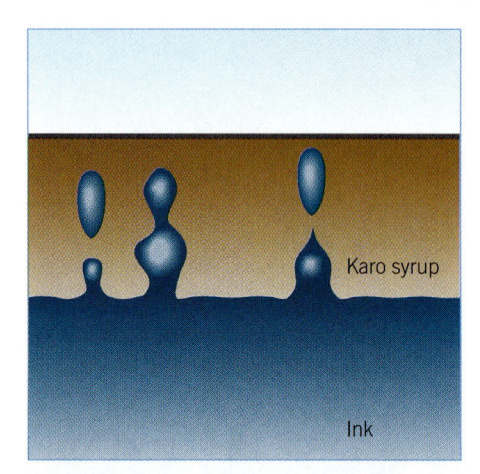

Figure 5.23 This experiment shows how density differences cause blobs of light liquid to rise through denser liquid. Geologists believe a similar process is at work in magma ascent.

Figure 5.24 Half Dome and upper Yosemite Valley, California. Glaciers carved and smoothed a giant granitic batholith, leaving behind these spectacular features.

Diapirs emplaced higher in the crust (5–15 kilometers) encounter host rocks that are cooler and more rigid. At this level, it appears that they simply push aside the host rock. Therefore, contacts between pluton and host rocks are sharper than at lower depths, and the zone of mixing is not as extensive. Diapirs emplaced still higher in the crust (0–5 kilometers) encounter rocks at relatively lower pressures and temperatures. There, the cold, brittle crust tends to fracture. Consequently, the diapir rises by a combination of forceful injection, the shattering and cracking of host rock, and *stoping.* Stoping occurs when the roof of the en-closing rock breaks into huge chunks, which are assimilated as the magma eats its way higher into the crust. These fragments, called *xenoliths,* become stranded like islands within the igneous rock near the contact (Figure 5.25).

Extensive fracturing of the shallow host rocks provides avenues for the magma to spread out into numerous channels and form dikes, sills, and laccoliths. At this level in the crust, the magma is under low pressure; and portions of it, fed from below, may be capable of bursting to the surface. The eruptions that result build volcanoes and other features (the subject of the next chapter). The high silica content of subduction zone magmas can make them extremely viscous, and their high viscosity prevents gradual escape of gases, so that pressures build until the magma erupts with great violence. As we will see in the next chapter, this condition makes subduction zone volcanoes extremely dangerous.

Figure 5.25 Basaltic xenoliths within granodiorite, Scotland.

Mineral Concentration Through Divergent Boundary Processes

Mid-ocean ridge hydrothermal activity plays a major role in the concentration of ore deposits in mid-ocean rift valleys. An ore is the naturally ocurring material from which an economically valuable mineral can be profitably extracted. Rift valley fractures are, in effect, an elaborate plumbing system, similar to the kind used in a large apartment house; the magma beneath the rift valley plays the role of basement boiler. Cold water enters the fractures and percolates deep into the crust, where it is superheated by the magma. The hot water then expands, rises through the fractures, and escapes to the ocean floor as metal-rich black smokers (Figure 5.26).

The brines that vent to the surface precipitate metals in an oxygen-rich environment, forming oxides, hydroxides, and silicate ore minerals. These minerals build the chimneys through which the brines escape. Often, the metals may precipitate directly around a small rock fragment, or shark's tooth, to form potato-shaped manganese nodules. This name is misleading, because manganese nodules also contain copper, iron, cobalt, nickel, and chromium (Figure 5.27). Billions of them are strewn throughout the sea floor. Their potential value is incalculable, and the nations of the world are busily working out treaties that will allow for their mining.

Mineral Concentration at Subduction Zones

The metals deposited by hydrothermal solutions upon or within newly created oceanic crust at divergent boundaries are swept conveyor belt style toward subduction zones. Though many of the metals are cycled back into the mantle within these zones, huge slices of metal-rich oceanic crust are sheared off the descending plates and shoved onto the bordering landmasses (see Chapter 17). These bodies of oceanic crust emplaced on land are called *ophiolites*. The island of Cyprus is an ophiolite formed from oceanic crust uplifted by the collision of the African and Eurasian plates. The name *Cyprus* is derived from *kyprus,* the ancient Greek word for copper—an appropriate name, considering the rich copper deposits of that island.

Figure 5.26 A plume of hot water issuing from a submarine chimney.

Subduction zones are also environments conducive to the development of hydrothermal circulation systems. The plates may partially melt as they descend into the mantle and release the metals that were concentrated in the upper oceanic crust by hydrothermal fluids at the mid-ocean rift valleys. At an ocean-continent plate boundary, subduction may also trigger melting of the overlying granitic crust. We know from the frequent earthquakes in the vicinity of subduction zones that the rocks overlying the subducting plate are highly fractured. These fractures provide avenues for underground water to penetrate the crust, absorb heat from the magma, and rise to the surface. They also provide the means for seawater trapped within the descending plates to escape. These hot waters react with crustal rocks and rocks of the upper mantle to dissolve metals and reprecipitate them within the crust, either concentrated as thick veins or scattered as *disseminated deposits.*

Figure 5.27 A large field of metal-rich manganese nodules on the floor of the northeast Atlantic Ocean. (See Figure 7.22 on page 166 for a close-up view of a nodule.)

Many valuable mineral deposits are formed through direct precipitation within the main body of the magma as it crystallizes to form igneous rock. Recall that a magma is an extremely complex silicate melt in which minerals crystallize over a broad range of temperatures. In many mafic magmas, metals such as nickel, chromium, tungsten, platinum, and cobalt crystallize at higher temperatures than most of the silicate minerals. Because they are denser than the melt, they settle to the bottom of the magma chamber, where they collect in thick, easily minable layers. This is the origin of many of the metals of the fabulous *Bushveld Complex* of South Africa, one of the richest deposits of rare, heavy metals on the Earth. The smaller but nevertheless important Stillwater Complex of Montana was formed in a similar manner. Each is of Precambrian age, and their exact origins are obscure. But association of the metals with mafic rocks indicates that they came from the mantle. Perhaps they crystallized at a rift zone or from mantle plume magmas that intruded the ancient crust. Valuable metal deposits are also concentrated in silica-rich magmas derived from the continental crust bordering subduction zones. The great Chilean copper deposits, for example, are closely associated with the andesite porphyries and granodiorites of the Andes.

Late-stage residual fluids of granitic magmas are an important source of mineral deposits. They are rich in silica, potassium, and rare elements that do not fit easily into the crystal structure of the major minerals composing the igneous rock mass. As the rock continues to cool and contract, the residual fluids are mobilized and pierce both the igneous rock and the surrounding host rock, crystallizing to a coarse-grained pegmatite. Some pegmatites contain deposits of gold, silver, uranium, lithium, bismuth, tungsten, tin, and antimony. In rare instances, protected cavities within the pegmatites contain beautiful gem minerals: varieties of beryl (emerald), topaz, garnet, and tourmaline.

The 700,000 year-old Galeras volcano in the Colombian Andes spews half a kilogram of gold into the air each day and may be depositing 20 kilograms of gold a year into the rocks that line the crater.

STUDY OUTLINE

The oceanic crust and most of the continental crust are igneous in origin.

I. IGNEOUS ROCK BODIES
  A. Igneous rocks begin as **magmas** generated from the melting of mantle and crustal rocks. Most magmas rise, intrude the crust, and crystallize to form rock bodies, or **plutons.** Some magmas escape to the surface as lava. Eruptions of lava can build conical volcanoes or thick lava plateaus.
  B. Plutons are classified on the basis of their relation to the layering of the host rocks that they intrude.
   1. **Concordant** plutons form parallel to the layering of the host rock either as flat, tabular **sills** or as mushroom-shaped **laccoliths.**
   2. **Discordant** plutons cut across the layering of the host rocks.
      a. Tabular **dikes** cut jaggedly through layers of country rock.
      b. Cylindrical pipes fill volcanic vents.
      c. Radiating dikes or concentric ring dikes may encircle these pipes.
      d. Diatremes are pipelike intrusions in which diamonds can be found.
      e. **Batholiths** are enormous granite plutons.

II. THE FORMATION OF ROCKS FROM MAGMA. A magma is a complex mixture of molten silicate rock material, dissolved gases, solid minerals, and chunks of country rock.
  A. The large mineral crystals of **intrusive** rocks form when the magma cools slowly, undisturbed. Such conditions are found only at depth. The small crystals of **extrusive** rocks form under conditions of rapid, disturbed cooling at the surface.
  B. There are five major types of igneous rock textures:
   1. **phaneritic**—coarse-grained, large crystals
   2. **aphanitic**—fine-grained, small crystals
   3. **glassy**—smooth texture, no crystals
   4. **porphyritic**—large crystals embedded in an aphanitic matrix
   5. **pyroclastic**—cemented lava fragments
  C. Igneous rocks can be grouped by mineral content into four families:
   1. the peridotite family, consisting primarily of olivine and pyroxene and found in the upper mantle
   2. the **basalt**-gabbro family, consisting primarily of pyroxene and calcium-rich plagioclase; the components of oceanic crust
   3. the andesite-diorite family, consisting primarily of amphibole and sodium-rich plagioclase; associated with subduction zone volcanoes
   4. the **granite**-rhyolite family, consisting of potassium feldspar, sodium-rich plagioclase, and quartz; the main components of the continental crust
  D. Igneous rocks arrayed on a classification chart display a number of gradational changes, from the peridotite toward the granitic end:
   1. The percentages of the ferromagnesian minerals decrease, whereas the percentages of potassium feldspar, sodium plagioclase, and quartz increase.
   2. The percentages of silica, sodium, and potassium increase, whereas the percentages of iron, magnesium, and calcium decrease.
   3. The density of the rock decreases.
   4. The temperature of magma crystallization decreases.
  E. The microscopic analysis of mineral textures and experimental work, summarized in **Bowen's reaction series,** bear out the general sequence of crystallization suggested in the rock classification chart.
   1. Reaction rims in the ferromagnesian minerals suggest a discontinuous reaction series. Zoning in the plagioclase feldspars suggests a continuous reaction. Quartz crystallizes last.

2. Through fractionation, a single basaltic magma may produce many kinds of igneous rocks.
3. Fractionation mechanisms in which early-formed minerals are separated from the remaining liquid include gravitational settling and flow segregation.

III. **THE GEOLOGIC SETTINGS OF IGNEOUS ACTIVITY.** Each setting is characterized by magmas of distinctive composition and by distinctive styles of intrusive and extrusive activity.

   A. At divergent boundaries, basaltic magma erupts and builds the layered igneous rock structure of the oceanic crust.

   B. Where oceanic and continental crust converge, granitic magmas are generated. The high silica and gaseous content of subduction zone volcanoes causes them to erupt violently.

   C. Hot spots beneath oceanic crust generate basaltic magmas that build huge, relatively peaceful volcanoes, such as those of Hawaii.

   D. In subduction zones, water expelled from the descending plate may initiate partial melting of mantle rock, generating granitic magmas. They rise in the crust as diapirs to form batholiths.

IV. **MINERAL CONCENTRATION AND IGNEOUS ACTIVITY** Hydrothermal activity plays a major role in the concentration of valuable mineral-bearing ore deposits at divergent boundaries and subduction zones.

STUDY TERMS

aphanitic (p. 112)
basalt (p. 114)
batholith (p. 110)
Bowen's reaction series (p. 117)
concordant (p. 110)
dike (p. 110)
discordant (p. 110)
extrusive (p. 112)
glassy (p. 112)

granite (p. 115)
intrusive (p. 112)
laccolith (p. 110)
magma (p. 109)
phaneritic (p. 112)
pluton (p. 110)
porphyritic (p. 112)
pyroclastic (p. 113)
sill (p. 110)

THOUGHT QUESTIONS

1. Distinguish between concordant and discordant plutons.
2. Describe the evidence, viewed under the microscope, that supports Bowen's reaction series.
3. Describe the compositional changes that occur across the spectrum of igneous rocks from peridotite to gabbro, diorite, and granite.
4. What is the origin of magma beneath the mid-ocean ridges? Explain the process by which this magma cools to form the four-layered structure of the oceanic crust.
5. How do the magmas produced beneath subduction zones differ from magmas produced beneath the mid-ocean ridges? What processes may generate subduction zone magmas?
6. Explain the origins of batholiths and how they are emplaced in the crust.

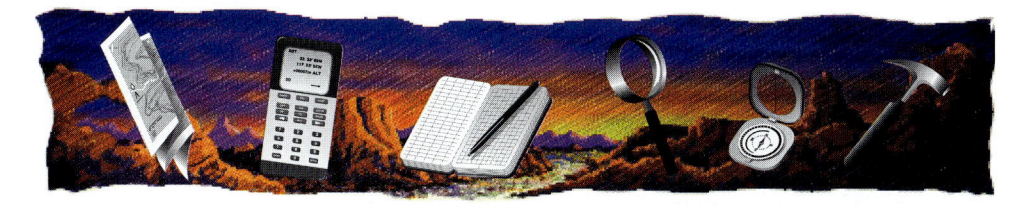

Chapter 6

The Anatomy of a Volcano

Tectonic Settings of Volcanism

The Mechanics of a Volcanic Eruption

The Materials of Volcanic Eruptions
- Lavas
  - Basaltic Lava Flows
  - Silicic Lava Flows
- Pyroclastic Materials
- Volcanic Gases

Volcanic Structures and Eruptive Styles
- Basaltic Volcanoes
  - Shield Volcanoes
  - Cinder Cones
  - Fissure Eruptions and Lava Plateaus
- Andesitic and Silicic Volcanoes
  - The Eruption of Mount Saint Helens
  - Ash Flows

Forecasting Eruptions

Volcanism and Climate

Constructive Aspects of Volcanism

Volcanism

The Klamath Native Americans have lived in northern California and Oregon longer than the redwoods, and each generation has passed on the following story to the next.

> Long ago, the Chief of the Below World, who resided within a mountain called Lao Yaina, became enraged when a young maiden declined to marry him. He swore he would seek revenge by annihilating her people with the Curse of Fire, and he rose to the summit of Lao Yaina, flames streaming from his mouth. As the mountain trembled he hurled molten rocks through the air and unleashed a hot rain upon the forests that burned the trees to charcoal stumps. The people prayed for deliverance to the Chief of the Above World, who drove the Raging One back inside Lao Yaina. The top of the mountain sank behind him as he retreated into his subterranean lair. The next morning, the high peak of Lao Yaina had vanished.*

Geologists and anthropologists believe this legend recounts an enormous volcanic eruption that occurred in the Cascade Range of the Pacific Northwest. Lao Yaina is almost certainly Oregon's Mount Mazama, in whose collapsed center now sits the magnificent Crater Lake (Figure 6.1). The huge quantities of debris found along the flanks of the decapitated **volcanic cone** and in the surrounding region are records of the eruption. A charred moccasin was also found in the bottom layers of the debris. Radiometric dating of this artifact and of fragments of charred wood has established that the eruption occurred about 7000 years ago. Thus the legend constitutes an unbroken 7000-year-old verbal chain that links an ancient people with their modern descendants.

From projections across Crater Lake, geologists estimate that the summit of the volcano before the eruption stood 3700 meters above sea level and 1200 meters above the

volcanic cone
An accumulation of lava and/or pyroclastics around a volcanic vent.

---

*Adapted from Lawrence R. Kittleman's interesting article "Tephra," *Scientific American*, December 1979.

◂ A lava eruption on Mauna Loa. The Hawaiian Islands are constructed of innumerable flows of this sort.

Figure 6.1 Oregon's Crater Lake fills the caldera of the former Mount Mazama, whose summit was destroyed during an explosive eruption about 7000 years ago. In the center of the lake is Wizard Island, a small volcanic cone that formed within the crater around the tenth century A.D. Geologists have reconstructed the history of the Mount Mazama eruption by analyzing the layers of lava and debris ejected from the volcano.

present rim of the cliff encircling the lake. A thick deposit of thinly layered ash at the base of the volcano provides evidence that the first products of the eruption settled out of the air. Above these ash layers lies a crude jumble of pumice, ash, and shattered rock. From observations of modern volcanoes, we know that some of these sediments were carried by a ground-hugging fiery cloud that surged down the mountainside; other sediments crept slowly as viscous masses of mud. In all, 30 cubic kilometers of debris were ejected from the volcano, which is about a hundred times the volume ejected by the Mount Saint Helens, Washington, eruption of May 18, 1980.

Deprived of support from below, the summit of Mount Mazama collapsed in a heap along steeply dipping circular faults, leaving in its center a crater 9 kilometers in diameter. About the tenth century A.D., a fresh eruption occurred within the crater and built a small volcanic cone. Today, the cone is called Wizard Island of Crater Lake. The lake itself covers 50 square kilometers and is 600 meters deep in places. Sheer cliffs tower 150–600 meters above the lake.

We have presented two contrasting descriptions of the same event: one ancient and metaphorical, the other modern and scientific. The event is no less awesome in either version, but let us now pursue the scientific explanation of volcanoes and their important contributions to the geology of the Earth's crust.

volcano
A vent in the surface of the Earth through which magma, gases, and rock fragments erupt; also the term for the landform that develops around the vent.

vent
A conduit through which magma rises to the surface.

The Anatomy of a Volcano

The term **volcano** stems from *Vulcan,* the Roman god of fire. It refers to an outlet to the Earth's surface through which magma, gases, and fragments erupt and to the commonly conical landform constructed of these materials. Usually, the outlet or **vent** is cylindrical, having been shaped by swirling gases and rock fragments escaping the

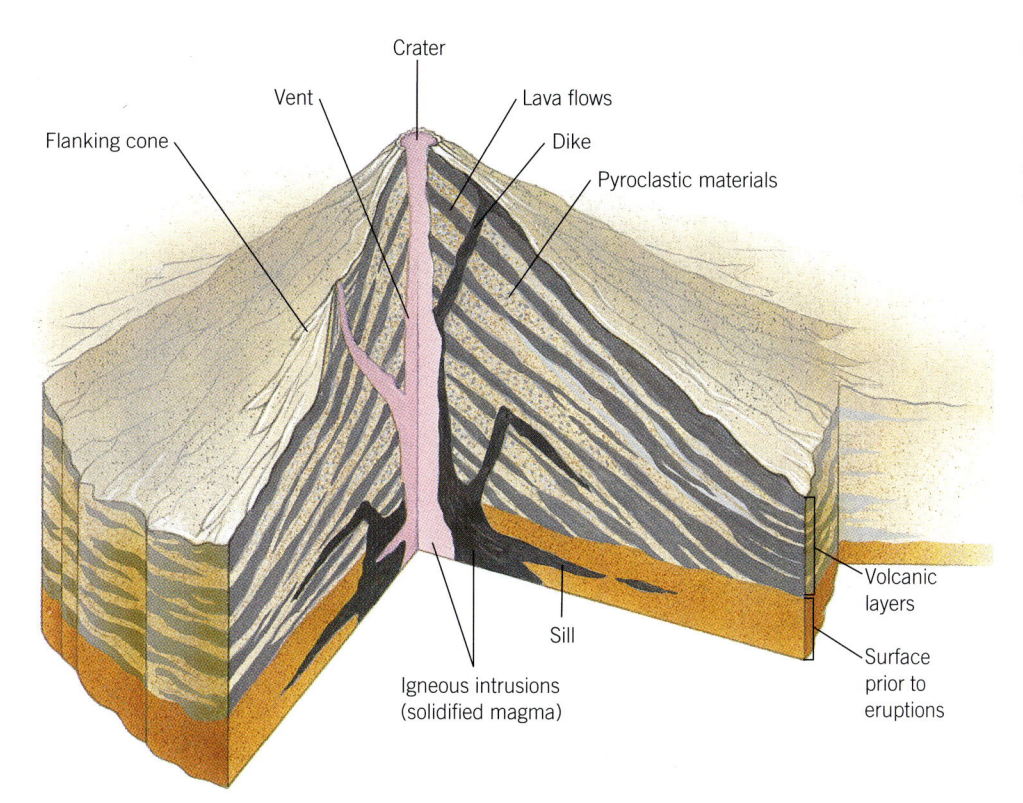

Figure 6.2 The internal skeleton of a volcano consists of a central vent and a network of dikes and sills filled with congealed lava. The cone itself is constructed of layers of lava and volcanic debris.

magma. During an eruption, most of the fragmental material accumulates close to the vent, as does the magma that pours out onto the surface as **lava**. Successive eruptions build the classic cone-shaped structure most people associate with the word *volcano* (Figure 6.2). The main pipe is generally located near the summit of the cone, where it opens to a bowl-shaped **crater,** or a much larger **caldera.** The crater can form from a violent eruption that blows the summit away or from the gentle piling of material on all sides of the vent. However, most craters are the result of collapse following an eruption, as in Mount Mazama. Large volcanoes are commonly fed through a network of vents that issue from their sides and create flanking cones. This elaborate plumbing system is filled between eruptions with congealed magma that provides the volcano with an internal skeleton. In active volcanoes, this skeleton serves as a pressure cap on the magma below.

lava
Magma that flows out onto the surface of the Earth; also refers to the rock body formed after the magma cools.

crater
A circular depression; a volcanic crater contains the vent or vents of the volcano.

caldera
A volcanic crater larger than 1 kilometer in diameter, usually formed by explosion or collapse.

Tectonic Settings of Volcanism

There are some 600 active volcanoes on the continents and islands of the world (Figure 6.3). That is, 600 have been observed by human beings to erupt during recorded history. Over two-thirds of these volcanoes constitute the *Ring of Fire* that borders the subduction zones of the Pacific. To the 600, add several thousand volcanoes that have not erupted in recorded history but whose lightly eroded cones indicate that they are of geologically recent origin and may spring to life again. Now add on the submarine volcanoes; 50,000 have been identified on the floor of the Pacific Ocean alone! How many are active is difficult to say, but it is obvious that volcanic eruptions, frequently presented in the media as exotic events, are instead among the most common phenomena that shape the Earth's crust.

Most volcanic activity, or *volcanism,* occurs along diverging and converging plate boundaries—that is, along the rift valleys of the mid-ocean ridges and along the island arcs and continental margins that border subduction zones. However, the Earth's dynamism is not exclusively expressed by volcanism at plate boundaries. Hot mantle plumes rising from the deep interior can penetrate the crust far from plate boundaries. Intraplate hot spots result wherever they force or melt their way to the surface.

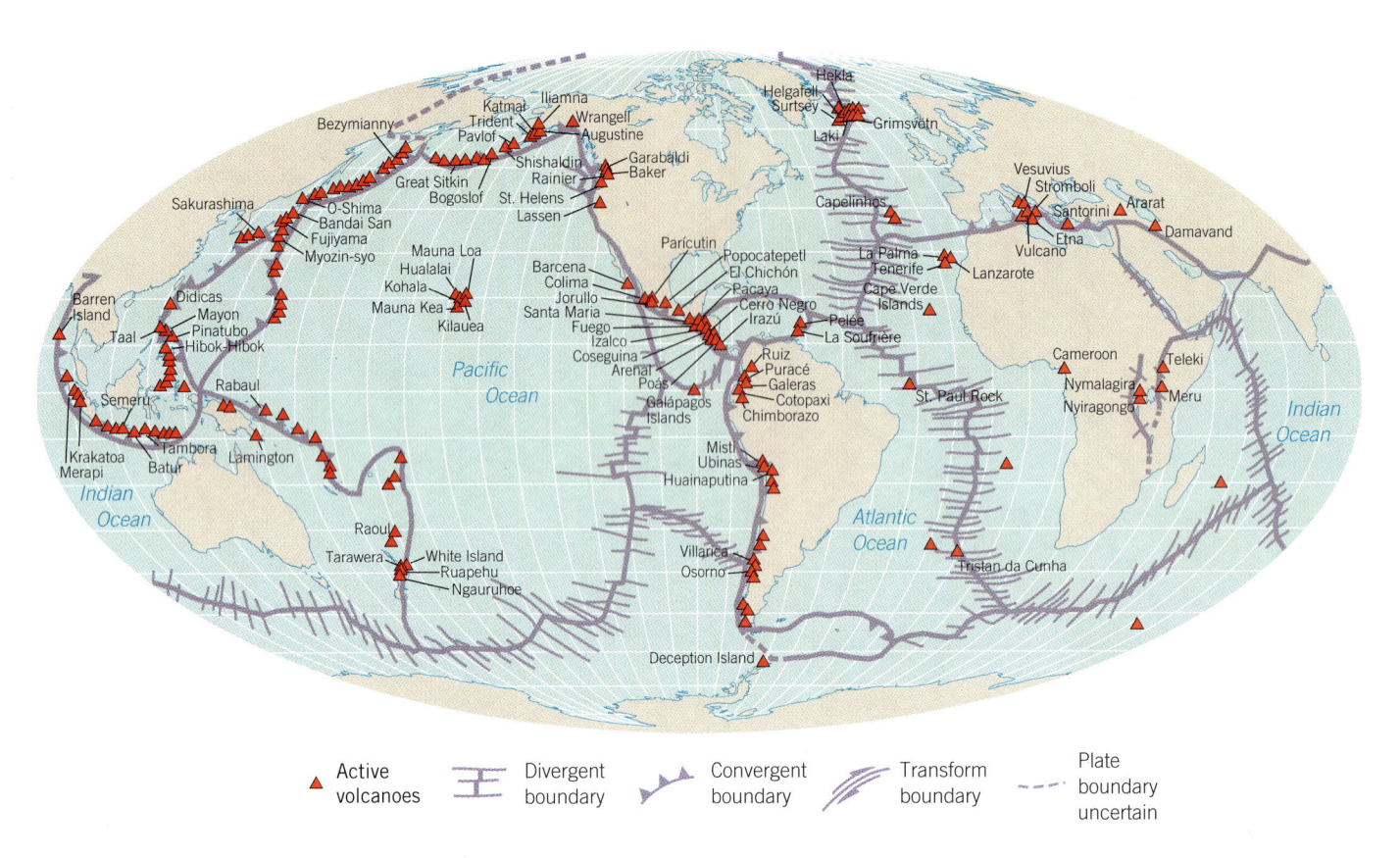

Figure 6.3 Most of the active above-sea-level volcanoes of the world are located at or near plate boundaries. They are so numerous along the margins of the Pacific Ocean that the region is referred to as the *Ring of Fire*.

pillow lava
A term applied to lavas of ovoid or pillow shape.

phreatic eruption
A volcanic eruption of steam, mud, or ash initiated by the contact of water with magma or hot rock.

The Mechanics of a Volcanic Eruption

The magma that fuels a volcanic eruption is generated by the melting of solid rock in the upper mantle or base of the crust. Because the molten liquid is hotter and less dense than the surrounding rock, it migrates upward through fractures. Prior to eruption, it commonly collects in a large reservoir, or *magma chamber*, a few kilometers or so beneath the surface. The ascent of the magma from the chamber into the volcano proper is marked by earthquake tremors. They occur as the rising magma weakens fractures, melts the overlying rock, and moves through the newly created space.

Picture suddenly pulling the cap on a bottle of warm soda you have just shaken. Carbon dioxide bubbles surge rapidly to the surface and out the neck of the bottle, carrying the liquid along with them. Volcanologists believe a similar combination of pressure reduction and escaping gases triggers volcanic eruptions. The pressure is relieved by the fracturing and shattering of sealed vents at the summit as the magma migrates upward. The gases that are then released separate and initiate the actual eruption.

Magma viscosity plays a key role in the style of an eruption because it determines how the gases separate from the melt. A highly viscous silicic magma, for example, retards the expansion and passage of the gas through the liquid, so that pressure builds within the magma. Also, the gas escapes more slowly, so most of it is retained as the magma rises higher in the vent beneath the summit. Thus when separation does occur, it is likely to be rapid and the resulting eruption explosive (Figure 6.4a). In low-viscosity basaltic magmas, gases escape uniformly and gradually, so that eruptions are relatively peaceful (Figure 6.4b).

Conditions at the time of eruption as well as the materials the magma encounters during ascent also influence the character of a volcanic eruption. Nowhere is this better illustrated than in typical sea-floor lava flows. Those that occur at great depths are seldom explosive, because the pressure of overlying water prevents gases from escaping the lava. Therefore, the lava flows smoothly and peacefully, and congeals to form ovoid **pillow lava** (Figure 6.5). On the other hand, **phreatic eruptions** involve the mixing of

Figure 6.4 Gas content and magma viscosity affect a volcano's eruptive style. (a) Explosive eruption. Viscous magma retains gases under great pressure. When the magma plug rises higher in the vent, pressure is lowered and the gases are released suddenly and violently. (b) Peaceful eruption. As fluid magma flows from the vent, gases can escape gradually. Note also that how a volcano erupts affects the shape of its cone: the debris of explosive eruptions fall near the vent, forming a steep cone, but fluid lava flows away from the vent, forming a low, broad cone.

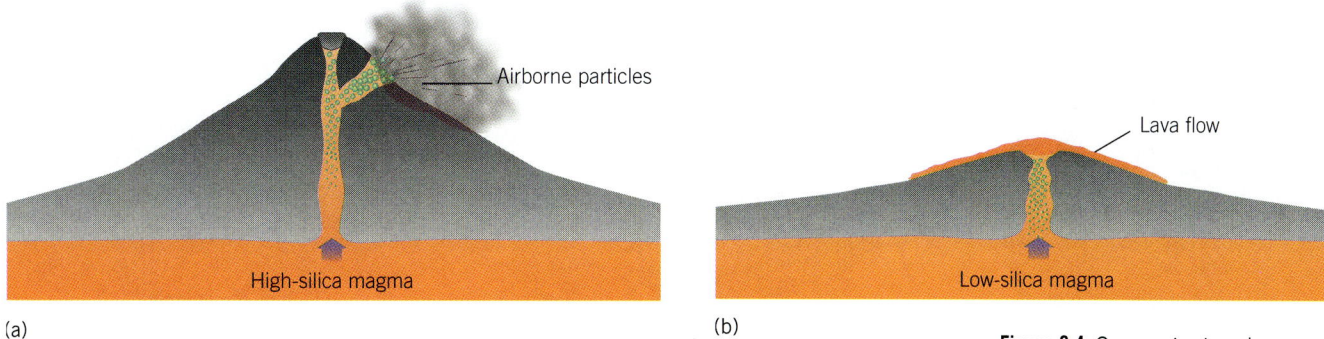

cold water and hot magma at shallow depths and are extremely explosive. One that occurred in 1963 created the new island of Surtsey off the coast of Iceland. The combination of hot magma and cold water that seeped into the vent generated steam of great propellant power (Figure 6.6). Two- and 3-ton lava bombs were shot hundreds of meters into the air for a period of weeks.

The Materials of Volcanic Eruptions

Volcanoes release three kinds of materials: lavas, gases, and fiery fragments. We will briefly describe these materials and the rocks that form from the lavas and fragments.

Lavas

In Chapter 5, we defined the two classes of igneous rocks: those that form from magmas that intrude country rock and cool within the Earth's crust, and those that form from lavas that erupt onto the Earth's surface. The two rock classes exhibit textural features reflective of their cooling histories. Intrusive rocks have coarse textures, because

Figure 6.5 This basaltic pillow lava, exposed on the face of a quarry wall, formed on the ocean floor. The rounded shapes were created when cold seawater came in contact with the molten lava.

Figure 6.6 A series of phreatic eruptions off the Icelandic coast in 1963 created the island of Surtsey. Steam and volcanic ash were propelled to great heights as cold seawater mixed with hot lava in the shallow volcanic vent.

(a) (b) (c)

Figure 6.7 Common textures of volcanic glasses. (a) Obsidian exhibits smooth texture and conchoidal fracture. (b) Pumice is composed of innumerable glassy threads and tiny air holes. (c) Scoria superficially resembles a sponge, with large air holes.

the magmas from which they were derived cooled slowly, allowing crystals to grow large. In contrast, extrusive rocks of the same composition exhibit fine-grained texture because the lavas cooled rapidly. The hot lava congealed when it met the cold air, and the flat shape of the lava flows exposed a greater surface area to cooling. Consequently, there was only time for microscopic crystals to form.

Mafic lavas form basalt, those of intermediate silica content form andesite, and silicic lavas (the equivalents of granodiorite and granite, respectively) form dacite and rhyolite. Though all lavas cool rapidly on the surface, differences in composition combined with widely fluctuating physical conditions create varied rock textures. If a lava loses heat too rapidly to form minerals, it may cool to a smooth glass. An example is *obsidian*, which lacks the organized internal atomic arrangement of a true mineral (Figure 6.7). Obsidian is usually a common feature of silicic lavas, whose high viscosities tend to prevent the migration of ions to crystal seeds within the melt. More commonly, volcanic glass is pockmarked with bubble cavities formed by escaping gases that failed to collapse in the viscous lava. Basaltic lavas exhibit large bubbles, which impart a spongy-looking texture to the cinderlike *scoria*. In silicic lavas, the bubbles are usually extremely numerous; and the light, frothy-looking, abrasive *pumice* is the result (see Figure 6.7).

Basaltic Lava Flows

Because of its relatively low silica content and high temperature, basaltic lava flows so easily from the vent that it is often referred to as runny lava. Individual flows have been traced for 200 kilometers in the Columbia Plateau of Washington and Oregon. When a thick basalt flow cools, it shrinks and develops **columnar joints.** The joints are a network of intersecting cracks that extend down the entire thickness of the flow and divide it into six-sided pillars (Figure 6.8).

Geologists distinguish between two kinds of basaltic lava flows, which bear Hawaiian names. **Pahoehoe lava** (pronounced pa-hoy-hoy) is billowy and undulating with a smooth continuous skin, but beneath this skin lies molten lava (Figure 6.9). **Aa lava** (pronounced ah′-ah) is the cooling viscous material that commonly forms at the front and sides of a river of pahoehoe lava (see Figure 6.9). It has a rough surface of blocks and fragments; partly congealed, red-hot lava lies beneath the rubble. Pahoehoe and aa lava are identical in composition, but aa lava is the more viscous of the two, because gas has escaped from it.

Silicic Lava Flows

Lavas of very high silica content, being more viscous than basaltic lavas, generally do not flow far from the vent. Those that crystallize into aphanitic rock form rhyolite, the extrusive equivalent of granite. However, highly viscous flows cool to form obsidian. In the absence of explosive gases, the viscous lava may extrude from the vent—like toothpaste from a tube—forming a **lava dome,** or it may ooze a short distance down the flanks of the volcano (Figure 6.10).

columnar joints
Cracks that form as the result of contraction during the cooling of lava flows, dividing the lava into columns.

pahoehoe lava
A lava flow with a smooth, ropy surface.

aa lava
A lava flow with a rough, jagged surface.

lava dome
A convex structure of solidified lava extruded from the vent of a volcanic crater.

Figure 6.8 Columnar jointing in the 60-million-year-old basaltic lavas of the Giant's Causeway, Northern Ireland. The parallel six-sided pillars were formed as the cooling lava contracted.

Pyroclastic Materials

The ancient Greek word for "fire" is *pyro,* and the word for "broken" or "fragment" is *klastos,* so the word *pyroclastic* accurately describes the materials that are expelled during the explosive phase of an eruption. Pyroclasts may be as small as dust particles or as large as trucks. They are classified according to size.

Ash, less than 2 millimeters in diameter, is not the residue of fire, as its name might imply. It is a tiny droplet of lava, composed of glass and mineral crystals, that cools in midair. *Cinders* and *lapilli* are pebble-sized fragments (2–64 millimeters) of the same composition as ash.

Blocks (defined as greater than 64 millimeters) are ejected as solid fragments. *Bombs* are the same size as blocks but are sent into the air in a semimolten state and are molded while in transit. Most bombs acquire the streamlined form that their name suggests. Upon

Figure 6.9 Dissolved water and gases can cause lavas of essentially the same composition to differ markedly in appearance and flow properties. Aa lava, deficient in dissolved gases, contains solid blocks of material and flows stiffly. Pahoehoe lava, richer in dissolved gases, flows smoothly and has a billowy or ropy surface.

Figure 6.10 Mount Saint Helens, Washington, in 1985. The large dome growing in the central crater is made of viscous silica that extruded from the volcano following the explosive 1980 eruption. Lava domes are a common late-stage feature of volcanoes having high silica content.

landing, they may flatten like pizza dough. The enormity of some of the bombs that are propelled into the air during especially violent eruptions defies the imagination. Mount Cotopaxi of Ecuador, in full eruption, hurled a 200-ton bomb 14 kilometers from the vent! Pyroclastic fragments cemented or welded together form **pyroclastic rocks.** Cemented ash and lapilli-sized fragments form *tuff* (Figure 6.11). If bombs, blocks, or other coarse fragments are incorporated within the tuff matrix, the rock is called *volcanic breccia*.

pyroclastic rock
A rock of any size formed from the cementation or welding of volcanic fragments.

Pyroclastic deposits ejected from a volcanic vent reach their final resting place by three means: air fall, surge, and mudflow. Air-fall deposits, or **tephra,** are fragments that settle out of the air; as a rule, they are well sorted. Large fragments land close to the vent, whereas finer materials are carried by winds and settle farther from the vent. The finest materials, volcanic dust, may actually circle the globe.

tephra
A general term that refers to all airborne pyroclastic debris.

Surge deposits are ground-hugging pyroclastic materials propelled down the slopes of some volcanoes on a bed of hot steam. A **nuée ardente** (French for "glowing cloud") is a mixture of hot incandescent gases and fragments that can surge at speeds exceeding 400 kilometers per hour (Figure 6.12). Upon cooling, the gases escape and the fragments fuse together to form *ignimbrite,* or *welded tuff,* a rock as hard and massive as concrete.

nuée ardente
A turbulent, ground-hugging, gaseous cloud erupted from a volcano.

Pyroclastic materials can also be transported as a muddy **lahar,** which is slower moving than a nuée ardente but can nevertheless outrun the inhabitants of a village downslope. The tops of large, steep volcanic cones are frequently covered with ice; others have well-forested slopes fed by abundant rainwater. Heat from the volcano combined with the shaking of the slopes during eruption mobilizes the waterlogged mixture of mud, soil, fragmented lava, and gases that pours rapidly into surrounding valleys.

lahar
A mudflow of volcanic material.

Volcanic Gases

fumarole
A vent or ground opening that spews volcanic fumes or vapors.

Volcanic gases dissolved in the magma under high pressure are released when the magma rises close to the surface. Carbon dioxide, water vapor, sulfur dioxide, hydrogen sulfide, hydrogen chloride, and nitrogen are the most common of these gases. Whereas lava accumulates relatively close to the vent, these gases, or *volatiles,* are dissipated into the atmosphere. Later in the chapter, we will discuss the possibility that carbon dioxide and fine sulfur dioxide droplets sent out during eruptions may markedly affect the Earth's climate.

Gases escape from lava in ways other than large eruptions. **Fumaroles** are vents that release hot gases (100 °C–1000 °C) either late in the history of a volcanic eruption or during the beginning phase. Some fumaroles emit

Figure 6.11 A pyroclastic rock. Welded tuff (or ignimbrite), composed of cemented particles of volcanic ash.

Figure 6.12 A nuée ardente photographed during the 1980 Mount Saint Helens eruption. Because the hot steam reduces friction, these ground-hugging, fiery clouds can move down a mountainside at speeds of hundreds of kilometers per hour.

gases that have directly escaped the magma; others—like those that once existed in the Valley of the Ten Thousand Smokes (Alaska)—release steam generated from water heated by buried yet still-hot ash deposits.

Volcanic Structures and Eruptive Styles

As we have seen, there is a direct connection between magma composition and the eruptive style of a volcano. This connection also extends to the form and structure of the volcanic cone.

Basaltic Volcanoes

Mid-ocean ridges. Oceanic hot spots. Certain continental fracture zones. These tectonic settings display broadly similar volcanic structures because they tap sources of extremely fluid basaltic magma derived from the mantle. The major difference among the structures is related to the shape of the vents through which the magma escapes to the surface. For example, volcanic cones are constructed around relatively localized and circular vents, whereas flat lava plateaus are built of lava that issues from a regional system of linear fissures in the crust.

Shield Volcanoes

Basaltic eruptions that occur on volcanic islands tend to be relatively peaceful. Mild explosiveness is confined to *lava fountains*—high arching sprays caused by escaping gas within the crater, or fissure, that can reach heights of 400 meters. During these displays, the eruptions thunder with a sound much like smashed crockery, as if a giant chef in a huge kitchen has gone mad. Following the display, a sheet of shimmering 1100 °C lava pours smoothly down broad mountain slopes at speeds of roughly 40 kilometers per hour, engulfing and smothering all in its path. An average flow forms a layer 10 kilometers long, 300 meters wide, and 12 meters thick and takes over three years to crystallize into a still formidably hot (750 °C) rock. Thousands of eruptions of this sort build **shield volcanoes.** Named for their broad, low profiles, shield volcanoes have slopes of less than 5 degrees (Figure 6.13). They are typical of mid-ocean ridge and oceanic hot-spot settings.

shield volcano
A broad, low-profile volcanic cone, commonly composed of basaltic flows.

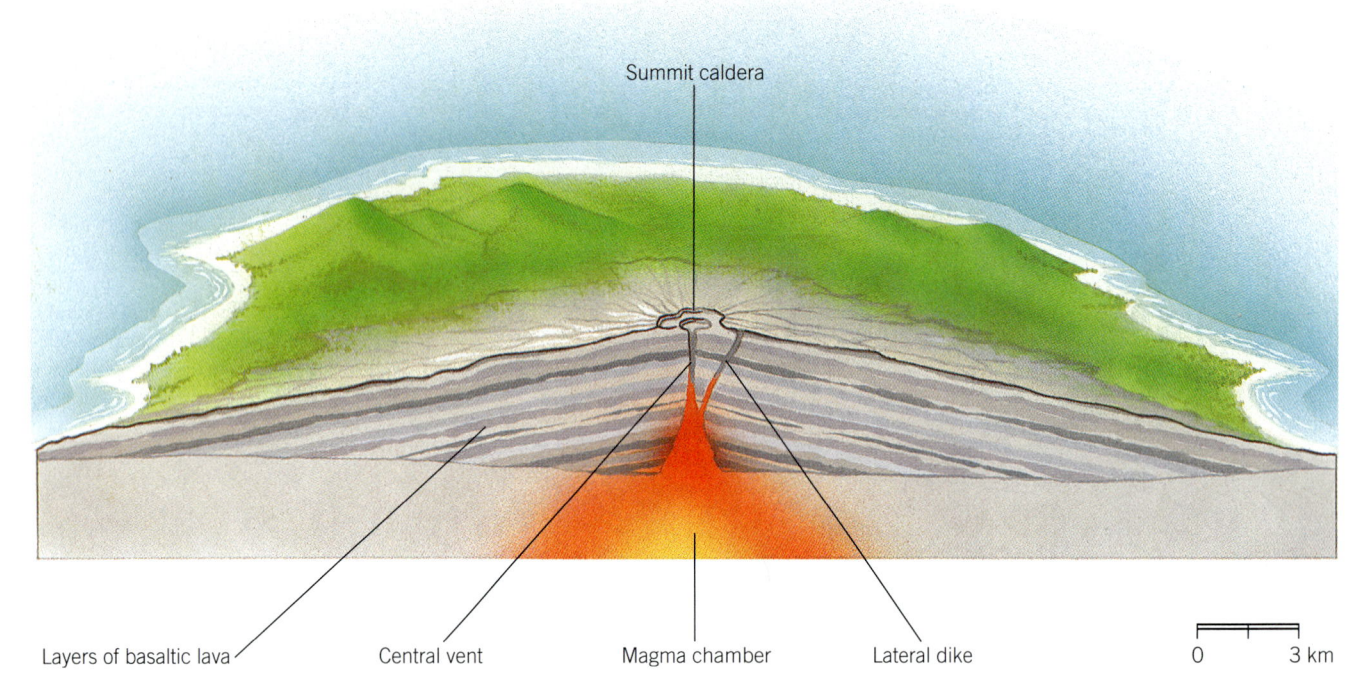

Figure 6.13 A typical shield volcano is a broad, low cone built of innumerable lava flows.

Labels: Summit caldera; Layers of basaltic lava; Central vent; Magma chamber; Lateral dike; 0 – 3 km

Located over the most active hot spot on earth, the island of Hawaii consists of five merging shield volcanoes. Mauna Loa, the world's largest active volcano, is 200 kilometers wide and stands 4170 meters above sea level. That does not tell the whole story, however, because its base rests on the Pacific Ocean floor, 5000 meters below sea level. So the full height of Mauna Loa is about 18.6 kilometers, taller than Mount Everest.

Kilauea is Hawaii's most active volcano. Earthquake analysis indicates that magma is generated by partial melting of mantle rock at a depth of approximately 50 kilometers. Prior to an eruption, magma rises to an accumulation chamber 1600–6400 meters beneath the summit. Lava escapes not only through the summit but also through fracture zones that radiate from the summit (Figure 6.14).

Figure 6.14 A fissure eruption along the eastern rift zone of Kilauea, Hawaii.

Figure 6.15 This cinder cone in Lassen Volcanic National Park, California, displays a graceful, symmetrical form.

Figure 6.16 Repeated basaltic lava flows have built the Columbia Plateau of Washington and Oregon.

Cinder Cones

Cinder cones are steep-sided, beautifully symmetrical volcanoes built entirely of pyroclastic debris (Figure 6.15). Many cinder cones are the product of basaltic eruptions and can be found as satellite cones on the flanks of Hawaiian volcanoes, although they are also common on the continents. Cinder cones are exceptions to the general tendency for pyroclastic materials to be of high silica content. The reason is obscure but may be related to gas concentrations within the magma chamber. Seldom greater than 500 meters in height, they are minuscule in comparison to the great Hawaiian shield volcanoes or the subduction zone volcanoes of the Cascade or Andes ranges.

Fissure Eruptions and Lava Plateaus

If mafic magma issues from long fractures in the crust, rather than from a localized cylindrical vent, its fluidity will enable it to spread evenly as a **fissure eruption.** The lava behaves much like a stream overflowing its banks—hence its name **flood basalt.** Flow upon flow stacks up in thick accumulations called **lava plateaus.** The volume of the Columbia Plateau in Oregon and Washington, built out of hundreds of individual lava flows, is estimated at 200,000 cubic kilometers (Figure 6.16). Most of the lavas are about 15 million years old, but some of the eruptions have occurred as recently as 6 million years ago.

The Columbia Plateau is dwarfed by other continental lava plateaus, such as Russia's Siberian traps and India's Deccan traps. The term *trap* is derived from the Swedish *trappa,* for "stairs," and it refers to the pattern made by the eroded flat layers. In all, they constitute the greatest concentrations of volcanic material on the continents, and plateaus of similar size are found in the ocean basins. Most occur far from plate boundaries and appear as huge volumes of material in a localized area. This observation leads geologists to wonder whether there is a connection between flood basalts and the mantle plumes that create hot spots.

cinder cone
A steep-sided volcano formed by the accumulation of ash, cinders, and other debris close to the vent.

fissure eruption
A volcanic eruption through a long fracture rather than a central vent.

flood basalt
Highly fluid basaltic lava produced during a fissure eruption.

lava plateau
An elevated, flat-topped region composed of a thick succession of horizontal lava flows.

Figure 6.17 Mount Rainier, Washington, compared to Mauna Loa, Hawaii. Although the steep, composite cones of continental subduction zone volcanoes appear huge and rise to great heights, they are in fact dwarfed in comparison to oceanic shield volcanoes.

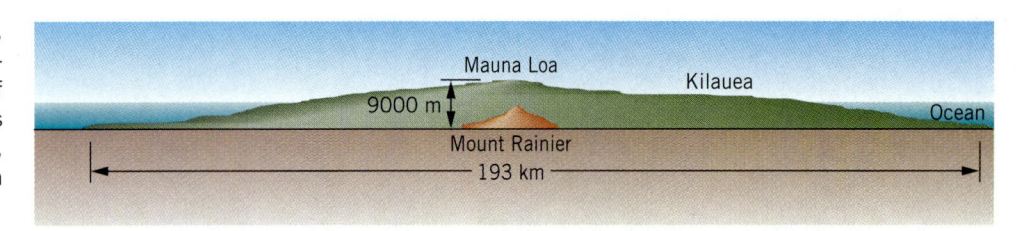

stratovolcano (composite cone)
A volcanic cone consisting of alternating layers of pyroclastic deposits and lava.

Andesitic and Silicic Volcanoes

The volcanic chains that border subduction zones, whether continental or island arcs, are fed by magmas of intermediate to high silica composition. The high viscosity and gas content of these magmas cause them to erupt explosively and, as a result, form volcanoes whose slopes and structures contrast markedly with those of shield volcanoes.

The high viscosity of subduction zone magmas traps gas bubbles and prevents them from expanding—a condition that intensifies as the magma continues to cool and rise close to the surface. Pressure within the magma may then build to such a level that the gases escape in an explosive eruption. Clots of molten magma and fragments of ancient volcanic rock are propelled into the air along with the gas. Following the escape of the pent-up gases, the eruption may enter a quiet phase, during which viscous magma oozes out the vent. The lava also congeals in the fissures and side vents that radiate from the central vent.

Many such cycles of violence and relative peace build a **composite cone** or **stratovolcano,** consisting of alternating layers of pyroclastic debris and viscous lava (Figure 6.17). Composite cones are much smaller in diameter than shield volcanoes, but they have steeper profiles and form some of the world's most majestic mountain peaks: Japan's Mount Fujiyama, Mount Mayon in the Philippines, Mount Cotopaxi of Ecuador, Italy's Mount Vesuvius, Indonesia's Krakatoa, and Mount Rainier of Washington, to name a few.

Subduction zone magmas of extremely high silica content intensify the cycle of eruption just described. Because these magmas are extremely viscous, they prevent the escape of gases until they rise within a few kilometers of the surface. When the gases are finally released, they are under extreme pressure. If their path to the top of the vent is blocked, they may instead blast through the sides of the volcano. A cataclysmic eruption then ensues.

Combine the explosiveness of a subduction zone volcano with the proximity of a population center and you have a formula that may result in huge losses of life. In 1902, Mount Pelée, on the island of Martinique in the West Indies, sent out threatening signals. For two weeks, minor flows and poisonous gas were observed escaping the vent. The emissions were strong enough to kill livestock and make people sick. The population of the nearby port city of Saint Pierre was terrified, but instead of evacuating the town, local politicians kept them there to ensure that they would vote in an upcoming election. Suddenly, the side of Mount Pelée blew open, triggering a nuée ardente. The surge, moving at 150 kilometers per hour, reached the nearby port city of Saint Pierre in a matter of minutes. Except for a badly burned shoemaker and a convict confined to an underground cell, the entire population of the city—approximately 30,000 people—perished in a matter of minutes.

The Eruption of Mount Saint Helens

Mount Saint Helens is a moderate-sized volcano in the Cascade Range. Its highly silicic dacite magma is derived from the subduction of the Juan de Fuca plate beneath the North American plate. In May 1980, it became the most publicized and best studied volcanic eruption of its kind (Figure 6.18).

Seconds before the eruption, seismographs deployed around the volcano had registered a 5.1 magnitude earthquake centered 1.6 kilometers below the north flank of the cone. The tremor triggered the collapse of the summit, which in turn produced the largest landslide-debris avalanche recorded in historic time. The rocks and fragments were waterlogged with melted snow, groundwater, and hot steam ejected from the volcano. At speeds of 200 kilometers per hour, a 4-cubic-kilometer of mass of rocks and mud moved from the north slope 28 kilometers to the Toutle River.

Collapse of the north wall pulled the plug on the hot gases and steam contained in the magma reservoir. In the instant following the landslide, a lateral blast of ash and steam blew out the north slope of the mountain. Traveling at close to the speed of sound, it was strong enough to rip trees 2 meters in diameter up by their roots within a few kilometers of the crater. At a distance of up to 30 kilometers from the summit, prime Douglas firs were snapped like matchsticks. A truck was found overturned 26 kilometers away, the plastic parts melted. In all, the blast devastated 590 square kilometers of countryside.

A cloud of ash and steam expanded into a mushroom cloud 25 kilometers high. Within hours, 10 centimeters of ash blanketed Yakima, Washington, 140 kilometers to the east. Spokane, Washington, 500 kilometers to the northeast, was so darkened that the automatic street lamps turned on and remained lit all day. Enough ash fell on Montana to spoil crops, and thin wisps of dust were traced across the country.

For five hours after the blast, a series of ground-hugging nuées ardente poured through the gaping hole in the north flank at speeds of 130 kilometers per hour, carrying ash and pumice downhill. At the same time, the glaciers of Mount Saint Helens melted and combined with centuries-old soil, rock fragments, and organic debris high up on the mountain. This dense fluid mass formed lahars that flowed down into the surrounding valleys.

The eruption took 400 meters off the top of the mountain and left a crater 750 meters deep (Figure 6.19). Less forceful eruptions have followed, but it seems that

Figure 6.18 The May 18, 1980, eruption of Mount Saint Helens, seen here at 2:44 P.M., sent a mushroom cloud of dust and debris high into the air.

Figure 6.19 The enormous crater left behind by the May 1980 eruption, which blew out the top and north side of the mountain.

Figure 6.20 Caldera eruptions such as those that occurred repeatedly in Yellowstone, Wyoming, and the Valles Caldera, New Mexico, are perhaps the most dangerous of all eruptions. Slowly rising plumes of viscous magma cause the crust to sink along circular fractures; at the same time, high-silica ash particles are released through the fractures. The ash forms a suffocating cloud that spreads over wide regions surrounding the caldera.

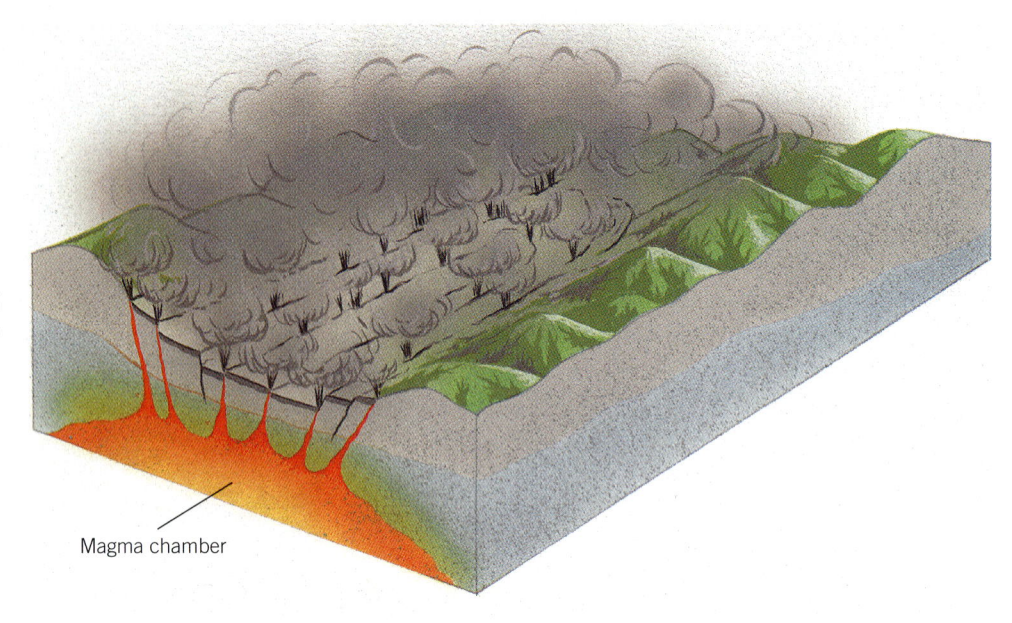

sufficient gas has been released to allow the magma to rise slowly rather than explosively. In the years since the eruption, the magma has been quietly building a 300-meter lava dome above the central vent (see Figure 6.10).

Ash Flows

Some of the largest and most explosive eruptions have occurred in continental regions far from plate boundaries. These eruptions have not built conventional cones but, rather, immense sunken craters surrounded by ash fields (Figure 6.20). Yellowstone, Wyoming; the vast ash field of the San Juan Mountains in Colorado; the Valles Caldera of New Mexico; and the Long Valley–Mammoth Lakes region of California were formed in this manner.

Yellowstone National Park is situated over the Yellowstone hot spot, a rising mantle plume that initiated the slow melting of granitic continental crust. The park is named for the color of the rhyolite lavas that characterize the region. Although no volcanic cone is present, geologists have been able to trace the outlines of three sunken calderas in the park and surrounding regions. From the volumes of pyroclastics ejected, they have concluded that these eruptions were far greater than those of typical subduction zone volcanoes.

Forecasting Eruptions

Krakatoa, the great Indonesian island volcano that erupted August 29, 1883, unleashed huge tsunamis that killed over 36,000 people and set off a cannonlike roar heard as far away as central Australia. It is a sobering thought that this awesome volcano, whose fine dust turned sunsets red for months around the world and noticeably reduced global temperatures up to a year later, erupted unexpectedly after two centuries of quiescence. A century later, the science of volcanology had progressed to the level that the forecasts and warning bulletins of the Mount Saint Helens eruption saved many lives. The mountain and surrounding regions were cordoned off and emergency teams were alerted well in advance. Most of the 60 people who died in the eruption had received adequate warning to leave the area but chose to stay instead. As with earthquakes, however, there is a significant difference between forecasting and predicting the exact moment of an event.

Many steps are involved in forecasting volcanic eruptions. To begin, geologists draw upon their background knowledge of the broad geologic settings in which volcanoes erupt. Next, field studies, combined with radiometric dating in laboratories, help geologists identify which volcanoes have been most active in the past, their frequency of eruption, and the type of activity they most likely can expect of them. For example, as we described at the beginning of the chapter, surveys of the volume and type of materials deposited on the flanks of a volcano help geologists to determine the power and explosiveness of its eruptions. Dating these materials establishes a volcano's rhythm—whether it erupts on the average of once each century or once every thousand years. The distribution of ejected material and its relation to local topography, such as valleys and ridges, suggests where material is likely to accumulate and, therefore, helps geologists to identify potential hazards. Work of this kind enabled geologists Dwight R. Crandell and Donald R. Mullineaux of the USGS to conclude, five years prior to the 1980 eruption, that Mount Saint Helens could erupt explosively within a few years. Studies such as theirs are not designed for short-range prediction but are useful in regional planning and in establishing research priorities.

Having decided that a volcano may erupt in the foreseeable future, geologists establish a monitoring system, including seismographs placed strategically around the flanks of the volcano and toward the summit to track the location, severity, and frequency of earthquakes and *tiltmeters* to detect ground swelling or other distortions.

Ground tilting and earthquakes are only two of the signs geologists monitor in the effort to forecast volcanoes. Other signs include changes in the composition and quantities of gases escaping the crater; changes in the heat escaping the crater; changes in groundwater composition and temperature; and changes in related igneous activity—for example, the appearance of geysers, hot springs, or steam vents (fumaroles) in the region. By integrating this information, geologists hope to improve their forecasting abilities.

Volcanism and Climate

The atmosphere plays a dual role in regulating the Earth's temperature. On the one hand, minute dust particles and acid droplets in the upper atmosphere reflect incoming sunlight, which lowers the Earth's surface temperature. On the other hand, carbon dioxide and other gases in the lower atmosphere absorb the heat radiated from Earth's surface, which has a warming effect. Volcanoes release all these temperature-regulating substances. For this reason, many scientists suspect that the rate of volcanism may strongly influence the Earth's climate.

The June 16, 1991, eruption of Mount Pinatubo clearly illustrates the cooling effect initiated by the venting of huge quantities of dust and sulfur dioxide into the upper atmosphere. Sulfur dioxide readily combines with water vapor to produce sulfuric acid droplets, which are very effective in scattering sunlight. As the particles and droplets spread around the globe, average world-wide temperatures dropped by more than 1 °C and did not recover for the next two years (Figure 6.21). Similar temperature declines have been noted following most other major eruptions involving fine pyroclastic dust. One of the best known followed the Mount Tambora, Indonesia, eruption of 1815. The following year was so cold that it snowed in New England in July, and 1816 came to be known in North America and Europe as "the year without a summer."

Carbon dioxide's role in regulating Earth's temperature is well known, for it is a key component of the **greenhouse effect.** The hot sun sends out light waves that the relatively cold surface of the Earth absorbs and converts to infrared radiation. Carbon dioxide, along with water and methane, acts as a filter. It allows sunlight to pass through the atmosphere but absorbs some of the infrared energy that rises from the Earth's surface. The infrared energy eventually escapes to space, but in the delay, the Earth's lower atmosphere and surface are warmed. The greater the concentration of carbon dioxide in the atmosphere, the more heat will be trapped and the higher Earth's temperature will be.

greenhouse effect
The warming of the Earth's atmosphere through the presence of atmospheric gases that absorb and re-radiate the heat rising from the surface.

Figure 6.21 Measurement of the thickness of the dust blanket from the Mount Pinatubo eruption of 1991. The dust rose into the atmosphere and lowered average worldwide temperatures nearly two degrees for almost three years. (Courtesy of the National Oceanic and Atmospheric Administration.)

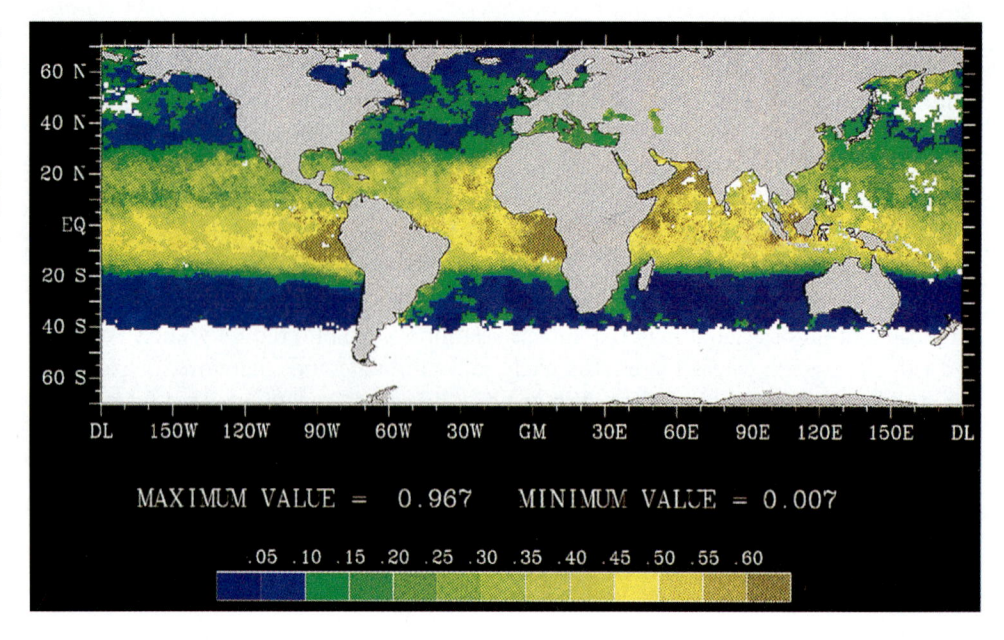

Some researchers see volcanism as the cause of the Earth's cooling whereas others see it as the cause of Earth's warming. These views are not necessarily contradictory, since the cooling effects of a single eruption commonly last 2 to 5 years, whereas carbon dioxide lingers far longer in the atmosphere. Carbon dioxide release, therefore, should have the greater cumulative effect, and we should expect long-term warming. However, intense volcanism sustained over long time-periods might release enough acid droplets and dust to cause catastrophic declines in sunlight and temperatures. Some geologists have, in fact, suggested this mechanism as a cause of ice ages and of the sudden mass extinctions of plant and animal life that have occurred at irregular intervals throughout geologic time. The causal relationships between volcanism and climate are obviously complex; investigating them is a major goal of the geosciences.

Constructive Aspects of Volcanism

In closing, let us mention some of the constructive aspects of volcanoes.

- They are major landforms of the Earth's surface; without volcanoes, there would be no Hawaii or Iceland.
- A volcano may temporarily devastate the ecosystem that surrounds it, but it provides rich soils that strengthen the ecosystem that returns (Figure 6.22). Soils of volcanoes and lava plateaus are among the most fertile on the Earth.
- The gases and particles that volcanoes emit are major regulators of climate. Most geologists agree that the oceans and atmosphere are the result of volcanism that occurred during the early stages of the Earth's formation.
- Valuable metals such as copper, gold, and mercury are brought close to the Earth's surface through volcanic vents and fractures.
- The near-surface magma bodies that supply volcanoes have enormous energy potential. The USGS estimates that geothermal heat of magma bodies within 10 kilometers of the Earth's surface in the continental United States could supply between 800 to 8000 times the energy the nation consumes each year! Iceland, New Zealand, and San Francisco have relied on geothermal power for years.
- Volcanoes are a priceless asset. Whether in repose or in eruption, they are beautiful.

Volcanism

Figure 6.22 The destructive effects of volcanism are short-lived, but the benefits are long-term. Here we see life reestablishing itself on the devastated flanks of Mount Saint Helens soon after the great eruption.

STUDY OUTLINE

The eruption of Mount Mazama led to the formation of Crater Lake, Oregon. A **volcanic cone** in the center of Crater Lake is evidence of a later eruption.

I. **THE ANATOMY OF A VOLCANO.** A **volcano** is a **vent** through which magma escapes to the surface as **lava.** The flows of successive eruptions build the cone-shaped structure; collapse afterward leaves a **caldera** or **crater** at its center. Between eruptions, the magma hardens in the vents to form the volcano's skeleton.

II. **TECTONIC SETTINGS OF VOLCANISM**
   A. Most volcanism occurs along diverging and converging plate boundaries, but there are also intraplate hot spots, caused by mantle plumes.
   B. Plate interactions are the foundation of the following causative chain:
      1. The type of plate boundary controls the composition of the magma.
      2. Composition, temperature, and gas content control magma viscosity.
      3. The magma viscosity controls the degree of explosiveness of the eruption.
      4. The degree of explosiveness controls the shape and form of the volcano.
   C. This chain results in the following general relationships:
      1. Divergent plate boundaries yield low-viscosity basaltic magmas.
      2. Convergent boundaries yield high-viscosity andesitic, dacitic, and rhyolitic magmas.

III. **THE MECHANICS OF A VOLCANIC ERUPTION.** Pressure reduction inside the vent and escaping gases trigger volcanic eruptions.
   A. The higher the silica content of the magma, the greater the viscosity. Viscous, high-silica magma retards gas expansion, which increases pressure, resulting in an explosive eruption. In low-silica, mafic magmas, gases escape easily, and eruptions are relatively peaceful.
   B. External conditions also affect the style of eruption. Peaceful sea-floor eruptions form **pillow lava.** The mixing of cold water and hot magma causes violent **phreatic eruptions.**

IV. **THE MATERIALS OF VOLCANIC ERUPTIONS.** Lava, gases, and fiery fragments, when cooled, form extrusive igneous rocks with fine-grained, or aphanitic, textures.
   A. Basalt, andesite, dacite, and rhyolite, in order of increasing silica content, are the four major rock types formed from lava.
      1. If a lava cools rapidly, it forms a smooth glass, obsidian. Pumice is the end product of bubbly silicic lava.
      2. Mafic, or basaltic, lavas flow easily away from the vent. A thick basaltic flow cools and develops parallel six-sided pillars, or **columnar joints. Pahoehoe lava** is billowy and ropelike. **Aa lava** is a rough surface of blocks and fragments.
      3. Silicic lava is viscous and remains close to the vent or extrudes from it as a **lava dome.**
   B. Pyroclastic materials come in different sizes, from **ash,** minute droplets of lava, to blocks and bombs that weigh many tons. Fragments welded together form **pyroclastic rocks.**
   C. Pyroclastic materials can be transported in three ways: as airborne pyroclastic debris (**tephra**), as a fiery cloud (**nuée ardente**), or as a mudflow (**lahar**).
   D. Volcanic gases, such as carbon dioxide, are expelled during a volcanic eruption. **Fumaroles** are vents through which gases or steam escape.

V. **VOLCANIC STRUCTURES AND ERUPTIVE STYLES**
   A. Low-viscosity, basaltic lava flows build broad, low **shield volcanoes,** such as those of the Hawaiian Islands, that erupt relatively peacefully. Exceptions are the pyroclastic eruptions of steep-sided **cinder cones.** Basaltic magma may also issue from crustal fractures in **fissure eruptions.** Cooled lavas form **flood basalts,** which build up over time into **lava plateaus.**
   B. Viscous andesitic, dacitic, and rhyolitic lavas cause explosive eruptions and build steep **stratovolcanoes,** or **composite cones.** Explosive intraplate eruptions build broad, sunken calderas surrounded by ash fields.
   C. The violent eruption in 1980 of Mount Saint Helens of the Cascade Range was typical for a moderate-sized, subduction zone volcano; the blast devastated 590 square kilometers of countryside.

VI. **FORECASTING ERUPTIONS**
   A. Dating volcanic deposits radiometrically helps geologists determine the eruption frequency of a volcano and the distribution of ejected material helps them to identify the hazardous zone.
   B. Geologists monitor ground swelling and tilting, seismic patterns, and gas emissions to forecast eruptions.

VII. VOLCANISM AND CLIMATE

A. In the **greenhouse effect,** carbon dioxide and other gases released by volcanoes absorb and reflect back the heat radiated from the Earth's surface, thus warming it.

B. However, the dust particles and acid droplets released by a volcanic eruption reflect incoming sunlight, resulting in a short-term cooling effect. Some geologists have suggested that prolonged volcanism over long periods may have cooled the Earth enough to cause the ice ages and mass extinctions of the past.

VIII. CONSTRUCTIVE ASPECTS OF VOLCANOES.
Volcanoes are major landforms. They are the source of rich soils, mineral deposits, and atmospheric gases, and their geothermal heat is a potential source of energy.

Study Terms

aa lava (p. 134)
caldera (p. 131)
cinder cone (p. 139)
columnar joints (p. 134)
crater (p. 131)
fissure eruption (p. 139)
flood basalt (p. 139)
fumarole (p. 136)
greenhouse effect (p. 143)
lahar (p. 136)
lava (p. 131)
lava dome (p. 134)

lava plateau (p. 139)
nuée ardente (p. 136)
pahoehoe lava (p. 134)
phreatic eruption (p. 132)
pillow lava (p. 132)
pyroclastic rock (p. 136)
shield volcano (p. 137)
stratovolcano (composite cone) (p. 140)
tephra (p. 136)
vent (p. 130)
volcanic cone (p. 129)
volcano (p. 130)

Thought Questions

1. How do geologists gauge the magnitude of an eruption that no one has witnessed?
2. Explain the relationship between a magma's silica content and the explosiveness of an eruption. What other factors may influence explosiveness? Which magmas tend to be most explosive? Which least explosive?
3. How does the theory of plate tectonics explain the Pacific's Ring of Fire?
4. Describe the mechanics of a violent volcanic eruption.
5. What kind of volcanic cone is associated with basaltic magma? How is its profile related to composition and eruptive style? What kind of volcanic cone is associated with a magma of high silica content? How is its profile related to composition and eruptive style?
6. Why are deep marine eruptions peaceful and shallow marine eruptions violent?
7. Account for the differences in form of lava plateaus and shield volcanoes.
8. What kinds of geological evidence might you search for to support the hypothesis that volcanism has significantly affected the Earth's climate in the past?

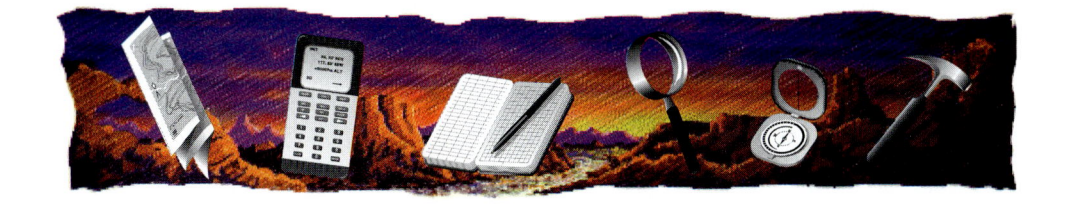

Chapter 7

Weathering, Transport, and Deposition of Sediments
   Weathering
   Transport and Deposition

Lithification

Classification of Sedimentary Rocks
   Detrital Rocks
      Coarse Detrital Rocks
      Fine Detrital Rocks
   Precipitated Sedimentary Rocks
      The Limestones

      Evaporites
      Chert
      Coal

Sedimentary Structures
   Fossils and Minor Structures

Oceanic Sedimentary Environments
   The Continental Margins
   The Deep-Ocean Environment
   Plate Tectonics and Ocean Sediments

Energy Resources in Sedimentary Rocks
   Coal
   Petroleum
      Oil Traps
      Oil Shales and Oil Sands
      Plate Tectonics and Petroleum
      The Petroleum Supply

Sedimentary Rocks

Sediments are fragmented particles weathered from preexisting rocks and transported and deposited by water, wind, and ice. Constituting only 7 percent of the crust by volume but covering 75 percent of its surface area, sediments and sedimentary rocks amount to little more than a thin coating on the hard igneous and metamorphic crust. Yet they are important for a variety of reasons.

Sediments and sedimentary rocks are the only Earth materials deposited at or near the surface under everyday conditions. As such, they contain the entire fossil record of life on the Earth. In them are recorded the composition, climate, and topography of former landmasses, as well as the physical, chemical, and biological conditions of oceans that no longer exist. Without sediments and sedimentary rocks, we would have virtually no idea of the geography of the past, the nature of the Earth's past environment, or the life that inhabited that environment.

Sediments also have significant economic importance. Virtually every material useful to civilization is mined from them: groundwater, gold, copper, zinc, iron, lead, diamonds, limestone, sand, gravel, and clay. Oil, gas, and coal are exclusive to sedimentary rocks; in fact, coal *is* a sedimentary rock.

sediment
Particles that have been mechanically transported by water, wind, or ice, or chemically precipitated from solution, or secreted by organisms, and deposited in loose layers on the Earth's surface.

◀ Layering is an important feature of sedimentary rocks, as illustrated by this sandstone exposure in the Paris Canyon Wilderness Area of Northern Arizona.

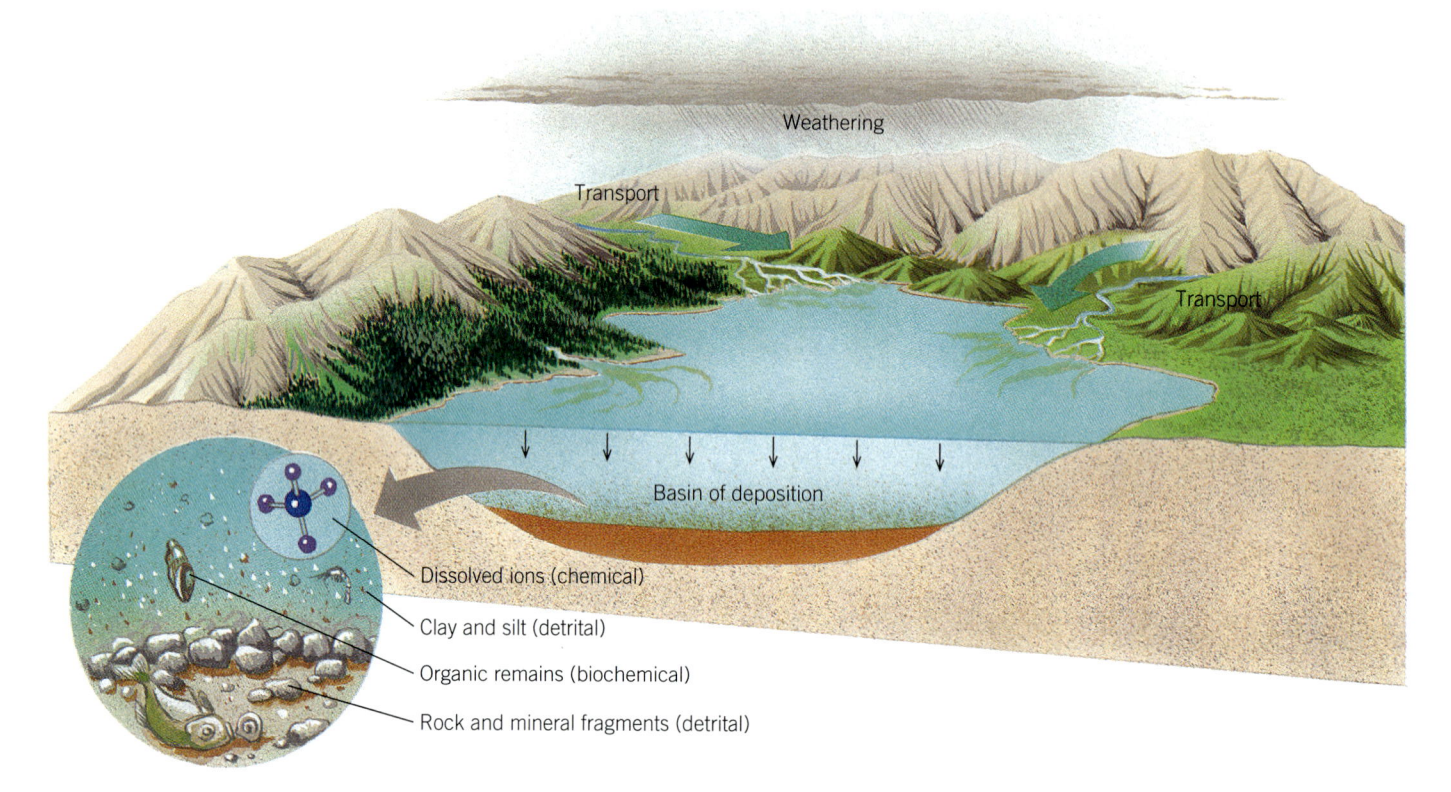

Figure 7.1 The components of a typical sediment are detrital, organic (biochemical), and chemical in origin. All result directly or indirectly from weathering, transport, and depositional processes.

detrital sediment
Fragments derived from the weathering of rocks, transported by water, wind, or ice, and deposited in loose layers on the Earth's surface.

chemical sediment
A sediment composed of particles precipitated directly from water.

biochemical sediment
A sediment precipitated directly or indirectly by the activities of organisms.

weathering
The physical and chemical alteration of rocks exposed to the atmospheric influences on the Earth's surface.

deposition
The gravitational settling of rock-forming materials by such natural agents as water, wind, or ice.

Weathering, Transport, and Deposition of Sediments

Imagine a low-lying area of the Earth's surface that acts as a receptacle or basin for sediments, such as the ocean floor or a lake (Figure 7.1). What kinds of sediments can be found there? The most readily identifiable are the fragmental or **detrital sediments,** the land-derived gravel, sand, silt, and clay brought to the basin by streams or, perhaps, by winds or glacial ice. A typical sedimentary basin also contains precipitated particles formed from ions that were transported in solution. These sediments are **chemical** in origin if precipitated directly from the water, or are **biochemical** in origin if precipitated as a result of organic processes. Often, a single sedimentary deposit will contain fractions of all three sediment types: detrital, chemical, and biochemical.

The raw materials that compose both detrital and precipitated sediments are derived from three processes: **weathering,** the physical and chemical breakdown of exposed surface rocks; transport of the weathered materials to the basin; and **deposition** in the basin.

Weathering

Weathering causes both mechanical disintegration and chemical decomposition of bedrock. In mechanical disintegration, bedrock is broken into small fragments or into individual mineral grains, but the composition of the materials remains unchanged. It occurs when water that has seeped into rock fractures and soil pores freezes, expands, and wedges them apart; when temperature variations cause minerals to expand and contract unevenly; when release of pressure causes rock fractures to widen; and when roots get wedged between rocks.

In chemical weathering, water and dissolved ions react with solid rock and mineral fragments to produce materials of fundamentally different composition. Hydrolysis is one of the many intricate chemical changes that occurs. It involves the reaction of water—in particular, the hydrogen ions in water—with minerals. Although hydrolysis occurs slowly in pure water, in nature the reaction is greatly accelerated by the presence of dissolved carbon dioxide, which is derived from the respiration and decay of organisms in the soil. Water and carbon dioxide then combine to form carbonic acid, which adds chemical potency to the solution by increasing the number of hydrogen ions present.

In the hydrolysis of feldspar, the most abundant mineral of the crust, carbonic acid and water alter the mineral to insoluble clay residues and dissolved ions (potassium, sodium, calcium, and silica). The clay residues are transported as solid particles, and the ions are carried in solution. This pattern of hydrolysis holds for all silicate minerals, although the weathering products vary with the precise composition of the minerals. Feldspar, quartz, and a small percentage of mica or amphibole are the major constituents of granite—the igneous rock that forms the bulk of the continental crust. The weathering products of these minerals—quartz and feldspar fragments, clay, mica, magnetite, and dissolved ions—all end up, one way or another, in sedimentary rocks.

Note: As you will see in Chapter 10, Weathering, Soils, and Mass Wasting, the production of raw materials for sedimentary rocks is only one aspect of weathering.

Transport and Deposition

Weathering prepares materials derived from bedrock for transport. As we saw in Chapter 1, it is the first step in the gravity-driven process of erosion that acts to level the continents. Erosion may act directly under the influence of gravity, as in landslides and rockfalls, or it may involve agents such as streams, glaciers, winds, and ocean waves and currents, all of which are capable of transporting sediments great distances from their sources. Streams carry the greatest sediment load by far. They deposit sediments in a variety of environments, where they are buried by later deposits and converted to sedimentary rocks. These environments include upland basins, continental margins, coastal plains, and the deep-ocean floor. Each of these broad environments is divided into myriad subenvironments. We will discuss them where appropriate throughout this chapter and the book.

Lithification

Lithification is the conversion of sediment to sedimentary rock, and **compaction** is the first step in the process. Successive layers of sediment exert pressure on the sediment beneath, squeezing grains tightly together. In this manner, the underlying sediment may be reduced to a quarter of its original volume. Nevertheless, spaces remain between the coarse grains, and the spaces may be empty or filled with a **matrix.** The composition of the matrix determines the path that lithification follows. If the matrix is composed of clay and silt, the heating and drying induced by compaction will harden the matrix to a bricklike mass that embeds the larger fragments. However, if the matrix is filled with groundwater or saltwater saturated with silica, calcium carbonate, or iron oxides, these compounds will precipitate and bind the grains together in a process called **cementation** (Figure 7.2). During **recrystallization,** the mineral grains themselves act as cementing agents. Under pressure, the grains dissolve slightly at points of contact and then recrystallize in the voids between the original fragments. The matrix of a sediment need not fill all voids in order to bind the fragments. Many sedimentary rocks contain a large percentage of empty space.

lithification
The conversion of sediment into rock through such processes as compaction, cementation, and recrystallization.

compaction
Reduction in volume of sediments resulting from the weight of newly deposited sediments above.

matrix
The fine-grained material surrounding larger grains in a rock.

cementation
The process by which precipitates bind together the grains of a sediment, converting it into rock.

recrystallization
The formation of new crystalline mineral grains in a rock.

Classification of Sedimentary Rocks

As with sediments, most sedimentary rocks may be divided into three broad categories: detrital, chemical, and biochemical (Table 7.1). The categories emphasize the origins of the rock components. The rocks are further classified within each of these categories on

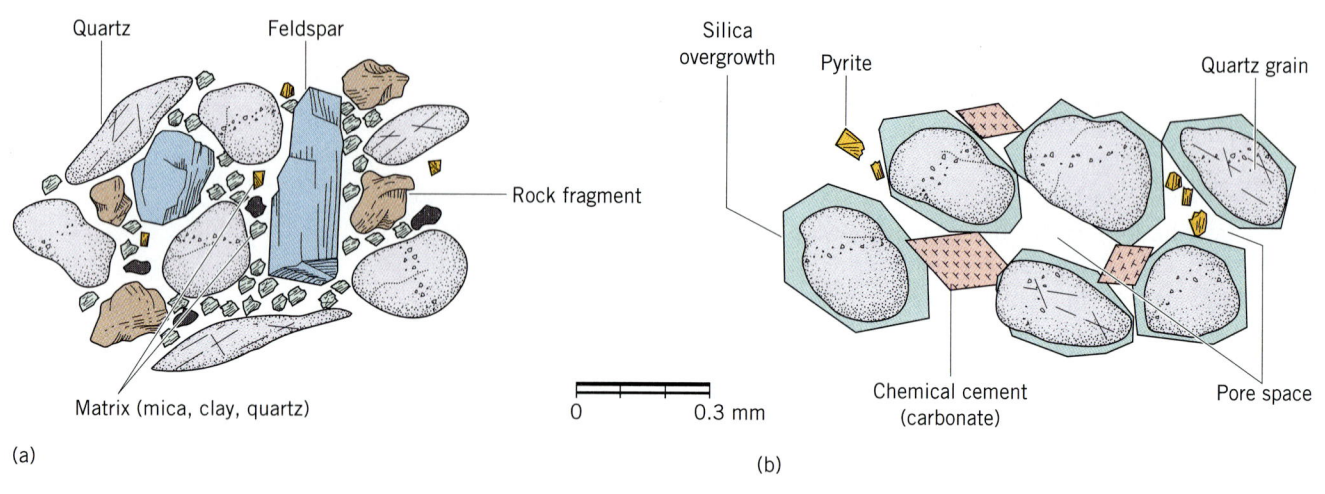

Figure 7.2 The grains of a typical detrital rock consist of a mixture of minerals and rock fragments. The grains may be bound together by (a) a clay and silt matrix or by (b) chemical cements.

the basis of texture and mineral content. Texture, as you may recall from our discussion of igneous rocks, refers to the size, shape, and arrangement of the components of the rock. This definition holds for sedimentary rocks also.

Detrital Rocks

As stated earlier, detrital rocks consist mainly of fragments weathered from preexisting rocks. The ancient Greek word for fragment is *klastos;* hence, detrital rocks are described as having **clastic texture.** However, clastic texture is so closely associated with detrital rocks that the term *clastic* is often used interchangeably with *detrital.* There are a number of characteristics a geologist looks for in analyzing the texture of a detrital rock. Among the most important are particle size, degree of sorting, and particle shape. We shall see that each has an interesting story to tell.

Perhaps the most obvious detrital textural characteristic is *particle size;* it is defined by the Wentworth scale (Table 7.2), which is the basis for classifying the common detrital rock types. Particle sizes range from boulders, cobbles, and pebbles down to sand, silt, and clay. Deposits consisting of sand or larger sizes are referred to as *coarse* sediments, whereas silt and clay are *fine* sediments. A rock formed largely of gravel is a *conglomerate* if the particles are rounded or a *breccia* if they are angular. A *sandstone* is formed of sand-sized particles and a *siltstone* of silt-sized particles. *Shale* and *mudstone*

> **clastic texture**
> Texture of a rock composed mainly of fragments of other rocks and minerals; most commonly used to describe detrital rocks.

Table 7.1 Classification of Common Sedimentary Rocks

Fragmental	Precipitated	
Detrital (classified by grain size)	Chemical	Biochemical
Breccia	Evaporites	Limestone
Conglomerate	Rock salt	Micrite
Sandstone	Gypsum	Chalk
Siltstone	Limestone	Coquina
Shale	Oolitic limestone	Reef rock
Mudstone		Chert (chalcedony)
		Coal

Sedimentary Rocks

Table 7.2 The Wentworth Scale for Classifying Sediments by Size

Diameter (mm)	Particle	Rock
	Gravel	
Above 256	Boulder	Conglomerate and breccia
64–256	Cobble	
4–64	Pebble	
2–4	Granule	
	Sand	
$\frac{1}{2}-2$	Coarse sand	Sandstone
$\frac{1}{4}-\frac{1}{2}$	Medium sand	
$\frac{1}{8}-\frac{1}{4}$	Fine sand	
$\frac{1}{16}-\frac{1}{8}$	Very fine sand	
	Mud	
$\frac{1}{256}-\frac{1}{16}$	Silt	Siltstone, shale, and mudstone
$\frac{1}{2048}-\frac{1}{256}$	Clay	

are composed of a mixture of the finest particles, clay and silt. The difference between them is that shale splits into thin layers, whereas mudstone is thickly layered and breaks into massive chunks.

Sorting refers to the range of particle sizes within the sediment and is another useful indicator of the differences between detrital rocks. Figure 7.3(a) shows a well-sorted sediment; its particles fall within a narrow range of sizes. It was taken from a typical beach, where it had been worked over by pummeling waves and currents. Contrast it with Figure 7.3(b), a poorly sorted sediment with a wide size range. It is *glacial till*, a sediment deposited directly by the ice of a melting glacier. These samples illustrate that sorting is related to the mode of sediment transport and deposition.

Particle shape tells a similar story, for whatever the shape of a fragment at its source, it will be modified during transport to the depositional site. Two measures of shape are the *sphericity* and *roundness* of the fragment. The first is a measure of the degree to

sorting
The process by which the agents of transportation (principally running water) separate sediments according to shape, size, and density.

Figure 7.3 The degree of sorting is closely related to the transport history of a sediment. (a) Well-sorted beach sand transported by waves and currents. The finer particles have been sifted out. (b) Poorly sorted glacial till is deposited directly by melted ice; large and small particles had no opportunity to separate during transport.

which the particle approaches a spherical shape during transport, whereas roundness refers to the degree to which the sharp corners and edges of the fragment have been worn down or abraded.

Coarse Detrital Rocks

Detrital rocks are also classified by composition, and the composition is determined mainly by the degree of weathering at the source. Those minerals with the greatest resistance to weathering have the greatest chance of showing up in detrital sedimentary rocks. Therefore, quartz and clay are the chief constituents of detrital rocks. Appreciable amounts of feldspar, mica, and rock fragments are present in those rocks derived from a source region where weathering has been incomplete. Most detrital rocks also contain a small percentage of heavy minerals such as magnetite or other metals or metal oxides.

The mineralogy and texture of detrital rocks are the products of the composition of the source rocks and of their weathering, transport, and depositional history. The following simplified account describes the connections between these factors.

Consider the mountainous region shown in Figure 7.4. Erosion is rapid on the steep slopes, leaving scant time for mechanical and chemical weathering to work to completion. As a result, the sediments deposited in the *talus slopes* at the base of the mountains are likely to be poorly sorted and angular, with a high percentage of coarse fragments embedded in a matrix of clay and silt. The sediments that are buried and lithified here will form breccia (Figure 7.5).

The streams that emerge from narrow canyons of the mountainous region spread sediment as *alluvial fans* on the adjacent valley floors. Sorting is somewhat improved, and abrasion has rounded the fragments considerably. If fragments remain pebble-sized but are rounded, then the rock that results will be a conglomerate (Figure 7.6). Finer fragments are transported farther and more rapidly than coarse ones, so that sand-sized sediments are predominant at the outer fringes of the alluvial fans and in channels farther downstream. The type of sandstone that forms will depend on the composition of the source rock. Granitic mountains will yield granitic weathering products: sand-sized

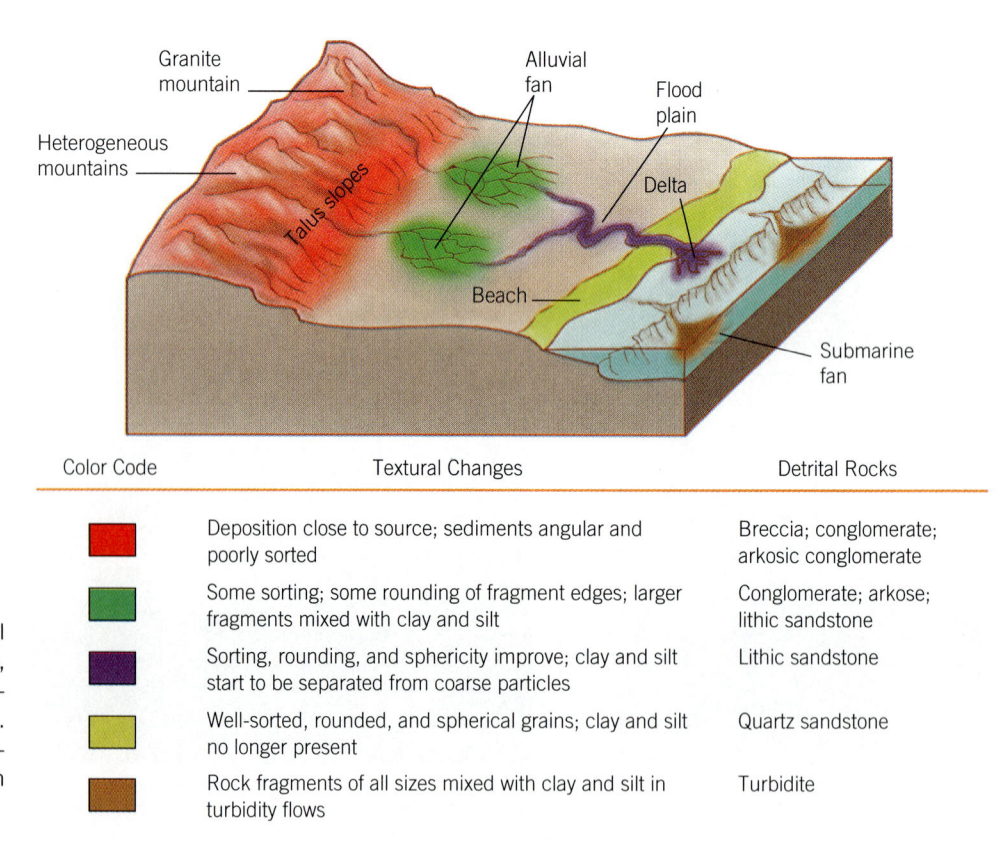

Figure 7.4 Some common depositional environments of detrital rocks. Sorting, rounding, and sphericity generally improve with distance from the source. Turbidites are a major exception because fragment sizes are remixed in turbidity flows.

Color Code	Textural Changes	Detrital Rocks
(red)	Deposition close to source; sediments angular and poorly sorted	Breccia; conglomerate; arkosic conglomerate
(green)	Some sorting; some rounding of fragment edges; larger fragments mixed with clay and silt	Conglomerate; arkose; lithic sandstone
(purple)	Sorting, rounding, and sphericity improve; clay and silt start to be separated from coarse particles	Lithic sandstone
(yellow)	Well-sorted, rounded, and spherical grains; clay and silt no longer present	Quartz sandstone
(brown)	Rock fragments of all sizes mixed with clay and silt in turbidity flows	Turbidite

Figure 7.5 Close-up of a typical breccia displaying angular fragments in a fine-grained matrix.

Figure 7.6 A quartz conglomerate. Compare the rounded pebbles in this sample with the breccia in Figure 7.5.

fragments of granite and grains of feldspar, quartz, and clay. A sandstone of this composition is called *arkose* (Figure 7.7). On the other hand, if the mountains consist of many rock types, the *lithic sandstone* derived from them will contain fragments reflective of the varied source region, in addition to quartz and clay. In both arkose and lithic sandstones, sorting is generally poor, with coarser fragments embedded in a clay and silt matrix. However, the lithic sandstone deposited near sea level in deltas and flood plains or on the continental shelf is better sorted, better rounded, and cleaner than its upstream cousin.

Much of the sediment that streams carry down from the source region will sooner or later reach the beach. As waves wash the grains back and forth across the beach and nearby sea bottom countless times, the fine particles are sifted out and the coarser fraction remains in the beach zone. For this reason, beach sands are commonly well-sorted sediments. The only common source mineral that is chemically and physically tough enough to survive the relentless attack is quartz (Figure 7.8). When cemented, these well-sorted and well-rounded quartz fragments form *quartz sandstone* (Figure 7.9). Where the wave action is very intense, the relatively scarce gravels are left behind on the beach and the coarse sands are carried offshore. The well-rounded beach gravels then form a *quartz conglomerate*.

Figure 7.7 Close-up of a typical arkose. The high percentage of pink feldspar, plus the angularity and coarse size of the fragments, suggest that the rock was deposited close to its granitic source region.

Figure 7.8 A detrital sediment is subjected to intensive abrasion, solution, and sorting during transport. The clay and silt matrix is sifted from the larger fragments, the fragments are rounded, and weak minerals are pulverized or dissolved. The product is commonly a well-sorted and -rounded quartz sandstone.

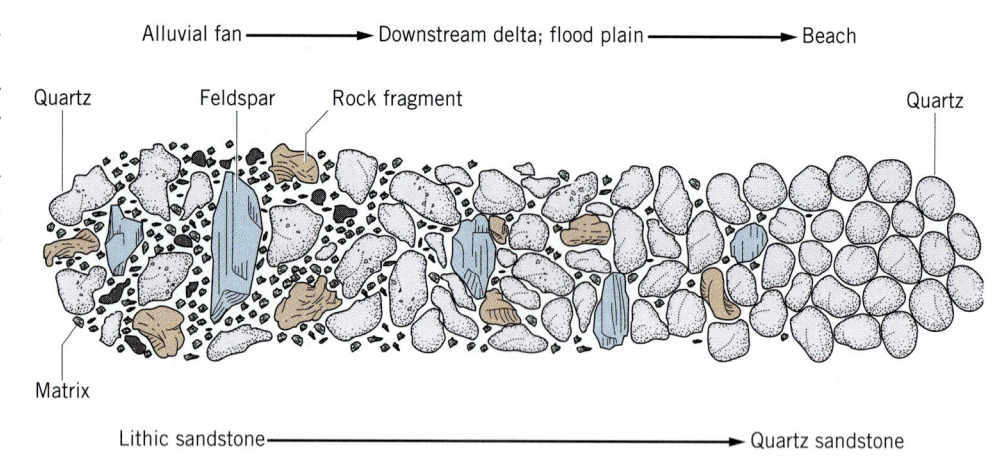

Turbidite Deposits The sediments brought to the ocean by streams enter a new realm, where they are subject to submarine transport. Often, the sand and silt that would otherwise form lithic or quartz sandstone are swept by currents and waves to delta fronts and to the edge of the continental shelf, where they are deposited precariously on underwater slopes. Any disturbance—an earthquake, a storm, or simply too great a supply of sediment—may send dense flows of clay, mud, sand, and gravel down submarine canyons. The flows, which hug the sea bottom, emerge from the canyons, lose power, and spread their sediment as *submarine fans*. The sediments (and rocks) that result from these flows are called *turbidites*. They are poorly sorted, with coarse fragments embedded in a dense clay and silt matrix.

Fine Detrital Rocks

Fine-textured sediment consists of minute clay minerals derived from the weathering of feldspar and tiny particles of quartz powder derived from the abrasion of larger quartz fragments. They tend to be poorly sorted because nature's sorting mechanisms are not sensitive to differences between extremely light, fine particles. The rounding process is also less effective, because the particles are so small that they are protected from abrasion and collision by films of water.

Because of their low mass, clay and silt particles do not sink easily. Slight turbulence can keep them in suspension nearly indefinitely, just as air turbulence can keep dust aloft. However, clay particles have surface charge and attract one another. When large clumps form, they sink.

Figure 7.9 Close-up of a typical quartz sandstone. Note the similarity to the loose beach sand in Figure 7.3(a).

Figure 7.10 Alternating beds of shale, mudstone, and sandstone in the Moenkopi Formation, Arizona. Notice the differences in the layering and textures of these rocks.

Clays and silts are deposited in calm waters, which can occur in a great variety of settings: lakes, flood plains, deltas, marshes and estuaries, the continental margin, and the deep ocean. Where their deposition occurs along with coarse sediments, the clay and silt form the matrix of the coarse detrital rock. Where the coarse sediments are absent, the clay and silt particles form shale and mudstone. The difference between shale and mudstone is connected to the marine life on the sea floor. Where burrowing organisms stir the sediment, mudstone tends to form. In the absence of such organisms, thinly layered shale forms (Figure 7.10).

Precipitated Sedimentary Rocks

Precipitated sedimentary rocks can be divided into two categories: chemical rocks, which form through direct chemical or physical means, and biochemical rocks, which form through the work of organisms in the water. Chemical precipitation is initiated by the evaporation of water within which ions are dissolved, by the addition of excess ions, or by temperature and pressure changes that affect the capacity of the water to hold the ions in solution. Biochemical precipitation is initiated by organisms through metabolic activities. These activities alter the pressure, temperature, and concentration of ions in water. Most of the organisms use these ions to synthesize shell matter.

The Limestones

Composed of calcium carbonate, *limestone* is the most common of the precipitated rocks. The majority of these rocks are biochemical in origin, having been derived from shell matter. Experiments prove that increasing water temperature and lowering pressure causes calcium carbonate to precipitate. Thus carbonate sediments are directly precipitated in the warm, shallow waters of tropical and subtropical seas but are rarely precipitated in great quantity at colder, higher latitudes or in very deep water.

Figure 7.11 Close-up of coquina, a limestone composed of large cemented shell fragments.

As with detrital sediments, shells and mineral matter are subject to abrasion and mechanical transport. Limestones are deposited in a wide range of marine environments. Thus their texture and composition are reflective of their transport and depositional history.

Biochemical Limestones Shells are the raw materials of biochemical rocks. For example, soft and fine-textured *chalk* consists of the shells of billions of microscopic organisms that settled in shallow water. So plentiful was it 135 to 65 million years ago that the entire geologic period was named for it—the Cretaceous (*creta* is Latin for "chalk").

What happens to shells after the organism dies largely determines the properties of the resulting rocks. They may remain intact, or waves and currents may break them into large fragments, carbonate sand, and limy mud. Their degree of sorting depends largely on the intensity and duration of wave and current action. *Coquina* is composed of large, poorly cemented shell fragments (Figure 7.11). Limestones made of shells that have been subjected to intensive abrasion in the beach zone display properties similar to those of quartz sandstones formed in the same environment. The ground-up shells are sand-sized, are well sorted, and have a clear, chemically precipitated matrix.

Now consider the other environmental extreme, that of deposition in water devoid of current and wave action, such as a tidal flat or quiet lake. Under these circumstances, fine particles settle out of the water and are deposited as lime mud. Limestones having this fine texture and composition are called *micrite*—a name that calls attention to the microscopic size of the particles. One famous micrite deposit, the Solenhofen limestone of Germany, was the site of a large lake 150 million years ago. The lake was so still, and the sediments so fine, that the remains of dead animals and plants that settled to the bottom are preserved to the most delicate detail. The Solenhofen contains one of the world's foremost "library" of flying reptiles and ancestral birds, including *Archaeopteryx,* which is considered by some paleontologists to be the first bird (Figure 7.19, page 163).

Most limestones are deposited in an environment between these energy extremes. They consist of a mixture of intact and fragmented shells embedded in a micrite matrix. In this respect, they resemble immature lithic sandstones, which consist of large rock fragments embedded in a clay matrix.

Coral Reefs Biochemical sediments predominate where there is a scarcity of land-derived sediments and where waters are warm and teeming with life. In these regions, shells of marine organisms deposited on the continental margins form *carbonate shelves.* Where warm ocean currents sweep abundant nutrients to shallow waters, conditions are ideal for growth of coral *reefs.* Anchored to the mainland or an island, the reefs are built of the myriad skeletal secretions of tiny colonial coral. The coral depends upon microscopic algae lodged in its gullet to help it digest food; because algae need sunlight for photosynthesis, the live coral must grow close to sea level, where light can penetrate.

Australia's Great Barrier Reef, 1500 kilometers long, is the largest living coral reef. A buried reef of similar dimensions, 150 million years old, runs several hundred kilometers offshore along the length of the eastern United States. It acted as a massive trap to sediments brought in from the continent and, in this way, was instrumental in building the continental shelf. Probably the greatest exposed fossil reef, the 220-million-year-old El Capitan of the Guadalupe Mountains, is currently stranded in arid West Texas (Figure 7.12). You can visit this spectacular structure and surrounding region and see all the features described above. Once you understand the origin of the rocks, you can almost hear the waves crashing in the desert.

Chemical Limestones One interesting environment where limestone is directly precipitated is among the lagoons and shallow reefs of the Grand Bahama Banks. There we find deposits of tiny caviar-sized particles called *ooids*. Sectioning and observing an ooid under the microscope, we see that it consists of a minute sand-grain nucleus, around which are wrapped layers of calcium carbonate. Geologists reason that the calcite layers precipitated around the sand grain as it was tossed about by the waves in this high-energy environment. *Oolitic limestones* are well sorted; the grains are held together by clear chemical cement (Figure 7.13).

Evaporites

Many **evaporites** are precipitated within the semienclosed basins of arid climates. In a typical setting, water within the basin is partially trapped by a shallow ledge or barrier, which closes off easy circulation with the open ocean. Heavy evaporation increases the salinity of the water in the basin to the point that it collects at the bottom as a dense brine from which the evaporites are deposited. The dense water creeps along the bottom and out into the open ocean, while the less saline water of the open ocean seeps into the basin to replace the salty water that escaped. It too becomes highly saline as it evaporates.

evaporite
Deposit from the evaporation of aqueous solutions. The most common example is rock salt.

Figure 7.12 El Capitan, an ancient reef complex that is now part of the Guadalupe Mountains of west Texas and New Mexico. The steep cliffs are remnants of a massive reef core; the forereef and deep basin strata are in the foreground.

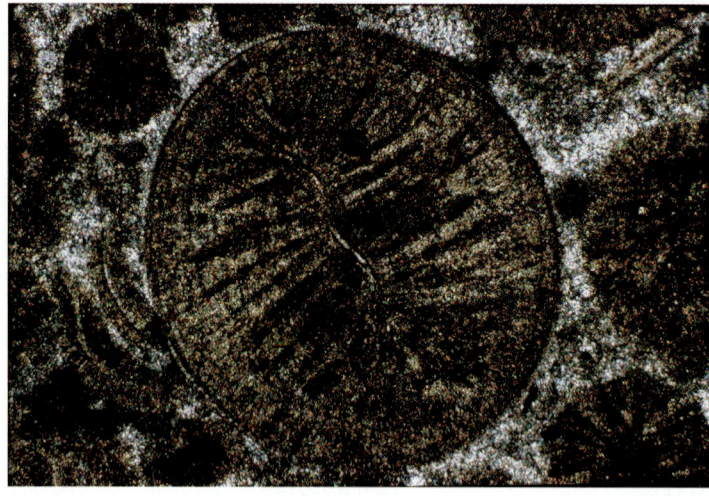

Figure 7.13 (a) Close-up of an oolitic limestone. (b) Photomicrograph of an ooid, displaying its delicate concentric structure.

If evaporation is extreme, salts precipitate in a bull's-eye pattern, with the least soluble on the fringes of the basin and the most soluble at the center. The sequence runs from carbonates (certain limestones and related rocks) to sulfates (gypsum) to halides (halite).

Thick evaporite deposits are spread over regions of New York, Pennsylvania, Ohio, Michigan, and Ontario. They are records of a shallow sea that covered the region some 400 million years ago. There are also evaporites of freshwater origin; the famous Bonneville Salt Flats of western Utah are the remains of an ancient lake that dried up; the Great Salt Lake is its descendant.

Chert

Some organisms use silica rather than calcium carbonate for their shells. Most biochemical silica is extracted from seawater by tiny single-celled *radiolaria* and by *diatoms,* an important single-celled alga. These shells rain on the ocean bottom and form layers of deep-sea ooze. Silica precipitates under the very same conditions that inhibit calcium carbonate precipitation, and oozes of this composition are found in polar seas and in regions where cold water rises to the surface from great depth. Upon lithification, the oozes form bedded *chert* (Figure 7.14). However, organisms with silica shells also exist in warm waters. If the shells are deposited along with abundant calcium carbonate, they will form impurities in limestone sediment and collect as nodules and geodes (structures we will soon discuss).

Coal

Coal is generally classified as a biochemical rock, although it is not precipitated by organisms but, rather, consists of the compressed, altered remains of vegetation. The typical environment where coal forms is a stagnant swamp. Because coal is highly combustible, it is a valuable source of energy. As such, it is discussed more fully later in this chapter.

Sedimentary Structures

No matter how they are transported, sediments ultimately settle to the sea bottom, lake bed, flood plain, and so on, under the influence of gravity. But sediments are never deposited continually in these environments. Streams overflow their banks onto the flood plain only seasonally, and shifting currents may prevent deposition on the sea bottom for a time. Temporary interruptions—which may vary from days to tens of thou-

Figure 7.14 A bedded chert from the Franciscan formation, San Francisco, California. Deposits like these are believed to be derived from minute silica shells of microorganisms that were altered by heat and pressure after burial.

sands of years—allow earlier deposits to begin to compact before fresh sediments cover them. The compacted sediments form cohesive layers, known as **beds** or **strata.** The surface of the bed, called the **bedding plane,** lies parallel to the depositional surface and is the boundary between layers. **Bedding** is the most prominent structural characteristic of any sedimentary deposit and is an integral part of the depositional process.

Various types of bedding give clues to the environments in which they formed. In **graded bedding,** each layer displays a change in particle size—coarse to fine—from the bottom to the top of the bed (Figure 7.15). This grading indicates that the velocity of the current carrying the sediment diminished. This condition occurs in submarine flows that deposit turbidites. Each turbidite bed represents a single flow event.

Cross-bedding refers to thin beds inclined at an angle to the main bed. Only the deposition of sediment carried by a moving stream of water or air can cause such a feature. Figure 7.16 shows a typical cross-bedded sandstone deposit, an ancient windblown sand dune. The cross-beds were formed as sand was blown up the gentle windward slope and rolled down the steeper lee slope. Shifts in the inclination of the cross-beds mark

beds (strata)
Visually distinguishable layers of sedimentary rock.

bedding plane
The plane that marks the boundary of each layer of stratified rock.

bedding
The stratification or layering of sedimentary rock.

graded bedding
Stratification in which particle size changes from coarse to fine from the bottom to the top of each bed.

cross-bedding
Thin strata laid down by currents of wind or water at an oblique angle to the main bed.

Figure 7.15 Graded bedding in a typical turbidite deposit. The layers were deposited during separate flow events by slurries of gravel and silt that hugged the continental slope and spread over the deep-ocean floor. The coarse fragments settled to the bottom of the bed as the slurry lost energy.

Figure 7.16 A spectacular 150-million-year-old example of a cross-bedded sandstone in what is now northern Arizona. These ancient sand dunes were probably formed in a coastal desert.

ripple mark
A corrugated form displayed in sedimentary rocks caused by currents of air or water that moved over the sediments prior to lithification.

mud crack
Polygonal crack formed by the drying and shrinking of mud, silt, or clay.

shifts in the direction of the wind that flowed over the sand dune; the average direction of cross-bedding reflects the average direction of the prevailing wind. The same principle applies to stream cross-beds and is a useful tool for interpreting the source region of the sediment. Cross-bedding can occur on a small or large scale—from fine layering within a single bed to the thick complex structures of huge deltas that form at the mouths of rivers entering calm portions of the ocean or lakes.

Ripple marks are formed by currents and waves that disturb the soft bottom where sediments have settled. The symmetrical ripple marks in Figure 7.17 indicate circular wave motion of the kind seen at beaches. Asymmetrical ripple marks result from currents flowing in one direction, such as stream and long-shore currents. Where sediment is deposited in water and then exposed to hot sun, **mud cracks** are caused by shrinkage when the sediment dries up (Figure 7.18). Tidal flats and seasonal lake beds are typical environments where mud cracks can be observed.

Figure 7.17 These ripple marks were formed by waves along a beach some 200 million years ago.

Figure 7.18 Mud cracks formed by the drying of sediment along the Colorado River.

Sedimentary Rocks 163

Figure 7.19 Some modes of fossilization.

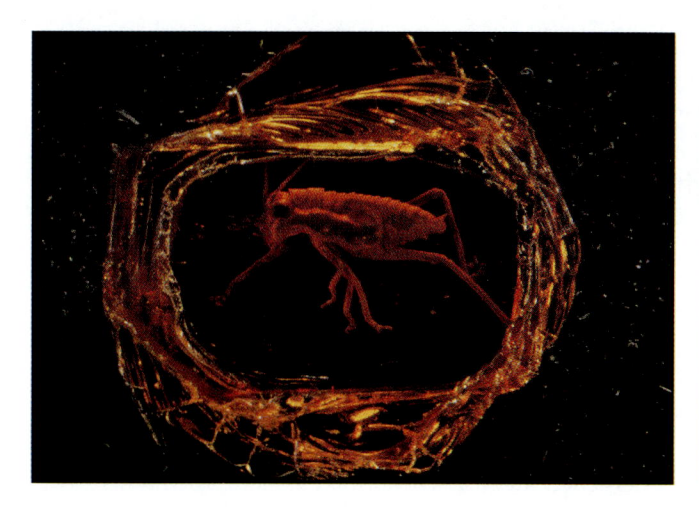

(a) *Direct preservation.* Forty million years ago, this grasshopper was trapped in a globule of tree sap that later hardened to amber.

(b) *Unaltered hard parts.* These rocks in Dinosaur National Monument, Utah, contain dinosaur bones.

(c) *Carbonization.* This 350-million-year-old seed fern has been altered to carbon, but so delicately that its structure is preserved.

(d) *Petrification.* The organic matter of this 200-million-year-old log has been replaced by silica, but the original tree-ring structure and bark texture have been preserved.

(e) *Molds and casts.* A mold is an impression of an organic structure, most commonly a shell. Sediment fills the mold, forming a cast in the shape of the organism. This near-perfect cast of *Archaeopteryx,* the oldest known bird fossil, was found in the 145-million-year-old Solenhofen Limestone, Germany.

(f) *Tracks and burrows.* Organisms that live in or pass over sediments often leave behind fossil tracks or burrows. These vertical worm burrows, filled with mineral cements, were made 600 million years ago in muds that are now part of the sedimentary rocks of Scotland.

Fossils and Minor Structures

fossil
The remains, trace, or imprint of a plant or animal preserved in rocks.

Many sediments contain **fossils,** the remains or traces of past life. Figure 7.19 (on the preceding page) illustrates the common modes of fossilization. Some are simple bedding-plane features, such as worm burrows, plant imprints, and animal tracks. Others are preserved bones, shells, and other organic matter. Typically, they are either recrystallized shells or molds and casts of these shells. The orientation of the shells can be indicative of sea-bottom conditions when they settled. For example, shells arranged haphazardly indicate a calm sea bottom, for the slightest current or wave action will flip shells concave sides down.

Nodules and *geodes* commonly occur as masses of silica within limestone (Figure 7.20). The nodules lack an internal structure, but the geodes consist of cavities lined with silica, which serves as the base for beautiful crystals that grow inward toward the center. Geologists think that the silica was deposited throughout the limestone first, and circulating groundwater later concentrated it within cavities. There the silica dehydrated and formed gels that gradually hardened into a massive, waxy-looking rock called *chalcedony,* a variety of chert. The quartz crystals that form the chalcedony are so minute that special X-ray techniques are needed to detect them. Chalcedony forms the lining of the hollow geodes and the entire mass of the nodules. In some locations, such as the upper Mississippi valley, geodes and nodules weather out of the soft limestone bluffs that border streams and, like so many apples, lie scattered in the streambeds—to the delight of mineral collectors.

Concretions are rounded, solid bodies found within shale, sandstone, and limestone beds. Their composition is variable—silica, iron oxide, and carbonate material. The concretions originate when material precipitates around a rock fragment, shell particle, or other impurity and then hardens; finally, the entire conglomeration is cemented together.

Oceanic Sedimentary Environments

Although most sediment is deposited on the continental margins, appreciable quantities reach the ocean floor. These sediments preserve the tectonic, environmental, and biological history of the deep-ocean basins.

Figure 7.20 A geode is formed by deposits of silica that line the cavities in limestones and other rocks. This geode contains amethyst, a variety of quartz.

Figure 7.21 (a) Microscopic shells of foraminifera, magnified 50 times. Most calcareous oozes are composed of these single-celled amoebalike organisms. (b) Microscopic shells of diatoms, magnified 100 times. Most siliceous oozes are composed of these single-celled algae.

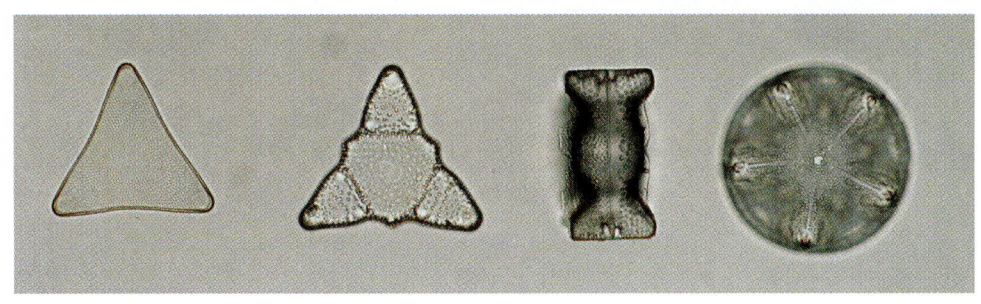

(a)

(b)

The Continental Margins

The east coast of North America is a good example of a passive margin, having been tectonically stable for several hundred million years. Detrital sediments predominate; indeed, in some places, the continental shelf seems to have been constructed from overlapping deltas that grew outward from the coast. As we go farther from shore, the sediments are finer and the continental slope is marked by silts and clays, in contrast to the coarse sands of the shelf.

Moving farther out to sea, we observe large ripple marks and dune structures in sand and silt. Evidently, sediments are being moved along the bottom toward the shelf edge. There, strong turbidity currents that hug the ocean bottom carry sediments down the continental slope and deposit their load on the continental rise. The turbidite beds then mingle with, and gradually give way to, the fine clay and microscopic shells that settled out onto the deep-ocean floor. These deep-ocean sediments are thousands of meters thick close to the continental margins and form the nearly flat *abyssal plains* of the deep-ocean basin that have buried the basaltic oceanic crust.

To understand depositional patterns on active margins, let us examine the west coast of South America. The Andes Mountain chain extends along the western edge of the continent and the deep Peru-Chile trench, located a few kilometers offshore, runs parallel to the coast. Oceanic crust is currently being subducted beneath that trench, and earthquake and volcanic activity is virtually unceasing. Under these circumstances, the continental shelf and slope are extremely narrow. Sediments are washed down into the trench, where they are trapped. Deep trenches rim much of the Pacific basin, in contrast to the passive Atlantic margin, which lacks bordering trenches. Thus turbidites in the active Pacific margin generally do not spread as far out to sea.

The Deep-Ocean Environment

Although deep-ocean sediments contain turbidites that have flowed down the continental slope and onto the abyssal plains, most sediments on the ocean bottom got there as the result of a steady drizzle of particles from the ocean surface. The particles are mainly of two kinds: *brown clay* and *deep-sea oozes*—the latter composed of the microscopic shells of minute organisms (Figure 7.21). Because the drizzle is exceedingly slow and fine, deep-ocean sediments accumulate at the rate of about 1 millimeter per thousand years. The distribution of brown clay and oozes over the vast expanse of ocean

Figure 7.22 A manganese nodule, split open to display its internal structure.

floor is largely controlled by variations in ocean depth, by deep-ocean circulation, by the kinds of organisms that thrive on the surface, and by movements of the lithospheric plates. The deep-sea oozes are classified on the basis of their composition. *Siliceous ooze* is derived mainly from organisms whose shells are composed of silica, and *calcareous ooze* mainly from the shells of organisms made of calcium carbonate.

Brown clay, the most common deep-ocean sediment, comes from a variety of sources. Most of it is derived from the weathering of crustal rocks and is brought to the ocean by streams. Being very fine, the clay particles float to the open ocean far from shore before they settle. Windblown particles distinctive of the Sahara Desert have been traced far out into the Atlantic Ocean off the West African coast. Minute, but measurable, quantities of meteoric dust, formed when meteors burn through the Earth's atmosphere, have been found mixed with the brown clay of the North Pacific. In contrast to carbonate ooze, brown clay is found at all depths.

Manganese nodules are deep-ocean deposits directly precipitated from seawater (Figure 7.22). They consist of concentric layers rich in copper, cobalt, and nickel, as well as manganese wrapped around a hard rock nucleus. Geologists think that the nodules precipitated as a result of reactions between cold seawater and hot fluids rising from fractures along the mid-ocean ridges. Though strewn throughout the ocean floor, they are concentrated mostly in the vicinity of mid-ocean ridges. Probably their wide distribution is at least partly explainable by the slow spreading of oceanic crust away from the ridges.

Plate Tectonics and Ocean Sediments

The migration of the oceanic crust and sediment away from the mid-ocean ridge crest brings us to an interesting point: since the oceanic crust is not permanent—being created at the ridges and destroyed at the trenches—the deep-ocean sediment deposited on it is not permanent either, at least not in its original condition. What happens to these former ocean sediments? Paradoxically, some are preserved on the continents. One way that deep-ocean sediment can be preserved is by being sheared off the oceanic crust as the crust descends into the mantle beneath deep-sea trenches. The sediment, together with slivers of oceanic crust, is highly deformed and slightly metamorphosed. This material, called a *mélange,* is often uplifted and thrust onto the landmasses bordering the trenches. The rocks of San Francisco are considered a classic example of a mélange. Other sediments become detached from oceanic crust during plate collisions and become attached to the continents as *accreted terranes.* Some of these terranes leave an excellent record of the ancient ocean floor. But, overall, the record of past deep-ocean sediments is confused and fragmented.

The most complete record of oceans past comes from the deposits of shallow seas that invaded the continents repeatedly over geologic time. These deposits form the relatively thin veneer that covers the igneous and metamorphic rock of the ancient, gnarled continents.

Energy Resources in Sedimentary Rocks

Coal, petroleum, and natural gas are derived from the partially decomposed remains of ancient organisms preserved within sedimentary rocks; that is why these hydrocarbon resources are referred to as **fossil fuels.** They are residues of the carbon-oxygen cycle. At the heart of the cycle is photosynthesis, whereby plants, using sunlight, extract carbon from the carbon dioxide in the atmosphere and incorporate the carbon into their living tissues as organic (hydrocarbon) molecules. The plants also release oxygen to the atmosphere as a by-product of photosynthesis, and this oxygen is extracted from the atmosphere by the respiration of bacteria and animals. During respiration (and during the decay of organic wastes), the oxygen recombines with the carbon of the organisms to form carbon dioxide, which is then released back to the atmosphere, where it is used again in photosynthesis. This process of cycling carbon into and out of the atmosphere through photosynthesis and respiration/decay is remarkably efficient, but not perfect. Over millions of years, a tiny fraction of carbon in the form of organic matter is diverted and buried along with other sediments. Depending upon the prevailing geological conditions, this organic matter is converted either to coal or to oil and natural gas.

fossil fuel
Any hydrocarbon that can be used for fuel.

Coal

Coal is a brown to black rocklike substance composed of the altered remains of ancient plants that thrived in freshwater and brackish swamps. The swamps were located in protected portions of coastal plains, estuaries, lagoons, deltas, and similar low-lying regions. Water circulation was often restricted in these environments, and the oxygen at the bottom was depleted, a necessary condition for the preservation of plant matter. The Great Dismal Swamp in Virginia is probably a modern equivalent of these ancient swamps.

coal
A combustible carbonaceous rock formed from the compaction of altered plant remains.

The transformation of dead plant matter to coal involves bacterial action, which partially decomposes the mosses, ferns, trees, and other plants into a dark brown to black residue called *peat,* the precursor of most coal. Subsequent stages are triggered by burial under younger sediment (Figure 7.23). When the peat is compressed to about a tenth of its original thickness and then heated, the volatiles are driven off, increasing the carbon content. Gradually, the peat is converted to coal, which is ranked as *lignite, bituminous,* and *anthracite* on the basis of increasing carbon content.

More than mere burial is generally required to produce anthracite; it tends to occur in regions of moderately folded and compressed rock strata (such as eastern Pennsylvania).

The swampy conditions that favor the formation of coal were prevalent over broad expanses of the east-central United States and central Europe during the Carboniferous period, some 300 to 350 million years ago, when the continents were near the equator and were subjected to shallow invasions and retreats of the sea. Thus swamps were repeatedly created and destroyed, which led to the formation of layer upon layer of coal between flood-plain sandstones and shales.

The United States is fortunate. When North America and Europe separated some 90 million years ago, the broad area that later became the United States retained a huge share of coal. Millions of years later, during the Triassic period, this stock was augmented by enormous deposits of lignite and bituminous coal that blanket the western plains and the Rocky Mountains states. Today, most of the nation's coal is mined in these western states, in part because of the low sulfur content and because it is found in surface or near-surface locations where it is easily extracted by open-pit or strip-mining methods.

(a) (b) (c)

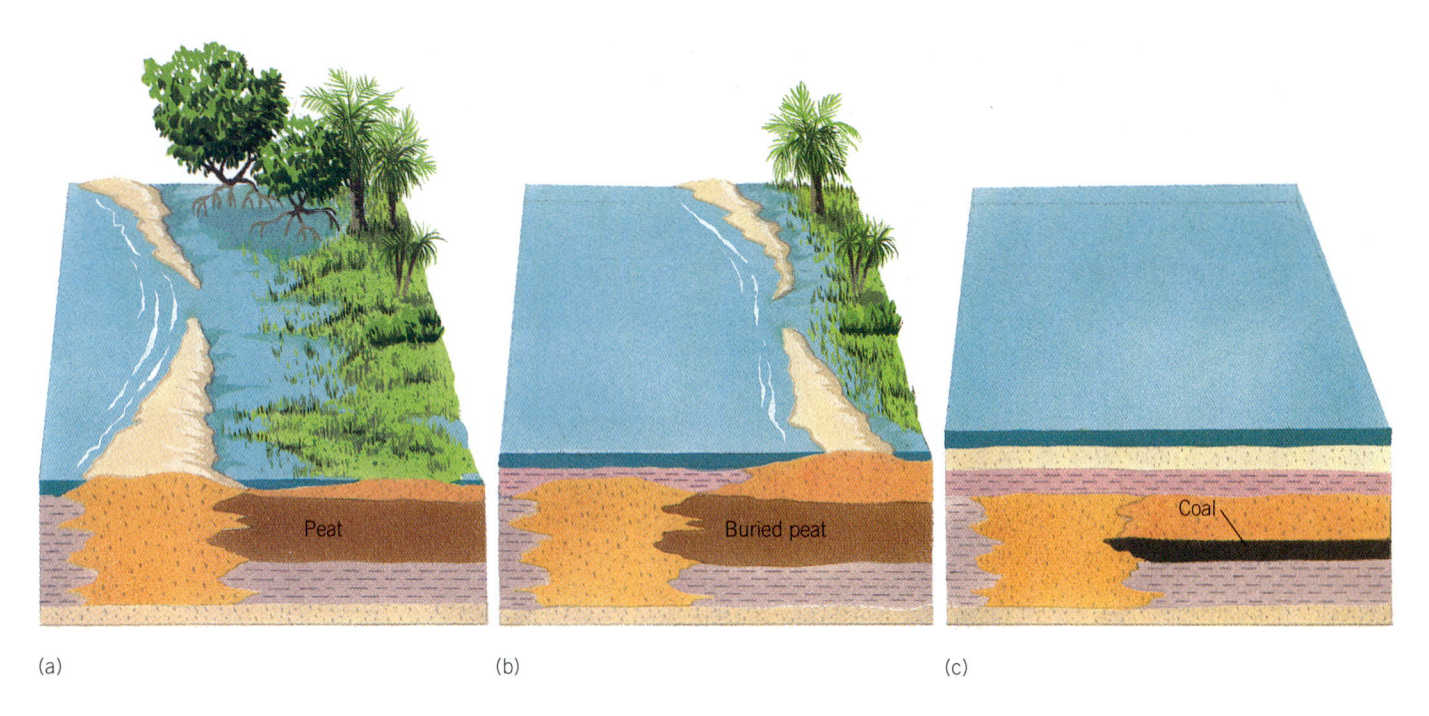

Figure 7.23 Stages in the burial and compaction of swamp plant matter leading to its transformation into coal.

petroleum
A naturally occurring liquid of complex hydrocarbon composition. (May include natural gas.)

Petroleum

Stripped to its Greek root, **petroleum** means "rock oil." It is a complex tangle of hydrocarbon compounds found almost exclusively in sedimentary rocks. Most known oil deposits were probably generated either in shallow basins within the continents, which at times had been invaded by the sea, or in depressions that lay offshore, within the continental shelves. Such modern-day basins as the Gulf of Mexico and the Orinoco River delta of South America are believed by petroleum geologists to be typical oil-forming environments. These basins are receptacles of three kinds of sediment: (1) detrital sands, clays, and silts washed in from the continents; (2) chemically and biochemically precipitated coral reefs, shell limestones, salts, gypsums, and cherts; and (3) the remains of organisms brought in by streams or that lived in the basins and collected on the sea bottom.

The organic matter is preserved in oxygen-poor portions of the basins, where decay is at a minimum, such as shallow lagoons or deltas closed off by barriers that restrict circulation. The oxygen in the water is soon depleted in this stagnant environment as organic remains accumulate in the muds. The organic matter is then slowly broken down into hydrocarbon compounds (petroleum) within these muds. Bacterial action plays an important role at first; but as the organic remains are covered by thick layers of younger sediments, the effects of compaction and the elevated temperatures that accompany burial become dominant.

When first deposited, the typical organic muds within which oil is generated consist of about 75 percent water. Compacted by burial beneath younger sediments to depths of 2000 meters, that same mud, now converted to shale, holds about 10 percent water. The oil and natural gas squeezed out with the water are driven from their source sediment or rock to regions of low hydrostatic pressure, where they accumulate. This primary migration is generally upward, toward the surface, but the path is also controlled by the porosity and permeability of the sedimentary rocks. Eventually, the water, oil, and gas will become concentrated in rocks with large, continuous pore spaces. Well-sorted beach sandstones, fragmented limestone, and reef deposits made of the skeletal framework of ancient corals make good **reservoir rocks.** If impermeable barriers lie above and below such reservoirs, preventing escape, the water, oil, and gas will be trapped.

Lest anyone think migration and accumulation in a reservoir is rapid, remember that the oil and water must trickle through minute spaces in which surface tension and friction exert tremendous retarding forces. A drop of oil and water probably moves from 2 to 10 centimeters per year. At this rate, it takes millions of years for a large oil field to form, whereas the oil field itself may be depleted in a matter of decades—which is why we call oil a nonrenewable resource.

reservoir rock
Any porous and permeable rock that could contain oil or natural gas deposits.

Oil Traps

Having become concentrated in reservoir rock during the primary migration stage, oil continues to migrate within the reservoir itself. This latter journey is called secondary migration. As in primary migration, the oil is driven from regions of high to low pressure, where it occupies the pore spaces between sediment grains, forming a *pool.* The location of the pool is controlled by an **oil trap,** a set of structural and/or stratigraphic conditions that prevent further migration of the oil. Often, a number of pools and traps occur in close proximity, related by similar structural and stratigraphic conditions, forming an **oil field.**

In a typical structural trap, the fluids are prevented from escaping by *cap rock,* impermeable strata that seal the oil-bearing reservoir rock. The fluids that accumulate in the reservoir separate according to density: natural gas (if present) on top, oil in the center, water below (Figure 7.24). A similar pattern exists in all oil traps.

oil trap
A structural or stratigraphic arrangement within a sedimentary basin that allows oil to accumulate and prevents its escape.

oil field
A geologically and spatially related feature containing two or more oil accumulations.

Oil Shales and Oil Sands

Nearly all of the world's oil of current commercial value is pumped as liquid from buried pools. *Oil shales* are fine-grained rocks consisting of clay minerals and silt-sized quartz grains. The spaces between grains are microscopic, making the rock highly impermeable. Trapped within these spaces is a solid, combustible hydrocarbon substance called *kerogen.* When heated, the kerogen breaks down and yields oil. We may think of oil shales as source rocks from which oil has not migrated.

Oil shale, like most hydrocarbon deposits, forms in stagnant lakes, in oxygen-deficient portions of shallow deltaic marshes and lagoons, or in the deeper oceans. These conditions preserve the accumulation of organic matter in the bottom muds. Typical are the oil shales of the Green River formation of Colorado, Utah, and Wyoming. Though variable in composition, they constitute the world's largest single oil shale accumulation, holding the equivalent of 1.8 trillion barrels of oil, or about twice the world's reserves of conventional oil.

Oil sands are sediments impregnated with viscous, asphaltlike crude oil. The oil probably seeped upward through porous reservoir rock to the surface, as in the famous La Brea tar pits of Los Angeles, California. Probably the world's best known tar sand deposits are located in Alberta, Canada. It appears that the tars, of Cretaceous age, were derived from oil trapped in a far more ancient Devonian reef. The reef was exposed by erosion, and much of the oil evaporated. The asphalt residue was then carried away and deposited with conventional sands.

Plate Tectonics and Petroleum

In Chapter 5, Igneous Activity, we described the close connection between plate tectonics and the concentration of valuable metals in the crust. We find a similar connection between plate tectonics and petroleum accumulations.

At the embryonic stage in the development of an ocean basin, seawater circulation is restricted in the rift valley, which leads to deposition of thick salt deposits and the

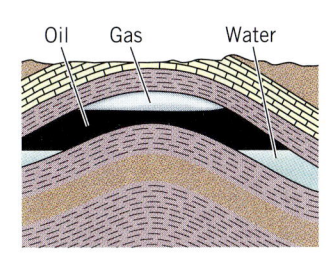

Figure 7.24 Common configuration of a structural oil trap.

Table 7.3 Estimates of Current and Future World Oil Supply (in billions of barrels)

	United States	World
Production (through 1990)	55	650
Reserves (1991)	25	~950
Undiscovered recoverable oil	~40	~500
Anticipated reserve growth	~20	Unknown
Total recoverable oil	~140	~2100
Total oil left for the future	~85	~1550

accumulation of organic remains in stratigraphic traps. In later stages of ocean basin development, the creation of new oceanic crust and sea-floor spreading cause the continents to be widely separated. Because the salt layers rest on old oceanic crust moved aside by sea-floor spreading, they are found along the flanks of the continents, far from the rift valley where they were precipitated. Rivers flowing down from the continents to the sea deposit thick wedges of sediments along the continental margins, burying the ancient salt layers. The salt is less dense than the overlying sediment and easily deforms in response to pressure. As sediments accumulate above it, the entire basin compacts. The salt is mobilized into streamlined diapirs, similar in shape to the granitic magmas that rise in the crust. The diapirs rise and pierce the sediment as salt domes. The salt domes act as traps for petroleum derived from the organic-rich sediments of the original rift zone and from the sediments that accumulate on the continental margin. The oil deposits of the Gulf of Mexico and off the Nigerian coast formed in this manner as a result of the widening of the Atlantic Ocean.

Major oil fields are also found in those regions where plates collide—near subduction zones that border continents and the marginal seas between island arcs and continents. Oil-bearing sediments accumulate in the tectonically active basins of these regions, where conditions favor the development of many structural and stratigraphic traps.

The Petroleum Supply

In view of the importance of economic resources, it is fair to ask whether we are in danger of running out of them. Independent information on the abundance of remaining oil resources and reserves is often difficult to obtain. Much of it is held confidentially by oil companies that have sponsored explorations, and they are not about to release it to competitors; in some nations, information concerning oil reserves is a state secret. Also, there is much debate over the scientific and statistical methods of measuring resources and reserves. Therefore, it is not surprising that estimates of the abundance of recoverable oil vary widely. The estimate of three U.S. Geological Survey specialists is shown in Table 7.3. According to them, the world has about a 70-year supply of oil at current consumption rates—much less if we consider population growth and expanding economies, much more if stringent conservation measures are taken.

Study Outline

Sediments are particles that have been transported mechanically or in solution by water or air currents and deposited in layers.

I. **WEATHERING, TRANSPORT, AND DEPOSITION OF SEDIMENTS**
   A. **Detrital sediments** are fragments weathered from rocks. **Chemical** and **biochemical sediments** are composed of ions that were weathered from rock,

transported in solution, and precipitated either directly (chemical) or indirectly with the aid of organisms (biochemical).

B. **Weathering** causes the mechanical disintegration and chemical decomposition of rock.

C. Streams, the main agents of transport and **deposition,** carry sediments to lakes, flood plains, deltas, beaches, and the ocean floor.

II. **LITHIFICATION.** There are three components to **lithification,** the process of converting sediments to sedimentary rocks.

   A. In **compaction,** layers are compressed by the weight of overlying layers.

   B. In **cementation,** mineral grains are held together by a **matrix** of precipitated ions.

   C. In **recrystallization,** dissolved edges bind neighboring mineral grains.

III. **CLASSIFICATION OF SEDIMENTARY ROCKS**

   A. Detrital, chemical, and biochemical sedimentary rocks are further classified by texture and mineral content.

   B. Detrital rocks display **clastic** (fragmented) **texture** and vary in the degree of **sorting** and rounding. *Conglomerates* have large, rounded fragments; *breccias* have large, angular fragments; *sandstones* are formed of sand-sized fragments; and *shale* is composed of the finest particles (clay and silt).

   C. Coarse detrital rocks are the products of the composition of the source rocks and of their weathering, transport, and depositional history. As a rule, the farther from the source rock, the finer, better sorted, and better rounded the sediments. *Turbidites* are detrital sediments that are subjected to further transport and deposition by undersea currents.

   D. There are two categories of precipitated sedimentary rocks: those that are precipitated by chemical means and those that are precipitated biochemically through the aid of marine organisms.
     1. The *limestones* are the most common of the precipitated rocks.
       a. Biochemical limestones are made of the shells of organisms.
       b. Coral reefs are formed from the skeletal remains of marine organisms.
       c. Oolitic limestones are precipitated in shallow waters.
     2. **Evaporites** are formed from salts that settle out of the evaporating waters of semienclosed basins in arid climates.

IV. **SEDIMENTARY STRUCTURES**

   A. **Bedding** is the predominant structure of sedimentary rocks. The parallel layers are called **beds** or **strata.** The surface of each bed is the **bedding plane.**
     1. In **graded bedding,** each layer displays coarse to fine particles from the bottom to the top.
     2. **Cross-bedding** is formed when a new bed is laid down at an oblique angle to the main bed.
     3. **Ripple marks** are formed when bedding is disturbed by currents or waves. **Mud cracks** form when the bed is exposed to air and dried.

   B. Other sedimentary structures include **fossil** remains, nodules and geodes, and concretions.

V. **OCEANIC SEDIMENTARY ENVIRONMENTS**

   A. On passive continental margins, detrital sediments are deposited such that the largest fragments are near the shore and the finest are far out on the continental slope. Active continental margins are short and steep, and many sediments are subducted into the trenches.

B. In the deep-ocean environment, land-derived turbidites, siliceous and calcareous oozes, brown clay, and manganese nodules accumulate.

C. Sediments deposited on the ocean floor are redistributed by the movement of the oceanic crust. At the trenches, some sediments are subducted, some are thrust up onto the continents as mélanges, and some are scraped off onto the continents as accreted terranes.

VI. **ENERGY RESOURCES IN SEDIMENTARY ROCKS** Petroleum, coal, and natural gas are known as **fossil fuels** because they are composed of the partially decomposed remains of once-living organisms. When oxidized, these organic compounds release heat.

A. **Coal** forms by the burial of plant matter in swampy environments that occur in coastal plains, lagoons, and deltas. Most of the coal found in the United States and Europe formed under swampy conditions caused by tectonic plate collisions during the Carboniferous period.

B. **Petroleum** is a fluid composed of hydrocarbon compounds found almost exclusively in sedimentary rocks. During generation, the organic matter accumulates in an oxygen-deficient part of a sedimentary basin, where it is buried and compacted.
1. The oil is driven by pressure to regions of low pressure within the basin, where it is concentrated in a **reservoir rock**. An **oil trap** prohibits further migration and the oil accumulates in a pool. Impermeable *cap rock* seals the reservoir rock. A group of traps and pools constitutes an **oil field**.
2. *Oil shales* are fine-grained deposits of kerogen, a hydrocarbon compound that, when heated, breaks down to form oil. *Oil sands* are sediments impregnated with asphalt.
3. Processes at convergent, divergent, and transform boundaries provide both the environmental settings conducive to the formation of oil and the structural traps that hold the oil pools.

Study Terms

bedding (p. 161)
bedding plane (p. 161)
beds (p. 161)
biochemical sediment (p. 150)
cementation (p. 151)
chemical sediment (p. 150)
clastic texture (p. 152)
coal (p. 167)
compaction (p. 151)
cross-bedding (p. 161)
deposition (p. 150)
detrital sediment (p. 150)
evaporite (p. 159)
fossil (p. 164)
fossil fuel (p. 167)

graded bedding (p. 161)
lithification (p. 151)
matrix (p. 151)
mud crack (p. 162)
oil field (p. 169)
oil trap (p. 169)
petroleum (p. 168)
recrystallization (p. 151)
reservoir rock (p. 169)
ripple mark (p. 162)
sediment (p. 149)
sorting (p. 153)
strata (p. 161)
weathering (p. 150)

Thought Questions

1. Describe the evidence that would convince you that a given sequence of sedimentary rocks was deposited in each of the following environments:
   a. on land, close to a mountain range
   b. on an ocean beach
   c. on the ocean floor, far from shore
2. What rock types might you associate with the environments listed in question 1?
3. Describe the changes you would expect to see in a detrital sediment as it moves downstream from source to ocean.
4. How do the textures of quartz sandstone and glacial till differ? Describe the differences in terms of fragment size, degree of sorting, and particle shape. Explain the differences in terms of weathering, erosion, transport, and depositional history.
5. Imagine mountain ranges A and B, each composed of identical granite. Range A is located in an arid climate and range B in a humid climate. What differences would you expect in the weathering products yielded by each mountain range?
6. What features of a turbidite indicate its mode of deposition?
7. In what ways are coquina and chalk similar? In what ways are they different?
8. What conditions favor the growth of large coral reefs?
9. Deep-ocean sediments are seldom found in an undisturbed state on the continents. Explain.
10. What conditions favor the formation of coal?
11. Why are fossil fuels found in sedimentary rocks rather than in igneous or metamorphic rocks?

Chapter 8

The Agents of Metamorphism
- Heat
- Pressure
- Chemically Active Fluids

Types of Metamorphism

Metamorphic Rocks
- The Minerals of Metamorphic Rocks
- Textures of Metamorphic Rocks
- Classification of Metamorphic Rocks
  - Foliated Rocks
  - Nonfoliated Rocks
- The Field Occurrence of Metamorphic Rocks
  - Contact Metamorphic Rocks
  - Regional Metamorphic Rocks

Metamorphism and Plate Tectonics
- Metamorphism of the Sea Floor
- Subduction Zone Metamorphism

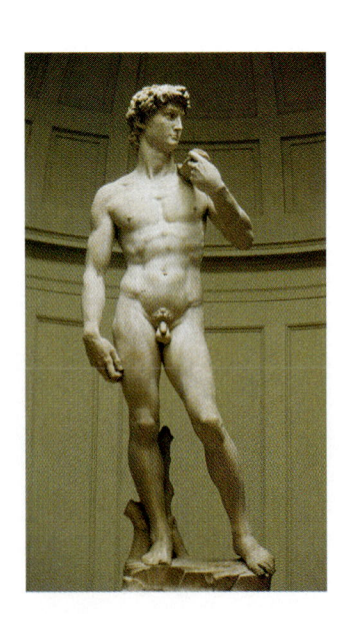

Metamorphic Rocks

The rock that became Michelangelo's magnificent sculpture *David* began life millions of years ago as limy mud on the bottom of a shallow sea that occupied a portion of what is now northern Italy. The mud was exceedingly fine, with few clay particles, quartz grains, or iron oxide compounds—what we call impurities. Upon burial, the mud was compacted and cemented into a dense, even-textured limestone. At that time, Italy was being squeezed against southern Europe to produce the Italian Alps, so the limestone deposit was subjected to much compression as well as heat from igneous intrusions in the region, and it recrystallized into a coarser-grained rock called marble. Prolonged erosion later exposed this marble, which was remarkably pure in color, texture, and composition and had few structural defects. For those reasons, architects and sculptors favored it for building ornaments and statues, and the marble was quarried for these purposes. Legend has it that Michelangelo took the particular block of marble that later became the *David* as a challenge. The block had been rejected by other sculptors because they considered it too thin and flawed for use.

This chapter describes metamorphism, the set of processes by which preexisting rocks are changed by the pressures, temperatures, and chemical conditions that prevail deep within the Earth. These changes produce rocks of markedly different mineralogy, texture, and structure—that is, metamorphic rocks.

Rocks remain in the solid state during metamorphism. What is mainly involved is change of form—a fact made clear by the Greek roots of the term: *meta*, "after" or "beyond," and *morphe*, "form." Metamorphism occurs at pressures and temperatures well above those prevailing at the Earth's surface but below those that cause melting. In most metamorphic transformations, the gross chemical composition of the rock changes

◀ A marble quarry in the Italian Alps (right). From a quarry similar to this one, Michelangelo chose the block of metamorphic rock from which he sculpted his *David* (left).

175

Figure 8.1 The pressure-temperature range of metamorphism.

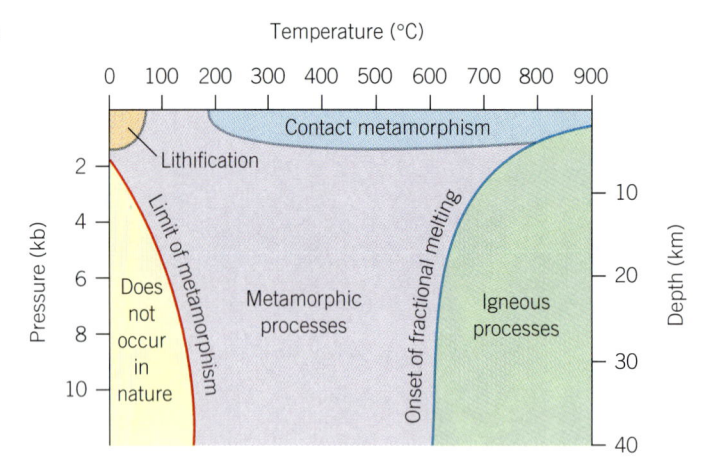

little, although reactions between minerals of the original rock may produce entirely new minerals.

The realm of metamorphism is inaccessible places: in the deep cores of mountain ranges, in the rocks surrounding igneous intrusions, in the fractured crust beneath mid-ocean ridges. In contrast to some igneous and sedimentary rocks, no metamorphic rock has ever been directly observed in the process of formation. It is the geologist's task to determine as closely as possible the nature of the preexisting parent rocks and the specific conditions responsible for their metamorphism. There are two components of this work:

1. Field studies of metamorphic terrains exposed through the erosion of overlying rock. These studies establish the mineralogical and structural features of the various metamorphic rocks and their variation in space and time.
2. Laboratory experiments on natural and synthetic rocks and minerals that attempt to simulate the conditions of metamorphism within the Earth.

Metamorphic rocks are found mainly in the core regions of the continents and mountain belts. Where exposed, they constitute 25 percent of the surface area of the continental crust. However, greater volumes of metamorphic rocks are blanketed by much younger sedimentary rock.

The Agents of Metamorphism

Heat, pressure, and chemically active fluids are the agents responsible for metamorphism. It will aid our understanding of the process to consider briefly the role played by each of them (Figure 8.1).

Heat

Heat is the energy source that powers metamorphic reactions. It raises the temperature of the rocks, which means that it causes the ions of the minerals to vibrate more rapidly. Chemical bonds are broken and the ions are forced to realign in combinations more in keeping with the high-energy environment. Thus new minerals are created as old ones are destroyed. In addition, heat mobilizes hot water and gases within the rock. The fluids act as a medium of exchange, dissolving and precipitating elements within the rock mass and, in general, accelerating the reaction rates. Heat also affects the mechanical properties of the rock by softening it and allowing it to deform plastically. Thus heat prepares the rock for the imprinting of characteristic metamorphic textures and structures.

Pressure

A body subject to pressure, in the form of a single force or a number of forces acting on it at the same time, is in a state of stress. If the forces are applied equally in all directions, then the body is said to be in a state of **lithostatic stress** (Figure 8.2a). In the Earth, this kind of stress is caused by the pressure of the overlying rocks.

Lithostatic stress squeezes rocks together and reduces their volume. Thus reactions in which the atoms of minerals recombine into denser forms are favored. We have mentioned two such reactions in previous chapters: the conversion of graphite to diamond in the upper mantle, and the conversion, through a series of steps, of olivine to perovskite in the transition zone of the mantle. High lithostatic stress, in conjunction with high temperature, also prohibits rocks from fracturing; instead, they deform plastically.

Although the degree of lithostatic stress, along with temperature, defines the general environment of metamorphism, nonuniform **compressive stress** is more directly responsible for many of the features associated with metamorphic rocks. For example, viselike pressure is exerted on rocks trapped between colliding lithospheric plates. Most rocks that are metamorphosed in this setting develop a pronounced layering, or **foliation,** perpendicular to the direction of maximum compression (Figure 8.2b).

Chemically Active Fluids

Fluids, especially water charged with dissolved gases, greatly accelerate chemical reactions because they have the ability to dissolve ions in regions of high pressure and temperature and to precipitate them in regions of lower pressure and temperature. Thus they are able to decompose old minerals and reconstitute new ones. These abilities increase enormously with depth, where the elevated pressures and temperatures convert the water to a superheated state, in which it is able to dissolve huge quantities of matter that are not ordinarily soluble at low temperatures and pressures. For example, quartz rarely dissolves on the Earth's surface, but it dissolves fairly easily in the presence of superheated hot water and dissolved gases. The superheated fluid also has great penetrative and propellant powers. Dissolved elements are commonly carried many kilometers from their source and disseminated throughout the rock mass.

The sources of the chemically active fluids that facilitate metamorphism are quite varied. The clays, the micas, and the amphiboles are among the many minerals that hold water chemically in the form of the hydroxide (OH^-) ion. These minerals release water if they are heated and subjected to high pressures. In addition, water is often held physically in the pore spaces and fractures of shallow rocks. The water in some sedimentary rocks may have originated as seawater trapped in the sediment at the time of deposition. In other instances, water is introduced from outside the rock body: from magma in the crust, from seawater that has penetrated fractured oceanic crust, or from groundwater that has filtered down from the Earth's surface. The dissolved gases and ions of metamorphic fluids are also derived from a

lithostatic stress
The uniform stress in the Earth's crust, caused by the weight of the overlying rocks.

compressive stress
The stress generated by forces directed toward one another on opposite sides of a real or imaginary plane.

foliation
The arrangement of a rock in parallel planes or layers; in metamorphic rocks, caused by parallel alignment of the minerals.

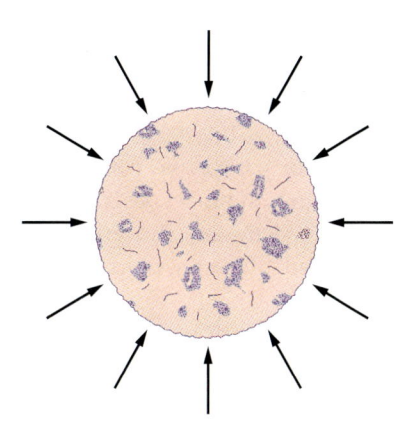

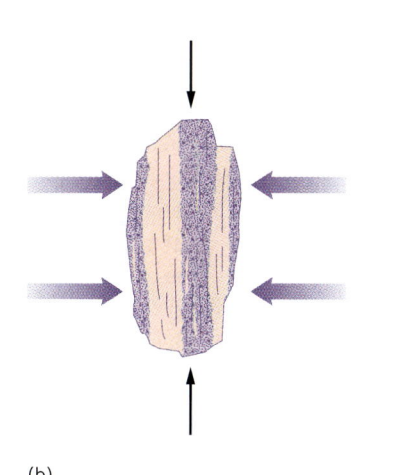

Figure 8.2 (a) Lithostatic stress results from pressure exerted equally in all directions. (b) Nonuniform compressive stress results in the parallel alignment of mineral grains, or foliation, a characteristic feature of many metamorphic rocks.

Figure 8.3 An outcrop of well-foliated and highly contorted metamorphic rock (gneiss), Painted Canyon, Mecca Hills, California. The pronounced layering reflects the segregation of minerals into light bands of quartz and feldspar and dark bands of biotite and hornblende. The structure, texture, and mineralogy of this rock are typical of regional metamorphism.

number of sources. Most are released from minerals subjected to high pressure and temperature. For example, carbon dioxide may come from the breakdown of calcite and shell matter.

Types of Metamorphism

Geologists classify the various types of metamorphism on the basis of the geologic processes responsible for them. **Regional metamorphism** is the most widely occurring type, and it is the type most commonly associated with the term *metamorphism*. It is induced in preexisting rocks during subduction and the plate collisions that produce fold-mountain ranges. In these mountain-building, or *orogenic*, environments, the rocks are invaded by the heat and fluids that escape from magmas rising from subducting plates, and they are subjected to enormous compressive stresses caused by the plates' squeezing together. Most regionally metamorphosed rocks exhibit a pronounced foliation developed in response to this stress (Figure 8.3). They occur as linear belts in the deeply eroded interiors of mountain ranges and are widespread in the core regions of the continents.

Contact metamorphism occurs where magma alters the cold country rock that surrounds an igneous intrusion or underlies a lava flow. The rock is baked, recrystallized, or otherwise changed through reactions driven by the infusion of heat and by the fluids mobilized within the magma and the rock itself.

Shear metamorphism typically occurs within active fault zones where one block of the crust moves with respect to another. High in the crust, where the lithostatic stress is low, rocks in the vicinity of the fault are subject to intensive grinding and are broken into rolled or rotated fragments, which are further pulverized to a fine powder. At deeper levels, the rocks are also subject to a great deal of plastic deformation and recrystallization. Faults and shear zones are generated along convergent, divergent, and transform boundaries. Therefore, shear metamorphism occurs in these settings.

Burial metamorphism is the change induced by the weight of overlying rocks. It occurs in the thick deposits of sedimentary basins at relatively low temperatures and is basically a continuation of the same processes that have converted the sediments into sedimentary rocks. These processes include the growth and hardening of the clay matrix, cementation, precipitation in pore spaces, and the dissolving and recrystallization of mineral grains at points of contact.

regional metamorphism
Metamorphism of an extensive area of the crust; generally associated with intensive compression and mountain building.

contact metamorphism
The transformation of rocks caused by heat escaping from an igneous intrusion or lava flow.

shear metamorphism
The transformation of rocks within the shear zone associated with active fault movement; mainly involves grinding and pulverizing high in the crust and recrystallization at deeper levels.

burial metamorphism
Metamorphism that results in response to the pressure exerted by the weight of the overlying rock.

Table 8.1 Mineral Content of Metamorphic Rocks

Category	Mineral
Minerals that crystallize from magma but also may form under metamorphic conditions	Quartz, feldspar, biotite, amphibole
Minerals most commonly, but not exclusively, formed under metamorphic conditions	Sillimanite, chlorite, garnet
Minerals that form only during metamorphism	Kyanite, andalusite, staurolite, talc

Shock metamorphism results from meteoroid impacts. Enormous pressures and temperatures are generated—and dissipated—in the short time interval during and immediately after the impact. Some rocks are fractured and pulverized; others are melted into droplets and fused into tiny glass beads; still others are recrystallized or converted to denser forms. Although meteoroid impacts are difficult to find on the Earth's surface, those that are well preserved get much attention because of their suspected links to mass extinctions of the dinosaurs and other species. Evidence of shock metamorphism plays a prominent role in the identification of meteoroid impact sites. Meteoroid impacts and the resultant shock metamorphic features are well preserved on the Moon, Mercury, and Mars, which lack the Earth's efficient erosion and plate tectonic recycling systems.

Hydrothermal metamorphism is triggered by water that percolates through fractures above shallow magma chambers. The magma directly or indirectly heats the water, which, in turn, alters the host rock. Hydrothermal reactions of this type are extremely important in altering the highly fractured and porous basalt lavas of mid-ocean ridge rift valleys.

shock metamorphism
Changes in rock and minerals caused by shock waves from high-velocity impacts, mainly from meteoroids.

hydrothermal metamorphism
The transformation of rock through the action of high-temperature solutions.

Metamorphic Rocks

Metamorphic rocks contain information concerning conditions *within* the crust, just as sedimentary rocks contain information concerning the *surface* of the crust. Their textures and internal structures are a consequence of the deep crustal forces that have molded them. Their mineral content is a consequence of the pressures and temperatures that prevailed within the crust at the time of metamorphism (Table 8.1). In this sense, metamorphic minerals are sensitive pressure gauges and geological thermometers.

The Minerals of Metamorphic Rocks

Because rocks remain solid during metamorphism, nearly all reactions occur between minerals that have not melted. They are called *solid-state* reactions, in which two minerals normally stable at high temperatures react when they come in contact. In some solid-state reactions, ions diffuse across grain boundaries, recombine in new arrangements, and form a third mineral that grows at the expense of the original two. In other reactions, complex minerals may break down by solid diffusion into simpler forms.

Metamorphism often dries out rocks. Diffusion reactions commonly convert *hydrous* (water-rich) minerals to simpler *anhydrous* (water-free) ones.

Diffusion reactions cause the growth of minerals through recrystallization, as in the conversion of limestone into marble. In this instance, the tiny calcite crystals of the limestone are transformed into the large, densely compacted crystals that characterize marble. Although technically the mineral composition of the rock remains unchanged, the original calcite structure of the limestone is obliterated, and the marble acquires a coarse texture. In the process, sedimentary features such as fossils are destroyed.

Textures of Metamorphic Rocks

The textures of metamorphic rocks—that is, the size, shape, and arrangement of their mineral grains—bear the stamp of the pressure-temperature environment in which recrystallization, mineral growth, and other reactions occur.

When rocks composed of granular minerals, such as quartz or calcite, are subjected to lithostatic stress, the minerals display no preferred direction of growth. As a consequence, the rocks develop a mosaic-tile pattern of interlocking equidimensional mineral grains—an *equigranular texture*. Rocks of this texture have a uniform, nonlayered appearance, as can be seen in many marbles and quartzites.

Most rocks subjected to the compressive stress of regional metamorphism display prominent parallelism of mineral grains. Minerals are sheared, rotated, and realigned at right angles to the direction of maximum compression. This action—plus growth of the platy minerals such as clay, mica, and chlorite—imparts a pronounced layering or foliation to the rock. Bladed minerals such as kyanite and sillimanite also line up with their long axes in the same direction, like swarms of parallel arrows. Marble and quartzite, composed of equidimensional calcite and quartz grains, respectively, tend to retain a superficially equidimensional texture even when subject to regional metamorphism. However, close inspection generally reveals subtle parallelism of structures even in these rocks.

Classification of Metamorphic Rocks

Metamorphic rocks, which are classified on the basis of texture and mineral content, fall easily into two categories: foliated and nonfoliated.

Foliated Rocks

Increasing metamorphic intensity caused by elevated temperatures and pressures often leads to an increase in grain size and a general coarsening of the texture of foliated rocks. Various terms describe these changes. *Slaty cleavage* refers to the foliation of exceedingly fine-grained rocks whose platy minerals are of microscopic size. *Schistosity,* on the other hand, refers to the foliation of coarser-grained rocks with highly visible minerals. High metamorphic intensity may also produce banded *gneissic* rocks, which display pronounced mineral segregation, or banding. Layers of dark platy or elongated minerals like biotite, mica, and hornblende alternate with light layers of feldspar and quartz. *Slate, gneiss,* and *schist,* then, are among the most common foliated rocks (Figures 8.4, 8.5, and 8.6). At higher levels of metamorphic intensity, the melting points of some quartz

Figure 8.4 A hand specimen of slate displaying closely spaced layering or slaty cleavage.

Figure 8.5 A hand specimen of biotite gneiss. Notice that the minerals are segregated into light and dark bands.

Metamorphic Rocks 181

(a) (b)

Figure 8.6 (a) A hand specimen of mica schist showing coarse texture and foliation. (b) The foliation is well displayed in this photomicrograph. The dark rounded minerals are garnet; the parallel bright-colored minerals, biotite; and the gray minerals, quartz.

and feldspar minerals are reached, and they form granitic fluids. The result upon cooling is *migmatite,* a complex mixture of banded gneiss and granite.

These rock names are generally made more specific by placing the names of key minerals in front of them: mica schist, garnet schist, sillimanite schist, hornblende gneiss, and so on. Some metamorphic rocks are referred to by color as a shorthand reference to their mineral content. *Greenschist,* generally the product of the metamorphism of basalt or gabbro, owes its color to the preponderance of chlorite and epidote (Figure 8.7). *Blueschist* derives its color from the metamorphic amphibole glaucophane. Hornblende schist is the higher-grade equivalent of greenschist.

Rocks subjected to shallow shear metamorphism are transformed to *fault breccia* (Figure 8.8). The rock consists of angular, fragmented mineral grains resulting from mechanical grinding and pulverization. Deeper in the fault zone, where temperatures are higher, rocks are subject to plastic deformation and are further ground down to a fine-grained, well-foliated *mylonite.* A compact rock with a streaky or banded structure, it is the product of extreme granulation and recrystallization (Figure 8.9).

Figure 8.7 Greenschist, which contains shiny green chlorite, green amphibole, quartz, and feldspar.

Figure 8.8 Fault breccia in the Grapevine Mountains, Death Valley National Monument, California. Notice the broken, sheared fragments—the products of mechanical grinding within the fine-grained matrix.

Figure 8.9 The transformation of granite into mylonite in the San Gabriel Mountains of California. Part (a) shows an undeformed granite at the periphery of a fault zone, while parts (b–d) show how the granite changes by degree into a fine-grained, well-foliated end-product.

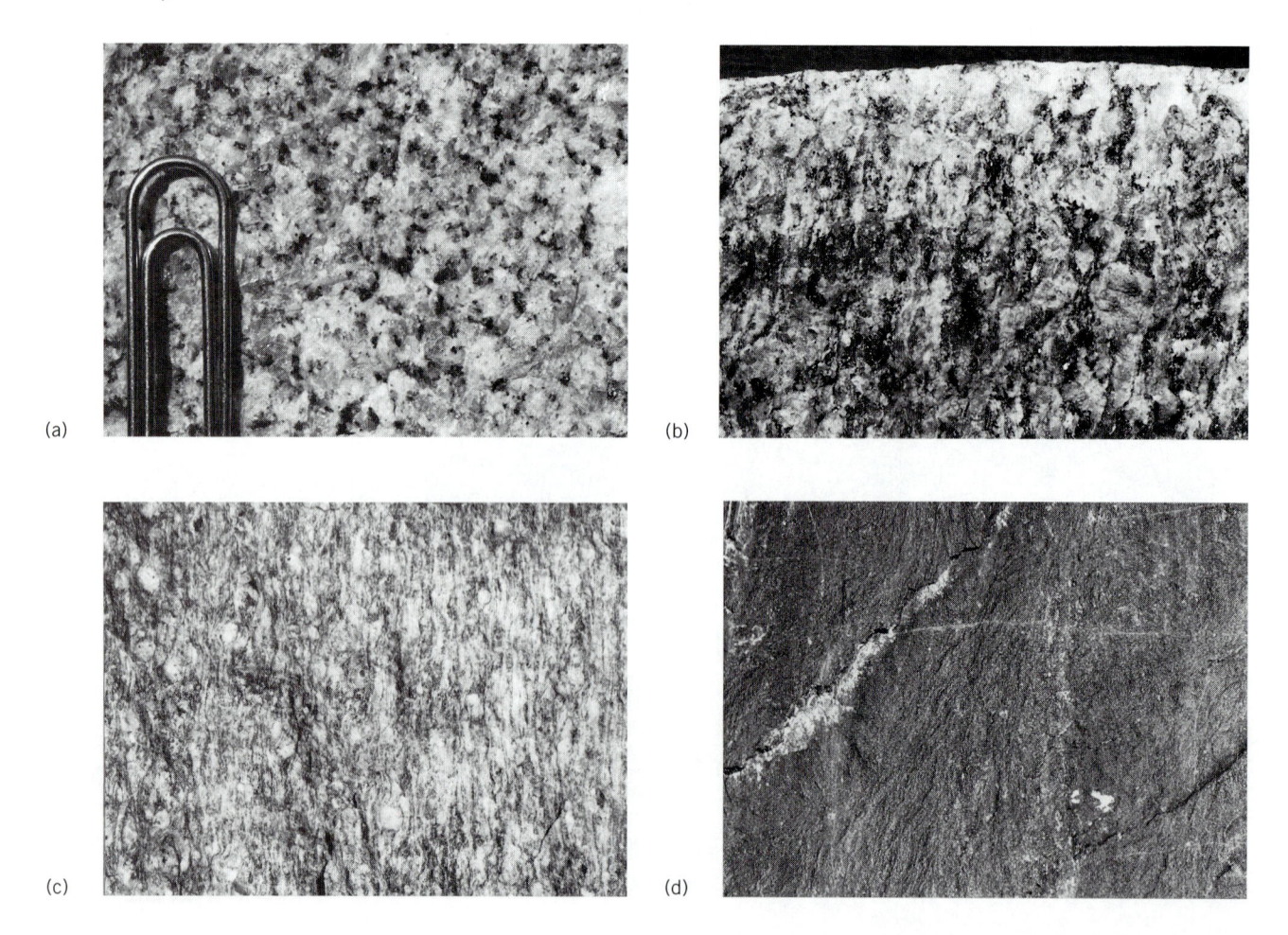

(a)

(b)

(c)

(d)

Figure 8.10 Hand specimen of a typical quartzite.

Nonfoliated Rocks

As described earlier, the nonfoliated rocks generally have an equigranular texture. They are distinguished mainly by their composition. *Quartzite* and *marble,* the most common types of nonfoliated rocks, are composed of interlocking quartz and calcite grains, respectively (Figures 8.10 and 8.11). They are the products of the metamorphism of sandstone and limestone. Greenstone, like the foliated amphibolite, is derived from the metamorphism of basalt or gabbro. It is composed of chlorite and epidote and is the low-grade, nonfoliated equivalent of greenschist.

Although equigranular textures generally imply mineral growth under lithostatic stress, heat is needed to trigger the reactions. Nearby magmas are the obvious heat source. Fieldwork bears out the fact that most equigranular textures are associated with rocks altered in contact zones that surround igneous intrusions. These rocks are given the special name *hornfels.*

The Field Occurrence of Metamorphic Rocks

Like living things, metamorphic rocks are best appreciated in their natural habitat. In the following descriptions of common metamorphic rocks as they appear in the field, the textural, structural, and regional properties of the rocks will be emphasized more than their mineralogy.

Figure 8.11 (a) A coarse-grained pink marble from Tate, Georgia. (b) This photomicrograph (×40) of marble, taken under polarized light, displays a mosaic of interlocking mineral grains (calcite). This texture is typical of nonfoliated rocks.

(a)   (b)

aureole
An halolike region of contact metamorphism surrounding an igneous intrusion.

Contact Metamorphic Rocks

Surrounding an igneous intrusion like a halo is a zone of transformed country rock called an **aureole** (Figure 8.12). In this display of contact metamorphism, the transformation takes the form of textural and mineralogical changes that grow more pronounced toward the igneous rock, so there is little doubt that heat and, in some instances, fluids escaping the magma were the agents responsible for the changes.

The extent and intensity of contact metamorphism depend upon a number of factors: the quantity, temperature, and composition of the magma; and the composition, rigidity, and permeability of the country rock. In general, the effects are most conspicuous in shaley limestone. The effects are not as great in sandstones, which are composed of heat-resistant quartz. Igneous rocks that have crystallized at high temperatures are least affected.

The best place to view the effects of contact metamorphism is in the field. There we can walk out and see the changes in the country rock firsthand, and take samples back to the laboratory to examine under the microscope. Figure 8.12 shows a typical contact aureole and the changes that can be traced from unaltered shale to a granitic intrusion.

In some instances of contact metamorphism, valuable residues, such as magnetite iron ore, are left behind. Most contact-metamorphic deposits are formed by the introduction of fluids as well as heat from the magma. Metal ions in the fluids react with and replace the ions present in the minerals of the host rock. The reaction is most effective when magmas intrude impure limestone, causing a recrystallization of calcite into a coarse-grained marble and the formation of many other minerals (Figure 8.13). Frequently, these minerals are of gem quality; the finest rubies and sapphires, found in the Magok region of Burma, are mined from aureoles and from nearby gravels weathered from the aureoles. Rocks subjected to the process also yield iron, zinc, copper, and tungsten.

Regional Metamorphic Rocks

Fold-mountain belts are commonly thousands of kilometers long and hundreds of kilometers wide. They are composed mainly of sedimentary rocks that were originally deposited as horizontal strata on continental margins, which in places have attained thicknesses greater than 10 kilometers. The mountains are formed where lithospheric plates collide. Strata caught between the plates are subjected to compressive stress and are deformed into a complex series of folds and faults. As streams strip the upper strata away, lessening the load on the mantle, the deep central cores of the mountains rise, and the streams cut into these as well. Eventually, the metamorphic core of the mountain belt is exposed. Figure 8.14 traces the changes in regionally metamorphosed rocks, concentrating on the transformation of shale, the most abundant sedimentary rock exposed on the Earth's surface.

Proceeding from the periphery of the mountain belt toward the core, we can often trace the transformation of sedimentary rocks into metamorphic rocks of increasing complexity. Here we are in the realm of the most widespread of metamorphic rocks, those formed as a consequence of deep burial, elevated temperature, and compressional stress.

Starting at the fringe of the mountain belt (point A in Figure 8.14), we note first that the folded rocks of the mountain belt have been squeezed in the direction of maximum compression and stretched in the direction of least compression (Figure 8.15). We note, too, that distinct layering has developed at right angles to the direction of maximum compression. These layers, called slaty cleavage (Figure 8.16), are especially well displayed in the shales. Cleavage consists of a series of closely spaced parallel planes, along which microscopic grains of mica and chlorite (derived from clay) have grown in response to stress. As the slaty cleavage grows more prominent in the shale, bedding

Metamorphic Rocks 185

Figure 8.12 A granitic intrusion into shaley country rock forms a contact aureole of metamorphosed rock. Walking outward from point Z to point Y to point X, we can observe that the intensity of metamorphism, as manifested by textural and mineralogical changes in the country rock, decreases with distance from the granite. (From: *Igneous and Metamorphic Petrology* by Best © 1982 by W. H. Freeman and Company. Used with permission.)

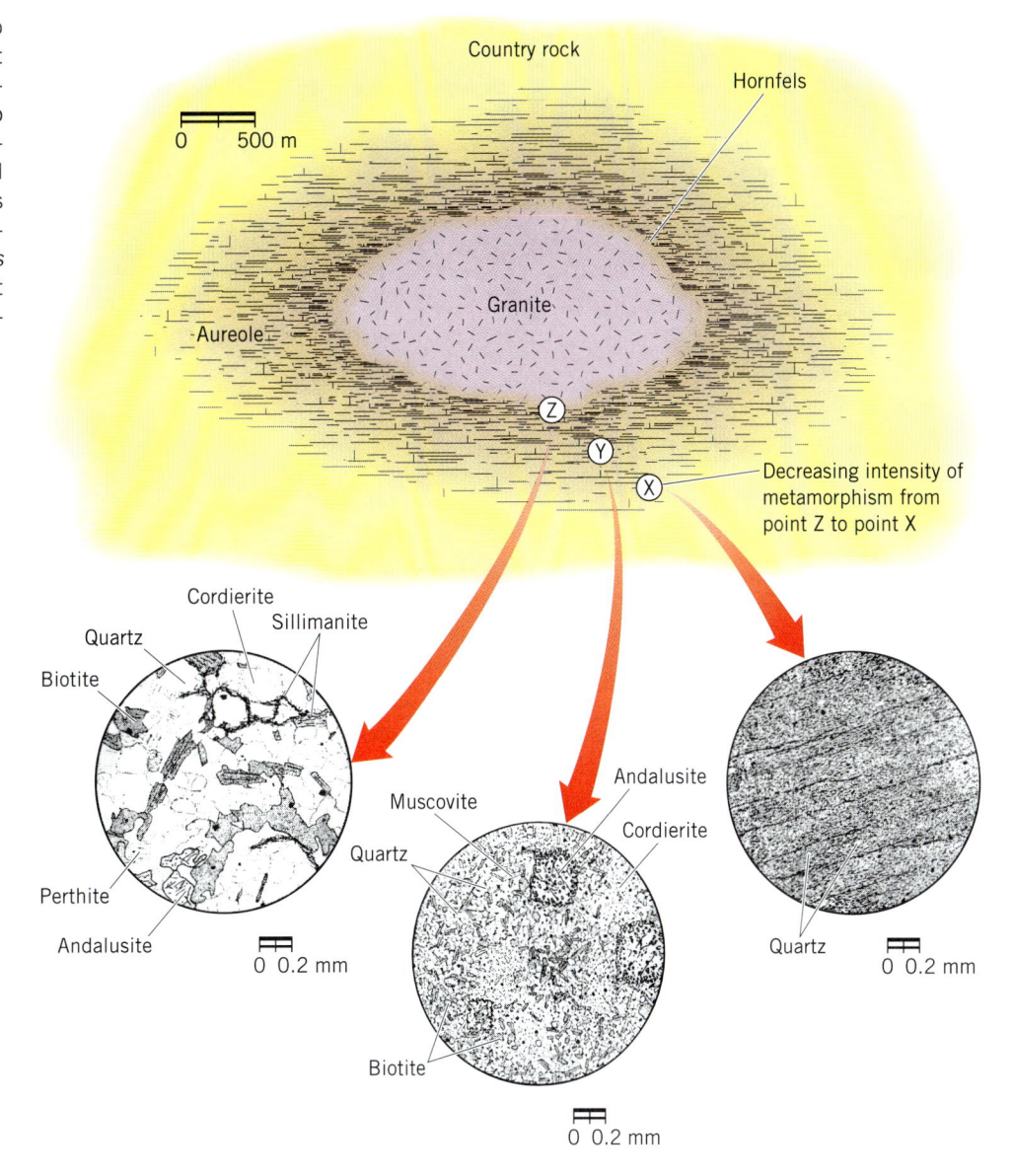

Figure 8.13 (a) An ore may form adjacent to an igneous intrusion in host rocks that are subjected to contact metamorphism. (b) An ore may also occur in disseminated form as fluids rise, escape from a magma, and react with host rocks. The ore is then deposited in minute fractures and faults. In both processes, some rocks, such as limestone, are more reactive than others.

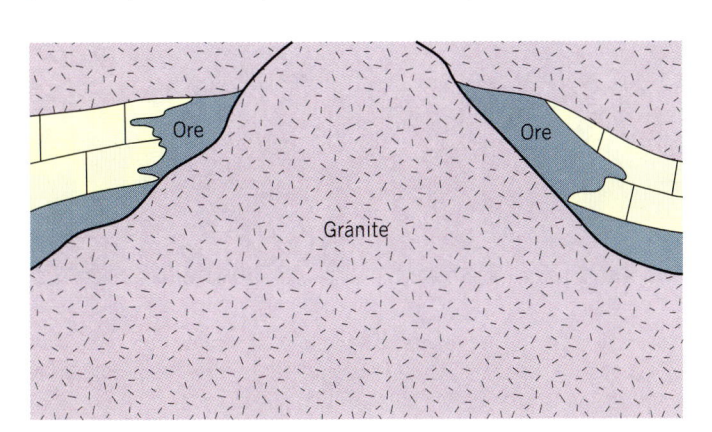

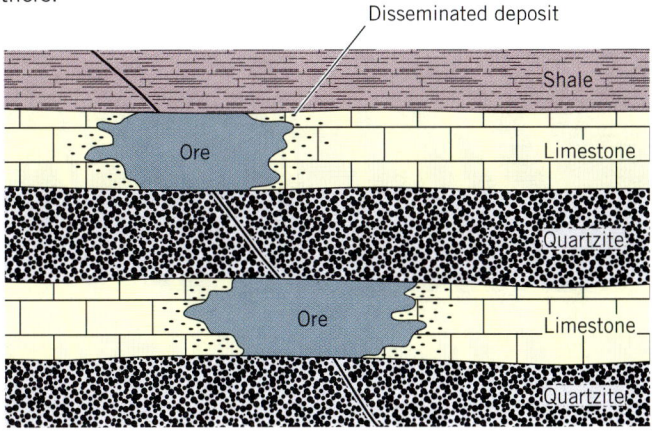

(a)

(b)

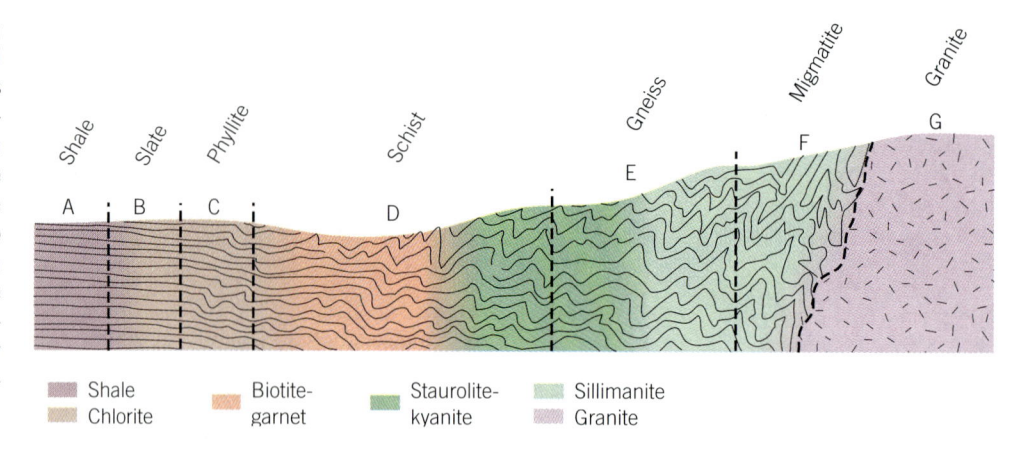

Figure 8.14 A schematic representation of the textural and mineralogical changes that occur in shale when it is subjected to increasing regional metamorphism. These changes can be seen when we trace the rock toward the eroded core of a mountain belt. The shale is transformed by degrees into slate, phyllite, schist, and gneiss. With increasing heat and pressure, the gneiss partially melts, forming a mixture of granite and gneiss, or migmatite. Finally, the rock melts completely, forming granite.

- Shale
- Chlorite
- Biotite-garnet
- Staurolite-kyanite
- Sillimanite
- Granite

planes and other features of the shale grow faint. We have now traced the original sedimentary shale into slate, a metamorphic rock (point B in Figure 8.14).

Tracing the slate deeper into the core of the mountain range, we note that the features of the original sedimentary rock are mostly obliterated. We have passed from slate to phyllite (Figure 8.17). Phyllite, in turn, grades into coarser-textured schist (point C to point D in Figure 8.14). The coarse texture is due to the growth and parallel alignment of constituent minerals, such as mica and chlorite. Quartz also is prominent.

Most schists are intricately folded (Figure 8.18). The intricate fold pattern, transferred to virtually every segment of the rock, is clear evidence that in this region of great metamorphic intensity, the rocks have deformed plastically in response to compressive stress and elevated temperatures.

Farther into the mountain core, we find that the schist grades into gneiss, which is similar to granite in mineral content: mainly quartz, feldspar, and a dark mineral, either biotite, pyroxene, or hornblende (point E in Figure 8.14). Dark and light minerals are segregated into a new type of foliation having distinct alternating bands.

Here in the center of the vise, the core of the mountain belt (point F in Figure 8.14), the gneiss gives way to a transitional zone of intricately mixed granitic and metamorphic rock, a migmatite (Figure 8.19). The migmatite, in turn, grades into a granite (point G in Figure 8.14). Evidently, temperatures inside the Earth reached the melting point of the

Figure 8.15 These elongated pebbles, found at the periphery of a fold-mountain belt, illustrate the role of compression in altering the original character of sedimentary rock.

Figure 8.16 Slaty cleavage in a shale formation. Notice how the cleavage planes intersect the bedding planes, which have been folded by compression. Growth of minerals along the cleavage planes is responsible for much of the foliation displayed by regionally metamorphosed rocks.

Figure 8.17 (a) A phyllite hand specimen showing pronounced foliation and silky sheen caused by the growth of mica. (b) This enlarged photomicrograph shows that the foliation of phyllite is caused by the alignment of minute mica and quartz grains.

metamorphic rock and transformed it to magma, which cooled to form granite. In other words, our journey from the fringes of the mountain belt, where we encountered sedimentary rock, to the core of the mountain belt, where we encountered the evidence that metamorphic rock melted to form granite, has been a journey around the rock cycle.

As we trace shale from the outer fringes of the metamorphic terrain toward the granite, we observe, in addition to textural and structural changes, systematic changes in the mineral content of the rocks. Certain minerals, such as chlorite, fade out and are superseded by others, such as biotite and garnet. These **index minerals** are formed by reactions with preexisting minerals at specific pressures and temperatures. They are therefore especially sensitive measures of *metamorphic gradze,* the intensity of metamorphism.

If we map the first appearance of an index mineral in a region, we get an *isograd,* a line joining points of equal metamorphic grade. In the areas between successive isograds, the **metamorphic zones,** the mineral content of the rock remains constant. Proceeding from low-grade slate through the high-grade schist and gneiss toward the granite, we encounter the chlorite, biotite, garnet, kyanite, and sillimanite zones.

index mineral
A mineral that characterizes a given intensity of metamorphism, having developed under a specific range of temperature and pressure conditions.

metamorphic zone
An area of equal metamorphic intensity between isograds in which the mineral content of rocks remains constant.

Figure 8.18 Intricate folds displayed in schist. These same folds are often reproduced on a regional scale.

Figure 8.19 A typical migmatite outcrop displaying a mixture of banded gneiss and granite.

Metamorphism and Plate Tectonics

Specific metamorphic rocks are closely associated with specific plate boundaries, in much the same way that specific kinds of igneous rocks and volcanoes are closely associated with these very same boundaries. Each type of plate boundary generates distinctive pressure-temperature-fluid conditions. Figure 8.20 is a plate tectonics model of the crust and upper mantle, with the principal sites of metamorphism labeled. Hydrothermal metamorphism occurs at divergent boundaries (mid-ocean ridges). Regional metamorphism occurs at subduction zones and in the mountain belts commonly associated with them. Within the subduction zone, rocks attached to the cold descending plate are subjected to relatively low-temperature but high-pressure metamorphism. In the bordering mountain ranges, the rocks of the continent or island arc are subjected to heating by rising magmas derived from the action of the subducting plate. Thus they are subjected to high-temperature but low-pressure metamorphism. Shear metamorphism occurs in the fault zones of convergent, transform, and divergent boundaries.

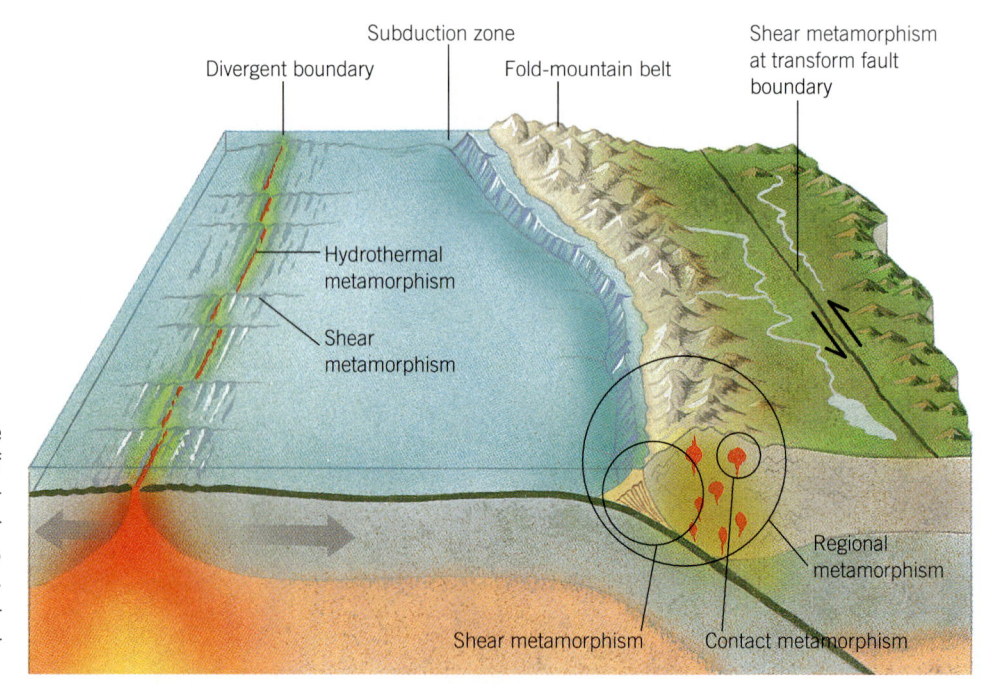

Figure 8.20 The relationship to plate boundaries of the various types of metamorphism discussed in this chapter. Some types, such as shear metamorphism, can occur at convergent, divergent, or transform boundaries. Others, such as regional metamorphism, are confined to a single boundary type (convergent).

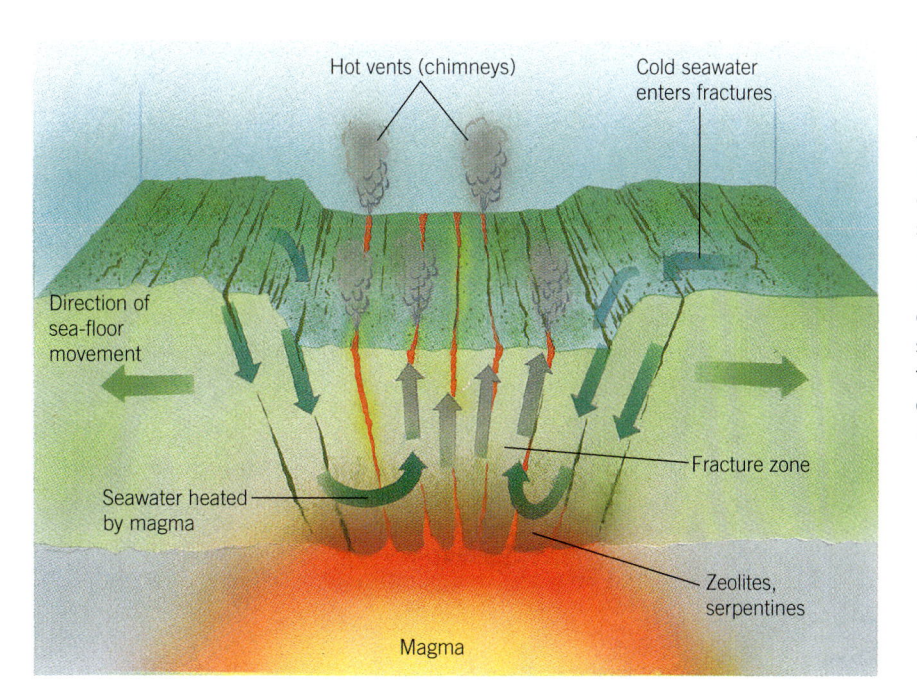

Figure 8.21 The environment of sub-sea-floor (hydrothermal) metamorphism. Fractures that develop within the central rift valley of the mid-ocean ridge act as passageways for seawater circulating within the crust. The cold seawater is heated by hot magma beneath the ridge crest. In the exchange between rock and water, the basaltic crust is converted to hydrous rocks, such as serpentinite, and metals extracted from the basalt by the water are concentrated high in the crust and near hot vents.

Metamorphism of the Sea Floor

Zones of intensive fractures develop as the oceanic crust slowly separates and spreads away from the mid-ocean ridge crests (Figure 8.21). The fractures penetrate deep into the crust and provide avenues for the intermittent rise of the magma lurking beneath the crests. The fractures are also an effective plumbing system. Cold seawater seeps deep into the crust along one set of fractures, where it is heated by the magma. It is then returned to the surface through another set of fractures that serve as hot-water vents.

The combination of plentiful seawater from above and heat supplied from below acts as a potent agent for the hydrothermal metamorphism of oceanic crust, during which water-free minerals are transformed into hydrous, or water-bearing, forms. Dredge and core samples recovered from drilling reveal chlorite-rich greenstone, converted from basalt, and serpentinite, the soft, dark green rock composed of fibrous serpentine minerals derived from the conversion of peridotite. The alteration is most thorough along fracture linings where water penetrates. Microscopic analysis reveals that serpentine is derived from the alteration of olivine. Though sub-sea-floor metamorphism occurs mainly along the mid-ocean ridges, sea-floor spreading carries the rocks formed by this process far across the ocean floor and eventually down into the mantle at subduction zones.

The complex reactions between seawater and oceanic crust lead to the formation of many other minerals, including valuable ore-grade metal deposits and *zeolites*. The latter are aluminum silicates whose crystal structures contain large quantities of water easily expelled when heated. They have extensive industrial use as water softeners, gas adsorbents, and drying agents.

Subduction Zone Metamorphism

The subduction zones beneath the deep-sea trenches are regions of relatively high pressure but low temperature. Temperatures are depressed by the cold slabs of oceanic crust that descend into the mantle. However, intense pressure and shearing stress are generated as the subducting plate grinds against the bordering plate. Scraped off the descending plate as it passes beneath the trench are deep-ocean sediments, turbidite deposits, and slices of oceanic crust (called *ophiolite*). These materials are plastered onto the bordering plate as a chaotic, wedge-shaped mixture called a *mélange*. The rocks of the mélange are generally of the high-pressure–low-temperature greenschist and blueschist facies. But deep within the subduction zone beneath the trenches, the descending plate is subjected to the very-high-pressure conditions of the upper mantle. The basalt slab is converted to *eclogite,* a colorful rock composed of red garnet and green pyroxene but of the same chemical composition as basalt.

The descending plates also play a role in generating magmas that rise as diapirs and intrude the folded, mountainous sedimentary rocks of the continents and island arcs. The diapirs are, in effect, concentrated sources of heat delivered to the continents and arcs. This heat powers the high-temperature reactions of regional metamorphism that produce migmatites, sillimanite gneisses, and schists. These rocks pass into lower-grade phyllites and slates at the fringes of the mountain belt.

Study Outline

Metamorphism is the process by which the mineralogy, texture, and structure of solid rocks are changed by the pressures, temperatures, and chemical conditions that prevail deep within the Earth.

I. THE AGENTS OF METAMORPHISM
  A. Heat softens rocks and breaks ionic bonds.

  B. **Lithostatic stress** reduces the rocks' volume; nonuniform **compressive stress** causes parallel layering, or **foliation.**

  C. Chemically active fluids dissolve and disseminate elements throughout the rock mass.

II. TYPES OF METAMORPHISM
  A. **Regional metamorphism** is induced in preexisting rocks during subduction and the plate collisions that produce fold-mountain ranges.

  B. **Contact metamorphism** is the result of the baking and recrystallization of country rock by igneous intrusions. Some instances of contact metamorphism leave behind valuable ores.

  C. **Shear metamorphism** grinds and deforms rocks at transform boundaries and other faults.

  D. **Burial metamorphism** results from the pressure of overlying rock on buried sedimentary rock layers.

  E. **Shock metamorphism** results from meteoroid impact.

  F. **Hydrothermal metamorphism** alters rocks through the introduction of hot fluids.

III. METAMORPHIC ROCKS
  A. Because rocks remain solid during metamorphism, nearly all reactions are *solid-state,* in which ions recombine to form new minerals. Reactions often involve the loss of water.

  B. Rocks with platy or elongated minerals display foliation in response to nonuniform compressive stress. Rocks with equigranular minerals display non-foliated texture.

  C. Foliated rocks are classified on the basis of grain size and presence of banding. They range from the fine-grained *slate* to the coarser *schist* and the banded *gneiss. Migmatite* is a mixture of granite and gneiss derived from partial melting of the gneiss.

1. The products of shear metamorphism are also commonly foliated. They include coarse, angular *fault breccia* and fine-grained mylonite.
2. Nonfoliated rocks are classified by composition. *Quartzite* is the product of the metamorphism of sandstone; *marble*, of limestone.
3. Rocks altered by igneous intrusions are called *hornfels*.

D. The field occurrence of metamorphic rocks follows a specific pattern.
   1. In contact metamorphism, an **aureole** of altered country rock surrounds an igneous intrusion. Textural and mineralogical changes grow more pronounced closer to the center, illustrating that the heat and fluids of the magma caused the changes.
   2. Regional metamorphic rocks are formed by deep burial, elevated temperature, and compressive stresses at convergent plate boundaries. The transformation involves a coarsening of texture, the growth of new minerals, and the development of foliation.

E. **Metamorphic zones** are areas where the mineral content of the rocks is constant. **Index minerals,** formed by reactions at specific pressures and temperatures, are measures of metamorphic intensity.

IV. METAMORPHISM AND PLATE TECTONICS
A. Conditions at plate boundaries commonly control the kinds of metamorphism that occur in the crust.
B. Hydrothermal reactions occur at mid-ocean ridges, regional metamorphism occurs at subduction zones, and shear metamorphism occurs in the fault zones of all plate boundaries.

STUDY TERMS

aureole (p. 184)
burial metamorphism (p. 178)
compressive stress (p. 177)
contact metamorphism (p. 178)
foliation (p. 177)
hydrothermal metamorphism (p. 179)

index mineral (p. 187)
lithostatic stress (p. 177)
metamorphic zone (p. 187)
regional metamorphism (p. 178)
shear metamorphism (p. 178)
shock metamorphism (p. 179)

THOUGHT QUESTIONS

1. If increasing pressure and temperature can transform a shale into a gneiss, do you think it is possible to convert a gneiss into a shale by decreasing temperature and pressure? Why or why not?
2. Suppose you were to trace shale deposits interbedded with basaltic lava into a region of steadily increasing metamorphic intensity. Furthermore, suppose you found that the shale had been transformed by regional metamorphism into sillimanite schist. Describe the mineral changes that occurred in the basalt.
3. How does the introduction of fluids and heat help to form ore deposits during contact metamorphism?
4. Describe the plate boundary interactions that trigger different types of metamorphism.

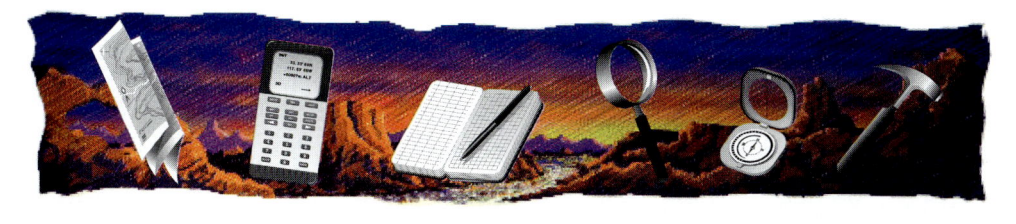

Chapter 9

Stress and Rock Deformation
　The Response of Rocks to Stress

The Map Depiction of Planar Features

Folds
　Plunging Folds
　Monoclines, Domes, and Basins

Fractures
　Joints
　Faults

Strike-Slip Faults
Normal Faults
Thrust Faults

Impact Craters
　The Structure of an Impact Crater
　Frequency of Impact

Rock Deformation

The evidence of Earth's past dynamism is preserved in the deformational features of rocks, in the contorted strata of mountain belts such as the Alps and Appalachians, in broad regional uplifts such as the Colorado and Tibetan plateaus, and in great faults such as the San Andreas of California. Most of these features are formed by the interaction of the Earth's lithospheric plates—namely, collision and subduction at convergent margins, rifting at divergent margins, and slide-by motions at transform margins.

This chapter describes the major deformational features of rocks: folds, faults, and joints, all of which are important to structural geologists because they are key elements in interpreting the architectural history of the crust.

Stress and Rock Deformation

Consider the three blocks of rock in Figure 9.1. To analyze the effect of the forces on these rocks, let us imagine an internal plane that divides the rocks as shown in the figure. Each force exerts **stress,** which is defined as the magnitude of the force applied divided by the area of the plane. The forces in Figure 9.1(a) exert **compressive stress,** because they tend to squeeze the rock. Typically, compression shortens, thickens, or buckles rocks. The forces in Figure 9.1(b) exert **tensile stress,** because they tend to pull the rock apart. Tension elongates and thins the rock. The forces in Figure 9.1(c) exert **shear stress,** because in acting parallel to the internal plane, they distort the shape of the rock. When subjected to shear stress, a cube is distorted to a rhomboid shape, and a sphere is made elliptical.

stress
The force applied to a plane divided by the area of the plane.

compressive stress
The stress generated by forces directed toward one another on opposite sides of a real or imaginary plane.

tensile stress
The stress generated by forces directed away from one another on opposite sides of a real or imaginary plane.

shear stress
Stress (force per unit area) that acts parallel to a (fault) plane and tends to cause the rocks on either side of the plane to slide by one another.

◀ Folded Paleozoic strata of the Northern Rockies near Borah Park, Idaho. Similar structures are typical of fold-mountain belts throughout the world.

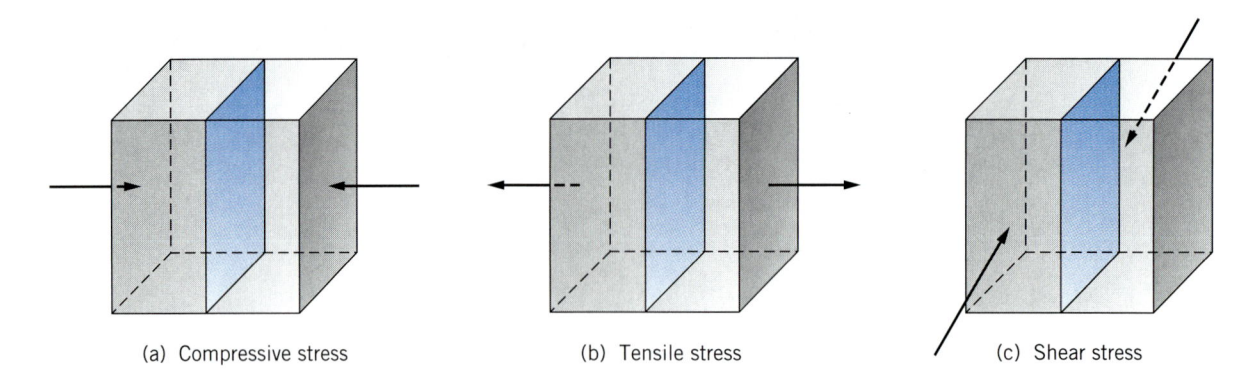

(a) Compressive stress (b) Tensile stress (c) Shear stress

Figure 9.1 The stress on an object is defined as the applied force divided by the area over which the force acts. The three basic types of stress are compressive, tensile, and shear.

The blocks of rock in Figure 9.1 are subjected to unequal stress. The block in Figure 9.2, on the other hand, is subjected to equal stress in all directions. As mentioned in Chapter 8, such uniformly distributed lithostatic stress (or pressure) occurs in the Earth's mantle under many kilometers of rock.

The change in volume or shape of an object that results from stress is called **strain;** the particular strain caused by unevenly applied stress is expressed by a change in the object's shape. An increase in lithostatic stress, on the other hand, leads to volume reduction.

The Response of Rocks to Stress

Experiments have shown that the response of rocks subjected to stress takes three forms: elastic, ductile (or plastic), and brittle. The response in a given instance depends upon the type of stress applied, the temperature and pressure conditions, and the mechanical properties of the rock.

If stress is light, a rock will deform in direct proportion to the stress; then when stress is removed, the rock will return to its original shape. This is an *elastic response.* However, above a certain level of stress, called the **elastic limit,** the rock may deform permanently in a *ductile response* (Figure 9.3). Strain is no longer in proportion to stress, and the rock will not return to its original shape once the stress is removed. On the other hand, stress continued beyond the elastic limit may cause a *brittle response,* in which minute cracks develop and the rock ultimately ruptures.

Experiments in rock deformation aid our understanding of the behavior of rocks subject to stresses set in motion by geological processes. The propagation of seismic waves through rocks is an elastic response. The waves are vibrations—that is, small distortions or strains triggered by stresses well below the elastic limit of the rock. The rock deforms in response to the stress but returns to its original shape once the stress is removed. Slamming the palm of your hand on a table will trigger a similar response in the table.

Ductile response is expressed by **folds,** which are permanent wavelike distortions in rocks. The stresses that created the folds may have ended millions or billions of years ago, yet the structures remain, frozen in time. In Chapter 8, we saw that in metamorphism, the shapes of pebbles, fossils, and other features are often distorted in folded rocks. Still further evidence of stress is the thickening and thinning of strata within the wavelike fold itself.

The brittle response to stress is expressed by fractures. **Faults** are fractures in the crust along which the rocks on either side have been offset with respect to one another. Where such movement is absent, the fracture is called a **joint.**

strain
The result of stress applied to a body, causing the deformation of its shape and/or a change of volume.

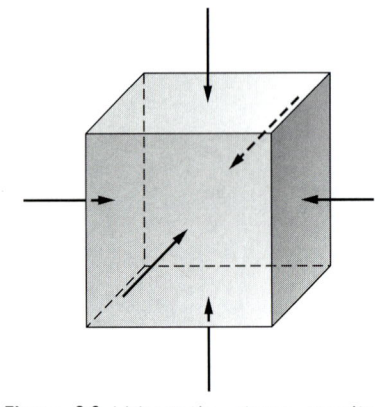

Figure 9.2 Lithostatic stress results when compressive stress is applied equally in all directions.

elastic limit
The maximum amount of stress a material can withstand before it deforms permanently.

fold
Permanent wavelike deformation in layered rock or sediment.

fault
A fracture in bedrock along which rocks on one side have moved relative to the other side.

joint
A fracture in a rock, without noticeable movement along the plane of fracture.

Rock Deformation 195

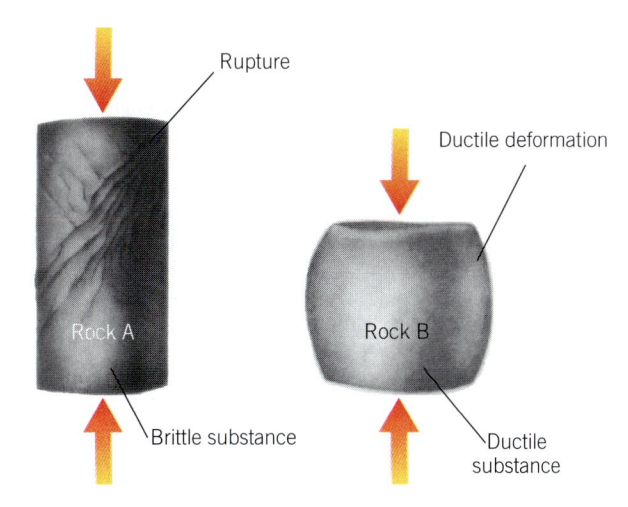

Figure 9.3 Two common responses to compressive stress. In a brittle response, rock A ruptures just beyond its elastic limit. In a ductile (or plastic) response, rock B is permanently deformed beyond the elastic limit. (Adapted from Marland P. Billings, *Structural Geology,* 2d ed., Englewood Cliffs, N.J.: Prentice-Hall, 1954.)

The Map Depiction of Planar Features

Suppose that while doing fieldwork in an unexplored region, we encounter the eroded, upturned strata of Figure 9.4. Because the layers were originally deposited parallel to the Earth's surface and are now inclined with respect to that surface, it is obvious that they have been disturbed in some way. Perhaps they are part of a fold or were tilted by a fault. One way to find that out is to measure the attitude, or orientation, of the layers with respect to the Earth's surface.

Two features, called the **strike** and the **dip** of the strata, are useful in this regard. Refer again to Figure 9.4; we see that the intersection of an imaginary horizontal plane with the bedding plane of one of the layers forms a straight line. The compass direction of that line is the strike of the layer or bed. It gives the trend of the layer on the Earth's surface. We also see that the bedding-plane surface forms an acute angle with the horizontal, if measured in a vertical plane perpendicular to the strike. This angle is called the dip of the bed. Notice that the direction of the dip is always at right angles to the strike. For example, the strike of the beds in Figure 9.4 is northeast, but the dip direction is 90° away, to the southeast. For this reason, strike and dip are depicted on maps, which enables us to immediately grasp the orientation of the layer with respect to the Earth's surface.

strike
The direction of the line formed by the intersection of a horizontal plane with a bedding or fault plane.

dip
The angle formed by the intersection of a bedding or fault plane and the horizontal plane; measured in a vertical plane perpendicular to the strike.

Note about strike and dip: *The symbol used by geologists to represent the strike and dip of a plane looks like this:*

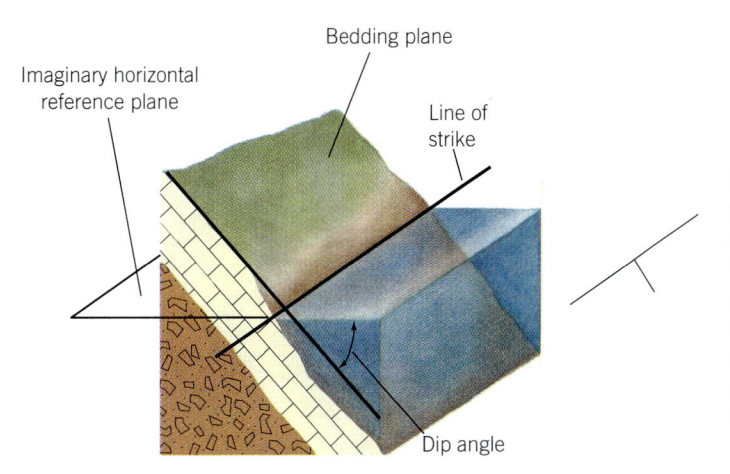

Figure 9.4 The strike of a rock layer is the direction of the line formed by the intersection of the horizontal plane and the bedding plane. The dip of the layer is measured in a vertical plane perpendicular to the strike. A strike-and-dip symbol (seen here to the right of the cross section) is used to represent the relationship between the two measurements.

CHAPTER 9

Folds

Folds are permanent wavelike bends or distortions in the planar features of rocks, such as in sedimentary strata and the foliation of metamorphic rocks. In three dimensions, they resemble the furrows in a rug and are caused by much the same thing: the buckling of originally horizontal layers due to compression.

Fold anatomy is best viewed in a block diagram. The surface of the block may be drawn either topographically or as a flat geologic map, and the sides drawn as cross sections. The block diagram of Figure 9.5 depicts a series of connected folds as they are commonly encountered in the field—with the top layers peeled away by erosion to expose the underlying strata.

The first thing to notice is that a fold may be divided more or less evenly by an imaginary surface called an *axial plane*. The two parts of the fold so divided, consisting of strata or foliated metamorphic rock, are called the *limbs* of the fold. Where the limbs intersect the axial plane, they form a line that can be considered the *fold axis*.

The block diagram of Figure 9.5 illustrates two types of folds, an **anticline** and a **syncline**. The limbs of the anticline dip away from the axial plane, giving the fold a convex-up orientation. Because of this orientation, erosion has exposed the oldest strata in the center of the anticline and progressively younger strata on the limbs. Relationships are reversed in the syncline. Limbs dip toward the axial plane, giving the fold a concave-up orientation, so that the youngest strata are exposed in the center of the fold. Since anticlines and synclines commonly alternate in the manner shown in Figure 9.5, the exposed strata of a series of folds alternate in a repetitive sequence.

Anticlines and synclines may occur in all varieties of attitudes with respect to the Earth's surface. Symmetrical folds have vertical axial planes and limbs that dip oppositely but at equal angles (see the photo at the beginning of the chapter). Asymmetrical folds have tilted or inclined axial planes and limbs that dip at unequal angles (Figure 9.6). In **overturned folds,** the limbs dip in the same direction rather than oppositely, as is usually the case. For this arrangement to occur, one limb must be rotated so that the older strata lie above the younger strata, in a reversal of the normal sequence. In regions of intense folding, the axial planes are themselves rotated until they are horizontal and the folds are pushed over on their sides, thus producing **recumbent folds.** Huge recumbent folds many kilometers in amplitude are key structural features of the Alps and other mountain ranges (Figure 9.7).

Plunging Folds

The folds we have discussed so far have axes that lie parallel to the Earth's surface—that is, horizontally. As a result, the outcrop pattern of eroded strata runs parallel to the strike of the fold axis. By contrast, a **plunging fold** is one whose axis is inclined to the Earth's

---

anticline
A fold with a core of stratigraphically older rocks; usually its convex side is upward.

syncline
A fold with a core of stratigraphically younger rocks; usually its concave side is upward.

overturned fold
A fold in which one limb has rotated past the perpendicular, so that both limbs dip in the same direction.

recumbent fold
An overturned fold whose axial plane is horizontal.

plunging fold
A fold whose axis is inclined rather than horizontal.

Figure 9.5 Anticlines and synclines in eroded strata. In an anticline, the limbs dip away from the axial plane. Progressively older strata are exposed toward the axis of the fold. These relationships are reversed in the syncline; the limbs dip toward the axial plane and progressively younger strata are exposed toward the fold axis.

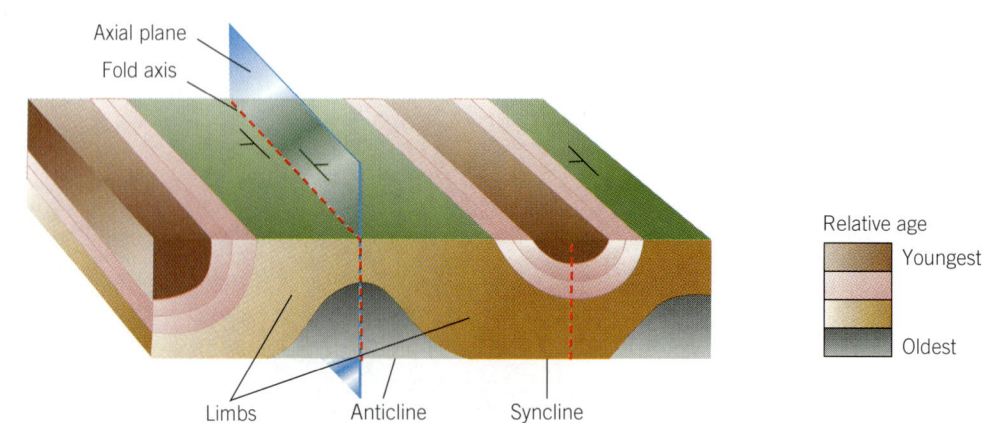

Rock Deformation 197

Figure 9.6 Asymmetrical folds near Palmdale, California, in the vicinity of the San Andreas fault. The axial planes are inclined, dipping to the right. Notice also that the strata on the opposite sides of the axial planes dip at different angles.

surface. As a consequence, the outcrop pattern of the eroded strata converges toward what is called the "nose" of the fold (Figure 9.8). Because anticlines and synclines are folded in the opposite sense, the noses of the folds are oppositely directed. Anticlines close in the direction of plunge, and synclines close in the opposite direction. Because plunging anticlines and synclines interconnect in elaborate fold systems, the exposed strata follow a wavy pattern.

The plunging fold system, combined with the effects of erosion, dictates the topography of many mountain belts. Resistant strata stand out as ridges; valleys are carved out of weaker layers. Together, they form a zigzag ridge-and-valley topography that is characteristic of mountain belts. The Zagros Mountains of Iran (see Figure 9.8) are a young mountain belt. As might be expected, the highest elevations belong to the anticlines, which are convex-up folds. However, deep erosion may produce a topographic

Figure 9.7 A recumbent fold is an extreme type of overturned fold in which the axial plane is horizontal. Notice how the stratigraphic sequence is reversed in recumbent folds: Older strata overlie younger strata.

Figure 9.8 (a) A plunging anticline in the Zagros Mountains of Iran; the fold limbs converge in the direction of plunge. (b) The axis of the fold is inclined with respect to the Earth's surface.

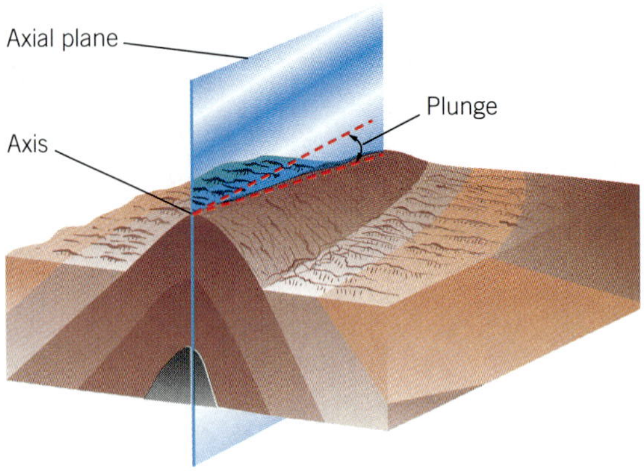

inversion, an interesting condition where the surface relief is out of phase with the underlying structure. Remember that anticlinal crests may be heavily fractured during folding, because the crests of convex-up structures are subject to tension. Thus they are easily eroded. Figure 9.9 shows a later stage of erosion in the Appalachians, a much older mountain belt than the Zagros. Anticlines are now the sites of deep valleys, while synclines form the high ground, standing above the anticlines as broad, flat-topped ridges.

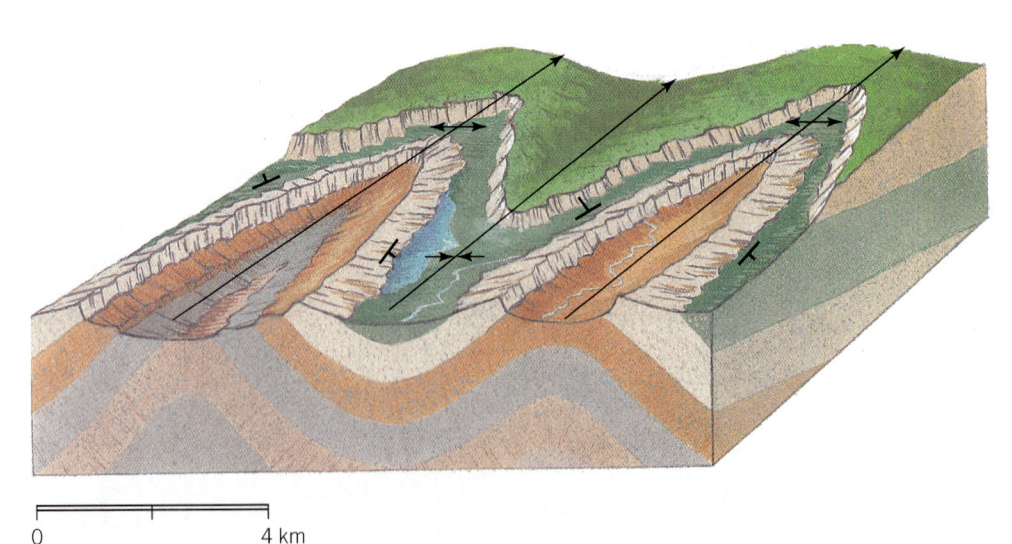

Figure 9.9 The relationship between plunging anticlines and synclines. The "nose" of the anticline points in the direction of plunge, whereas the nose of the syncline points in the opposite direction. Also, it is common for deep valleys to erode in the cores of anticlines, leaving the synclines standing above them.

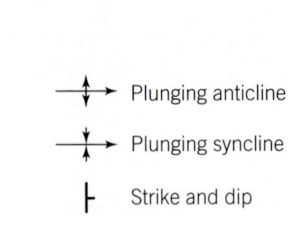

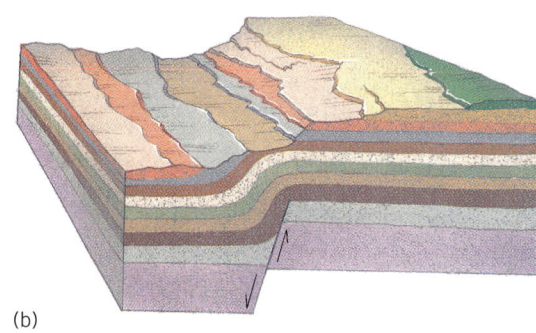

Figure 9.10 (a) A single flexure, or monocline, in the Cambrian sandstone of the Bighorn Mountains of Wyoming. (b) The flexure was formed by faulting in the rigid basement rocks beneath the sandstone strata.

Monoclines, Domes, and Basins

Not all folds form the repetitive pattern of anticlines and synclines that distinguish fold-mountain belts. Three important types that do not fit this pattern are monoclines, domes, and basins. A **monocline** is a single flexure, or bend, in otherwise horizontal rock strata. **Domes** and **basins** are rounded versions of anticlines and synclines.

Monoclines are commonly formed by faults that cut brittle strata at depth but do not break through younger strata above (Figure 9.10). The older rocks are displaced along the fault plane, whereas the more ductile, younger rocks merely fold into the monocline. They are common structures of the Colorado Plateau, where strata deposited in shallow water have been uplifted thousands of meters without intense deformation. Monoclines are also formed by compression in the manner of anticlines.

Figure 9.11 shows a typical dome, in which progressively younger strata dip away symmetrically from an older central core. Notice the topographic inversion—a valley

monocline
A sudden steepening in an otherwise gently dipping strata.

dome
An anticlinal circular structure with rocks dipping gently away from the center.

basin
A synclinal circular structure with rocks dipping gently toward the center.

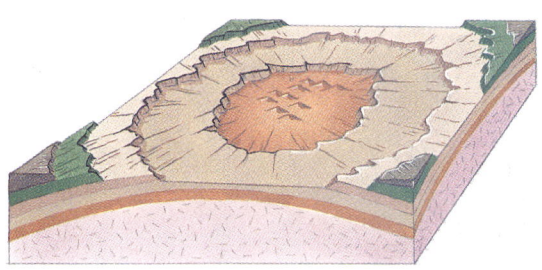

Figure 9.11 (a) Upheaval Dome, near the Utah Canyonlands. Notice how the strata dip symmetrically away from the up-arched center. (b) In a topographic inversion, a valley has been eroded in the core of the dome.

occupies the center of the dome because the apex of the fold has been subjected to deep erosion. This observation illustrates a general rule: One should not confuse underlying rock structure with topography—that is, the general configuration of the land surface. Although rock structure influences topography, the two are not the same. Topography is also a function of the resistance of rocks to erosion in a given climate, the age of the particular features, and many other factors.

The Black Hills of South Dakota are considered a classic dome. There, a Precambrian granite core is encircled by progressively younger strata that dip away from the core at relatively high angles. Whereas its topography is rugged, most domes and basins of the interior of the continent are far subtler structures, detectable only on a broad scale. For example, the rock strata that underlie the state of Michigan appear flat at any given locality. But when their outcrop patterns are studied on a regional scale, we see that Michigan is, in fact, a basin. Observe in Figure 9.12 the circular outcrop pattern of the formation, youngest toward the center.

The domes and basins of the midcontinent region are far from plate boundaries, and their origin is a matter of debate. Perhaps they are the distant ripples of plate collisions; perhaps they result from vertical faults unrelated to present plate collisions; or perhaps they are related to irregularities in the underlying basement complex.

Fractures

Suppose, while standing, you were to hold a heavy suitcase in each hand. Plainly, your torso would be subjected to compression at the same time that your arms were subjected to tension. The parts of your body would respond according to their orientation with respect to the external forces applied. Rocks respond in similar fashion. When a rock's tolerance to opposing stresses is exceeded, the rock will fracture.

Joints

Natural fractures in rocks are called joints, and a group of parallel fractures is a *joint set*. Because joint sets bear a definite orientation to the applied force, studies of them help us understand regional topography as well as regional stress patterns. For example, we can see in Bryce Canyon (Figure 9.13) that joint sets intersecting at right angles divide the soft strata into rectangular blocks.

Faults

For most purposes, the classification of faults is based on the nature of the relative movement along the fault plane. Consider Figure 9.14a, which is a block diagram showing a fault plane that dips at an angle of approximately 60°. Imagine you are a miner working in a shaft parallel to that plane. To use terms that were coined by Welsh coal miners in the nineteenth century, the block upon which you stand is the *footwall* (side A), and the block slanting diagonally overhead is the *hanging wall* (side B).

A **normal fault** is one in which the hanging wall has moved down with respect to the footwall (Figure 9.14b). Conversely, a **reverse fault** is one in which the hanging wall has moved up relative to the footwall (Figure 9.14c). In either case, displacement is along the dip of the slanting fault plane.

The normal and reverse faults in Figures 9.14(b–c) both dip at high angles (greater than 45°). However, there is an extremely important type of reverse fault in which the fault plane dips at less than 45° and most often at less than 15°. It is called a **thrust fault** because hanging-wall rocks appear to have been driven over the footwall rocks (Figure 9.14d).

Relative movement of the hanging wall in normal, reverse, and thrust faults is either up or down the fault plane. For this reason, all are referred to as **dip-slip faults.** In a **strike-slip fault,** however, displacement is horizontal—parallel to the strike of the fault plane (Figure 9.14e).

---

normal fault
A fault in which the hanging wall has been moved downward relative to the footwall.

reverse fault
A fault in which the hanging wall has been raised relative to the footwall.

thrust fault
A low-angle reverse fault that generally dips at about 15° or less.

dip-slip fault
A fault in which the movement is parallel to the dip of the fault plane.

strike-slip fault
A fault in which the movement is parallel to the strike of the fault plane.

Rock Deformation 201

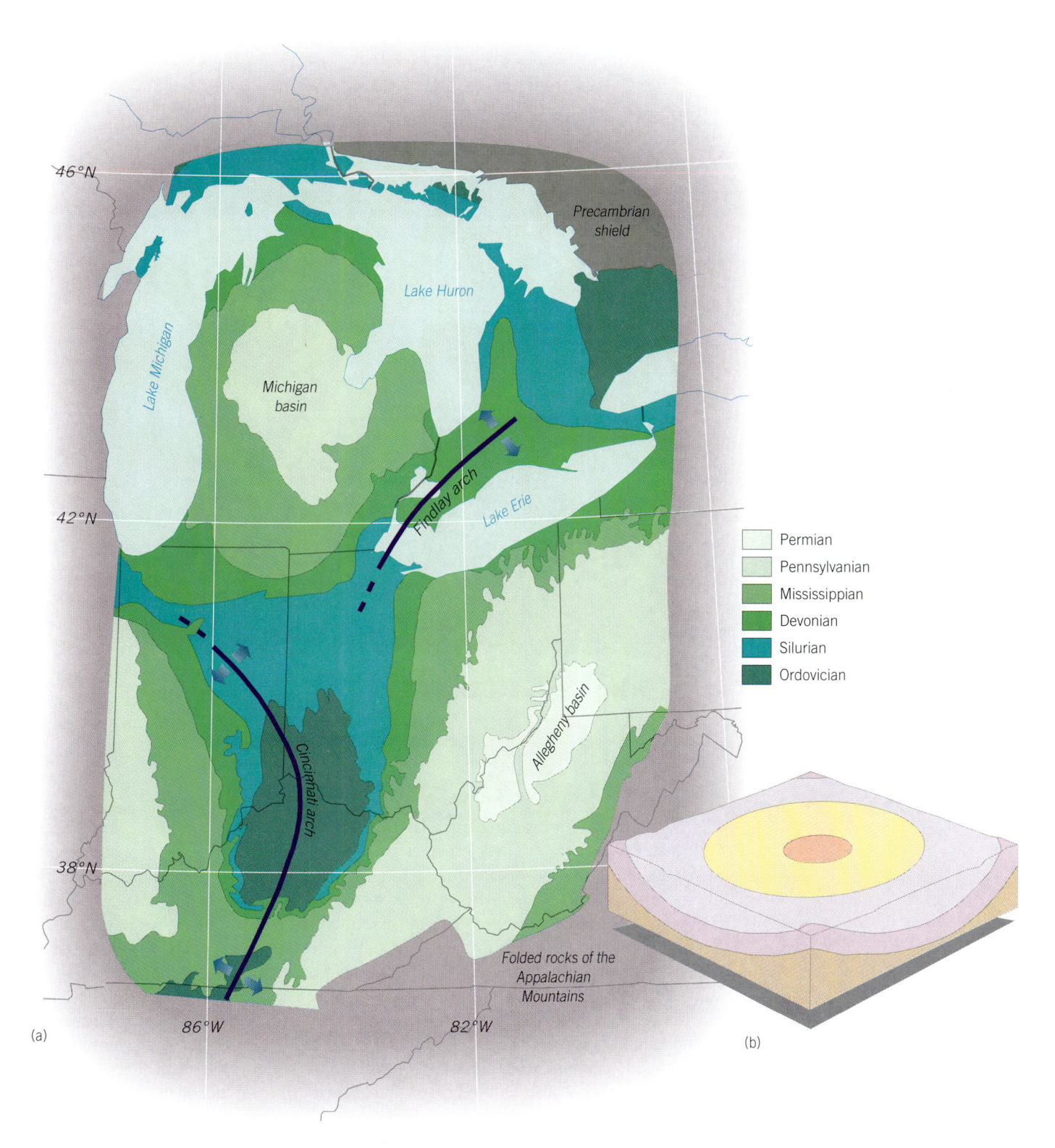

Figure 9.12 (a) The mid-region of the North American continent consists of a series of broad, gentle domes and basins. (b) The strata of the Michigan basin dip so gently that the structure is discernible only through mapping on a regional scale. Notice the circular pattern of the rock formations, with the youngest formation at the center of the bull's-eye.

Figure 9.13 (a) These striking columns or "hoodoos" of Bryce Canyon, Utah, were formed from joint sets that intersected at right angles, dividing the soft limestone into rectangular blocks. (b) Weathering and erosion have widened the joints and rounded the columns.

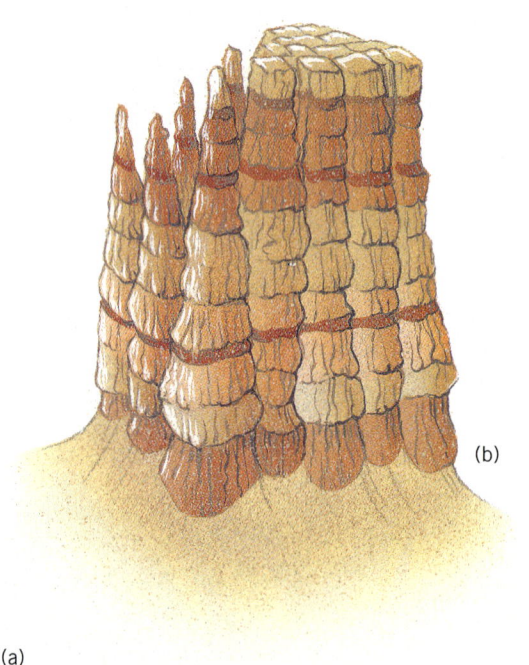

(a)

(b)

Strike-Slip Faults

The transform faults resulting from lateral tectonic plate movement are all strike-slip faults. Many of them connect other large-scale structural features—for example, the offset segments of mid-ocean ridges. Indeed, many strike-slip faults are formed by the movement of the crust between these offset segments. The San Andreas is a right-lateral strike-slip fault, meaning that displacement occurs to our right as we face the fault from either side. Fences, roads, and similar features were offset 3 to 4 meters in the right-lateral sense during the 1906 San Francisco earthquake.

Not all large strike-slip faults are located at plate boundaries, however. The enormous Altyn Tagh fault of China is one of many left-lateral faults that extend hundreds of kilometers across the Tibetan Plateau, the vast region north of the convergent boundary between the Indian and Eurasian plates. These faults result from horizontal displacement in the plateau as the Indian plate pushes into the Eurasian plate, like a fist jambing into a giant pizza dough.

Normal Faults

Normal faults result from nonuniform tensile stresses that pull apart the crust, as occurs in the rift zones of the Earth's lithospheric plates. As the crust is stretched, some fault blocks drop relative to others along these planes to fill the excess space. Normal faults are the dominant structural units of Nevada, portions of its neighboring states, and northern Mexico. The region is referred to as the Basin and Range Province, a name that emphasizes the topography produced by these faults. The ranges are massive upthrown blocks called **horsts;** the basins are called **grabens** (Figure 9.15).

horst
An uplifted block bounded by normal faults on its long sides.

graben
An elongated, depressed block bounded by normal faults on its long sides.

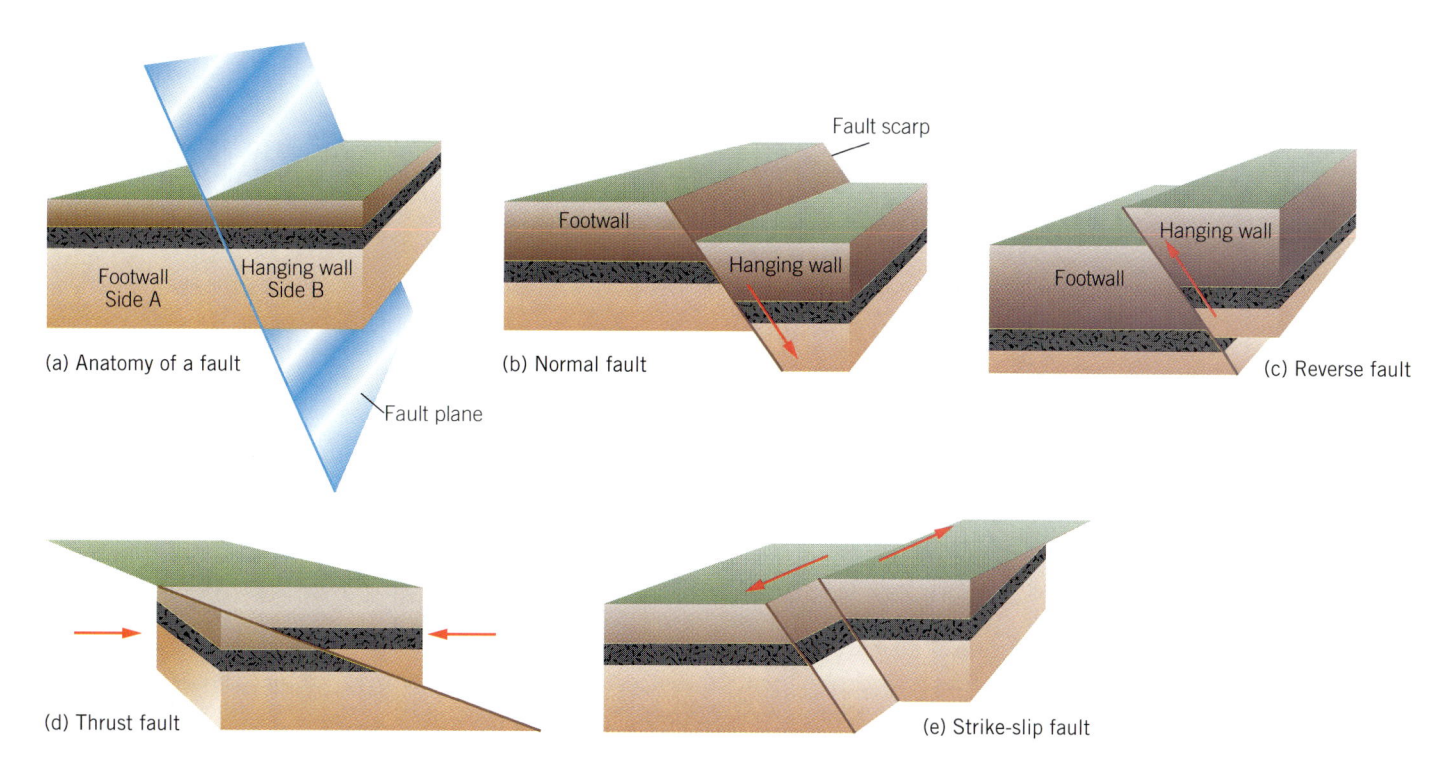

Figure 9.14 Some common types of faults and their associated terminology. Relative motion of the opposing blocks on either side of the fault plane determines whether a fault is classified as normal, reverse, or strike-slip. A thrust fault is a low-angle reverse fault.

Thrust Faults

Thrust faults occur in regions subject to compressive stress, such as along colliding plate boundaries. Typically, such faults are an integral part of fold-mountain belts and, like them, are evidence of crustal shortening. Thrust faults tend to drive older rocks over younger ones. The cross section of the Canadian Rockies in Figure 9.16 illustrates that thrust faults often occur as thin sheets, in which fault planes overlap one another in series, like so many stacked shingles. Notice how the faults cause omission and out-of-sequence placement of the strata.

Figure 9.17 shows that thrust faults are capable of displacing strata great distances. Chief Mountain in Montana is an isolated remnant of a thrust fault separated by erosion from the main mass of the thrust sheet. The top of Chief Mountain consists of Precambrian strata about 1 billion years old. It rests on Cretaceous strata only 65 million years of age. It has been estimated that the Precambrian strata were driven a minimum of 25 kilometers. Some geologists believe the combination of folding and faulting is evidence that the crust has been shortened hundreds of kilometers in fold-mountain systems.

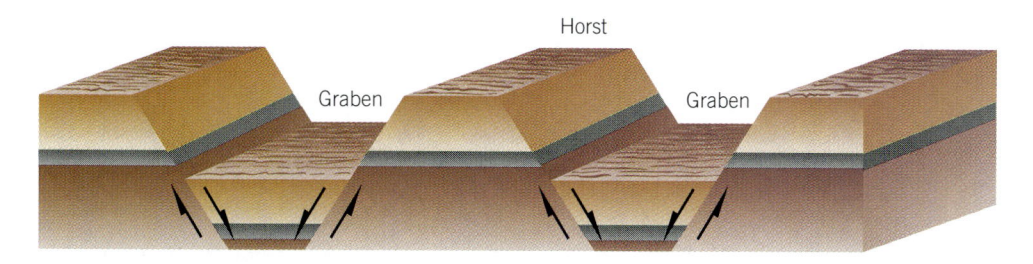

Figure 9.15 In the Basin and Range Province of the western United States, the crust is being stretched, causing some blocks (grabens) to drop relative to others (horsts) along normal fault planes.

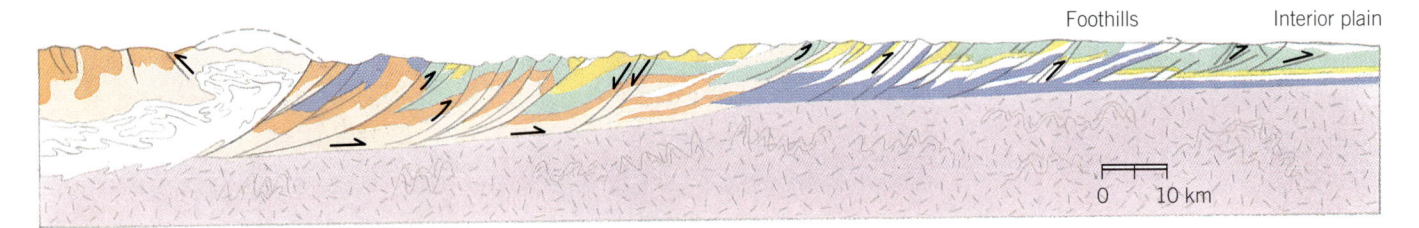

Figure 9.16 A cross section of the Canadian Rockies, which consist of a series of shallow, overlapping thrust sheets. For rocks to be driven over one another in this manner, considerable crustal shortening must have occurred. The westward dip of the thrust sheets is evidence that the compression drove the rocks from west to east.

Figure 9.17 Crustal shortening in the Northern Rockies as illustrated by Chief Mountain, of Glacier National Park, Montana. Precambrian rocks of the mountain have been driven up to 25 kilometers to the east along the Lewis thrust fault, and have come to rest upon Cretaceous rocks. The mountain is called an outlier because erosion has separated it from the main mass of Precambrian rocks to the west.

Impact Craters

Though Thomas Jefferson was open-minded about many subjects, the existence of meteoroids was not one of them. "Do we believe our common sense or the word of some Yankee professors?" was his response to a report that two Yale University researchers had retrieved rock fragments that had fallen from the sky over Connecticut. The observations of the Yankee professors were, of course, correct.

Most meteoroids originate in the asteroid belt located between Mars and Jupiter. Orbiting the Sun within the belt are more than 1 million rock and iron fragments with diameters that range from a few millimeters to a few thousand kilometers. A *meteoroid* is formed when a collision sends a fragment of one of these bodies on an eccentric path. Sometimes the path of a meteoroid intercepts the Earth's gravitational field. A *meteorite* is the rocky remains of a meteoroid that reached the Earth's surface.

Comets are another source of meteorites. Originating in the Oort Cloud at the outer reaches of the Solar System, these fragments of rock, iron, and ice sometimes pass close to the Sun, break up, and leave a trail of debris behind in the inner Solar System.

Probably millions of meteoroid particles shower the Earth every 24 hours. Most burn out in the atmosphere, and are observed as *meteors* or "shooting stars" in the night sky. Occasionally, boulder-sized meteoroids make it through the atmosphere; although they can flatten a car on impact, they have been slowed considerably by the time they reach the Earth's surface. Meteoroids large enough to blast impact craters are a different story, because the atmosphere has little effect on them. To survive the Earth's atmosphere and crash into the planet's surface with enough force to form a crater, a meteoroid must weigh at least 320 metric tons and travel at 15 to 30 kilometers per second—roughly 50 to 100 times the speed of sound. The rule of thumb is that a meteoroid will blast a hole roughly ten times wider than its diameter, although actual craters may vary from this estimate by as much as a factor of ten.

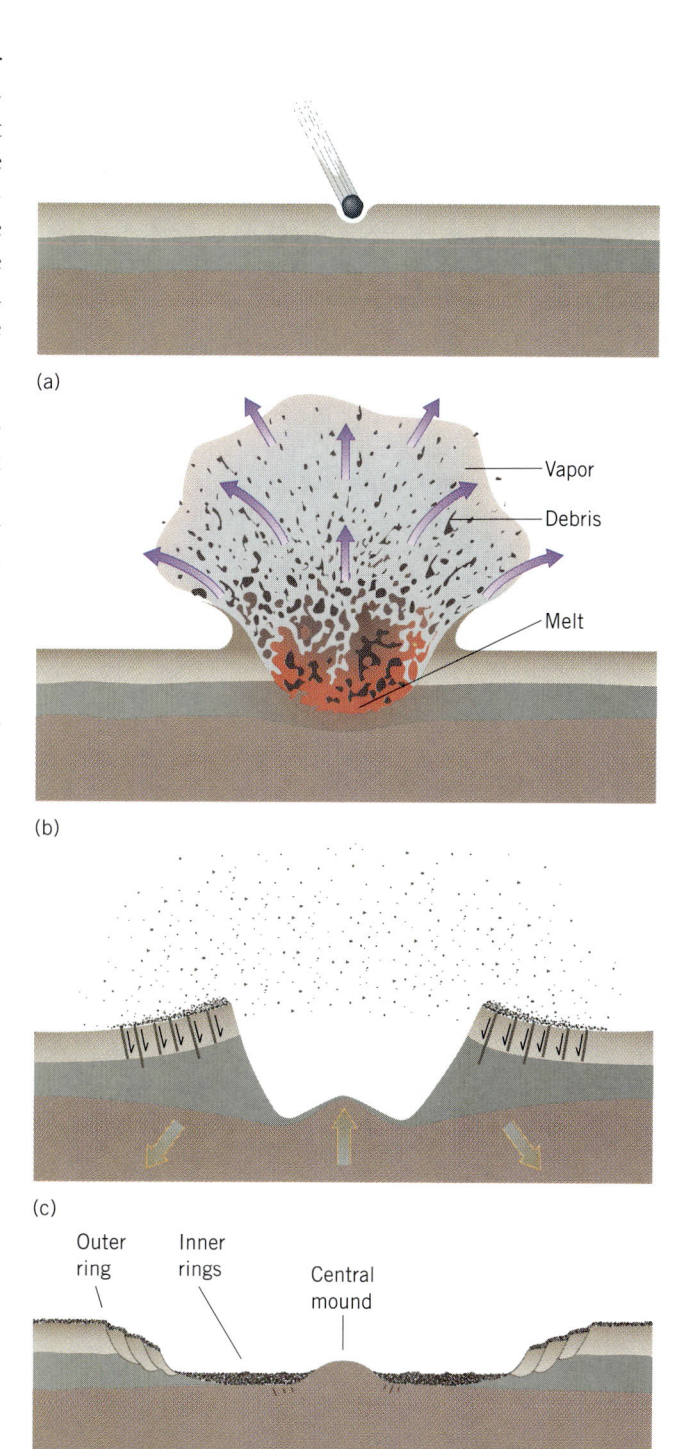

Figure 9.18 Impact of a meteoroid. (a) An incoming meteoroid heavier than 320 tons may be moving as fast as 100,000 kilometers per hour. (b) The impact shock causes such high temperatures and pressures that most of the meteoroid and crater rock are vaporized and melted. (c) The release wave following the shock wave causes the center of the floor in the transient crater to rise. (d) The fractured walls fail and slide into the crater, creating a wider and shallower final crater.

The Structure of an Impact Crater

Although no one has ever witnessed an impact-cratering event firsthand, the landforms created by an impact are unmistakable (Figure 9.18). Most distinctive is the circular pit of the crater, which is bounded by an *outer ring* consisting of steep walls and an elevated *crater rim* that dips gently away from the center of the pit. The floor of the crater itself contains one or more circular fractures, or *inner rings,* that encircle the *core* of the crater. The core consists of a *central mound,* or peak, of elevated crater floor.

As shown in parts (a) and (b) of Figure 9.18, the inner ring fractures are generated instantaneously by the enormous shock wave that spreads from the point of impact. In part (c), rebound from this wave causes the floor of the crater to rise, forming the central mound. A considerably longer interval of readjustment follows these rapid events. The crater walls stabilize by slumpage and other forms of mass wasting, which widens the crater and reduces its steepness. As indicated in part (d), the final circumference of the crater constitutes its outer ring.

A large-scale impact leaves behind other compelling evidence as well:

- *Ejecta,* consisting of particularized bedrock blasted out of the crater, are strewn about the surrounding region.
- *A zone of fractured and pulverized rock* may extend for hundreds of meters beneath the crater floor.
- *Meteorite fragments* are identifiable as chunks of iron and nickel or rocks of characteristic mineral assemblages that are coated with a remelted and glazed crust.
- *Distinctive minerals, melted rock, and shock structures* that form only during high-velocity impact (i.e., certain forms of quartz, glassy beads, and metal droplets) may be present.
- *Rare elements* may be found in the ejecta—for example, an iridium content that matches meteorites but no other materials formed on the surface of the Earth.

Frequency of Impact

Analysis of the frequency of impacts on our closest neighbor, the Moon, suggests that the Earth has suffered approximately 2400 large impacts over the past 3 billion years, with 720 of them occuring on the continents. Some 150 impact craters have been dis-

Figure 9.19 Barringer (Meteor) Crater of northern Arizona.

covered on the continents thus far. It will be difficult to identify more than an additional few hundred, however, because many—perhaps most—have been destroyed by erosion and crustal recycling. Because of the prevailing dry climate, the Barringer (or Meteor) Crater of northern Arizona is probably the world's best preserved impact structure (Figure 9.19).

Many of the craters discovered so far have been implicated in the formation of a number of important geological features and historical events. For example, the inner and outer rings of an 85-kilometer-wide crater lie beneath Chesapeake Bay and the Delmarva Peninsula. At the center of the bull's eye lies a thick zone of fractured rock, churned-up sediments, and telltale shocked quartz grains. The 4-kilometer-wide meteoroid was roughly the size of the largest fragment of the Shoemaker-Levi 9 comet, which entered Jupiter's atmosphere in 1994. The meteoroid struck the Earth about 35 million years ago, and its impact probably created the depression in the continental shelf that later became Chesapeake Bay.

A large, 65-million-year-old crater buried beneath the Yucatan Peninsula of Mexico may be the most intensively studied impact crater ever found. Its age and placement support the theory that its occurrence caused the extinction of the dinosaurs and many other life forms at the close of the Cretaceous period 65 million years ago. According to this theory, the extinctions were caused not by the force of the impact itself, but by secondary effects. First the atmosphere was heated to a boil by the fiery dust clouds and forest fires ignited by the impact. It was then cooled as the dust circled the Earth and blocked the sunlight. These effects, plus the torrential acid rains that followed, disrupted the Earth's ecosystems to the point where the mass extinctions occurred in a matter of months.

STUDY OUTLINE

The evidence of Earth's past dynamism is preserved in the deformational features of rocks: folds, faults, and joints.

I. STRESS AND ROCK DEFORMATION
  A. **Stress** is the magnitude of the force on a rock divided by the area of the plane. **Compressive stress** squeezes the rock from two opposite directions, **tensile stress** pulls it apart, and **shear stress** distorts its shape. Lithostatic stress applies equal forces in all directions.

  B. The change in a rock's volume or shape that results from stress is called **strain.**

  C. Rocks exhibit three modes of response to stress:
  1. A rock exhibits *elastic response* if it returns to its original shape once the stress has been removed.
  2. Beyond the **elastic limit,** rocks can exhibit *ductile response,* in which the shape of the rock is permanently changed. The result is **folds,** which are wavelike distortions in the rock.
  3. In *brittle response* to stress, rocks fracture. **Faults** are fractures in which the rocks on either side have been offset. If there has been no movement along the plane of fracture, the fracture is called a **joint.**

II. THE MAP DEPICTION OF PLANAR FEATURES.
Strike is the compass direction of a line along the intersection of an imaginary horizontal plane and a bedding plane. The **dip** is the acute angle between the bedding plane and the horizontal plane, measured in the vertical plane.

III. FOLDS.
A fold may be divided into two parts, or limbs, by an imaginary surface called an *axial plane.* The line of intersection of the limbs and axial plane is the *fold axis.* An **anticline** is a fold with a convex-up orientation. A **syncline** is a fold with a concave-up orientation.

B. Folds may be symmetrical or asymmetrical. In **overturned folds,** a limb is rotated so that older strata lie above younger strata, in a reversal of the normal sequence. In a **recumbent fold,** the axial plane is horizontal.

C. The axis of a **plunging fold** is inclined; in a nonplunging fold, the axis is parallel to the Earth's surface.

D. A **monocline** is a single flexure, or bend, in otherwise horizontal rock strata. A **dome** is a circular flexure that dips gently away from the center; a **basin** is a circular flexure that dips gently toward the center.

IV. FRACTURES

A. Joint sets are parallel fractures that intersect at right angles and divide rock into rectangular blocks.

B. The classification of faults is based on the nature of the relative movement along the fault plane.
1. When a rock is dissected by a fault that is diagonal to the Earth's surface, the underlying block is the *footwall* and the overlying block is the *hanging wall.*
2. In **dip-slip faults,** movement is vertical along the dip of the fault plane. In a **normal fault,** the hanging wall has moved down with respect to the footwall. In a **reverse fault,** the hanging wall has moved up relative to the footwall. A **thrust fault** is a reverse fault in which the hanging wall overlaps the footwall.
3. In a **strike-slip fault,** displacement is horizontal—parallel to the strike of the fault plane.

V. THE TECTONIC SETTING AND TOPOGRAPHY OF FAULTS

A. Transform faults are strike-slip faults that result from movement of the crust between offset plate segments.

B. Normal faults result from tensile stresses that pull the crust apart at divergent plate boundaries. *Basin-and-range* topography is produced by normal faulting. The ranges are massive upthrown blocks called **horsts,** and the basins are called **grabens.**

C. Thrust faults are the result of compressive stress along convergent plate boundaries. They are an integral part of fold-mountain belts and are the result of crustal shortening.

VI. IMPACT CRATERS.
Impact craters are caused by the collision of meteoroids with the Earth's surface.

A. The structure of an impact crater consists of an outer ring, crater rim, inner rings, core, and central mound.

B. Impact cratering may have been responsible for a wide variety of events in Earth history. Such events include mass extinctions and the formation of deep basins, such as the one beneath Chesapeake Bay.

STUDY TERMS

anticline (p. 196)
basin (p. 199)
compressive stress (p. 193)
dip (p. 195)
dip-slip fault (p. 200)
dome (p. 199)

elastic limit (p. 194)
fault (p. 194)
fold (p. 194)
graben (p. 202)
horst (p. 202)
joint (p. 194)

monocline (p. 199)
normal fault (p. 200)
overturned fold (p. 196)
plunging fold (p. 196)
recumbent fold (p. 196)
reverse fault (p. 200)
shear stress (p. 193)

strain (p. 194)
stress (p. 193)
strike (p. 195)
strike-slip fault (p. 200)
syncline (p. 196)
tensile stress (p. 193)
thrust fault (p. 200)

Thought Questions

1. Explain the difference between stress and strain.
2. A single, externally applied force may cause many kinds of stress in a rock body. Explain.
3. Why do we call an anticline whose limbs dip in the same direction an overturned fold?
4. What kind of fault is associated with crustal shortening? With crustal extension? With crustal shear? Name three regions of the United States that display these features.
5. Give two examples of how faulting followed by erosion may influence the sequence of rock strata you encounter in the field.
6. Which feature of the oceanic crust resembles a graben? A strike-slip fault? A thrust fault?
7. Describe the distinguishing features of a meteoroid impact crater. Why are impact craters easy to observe on the Moon but difficult to find on the Earth?

Chapter 10

Weathering, Mass Wasting, and Gradation

Weathering Processes

Mechanical Weathering
   Frost Wedging
   Salt Crystallization
   Sheeting, Exfoliation, and Spheroidal Weathering
   Weathering Effects of Plants and Animals
Chemical Weathering
   Direct Solution
   Hydrolysis
   Oxidation
Rates of Weathering
Differential Weathering

Soil Formation

The Soil Profile
Nutrient Cycling
Climate and Soil Types
   Soils of Tropical, Humid Climates
   Soils of Arid and Semiarid Climates
   Soils of Cold to Temperate Climates

Mass Wasting

Falls
Slides
   Conditions Favoring Slides
   Slumps
Flows
Creep

Predicting and Preventing Mass-Wasting Disasters

Weathering, Soils, and Mass Wasting

In time the Rockies may tumble,
Gibraltar may crumble,
They're only made of clay,
But our love is here to stay.

—*George and Ira Gershwin*

Weathering, Mass Wasting, and Gradation

Allowing that George and Ira Gershwin were great songwriters, not geologists, they expressed the essence of an important fact: nothing exposed to the surface environment lasts forever. The Rock of Gibraltar does indeed crumble—although it is made of limestone rather than clay.

Weathering is the first stage in the vast redistribution process of **gradation,** in which erosion wears low the high portions of the land surface, and deposition builds up the low portions. In the course of gradation, the Earth's scenery is rearranged—indeed, to a large extent, created. Tributaries of the Mississippi River carve the valleys of the western slopes of the Appalachians and the eastern slopes of the Rockies. When it reaches the Gulf of Mexico, the Mississippi River constructs its delta out of the sediments excavated from these mountains.

gradation
The balance between erosion and deposition that maintains a general slope of equilibrium, trending toward sea level.

◀ Sugarloaf, the famous landmark of Rio de Janeiro, Brazil. The rounded form is the result of weathering, which causes thin concentric shells to peel away from the massive granitic rock.

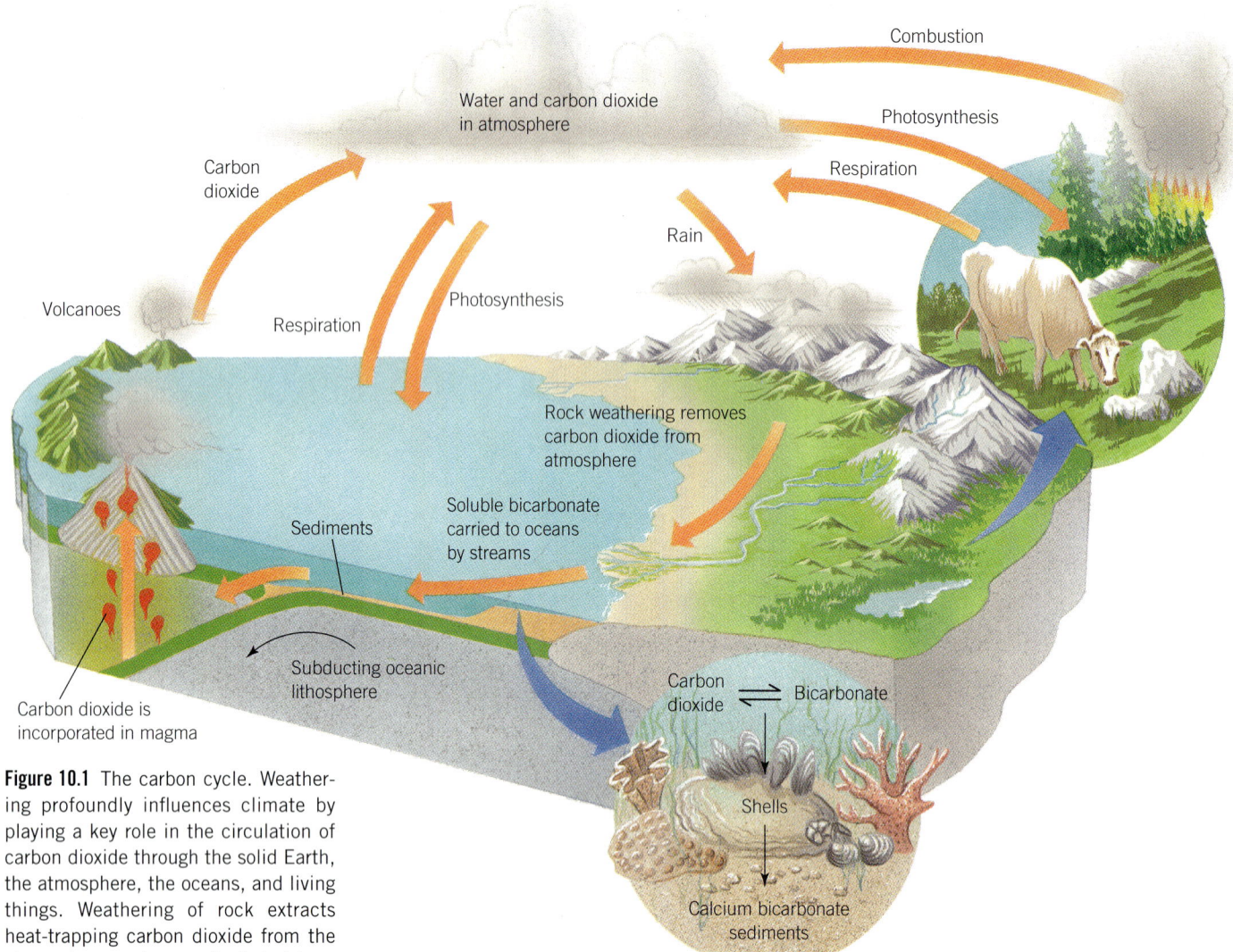

Figure 10.1 The carbon cycle. Weathering profoundly influences climate by playing a key role in the circulation of carbon dioxide through the solid Earth, the atmosphere, the oceans, and living things. Weathering of rock extracts heat-trapping carbon dioxide from the atmosphere and converts it to bicarbonate ions. Transported to the oceans by streams, the bicarbonate ions are temporarily stored in the calcium carbonate shells of corals, clams, microscopic plankton, and other marine organisms. Carbon dioxide is returned to the atmosphere via respiration, combustion, and subduction zone volcanism.

The process called weathering is so inconspicuous that its importance is often underestimated. Weathering creates the soil that makes plant life possible, and it releases to the ocean the nutrients necessary for the growth of marine algae. Algae and plants produce oxygen through photosynthesis; thus weathering is indirectly responsible for the air we breathe. Weathering also plays a key role in the carbon cycle—the circulation of carbon dioxide through the solid Earth, the atmosphere, the oceans, and living things (Figure 10.1). As we learned in Chapter 6, carbon dioxide regulates the Earth's temperature, for it is a key component of the greenhouse effect.

Whereas weathering causes the slow, steady wearing away of Earth materials fragment by fragment, mass wasting is the downslope movement of those materials caused by gravity. A loosened rock fragment skipping down the face of a sheer cliff is an example of mass wasting, but so is an entire mountainside crashing down on the village nestled at its base. Mass wasting is also an integral part of the process of gradation, along with that most important gradational agent, streams (which we will come to in the next chapter). Weathering, mass wasting, and streams are as closely linked to one another as the men who quarry rock are to those who truck it away.

Figure 10.2 illustrates that weathering, mass wasting, and slope erosion supply most of the sediment that streams carry away. Conversely, by downcutting deeply into the land, streams create the sloping surfaces that make mass wasting possible. Other grada-

Figure 10.2 The Yellowstone River occupies a small portion of the V-shaped valley it formed by downcutting into the land surface. The sloping valley walls are widening through mass wasting, which supplies most of the sediment that the stream transports.

tional agents also contribute to mass wasting by creating conditions of instability on slopes. Glacier, groundwater, breaking waves, and even the wind all work to undercut and oversteepen slopes.

Because most erosion occurs above sea level, and most deposition below, gradation works in the long term toward reducing the land surface of Earth to a featureless sea-level plain, with the planet seeming as monotonous as a billiard ball. Except in localized regions, however, this reduction has never happened, because the Earth is a dynamic planet. Tectonic, isostatic, and volcanic activity constantly uplift the surface and supply fresh material to the crust, so that the gradational agents of wind and water have never run out of work. We will see gradational processes at work in this chapter and in the five chapters that follow.

Weathering Processes

Two major kinds of weathering take place on the Earth's surface. In **mechanical weathering** (disintegration), rock and mineral fragments are reduced to small sizes by purely physical means, without change in composition. In **chemical weathering** (decomposition), chemical reactions fundamentally alter the original composition of the rock.

There is often a reciprocal relationship between mechanical and chemical weathering. Mechanical disintegration accelerates the process of chemical weathering by

mechanical weathering
The combination of physical processes that disintegrates a rock without chemical change.

chemical weathering
The surface process that decomposes rocks and minerals through chemical reactions.

Figure 10.3 By breaking a rock into smaller fragments, mechanical weathering increases the surface area that is subjected to chemical attack.

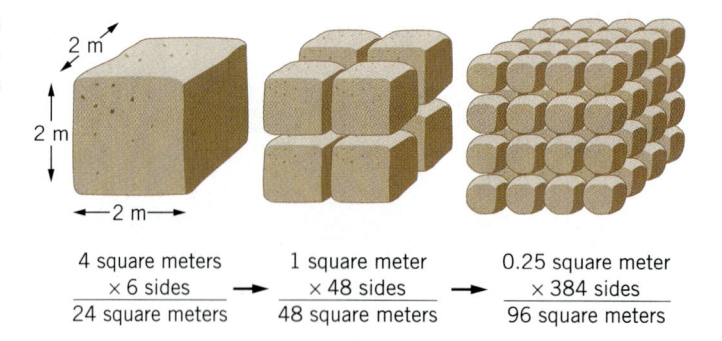

$$\frac{\text{4 square meters}}{\text{24 square meters}} \times 6 \text{ sides} \longrightarrow \frac{\text{1 square meter} \times 48 \text{ sides}}{\text{48 square meters}} \longrightarrow \frac{\text{0.25 square meter} \times 384 \text{ sides}}{\text{96 square meters}}$$

exposing more surface area of the rock to chemical attack (Figure 10.3). In decomposing minerals, chemical weathering weakens the internal fabric of rock masses, leaving them susceptible to mechanical disintegration.

Mechanical Weathering

Mechanical weathering is accomplished through a variety of processes, from the freezing and thawing of water to the invasion of plant roots. All of them work to break rocks into smaller pieces.

Frost Wedging

Those of us who live in northern climates are well aware that pipes burst when water freezes in them over the winter. The reason they burst is that water expands about 9 percent upon freezing, which exerts enormous pressure against the pipe walls. Ice plays a similar role in rock weathering in any climate that allows ice to melt and freeze again. Water seeps into rock fractures, expands as it changes to ice, and dislodges rocks. The process is called **frost wedging.** At high elevations in mountainous regions, it can result in an enormous *boulder field* (also called *felsenmeer,* or "sea of rocks"), as shown in Figure 10.4. If ice wedging takes place along cliff faces, the dislodged blocks tumble down to the base of the cliff, where they accumulate. The newly exposed rock of the cliff face is, of course, subject to similar attack; and in this way, the mountain slowly crumbles.

frost wedging
A process by which water seeps into rock joints, freezes, and wedges the rock apart.

Figure 10.4 These jagged-edged boulders in Finnish Lapland were formed through frost wedging, which shattered solid rock into smaller fragments.

Figure 10.5 Ruins of the Anasazi cliff dwellings at Mesa Verde National Park, Colorado, where Native Americans constructed their homes in caves that were weathered from the base of thick sedimentary rock formations. Salt crystallization played a major role in the weathering process.

Salt Crystallization

Salt crystallization is effective in causing sandstone to disintegrate in arid climates. Fractures and pore spaces allow water to trickle through the rock. As the water evaporates, minute salt crystals are precipitated. As the crystals grow larger over the years, they pry apart sand grains and widen fractures. Eventually, larger fragments and blocks of rock are wedged loose from the cliff. Most of the action takes place at the base of the cliff, where salt tends to accumulate. In time, caves as large as amphitheaters will develop there. Caves formed in this manner provided shelter for the cliff-dwelling Indians of the American Southwest and for peoples of other cultures the world over (Figure 10.5).

Sheeting, Exfoliation, and Spheroidal Weathering

Miners and sandhogs who dig deep tunnels in hard rock do not have the world's safest jobs. Besides worrying about cave-ins, floods, blasting accidents, fires, and ventilation hazards, they must also worry about particularly nasty and unpredictable rock bursts. Rock bursts occur when freshly excavated rock experiences a sudden release of pressure. The rock expands, or "pops," often with a loud noise. Fragments that split along fractures go flying through the air.

In nature, rocks are exposed gradually by erosion and mass wasting; as the underlying rocks slowly expand, the release of pressure causes **sheeting,** or fracturing parallel to the rock surface. In **exfoliation,** a related though not identical process, concentric plates or scales are stripped from the exposed surfaces of large rock masses. The exact mechanisms are not well understood, but they involve physical and chemical processes that widen fractures parallel to the exposed surface. Large exfoliated exposures of massive granite or similar rock form striking, rounded *exfoliation domes.* Stone Mountain, Georgia; Sugarloaf, Rio de Janeiro, Brazil; and the peaks of Yosemite, in the California Sierras, are classic examples of these features (see the photograph of Sugarloaf at the beginning of this chapter).

Spheroidal weathering is the peeling away of shells of decayed rock from isolated unaltered boulders and small rock outcrops. This type of weathering is caused by water that seeps into the fractures of jointed rocks and chemically weathers the minerals along the surfaces. Spheroidal weathering operates on a smaller scale than exfoliation but produces similar rounded onionskin layers of decayed rock (Figure 10.6). Weathering molds rocks into spherical shapes even if they are not subjected to peeling action. The points of rectangular fragments are attacked from three directions at once, whereas the edges are attacked from two and the faces of the rock from one. Thus the points and edges weather more rapidly than the sides. Eventually, a spherical shape is attained.

sheeting
A type of jointing parallel to the rock surface caused by pressure release. Similar to exfoliation.

exfoliation
The flaking or stripping of a rock body in concentric layers.

spheroidal weathering
The peeling of small rock bodies into onionlike layers.

Figure 10.6 Spheroidal weathering of loose granite boulders in the Algerian Sahara desert.

Weathering, Soils, and Mass Wasting 217

Figure 10.7 An eroded stream bank in the Lost Maples State Natural Area of central Texas exposes tree root systems in the process of disintegrating solid rock.

Weathering Effects of Plants and Animals

Living organisms accelerate rock disintegration. Figure 10.7 shows the roots of a tree wedging between and enlarging cracks in the surrounding rocks and, with the aid of water and dissolved elements, reducing the rocks to fine particles. The root systems of plants and trees also extract nutrients from soil and act as a medium of chemical exchange, a topic we shall discuss later in this chapter. Burrowing organisms churn the soil, which exposes fresh rock and mineral fragments and allows air and water to penetrate to greater depth. Bacteria and fungi cycle elements through the soil and are responsible for the production of acids that are important to chemical weathering. Humans accelerate weathering by digging, blasting, excavating, and, in a variety of ways, uncovering rock and sediments that would otherwise be protected.

Chemical Weathering

The vast majority of chemical-weathering processes involve three major reactions: (1) direct solution, (2) hydrolysis, and (3) oxidation. These are descriptive terms that geologists find useful and are not necessarily the formal definitions used by chemists.

Direct Solution

Direct solution is a reaction whereby mineral and rock materials are chemically attacked and dissolved. In pure water, reactions can be traced directly to the dipolar structure of

direct solution
The dissolving of rock or mineral materials.

Figure 10.8 The dissolution of halite. The positive hydrogen ends of the water molecules attract the negative chlorine ions and the negative oxygen ends attract the positive sodium ions.

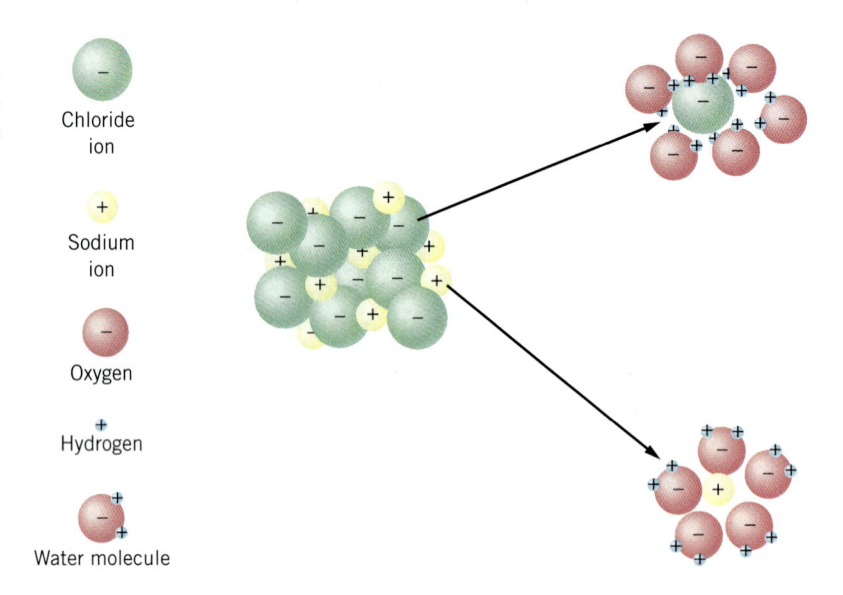

the water molecule (Figure 10.8). This structure allows the molecule to act as if it were a minute magnet whose north and south poles attract opposite charges and repel like charges. Figure 10.8 illustrates a well-known example: the dissolving of common salt (halite, NaCl). The positive hydrogen ends of water molecules dislodge the negative chlorine ions (Cl$^-$), and the negative oxygen ends dislodge the positive sodium ions (Na$^+$). Subjected to this attack, halite is thus dissolved. Direct solution works best on minerals in which the bonds between ions are relatively weaker than the attraction of these ions for the water dipole.

Hydrolysis

hydrolysis
The reaction between minerals, especially silicates, and water.

Hydrolysis is an exchange reaction involving minerals and water. Even pure water contains a small but measurable amount of free hydrogen (H$^+$) and hydroxide (OH)$^-$ ions that are shaken loose through random collisions of water molecules. These ions are potent agents able to replace mineral ions and force them into solution. In the process, the mineral's atomic structure is converted to other forms.

Although hydrolysis works in pure water, the reaction rate is accelerated enormously if other sources of hydrogen ions are also present. Dissolved carbon dioxide (CO$_2$) supplies these ions. It is present in the atmosphere and soil, being a product of the respiration and decay of animals and plants. The carbon dioxide combines with rain and groundwater to form carbonic acid (H$_2$CO$_3$) that instantly separates into positive hydrogen ions and the negative bicarbonate ion (HCO$_3^-$).

$$H_2O + CO_2 \rightarrow H_2CO_3 \rightarrow H^+ + HCO_3^-$$

Hydrolysis involving dissolved carbon dioxide is the fundamental weathering reaction of the silicate minerals, which compose well over 90 percent of the Earth's crust. As an example, consider the hydrolysis of potassium feldspar, a representative of the most common mineral group of the crust.

$$2\,KAlSi_3O_8 + 2(H^+ + 2\,HCO_3^-) + H_2O \rightarrow Al_2Si_2O_5(OH)_4 + [2\,K^+ + 2\,HCO_3^- + 4\,SiO_2]$$

Potassium feldspar + Carbonic acid ions + Water Yields Kaolinite (clay residue) + Potassium ion + Bicarbonate ion + Silica

[In solution]

First, note the overall reaction: potassium feldspar, in the presence of carbonic acid, is converted to an insoluble clay mineral. The reaction involves an exchange: the hydrogen ions (H$^+$) enter the clay, whereas the potassium ions (K$^+$) and silica (SiO$_2$) are released. The clay is incorporated in the soil and sedimentary rocks. The soluble potassium ions either react to form potassium bicarbonate (KHCO$_3$) or are taken up by plants, which use them as nutrients. The dissolved silica will either be carried away by groundwater or will precipitate as chert or flint.

This reaction follows the hydrolysis pattern of all silicate minerals: the mineral is converted to *insoluble residues* and *dissolved ions*. The composition of these weathering products depends mainly on the composition of the mineral under attack. For example, feldspar contains abundant aluminum, which is not soluble; thus weathering produces clay mineral residue. Olivine, on the other hand, contains no aluminum, and its iron is insoluble under weathering conditions. Consequently, iron compounds, rather than clay, become the insoluble residue. Quartz is another example of special note, because it is chemically and physically resistant. Hydrolysis has little effect on it, except in extremely hot, humid climates and only after thousands of years of exposure.

Oxidation

Oxidation is a chemical reaction that results when elements combine with oxygen. Oxidation reactions are probably the form of chemical weathering most familiar to us. They turn wrought iron to orange rust and copper pennies green. Iron that remains in the soil or on exposed rock surfaces also rusts: it combines with oxygen to form the red mineral hematite or with water to form the yellow mineral limonite. Iron oxides and hydroxides such as hematite and limonite are responsible for the bright colors of many soils.

In certain oxygen-poor environments, such as stagnant swamps, iron combines with sulfur to form pyrite. If, at a later time, pyrite is exposed to oxygen, either directly through erosion or by circulating groundwater, the oxygen will replace the sulfur. The released sulfur may combine with hydrogen ions in the water to produce hydrogen sulfide (H$_2$S) or, if dissolved oxygen is abundant, sulfuric acid (H$_2$SO$_4$). Both are potent weathering agents.

oxidation
The process by which an element combines with oxygen.

Rates of Weathering

Rocks of different compositions and origins weather at different rates. Figure 10.9 lists the common igneous-rock–forming minerals in order of their increasing resistance to weathering. The arrangement closely resembles Bowen's reaction series, which gives the

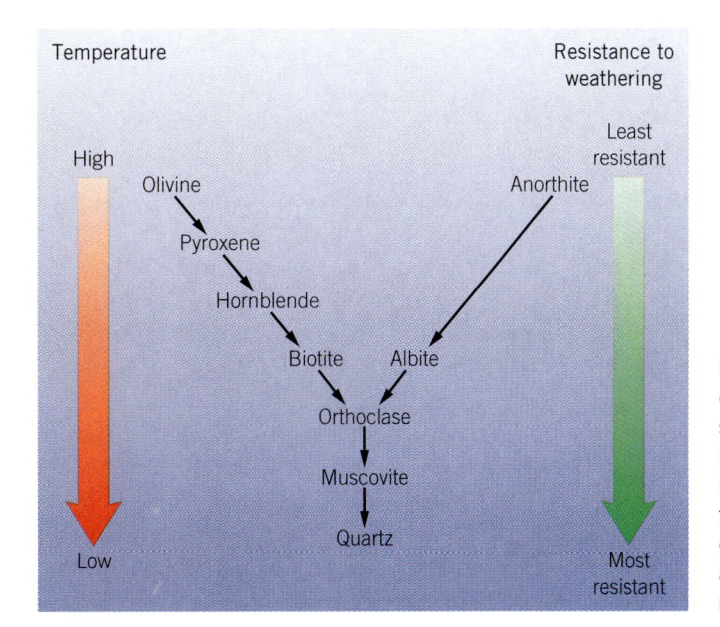

Figure 10.9 Minerals are weathered chemically in reverse order to the sequence of crystallization outlined in Bowen's reaction series (see Chapter 6). Minerals that crystallize at high temperatures are generally the most easily weathered; those that form at low temperatures are the most resistant.

Figure 10.10 Differential weathering of gravestones, displayed in a Long Island, New York, cemetery. The lettering is more decipherable on (a) the slate stone, dated 1701, than on (b) the limestone marker, dated 1816.

(a)

(b)

general sequence of mineral crystallization during the cooling of typical basaltic magmas. We see that olivine, the mineral that precipitates from magma at the highest temperature, also weathers the most rapidly. Quartz, which precipitates at the lowest temperature, weathers at the slowest rate. In general, those minerals that form closest to surface conditions are the most resistant.

We can see now why quartz and clay minerals are the most common of the loose materials we usually observe on the beach, in streambeds, and similar places—they are the most stable minerals at surface conditions. Quartz is the chief mineral of sandstone and conglomerate. In combination with clay derived from feldspar, it forms the bulk of the thick sediment deposits found on the Earth's continental margins. Remember that granite is the most common rock of the continental crust, and that it is composed mainly of feldspar and quartz. Most of the carpet of loose materials and sedimentary rocks that cover the Earth's surface is derived directly or indirectly from the weathering of granitic rocks.

Differential Weathering

Differential weathering refers to rocks that weather at different rates in the same climate. Graveyards are excellent places to compare the weathering rates of various kinds of rocks. The tombstones are dated, and we need only observe the condition of rocks of the same age. A walk through a New England cemetery, or a cemetery in any other suitably humid climate, is especially revealing (Figure 10.10). Limestone and marble tombstones weather most rapidly because of their susceptibility to the dissolved acids in rainwater. The inscriptions of stones whose age is greater than about two hundred years are usually indecipherable. By contrast, slate tombstones remain clear and readable. Granite tombstones hold up longer than limestone, but they lose their gloss within a century. The feldspars appear dull and discolored, and the rock surfaces are somewhat pitted, perhaps due to the dissolution of more soluble dark minerals like biotite or hornblende. But the blocks are otherwise in good shape.

Limestone and marble tombstones of the arid Southwest are in strikingly better condition than those of the same age in the rainy East. Again, climate is the key influence.

The sparse moisture inhibits chemical weathering, allowing the mechanical properties of the rock to determine how rapidly it will disintegrate. Because limestone blocks are relatively compact and largely homogeneous, they offer greater resistance to temperature change, salt crystallization, and similar mechanical processes than many sandstones and shales. This resistance translates to sheer, massive limestone cliffs in arid regions, just about the opposite of the retiring aspect that limestone assumes in humid climates.

Soil Formation

The loose rock material that nearly everywhere forms the land surface and covers the bedrock of the crust is called **regolith,** from the Greek *regos* ("blanket") and *lithos* ("stone"). Except for minor amounts of volcanic ash and infinitesimally small quantities of meteoric dust, all regolith is derived from the weathering of bedrock. **Soil** is regolith that has incorporated organic matter and is capable of supporting vegetation.

The soil of any given locality is a dynamic body. Its condition results from the interplay of climate, rock material, groundwater, organic activity, topography, and time. It is no wonder, then, that soils are difficult to study and classify. However, the overall patterns of soil development are more easily understood.

The condition of a typical soil is determined by additions, subtractions, rearrangements, and recyclings of materials. Materials are added to the soil by weathering that reduces bedrock to clay, silt, and sand and that releases soluble elements for uptake by plant roots. Materials are subtracted from and rearranged within the soil body by rainwater, which percolates through the soil on its way to the water table. **Leaching** is the process by which soluble elements are transported downward in solution from the upper levels. Some are redeposited at depth along with clay and silt carried in suspension. Materials are also recycled and retained within the soil by complex plant, animal, and bacterial communities that live within and above the soil. The net result of their activities is called *nutrient cycling.*

The Soil Profile

The downward percolation of water through soil produces a characteristic layering. Each layer is called a **soil horizon,** and the totality of these horizons constitutes the **soil profile** (Figure 10.11).

Just below the vegetative mat of partially decomposed plant matter, called the *O horizon,* is the first soil layer, called the *A horizon.* This is the zone from which positive

regolith
All the fragmented and unconsolidated material overlying bedrock.

soil
Unconsolidated material capable of supporting vegetation.

leaching
The transfer in solution of organic matter and other elements from upper to lower soil levels.

soil horizon
A soil layer with physical and chemical properties that differ from those of adjacent soil layers.

soil profile
A vertical cross section of soil that displays all soil horizons.

O horizon: Partly decomposed plant matter

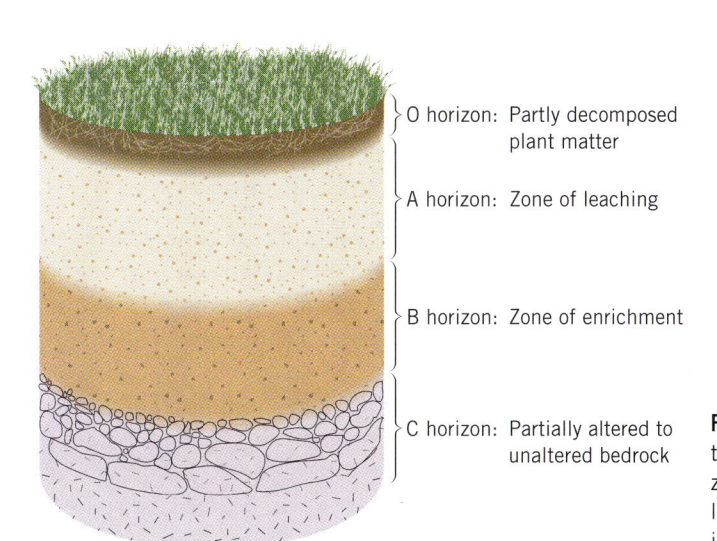

A horizon: Zone of leaching

B horizon: Zone of enrichment

C horizon: Partially altered to unaltered bedrock

Figure 10.11 A typical soil profile for temperate climates. The various horizons are the products of weathering, leaching, and redeposition by percolating groundwater.

ions are leached and clays are removed. Because of the leaching, the layer tends to be porous with a sandy or silty texture. At the surface, its color is dark because it mixes with partially decomposed organic matter; lower down, its color generally lightens. The *B horizon* marks the enriched zone in which the positive ions that are leached from the A horizon are precipitated, and the clays carried in suspension are deposited. The accumulation of clay may produce a dense structure that is hard, physically tough, and difficult to plow. "Hardpan" is the common name given to this layer. The B horizon is characteristically richly colored. If calcium accumulates, as in desert climates, it is white. If iron accumulates, as in humid climates, it is dark red or orange.

A zone of partially weathered bedrock, the *C horizon*, lies beneath this enriched layer. It is formed by direct reaction between the downward-percolating groundwater and the rock. The C horizon grades into fresh bedrock below and closely resembles it, but appearances are deceiving: the feldspars have been weathered to clay, and the "rock" crumbles easily. Drillers aptly describe this zone as "rotten rock." As long as conditions are unchanged, the A, B, and C horizons will continue to thicken at the expense of the bedrock beneath.

Nutrient Cycling

Leaching impoverishes the A and O horizons. Thus, over the long term, it would inhibit plant growth were it not for a countervailing process that replenishes the soil: nutrient cycling. Nutrients extracted from the soil are returned to it when plants die or shed leaves. Leaves and other plant matter are broken down by bacteria, which produce humus, a black to dark brown organic substance. Humus mixes with clay to provide numerous bonding sites for nutrients such as potassium, sodium, magnesium, iron, and calcium ions. Because the nutrients are weakly held, plant roots can easily extract them. Thus soils with rich clay-humus complexes are generally highly productive, whereas those lacking sufficient humus tend to be infertile. There is a practical application here for gardeners who bag and discard fallen leaves and lawn cuttings. This practice interferes with the nutrient cycle by not allowing the leaves to naturally decompose to soil nutrients—driving the gardeners to spend money on fertilizers to replace them.

A given plant species may cycle some nutrients and exclude others. For this reason, agricultural systems—especially those that rely heavily on a single crop not native to the region's soil—often impoverish the soil for other species, including other crops, by altering the nutrient balance.

Climate and Soil Types

The two major soil-forming processes, leaching and nutrient cycling, depend upon the availability of water in the form of precipitation and energy in the form of solar heat. Since precipitation and solar heat are the key elements of climate, most soils, given sufficient time, reflect regional climate. As climate varies, so do soil types.

Soils of Tropical, Humid Climates

Brick red *laterite* (from the Latin *later*, meaning "brick" or "tile") soils develop in hot, humid climates that support rain forest or savanna vegetation. The drenching rains and high temperatures of these climates promote extensive soil leaching. Left behind in the leached soil are oxides of iron and aluminum. The iron oxides produce the red color of laterite soils.

In some regions, these residual aluminum and iron oxides accumulate in commercially valuable concentrations. Bauxite, the chief aluminum ore, forms in this manner. Huge reserves of it are found in the deep laterite soils of Jamaica, among other places (Figure 10.12). The iron oxide ores of western Australia have formed in similar fashion. They constitute one of the world's richest iron deposits.

Thick laterite soils support the magnificent rain forests of the Amazon, Central America, equatorial Africa, and Southeast Asia. Some of the world's poorest and most densely populated nations are also located in or near these rain forests. It is one of nature's cruel ironies that these soils, capable of spawning luxuriant plant growth, will not generally support sustained agriculture or ranching; they are among the world's poorest soils for these purposes. High rainfall leaches the nutrients out of the A horizon. In addition, those nutrients that remain in the soil are snatched up rapidly by the fast-growing plants. We can readily predict the results of aggressive human intrusion into this environment. Clearing the land, planting crops, and allowing cattle to graze further impoverishes the soil, because it destroys the vegetative cover and short-circuits the recycling of nutrients in the soil by plant decay. It also accelerates leaching, erosion, and the development of hard laterite crust in the B horizon.

Figure 10.12 This bauxite, an aluminum ore, is the product of prolonged and intensive soil leaching in a humid tropical or subtropical climate.

Soils of Arid and Semiarid Climates

Soils formed in arid and semiarid climates are as different from laterites as deserts are from tropical rain forests. Whereas laterites are residues, the products of extensive leaching, arid soils result from insufficient leaching, so that they retain highly soluble nutrients that would be carried away under more humid conditions.

Two factors work in arid climates to keep nutrients from escaping the soil: **calcification** and natural vegetation. Calcification is the process of enriching the soil in highly alkaline elements, especially calcium compounds. It works in the following manner: downward-percolating rainwater is evaporated before it can pass through the soil; indeed, it may actually be drawn back up toward the surface by capillary action. Left behind in the B horizon are very soluble ions that are normally carried away in rainier regions: sodium, potassium, magnesium, and calcium. These ions impart a light, chalky color and texture to the B horizon. Due to evaporation and capillary action, these highly soluble ions are plentiful in the A horizon as well.

calcification
The encrusting of a soil with alkaline compounds, particularly calcium oxides.

Arid conditions may therefore favor the development of potentially fertile soils, because some of them contain large stores of nutrients and humus close to the surface in the A horizon. All that is needed is sufficient rainfall to support vegetation and leach away some of the alkaline elements. The proper blend of ingredients occurs in mid-continent regions of middle latitudes, where extremely arid conditions give way to a somewhat wetter climate capable of supporting grassland vegetation. These areas are the great grain-producing breadbasket regions of the world: the North American Great Plains, the Ukraine, Asia's Manchurian plains, and South America's pampas.

Salinization is a serious problem in lands irrigated for agriculture in arid climates. The irrigation water soaks into the ground and raises the level of underground water until it seeps to the surface and evaporates, leaving behind a white, powdery salt residue (Figure 10.13). Little of value can grow in soil like this; the salty soil kills plants and soil bacteria alike. Agricultural scientists estimate that each year, salinization ruins hundreds of thousands of acres the world over. For example, salinization threatens major portions of California's San Joaquin Valley, which produces billions of dollars worth of crops annually. In the past, salinization has literally destroyed entire civilizations; 4000-year-old records from Mesopotamia (modern Iraq), for instance, describe abandoned farmland and mass migration as the land salted over.

Soils of Cold to Temperate Climates

The soils of cold, subarctic climates support the great belt of pine forest that stretches across northern Canada and Europe. These pine forests have shaped the properties of the soils that support them by affecting the chemistry of the plentiful rainwater that filters through the fallen pine needles matting the forest floor. The water that filters through the acidic pine needles also becomes acidic. This water, in turn, leaches from the A horizon not only such soluble elements as sodium, calcium, potassium, and

Figure 10.13 Salt-ravaged agricultural land northwest of Victoria, Australia.

magnesium but also normally insoluble aluminum and iron oxides. The oxides are derived from the clay-humus complexes, which are decomposed by the acidic water. Left behind in the A horizon is a pale, ash-colored, silica-rich soil that is poor in nutrients. The pines have adapted by cycling few nutrients through the soil.

However, the clays, oxides, and organic matter leached from the A horizon are redeposited in the B horizon. They color that layer yellow, red, or black, which gives the entire profile a distinctive color banding. The soil-forming process we have described is called *podzolization*.

North of the pine forests, moisture and precipitation diminish and weathering is slow. Podzol soils grade into the poorly developed permafrost soils of the Arctic tundra. The ground surface is frozen much of the year, and nutrient cycles are greatly retarded. Only mosses, lichens, low shrubs, and other tenacious species can survive in this harsh environment.

To the south of the pine forest region, however, conditions are less restrictive. The vegetation changes to the mixed-forest type; trees and brush are more luxuriant; and roots sink deeper into the soil in search of greater quantities of nutrients. Nutrient cycling is more intense, and in general, the soils are richer than those farther north. Much of the mixed-forest soils of the mid-Atlantic states has been converted to highly productive farmland.

Mass Wasting

Mass movements of materials downhill take four general forms: fall, slide, flow, and creep. Because materials often exhibit more than one type of movement in their journey downslope, it is not always easy to classify a specific event.

Falls

Falls result when rock fragments are dislodged from a steep cliff face and fall unrestrained or when the face is undermined, causing it to collapse in a heap along vertical joints.

Earlier in this chapter we discussed mechanical- and chemical-weathering effects that can dislodge rock materials at high elevations: frost wedging, exfoliation, and salt crystal growth. Generally, these effects produce a steady, though scarcely noticeable,

fall
The free downward movement of detached rock fragments through the air from a cliff or steep slope.

Weathering, Soils, and Mass Wasting 225

Figure 10.14 Rockfall boulders stacked at the base of the sheer cliff of the Palisades, New Jersey. The cliff is divided by vertical joints or fractures into massive columns. Not visible in the photo is an easily weathered layer at the base of the cliff (the olivine zone). Erosion of this layer deprives the columns of support, which leads to their collapse.

cascade of fragments. Occasionally, however, weathering produces sudden, dramatic rockfalls. Figure 10.14 shows the results of collapse along the Palisades sill of New Jersey, which runs along the west bank of the Hudson River. The steep sill is composed of resistant diabase and is subdivided into massive columns by vertical joints. The joints are responsible for the steep, imposing appearance of the cliff. Because the thin zone of olivine-rich rock at the base of the sill weathers more easily than the rest of the sill, it leaves the huge vertical columns unsupported. As the joints widen, friction is diminished, and the rock columns collapse under their own weight. In this manner, the Palisades are being worn steadily back from the Hudson.

The fragments dislodged in falls collect at the base of the cliff, forming wedge-shaped accumulations called **talus** slopes (Figure 10.15). The angular fragments of the talus slopes come to rest at steep **angles of repose** (about 35°), which are maintained as fresh talus is added to the slopes.

The angle of repose of a natural slope depends upon the size of the individual particles, their angularity, the degree of sorting, and their wetness or dryness. Larger, more angular particles, such as gravel, tend to form slopes with steeper angles of repose than finer particles. Most natural slopes range from 25° to 40° in steepness.

Given sufficient time, the cliffs can choke in their own debris. More typically, however, fragments at the base of talus slopes are removed through stream erosion, which allows the slow downhill movement of fragments from the head to the base of the slopes. Thus, the form of talus slopes may remain constant, but over long time periods, the fragments that compose them change continuously.

Slides

The term **slide** generally refers to the mass movement of rock and soil downhill along a shear plane (Figure 10.16). More specifically, a **rockslide** is the movement of blocks of bedrock, and a **landslide** is the movement of a more or less consolidated mass of soil and rock fragments.

Rockslides and landslides may occur quite suddenly, although close inspection—usually after the fact—reveals plenty of warning signs. On October 1, 1963, a shepherd

talus
Coarse, angular rock fragments that collect at the base of the cliffs from which they were dislodged.

angle of repose
The maximum slope, or angle, at which a pile of loose material remains stable.

slide
The rapid downslope movement along shear planes of more or less consolidated rock or fragmented materials.

rockslide
The rapid downslope movement along shear planes of a mass consisting mostly of large chunks of bedrock.

landslide
The rapid downslope movement along shear planes of a mass of rock fragments and soil.

Figure 10.15 Talus slopes at the base of the Crowfoot Glacier in Canada. The slopes are composed of fragments weathered from the cliff face. The fragments were dislodged from loosely cemented sedimentary rocks, which accounts for their relatively small size. Contrast their size with the massive boulders at the base of the Palisades in Figure 10.14, which were derived from resistant igneous rock.

noticed that his flock of sheep refused to graze on the north slope of Mount Toc in the Italian Alps. Eight days later, 300 million cubic meters of rock and soil slid down the north slope at more than 90 kilometers per hour and plunged into the Vaiont Reservoir, which runs along the axis of a narrow valley (Figure 10.17). A chunk of mountainside nearly 600 meters high and 2.4 kilometers wide had collapsed and filled the reservoir with debris 150 meters above its former level. To cart the load away, a typical pickup truck would have had to make 500 million trips!

The landslide sent a huge blast of compressed air, water, and rock up the valley walls; the blast was strong enough to lift the roof off a house 250 meters above the reservoir. Like an enormous plunger, the slide also sent a wall of water 100 meters high over the top of the Vaiont Dam. There, it poured as a 65-meter wave to the Piave River 1.5 kilometers below the dam. The surging waters spread upstream and downstream along the Piave River for many kilometers and flooded the populated valley on either side. The lives of 2600 people were lost in an event that took about 7 minutes from start to finish.

Figure 10.16 Two kinds of slides. (a) A rockslide involves the movement of large blocks of bedrock along a shear plane. (b) A landslide involves more or less unconsolidated rock and soil, although the term is often used in a more general sense to include many types of mass wasting.

(a) Rockslide

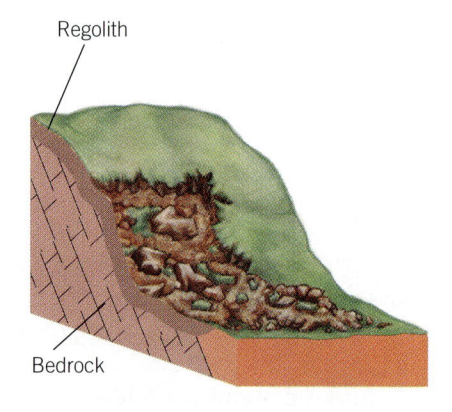

(b) Landslide

(a)

(b)

Figure 10.17 (a) This photo, taken shortly after the Vaiont Dam landslide disaster, shows the enormous quantity of debris that filled the reservoir. The debris acted as a plunger, forcing a wall of water to spill over the dam and flood the communities downstream. The crest of the dam, visible in the photo, stands approximately 265 meters above the valley floor. (b) This geological cross section of the reservoir behind the Vaiont Dam illustrates some of the conditions that led to the disaster: shear planes dipping steeply toward the valley; massive limestone formations interbedded with clay layers that served as weak support; and a rising water table that lowered friction along the slide surfaces.

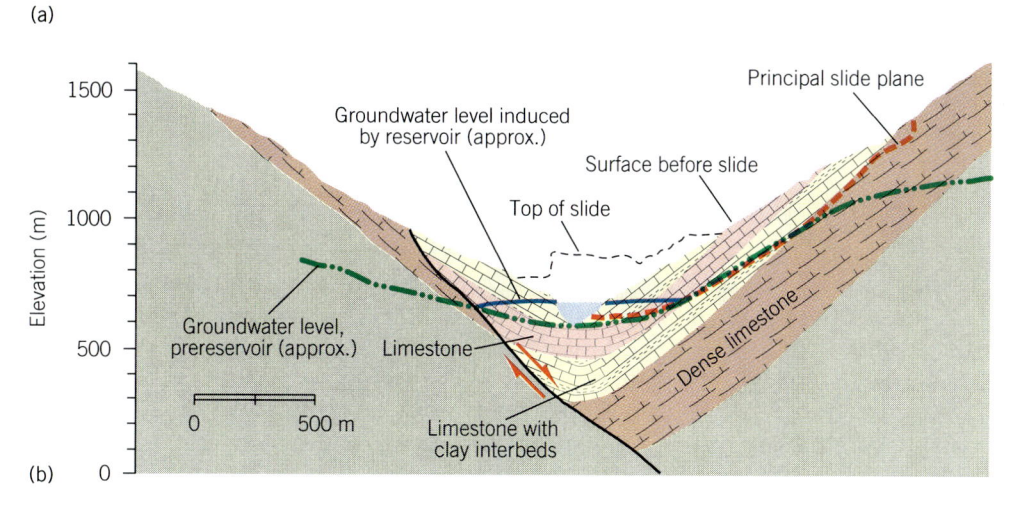

The Vaiont Dam was newly constructed at the time and one of the world's highest, at slightly more than 265 meters. It sustained no significant damage, despite the tremendous stresses placed upon it—which is ironic, since it was the construction of the dam that set the disaster in motion. First, the dam had been built in an area of unfavorable geologic structure. The bedding planes of the limestone and shale formations dipped steeply toward the valley of the reservoir, as did prominent fault planes within these rocks. The limestone and shale strata, weakened by fracturing and weathering, were subject to shear failure.

Second, the groundwater level had been raised by impounding the water behind the dam. The water infiltrated the bedding planes, fractures, and pore spaces of the rock formations, further lowering the shear strength of the rock. Conditions had worsened in the weeks preceding the disaster. Drenching rains saturated the soil and seeped into the bedrock, thus adding weight to the slope. In the three years prior to the heavy rains, the rate of soil creep along the north slope of Mount Toc had been a few centimeters per week, but it jumped to 25 centimeters per day in late September. The day before the slide, the creep rate had reached 1 meter per day as the mountainside began to slide along the shear planes.

These events were not lost on the engineers at the site. They suspected that disaster was imminent and sought to release water via the spillway of the dam. But the water released was more than offset by the rainwater washing down from the slopes. The Vaiont Reservoir and Dam had turned out to be a magnificently designed deathtrap.

Conditions Favoring Slides

The conditions that led to the Vaiont disaster are similar to those of most other slides:

1. *Steep slopes and adverse geologic structures.* Most slides involve either bedding or fracture planes that dip steeply downslope. This orientation provides shear planes over which rocks can slide. On the slopes on either side of the Vaiont Dam, the bedding planes of the limestone and shale formations, as well as the fault planes within them, dipped steeply toward the valley of the reservoir—a catastrophe just waiting to happen.
2. *Weak layers that support heavy loads.* Shale, especially when wet, lacks the shear strength of sandstone or granite. In the Vaiont Dam disaster, increased penetration of groundwater greased the skids of the shaley layers while adding weight to the slope.
3. *A triggering mechanism.* Slides are initiated by natural processes and human activities: earthquakes, oversteepened or undercut slopes, waterlogged rock and soil, and altered groundwater levels. In the Vaiont Dam disaster, unusually heavy rains lent enough additional weight to the north slope to pull a 600-meter-high chunk of the mountainside down in a heap.

Slumps

Not all slides involve the downslope movement of bedrock along bedding planes or preexisting fractures. A **slump** is a type of slide that occurs when intact blocks of rock or unconsolidated debris slide downward along concave planes. The blocks also tend to tilt backward during slump, so that rotation as well as downward motion takes place. Commonly, slumps occur on hills thickly blanketed with soil or on steep shoreline cliffs made of loose, poorly cemented materials.

The reason the glide planes of slumps are curved is not precisely known, but as with all slides, they are surfaces of shear failure. As in all mass movements, slumps are most common on slopes that are oversteepened, water-saturated, and undercut by streams, waves, or human activities. Earthquakes also set slumps in motion. Most of the damage that Anchorage, Alaska, sustained during the earthquake that nearly destroyed the city in 1964 has been attributed to slumps and other mass-wasting effects (Figure 10.18).

Precarious subsurface conditions exist beneath Anchorage. The city was built on an 18-meter layer of gravel supported by the thick Bootlegger Cove Clay. This water-soaked clay formation liquefied as the ground shook during the earthquake, a situation intensified by water pouring from broken water mains and pipes. The relatively stable gravel deposit and stiff upper portion of the clay slid over the liquefied clay beneath, toward Cook Inlet to the southwest. In the center of the city, the motion caused huge fault blocks to sink along fractures. At Turnagain Heights, the high bluff overlooking Cook Inlet, the ground split as the heights slid toward the inlet.

slump
Downslope movement of rock or loose debris along a concave plane.

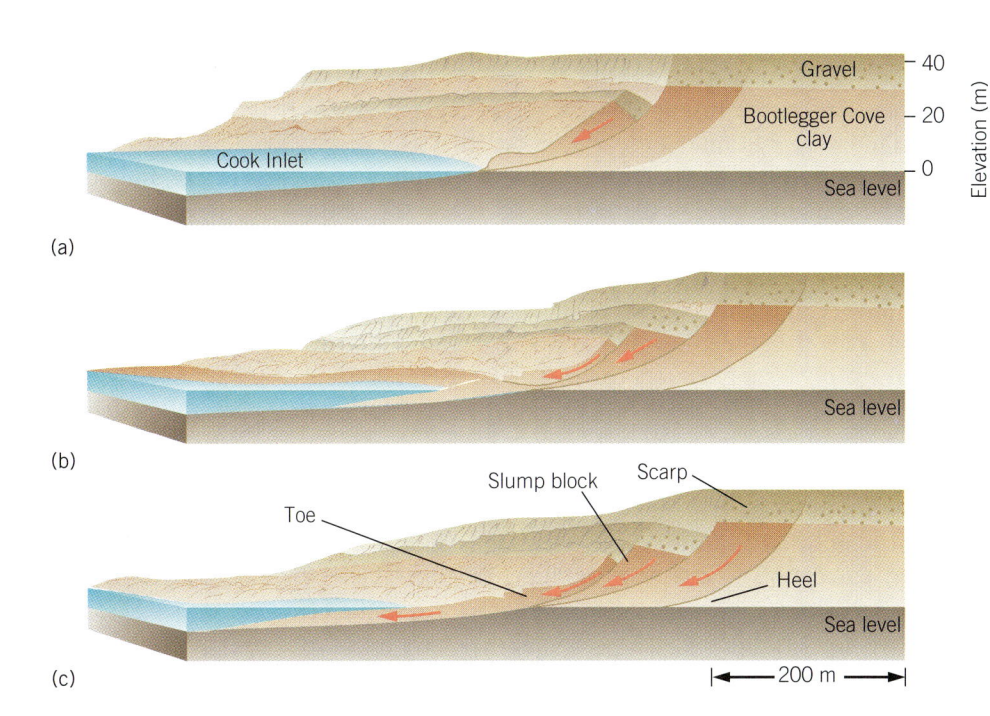

Figure 10.18 (a–c) Stages in the development of slump blocks in Turnagain Heights. The slumping was set in motion by the Anchorage, Alaska, earthquake of 1964. Notice the steep scarp, the curved slide plane, the heel and toe of the slide. All are characteristic of slump structures. (d) An aerial photograph of Turnagain Heights, showing the complex, curving fractures at the top of the scarp.

The cracked ground formed huge slump blocks that followed curved paths downslope (Figure 10.18a–c). Note the steep, curving *heel* of the slump and the *toe* of slope material that extends outward at the base. This arrangement reflects the balance of forces that brought the block to rest as it slid downward and rotated along the curved glide planes. Should the toe of the slump be removed—by natural erosion or by a poor road cut—the instability will trigger further slumpage.

Turnagain Heights was a fashionable suburb, but the slump blocks did not respect property; houses were severed and splintered. The panicked inhabitants were carried down in the dark on the slump blocks together with their shattered houses. It was sheer luck that few people were injured.

flow
Downslope movement of loose rock and soil as a viscous fluid mass.

earthflow
Downslope movement of a loose mass consisting mainly of rock fragments and soil in a semifluid state.

mudflow
Downslope movement of a fluidized mass of clay and other fine-grained materials.

Flows

A **flow** is a water-soaked mass of loose rock, soil, and sediment that tends to behave as a viscous fluid; that is, it more or less sticks together as it moves downhill. However, the characteristics of a given flow depend on its water content, its particle size, the degree of particle sorting, the angle of the hillside slope, and the width of the flow channel. All these features make flows difficult to classify, but there are three main types: earthflows, mudflows, and debris flows.

Earthflows commonly consist of water-saturated clay, shaley fragments, soil, and similar materials that slide downhill as a viscous mass. During downslope movement, an earthflow may be confined to a channel of some sort. Once it reaches the valley floor, it spreads more like stiff molasses than muddy water. For this reason, earthflows end in distinctly curved boundaries or lobes.

Mudflows are less viscous and consist of finer-grained materials than earthflows. Containing up to 60 percent water by weight, these dense, well-lubricated masses can move very rapidly. As you may recall from Chapter 6, snowcapped volcanic peaks are particularly susceptible to a type of mudflow called a lahar. The peaks frequently contain thick deposits of pyroclastic debris set at a high angle of repose on steep slopes. During a volcanic eruption, the snow around the crater melts and begins to flow quickly downslope. At the same time, the shaking of the mountain loosens debris, which mixes with the flowing water. Huge volumes of mud may run down the flanks of the volcano at high speeds and smother all below it.

In fact, much of the damage and death in the Mount Saint Helens, Washington, eruption of 1980 was caused by lahars rather than by fresh lava flows. Also, the lahars triggered by the eruption of Mount Pinatubo in 1991 devastated huge portions of the island of Luzon in the Philippines. Green farmland and thriving villages were smothered in hot mud, which cooled to an ugly gray mass, as hard as cement. A far greater disaster, in human terms, occurred on November 13, 1985, when a lahar descended the 5400-meter slope of Nevado del Ruiz, Colombia (Figure 10.19). Channeled within the valley of the Lagunilla River, it was 15 meters high and traveling 70 kilometers per hour by the time it reached the town of Armero, 48 kilometers from the summit. The inhabitants did not stand a chance. Approximately 25,000 people were buried alive in the mud.

Figure 10.19 In 1985, this mudflow (or lahar) poured down the flanks of the Nevada del Ruiz in Colombia at 70 kilometers per hour. The town of Armero and its 25,000 inhabitants were buried in 15 meters of mud.

Figure 10.20 The regional setting of the Yungay, Peru, debris flow of 1970. The scar in the face of Nevada de Huascarán at the right marks the site where a huge block of ice was dislodged from the mountain by a large earthquake. Upon impact, the snow and ice melted, thus setting in motion the debris flow that buried Yungay and neighboring towns along the Rio Santa at the base of the mountains.

Debris flows are defined as moving masses of rock and soil in which more than half the materials are coarser than sand size. Depending upon the viscosity of the mass, debris flows can vary in speed between 1 meter per year to hundreds of kilometers per hour. In 1970, Yungay, Peru, bore the brunt of what is probably this century's most disastrous debris flow. Figure 10.20 illustrates the chain of events set in motion by a large earthquake that occurred in the subduction zone bordering the Peruvian coast. The earthquake shook an enormous ice block from a glacier high up on the sheer face of 6540-meter-high Nevado de Huascarán—a mountain that looms above the valley where the city is located. The ice plunged 1000 meters before striking the ground. The collision set in motion 80 million cubic meters of rock, melted snow, and mud that roared down the valley of Rio Santa at 280 to 335 kilometers per hour. Well over 20,000 people—including the entire population of Yungay and the inhabitants of innumerable tiny villages—were killed.

Slumps, earthflows, mudflows, and their coarse-loaded cousins, debris flows, are common occurrences in arid climates. Lacking vegetation, the soil of hillside slopes in these climates is quickly saturated when beset by sudden, torrential rains. Of course, humans often unknowingly use their ingenuity to create these very same conditions in nonarid climates. People strip hillside slopes naked through deforestation, careless forest fires, or poor farming practices. They may then compound the problem by building houses and planting grass on the slopes, and watering the grass for hours on end with sprinklers. At some later date, they may express genuine surprise and consternation when the hillside runs away with their homes, their lawns, and all their possessions. Figure 10.21 shows some all-too-common ways that human activities can turn a stable slope into an unstable one.

debris flow
Downslope movement of a viscous mass of rock and soil particles, more than half of which are larger than a sand grain.

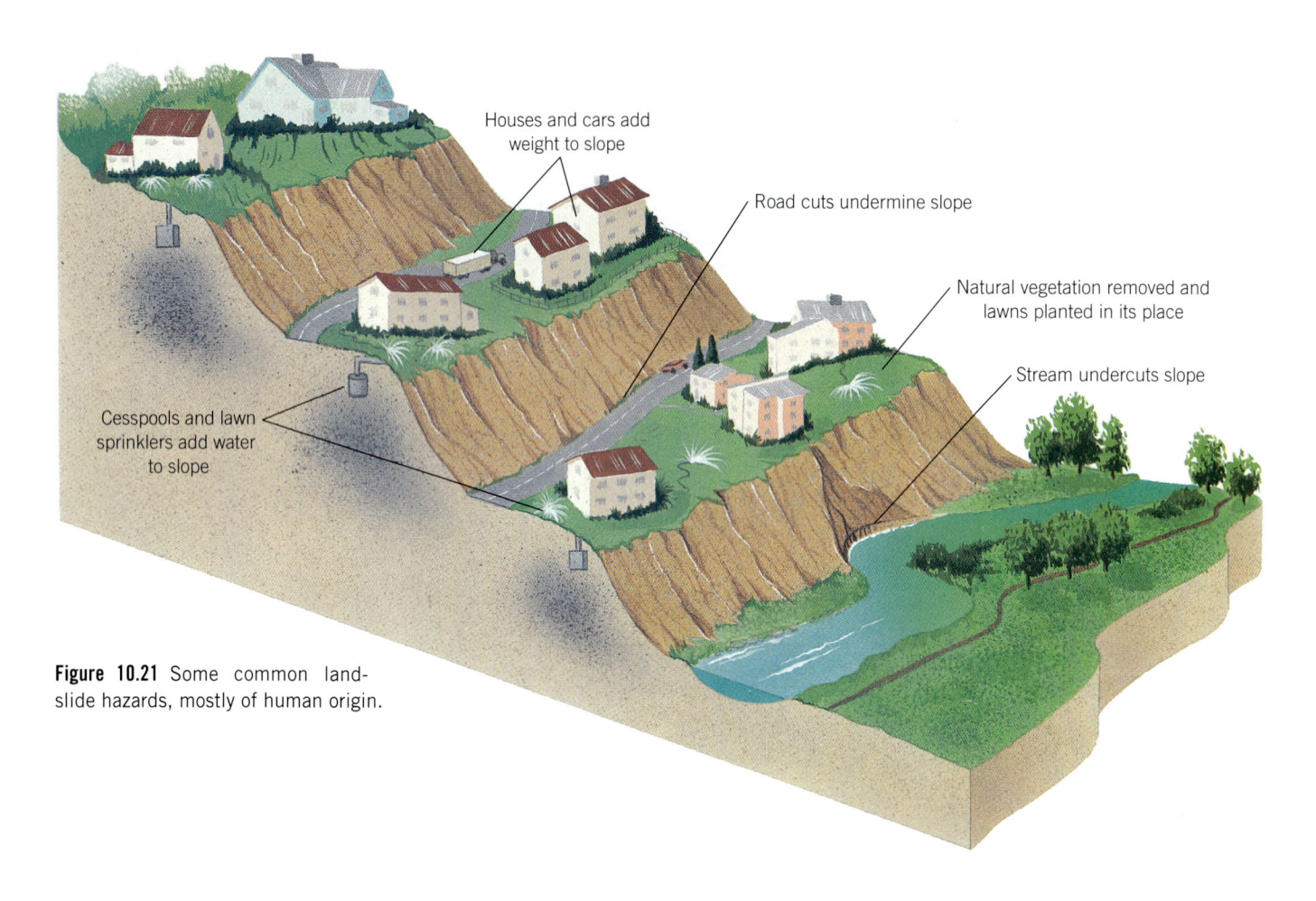

Figure 10.21 Some common landslide hazards, mostly of human origin.

Creep

creep
The imperceptibly slow downslope movement of soil and rock particles, mainly occurring in cold climates where water alternately freezes and thaws.

Creep is so slow that it can be measured in centimeters by the downward migration of fence posts, tombstones, and trees over decades or centuries (Figure 10.22). Yet on a worldwide basis, more material moves to the bottom of hills by creep than by any other form of mass wasting. The secret to understanding the enormous impact of creep is to grasp the cumulative impact of gravity acting over long time-periods. Whether it is the size of a sand grain or a boulder, a dislodged rock will roll downhill, not up. A soft, semifluid mass of soil will likewise sag downhill. Thus, nearly every event that loosens fragments or softens clay, silt, and sand contributes to downslope movement: the impact of grazing and burrowing animals, the splash of raindrops, the saturation effects of groundwater.

Creep processes are common at high latitudes and high elevations where water alternately freezes and thaws. As water seeps into the ground and freezes, its expansion lifts rock and soil at right angles to the sloping hillside surface. However, when the ice melts, materials are dropped vertically under direct influence of gravity. The result is a net downhill movement of the surface, scarcely noticeable but cumulatively important. The alternate wetting and drying of clay minerals, leading to expansion and contraction, produces similar downslope movements in all climates.

permafrost
A condition of permanently frozen soil or subsoil occurring in cold climates.

The freezing and thawing of water produces a number of striking landforms and other effects in cold climates. Most are associated with **permafrost,** a condition that develops when all the water within the pore spaces in regolith and rock remains frozen solid throughout the year. Permafrost covers 20 percent of the land area of the Earth, including 80 percent of Alaska. The frozen ground can reach depths greater than 600 meters in Alaska and Siberia. The thin soil blanket above the permafrost thaws each

Figure 10.22 Unrelenting hillside creep is buckling this stone wall on a Berkeley, California, street.

summer and becomes saturated, because the frozen ground beneath will not allow water to seep deeper into the regolith. This process is responsible for the frequently boglike, marshy conditions of arctic tundra landscape. Saturation is also responsible for **solifluction,** a form of creep in which the waterlogged soil slips slowly downhill and, like ripples in a rug, forms wavy *solifluction lobes* (Figure 10.23). Although the lobes give the appearance of great activity, solifluction is a slow process.

Permafrost presents difficult environmental and engineering problems that depend on soil type, regional drainage, local climate, and type of construction planned. At the root of most of these problems are the human-made disruptions in the natural regime that cause the subsurface ice to melt: construction that adds weight to the soil; the clearing of natural vegetation that alters the amount of solar heat absorbed by the soil; rumbling heavy machinery that compresses and disrupts the subsurface; and subsidiary roads that disturb the natural surface drainage. If a summer is unusually warm, the melting permafrost can cause extensive damage. The softened ground leads to land subsidence, settling of structures, lateral slippage of load supports, landslides, and flows. When the ground refreezes in the winter, it triggers massive frost heaving, which, in turn, causes further stress on structures.

The ecology of arctic tundra has come under increasing stress as oil and other resources have been discovered near and within the Arctic Circle and as populations have spread north. Though the survival problems of tropical rain forests have received a great deal of attention, comparatively little has been given to the perils facing the starkly beautiful tundra landscape, underlain by permafrost.

solifluction
The slow downslope movement of waterlogged soil caused by repeated freezing and thawing in cold climates.

Predicting and Preventing Mass-Wasting Disasters

Having discussed the ways in which soil, rock fragments, and boulders move from the top to the bottom of hills, it is reasonable that we should now discuss ways to prevent them from burying us. The superficially simple solution—get out of the way—is not

Figure 10.23 Solifluction lobes of the Alaskan tundra.

always possible, as the examples cited in this chapter and elsewhere illustrate. Assuming that it is possible to escape with one's skin intact, there is the further problem of preventing or minimizing property loss to roads, utilities, buildings, and dams. These structures cannot get out of the way. We also have to contend with the slower, more subtle forms of mass wasting, such as soil creep. While not immediate threats, they often spell disaster on the installment plan—the ruination of farms and other valuable property.

Recognition of unfavorable sites begins with a detailed geological survey of the region in which geologists search for all the conditions discussed in this chapter.

1. *The relationship of bedrock structure to topography.* Do the rock strata dip toward the sea, where wave erosion is active? Do they dip toward steep valleys, where streams undercut slopes?
2. *The condition of the bedrock.* Do massive rock and soil formations overlie strata of low shear strength? Is the rock highly fractured?
3. *The composition and steepness of hillside slopes.* Is the hill mostly bedrock? Or loose material? If the latter, what is the percentage of clay, sand, and boulders? Are these materials close to their angle of repose, or are the hillsides oversteepened? What is the shear strength of the materials on the slope? How will they be affected by heavy rains? Are the slopes vegetated or bare? Has there been evidence of recent slippage?
4. *The groundwater conditions that underlie the slopes.* How do such structures as reservoirs or factories affect them?
5. *The tectonic history of the region.* What is the likelihood that earthquakes of a given magnitude will occur? How are the rocks and soil at specific locations likely to respond to severe ground shaking?
6. *The location of roads, dams, bridges, buildings*—indeed, whole towns—with respect to potential hazards.

Geologists are, of course, primarily concerned with being able to predict mass-wasting disasters. Telltale signs include nearly parallel, curving cracks high on steep slopes or recent excavations, fresh springs, and acceleration of the soil creep rate. A slope on the move is generally wavy and has few deep-rooted trees. The creep rate of centimeters per year often changes to a gallop as a full-fledged landslide develops. Thus, if cracks develop in your hillside home, you should have an engineering geologist check out the site, although there may be other explanations.

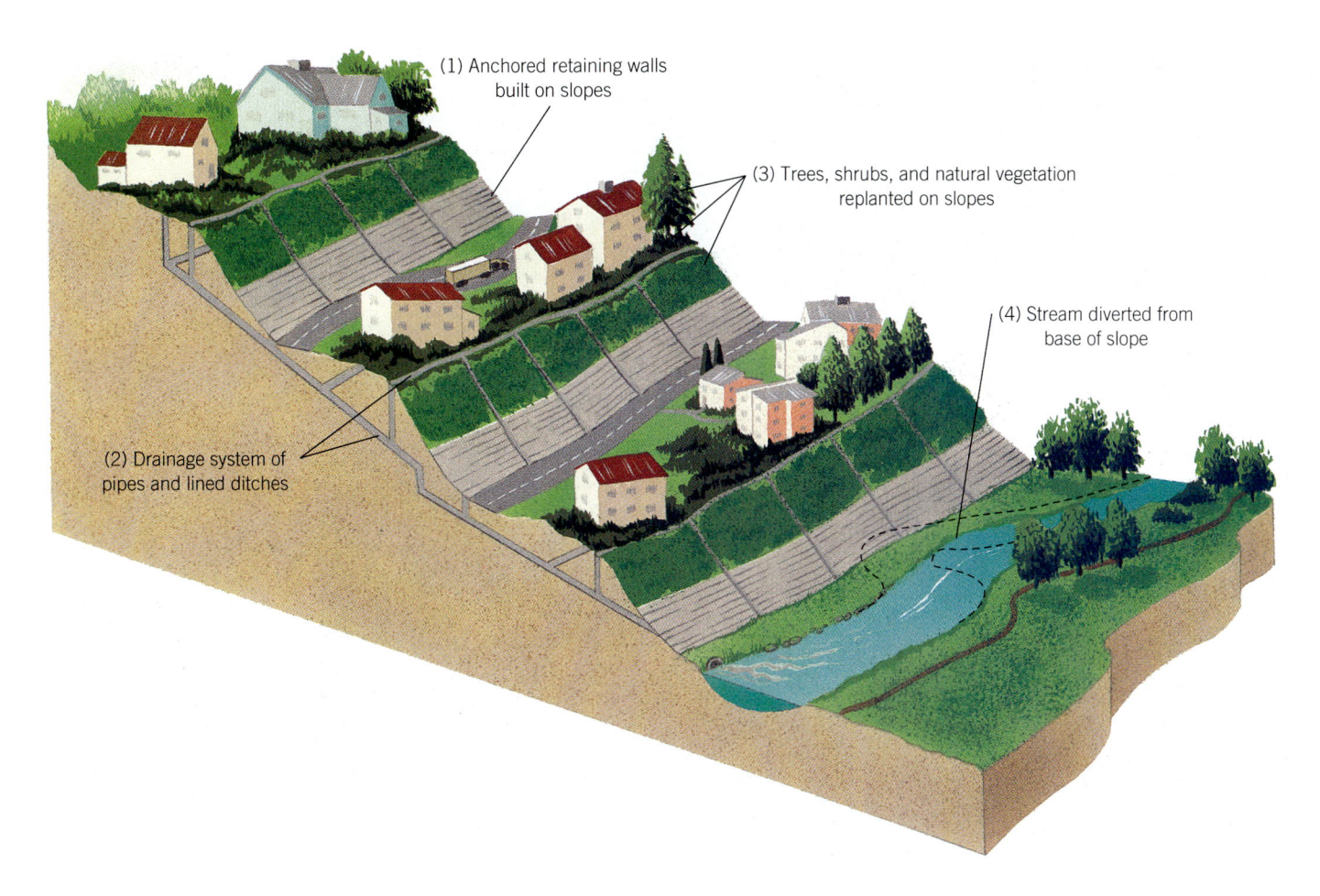

Once a potential landslide site has been spotted, how can the slope be modified so that it remains stable? Recall that on stable slopes, a balance exists between gravity and driving forces (which tend to pull material downslope) and frictional and cohesive resisting forces (which prevent movement). In theory, there are many ways to achieve stability (Figure 10.24). However, the engineering geologist must pick the most practical means available within the physical and financial constraints of the project, which is quite another matter. Also, some solutions are beyond the capacity of geologists and engineers. They require changes in zoning regulations and lifestyle habits. For example, if waterlogging is a problem, community waste may have to be treated in a separate facility rather than in septic tanks on the slope. Or perhaps lawn sprinkling will have to be rationed.

This discussion of mass-wasting control has focused on the hazards confronted in the United States and similar economically advanced nations. In parts of the world that are less developed economically, however, the prevention of slides and mudflows is an extremely serious matter affecting millions of people. Population experts forecast that by the year 2000, more than half of humankind will reside in huge metropolitan regions called "megacities." The most spectacular growth will occur in less-developed countries where recent migrants from the provinces frequently settle in the shanty towns that carpet the steep hillsides surrounding older central cities such as Lima, Peru; Rio de Janeiro, Brazil; or Tijuana, Mexico. These slums often lack sewers and plumbing and are bare of vegetation; a heavy rain can thus mobilize an entire hillside (Figure 10.25). As the mud and water pour downslope in torrents, they carry away virtually everything these impoverished people own. Preventing mudflows is a vital step toward improving their lives.

Figure 10.24 Steps commonly taken to prevent landslides on a developed hillside: (1) buttress slopes with retaining walls or bulkheads; (2) install a drainage system of pipes and ditches to carry water off the slope; (3) plant trees and shrubs to soak up water and retard erosion; and (4) divert stream channels from base hillside slopes.

Figure 10.25 The aftermath of a mudslide in Puerto Rico. Slides such as these are tragedies wherever they occur, but their impact on impoverished communities is especially devastating.

STUDY OUTLINE

I. **WEATHERING, MASS WASTING, AND GRADATION.** Weathering and mass wasting are the first and second steps in the vast redistribution process of **gradation,** in which erosion wears low the high portions of the land surface and deposition builds up the low portions.

II. **WEATHERING PROCESSES**
   A. **Mechanical weathering** reduces the size of rock and mineral fragments without changing their composition.
      1. Water may seep into rock cracks and pores and then freeze. As the ice expands, it pries the rock apart in a process called **frost wedging.**
      2. In arid climates, expanding salt crystals, precipitated from evaporating water in cracks and pores, pry the rocks apart.

3. In **sheeting,** caused by pressure release, and **exfoliation,** caused by physical and chemical processes, large rock exposures fracture and peel in layers. In **spheroidal weathering,** rocks and boulders peel in rounded, very thin layers.
4. Plant roots also penetrate rock cracks and pry them apart. Burrowing animals expose rock surfaces to the weather.

B. **Chemical weathering** increases the efficiency of mechanical weathering by weakening chemical bonds that hold rocks together.
  1. In **direct solution,** the positive and negative poles of water molecules dislodge negative and positive ions in a mineral, thus dissolving it.
  2. In **hydrolysis,** free hydrogen and hydroxide ions in water replace ions of silicate minerals and convert them to insoluble residues and dissolved ions.
  3. **Oxidation** (commonly known as rusting) results when elements (such as iron) combine with oxygen.

C. Rocks of different composition weather at different rates under the same climatic conditions.

III. SOIL FORMATION

A. Weathering disintegrates and decomposes rocks and minerals, thus rendering elements mobile. The loose rock material is called **regolith. Soil** is regolith that has incorporated organic matter and is capable of supporting vegetation.

B. Soil formation involves two main steps.
  1. Soluble elements needed by plants are released by the chemical weathering of the clays, silts, and sands that were mechanically weathered from bedrock.
  2. Soil materials are rearranged by rainwater that percolates through the soil and leaches soluble elements down through the soil levels.

C. **Leaching** produces soil layering, or the **soil profile.** Each layer is called a **soil horizon.**

D. Nutrient cycling through the soil is accomplished by complex interactions involving plants, animals, and bacterial communities in the soil.

E. Soil types vary with climate.
  1. *Laterite* soils develop in tropical, humid climates. They support tropical rain forests in humid climates but are ill-suited for agriculture.
  2. Because of minimal leaching by rainwater, soils of arid and semiarid climates retain nutrients through the process of **calcification.** Although arid-climate soils are fertile, salinization can ruin soils that are overirrigated.
  3. Nutrient cycling is retarded in arctic soils because the ground is frozen nearly all year.
  4. Soils of temperate-climate regions are fertile because of intense nutrient cycling.

IV. MASS WASTING

A. **Falls** involve the free fall of rock and soil fragments. They result from the undermining of sheer cliffs by weathering, erosion, wave action, glaciers, earthquakes, and human activity. The pile of dislodged fragments that collects at the base of a cliff is called **talus.**

B. When loose fragments form a pile, the steepness of its slope, or **angle of repose,** depends upon the size of the individual particles, their angularity, the degree of sorting, and whether the particles are wet or dry.

C. **Slides** involve the downslope movement along shear planes of blocks of rock (**rockslide**) or rock fragments and soil (**landslide**). Three conditions favor slides.
   1. Steep slopes and adverse geological structures—in particular, the downhill orientation of bedding planes and rock fractures;
   2. Weak layers—for instance, shale or loose soil—that support heavy loads; and
   3. Triggering mechanisms, which can be natural or human-made.

D. **Slumps** involve intact blocks of rock or unconsolidated debris that slide downward and tilt backward along concave planes. Slumps occur most frequently on hills with thick soil blankets and steep cliffs made of loose materials.

E. **Flows** involve the movement of rocks and soil as a fluid mixture. Flows behave differently depending on their water content, particle size, and sorting.
   1. Viscous **earthflows** commonly consist of water-saturated clay, shaley fragments, and soil.
   2. **Permafrost** is a condition of cold climates. The water in regolith and rocks remains frozen solid all year, but in summer, the soil layer thaws and becomes saturated. In **solifluction,** saturated soil slips slowly downhill, causing wavy solifluction lobes. Melting and refreezing cause *frost heaving*.
   3. **Mudflows** contain more water and are less viscous than earthflows.
   4. **Debris flows** are defined as masses of rock fragments, more than half of which are larger than sand grains.

F. **Creep** is a net downhill movement of rock and soil, scarcely noticeable but cumulatively important.

V. **PREDICTING AND PREVENTING MASS-WASTING DISASTERS**

A. To recognize unfavorable sites and imminent hazards, geologists analyze the condition of the bedrock, the steepness of hillside slopes, underlying groundwater conditions, the tectonic history of the region, and the location of human facilities.

B. To control mass-wasting hazards and achieve slope stability, engineers can adjust the steepness of the slope, regulate or divert excess water, or artificially buttress the slope.

Study Terms

angle of repose (p. 225)
calcification (p. 223)
chemical weathering (p. 213)
creep (p. 232)
debris flow (p. 231)
direct solution (p. 217)
earthflow (p. 230)
exfoliation (p. 216)
fall (p. 224)
flow (p. 230)
frost wedging (p. 214)
gradation (p. 211)
hydrolysis (p. 218)
landslide (p. 225)
leaching (p. 221)

mechanical weathering (p. 213)
mudflow (p. 230)
oxidation (p. 219)
permafrost (p. 232)
regolith (p. 221)
rockslide (p. 225)
sheeting (p. 216)
slide (p. 225)
slump (p. 228)
soil (p. 221)
soil horizon (p. 221)
soil profile (p. 221)
solifluction (p. 233)
spheroidal weathering (p. 216)
talus (p. 225)

Thought Questions

1. How does mechanical weathering aid chemical weathering? How does chemical weathering aid mechanical weathering?
2. Why is it not surprising that the order of chemical weathering in silicate rocks is the reverse of Bowen's reaction series?
3. Describe how leaching and nutrient cycling keep a soil in balance.
4. Why are tropical rain forest soils generally poor agricultural soils?
5. Distinguish among falls, slides, slumps, and creep.
6. Describe five ways that human activities contribute to mass wasting.
7. Describe three types of physical evidence that a geologist looks for in identifying a landslide hazard.
8. Describe three methods of stabilizing a slope.

Chapter 11

The Energy of Streams
    Stream Discharge
    Stream Flow and Channel Erosion
    Transport and Deposition

The Graded Stream
    Waterfalls

Two Contrasting Stream Types
    The Braided Stream
    The Meandering Stream

Mineral Deposits in Streams

The Flood Plain
    The Debate Over Flood Control

Deltas

Stream Rejuvenation

Drainage Patterns
    Reorganization of Drainage
    (Stream Piracy)

Streams

Each year, the Mississippi River transports about 300 million tons of North American sediment to the Gulf of Mexico. The river spreads the sediment over its flood plain and delta and in this manner enlarges the Gulf states. Thus the Mississippi simultaneously adds to and subtracts from North America as it carries out the three major functions of all streams: erosion, transport, and deposition of sediment. In the process, new landforms are created, old ones are destroyed, and the configurations of continents are altered. **Streams**—to the geologist, this term means channeled flows of any size—are, in fact, the chief agents responsible for these gradational changes on the Earth's surface.

As we saw in Chapter 1, streams are also key components of the hydrologic cycle (see Figure 1.5). Each year, 496,000 cubic kilometers of water are evaporated from the Earth's surface, and the same amount is returned. However, the oceans surrender more water to the atmosphere than they receive from it, and more water falls on the continents than is evaporated. This trend creates a water deficit in the oceans and a surplus on land. Streams are the instruments by which the surplus water that falls on the land is returned to the oceans.

A stream is best studied as a complex interdependent system—one that includes every branch, from the main trunk, to the **tributary streams** that feed it, to the most insignificant-looking hillside gully. The Mississippi system, for example, covers two-thirds of the United States; every drop of water that falls within this area drains into the branches that feed into the main trunk. The total area drained by the Mississippi constitutes its **drainage basin** (Figure 11.1).

stream
Flowing water within a channel of any size.

tributary stream
A stream that flows into a larger stream.

drainage basin
The total area that contributes water to a stream system.

◀ This turbulent segment of the Big Thompson River in the Rocky Mountains is cutting a narrow gorge through resistant bedrock. Eventually the rapids will be eliminated and the stream will establish a smooth channel.

Figure 11.1 The major drainage outlets of the United States and Canada. A large part of the continent is drained by the Mississippi and its tributaries.

A stream system is, in fact, an "organism" remarkably sensitive to its environment. It adapts to environmental change by adjusting its course, gradient, channel, and velocity to the volume of water and sediment it conducts to the sea. The nature of these adaptations is the major theme of this chapter.

The Energy of Streams

Energy is the capacity to do work. Lift a rock above a given reference plane, such as a table, and you have given the rock a certain potential energy relative to the plane. Three factors determine the potential energy of the rock: its mass, elevation, and gravity. The same is true of all objects, including the water that constitutes streams.

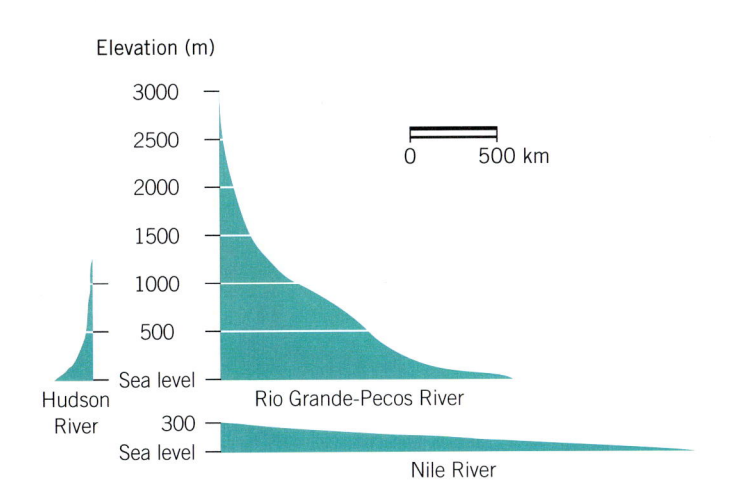

Figure 11.2 Some representative stream profiles drawn to the same scale. Although they vary a great deal in length and elevation at their sources, they all exhibit concave profiles. (Vertical scale is greatly exaggerated.)

Figure 11.3 The base levels encountered by a stream, shown in profile. All base levels are local (and temporary) except sea level.

As they flow downhill, streams convert their potential energy into motion. The velocity of the stream is governed by the steepness of its slope, or its gradient. All other factors being equal, the steeper the gradient, the more rapidly the stream flows. It is significant that all streams are steep at the source, or head, and taper to gentle slopes before they enter the sea. Thus when viewed in cross section throughout their entire length, they exhibit a concave, longitudinal profile, or **long profile**. Although their slopes are concave, stream profiles nevertheless vary from one another in form. Compare, for example, the steep gradient of the Hudson River with the nearly flat gradient of the Nile River shown in Figure 11.2.

A stream's profile is controlled by the lowest level to which it can erode its channel—its **base level** (Figure 11.3). For a stream that empties into a below–sea-level inland basin, such as the Jordanian Dead Sea, the basin acts as the base level. However, the ultimate destination of the overwhelming majority of streams is the sea. For this reason, sea level is considered their *ultimate base level*; lakes, dams, flat valleys, or stream junctions serve as *local base levels*. We will see that local base levels often exert greater control over streams than the more distant sea level.

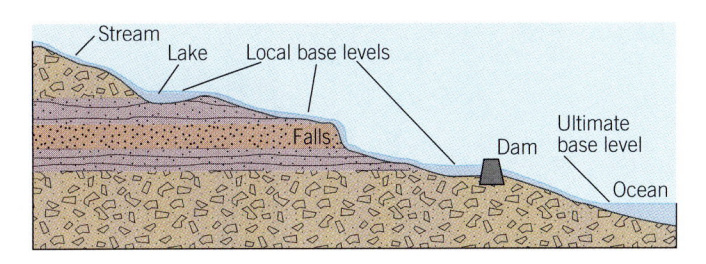

long profile
A cross section of a stream channel showing the gradient from the source to the mouth of the stream.

base level
The limiting level below which a stream cannot erode its bed.

Stream Discharge

The more water that flows through the channel in a given period, the greater the energy available to the stream. Therefore, an extremely important measure of a stream is its **discharge**, or the volume of water that flows past a monitoring station in a given time interval (usually 1 second). Discharge, measured in cubic meters per second, is the product of the cross-sectional area of the channel and the stream velocity (Figure 11.4).

At a given locality, stream velocity, channel width, and channel depth vary along with discharge, which is commonly controlled by seasonal rainfall patterns and snow melts. Because the main trunk of a stream system must transmit all the water funneled into it by its tributaries, discharge tends to increase in the downstream direction. To accommodate the added discharge, the stream channel changes downstream in much the same way that it changes at a specific location during the rainy season: stream velocity increases along with channel width and depth. These latter changes lower friction and facilitate flow, because a lesser proportion of the water is in contact with the channel surface.

discharge
The volume of water moving past a particular point in a stream over a given time interval; the product of the cross-sectional area of the channel and the stream velocity.

Stream Flow and Channel Erosion

If all the potential energy of a stream were converted to motion, most streams would rush to the ocean at the speed of a waterfall. This does not happen, however, because friction drains away most of the stream's energy. Friction is generated by water rubbing

Figure 11.4 (a) The shallow channel and (b) the deep channel have approximately the same cross-sectional area (30 m²). However, the deep channel conducts water more efficiently than the shallow channel does because less water comes in contact with the channel walls.

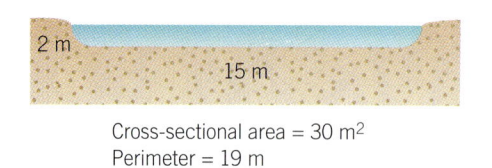

Cross-sectional area = 30 m²
Perimeter = 19 m

(a) Wide, shallow channel

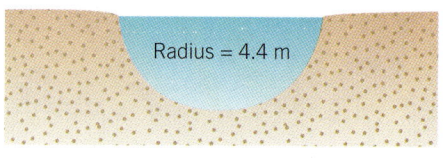

Cross-sectional area = 30 m²
Perimeter = 13.7 m

(b) Semicircular channel

against the sides and bottoms of the stream channel, by water particles colliding with one another, and by water dragging sediment along the streambed. It acts as an important brake on stream velocity and is also a key factor in channel erosion and sediment transport. The moving sand and gravel particles also contribute to the erosion of the stream channel. Their work is most evident where the particles, spinning rapidly in whirlpools, have scoured cylindrical *potholes* in a bedrock channel (Figure 11.5).

Transport and Deposition

Geologists refer to all of the materials transported by a stream as its load. Larger, heavier particles are dragged and skipped along the bottom as **bed load.** Fine particles carried within the mass of the stream make up the **suspended load,** and ions carried in solution are the **dissolved load.**

Most dissolved load is brought to streams by groundwater seepage into the channel. The ions in solution are derived from chemical weathering of rock and soil. Thus the dissolved load of a particular stream is directly related to the degree of chemical weathering in the drainage basin. Each year, streams in the 48 contiguous United States carry a dissolved load that amounts to an average of 35 percent of their total load.

Most of the solid load of a stream is probably transported in suspension. This load frequently colors the stream or makes it opaque. The Colorado River was named by Spanish explorers for the brilliant red silts it carried in suspension before its flow was regulated by twentieth-century dams.

Two useful measures of a stream's ability to transport solid load are its capacity and competence. **Capacity** refers to the *quantity* of sediment the stream can carry past an

bed load
Large or dense particles that are transported by streams on or immediately above the streambed.

suspended load
The fine particles carried in the mass of a stream and kept aloft by turbulence.

dissolved load
The portion of a stream load carried in solution.

capacity
The transporting ability of a stream as measured by the quantity of sediment that the stream carries in a given time interval.

Figure 11.5 These potholes in the stream channel were abraded by gravel and sand that were trapped in swirling eddies and whirlpools.

observation point in a given time span. Clearly, it is proportional to discharge—the water volume flowing through the channel. **Competence** refers to the maximum sediment *size* the stream can carry. The velocity of flow determines the competence of a stream: the higher the velocity, the greater the stream competence.

The stream velocity necessary to transport a particle in suspension may differ markedly from the one required to lift it from the streambed. Transport depends upon the *settling velocity* of the fragment, the rate at which it sinks through the water. If the stream's velocity is greater than this sinking rate, the fragment will stay in suspension; if less, it will sink.

Fine silt and clay are transported at such low settling velocities that, theoretically, they may remain in suspension indefinitely; for this reason, they are the major constituents of suspended load. Coarse fragments, in contrast, have such high settling velocities that stream turbulence cannot keep them in suspension for long and they are therefore transported as bed load. Heavier particles roll and slide along the bottom, but lighter bed load particles may travel by **saltation,** in which grains bounce and skip along the bottom. These modes of transport result in a natural sorting of sediments during their journey downstream. Fine suspended load moves with the velocity of the water, whereas the coarser bed load that is dragged along the bottom lags behind. In this manner, the percentage of suspended load increases downstream.

competence
The transporting ability of a stream, as measured by the largest-sized particles that it can carry.

saltation
Transport by streams in which bed load particles are moved in short skips and bounces.

The Graded Stream

As we have seen so far, the function of a stream is to transport all the water and sediment funneled through its channel. If given enough time, the slope and channel of the stream adjust so that transport is accomplished with maximum efficiency. A stream that has reached this state is said to be in a graded condition. In a **graded stream,** slope and channel are so adjusted that the stream has just enough energy to transport its load; there is no excess energy present to erode the channel, nor is there a deficiency of energy leading to deposition of sediment in the channel. This ideal state of equilibrium is rarely achieved and even more rarely maintained. Because the discharge and the velocity of streams are variable—especially over short time periods—streams are constantly either eroding their channels or depositing sediment. Over a sufficiently long time span, however, a stream can come close to a graded condition by striking a rough balance between erosion and deposition. In this respect, a graded stream is analogous to a region's climate: the weather may vary from day to day, but the overall pattern remains constant from year to year.

graded stream
A stream whose slope and channel are so adjusted that it has just enough energy to transport its load; no excess energy is present to erode the channel, nor is there a deficiency of energy that leads to deposition of sediment in the channel.

Waterfalls

Waterfalls have many causes, but no matter what their origins are, they share a common property: all are located at a *knickpoint,* a place where the stream profile steepens abruptly, causing a vertical drop. Because energy is concentrated there in the form of high water-velocity, abrasion, and turbulence, a waterfall is a temporary feature in the course of a stream. The falls wear back and reduce the knickpoint, causing it to retreat upstream until it merges with the smooth, concave profile of the stream, at which time the fall is destroyed (Figure 11.6).

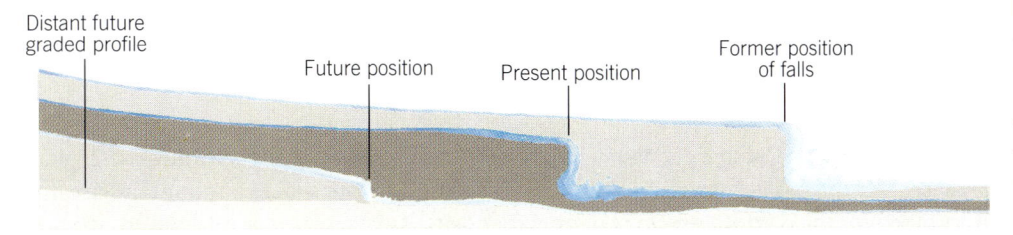

Figure 11.6 A waterfall is an irregularity of the stream profile and as such will eventually be destroyed. Energy is concentrated at the head, or knickpoint, of the falls. As the falls are worn back, their height is reduced until the irregularity vanishes and the smooth stream profile is restored.

(a)

(b)

Figure 11.7 (a) Niagara Falls. (b) Yosemite Falls.

Figure 11.7 shows two waterfalls, New York's Niagara Falls and California's Yosemite Falls. Nearly a kilometer wide and 50 meters high, Niagara Falls is capped by a thick, resistant layer of rock, the Lockport dolomite, beneath which are soft shales. At the base of the falls, turbulence is eroding the soft shales that support the dolomite, and it has worn the falls back 1 kilometer since its origin some 8000 to 10,000 years ago. In the near geological future—perhaps as little as 50,000 years from now—the falls will probably be worn back and reduced until it becomes just another smooth stretch of the Niagara River.

A narrow sliver compared with Niagara Falls, the 700-meter-high Yosemite Falls obviously has a long way to go before it is reduced to a quiet stretch of stream. This waterfall is cut in resistant granites and gneisses and empties into a deep valley oversteepened and carved by a large glacier that has since melted. In spite of these obstacles, the tendency still exists for Yosemite Falls to wear back the formidable cliffs and produce a smooth, concave profile. The process will simply require a long period of time, perhaps several million years.

Two Contrasting Stream Types

Graded streams are alike in the sense that, over a long time span, they transport their loads with no net channel erosion or deposition. However, they may differ from one another in channel characteristics and other features, because they attain the graded condition in response to differing requirements of sediment load and discharge. To illustrate this point, we will describe the channel, gradient, and landforms produced by two contrasting stream types: the braided stream and the meandering stream.

The Braided Stream

Braided streams are typically associated with the **alluvial fans** that accumulate at and spread from the base of dry mountain ranges. A fan develops where a mountain stream, no longer confined to its narrow channel, deposits its coarse sediment load in a large,

braided stream
A stream that divides into branching and intertwining subchannels separated by islands or sandbars.

alluvial fan
A gently sloping, fan-shaped mass of sediment deposited by a stream where it issues from a narrow canyon onto a plain or valley floor.

sloping, apronlike structure at the point of emergence onto the plain. Braided streams twist their way through the fan and carry sediment from the fan's apex to its fringes, thus enlarging it.

A braided stream derives its name from the way its main channel divides into a complex series of interwoven branching and merging subchannels (Figure 11.8). Tear-shaped islands as well as bars of sand and coarse sediment lie between the subchannels, giving the entire stream a braided appearance when viewed from above. Seemingly smothered by its own debris, the braided stream is, in fact, a highly competent earth-moving machine. The numerous shallow subchannels provide maximum contact between the coarse sediment and the meager amount of water available to transport it.

The Meandering Stream

As they approach base level and their gradients lower, most streams begin to **meander**—that is, their channels form wide loops in their course downstream. Like the undulations of a giant snake, the loops slowly migrate downstream as they wind back and forth across the nearly flat valley floor. Meandering streams may be as large as the Mississippi or as small as the Animas River, shown in Figure 11.9.

Although the typical meandering stream has a large suspended load, it also has more than adequate discharge to carry the load. Lacking sufficient coarse fragments to abrade and downcut the streambed and having, in most instances, a channel floor already close to base level, the stream expends its excess energy laterally as it winds back and forth across its valley. The outer channel wall is eroded through impact, slumping, and undercutting, thereby forming a steep **cutbank.** At the same time, the complementary process of deposition occurs opposite the cutbank. Sediment is funneled to the low-energy inner bank downstream, where it forms a **point bar.**

Because the meandering stream has a gentle gradient, maximum cutbank erosion occurs at the most downslope portion of the bend, slightly below the midline of the meander. This seemingly trivial effect has great consequence, because it forces the meanders to migrate downstream (Figure 11.10). The meanders do not move with equal speed, however. Some are delayed because of cutting through materials that are difficult to erode, such as compacted clays or sediments tightly bound by vegetation. The delay allows the upstream meander to catch up and break through the short segment of land

meander
As a verb: to flow in a winding, sinuous course. As a noun: a loop created as a stream winds back and forth across a valley.

cutbank
A steep slope caused by erosion of the outer bank of a meander.

point bar
A low, crescent-shaped deposit of sand and gravel developed on the inner bank of a meander where the water velocity is low.

Figure 11.8 A braided stream draining an alluvial fan in the Jago River, Alaska. Notice the sand bars among the twisting, branching stream channels.

Figure 11.9 The Animas River of Colorado. In this view we see many of the features of the meandering stream: broad flood plain, oxbow lakes, meander scars, and point bar deposits. Notice that the point bar gravels are slightly off-center from the meanders in the downstream direction.

meander cutoff
A new channel created when a stream takes the shorter route across the neck of a meander.

oxbow lake
A crescent-shaped, abandoned meander channel isolated from the stream channel by a meander cutoff and sedimentation.

between the loops. For a while, the channel will divide, one portion of the stream taking the long route, the other taking the **meander cutoff.** Eventually, the stream will be entirely diverted through the cutoff, which possesses the steeper gradient. The stranded portion of the meander will be isolated, forming an **oxbow lake.** In time, the lake will silt up, leaving only a *meander scar* (see Figure 11.10).

Mineral Deposits in Streams

Many important metals and valuable minerals are weathered from rock outcrops and are carried along by streams. They are physically resistant and do not dissolve readily, so they tend to survive the rigors of transport. Because they are dense, they accumulate wherever the transporting water loses velocity (Figure 11.11): in the point bars of meandering streams, in stream channel pools, and in the upper berm faces of beaches, where backwash lacks the power to transport them back to the ocean. Mechanical concentrations like these are called *placer deposits.*

The gold sifted from stream gravels and sands at Sutter's Mill in northern California in 1848 probably constitutes this nation's most famous placer deposit. In sparking the Gold Rush, it also sparked a major population push to the West Coast. By panning the streams that flowed west from the Sierras, prospectors knew that the source from which the gold had weathered—the elusive "mother lode"—lay to the east, within the mountain range. Indeed, the mother lode was later found to be gold disseminated in scattered quartz veins within the granites of the western Sierras.

With rare exceptions, those who invested in outfitting the prospectors in clothing, food, and rigging profited more than the prospectors themselves. The same cycle seems

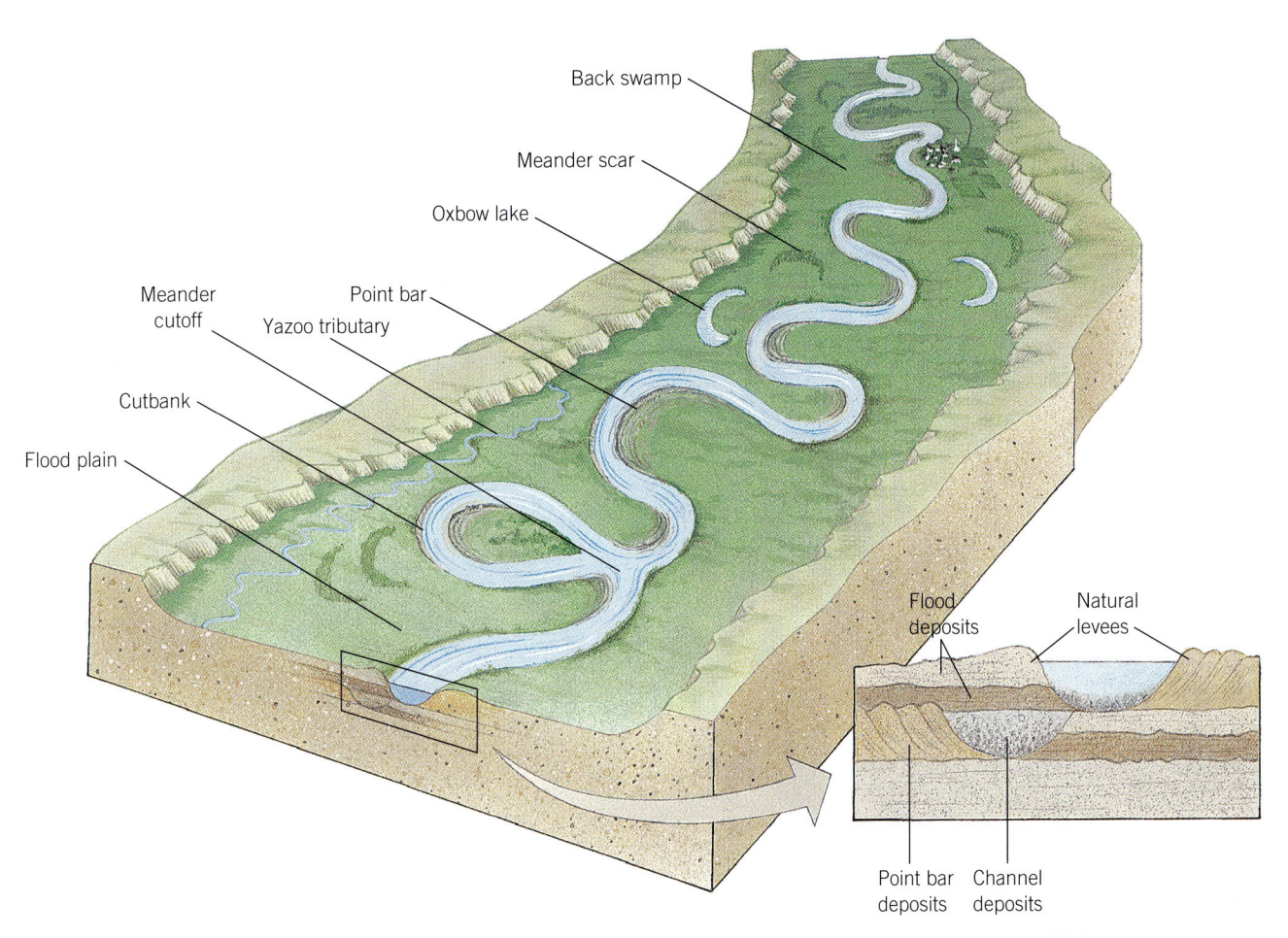

Figure 11.10 Major features of the meandering stream and flood plain. The valley is widened by lateral erosion at the same time that the flood plain is built up by complex depositional processes. Compare this diagram with Figure 11.9.

to be repeating itself in the Amazon basin of Brazil, where vast numbers of poor prospectors have flocked in search of gold. Most will have little to show for much hard work that damages the rain forest. This damage goes beyond clearing patches of the forest. The mercury used to extract the gold is contaminating streams throughout the Amazon basin, and it has become a major environmental problem.

Placer deposits are commonly mined for rich accumulations of platinum, tungsten, and tin. The diamond placers deposited on South African beaches alerted prospectors to the mother lode found inland: the dark, igneous, cylindrical kimberlite pipes of South Africa. Submerged stream and beach placers of the Atlantic continental shelf are potentially valuable metal resources, provided that ways can be found to mine them economically without disturbing the marine environment.

The Flood Plain

As the meandering stream winds back and forth across its valley, it simultaneously widens the valley through erosion and builds up the valley floor through deposition. Erosion occurs where the stream undercuts the valley walls, attacking one side and then the other as it pursues a winding course. Eventually, this lateral erosion wears back the hillside slopes and widens the valley floor. At the same time, the stream spreads channel and point-bar deposits across the valley floor. Periodically, when discharge becomes too great for the channel to handle, the stream overflows its banks. Layers of silt are deposited with each successive overflow, supplementing the channel deposits. Over thousands of years, this combined depositional process builds a thick sediment blanket, the **flood plain,** upon which the stream flows.

flood plain
Layers of sediment beds deposited across a valley as a stream periodically overflows its banks and meanders laterally.

Figure 11.11 Three areas of streams where dense placer deposits may accumulate.

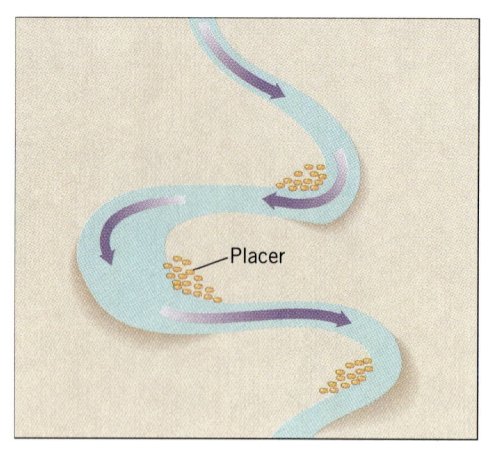

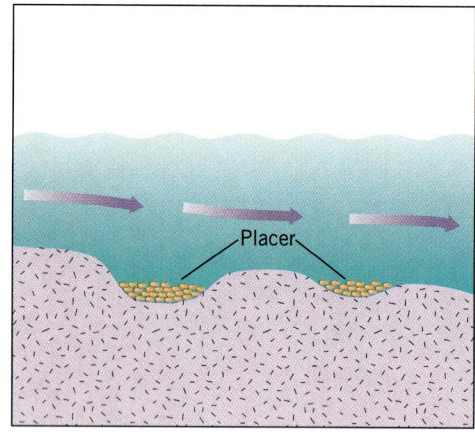

(a) Inside meander loops

(b) In potholes

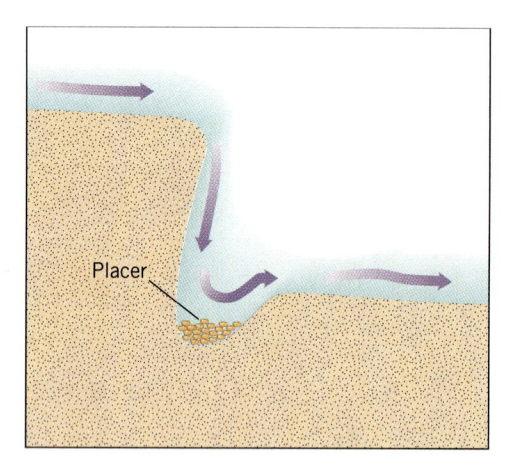

(c) Below waterfalls

During floods, streams deposit the greatest portion of their loads close to their banks. Over the years, thick wedges of sediment accumulate there (see Figure 11.10). These wedges are called **natural levees,** and they serve as partial barriers against future floods. The flood plain tapers gently from the natural levees toward the main valley walls, forming subtle depressions along the base of the valley walls. Drainage is often poor in these regions and vegetation is thick; swamps are common. A meandering tributary stream may wander for quite a distance through these depressions, blocked from joining the main stream by the levees. Such a stream is called a *yazoo* tributary, after the small river that parallels the Mississippi River and eventually joins it at Vicksburg.

natural levee
A ridge of sand and silt that parallels the stream channel and that is deposited over time, when the stream overflows its banks during floods.

The Debate Over Flood Control

Throughout the late spring and summer of 1993, a series of storm systems drenched the same midwestern states repeatedly. The subsurface was soon saturated and the runoff swelled the streams of the upper Mississippi and Missouri drainage basins.

A flood occurs when the discharge exceeds the capacity of the stream channel to contain the water flowing in it. Engineers can measure the magnitude of the resulting flood by surveying the percentage of the flood plain covered when the stream overflows its banks. How often we can expect floods of similar magnitude to occur determines the recurrence interval of the flood—whether it is a ten-year flood, a fifty-year flood, or so on. By this criterion, the recurrence interval of the 1993 Mississippi floods was on the order of centuries; the floods spread over portions of the upper Mississippi Valley that had not been covered since the time when recordkeeping had begun. However, the high discharge of the upper Mississippi River—11 times normal and 4 times the flow of Niagara Falls—was easily accommodated south of Saint Louis. There, the channel of the Mississippi River broadens to receive the discharge of the entire drainage basin, covering 31 states, and the contribution of the upper Mississippi had little impact.

Although the 1993 flooding of the Mississippi River wreaked havoc on flood-plain communities, it was a natural event, similar to others that occurred throughout the long history of the river. After all, without floods, flood plains would not exist. The economic loss, misery, and deaths from floods are of human origin—resulting from the population density and the concentration of human activities on the flood plain. In these terms, the floods of 1993 were of catastrophic proportions—48 lives and $12 billion lost (Figure 11.12). The catastrophe has renewed an old debate among scientists and the public at large over the effect that human interference has had on the natural flow of the Mississippi River system—and, by extension, other river systems. The main source of conflict involves the impact that the construction of 5800 kilometers of dikes and levees has had on the river.

Those who support the construction and maintenance of levees point to the development of vast tracts of land for agricultural, commercial, and residential purposes, all made possible by the levees. Those who disagree accuse their opponents of fostering a false sense of security among flood-plain residents and, more important, of intensifying flooding by constructing the levees.

How may a levee failure intensify flooding? When a stream's flow is constricted, the sediment it would normally spread across the flood plain is deposited instead in the channel. This deposition raises the level of the channel above the flood plain and thus requires engineers to add height to the levee. Indeed, heights of 12 meters or more have become common along the upper Mississippi. The river at or near flood stage roars downstream within its confined channel at high speed, and it works to destroy the levees by a variety of means (Figure 11.13). When the levee is breached, it is like a dam bursting; a high-pressure wave of water spreads out across the countryside at great speed. The erosive power of the floodwaters is therefore enhanced. In contrast, a stream that is allowed to flood naturally does so from the lower height of the stream bank, and its ability to erode is reduced. Furthermore, the levee and other structures retard the drainage of floodwaters back to the river channel. This leaves the land flooded longer than in a natural flood.

It is an underappreciated fact that flood plains are instruments of flood control, which they accomplish by storing and slowing floodwaters and by reducing floodwater heights. For this reason, the states hardest hit by the 1993 flood were the ones that had converted their wetland flood plains to intensive agriculture at the same time that they leveed and channeled the streams. Over the years, Missouri has lost 87 percent of its wetlands, Iowa 89 percent, and Illinois 85 percent. But the altered flood plains and stream channels are merely components of the much larger problem of the transformation of the entire drainage basin of the Mississippi. What was once water-absorbent forest and prairie now has been converted to runoff-enhancing farms and paved-over suburbs and cities. More rainwater goes directly to the stream channels and it gets there faster, thereby intensifying flood conditions.

Figure 11.12 The Mississippi River flooded this Illinois town in July 1993.

Figure 11.13 How a levee fails. Powerful flow tears away at the levee walls on the river side. Water under high pressure may seep beneath the levee embankment and boil up on the flood plain side, undermining the levee and causing it to collapse. Finally, the levee may collapse of its own weight as water saturates the structure and turns it to mud.

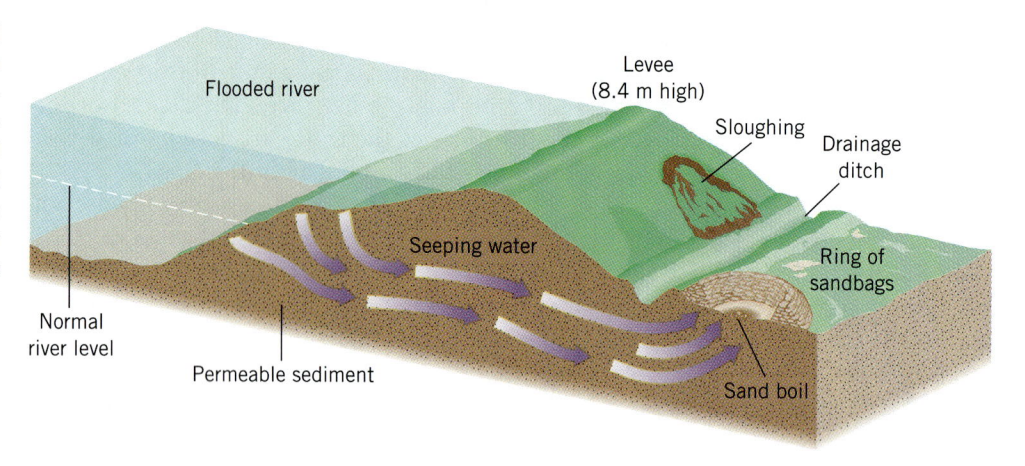

Most experts agree that the present system of flood control leaves much to be desired. But they also agree that there is no going back to a state of pristine nature. In 1991, a federal task force reported that, nationally, 17,000 communities occupied our nation's flood plains, which amounted to 10 million households and $390 billion in property. Having learned some painful lessons from major floods, states and communities across the nation are beginning to pursue "soft" alternatives to the "hard engineering" solutions of dikes and levees. They are preserving the few stretches of flood plains left within cities and suburbs, buying back additional wetlands from private owners, and passing zoning regulations that discourage development of flood-prone areas. Some of these open spaces serve as parks and farmland between floods. Some states are planning for unleveed stretches that allow the river elbowroom to flood naturally. This approach will probably save lives and money in the long run.

Deltas

When a stream enters an ocean or lake, its velocity is checked and much of its sediment load is deposited near the mouth of the stream. If the rate at which the sediment is supplied exceeds the capacity of the waves and currents to carry the load away, sediment will accumulate and form a more or less triangular structure, with the river mouth at the apex. The name of the structure, **delta,** is from the Greek letter delta, Δ. It was first used by Herodotus in the fifth century B.C. to describe the Nile River deposits in the Mediterranean Sea north of Cairo. Nevertheless, variations in local currents, wave action, and shoreline configurations prohibit many deltas from achieving the idealized triangular shape.

A delta derives its form from the branching **distributary** channels, which flow from and divide the main trunk of the stream, bringing sediment to the delta front. The mouth of a stream entering a quiet lake is a good place to observe the formation of a delta (Figure 11.14). As the stream enters the lake and deposits sediment, a broad-layered platform that slopes into deep water is gradually constructed. These layers are the *foreset beds* of the delta. This platform becomes a base upon which a complex mixture of river and shallow-bay sediments are deposited. They form the *topset beds* of the delta. At the same time, fine clays and silt are carried to deeper, calmer portions of the lake and settle to the bottom. These *bottomset beds* grade into the foreset deposits in the direction of the delta and merge imperceptibly with the nondeltaic sediments of the middle of the lake. The delta continues to grow as topset-foreset-bottomset deposits reach ever farther into the lake. These three structural units form distinctive *deltaic crossbeds*.

delta
A triangle-shaped landform built up over time where a stream enters a calmer body of water and deposits its sediment load.

distributary
One of a system of small channels that carry water and sediment from the mainstream channel and spread them over the delta surface.

Many deltas have built-in cycles of growth and retreat. During the growth phase, the delta pushes seaward for a time by depositing a thick wedge of sediment on the continental shelf. Coarse beach sands are thrown up by wave action on the delta front, while flood-plain and marsh sediments are deposited behind it. As the delta grows, the sediment compacts of its own weight, and the crust beneath the shelf may buckle slightly in response to the sediment load. At the same time, extension of the delta lessens the stream gradient, which leads to a decrease in sediment supply. These events trigger a gradual flooding of the delta surface and the onset of the retreat phase. As the sea washes in, the beaches are beaten backward, and the encroaching ocean spreads shallow marine muds over the flood plain and marshes. But the growth cycle of the delta is eventually renewed, because the marine invasion temporarily increases the stream gradient and therefore the capacity of the stream to deliver sediment.

Figure 11.15 shows that the Mississippi River has shifted outlets many times over the past 7000 years, first to one side of the broad deltaic plain, then to the other. Progressive outbuilding has overextended the modern delta to the point where the Atchafalaya River, which branches off the Mississippi above Baton Rouge, offers a much shorter route to the sea. Indeed, the flood history of the past few decades has proven that if the Mississippi had its way, it would shift channels and flow through the Atchafalaya. New Orleans would be bypassed, a new delta would be established at the mouth of the Atchafalaya River, and the present bird's foot would gradually be worn back. In fact, the present course of the Mississippi is maintained artificially through the investment of millions of dollars, intensive labor, and sophisticated engineering. The Atchafalaya River and surrounding lakes are used mainly as controlled safety valves to handle the excess discharge of the Mississippi River at high-flood stage and to limit flooding of New Orleans.

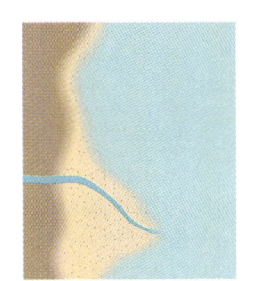

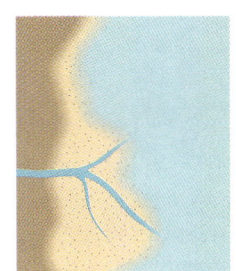

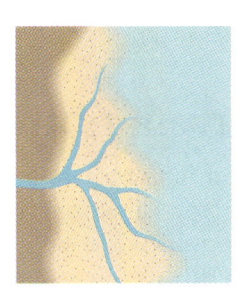

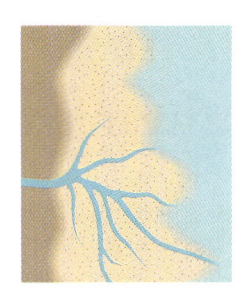

Figure 11.14 Stages in the development of a simple lake delta. In the ocean, strong waves and currents add complexity to this basic structure.

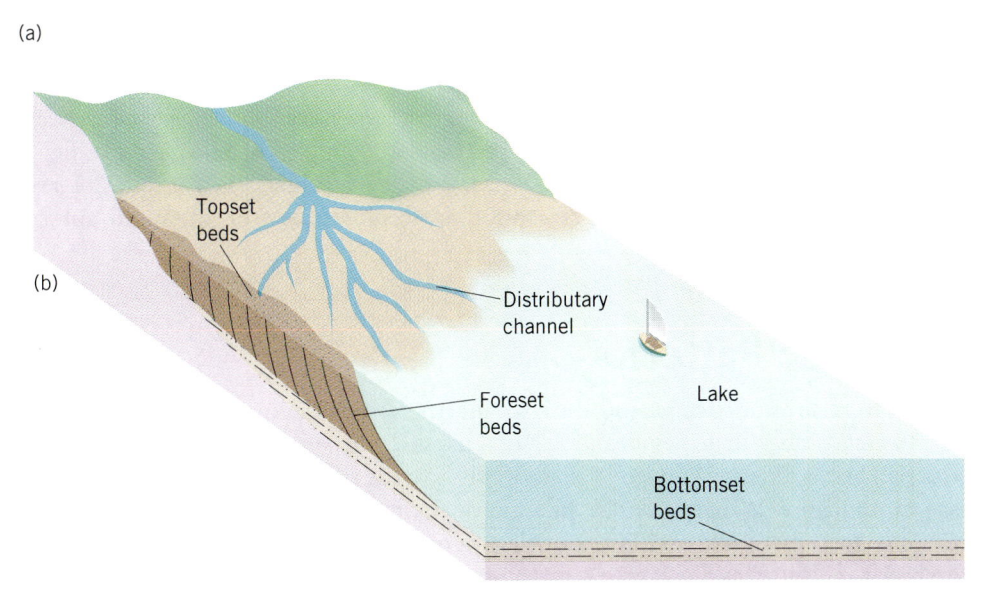

Figure 11.15 Some of the shifting outlets of the Mississippi River over the last few thousand years. Notice that the Atchafalaya River offers a shorter route to the Gulf of Mexico than the present course of the Mississippi does. Elaborate measures are employed to prevent the Mississippi from bypassing New Orleans, one of the nation's leading ports.

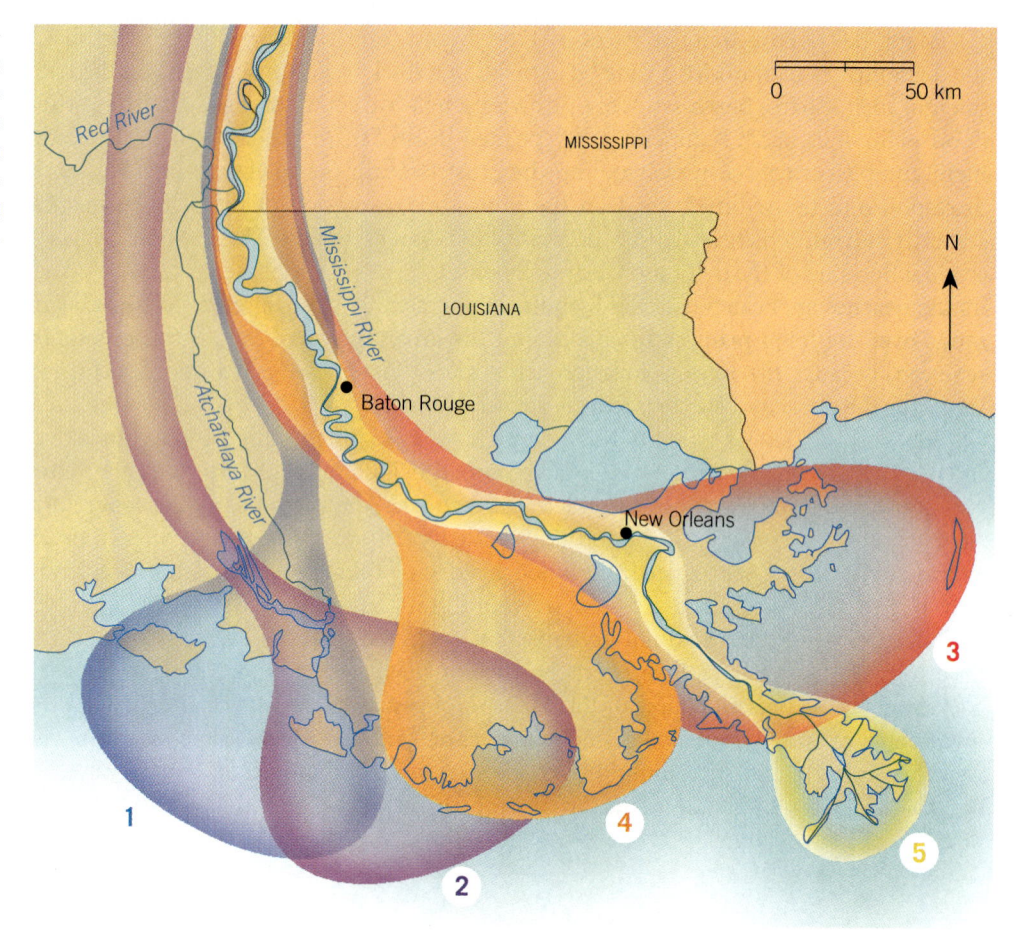

stream terrace
An elevated, shelflike surface upon which the stream formerly flowed; typically, ancient flood-plain remnants abandoned as the stream cut downward and established a course at a lower level.

incised meander
A deep, narrow, winding stream valley caused by the combination of regional uplift and stream downcutting.

Stream Rejuvenation

Streams that receive an infusion of energy from uplift, sea-level retreat or increased discharge due to long-term climatic changes are said to be rejuvenated. They put their fresh energy supply to work lowering the stream gradient. For a meandering stream that flows on a flood plain, the method of accomplishing this task is to cut deeply into the thick flood-plain sediment. Left behind on either side of the newly incised channel are **stream terraces,** the flat-topped remnants of the former flood plain. Complex terrace patterns commonly consist of many levels, like staircases. Rising above the main level of the flood plain, they are often well drained, have fertile soil, and usually escape flooding. Active tectonism may produce spectacular examples of rejuvenation. Figure 11.16 shows **incised meanders** along the course of the San Juan River of the Colorado plateau. The river, cutting rapidly downward, is keeping pace with the rapid uplift of this plateau.

Drainage Patterns

Viewed from the air, streams and their tributaries etch characteristic drainage patterns into the landscapes that they drain. A large pile of earth left uncovered around a construction site is an especially good place to observe the development of a drainage pattern in miniature. Watch carefully the way water runs downslope during a heavy rain. If rain comes down hard and fast enough, it soon exceeds the capacity of soil and sand to absorb moisture, and the excess water runs off the pile in thin sheets. Downslope,

Figure 11.16 The incised meanders of the San Juan River, Utah, are the result of stream rejuvenation. Rapid uplift has energized the stream, which has responded by downcutting its winding channel.

the sheets break into turbulent ribbons of water called *rills,* caused by local slope irregularities and slight differences in composition of the materials. The rills scour sediment and erode channels in the earth pile, creating small stream valleys. The hill crests separating neighboring valleys are the **stream divides,** and the areas encompassed within the divides are the drainage basins. Rainwater striking the pile within a given drainage basin descends downslope to the stream channel.

With time, an intricate *stream system* will evolve that consists of a main trunk and branching tributaries. Its drainage basin will then include all the drainage basins of streams that belong to the system. Thus all the water and sediment of the tributaries will be transported off the pile through the main stream. The stream system is in a constant state of adjustment. With each rain, the main stream cuts deeper into the pile, which forces the tributaries joining the main stream to follow suit. As the streams cut deeper into their channels, the heads of the tributaries gain the energy to eat farther back into the earth pile through **headward erosion** and to develop many branches of their own. Eventually, this system and competing systems drain the entire surface of the pile. In this manner, the pile is eroded and—given enough rainfall—is eventually flattened.

Drainage patterns to a large extent reflect underlying rock structures and differential erosion rates. In our example, the pile is essentially homogeneous and exhibits tree-like **dendritic drainage.** It is characteristic of regions underlain by materials of uniform composition or by horizontal layers whose properties do not change laterally (Figure 11.17a). Dendritic drainage is a pattern of random development based on rainfall and regional slope. However, much of the Earth's surface is underlain by rocks of variable composition and structure. Streams carve channels in belts of easily eroded rock and in zones of faulting and fracture, in which rocks have been mechanically crushed and weathered. Thus geologists are often able to identify underlying rock structures and compositions from inspection of drainage patterns shown in aerial photographs and maps.

Trellis drainage is characteristic of regions underlain by parallel belts of tilted strata, having differing resistances to erosion (Figure 11.17b). Main streams follow the regional slope and cut across the strata. Tributary streams then erode valleys in the weaker strata. The entire system resembles the pattern that vines make on a trellis. Trellis drainage is

stream divide
The boundary that separates adjacent stream valleys.

headward erosion
The lengthening of a stream or gully by erosion at the head of the stream valley.

dendritic drainage
A stream drainage pattern resembling a branching tree that is common to regions underlain by materials of uniform composition or horizontal layers.

trellis drainage
A drainage pattern in which the main stream cuts across the regional structure (typically folded or tilted strata), and tributaries follow parallel belts of weak strata that lie perpendicular to the main stream.

well displayed in the gently inclined strata of the Atlantic Coastal Plain and in the deeply eroded folds of the Appalachian Mountains of Pennsylvania. *Annular drainage* is similar to the trellis drainage of fold-mountain belts but is primarily associated with domes and basins (Figure 11.17c). The annular ring pattern develops because erosion of these structures produce circular rather then parallel rock exposures. Another pattern, organized like the spokes of a wheel, is the **radial drainage** common to the flanks of volcanoes (Figure 11.17d). It is formed by streams that flow outward in all directions from the cone summit. The stream valleys originate at the volcano base, and work upslope by means of headward erosion.

Often, fractures and faults form the major structural elements of massive rocks of homogeneous composition—commonly granite. These features may intersect at right angles and provide zones of weaknesses for streams to exploit. A pronounced **rectangular drainage** pattern develops where the stream system incises channels along the joint and fault planes (Figure 11.17e).

Reorganization of Drainage (Stream Piracy)

As they work backward into their upland regions through headward erosion, the drainage divides between the tributaries of the same and neighboring stream systems are narrowed. The stream having the steeper gradient possesses a distinct advantage over its neighbor in offering a shorter, more rapid route to base level. Eventually, the steeper stream may undercut the divide between the two and divert through its own channel the headwaters of the neighboring stream. Appropriately enough, the process is called **stream piracy** (Figure 11.18). Often, an abandoned valley or *wind gap* exists where the diverted headwater of the pirated stream used to flow.

Stream piracy is the way drainage systems adjust in order to achieve maximum energy efficiency in the course of transporting water and sediment to the sea. As tectonic and climatic changes occur, the stream adjusts accordingly, and stream piracy is one type of response.

---

radial drainage
A stream pattern that radiates like the spokes of a wheel from the cone summit of a volcano.

rectangular drainage
A drainage pattern commonly found in homogeneous igneous and metamorphic rocks in which tributary streams display right-angled bends that follow joints and faults.

stream piracy
The diversion of a stream into another stream that has a steeper gradient.

Figure 11.17 The relation of five common drainage patterns to underlying rock structure.

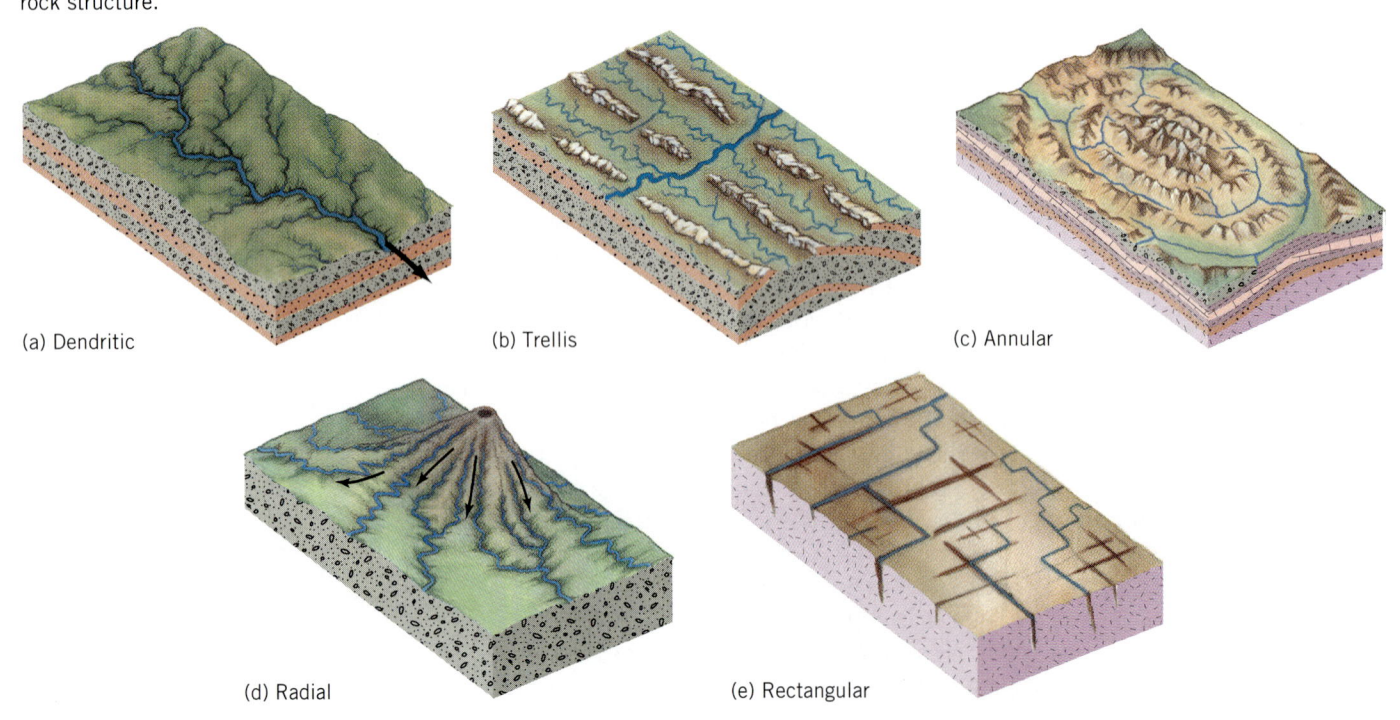

(a) Dendritic

(b) Trellis

(c) Annular

(d) Radial

(e) Rectangular

(a)   (b)   (c)

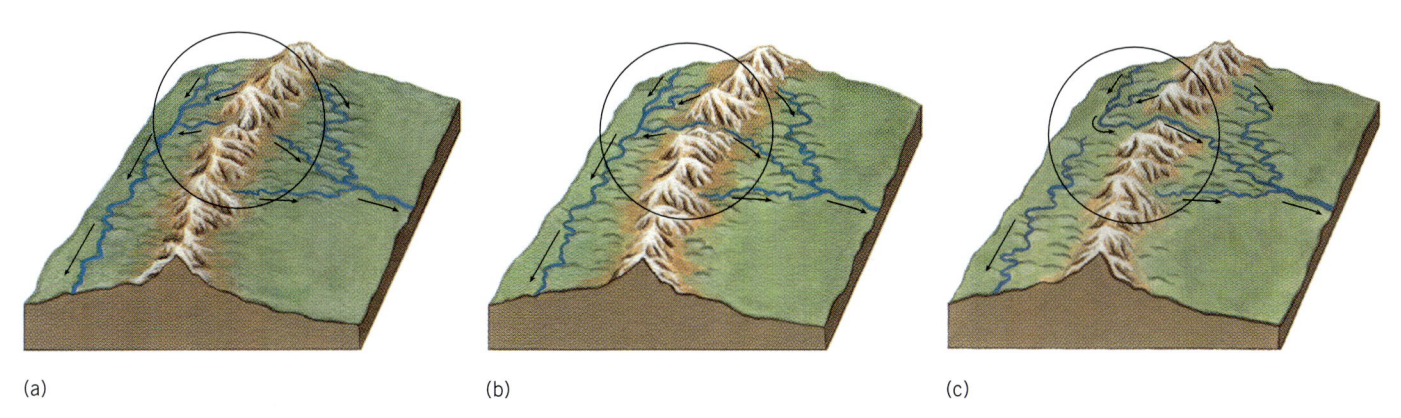

Figure 11.18 Drainage reorganization through stream piracy. (a) The streams flowing down the east side of the mountain range have steeper gradients than streams on the west side. (b) Streams on both sides of the mountain range cut into the mountains in headward erosion. (c) When their channels meet, the higher-gradient stream diverts or captures the headwaters of the stream with the gentler gradient.

STUDY OUTLINE

Streams are defined as channeled flows of any size; their three major functions are erosion, transport, and deposition of sediment. A stream system consists of the main trunk, **tributary streams,** and **drainage basin.**

I. **THE ENERGY OF STREAMS.** As streams flow downhill, they convert their potential energy into motion. The steeper the stream *gradient,* or slope, the more rapid the flow. All streams have concave **long profiles.** A stream cannot erode its channel below **base level.**
   A. As stream **discharge** increases, so do stream velocity, channel width, and channel depth.
   B. Friction drains away most of a stream's energy while eroding channel walls. Particles spinning in whirlpools may scour *potholes* in the channel.
   C. Materials are transported by streams in three ways: dissolved ions in **dissolved load,** fine particles in **suspended load,** and larger particles in **bed load.**
      1. **Capacity** is the amount of sediment that a stream can carry, and **competence** is the size of particles that it can carry.
      2. If a particle's *settling velocity* is low relative to stream velocity, it is transported in suspension. Heavier particles are rolled along the bottom; lighter ones are skipped along in **saltation.**

II. **THE GRADED STREAM.** A **graded stream** is one that has just enough energy to transport its load without eroding its channel or depositing sediment.
   A. Variable conditions are constantly throwing a stream out of grade; increased erosion and deposition act to return the stream to grade, as do changes in channel dimensions.
   B. Irregular features in the stream profile, such as waterfalls, are destroyed over time by channel erosion at high points and deposition at low points.

III. **TWO CONTRASTING STREAM TYPES**
   A. Where a stream must transport excessive bed load under low discharge, it develops a series of interwoven subchannels that give it a braided appearance. **Braided streams** commonly form in **alluvial fans** at the foot of desert mountains.

B. As they approach base level, most streams form **meanders,** or loops. A meandering stream has a large suspended load but adequate discharge to carry it.
   1. **Cutbanks** and **point bars** form on the outer and inner channel walls.
   2. Meanders tend to migrate downstream. The segment of land between the loops may be breached, forming a **meander cutoff.** The stranded portion of the meander forms an **oxbow lake.**

IV. **MINERAL DEPOSITS IN STREAMS.** *Placer deposits* are metals and valuable minerals that are weathered from rock outcrops, transported by streams, and deposited wherever the water loses velocity.

V. **THE FLOOD PLAIN**
  A. A meandering stream flows over a self-constructed **flood plain.**
  B. Deposits on the sides of the stream channel, called **natural levees,** serve as partial barriers against future floods.
  C. The debate over flood control continues.
   1. Flooding of the Mississippi River in 1993 was a natural event, but the devastating results may have been intensified by human interventions designed to protect against floods.
   2. Supporters of the levee system of flood control point to the economic benefits of flood protection; opponents claim that confining the river forces it to grow higher and flow faster, which increases flood severity when the levees are breached.

VI. **DELTAS**
  A. When a stream enters an ocean or lake, its sediment load is deposited near the mouth and accumulates in a triangular structure called a **delta.** The sediment, which is brought to the front by branching **distributary** channels, is deposited as foreset, topset, and bottomset beds.
  B. Many deltas have built-in cycles of growth and retreat. The Mississippi River, for example, has shifted outlets many times in the past.

VII. **STREAM REJUVENATION.** A stream that is revived through tectonic uplift or increased discharge will cut into its flood plain and leave behind **stream terraces** that mark the former level of the plain. Tectonic uplift and stream downcutting also form **incised meanders.**

VIII. **DRAINAGE PATTERNS**
  A. As thin sheets of water from heavy rainfall runs downslope, the water collects in turbulent ribbons called rills, which erode channels, creating small stream valleys. The crests between channels form **stream divides.** In time, channels join together to form tributaries and a main trunk. The tributaries develop branches of their own through **headward erosion.**
  B. Stream systems take the following patterns.
   1. **dendritic drainage:** a treelike pattern characteristic of horizontal strata of homogeneous rock
   2. **trellis drainage:** a vinelike pattern found in fold-mountain belts
   3. **radial drainage:** a pattern that radiates from a volcanic crater
   4. **rectangular drainage:** a pattern that forms along joint and fault planes
  C. **Stream piracy** occurs when a steeper stream undercuts a divide and diverts the waters of a neighboring stream through its own channel.

Study Terms

alluvial fan (p. 246)
base level (p. 243)
bed load (p. 244)
braided stream (p. 246)
capacity (p. 244)
competence (p. 245)
cutbank (p. 247)
delta (p. 252)
dendritic drainage (p. 255)
discharge (p. 243)
dissolved load (p. 244)
distributary (p. 252)
drainage basin (p. 241)
flood plain (p. 249)
graded stream (p. 245)
headward erosion (p. 255)
incised meander (p. 254)
long profile (p. 243)
meander (p. 247)
meander cutoff (p. 248)
natural levee (p. 250)
oxbow lake (p. 248)
point bar (p. 247)
radial drainage (p. 256)
rectangular drainage (p. 256)
saltation (p. 245)
stream (p. 241)
stream divide (p. 255)
stream piracy (p. 256)
stream terrace (p. 254)
suspended load (p. 244)
trellis drainage (p. 255)
tributary stream (p. 241)

Thought Questions

1. Use the terms *energy* and *base level* to explain why streams cannot erode their channels below sea level.
2. Waterfalls are temporary features in the life of a stream. Explain.
3. Describe the downstream changes in channel dimensions, discharge, and sediment load in a typical stream. How might they explain how a stream changes from a braided to a meandering pattern?
4. Explain what happens to a stream when it reaches a lake or ocean. You may want to include the following terms in your discussion: delta, channel, sediment, distributary, outlet, deltaic bed.
5. Describe how a straight stream becomes a meandering stream.
6. Describe the development of a stream system. You may want to include the following terms in your discussion: sheet, rill, channel, valley, divide, drainage basin, tributary, main stream.
7. Describe the extent of human interference with the Mississippi River system. In your opinion, has the interference been generally beneficial or harmful? What are some possible alternatives to present practices?

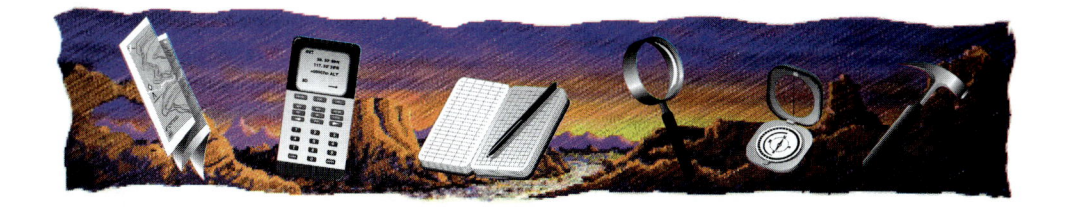

CHAPTER 12

Accumulation of Groundwater
   Porosity and Permeability
   Aquifers

Movement of Groundwater

Wells
   Artesian Wells
   Well Extraction Problems
      Land Subsidence
      Saltwater Encroachment
      Loss of Hydraulic Pressure

The Geologic Work of Groundwater
   Caves
   Karst Topography
   Geysers, Hot Springs, and Fumaroles

The Quality of Groundwater
   Groundwater Pollution

Groundwater

Underground water makes up about 95 percent of the world's supply of freshwater, exclusive of glacial ice (Table 12.1). To gauge the enormity of this hidden resource, consider that the volume of water trapped within a half mile of the continental surface is 40 times greater than the water residing in all rivers and lakes. However, these reserves are not inexhaustible, nor are they evenly distributed. Often, the demand is greatest where supply is least, such as in arid regions brought under agricultural cultivation.

Underground water—extracted at a rate of 340 billion liters per day—provides one-fifth of the water used in the United States. It is a fragile resource subject to contamination from agricultural, industrial, and municipal sources. Approximately 25 percent of the supply of underground water considered usable for drinking in the United States is contaminated. This percentage is likely to increase, because contamination has not been brought under control despite tougher cleanup laws.

This chapter describes how water accumulates underground and migrates through the subsurface; how it is extracted through wells; and how it does the geologically important work of forming caves, shaping the landscape, preserving fossils, and creating hot springs and geysers. The role of humans in polluting this valuable natural resource will also be examined.

Accumulation of Groundwater

Water that strikes the land surface as rain either evaporates, runs off directly as overland flow, or seeps into the ground. The amount of seepage in a given location depends

◀ Groundwater springs issuing from cliffs in Grand Canyon National Park, Arizona.

CHAPTER 12

Table 12.1 Quantities of Water Held in Storage

Location	Amount (km³)	Percent
Oceans	1,350,000,000	97.4
Glaciers	27,500,000	2.00
Groundwater	8,200,000	
Lakes and inland seas	205,000	0.6
Atmosphere	13,000	
Streams	1700	

zone of aeration
The subsurface zone between the water table and the surface; it retains little moisture except within the belt of soil moisture and the capillary fringe.

zone of saturation
A subsurface zone where groundwater accumulates and completely fills the voids in sediments and rocks.

groundwater
Underground water within the zone of saturation, below the water table.

water table
The surface marking the upper limit of the zone of saturation.

upon a number of factors: the degree of slope; the frequency, intensity, and duration of rainfall; the absorbing and transmitting capacity of the subsurface; and the vegetation present. The rainwater that does seep into the ground percolates beneath the surface through the interconnected spaces between soil, sediment, and rock particles of the **zone of aeration.** Eventually, the water reaches a depth where movement is retarded. There, it accumulates and completely fills the voids between solid materials, forming a **zone of saturation** (Figure 12.1). Many hydrogeologists distinguish between the general term *underground water,* which encompasses all water beneath the surface, and the more specific term **groundwater,** which refers to water in the zone of saturation.

The upper limit of the zone of saturation, called the **water table,** fluctuates with the seasons, rising during periods of high rainfall and sinking when rainfall is scarce. Thus the zone of saturation, a natural reservoir, plays a key role in regulating the regional water economy. By slowly releasing water to streams during dry spells and storing it during heavy rains, it lessens the severity of droughts and floods.

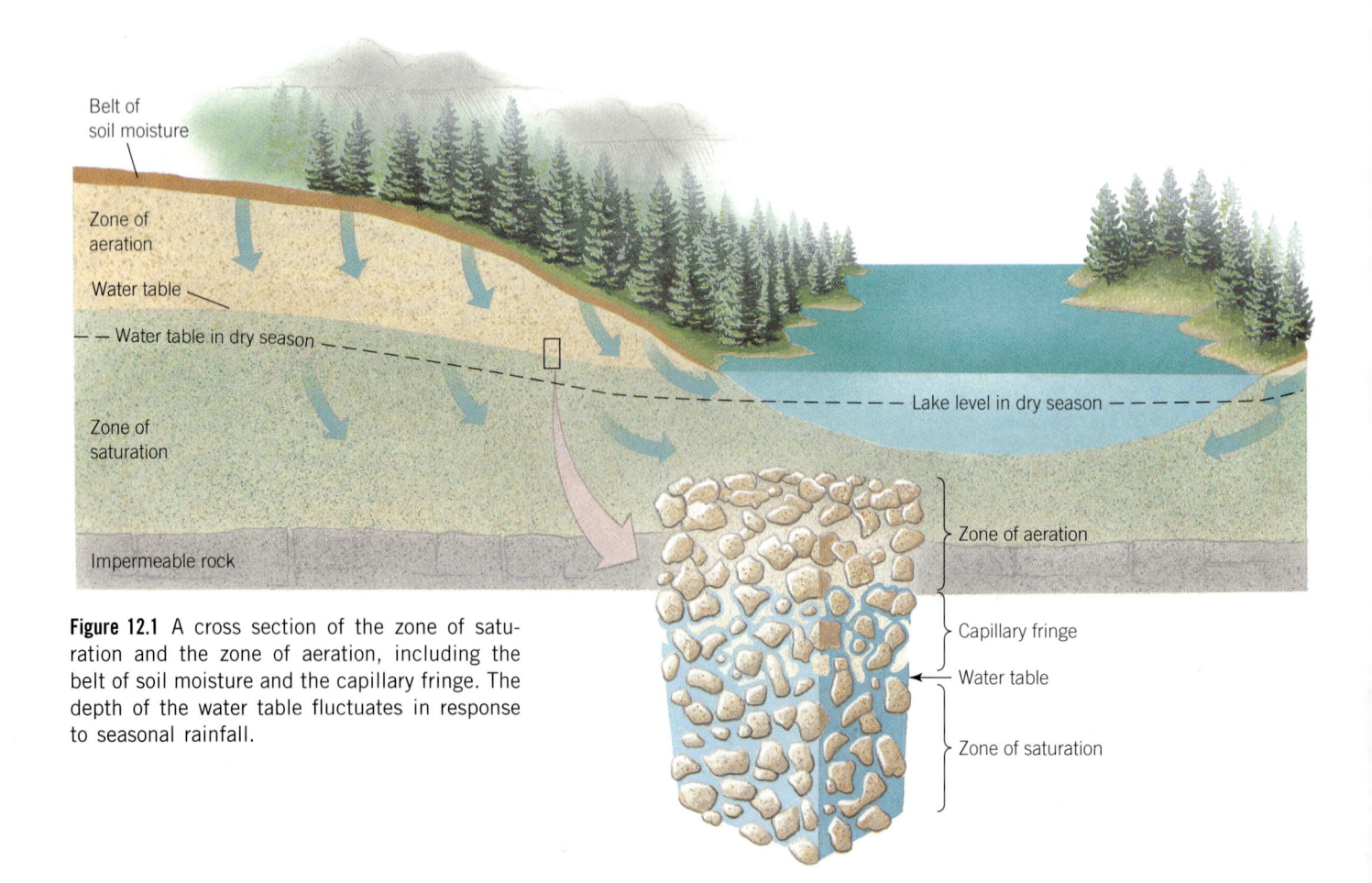

Figure 12.1 A cross section of the zone of saturation and the zone of aeration, including the belt of soil moisture and the capillary fringe. The depth of the water table fluctuates in response to seasonal rainfall.

Though largely dry, the zone of aeration does contain two significant accumulations of water: the **belt of soil moisture** directly beneath the surface and the capillary fringe just above the water table. The belt of soil moisture is formed when a small portion of the water that seeps beneath the surface is trapped in the upper few meters of soil by sand, silt, clay, and organic matter. Plants draw on the soil moisture in this thin belt; the presence of this water is the reason many species survive dry spells. But the soil is extremely loose and pockmarked with voids left by decayed roots and burrowing animals, which allows most rainwater to percolate down to the zone of saturation.

The **capillary fringe** is the wet halo that extends upward about a meter or so from the water table. It results from the adhesive forces between mineral grains and water molecules that work against gravity to draw the water up from the saturated zone into small openings between grains. You can observe the same kind of thing by dipping the edge of a napkin into a glass of water. The boundary between the wet and dry portions of the napkin will rise above the water surface.

belt of soil moisture
The thin layer of moisture just beneath the land surface; the uppermost subdivision of the zone of aeration.

capillary fringe
The thin belt just above the water table within which moisture is drawn up by surface tension and partly fills the voids in sediments and rocks.

Porosity and Permeability

If you fill a container with common sand and then pour water into the sand all the way to the top, you may be surprised to find out just how much water the sand absorbs—up to 25 percent of the volume of the container. What you have measured is the sand's **porosity**—that is, the ratio of voids to the total volume of the sample.

The porosity of loose material depends mainly on three factors. The first, *sorting*, is the degree of uniformity of particle size. A poorly sorted gravel and silt mixture has less open space than pure gravel (Figure 12.2a). The second factor, *packing*, is a measure of how tightly the particles are arranged; the tighter the arrangement, the less the porosity (Figure 12.2b). The third factor is the *shape* of the particles. For example, spherical particles leave more open space than nonspherical ones (Figure 12.2c).

Typical soil is extremely porous because of its tendency to form clumps and because of its root systems and animal burrows. A soil's degree of porosity is an important component of its fertility, which is one reason that farmers plow the soil before they plant. The porosity of consolidated sedimentary rocks is affected by the three factors listed above and by the material that cements the particles. Where particles are partially cemented, many open spaces may remain; however, if cement fills the voids between particles, then obviously porosity is much reduced. If the cementing agent is soluble, then groundwater will dissolve it. Hard, massive rocks—like granite, basalt, and certain kinds of limestone—would appear at first glance to lack any space for water. However, because these rocks are frequently fractured, they may hold great volumes of water. Basalt and other lavas are often quite porous owing to the holes left behind by escaping gas, lava tubes, and other features.

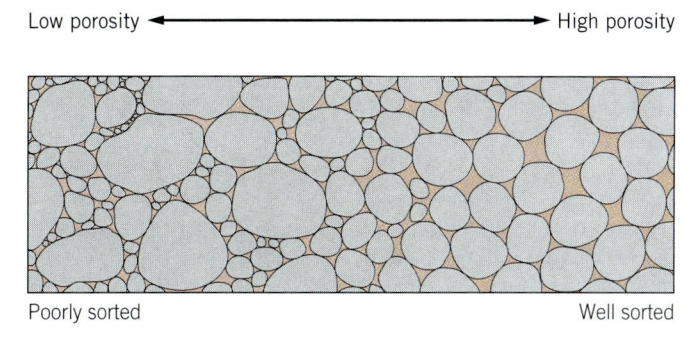

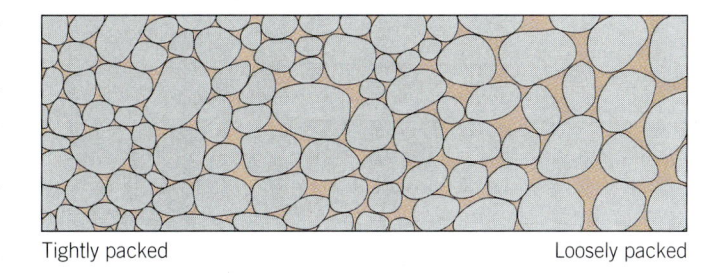

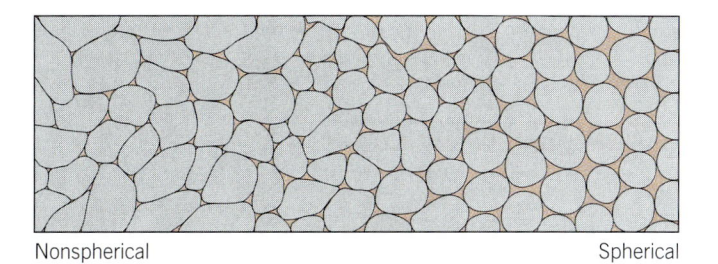

porosity
The ratio of open space to total volume of a subsurface material.

Low porosity ←————————→ High porosity

Poorly sorted   Well sorted
(a) Sorting

Tightly packed   Loosely packed
(b) Packing

Nonspherical   Spherical
(c) Particle shape

Figure 12.2 In general, the particles of highly porous (coarse) sediments are (a) well sorted, (b) loosely packed, and (c) spherical. These characteristics maximize the open space within the sediment.

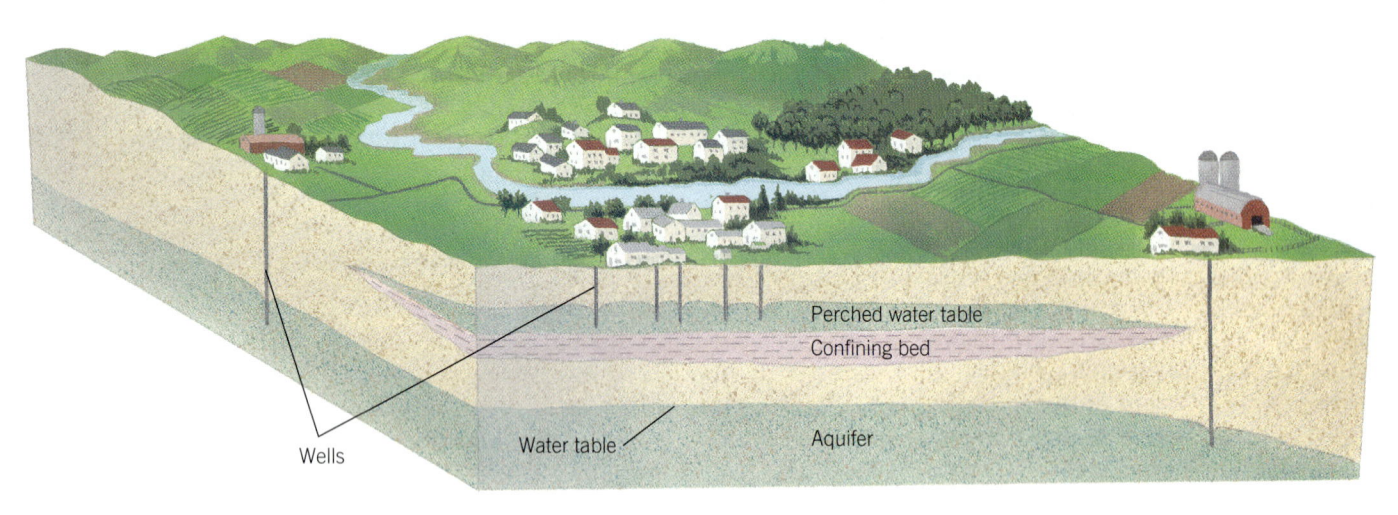

Figure 12.3 A lens of impermeable strata above the regional water table may trap groundwater as it percolates downward, thus creating a perched water table.

permeability
The capacity of a material to conduct water; measured by the volume of water that will move through a cross-sectional area in a given period of time, subject to a given pressure difference.

Knowledge of the **permeability** of sediments and rock samples—that is, the ease with which they transmit water—is also important to our understanding of groundwater accumulation and movement. Permeability is measured by the volume of water that will move through a given cross-sectional area of material in a given period of time, subject to a given pressure difference. Clay, for example, usually yields only 2 percent of its water. The balance is held within the small openings between grains and within the crystal structure of clay itself. By contrast, gravel yields most of the water that it holds.

Whereas a permeable substance must also be porous, porous substances are not necessarily permeable. For example, the volcanic rock pumice contains so many open spaces that it floats in water; the spaces are not connected, however, and so water cannot penetrate it. Thus permeability depends upon the size, arrangement, and interconnectedness of openings, as well as on porosity. Unfractured granite, shale, and clay are impermeable; most soils, cavernous limestone, gravel, and sandstones are relatively permeable.

The porosity and permeability of rock and loose material frequently vary horizontally and vertically, and these changes affect groundwater accumulation and movement. Figure 12.3 shows what happens when downward-percolating groundwater encounters an impermeable barrier, or **confining bed**—typically, a lens of clay within a sand or gravel deposit, or a shale layer within sandstone or conglomerate—before reaching the zone of saturation. Groundwater accumulates directly over the lens and forms a **perched water table** above the main zone of saturation. Perched water tables are often extensive, and wells sunk in them can extract large volumes of water.

confining bed
An impermeable layer adjacent to an aquifer that prevents escape of groundwater from the aquifer.

perched water table
The upper limit of a local groundwater body stranded above the regional water table by an impermeable layer of rock or sediment.

aquifer
A permeable body of subsurface sediment or fractured rock that conducts water.

Aquifers

Groundwater is contained within porous and permeable layers called **aquifers** (see Figure 12.3). Where rainwater is able to percolate directly from the surface to the zone of saturation, it is contained within an *unconfined aquifer,* so named because the water table can rise or fall freely. However, aquifers are often sealed between confining beds, and this arrangement prevents local rainwater from percolating down to it. The groundwater in these *confined aquifers* originates in another region where the aquifer is exposed at higher elevation and receives precipitation directly. The groundwater in a confined aquifer is generally under considerable pressure and often will rise to the surface when wells are sunk. Wells like these, in which groundwater rises above the level of the aquifer under its own pressure, are called **artesian wells.** We will return to the subject of wells later in the chapter.

artesian well
A well under sufficient hydrostatic pressure to force the water to rise above the top of the aquifer.

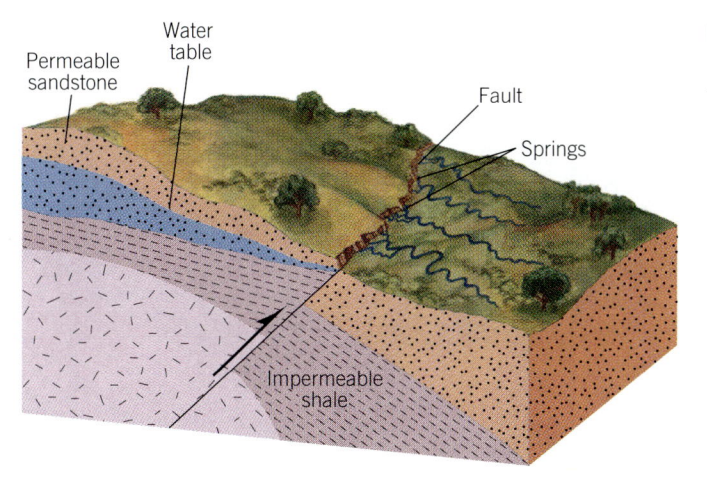

Figure 12.4 One of the numerous subsurface conditions that may lead to the occurrence of springs: faulted strata cause the water table to intersect the land surface.

Movement of Groundwater

The term *water table,* taken literally, can be misleading, for it implies that the top of the zone of saturation forms a level surface beneath the countryside. Instead, the water table follows the outline of the topography, although in a more subdued and less detailed fashion. It is generally higher beneath hills than it is beneath valleys.

In lowland areas, the water table may intersect the land surface, forming swamps and marshes. More typically, it intersects stream channels. Streams that gain groundwater in this manner are called **effluent** (or gaining) **streams.** In arid regions, the water table may lie beneath the stream channel, in which case the **influent** (or losing) **stream** feeds the water table. At higher elevations, the water table may not mirror the topography because of abrupt changes in rock permeability, faulting, or other factors. In these instances, the water table may intersect the land surface where groundwater issues directly onto the surface as a **spring** (Figure 12.4).

Like all continuous bodies of water, groundwater in the zone of saturation flows from those regions where the water stands high to those where it is low. The velocity of flow is determined by the permeability of the material through which it flows and by the gradient or slope of the water table—the steeper the gradient, the faster the flow.

The normal flow of groundwater is from **recharge areas** (the broad uplands between stream valleys that receive rainfall and soak up groundwater), through the zone of saturation, to **discharge areas** (commonly, streams, lakes, and marshes). In time, the configuration of the water table reflects the balance between the rate of recharge, the rate of discharge, and the porosity and permeability of the subsurface. If these factors remain fairly constant, so will the configuration of the water table.

Wells

Wells that extract water from an unconfined aquifer are known as *water-table wells.* As points of artificially created discharge, they depress the water table in the vicinity of the well. Thus a **cone of depression** is produced in the water table as water is drawn into the well from all directions and *drawdown,* the distance from the bottom of the cone to the water table before pumping, increases (Figure 12.5). Even a stable cone of depression will slowly widen with time, because as the water table is lowered, the groundwater that is extracted must come from greater distances. The widening cone means that the hydraulic gradient between the top and the bottom of the cone is being reduced. This is bad news, because energy must be supplied by the pump to compensate for energy loss from the declining gradient. Water then becomes increasingly expensive to pump from the ground.

effluent stream
A stream that intersects the water table and gains water from the zone of saturation.

influent stream
A stream that lies above the water table and loses water to the zone of saturation.

spring
A place where the land surface intersects the water table and groundwater seeps or flows naturally out of the ground.

recharge area
The area—mainly the broad upland between stream valleys—that receives precipitation and adds water to the zone of saturation.

discharge area
The area—primarily stream channels—that receives groundwater from the zone of saturation and conducts it away.

cone of depression
A concentric indentation in the water table that develops around a well.

Figure 12.5 Pumping leads to a general lowering of the water table, which forms cones of depression surrounding the wells. The water table falls during a pumping period and rises when pumping ceases. Unless water is put back into the ground, however, the level of the water table gradually declines.

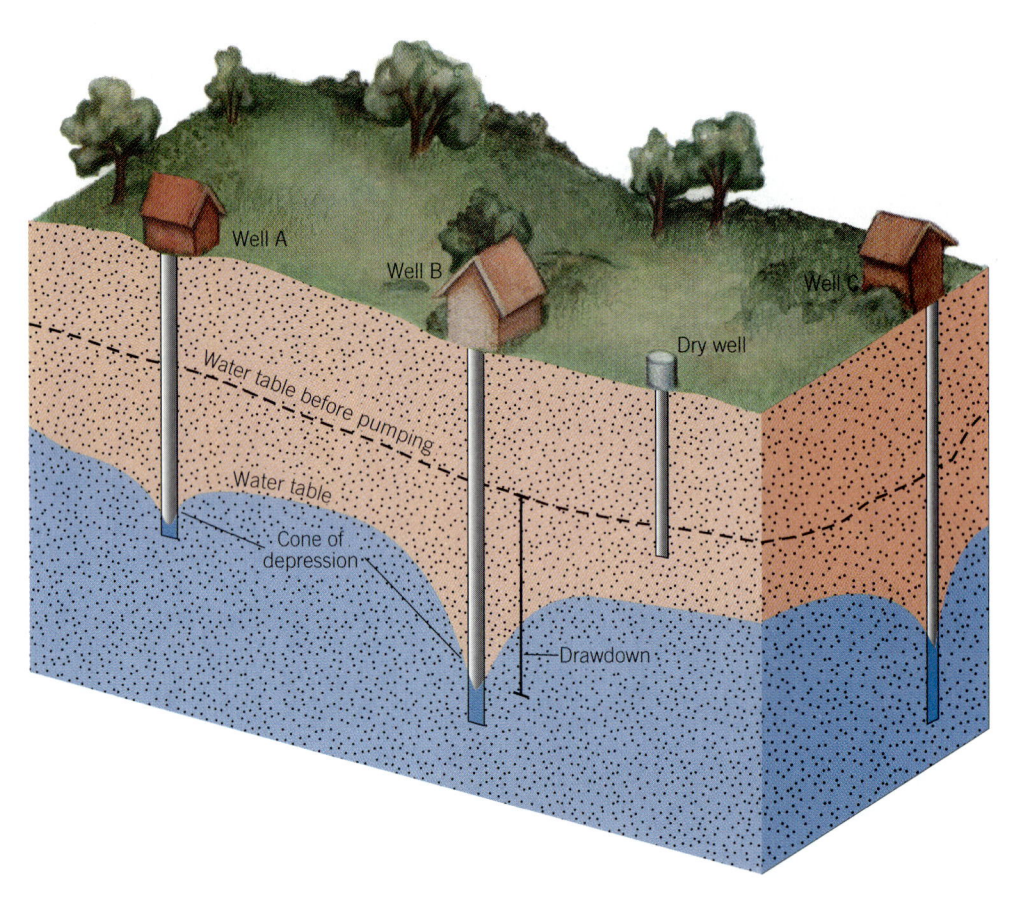

Long-term pumping at high rates from a large well can produce a cone of depression with a radius of 3 kilometers or more. Frequently, the cone may extend beneath neighboring properties, causing wells there to run dry. Lawsuits have been fought over this issue—water is the one thing that crops, livestock, factories, and people cannot do without. Where demand is high, many closely spaced wells are sunk. Cones of depression intersect and lead to a general lowering of the water table, to everyone's detriment.

Artesian Wells

Early in this century, railroad planners faced a serious problem: to find a reliable supply of water that could feed the huge steam-driven locomotives that moved goods and people across the arid South Dakota plains. Planners turned to the United States Geological Survey for help, and it is interesting to see how the USGS geologists solved the problem. They began by investigating the geology of the Black Hills dome of South Dakota. The Black Hills consist of a core of ancient igneous and metamorphic rocks encircled by a thick sequence of sedimentary strata. At one time, the strata covered the entire core region, but they have long since been eroded. As with all domal features, strata dip away from the core and are found at great depth far from the uplift.

Prominent in the sequence of strata is the Dakota sandstone, a porous, permeable formation that was deposited when a broad, shallow sea covered the western United States approximately 100 million years ago. In the vicinity of the Black Hills, the Dakota sandstone forms a thick ridge that is sandwiched between impermeable shales. On the basis of these relationships, survey geologists predicted that a 1000-meter-deep well drilled several hundred kilometers to the east—a location convenient to the railroad—would tap the Dakota sandstone. The prediction proved correct when water gushed from the new well.

The Dakota sandstone is the key element of a classic artesian well system—one in which well water rises above the level of the confined aquifer without the need for pumping (Figure 12.6). The Dakota sandstone is a porous and permeable aquifer with

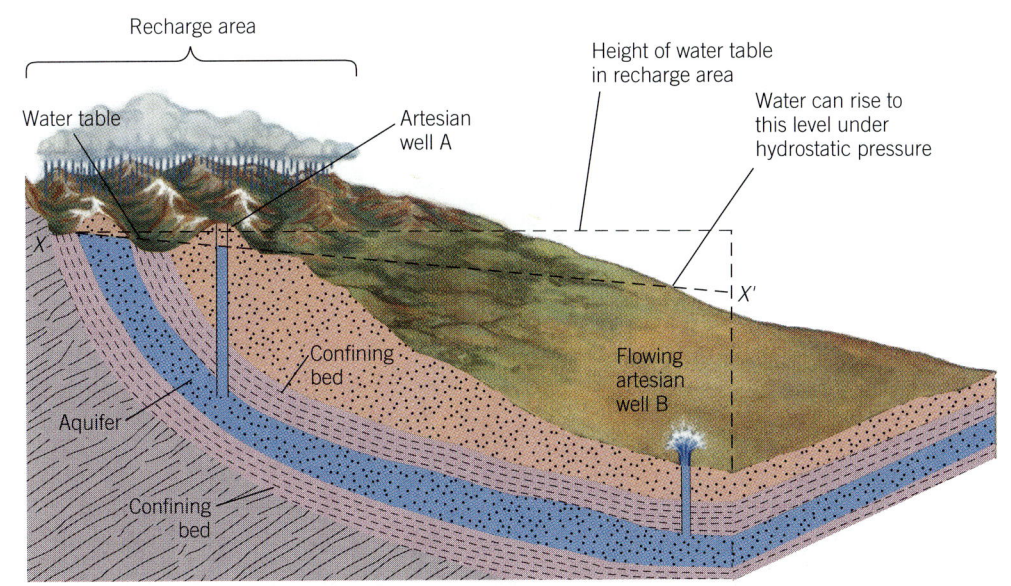

Figure 12.6 The conditions necessary for an artesian well in a confined aquifer. Well B is far below the height of the recharge area, so water gushes to the surface under hydrostatic pressure. Were it not for friction, the water at this location would rise to the level of the recharge area, but in actuality, it reaches level $X - X'$.

impermeable confining beds above and below the aquifer that prevent groundwater escape. The Black Hills recharge area receives water at a greater elevation than the outlet or discharge well. Therefore, water at the discharge well is under enough hydraulic pressure to rise toward the surface.

If it were not for friction, water spouting from wells drilled in the aquifer would rise to the same elevation as the aquifer at recharge. Because of friction, however, the level to which artesian well water is able to rise decreases with the distance of the well from the recharge area. The sloping line $(X - X')$ in Figure 12.6 indicates how this level changes with distance. If the land surface at a particular location lies beneath this line, water will reach the surface under its own power. Should the land surface lie above the line, water will rise only partially to the surface, and pumping is required to bring it up the rest of the way.

The Dakota sandstone and similar confined aquifers are the major water sources for numerous cities, towns, and farms of the Plains states. First tapped in the 1870s, they probably had more influence on the settlement of the American West than the covered wagon. Artesian systems also are important water sources in the upper midwestern states and in the humid Atlantic and Gulf states.

Well Extraction Problems

As we mentioned at the beginning of this chapter, groundwater reserves are not inexhaustible. Excessive extraction can lead to land subsidence, saltwater encroachment, and loss of hydraulic pressure.

Land Subsidence

Discharge from an artesian well triggers a pressure drop surrounding the well similar in form to the cone of depression in a water-table well. However, the drop in an artesian well is commonly more rapid and far deeper and extends over a wider area, because water in a confined aquifer plays an important structural role. The weight of overlying strata is supported by the framework of the particles that compose the aquifer and by the fluid pressure of the water in the pore spaces between particles. As the well draws water from the confined aquifer, the particles distort and pack together more tightly, squeezing water from the pore spaces as from a sponge. The framework can no longer support the load and begins to collapse, causing the land surface above to subside, or sink. Thus excessive extraction of water from a confined aquifer may lead to **subsidence** of the land. Subsidence is accelerated if there is additional surface load in the form of buildings, roads, and other structures.

subsidence
The sinking of an area of the Earth's surface upon compaction or dissolution of subsurface materials, often due to groundwater extraction.

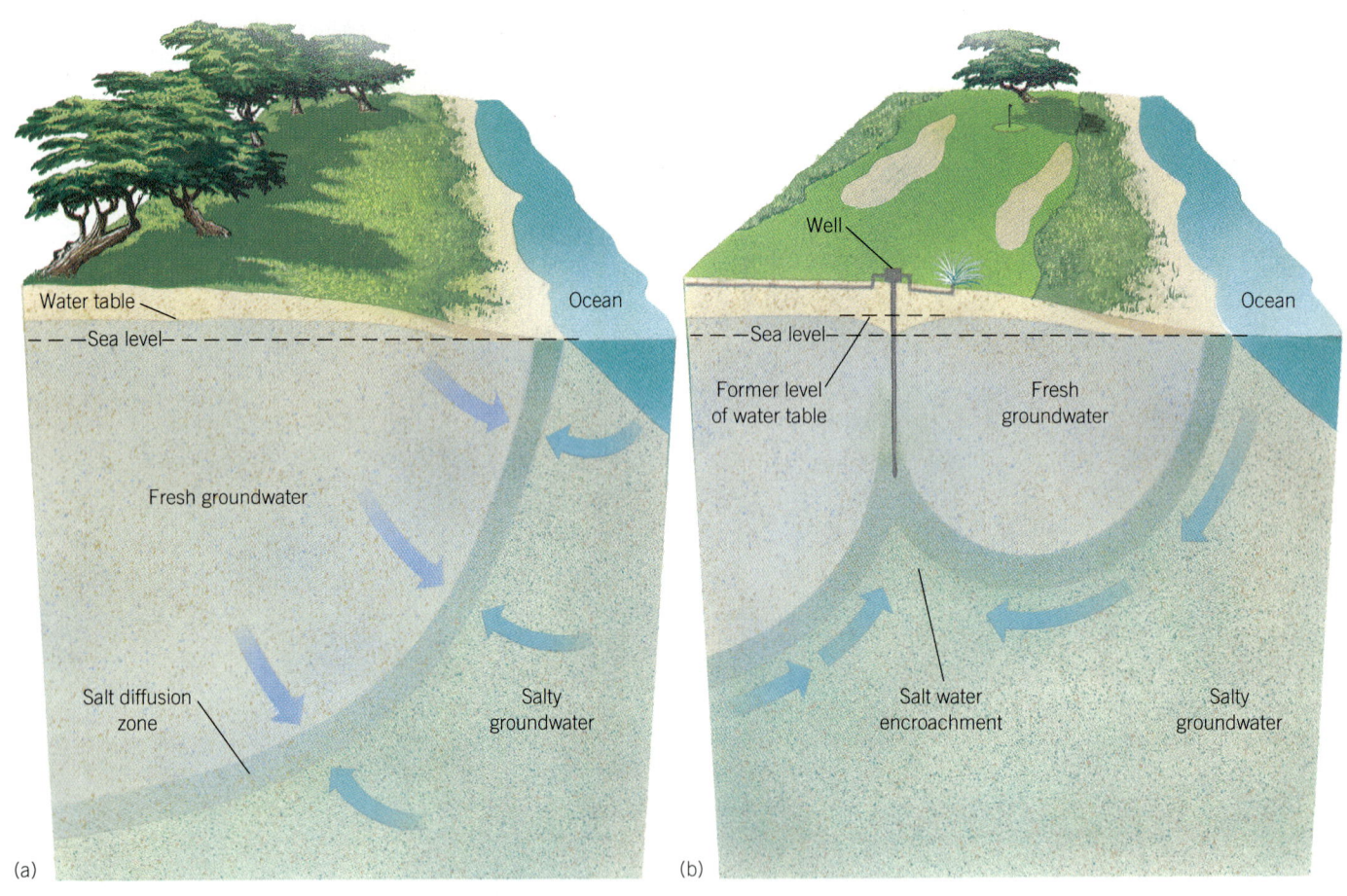

Figure 12.7 Saltwater encroachment and well contamination in a coastal region. (a) Because saltwater is denser than freshwater, it passes beneath and lifts normal groundwater. (b) A slight drawdown of freshwater due to pumping causes a great rise in the saltwater level beneath the well.

Subsidence may cause many problems. Subsurface utilities, such as water and drainage pipes, are commonly ruptured, and electrical installations are damaged. Foundations may shift, causing shear fractures in the supports, warped doors and windows, and other things that make buildings uninhabitable. Cracks can develop suddenly in roads, and roadbeds can sag. If subsidence occurs near coastal regions, the land surface may actually drop below sea level, which accelerates coastal erosion, shifts stream courses, and causes flooding.

Land subsidence can be seen in the San Joaquin Valley of California, where groundwater extraction to feed this rich agricultural region has led to a lowering of the land surface by 9 meters in 50 years.

Saltwater Encroachment

In coastal regions, groundwater in the zone of saturation floats on saltwater that has seeped in from the ocean and penetrated the strata. The density relationship between the two water bodies is such that for every 1 meter that the water table rises above sea level, the freshwater sinks 40 meters below sea level (Figure 12.7). Therefore, a 1-meter well drawdown of the water table results in a 40-meter rise of the saltwater beneath the well. Unrestrained pumping thus brings saltwater up into the well very quickly.

Artesian wells are particularly vulnerable to **saltwater encroachment,** because of the rapid, deep, and widespread pressure drops in confined aquifers. As saltwater works its way up in the aquifer, it contaminates wells ever farther inland; it is thus a major problem along

saltwater encroachment
The displacement in the zone of saturation of freshwater by saltwater.

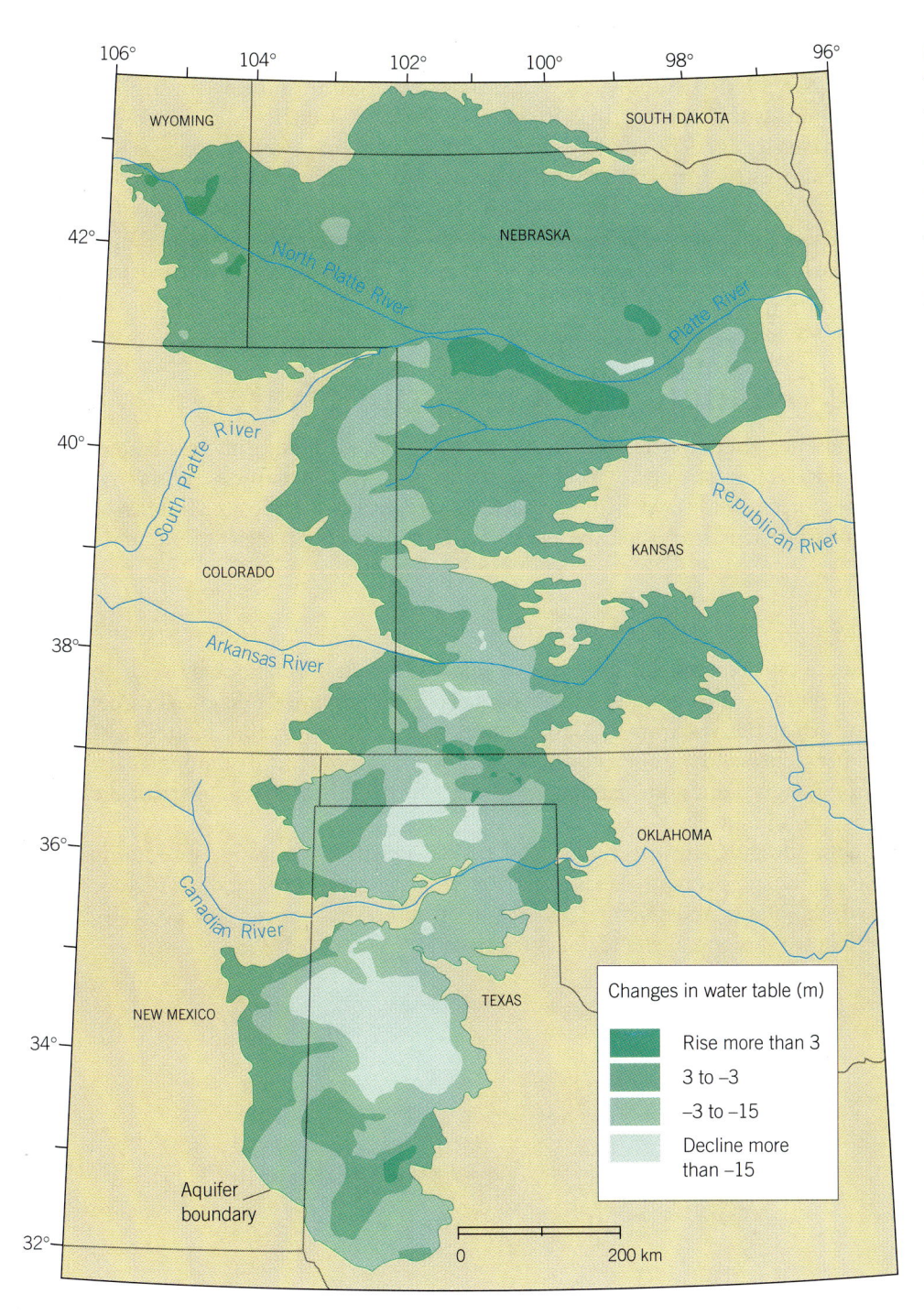

Figure 12.8 Changes in the level of the water table within the Ogallala aquifer over the past 30 years. Overpumping has led to a decline in hydraulic pressure throughout this rich agricultural region, which is felt most acutely in western Texas and the Texas-Oklahoma Panhandle. (Adapted from the 1982 *USGS Annual Report*.)

some reaches of the Atlantic Coastal Plain. Needless to say, saltwater contamination makes water useless for drinking and industrial purposes; it also ruins machinery and the wells themselves. It is not necessary for seawater to actually invade wells for contamination to occur; just the upward diffusion of salt is sometimes sufficient to contaminate freshwater.

Loss of Hydraulic Pressure

Serious pressure problems develop where confined aquifers are overtapped. A well-known case is the Ogallala aquifer, a 65-meter-thick, porous deposit that spreads beneath the arid Texas Panhandle, western Kansas, eastern Colorado, and much of Nebraska (Figure 12.8). The entire region underlain by the Ogallala aquifer is responsible for feeding 40 percent of the nation's cattle and generating $30 billion a year in agricultural products. The aquifer is vital to at least 5 million cultivated acres of the Texas Panhandle and New Mexico. Although the Ogallala is recharged to some extent by present-day rainfall, most of the water that it contains was added during the wetter climate of the previous ice age. The aquifer is like a bank account from which money is withdrawn but not deposited. Over 150,000 wells are currently tapping the Ogallala aquifer at 10 times the rate of recharge, and the zone of saturation is shrinking drastically. Should the practice continue, this essentially nonrenewable aquifer will be depleted in less than 30 years. Because no other source of water is available to replace it, the loss of the aquifer would have a disastrous effect on the economy and ecology of the region.

The Geologic Work of Groundwater

Groundwater is a powerful geologic agent. It creates such landforms as sinkholes, caves, and underground streams. It preserves the record of extinct forests, erupts as geysers and hot springs, and leaves behind diverse mineral and rock deposits. Groundwater does its most spectacular work on the thick limestone deposits of ancient seas.

Caves

cave
A natural cavity beneath the surface of the Earth, usually formed by the dissolution of limestone by groundwater; large caves are also called caverns.

Caves are open spaces dissolved out of thick limestone formations by acidic groundwater solutions. Groundwater usually contains a fair amount of dissolved carbon dioxide, which it derives from the atmosphere and decayed organic matter in the soil. Together, water and carbon dioxide form a weak carbonic acid solution that decomposes minerals. Limestone, composed mostly of calcium carbonate, is particularly vulnerable to acid attack.

Groundwater rich in carbonic acid trickles down to the water table and occupies the fractures and bedding planes of the limestone (Figure 12.9). As the limestone is dissolved, the planes and fractures enlarge in all directions, forming water-filled cavities that follow the slope of the regional water table.

Figure 12.9 The development of a limestone cave network begins at the top of the zone of saturation and follows the lowering of the regional water table.

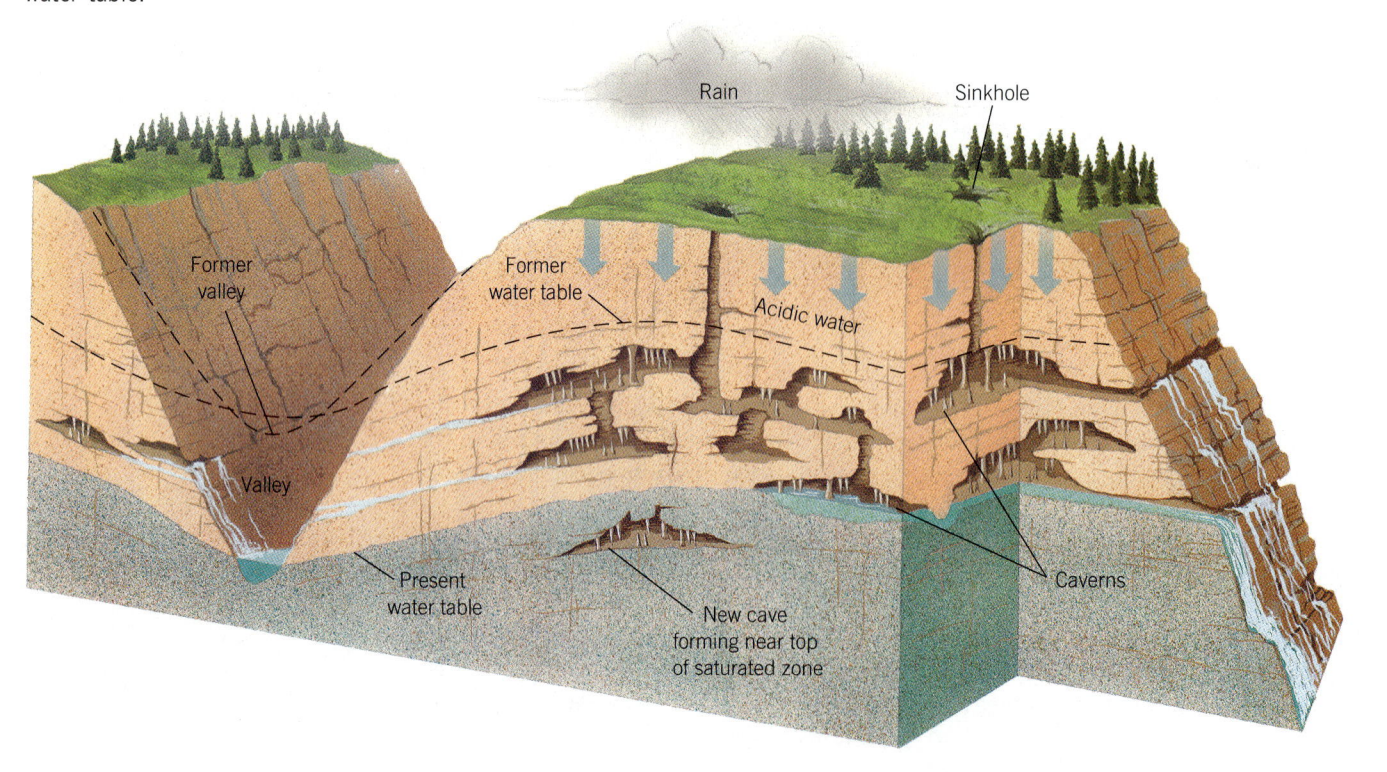

Figure 12.10 Prominent stalactites (ceiling) and stalagmites (floor) in a cave in New Mexico.

As stream valleys are lowered by erosion, the water table falls, because groundwater flow (like surface runoff) is controlled by the level of the stream that it enters. The drop in the water table exposes the solution cavities to the air and extends the solution process to greater depth. Over time, the falling regional base level leads to an intricate, many-storied network, complete with weirdly shaped passages and vast, unexpected rooms that delight explorers.

The lowered water table is also responsible for two of nature's delightful creations: *stalactites,* which hang from the roof of a cave, and *stalagmites,* which grow up from the floor (Figure 12.10). Their development begins with a vertical crack in the limestone cave roof. The carbon dioxide escapes when it reaches the open-air cave. Calcite is precipitated as the water drips through the crack and thin concentric layers accrete to the stalactite's surface as the water runs down over it. Meanwhile, the dripping groundwater falls to the floor of the cave and, as the water evaporates, a thick calcite mound accumulates and forms the stubby, wider stalagmite beneath the stalactite.

Underground caves that are stranded in the zone of aeration by water-table lowering are subject to weathering and erosion. Their roofs are unsteady and frequently collapse for lack of support. If collapse undermines the land surface, the land will subside, forming a large, circular depression, or **sinkhole.** Frequently, sinkholes fill with groundwater to form deep lakes. Numerous sinkholes give the countryside of Florida (as well as other parts of

Note: The terms stalactite *and* stalagmite *are easily confused; try associating the* c *in* stalactite *with* ceiling, *the* g *in* stalagmite *with* ground.

sinkhole
A circular surface depression that occurs when the roof of an underground cave collapses or dissolves.

Figure 12.11 A sinkhole, formed from the sudden collapse of limestone bedrock, near Orlando, Florida.

the southeastern United States) a pockmarked appearance. While they provide scenic and recreational benefits, as well as habitats for wildlife, they also cause major hazards for homeowners and engineers. It is difficult to tell when or where a sinkhole will suddenly open (Figure 12.11). Urban sprawl has increased groundwater pumping and a corresponding lowering of the water table, which has, in turn, accelerated the formation of sinkholes.

Karst Topography

The Karst region of Slovenia along the eastern shore of the Adriatic Sea is characterized by many of the features discussed so far: thick limestone deposits with numerous sinkholes and intricate cave networks extending thousands of meters below the surface. Their depth is evidence that in the past, sea level was much lower than it is at present. The entire subsurface has become so pockmarked with caves, sinkholes, and interconnected channels that streams no longer run on the surface but drain underground through the water table. The porous topography takes on an arid aspect, even though rainfall may be plentiful.

Regions of the world that have features similar to those just described are said to display **karst topography.** The Yucatan Peninsula and portions of Florida, Indiana, and Kentucky all exhibit relatively early stages of karst development, in which most of the land surface has not yet been eroded, despite the proliferation of sinkholes and caves. The Kwangsi Province of China is a classic example of late-stage karst topography (Figure 12.12). The once-thick limestone strata have been reduced to an eerie landscape of stacks and towers. Full-fledged karst topography takes millions of years to evolve.

Far from being a geological oddity, karst topography covers about 15 percent of the Earth's land surface. Because it is characterized by underground drainage, the potential for the spread of pollution is enormous. For instance, about 20 percent of the United States' freshwater drains through karst topography. Unfortunately, uncounted tons of domestic, agricultural, and industrial wastes have been pumped down into cave networks, converting them to sewers and polluting their waters, which were once symbols of purity. Bowling Green, Kentucky, for example, is situated on karst topography. *Time* magazine reported that on several occasions during the 1980s, benzene and other chem-

karst topography
Topography formed by the intensive dissolution of underlying limestone bedrock, featuring sinkholes, intricate cave networks, and diversion of surface drainage underground.

Figure 12.12 Late-stage karst topography, south of Auilin, China. The towers are the remnants of thick limestone formations dissolved away by groundwater.

ical fumes rose up from the caves, endangering homes and schools. Fortunately, cave specialists from Western Kentucky University were able to use their knowledge of groundwater flow to identify the sources of the pollution.

Geysers, Hot Springs, and Fumaroles

Chapter 8 described what happens when underground water comes in contact with hot rock or magma beneath the sea floor. When cold seawater trickles down through fractures along oceanic rift zones, it absorbs heat from the magma and rises back up to the ocean floor as hot-water chimneys. **Geysers** and related features are land analogies to these submarine chimneys (Figure 12.13). Most geysers erupt at intervals, although not necessarily with the clocklike efficiency of Old Faithful in Yellowstone National Park, Wyoming.

The eruptive cycle of a geyser begins as water filters down through fractures and passageways to a heat source at depth (Figure 12.14). The complex fracture network in effect forms a vessel or container. Water at the bottom that is in contact with the hot rock is under great pressure, which allows the water temperature to rise well above the normal 100 °C surface boiling point. Over time, the temperature reaches a critical point where the water at depth vaporizes to steam despite the pressure. The expansion causes water at the surface to spill over in a sort of preliminary event, reducing pressure on the entire water column, and allowing the superheated water in the column to vaporize to steam and erupt violently. After the pressure is released and the water column cleared, the water of the geyser

geyser
A periodic eruption of hot water and steam, caused by the heating of pressurized groundwater in a network of underground channels in contact with hot rock or magma.

Figure 12.13 A geyser in full eruption, Iceland.

trickles back down through the cracks to repeat the cycle. Left behind on the surface near the geyser vent is thick, encrusted silica called *geyserite*. Similar in composition to quartz, but lacking an organized internal structure, geyserite precipitates from the hot water.

Although geysers are the most spectacular manifestation of the interaction between groundwater and subsurface magma, there are others. Where spring waters have come in contact with magma or hot rock before issuing from the ground, they form **hot springs**—some warm enough to bathe in, others hot enough to boil eggs in. If magma-heated underground water intersects volcanic gases, the mixture may erupt through a vent in the surface as a steaming **fumarole.** Geysers, hot springs, and fumaroles are common in such volcanically active regions as New Zealand, Iceland, Alaska, northern California, Wyoming, and Montana. In many parts of the world, these features have been tapped as a source of geothermal energy. In fact, fumaroles in northern California supply much of the energy

hot spring
A spring whose waters have been heated above body temperature (36.7 °C) by hot rock or magma.

fumarole
A vent from which volcanic gases and water vapor escape.

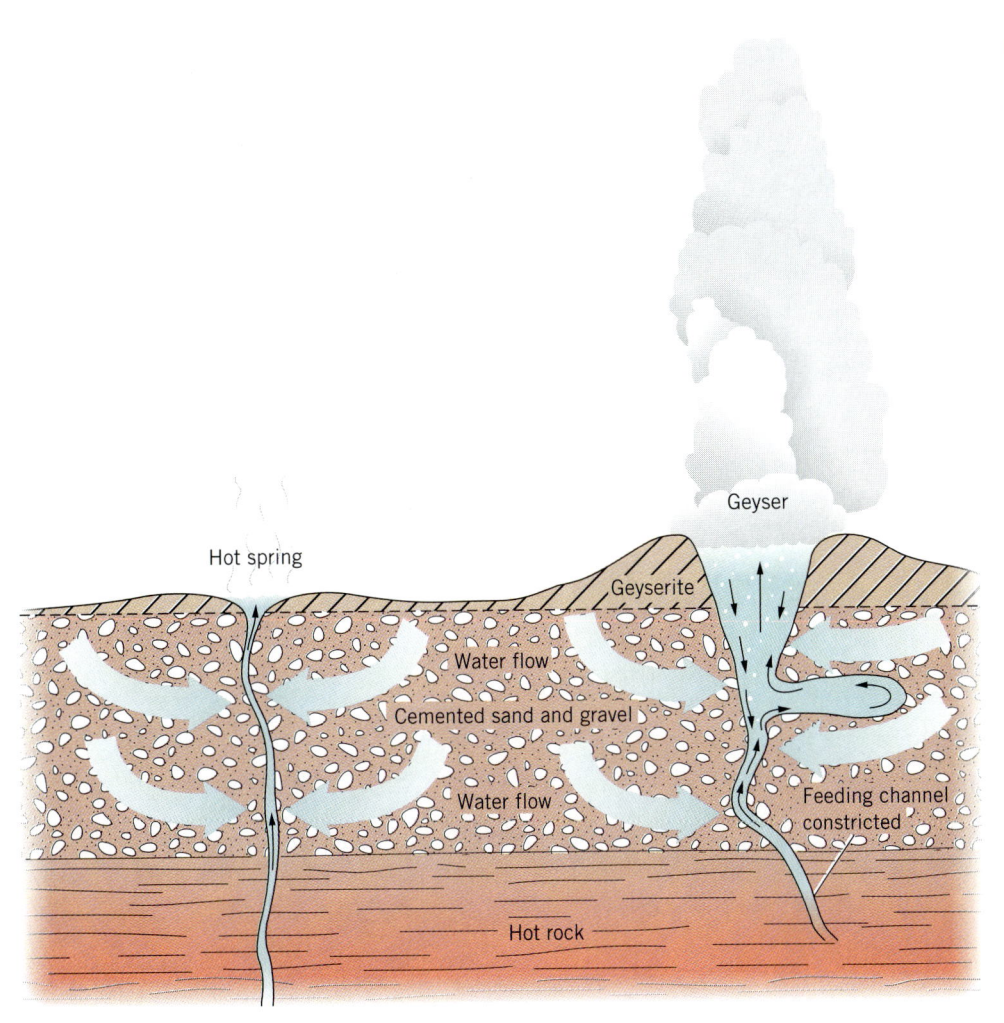

Figure 12.14 Two ways in which groundwater interacts with magma or hot rock: hot springs and geysers.

needed to run the utilities of San Francisco. Iceland and New Zealand also have been tapping subterranean heat for years; and recently, Hawaiian utility companies have begun to harness some of the enormous energy released by the island's active volcanoes.

The Quality of Groundwater

Totally pure water is found only in laboratories; underground water contains many naturally dissolved substances. More often than not, groundwater quality depends on the composition, the concentration, and (in some places) the interaction of these dissolved substances. The common characteristic of *hardness* is caused by relatively high concentrations of calcium, magnesium, and iron salts that are dissolved in water; they can impart an unpleasant taste to the water and prevent soap from lathering. Hard water also gunks up machinery; it forms scales in boilers and hot-water pipes.

The calcium and magnesium ions in hard water are derived mainly from the solution of limestone, dolomite, and gypsum and, in part, from the solution of metamorphic and igneous minerals like olivine and pyroxene. Regions with underlying carbonate and sulfate sediments tend to have hard water. Often, chemical treatments that extract or filter these ions from the water are required.

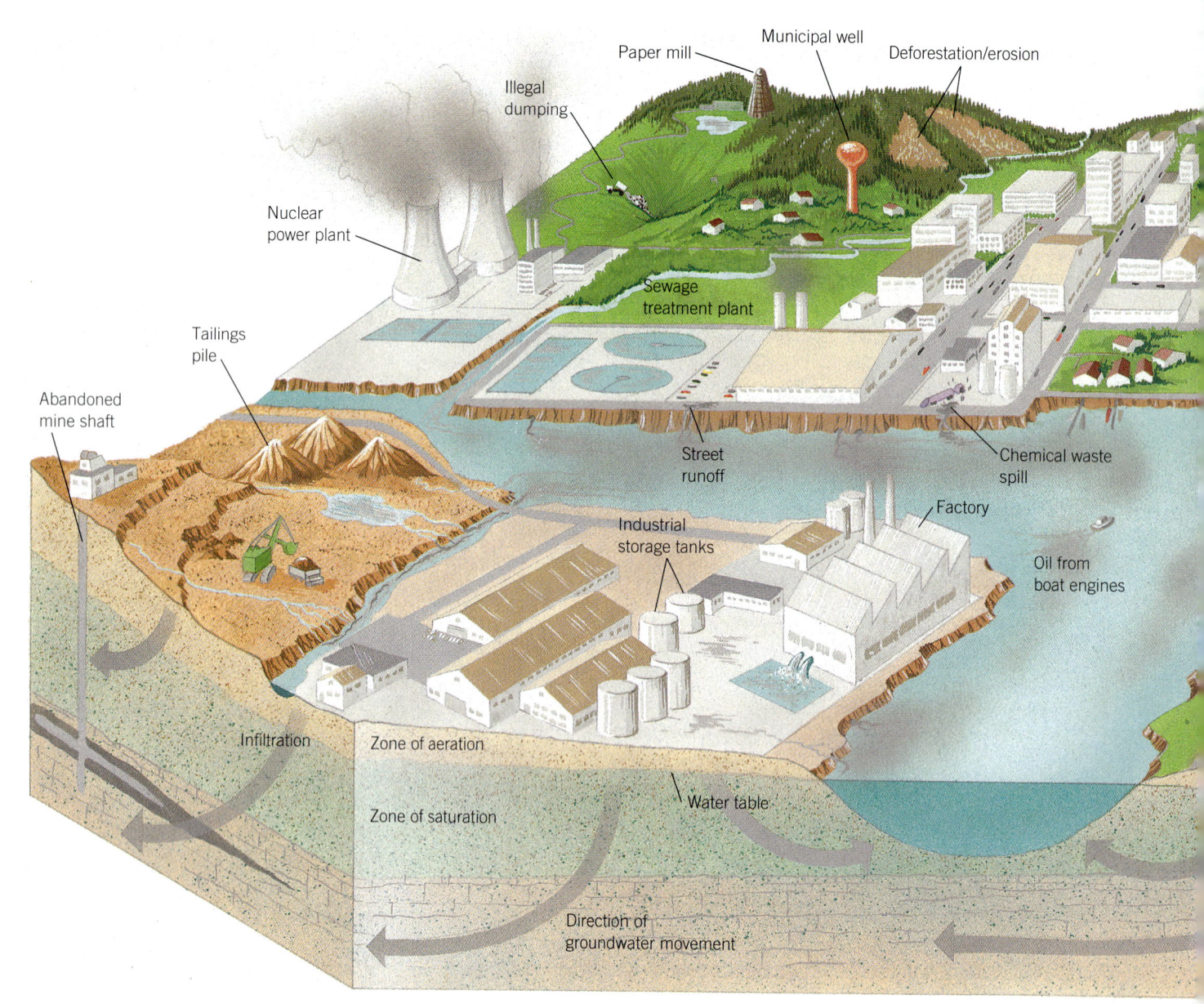

Figure 12.15 Some of the myriad sources of groundwater contamination. Sources such as crop dusting with pesticides spread contaminants over a wide surface area. Others, such as leaky industrial storage tanks, are localized point sources. Upon percolating through the subsurface, the pollutants migrate according to the regional slope of the water table. Streams, lakes, and wells are at risk.

Groundwater Pollution

Groundwater pollution is broadly defined as any deterioration in water quality. Saltwater contamination of freshwater aquifers that results from excess well pumping fits this definition. However, the main cause of groundwater pollution stems from waste disposal on or beneath the land surface. Industrial chemicals, radioactive wastes, heavy metals, pesticides, and other substances seep—or are purposely injected—into the ground. The main culprits are factories, nuclear power plants and weapons facilities, farms, landfills, buried gasoline tanks, and septic systems (Figure 12.15). As populations grow and as industrial and agricultural activities multiply, the extent and intensity of waste disposal also increase. Approximately 25 percent of our nation's aquifers are unusable, the percentage expanding to nearly 100 percent in some states.

Once polluted, a groundwater system may take centuries and incredible sums of money to purify. On the national scale, billions of dollars are already being spent on purification. A more serious problem is that while some contaminants are evident because they are obnoxious in taste, odor, color, or the like, others are silent killers. If these poisons go undetected, over the long run, they can cause disease in humans, animals, and plants. As with so many environmental hazards, it is often difficult to trace the exact cause of illness (or death) to its source.

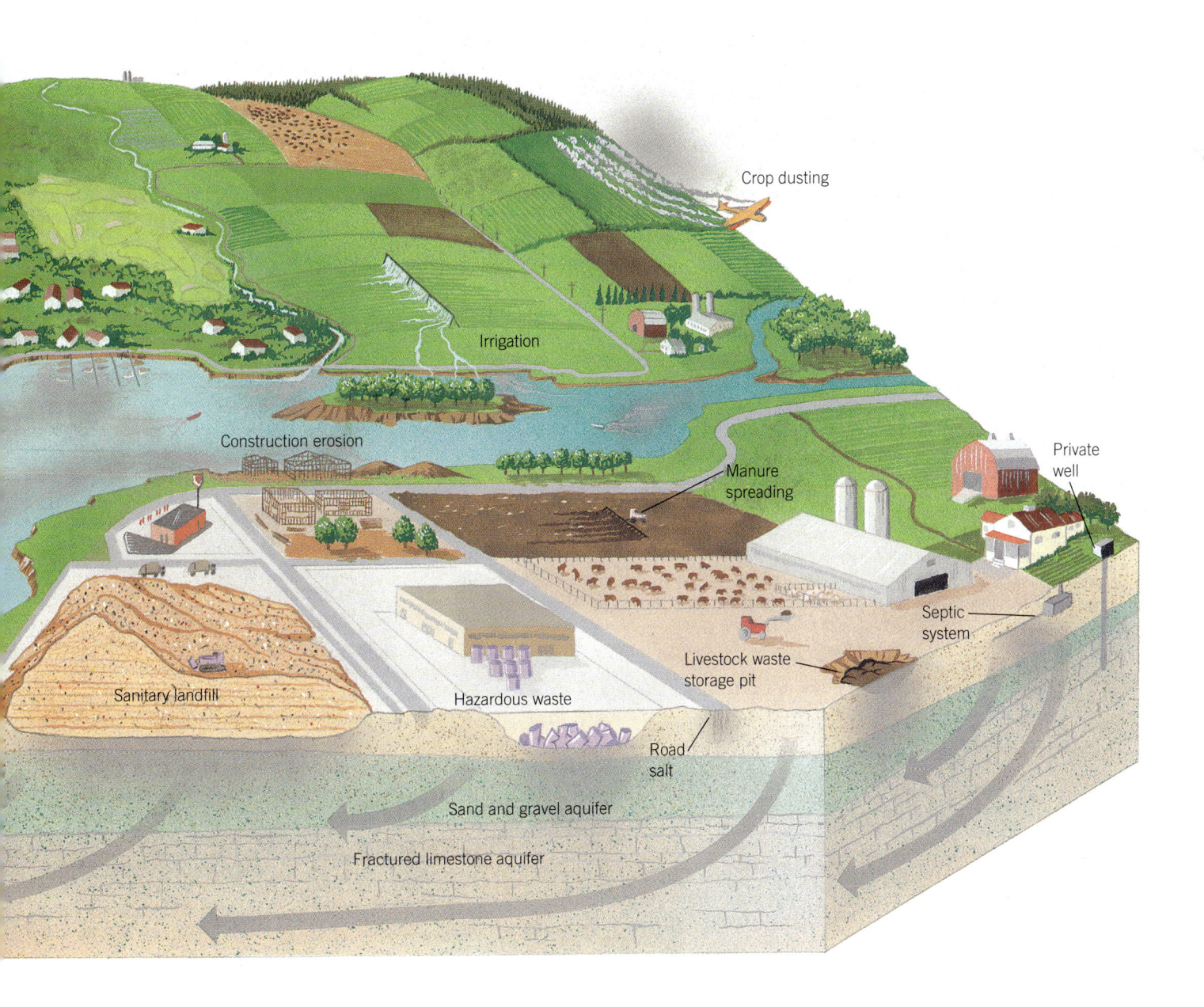

Figure 12.15 illustrates that pollutants are released to the subsurface from point sources—septic tanks, leaking gasoline storage drums, and industrial injection wells—and over broad areas, such as cropland that is sprayed with pesticides. Point-source and broad-area release occur in both urban and rural settings and have two kinds of effects on the environment: (1) local and usually short-term and (2) pervasive and usually long-term.

Take, for example, a typical point-source polluter: a septic tank located a short distance uphill from a well. Sinking a well in this setting is asking for trouble, since the pollutants will migrate toward the well. Nevertheless, the source of pollution can be pinpointed quickly, and the tank or the well can be moved. Some point-source problems

Table 12.2 Drinking Water Standards of the U.S. Environmental Protection Agency

Contaminant	Health Effects	Maximum Contaminant Level (mg/L)	Sources
Arsenic	Dermal and nervous system toxicity effects	0.05	Geological, pesticide residues, industrial waste, and smelter operations
Cadmium	Kidney effects	0.01	Geological, mining, and smelting
Lead	Nervous system damage; kidney effects; highly toxic to infants and pregnant women	0.05	Leaches from lead pipes and lead-based solder pipe joints
Mercury	Central nervous system disorders; kidney effects	0.002	Used in manufacture of paint, paper, vinyl chloride; geological
Nitrate	Methemoglobinemia ("blue-baby syndrome")	10	Fertilizer, sewage, feedlots, geological
Selenium	Gastrointestinal effects	0.01	Geological, mining

Source: U.S. Environmental Protection Agency, 1977.

are not as easily corrected because the pollutants migrate surprising distances. In addition, their slow rate of creep through the subsurface makes them difficult to locate in areas of intense land use.

Long-term pollution problems involve a slow decline in water quality and are often reflected by changes in community health. Unfortunately, water lives up to its reputation as the universal solvent where pollutants are concerned—virtually all are soluble in water, at least to some degree. Table 12.2 shows the maximum concentration allowed for a selection of inorganic contaminants in drinking water by the U.S. Environmental Protection Agency. Note that it does not take a high concentration for some elements to be considered harmful. Because pollutants migrate slowly, some problems are the harvest of mistakes planted decades earlier. Thus groundwater quality continues to deteriorate despite current enlightened pollution control and treatment policies.

Pollutants travel with groundwater, so their dispersion rates depend on the porosity and permeability of the subsurface. In sand or gravel, point pollutants spread in a widening cone, their concentration varying with time. Minute amounts arrive first, and amounts increase rapidly as the mass of groundwater arrives. Concentrations diminish slowly as the source is exhausted, but dispersion is intensified if leakage is continuous.

If given sufficient time, some natural processes will diminish groundwater pollution. Biodegradable compounds eventually break down, for instance, and radioactive elements with a short half-life eventually decay. Most important, the subsurface has a remarkable capacity to scrub groundwater clean. Sand and gravel are natural filters that remove harmful particles, and the oxygen-rich zone of aeration acts to decompose many potentially dangerous compounds. Clays are also very effective natural absorbers of metal ions and organic compounds that can be dissolved in groundwater. However, they are not permanent repositories of pollutants and can release them back into the groundwater. Also, the subsurface through which groundwater moves is of finite size, so its purifying capacity can be overloaded. Finally, there are some highly soluble industrial compounds, such as methyl mercury, for which there is no natural defense.

Groundwater

STUDY OUTLINE

Underground water supplies most of the world's usable freshwater.

I. **ACCUMULATION OF GROUNDWATER.** When rainwater seeps into the ground, it passes first through the **zone of aeration,** which contains little moisture except in the **belt of soil moisture,** directly beneath the surface. The water fills the voids in sediments and rocks, forming a **zone of saturation.**

  A. The term **groundwater** refers specifically to water in the zone of saturation, and the upper limit of that zone is the **water table.** The **capillary fringe,** where water rises up through capillary action, is located just above the water table.

  B. The **porosity** of loose materials depends on three factors: sorting, packing, and particle shape. The porosity of consolidated rock is also affected by the material that cements the particles and by the degree to which the rock is fractured.
  1. **Permeability** is a measure of the ease with which materials transmit water. It depends upon the material's porosity and upon the size, arrangement, and interconnectedness of openings.
  2. When groundwater cannot reach the zone of saturation because of an impermeable barrier, it forms a **perched water table.**

  C. **Aquifers** are porous and permeable strata that hold and release groundwater. In an unconfined aquifer, the water table can rise or fall freely. A confined aquifer, sealed between impermeable **confining beds,** receives water from a distant, unconfined source.

II. **MOVEMENT OF GROUNDWATER**

  A. The water table tends to mirror the topography. Where it intersects lowland surfaces, lakes or marshes are formed. Streams that gain water from the water table are called **effluent streams.** In arid regions, **influent streams** lose water to the water table. Due to faulting or other irregularities, the water table may be exposed and flow out as a **spring.**

  B. The flow of groundwater is from **recharge areas,** where precipitation replenishes the groundwater supply, to **discharge areas** (streams, lakes, and marshes).

III. **WELLS**

  A. Wells sunk in unconfined aquifers produce a **cone of depression** in the water table. Drawdown is the difference between the levels of the water table before and after pumping.

  B. An **artesian well** will flow without the aid of pumping if the recharge area lies above the elevation of the well. Because of friction, however, this effect decreases with the well's distance from the recharge area.

  C. Well extraction problems
  1. Extraction from a confined aquifer may lower its water pressure and cause its framework of particles to compact. The result may be land sinking, or **subsidence.**
  2. Excessive pumping of coastal plain aquifers may cause **saltwater encroachment.** A slight freshwater drawdown leads to a marked rise of saltwater into the well.

3. Overuse of a confined aquifer increases the expense of extraction because of falling hydraulic pressure. Excessive extraction can be offset somewhat by treating the water and returning it to the ground through recharge basins or wells.

IV. THE GEOLOGIC WORK OF GROUNDWATER

A. **Caves** are open spaces dissolved out of limestone or other soluble rock by acid solutions. They form at or beneath the water table and are enlarged as the regional water table is lowered.
   1. Groundwater and carbon dioxide combine to form carbonic acid, which decomposes limestone. In the process, the water table is lowered, leading over time to exposed cave networks. Water supersaturated with carbon dioxide that drips from cave ceilings precipitates calcite to form *stalactites* and *stalagmites.*
   2. When an underground cave collapses, it may form a circular depression, or **sinkhole,** on the surface.

B. **Karst topography** is formed by the intensive solution of underlying limestone bedrock. It features sinkholes, intricate cave networks, and diversion of surface drainage underground.

C. **Geysers** and **hot springs** occur when water filters down through fractures and contacts hot rock or magma. **Fumaroles** are vents through which volcanic gases and vapor escape. In some regions of the world, they are tapped as a source of geothermal energy.

V. THE QUALITY OF GROUNDWATER.
Groundwater quality depends upon the type and quantity of naturally dissolved substances.

A. Groundwater pollution—broadly defined as any deterioration in water quality—stems mainly from waste disposal. Once polluted, a groundwater system may take centuries to purify.

B. Point-source pollution is local and short-term; broad-area release is pervasive and long-term. A pollutant's migration is governed by its density and the porosity and permeability of the materials through which it moves.

STUDY TERMS

aquifer (p. 264)
artesian well (p. 264)
belt of soil moisture (p. 263)
capillary fringe (p. 263)
cave (p. 270)
cone of depression (p. 265)
confining bed (p. 264)
discharge area (p. 265)
effluent stream (p. 265)
fumarole (p. 274)
geyser (p. 273)
groundwater (p. 262)
hot spring (p. 274)

influent stream (p. 265)
karst topography (p. 272)
perched water table (p. 264)
permeability (p. 264)
porosity (p. 263)
recharge area (p. 265)
saltwater encroachment (p. 268)
sinkhole (p. 271)
spring (p. 265)
subsidence (p. 267)
water table (p. 262)
zone of aeration (p. 262)
zone of saturation (p. 262)

THOUGHT QUESTIONS

1. Explain how the zone of saturation helps regulate the regional water economy during a one-year period.
2. A porous substance is not necessarily permeable. Explain.
3. Diagram and label the conditions necessary for an artesian well.
4. Describe three major well extraction problems.
5. Explain the origin of a limestone cave.
6. Describe four features associated with karst topography.
7. Explain the similarity between sea-floor chimneys and geysers.
8. Distinguish between point-source and broad-area pollutants. Make a list of each type.

Chapter 13

The Formation of Glaciers
   The Movement of Glacial Ice
   The Glacial Budget

Erosional Features of Glaciers
   Ice-Sculpted Mountain Topography

Depositional Features of Glaciers
   Moraines
   Landforms of Stratified Drift

Glacial Lakes

Glacial and Interglacial Ages
   Causes of the Ice Ages
      Plate Tectonic Effects
      Variations in Earth's Orbital Geometry: The Milankovitch Theory
      Variations in Atmospheric Carbon Dioxide and Dust

Human-Induced Global Warming
   The Impact of Global Warming on Human Society

Glaciers and Climate

Most of the world's ice is confined to two locations: Greenland, in the Northern Hemisphere, and Antarctica, in the Southern Hemisphere. The Greenland ice sheet is nearly a kilometer and a half thick, and it covers an area roughly the size of California, Arizona, New Mexico, Nevada, Utah, and Colorado combined. The Antarctic ice sheet is even larger. Concealing a continent the size of the United States and Mexico, it is 7.5 times as extensive as the Greenland ice sheet.

Those giant deep-freeze lockers, the Antarctic and Greenland ice sheets, exert major control over the Earth's climate by extracting heat from the overlying air and surrounding ocean. For this reason, the ice sheets are closely monitored for climatic trends. Is the Earth getting warmer? If so, a 5 °C increase would melt the glaciers and raise sea level at least 65 meters—enough to drown the land occupied by two-thirds of the world's population. Is the Earth getting colder? If so, a 5 °C drop would return the world to where it was 18,000 years ago, when glaciers covered 30 percent of the Earth's land surface and drastically altered its geography and environment (Figure 13.1). Although this vast ice sheet was indeed impressive, there is clear evidence that its appearance was not a unique event. Glacial advances and retreats have been the pattern over at least the past 3 million years, and probably longer.

Evidence in support of the theory that ice sheets periodically covered large portions of the Earth's land surface comes, first, from observing today's glaciers in action and the geological features they produce and second, from observing that vast regions, now ice-free, contain deposits and landforms derived from glacial action. Glaciers have sculpted the mountain peaks of the Canadian Rockies, the Cascades, the Andes, the Alps, and the

◀ Glaciers of the Fairweather Range, Glacier Bay, Alaska.

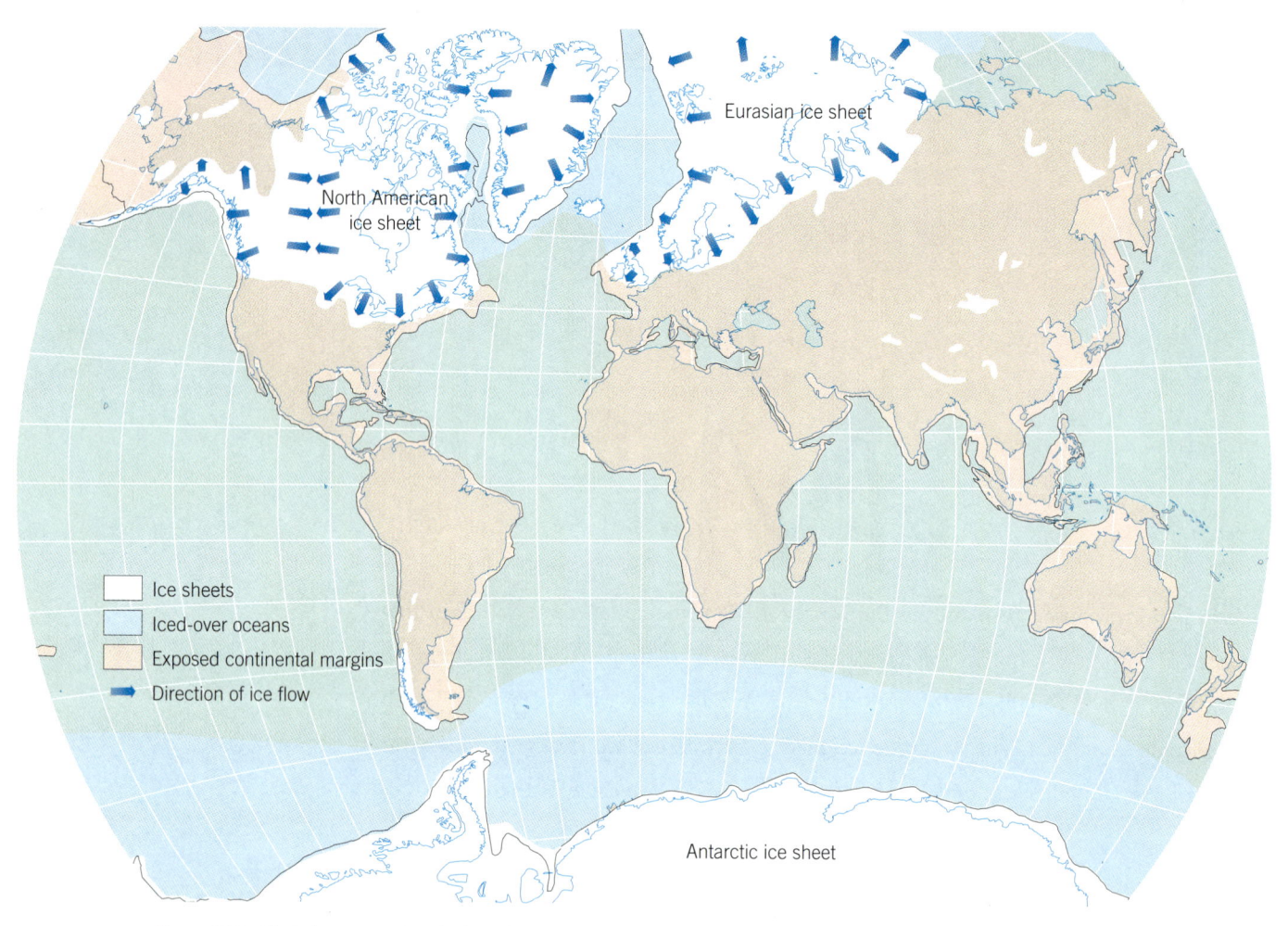

Figure 13.1 Global geography during the last glacial age. The growth of vast ice sheets lowered the worldwide sea level and exposed the continental shelves. Australia and Indonesia were one; Japan, Malaysia, and many Southeast Asian islands were joined to the mainland; and the Red, Black, and Caspian seas were dry. Humans migrated across the exposed Bering Strait from Asia to settle in North America.

Himalayas. The Vikings sailed out of the ice-carved fjords of Norway. Glaciers scoured and scraped clean the bedrock of eastern Canada and, upon melting, bequeathed a wilderness landscape of lakes, marshes, and rearranged stream courses. Also left behind by the melting ice were the stony soils of New England, Canada, and Northern Europe. Meltwater carrying fine silts and sands from the glaciers built the heavily farmed plains of Iowa, Central Europe, and China.

The Formation of Glaciers

A **glacier** is a large, slowly moving mass of ice. To understand how one forms, we must consider a key feature called the **snowline**—the altitude above which there is permanent snow. Above the snowline, temperatures are so low that snow forms into large snowfields; within these fields, the snow is converted to glacial ice. The conversion begins when the freshly fallen snow of these fields is changed through compaction into hard little ice pellets, or **firn.** The firn is buried under later snowfalls, and the added weight of the overlying snow subjects it to great pressure. Pressure is greatest where the grains make contact and least in the open spaces between the grains. The ice then melts along

glacier
A large ice mass formed on land by the compaction and recrystallization of snow that survives from year to year and shows evidence of present or past flow.

snowline
Altitude above which the snow is permanent.

firn
The pelletlike form assumed by snow in its transition to glacial ice.

Figure 13.2 Valley glaciers of the Wrangell Mountains in Alaska occupy former stream valleys. Note the rounded cirques at the heads of the glaciers and the knifelike arêtes that divide the glaciers. The numerous crevasses are caused by stress generated in the glaciers as they slide downslope.

the points of contact, and the water migrates to the spaces, where it refreezes. The process takes several decades or centuries, depending upon temperatures, but the net result is a dense, fused mass of interlocking ice crystals at depths exceeding 50 meters.

The snowline defines just where glaciers can form, and it is generally found at or near sea level toward the polar regions. There, sea-level temperatures year-round are low enough to permit the growth of ice masses of continental proportions that spread outward in all directions regardless of the topography. Ice masses of this magnitude, such as those presently covering Greenland and Antarctica, are called **continental glaciers,** or simply **ice sheets.**

The snowline rises toward the equator, although its exact elevation is influenced by such factors as sunlight, temperature, snowfall, and wind conditions. Away from the polar regions, glaciers are confined to high altitude, the mountains. Mountain glaciers, which are more formally called **alpine** or **valley glaciers,** occupy former mountain stream valleys (Figure 13.2). Alpine glaciers occur at all latitudes, including the equator, as in the case of Mount Kilimanjaro.

The Movement of Glacial Ice

As the ice in a valley glacier thickens by the addition of new snow, it begins to deform under its own weight. Ever so slowly, it flows, slips, and slides downhill. Geologists measure this motion by hammering a straight line of pegs into the ice across the valley and noting how the line changes in subsequent years (Figure 13.3). Over time, the line gradually deforms into a curve that bends downhill, which proves that ice velocity is greatest at the valley center and least along the valley walls. Clearly, friction with the valley walls and bottom retards ice flow, just as it retards stream flow.

Measurements like these show that the mechanics of glacial motion depend upon the response of the ice to stresses induced by gravitational and frictional forces. The upper 35 meters or so constitute the *zone of fracture*, in which the ice responds rigidly, storing stress until it cracks (see Figure 13.3). The stress results from the downhill pull of the ice at depth and from friction with the valley walls. It causes a series of fractures, or **crevasses,** to develop that extend in arcs across the width of the glacier. The crevasses open and close as the glacier creeps down over ledges in the valley floor.

Crevasses do not extend below about 35 meters, because at that depth, the ice deforms under its own weight in the **zone of plastic flow.** That is, rather than fracturing, the ice

ice sheet or **continental glacier**
A glacier more than 50,000 square kilometers in area that spreads in all directions, unconfined by underlying topography.

alpine glacier or **valley glacier**
A glacier in mountainous terrain that flows downslope in a valley previously eroded by a stream.

crevasse
A fracture in the rigid outer layer of a glacier caused by glacier movement.

zone of plastic flow
The inner mass of a glacier where the ice moves without fracturing.

changes shape permanently and continuously in response to pressure generated by the weight of the overlying ice. In this zone, the ice flows—but slowly (see Figure 13.3). The squeezing is greatest at the head of the glacier where the ice is thickest. Constrained by the valley walls and floor, the ice is forced to flow downhill toward the *glacial terminus* (or front), like toothpaste squeezed out of an open tube.

Motion in the zone of plastic flow can be measured by drilling holes greater than 35 meters deep into the ice and observing how the holes deform and bend with time (see Figure 13.3). Again, the overall similarity between ice motion and stream flow is striking. For example, just as water velocity increases with height above the streambed, so ice velocity increases with height above the valley floor. In each case, the flow velocity increases in a rising curve, reaching its maximum about halfway up.

Glacial motion seems to involve both sliding layers and melting: as the ice moves under pressure, it alternately melts and refreezes in the downhill direction. The melting and refreezing of the ice speed up movement in the zone of plastic flow. The process also supplies the water that collects on the valley floor and mixes with sand and gravel that has attached to the base of the glacier, creating a slippery slush. Like a greased skid, this slush allows the glacier to decouple from the valley floor and slide en masse downhill. This action is called **basal slip** (see Figure 13.3). On occasion, the decoupling is very sudden and dramatic; glaciers have been recorded moving downhill at up to 20 meters per day. Movement of this kind is called a **surge**.

The great ice sheets, which flow from thick interiors to thin coastal margins, are not as well understood as valley glaciers. Teams of geologists are currently measuring them and combing satellite images to advance our understanding of these complex masses. It appears that the Greenland and East Antarctic ice sheets, for the most part, move exceedingly slowly—on the order of centimeters per year in their interiors. Contributing to this slow movement are low regional slopes and low year-round temperatures, which makes the ice more rigid and binds it to bedrock. On the other hand, portions of the West Antarctic ice sheet have been observed to surge on the order of 3 meters per day.

The Glacial Budget

Glaciers can be said to operate like a business. They have income and expenditures and, over a period of time, show profits and losses. Income for a glacier means, of course, snow *accumulation* converted to firn and then to glacial ice. Expenditures mean ice loss through melting, evaporation of water that did not refreeze, and *sublimation*—that is, the passing of ice directly into the vapor state (which is the reason ice cubes tend to shrink in the freezer). Such expenditures go under the name of *ablation,* or wastage (Figure 13.4).

For a typical valley glacier, accumulation exceeds ablation above the snowline, that is, where the glacier forms. However, ablation exceeds accumulation below the snowline, because temperatures are higher at the lower elevations and melting is greater. If, over a given time period, net accumulation above the snowline exceeds net ablation beneath it, ice is produced faster than it is wasted. This is a profit or growth situation, and the glacial terminus will advance from its previous position. If the situation is reversed—that is, if ablation exceeds accumulation—the glacial front will retreat up the valley, although ice will continue to move from the head to the terminus.

basal slip
Movement in which a glacier decouples from the valley floor and slides downslope.

surge
A brief period of rapid glacial flow.

Figure 13.3 (page 287) The major structural units of a typical valley glacier. Note the various ways the glacier moves downslope: fractures that open and close near the surface, plastic flow at depth, and basal slip along the bottom. The rate of movement at the surface is greatest toward the center, as shown by the progression downslope of a line of stakes. Curving of drill holes indicates motion in the zone of plastic flow. The glacier slides downhill on flattened glide planes that stack up at the glacial terminus. A terminal moraine is deposited as the glacier melts, and the sediments deposited by meltwaters escaping the glacier form an outwash plain.

Glaciers and Climate 287

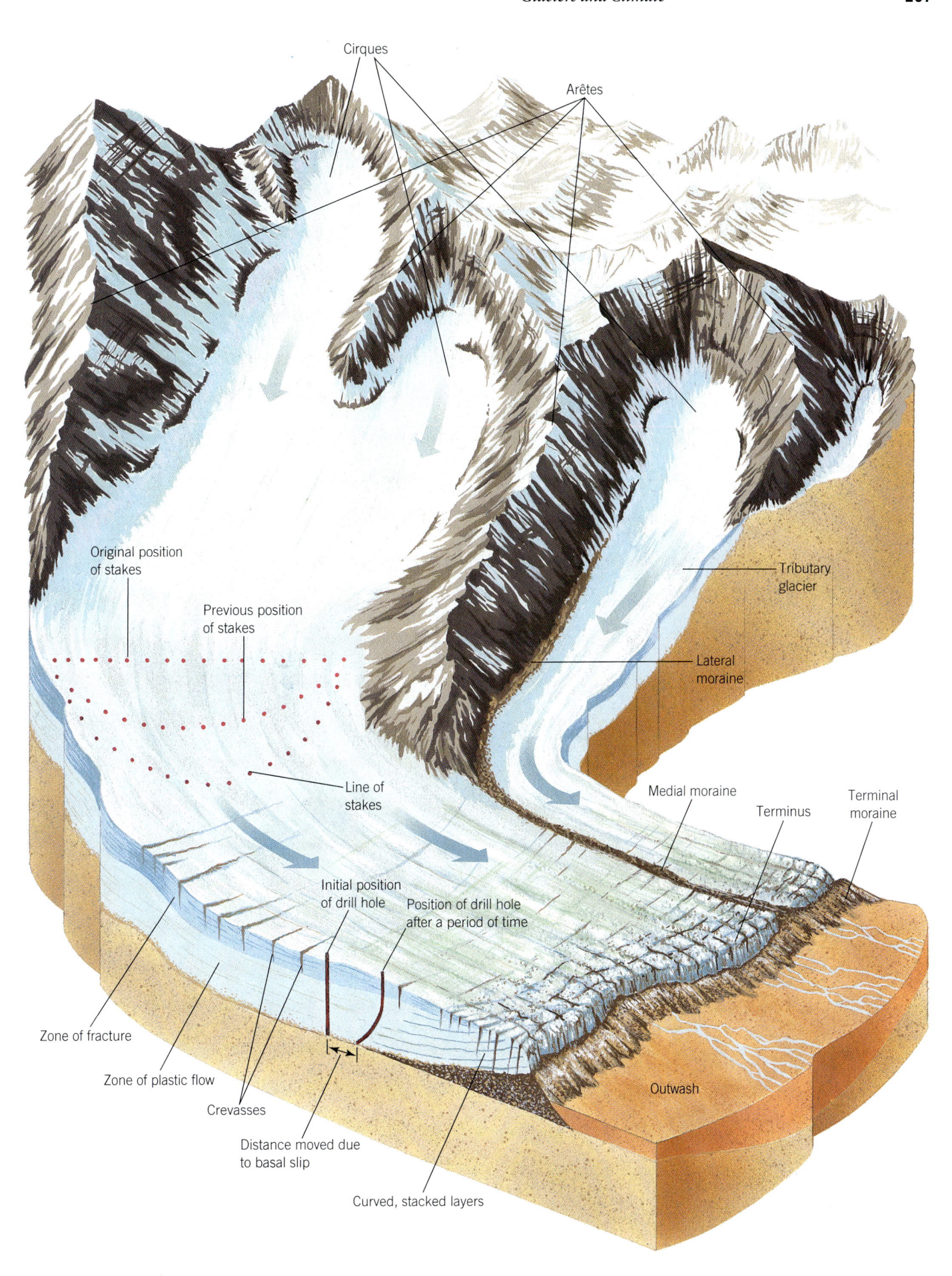

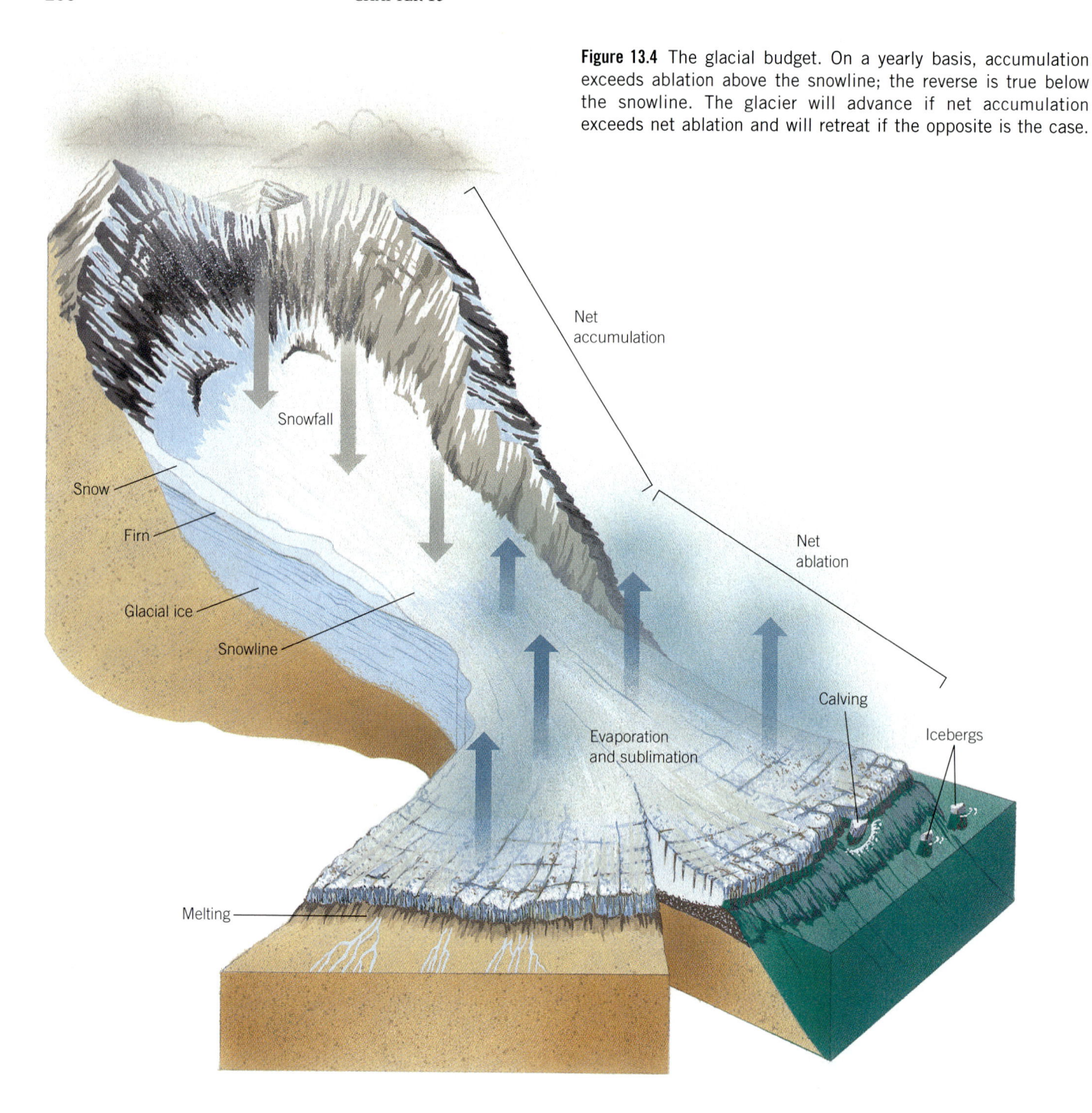

Figure 13.4 The glacial budget. On a yearly basis, accumulation exceeds ablation above the snowline; the reverse is true below the snowline. The glacier will advance if net accumulation exceeds net ablation and will retreat if the opposite is the case.

The Antarctic ice sheet continues to grow even though very little snow falls on it (just 5 centimeters per year); because it is so cold, there is almost no melting. However, this growth is offset by ice that is lost to the sea along the glacier margin. The glacial ice flows from the interior to the edges of the continent, where it extends far out to sea as broad, flat *ice shelves*. Attached to the land on one end, the shelves terminate in steep-walled ice cliffs—starkly beautiful sights. Ice shelves in the ocean behave like an ice cube in a glass of water, which floats with most of its volume submerged. The ice shelves grow until they lack a firm base of support on the seaward end. When they reach depths that allow 80 to 90 percent of their volume to sink underwater, the edges of the ice shelves calve—that is, they break into huge blocks and float out to sea as **icebergs** (Figure 13.5). Because melting is minimal on the vast Antarctic ice sheet, **calving** is the chief means by which the ice sheet is ablated.

iceberg
A block of glacial ice floating on a body of water, with 80 percent or more of its volume below the surface.

calving
The process through which blocks of ice break off from ice shelves and float out to sea as icebergs.

Figure 13.5 An iceberg calving from the face of the Margerie glacier in Glacier Bay National Park, Alaska.

Erosional Features of Glaciers

The geologic work of glaciers has a definite pattern, whether confined to mountain valleys or spread over continents. The ice erodes outward from its center of accumulation and transports the eroded material to the terminus, where it is deposited. As the melting glacier retreats toward its source, it leaves behind a countryside characterized by an intricate mixture of erosional and depositional features.

Two relatively simple processes are responsible for carving the wide variety of erosional features associated with glaciers. The first, **plucking,** is related to the fact that ice generally melts under pressure when squeezed against solid-rock surfaces. The meltwater trickles into fractures in the rock and refreezes. The water expands upon freezing, which exerts pressure and breaks off the rock fragments—some the size of a van, others the size of a sand grain. The rock fragments are dislodged (or plucked) and incorporated in the glacier as it slowly moves toward the ice front. The glacier then uses the fragments to grind away the surfaces it moves against. The process is called **abrasion,** and it is the second way that the ice erodes its surroundings.

Plucking is largely responsible for the characteristic signature carved at the head of valley glaciers, the **cirque** (see Figures 13.2 and 13.3). The plucked fragments serve as abrasives on the bottom and sides of the glacier, removing loose materials and gouging, polishing, and crushing bedrock. Figure 13.6 shows typical features of a rock outcrop subjected to glacial scouring. Note, first, that the outcrop has been scraped clean of soil. Parallel grooves and fine scratches, called **glacial striations,** were gouged in the outcrop by rock fragments attached to the bottom of the ice as the glacier slid over it. Finer silt and sand particles in the ice, mixed with slush, acted as sandpaper to give the rock its finer polish. The polish is smooth to the touch in the direction of ice movement but rough in the opposite direction, like planed wood. And like the fine sawdust created when wood is sanded, rock dust, or *rock flour,* was generated by the grinding and polishing of the rock surfaces and was washed away by streams of meltwater.

plucking
An erosion process in which meltwater trickles into rock fractures, refreezes, and expands, breaking off rock fragments.

abrasion
The mechanical wearing or grinding of rock surfaces by friction and impact of rock particles transported by wind, ice, waves, running water, or gravity.

cirque
A deep, curving, steep-walled depression scooped out of the bedrock at the head of a valley glacier.

glacial striations
Parallel grooves and scratches that were gouged in the bedrock by rock fragments attached to the bottom of the ice as the glacier slid over it.

Ice-Sculpted Mountain Topography

Glacial plucking and scouring produce spectacular alpine scenery. Unlike a stream, which physically occupies only a minute percentage of the valley it has carved, a

Figure 13.6 Glacial striations in a rock outcrop in the Van Horn Range in Alaska. Mapping striations on a regional scale enables geologists to trace the direction of ice movement.

arête
A divide between adjacent valleys, narrowed to a knife-edged ridge by the backward erosion of the valley walls.

horn
A steeply carved rock peak, formed by the walls of three or more cirques.

glacial trough
A U-shaped valley carved by a mountain glacier in what was previously a V-shaped stream valley.

hanging valley
A glacial valley, formed in what was previously a tributary stream valley that was cut off and left stranded at a higher elevation from the main glacial valley.

fjord
A deep, narrow arm of the sea, formed when a glacial trough that extended to the sea became submerged after melting of the ice.

drift
Glacial deposit of rock fragments.

glacier may fill the entire valley, so that erosion extends far up the valley walls (Figure 13.7).

Just as tributary streams feed water into the main stream, tributary glaciers feed ice into the main valley glacier. The ice originates at the headwaters of former streams, where, as we have seen, glacial plucking produces cirques. Expansion of the cirques narrows the divides between glaciers to knife-edged **arêtes.** In places, the glaciers expand their cirques headward to produce isolated peaks of rock, or **horns,** which stand like massive towers above the surrounding valleys.

As the ice flows and slides relentlessly downhill, the V-shaped stream valleys are converted into characteristic U-shaped **glacial troughs**—steep-sided, broad, and deep. In the process, the main glacier may amputate the former tributary stream valleys; it may also excavate a far deeper trough than the tributary glaciers dug. Thus when the glaciers melt, the tributary valleys are left stranded at a higher elevation than the main valley, forming spectacular **hanging valleys** that run along the upland margins of the main U-shaped glacial trough.

The rainwater that falls on this glaciated region will eventually erode these features and reestablish the dominance of stream erosion in shaping the landscape. Until then, running water accommodates itself to the reality of what the ice has accomplished. Cirques become cold, crystal-clear lakes, or *tarns.* Unevenly scoured bedrock forms the basins of *paternoster lakes* that are sometimes connected by a series of cascading waterfalls and rapids.

Mountain glaciers extending to a coast often cut their U-shaped valleys below sea level. When the ice in the valley finally melts, the sea invades these steep-sided troughs, forming **fjords**. These deep, narrow arms of the sea are prominent features of high-latitude seacoasts bordered by glaciated mountain ranges: the coasts of Norway, Greenland, Alaska, and British Columbia in the Northern Hemisphere, and Antarctica, Chile, and New Zealand in the Southern Hemisphere. Aside from being objects of scenic splendor, many fjords make great natural harbors.

Depositional Features of Glaciers

Glacial deposits are called **drift**—a term held over from the time, not so many years ago, that they were believed to have been laid down during Noah's Flood. The two vehicles for depositing drift are meltwater and ice. Meltwater allows sediment to separate

Glaciers and Climate 291

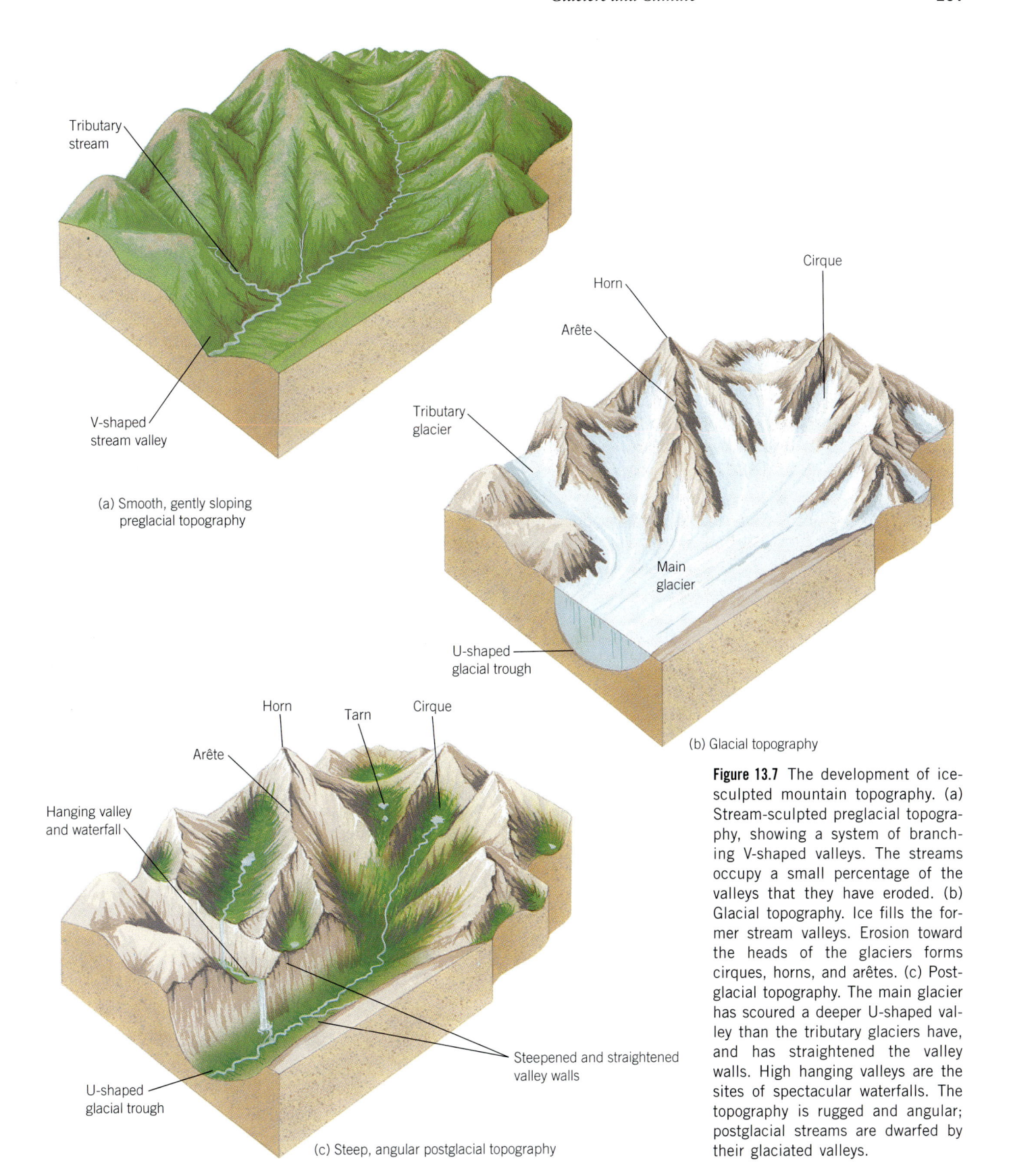

(a) Smooth, gently sloping preglacial topography

(b) Glacial topography

(c) Steep, angular postglacial topography

Figure 13.7 The development of ice-sculpted mountain topography. (a) Stream-sculpted preglacial topography, showing a system of branching V-shaped valleys. The streams occupy a small percentage of the valleys that they have eroded. (b) Glacial topography. Ice fills the former stream valleys. Erosion toward the heads of the glaciers forms cirques, horns, and arêtes. (c) Postglacial topography. The main glacier has scoured a deeper U-shaped valley than the tributary glaciers have, and has straightened the valley walls. High hanging valleys are the sites of spectacular waterfalls. The topography is rugged and angular; postglacial streams are dwarfed by their glaciated valleys.

stratified drift
Deposits left by streams or pools of glacial meltwater that are sorted and layered according to size.

till
Unsorted, unlayered drift deposited directly by glacial ice.

moraine
A landform composed of till left behind by a retreating glacier.

terminal moraine
A thick pile of till deposited at the line of maximum glacial advance.

ground moraine
A rough blanket of till that accumulated under the glacier, exposed as the terminus recedes toward the head of the glacier.

recessional moraine
A lateral or end moraine accumulated during a pause in a glacier's retreat.

according to size. Hence, the resulting deposits are sorted and layered, forming **stratified drift.** Sediments deposited directly by ice, called **till,** are neither sorted nor stratified, because all fragments, from the fine clay to boulders the size of trucks, are locked in place within the moving ice and dumped wherever the ice happens to melt.

Moraines

Moraines are landforms composed of till (see Figures 13.3 and 13.10). Like other sedimentary landforms—sand dunes, beaches, deltas—a moraine reflects the manner in which the sediment was brought to the depositional site and the environment in which the sediment was deposited.

Let us consider the situation at the front of a glacier whose budget is in balance—which is where ice accumulation equals wastage. Because ice is being produced at the same rate as it is being destroyed, the glacier can neither advance nor retreat. But as we have seen, the ice *within* the glacier continues to move to the front, where it melts. Deposited at the front of the glacier is the debris, or till, that the ice brought with it. Should the ice remain stationary for a number of centuries, a thick pile of till called an *end moraine* will accumulate. If the stationary front is, in fact, at the line of maximum glacial advance, then its end moraine is called a **terminal moraine.**

Now let us assume that the budget of this glacier becomes unbalanced for several centuries by a warming climate, during which time melting exceeds ice accumulation. In this case, the ice margin steadily recedes toward the head of the glacier, leaving a rough blanket of till called a **ground moraine,** the most common and widespread type of glacial deposit. If the glacial budget temporarily comes into balance during the long period of retreat, it may produce another stationary ice front, along which a **recessional moraine** will accumulate. Depending on the number of times the ice stalls, a series of recessional-moraine ridges may form parallel to the terminal moraine of the glacier. End and ground moraines are features of both continental and mountain glaciers.

Figure 13.8 A glacial erratic in Tripod Rock Reserve, New Jersey.

Figure 13.9 A drumlin field along Copper River, Alaska.

A valley glacier, however, abrades and plucks the valley walls as well as the valley floor. This oversteepens the walls, causing landslides, slumps, and rockfalls. As this debris collects at the sides of the glacier, it forms a **lateral moraine.** Should lateral moraines merge, as when two tributary glaciers enter the main valley, they form a **medial moraine.** These moraines ordinarily are not well preserved because they are destroyed by meltwater during glacial retreat.

When a glacier melts, it leaves boulders strewn across the glaciated landscape. We know that they are ice deposits because they are too large to have been transported by running water, even a huge flood. Called **glacial erratics** (Figure 13.8), they commonly do not match the underlying bedrock, which makes them useful for what they reveal about the direction and mechanics of ice movement. The way to use them is to note carefully their distinctive features—mineral content, fossils, texture, color—and trace them boulder to boulder back to their bedrock source. In many instances, that source is local, perhaps a kilometer or two away. In other instances, ice carried these boulders hundreds of kilometers. For example, boulders of granite from Canada, weighing several tons, today rest peacefully on the limestone plains of Iowa.

In places, the ground moraine forms smoothly rounded, elongated hills called **drumlins** (Figure 13.9), whose long axes parallel the direction of ice flow. Their origin is uncertain. They may have been molded by the base of the glacier moving over previously deposited moraines. Averaging about 30 meters high and 1500 meters in length, most drumlins appear in groups or swarms. From the air, a group resembles a school of whales, their broad backs breaking the water surface. Some ten thousand drumlins are found in western New York State alone. They are also locally prominent landscape features of Wisconsin, Minnesota, Ontario, and New England. While not one of the Earth's major landforms, the modest drumlin has had its rendezvous with history: Bunker Hill, where in 1775 Boston citizens fired on British soldiers in a battle that sparked the American Revolution, is a drumlin.

lateral moraine
A long, narrow mound of till that lies perpendicular to the glacier front, formed by plucking and rockfalls along the valley walls.

medial moraine
The joining of lateral moraines where two valley glaciers merge.

glacial erratic
A large rock fragment or boulder, carried by glacial ice away from its place of origin and usually deposited on bedrock of a different type.

drumlin
A rounded, low, and elongated hill of compacted till, with its blunter end pointing upstream and its tapering end pointing in the direction of ice flow.

Landforms of Stratified Drift

The landforms of stratified drift are composed of glacial meltwater deposits. These waters issue from continental and valley glaciers, so it is no surprise that similar

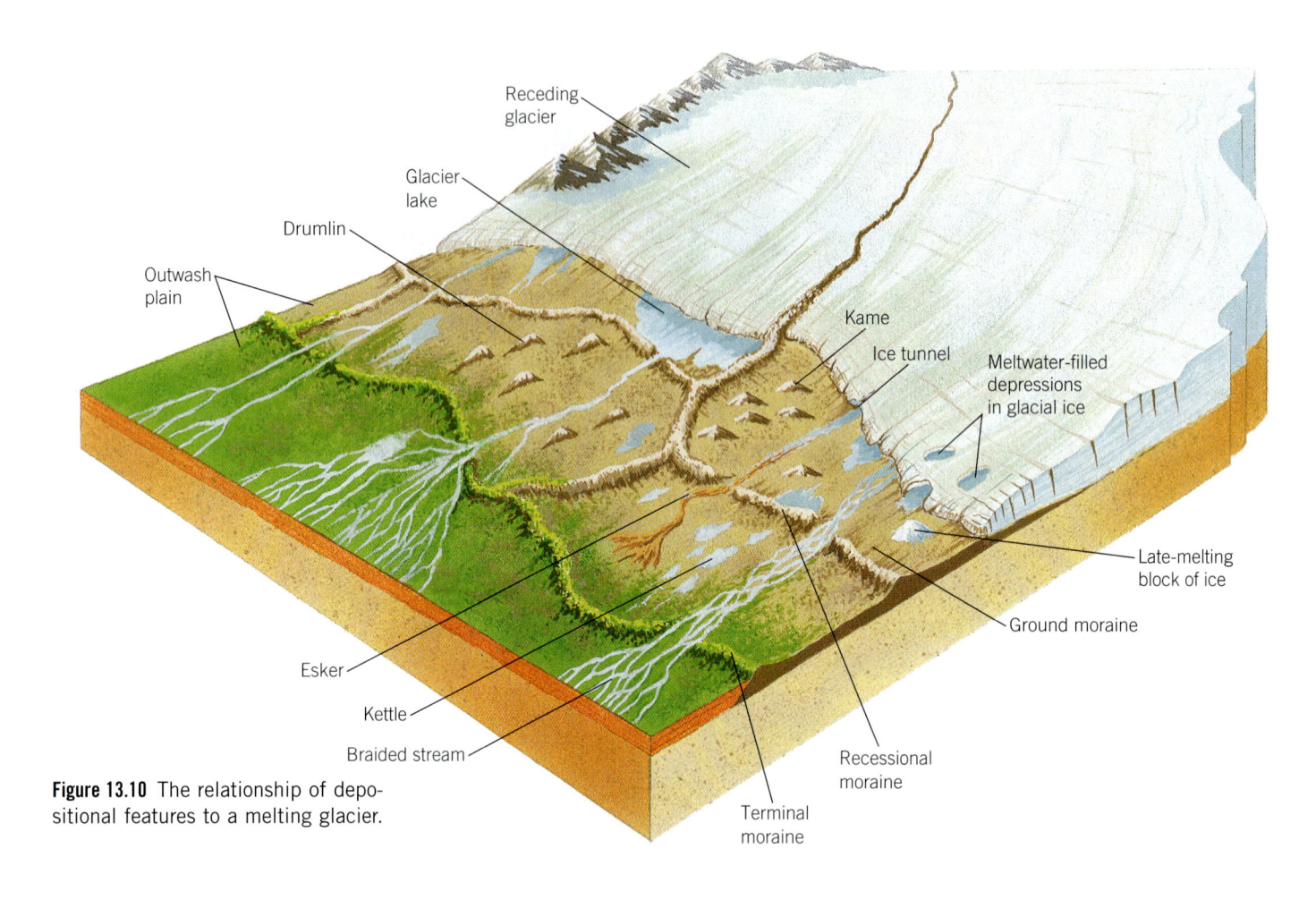

Figure 13.10 The relationship of depositional features to a melting glacier.

outwash plain
A broad, gently sloping sheet of stratified sand and gravel sediment deposited by meltwater streaming out of the front of a glacier.

esker
A snakelike ridge, up to 500 kilometers or longer, of roughly stratified gravel and sand, left behind when glacial ice melted.

landforms are common to both—though they may differ in size and frequency of occurrence (Figure 13.10).

The meltwater issuing from a glacial terminus carries a choking load of coarse bed load sediment. Unable to move this load all at once, the stream deposits most of it within the channel and assumes a braided pattern (see Figure 13.10). As discussed in Chapter 11, this is the most efficient way to move coarse bed load in a low-volume stream. Over time, the stream's constantly shifting channels will deposit sediment in front of the retreating glacier, forming a broad, flat **outwash plain** (see Figure 13.10). As a valley glacier retreats, it leaves behind a long, narrow body of outwash called a *valley train*. Like all stream deposits, outwash plain and valley train sediments are sorted and layered. Coarse gravels are found closest to the glacier front, sands and finer materials farther away. Because most of the load is deposited closest to the glacier, outwash plain deposits thin as they slope away from the terminal moraine. As may be expected, the outwash plains of ice sheets are far more extensive than those of valley glaciers.

If we trace outwash plains back toward the source of the glacier, we find that they often connect with long, snakelike winding ridges called **eskers** (see Figure 13.10). The eskers of continental glaciers may reach impressive dimensions: over 500 kilometers long, including gaps, with heights of 300 meters. Typically, however, they are far more modest, having lengths of a few kilometers and heights of perhaps 10 meters. That eskers are composed of poorly stratified gravel and sand and that they connect to outwash plains suggest that they were formed by streams escaping through tunnels at the base of the melting glacier.

There are many other opportunities for meltwater to deposit sediment within and along the margins of a melting ice sheet or valley glacier. As the ice recedes, its sediment forms mounds, knobs, or short ridges above the surrounding countryside. These

Figure 13.11 Kame and kettle topography south of Big Delta, Alaska. The mounds are kames; the lakes are kettles. This topography typically forms over a broad zone at the terminus of a retreating ice sheet.

elevated features are collectively called **kames,** from the Scottish word meaning "steep-sided monument." *Kame terraces* are long, narrow ridges of sediment that were deposited by streams that flowed between the ice and the walls of valley glaciers.

Kettles are depressions in moraines and outwash plains formed when blocks of late-melting ice were isolated from the retreating glacier. They often become rounded and secluded lakes. Henry David Thoreau's Walden Pond in Massachusetts is such a feature. Pitted *kame-and-kettle topography* is especially common along the ice margin, with its array of mounds, ridges, depressions, and small lakes (Figure 13.11).

Glacial Lakes

A resource of inestimable value, the Great Lakes contain over 30 percent of all the fresh liquid water on the Earth's surface. It is perhaps difficult to imagine a feature of such magnitude being a mere ten thousand or so years old. However, the Great Lakes basins were filled by the North American ice sheet. As the ice retreated, the streams draining the Great Lakes flowed first to the Gulf via the Mississippi, then to the Atlantic via the Hudson, before finally flowing out to the Atlantic via the Saint Lawrence River.

The Great Lakes, Finger Lakes, and others, though formed by glaciers, survive because they are regional low points fed by present-day streams. But many glacial lakes have not survived because the ice sheet formed their margins, either in whole or in part; when the ice melted, the lakes drained. Left behind were lakebed sediments, as well as ancient beaches that defined their shorelines. Glacial Lake Agassiz, which spread over enormous areas of central Canada, Minnesota, and North Dakota, was the most extensive of the glacial lakes, larger than any of the Great Lakes. Its legacy is some of the richest farmland on the continent.

The climate that produced the ice sheets was also responsible for the formation of lakes farther south. Dry regions, such as Utah, Nevada, and Southern California, were in fact heavily forested or had lush grassland supporting a fauna whose abundance rivaled the plains of Africa. The melting of valley glaciers—or simply abundant rainfall—fed water into the lowlands forming large **pluvial lakes.** Lake Bonneville, Utah, of which Great Salt Lake is a puny descendant, was one of them. Its shorelines are visible in the mountains high above the Bonneville Salt Flat, where the lake evaporated as the climate turned arid.

kame
A mound or short ridge of stratified sand and gravel deposited by water streaming under or trapped within glacial ice; from the Scottish word meaning "steep-sided monument."

kettle
A depression in a moraine or outwash plain formed when a large block of ice was isolated from the retreating glacier and buried in the drift, melting afterward.

pluvial lake
A body of water in a nonglaciated region formed by abundant rainfall due to the cooler climate caused by the growth of ice sheets.

Glacial and Interglacial Ages

Northern Hemisphere glaciation began at least 3 million years ago during the Pliocene epoch. It reached its maximum extent in the period most associated with glaciation, the Pleistocene epoch, and continues to this day. Each ice advance, or **glacial age,** has been followed by a period of ice retreat, or **interglacial age.** Right now, we are living in the Holocene interglacial age. Geologists have arbitrarily set its beginning at the onset of the rapid sea-level rise that resulted from melting of the last ice sheet about 11,000 years ago.

glacial age
An interval of geologic time in which glacial ice sheets advance.

interglacial age
An interval of geologic time in which glacial ice sheets retreat.

Causes of the Ice Ages

Ice ages are an expression of climate, and much of what we know about them comes from the study of past and present climates. The climatic conditions that have led to glacial and interglacial ages are the result of a multitude of factors. The three major factors we discuss are:

1. the movement of the Earth's tectonic plates and associated volcanic and crustal uplift
2. long-term cyclical variations in solar energy received by the Earth
3. the condition of the Earth's atmosphere—most importantly, its carbon dioxide and dust content

Plate Tectonic Effects

The rocks and sediments of the mid-Cretaceous period, 120 million years ago, preserve evidence of a world warmer by 15 °C than today's—a world in which dinosaurs spread throughout the globe feeding on lush tropical vegetation or on animals who fed on it. Two effects of the movement of the Earth's lithospheric plates are responsible for the pronounced decline in average world temperature since the Cretaceous period, a decline that made the advent of the ice ages possible. The first is the slow drift of the continents toward high latitudes. The second is the general elevation of the continents due to volcanism and mountain building.

Present ice sheets are concentrated in, and have spread from, high latitudes. High latitudes are favored because temperatures generally decline toward the poles. However, continents must also be present in high latitudes to support the ice, and as we learned in Chapter 1, the continents have been drifting toward the poles since the breakup of Pangaea some 180 million years ago (see Figure 1.15).

The Cenozoic era (our era) has been a time of active tectonism. The enormous uplifts of the Himalayas, the Tibetan Plateau, and portions of the Rockies and Andes mountains have occurred during this time frame. Yet even slight crustal uplift can raise elevations enough to lower temperatures by a few degrees in surrounding regions. Uplifts far from ice sheets may affect climate in subtler ways. For example, they may alter global wind patterns, forcing frigid arctic air farther south.

Clearly the drift of the continents to high latitudes, combined with crustal uplift and mountain building—both of which are controlled by the movement of the Earth's lithospheric plates—have contributed heavily to a lowering of worldwide temperatures during the past 120 million years. However, the plates move too slowly to cause periodic glacial and interglacial ages. Neither is there any evidence that they oscillated back and forth to cause the alternate warming and cooling necessary to trigger advances and retreats of the ice. What mechanism could be responsible for these rhythmical changes?

Variations in Earth's Orbital Geometry: The Milankovitch Theory

One mechanism that might trigger glacial advances and retreats is cyclical variations of the Earth's orbit and rotational axis that control the amount of solar energy reaching Earth's high latitudes. The ice would advance when incoming solar heat is reduced and

Glaciers and Climate   297

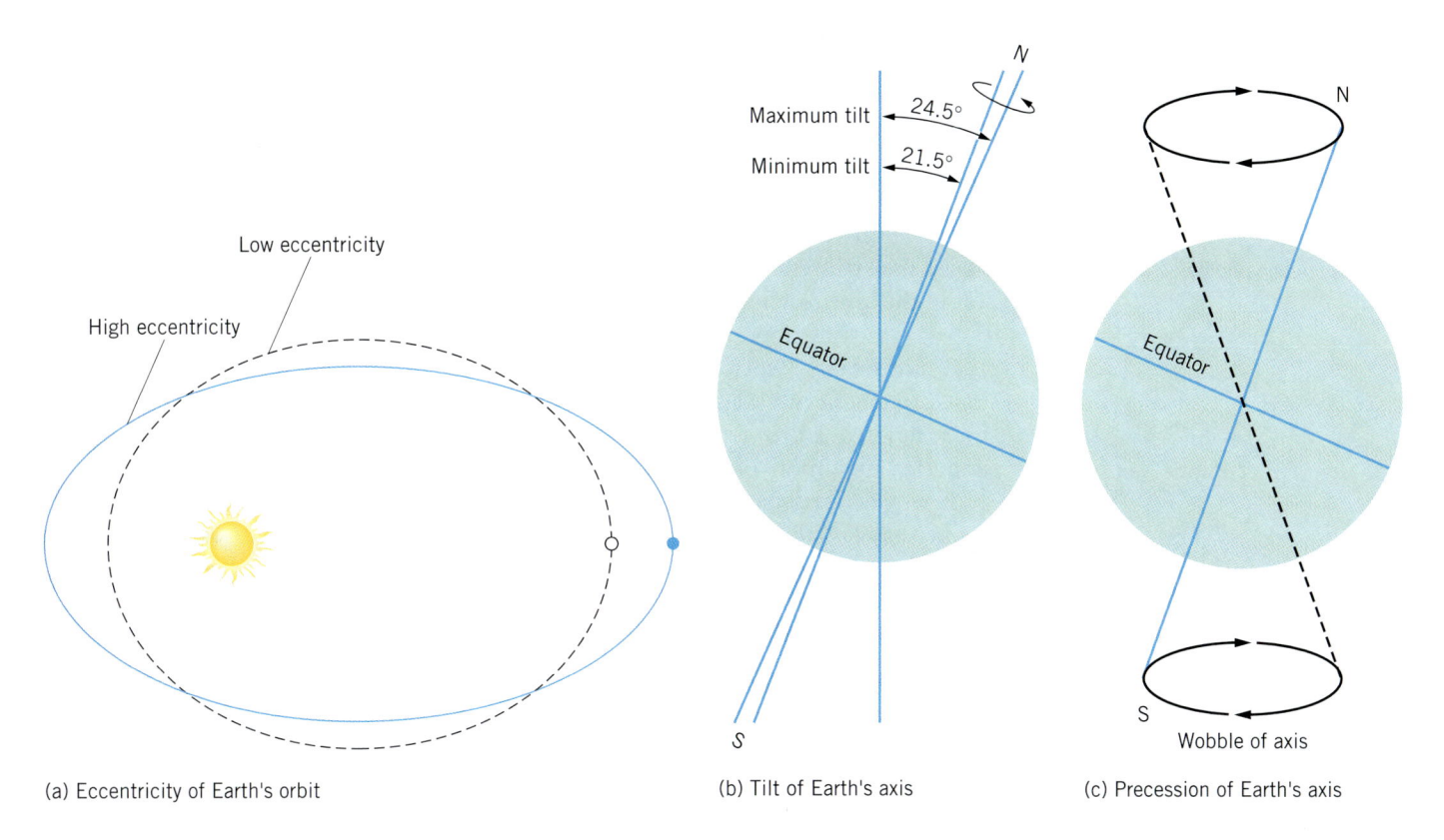

(a) Eccentricity of Earth's orbit

(b) Tilt of Earth's axis

(c) Precession of Earth's axis

retreat when these latitudes are warmed again. Most glacial geologists agree that a sustained drop of about 3 °C in average annual temperatures is sufficient to trigger an ice advance. This is the approximate difference in mean temperature between Boston and Washington, D.C. The entire difference in average temperatures between peak glacial and interglacial ages is probably on the order of 5 °C.

We are already familiar with one form of cyclical variation of solar heat over the Earth's surface that is controlled by orbital geometry—the seasons. Summer occurs in the Northern Hemisphere when, because of the tilt of the Earth's axis with respect to its orbit about the Sun, the Sun's rays strike most directly above the equator. Winter occurs when the rays strike most directly south of the equator. (Of course, the timing of the seasons is reversed in the Southern Hemisphere.) In the 1920s and 1930s, Serbian astronomer and meteorologist Milutin Milankovitch (1879–1958) proposed a concept similar to the seasons to explain ice ages. He suggested that other, more subtle orbital characteristics were responsible for slight, longer-term variations in the distribution of solar heat over the Earth's surface—variations sufficient to control ice advances and retreats. These variations are caused by the gravitational tug on the Earth by the other planets and the Moon as the Earth spins like a top and revolves around the Sun. Three gravitational effects are evident.

1. *Eccentricity of the Earth's orbit.* Figure 13.12(a) shows that the Earth periodically departs from a nearly circular orbit to a more elliptical one. Maximum departure occurs about once every 100,000 years. This effect periodically brings the Earth closer to and farther away from the Sun than at present and causes a slight ebb and flow of the heat received by the Earth over time.
2. *Tilt of the Earth's axis.* The Earth's rotational axis also tilts with respect to the plane of the Earth's orbit about the Sun. At present, the angle of tilt is 23.5°, but it varies (nods) between 21.5° and 24.5° every 40,000 years or so. The greater the angle of tilt, the more sunlight that reaches the polar regions in summer and the less in winter (Figure 13.12b).
3. *Precession of the Earth's axis.* In addition, the Earth's axis precesses, or wobbles, slightly, repeating each wobble about every 26,000 years. This precession

Figure 13.12 Three periodic variations in the Earth's orbital and spin characteristics contribute to the Milankovitch cycle. (a) Orbital eccentricity. (b) Axial tilt. (c) Axial precession. An ice age may be initiated when these motions coincide periodically, thus reducing the solar energy that reaches the Northern Hemisphere during summers.

changes the date when the Earth is closest to the Sun and, therefore, varies slightly the intensity of the sunlight striking polar regions in summer and winter (Figure 13.12c).

Milankovitch proposed that these cyclical motions periodically reinforce one another. When the combined effects warm the Northern Hemisphere slightly, they trigger an ice retreat. When they combine to reduce temperatures in the Northern Hemisphere, they trigger an ice age. Apparently, an ice advance correlates with a slight reduction in the heat reaching northern latitudes during the summer months, which leads to less snow melting and an overall ice surplus.

Unfortunately, when Milankovitch first proposed his hypothesis, it suffered a fate similar to that of many advanced ideas; seemingly impossible to prove or disprove, it was largely ignored. However, attitudes changed decades later when studies of deep-sea cores supported the timing of glacial advances and retreats predicted by Milankovitch.

That glacial advances and retreats correlate with variations in solar heat as predicted by the Milankovitch cycle is indeed an important contribution to our understanding of the origin of glaciers. But it is far from a complete explanation, for the cyclical heat variations are too weak to induce the temperature extremes necessary to cause glacial and interglacial ages. Other factors must serve to amplify the effect.

Variations in Atmospheric Carbon Dioxide and Dust

Earth's temperature is very sensitive to the percentage of atmospheric carbon dioxide present, because the carbon dioxide traps the heat rising from the Earth's surface. In this respect, the carbon dioxide acts like a blanket that warms you by slowing the dissipation of the heat your body radiates. This well-known role of carbon dioxide and other gases in regulating the Earth's temperature is called the greenhouse effect. Sunlight is able to penetrate the atmosphere to warm the Earth's surface, but carbon dioxide delays the escape of infrared radiation (that is, heat) rising from the surface. The greater the concentration of carbon dioxide, the more heat that is trapped and the higher the Earth's temperature will be; the lesser the carbon dioxide content, the lower the temperature.

Because the greenhouse effect is known to regulate present-day climates, the carbon dioxide content of ice cores drilled in the Greenland and Antarctic glaciers constitutes highly significant evidence of past climatic changes. Data from the cores suggest that a strong correlation exists between variations in atmospheric carbon dioxide content, glacial and interglacial ages, and the Milankovitch cycle. The low temperatures of

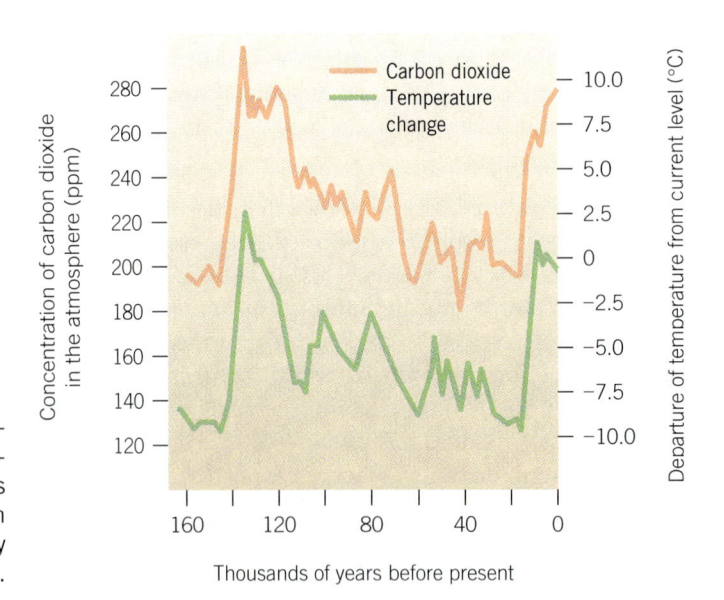

Figure 13.13 Variations in atmospheric carbon dioxide and temperature over the past 160,000 years are closely matched. This correlation suggests that carbon dioxide is a key regulator of the Earth's temperature.

the ice ages correlate with low carbon dioxide—over 30 percent less than the present amount. The high temperatures of interglacial ages show correspondingly high carbon dioxide content. In addition, the data show that over the past 150,000 or more years, carbon dioxide buildups have preceded the onset of global warming, whereas declines have signaled cooling periods (Figure 13.13).

Atmospheric Dust   Greenland and Antarctic ice cores contain other important information. Trapped in their layers are dust particles that had settled out of the atmosphere at the time of deposition. Again, the correlation with the Milankovitch cycle is striking. Low dust volumes match warming periods; high dust volumes match cooling periods. Dust particles in the upper atmosphere—most importantly, sulfate droplets—both reflect and absorb sunlight and therefore cause worldwide temperature reductions. As discussed in Chapter 6, volcanic eruptions are the chief source of these particles.

Human-Induced Global Warming

Is the Earth warming or cooling? This is an interesting question aside from practical considerations, because the answer depends upon the natural contributions to climate we have been discussing so far and the "unnatural" contributions made by industrial civilization. The latter effect is widely known today as *human-induced global warming*.

Taking the long view, there is little doubt that the Earth will grow increasingly colder and drier. We are currently in a warm, wet interglacial age that began approximately 6000 years ago. However, judging from the length of previous interglacial ages and the Earth's present position in the Milankovitch cycle, temperatures will probably decline within the next few thousand years.

Over the short run (that is, the past 150 years), worldwide temperatures have been on the rise. Geoscientists generally agree that the short-term temperature spurt is human-induced and may cause an average temperature increase of 2 to 5 °C by the year 2050. They trace global warming to two types of human activities.

The first category includes activities that accelerate emissions of greenhouse gases: *carbon dioxide* from cars, factories, electrical power plants, and burning of forests; *methane* from natural gas wells, cattle, and rice cultivation; *nitrous oxide* from agriculture; and *chlorofluorocarbons* (CFCs) from refrigerants and other sources. As these gases are added to the atmosphere, they trap a higher percentage of the heat radiated from the Earth's surface.

The second type of activity contributing to global warming is the ongoing destruction of the world's great forests and other ecosystems, which decreases the amount of carbon dioxide that is extracted from the air via photosynthesis. Most geoscientists reason that the average global temperature is rising because heat-trapping gases are being added to the atmosphere more rapidly than they are being extracted.

While most scientists are becoming alarmed at the prospect of global warming, a minority continues to wonder what the fuss is about. They remind us that natural climatic fluctuations lasted for centuries in the recent past; yet the human race survived prolonged periods of both relative warmth and relative cold. For example, about 900 years ago, the climate in northern Europe during the Medieval Warm Period (1100–1375) was milder than it is today. Writers of that time described a Britain so mild that wine grapes flourished and a Greenland so warm that the Vikings farmed the now-frigid coast. A century or two later, during the Little Ice Age (1550–1850), Europe became so cold that the British held their winter carnivals on the frozen Thames River and the Greenland ice sheet extended back toward the coast.

Although an hysterical response to the evidence of global warming may not be called for, neither is complacency. During the Medieval Warm Period, the world's population probably numbered fewer than 600 million people. Today's population is closer to 6 billion, and is welded together by a technological civilization infinitely more complex than it was centuries ago. The following are some of the changes we can reasonably anticipate as a result of global warming:

- We can expect a higher degree of weather variability than at present. As more water vapor and energy are transferred to the atmosphere, hurricanes and other storms are likely to grow more violent. Also, global warming is likely to increase the seasonal contrasts between land and sea temperatures. Midcontinental droughts and heat waves will intensify in summer; in winter, a reverse cooling effect will bring more frigid, longer-lasting cold spells.
- The highest latitudes are expected to receive the sharpest temperature increases, which, in turn, may accelerate the melting of the ice caps. The degree of melting remains a point of disagreement among scientists. Indeed, some think that the glaciers will actually grow for a time because of increased precipitation. However, if sea level should rise at close to current rates, as the Intergovernmental Panel on Climate Change (IPCC) has estimated, the oceans will rise between 0.5 and 1.5 meters over the next century. Some extreme estimates predict a sea-level rise of 6 meters.
- Temperature increases may also cause the North American and Eurasian rain belts to shift significantly northward. The reason for the shift is that most of the rain is generated when warm moist air from the subtropics clashes with dry, cold polar air along constantly fluctuating boundaries or *fronts*. In North America, the current "line of battle" extends across the northern Plains States; with global warming, the warm and cold air masses are likely to clash at higher latitudes, moving well into Canada. If the rain belts migrate northward, we can expect the Plains States to become hotter and drier.

    As with most rapid change, there will be both winners and losers to the latitudinal creep of the warm climatic belts. The humid tropics and subtropics are expected to broaden, bringing rain and renewed fertility to the north African deserts and the surrounding semiarid zones, which today are periodically famine-stricken (see Chapter 14, Deserts and Winds).
- We can also expect major ecological changes. Global warming will endanger any species unable to evolve swiftly in the face of rapid environmental change. Animals that have long life spans, relatively few offspring, and specialized food and habitat requirements—most large land mammals, for example—will not be able to adapt to such rapid change and may die out. On the other hand, rodents and insects, being omnivorous and highly adaptable, will probably prosper.

The Impact of Global Warming on Human Society

The physical and biological effects of global warming will have major economic and social implications for that most prolific and adaptable of large mammals, Homo sapiens. The changes will affect two important domains: agricultural production and coastal habitation.

The agricultural potential of a region depends upon its climate, soil type, and soil condition. Many regions of the continents are unsuitable for anything beyond small-scale farming, while other regions have become worldwide suppliers of their agricultural produce. The United States, for example, is the world's largest grain exporter, and most of the grain is grown in the midwestern Plains States.

If the rains that now water the grain fields of the Plains States shift northward, will Canada, the recipient of the shifting rains, eventually be able to take up the agricultural slack? Probably not. Much of the soil of northern Canada is underlain by stony glacial deposits and is thin and poor. It is doubtful that the region could ever match the productivity of the present corn and wheat belts, where glacial outwash nourished by thousands of years of rainy climate has become the thick, rich soil blanket of the midwestern United States.

The prediction of a 1-meter sea-level rise doesn't seem impressive until you consider that as much as 70 percent of the world's population lives along coasts. A sea-level rise of 1.0 to 1.5 meters would displace tens of millions of people, contaminate the freshwater wells and storage facilities of coastal communities, and intensify inland flooding during storms. The United States alone would lose 66,300 square kilometers of coast. The most extreme estimates place the sea-level rise at a catastrophic 6 meters. In that case, nearly all the world's coastal cities, farming belts, deltas, and flood plains would drown.

Global warming will facilitate the spread of a host of nasty tropical diseases to higher latitudes and altitudes. For example, the mosquito that carries malaria will be able to migrate to regions currently too cold for it to survive. IPCC scientists estimate that an average global temperature increase of approximately 3 °C will add 50 to 80 million malaria cases per year to the 300 million per year already occurring.

Finally, we must keep in mind that short-term global warming is a problem caused by human economic behavior. Scientists can only explain the causes and recommend solutions—that is, lower the rate of greenhouse gas emissions and prohibit further destruction of the world's rain forests. How to engage the cooperation of all economic interests and institute these solutions among all the nations of the world is a political, rather than a scientific, issue.

STUDY OUTLINE

Glacial ice covers 10 percent of the Earth's land surface, with most of it concentrated in the Greenland and Antarctic ice sheets. Glacial advances and retreats, accompanied by profound climatic fluctuations, have been the pattern for at least 3 million years. Glacial advances cause the sea level to fall, which exposes the continental shelves. Glacial retreats cause the sea level to rise, which floods the coastal regions of the continents. Glaciers are important agents of erosion, transport, and deposition.

I. **THE FORMATION OF GLACIERS**

A. A **glacier** is an ice mass that survives from year to year. Its formation begins with the compaction of snow into ice pellets called **firn**. The **snowline** is the elevation above which there is permanent snow. Glaciers of continental proportions are called **continental glaciers**, or **ice sheets**. Away from the polar regions, in areas of high elevation, **alpine** or **valley glaciers** occupy former mountain stream valleys.

B. As an alpine glacier flows downhill, ice velocity is greatest at the valley center and least along the valley walls.
   1. The top rigid layer cracks under the stress of movement, forming **crevasses**. Below, in the **zone of plastic flow,** the ice flows under the pressure of its own weight toward the *glacial terminus*. Melting along the valley floor allows the entire glacier to slide downhill in **basal slip** or in more rapid **surges.**
   2. Because ice sheets form at high latitudes, they are colder and more rigid than alpine glaciers. Ice sheets flow slowly outward in all directions from the centers of accumulation.

C. Like a business that has income and expenditures, the glacial budget is a balance between *accumulation* of snow and *ablation*—water loss through melting, evaporation, and **calving.**
   1. If accumulation exceeds ablation, the glacier grows and advances. If ablation exceeds accumulation, it shrinks and retreats.
   2. When a glacier reaches the sea, it extends outward as an *ice shelf,* where chunks break off, or calve, forming **icebergs.**

II. **EROSIONAL FEATURES OF GLACIERS.** Glaciers erode rock by **plucking,** in which ice melts, seeps into rock fractures, refreezes, and breaks off rock fragments. Through **abrasion,** the ice polishes and grinds bedrock.

   A. A mountain glacial system fills the former stream valley and all its tributaries, eroding V-shaped stream valleys into U-shaped **glacial troughs.** The main glacier may cut off the tributary glacier valleys, forming **hanging valleys.** *Tarns* and *paternoster lakes* fill the cirques and depressions left by a glacier in the valley floor. **Fjords** are seawater-filled glacial troughs that extend to the coast.

   B. Bowl-shaped **cirques** are formed by the **plucking** of a valley glacier headwall. Between cirques are knife-edged **arêtes** and rock peaks called **horns.**

   C. Parallel grooves in rock, caused by abrasion, are called **glacial striations.** The fine debris of abrasion is called *rock flour.*

III. **DEPOSITIONAL FEATURES OF GLACIERS.** There are two types of glacial deposits, or **drift: stratified drift** and **till.**

   A. Till consists of unsorted, unstratified rock fragments (including boulder-sized **glacial erratics**) that are transported by glacial ice and dropped at the terminus in mounds called **moraines. Terminal** and **recessional moraines** form at the foot of the glacier; **lateral** and **medial moraines** are formed along the sides. A **ground moraine** is a blanket of till left behind by a receding glacier. In places, the ground moraines of continental glaciers feature rounded, elongated hills called **drumlins.**

   B. Meltwater flowing in braided streams from the glacial terminus deposits sorted and layered **stratified drift,** leaving landforms such as broad, flat **outwash plains** and snakelike ridges called **eskers.** Other stratified-drift landforms are mounds called **kames** and depressions called **kettles.**

IV. **GLACIAL LAKES**

   A. The last North American ice sheet excavated the Great Lakes and other new lakes.

   B. The climate that caused the ice sheets was also responsible for large **pluvial lakes,** formed in regions that today are arid.

V. **GLACIAL AND INTERGLACIAL AGES**

   A. Great ice sheets have occurred ten or more times during the past 2 to 3 million years. The most recent interval of glaciation began during the Pleistocene epoch. Each ice advance, or **glacial age,** has been followed by a period of ice retreat, or **interglacial age.**

   B. The following factors have led to glacial and interglacial climate conditions:
      1. movement of the Earth's tectonic plates from low to high latitudes and the associated elevation of the continents due to volcanism and mountain building;
      2. the Milankovitch cycle: long-term variations in solar energy received by the Earth; and
      3. the condition of the Earth's atmosphere—most importantly, its carbon dioxide and volcanic dust content.

VI. **HUMAN-INDUCED GLOBAL WARMING.** Future climate trends will depend on natural and human-induced effects.

   A. Temperatures in the long-range future will probably decline; however, over the short run, worldwide temperatures have been on the rise.

B. Many experts now agree that dangerous global warming is being caused by two areas of human activity: increased emissions of greenhouse gases, such as carbon dioxide, and deforestation.

Study Terms

abrasion (p. 289)
alpine glacier (p. 285)
arête (p. 290)
basal slip (p. 286)
calving (p. 288)
cirque (p. 289)
continental glacier (p. 285)
crevasse (p. 285)
drift (p. 290)
drumlin (p. 293)
esker (p. 294)
firn (p. 284)
fjord (p. 290)
glacial age (p. 296)
glacial erratic (p. 293)
glacial striations (p. 289)
glacial trough (p. 290)
glacier (p. 284)
ground moraine (p. 292)
hanging valley (p. 290)
horn (p. 290)
iceberg (p. 288)
ice sheet (p. 285)
interglacial age (p. 296)
kame (p. 295)
kettle (p. 295)
lateral moraine (p. 293)
medial moraine (p. 293)
moraine (p. 292)
outwash plain (p. 294)
plucking (p. 289)
pluvial lake (p. 295)
recessional moraine (p. 292)
snowline (p. 284)
stratified drift (p. 292)
surge (p. 286)
terminal moraine (p. 292)
till (p. 292)
valley glacier (p. 285)
zone of plastic flow (p. 285)

Thought Questions

1. Explain how it is possible for ice to advance from the head to the terminus of the glacier even as the glacier retreats.
2. What factors affect the budget of a glacier? Under what circumstances will a glacier advance or retreat?
3. Assume that sometime in the future, all the world's glaciers will have melted. If you were living at that time, what evidence preserved on land would suggest that ice sheets had once existed?
4. Name and explain at least three factors that may be involved in climate regulation.
5. Why do many scientists suspect that humans are contributing to global warming?
6. Consider the economic and environmental impact of global warming on your current home town. In what ways do you think the weather and climate will change? How will those changes impact on the local flora and fauna? How may global warming impact on your livelihood?

Chapter 14

The Formation of Deserts
   Winds

Weathering and Erosion in Arid Climates
   Structural Influences on Desert Topography
      The Colorado Plateau
      Basin-and-Range Topography

The Geological Work of Wind in Arid Climates
   Features of Wind Erosion
   Features of Wind Deposition

   Loess Deposits
   Sand Dunes

Water Resources in Arid Climates

On the Fringe of the Desert: The Sahel
   The Contribution of Climate
   The Human Contribution

Deserts and Winds

Sand and deserts go together in the popular imagination. Most desert regions, however, are not covered by windblown sand. Running water—and not wind—is the chief gradational agent of deserts, and great sand dunes, though certainly impressive, occur only under specialized conditions. Perhaps the most impressive fact about deserts is that they occupy 19 percent of the land surface of the Earth. Semidesert steppe regions account for another 15 percent. Far from being exotic, either arid or semiarid conditions are common to one-third of all land (Figure 14.1).

Another popular myth to dispel is that all deserts are hot. Deserts are defined by lack of precipitation, not by temperature. Temperatures in the Sahara Desert, which can rise to well over 40 °C during the day, can easily drop below 4 °C at night. Midlatitude deserts and steppes can get brutally cold in winter—ask a Wyoming rancher or read the accounts of soldiers who fought on the Russian front during World War II. Because deserts are defined on the basis of annual precipitation, portions of Antarctica, Greenland, Siberia, northern Canada, and Alaska must be included. We do not usually think of these regions as deserts, though, because of their low temperatures and the effects of the ice sheets that either still cover them or have recently receded. Having discussed these regions in Chapter 13, we devote this chapter to the deserts and semiarid regions that fit the more conventional meaning of the term. We will look at the origins of deserts, the topographic features of deserts, the work of desert winds, desert water resources, and the human contributions to desertlike conditions—a serious problem, as we shall see.

◀ A striking desert scene at the foot of the Black Mountains, Death Valley, California. The colors are caused by the oxidation of mineral-laden groundwater seeping to the surface.

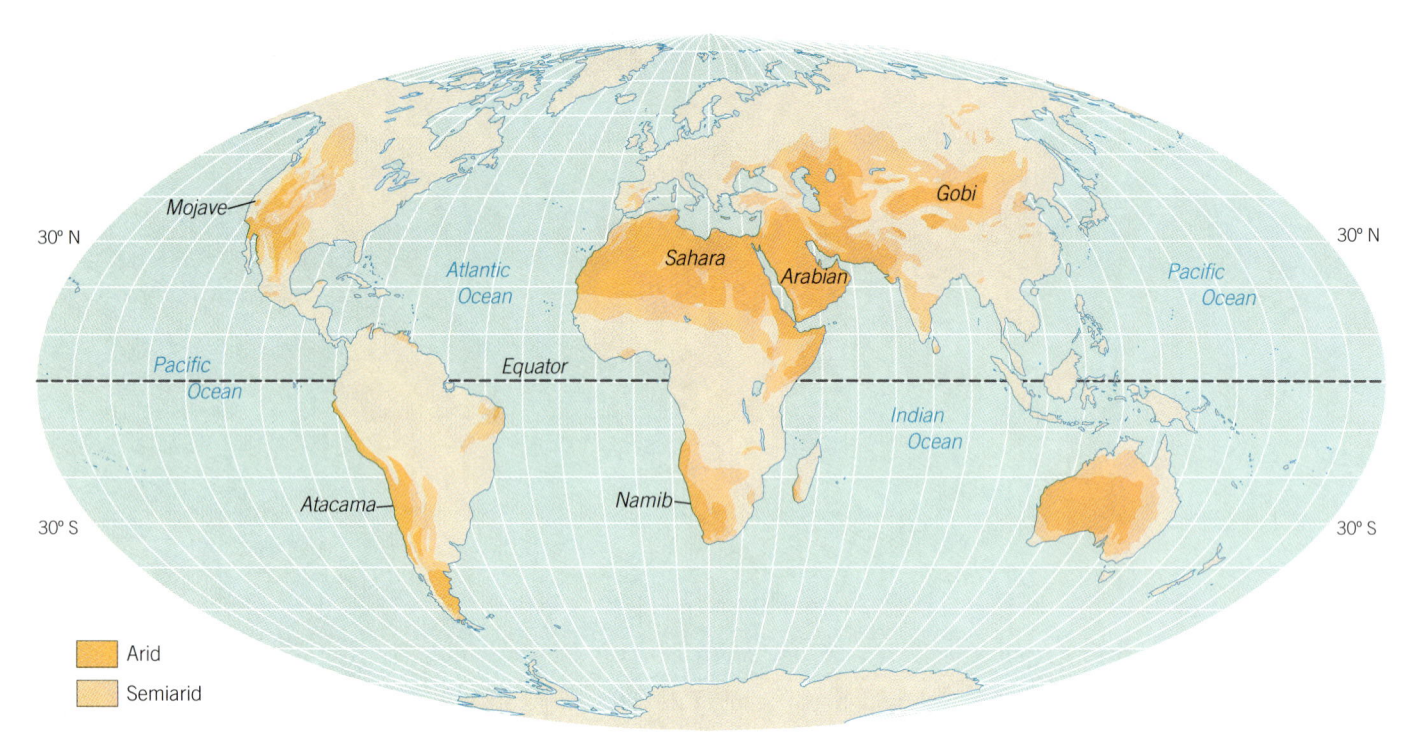

Figure 14.1 The worldwide distribution of arid and semiarid lands. Notice the concentration of deserts within belts approximately 30° north and 30° south of the equator.

desert
A region with an average annual precipitation of less than 25 centimeters, too dry to support more than sparse vegetation. (Climatologists define deserts more complexly on the basis of temperature-precipitation-evaporation ratios.)

The Formation of Deserts

Lack of abundant, consistent rainfall (less than 25 centimeters per year) is the irreducible feature of all **deserts** and is responsible for all that is unique to them—including the fact that the land surface is too dry to support more than a sparse plant cover. Three factors, acting separately or in combination, create the arid conditions. The first concerns the placement of the continents with respect to the worldwide wind system; the second concerns the influence of topography on rainfall patterns; and the third concerns the proximity of the land to certain cold ocean currents.

Winds

The origin of deserts is intimately connected to the planetary wind system, which controls the delivery of moisture to the continents. Winds are air movements and are controlled, either directly or indirectly, by the uneven distribution of the Sun's rays, which strike the surface most directly at the equator and least directly at the poles. Thus the equator receives more heat than the poles, and temperatures generally decline poleward—although seasonal variations, the location of landmasses, and ocean circulation complicate matters.

Though our intuition may tell us otherwise, the atmosphere does not draw its heat directly from the Sun. Rather, the Sun's rays penetrate to the Earth's surface, where they are absorbed; the warmed Earth *then* heats the atmosphere from below. That is why air temperatures decline with altitude. The air in contact with the surface of the Earth at the equator is warmed more than air over other regions; as it is warmed, it expands, becomes less dense, and rises. This warm air holds huge quantities of moisture, because a large portion of the solar heat that strikes the equator is used to evaporate water from the ocean. As the rising air cools, however, it releases the moisture as rain. Thus the vast rain belts of the world are located at the equator, as are the tropical rain forests.

As air at the equator continues to rise, it eventually hits a "lid" of even lighter air, known as the stratosphere. Unable to rise farther, it spreads—one loop heading north,

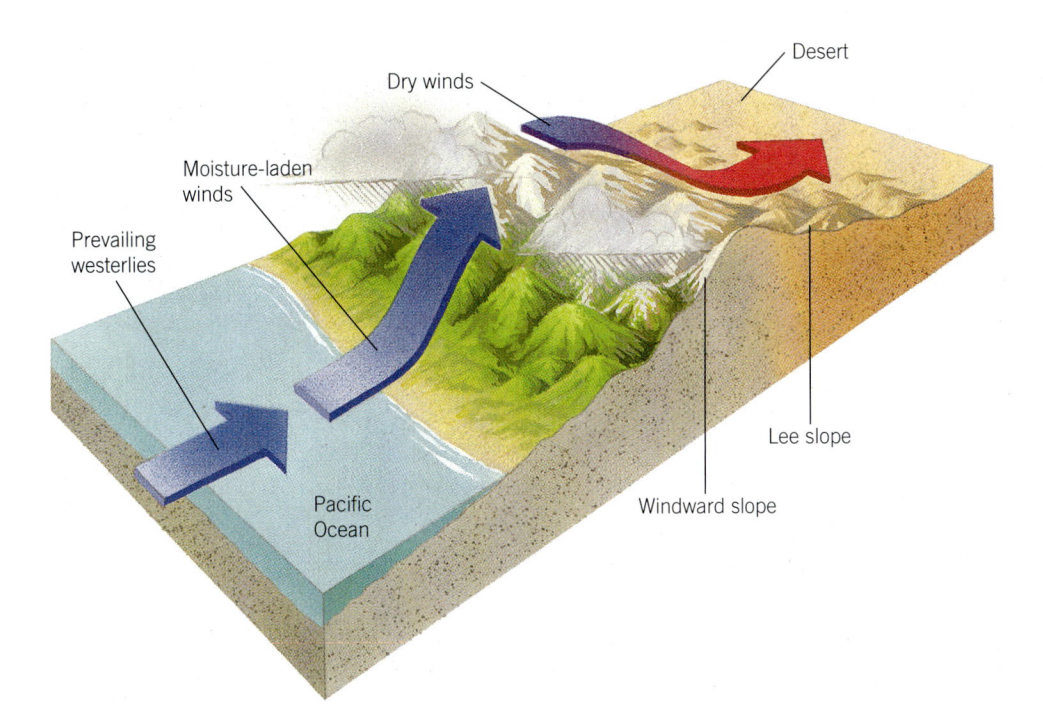

Figure 14.2 Rain shadow deserts are formed on the leeward side of mountain ranges. Rising winds release their moisture on the windward slopes, causing these slopes to be heavily forested. Descending winds on the lee slopes of the range are dry.

the other south. At about 30° latitude north and south of the equator, the air has cooled sufficiently to sink and pile up in huge masses. As they pile up, they compress and therefore warm the air at the surface, which increases the air's capacity to hold moisture; so water is evaporated from the surface. As a consequence, rainfall is generally sparse along the subtropical zones, centered at approximately 30° north and south of the equator. If you look at Figure 14.1 again, you will see that the world's great subtropical deserts are located at these latitudes.

The planetary wind belt is also instrumental in forming **rain shadow** deserts (Figure 14.2). For example, the deserts of the American West form where moisture-laden winds coming off the Pacific Ocean find their eastward progress blocked by a series of high mountain ranges. The winds cool as they rise over the peaks; their capacity to hold moisture is thus diminished, and they drop rain on the western windward slopes. The now-dry winds are heated as they descend the eastern lee slopes of the mountains. Thus their capacity to absorb moisture is increased. The combination of having little moisture to give and an increased capacity to hold it ensures that deserts will develop east of the mountains. One need only compare dry Nevada with wet northern California to see the influence of the Sierra Nevada Mountains on the creation of deserts. Farther north, the Olympic Peninsula and the west coast of Washington constitute the rainiest portion of the United States, whereas central Washington is desert. The latter is in the rain shadow of the Olympic-Cascade mountain ranges.

The creation of deserts does not always involve low latitudes or mountain ranges adjacent to the ocean. For example, the great Gobi Desert of China is located at midlatitude in a deep continental basin within the vast interior of Asia. Winds there are simply exhausted of moisture before reaching the basin, having passed over the Himalayas, the mountains of Afghanistan, and the Tibetan Plateau.

The planetary wind system also plays a major role in forcing the ocean currents to flow in enormous circulation loops. These currents deliver a huge supply of heat from the equator to high latitudes. Most of the heat transfer occurs along the western boundaries of the oceans, and by the time the currents reach the eastern boundaries, they are cold. Those that run parallel to the western coasts of South America and Africa are responsible for the coastal deserts there. The cold Benguela current runs directly

rain shadow
A dry region on the lee side of a mountain range where rainfall is significantly less than on the windward side; the region is said to be "in the rain shadow" of the mountain range.

Figure 14.3 The Namib coastal desert of western Africa; the cold Benguela current is just offshore.

offshore of the Namib Desert, for example (Figure 14.3). The cold water cools the air above it, causing fog and lending the illusion of plentiful rainfall to follow. The cool air, however, contains very little moisture. The cold water causes the air mass to sink and spread, and as it sinks, it warms up. The warm air, with its increased capacity to hold moisture, dissipates the fog, leaving the land dry.

Weathering and Erosion in Arid Climates

Sparse rainfall has an impact on every aspect of the erosion process in desert regions. First, the rate of chemical weathering is greatly retarded, and many rocks and minerals are not decomposed to the extent common in temperate and humid climates. Instead, mechanical weathering and incomplete chemical weathering produce a blanket of loose, porous material that covers the bedrock. Water trickles through these partially decomposed rocks and minerals during infrequent rains and dissolves soluble elements. However, water also evaporates rapidly in the desert, and these elements are precipitated near the surface rather than carried away, as in more humid climates. The result is highly alkaline soil, often colored pink, white, and buff. In the extreme, hard white masses of caliche, a calcium carbonate precipitate, accumulate at or near the surface. With the aid of bacterial action, loose fragments and bedrock surfaces may also acquire a strikingly dark-colored coating of *desert varnish*, an encrustation of clay, manganese, and iron oxide (Figure 14.4). The alkaline, granular nature of the soil, combined with the lack of water, accounts for the sparse, widely spaced vegetation of deserts. Sparse vegetation in turn greatly accelerates erosion and is thus responsible for many features of desert landscapes.

Figure 14.4 Dark coatings of desert varnish on gravels.

Another factor that affects erosion in arid regions is the nature of the water flow. Although some of the great rivers of the world cross deserts, most of them receive the greater part of their water from regions of high rainfall. The headwaters of the Nile, for example, originate near the humid equator. The Colorado River is nourished from melting snows north of the Colorado Plateau. However, streams that originate within deserts leak water to the water table rather than being fed by it. Furthermore, there is little vegetation in the drainage basins to absorb water and release it slowly to streams, so desert streams must rely almost solely on the rainfall that occurs at widely spaced and irregular intervals. In addition, the high evaporation rate returns a great deal of water to the atmosphere soon after it falls. Under such conditions, these **intermittent streams** conduct water during periods of high rainfall and are left dry during intervening periods.

The combination of sparse vegetation and sporadic water flow allows slopes in arid regions to be eroded steeply and rapidly. Even mild downpours cause extensive erosion, and the brief but torrential rains, common to many deserts, trigger flash floods that transport huge quantities of sediment. Extensive and rapid erosion accounts for the steep, angular, and generally rugged-looking topography of deserts. Resistant strata, such as well-cemented quartzite, basaltic lava flows, and massive limestone, form steep cliffs.

intermittent stream
A stream whose channel conducts water during periods of high rainfall and is dry during intervening periods.

Easily eroded shales and weak sandstones form gentler slopes. Cliffs with alternating strata of varying composition display varying slope angles.

Eroded fragments that collect at the base of desert cliffs are not colonized by vegetation, so they too are easily removed by the intermittent slope wash and do not accumulate to the same extent as they would in humid climates. Thus the hillside slopes of arid regions retreat at relatively constant angles. In contrast, the slopes in humid regions become gentler as they are eroded. Vegetation lessens the effect of slope wash by soaking up water, and fragments remain close to the base of the hill, which reduces the hill's angle of slope. We can think of slopes in arid regions as being worn back and slopes in humid climates as being worn down.

Structural Influences on Desert Topography

Deserts exhibit a wide variety of erosional and depositional features that are dependent upon the bedrock structure of the region and upon the tectonic activity currently at work. As illustrations, we will discuss two adjacent rain shadow deserts of the American West: the Colorado Plateau and the Basin and Range Province of Nevada (Figure 14.5). Each is the product of complex interactions involving the North American and Pacific plates.

The Colorado Plateau

The Colorado Plateau is an elevated region of relatively flat land, with a steep drop to lower ground on its western and southern borders. It is composed of sedimentary rock thousands of meters thick, and it is the nearly horizontal layering of this strata that is responsible for the flatness of the plateau. The entire region reminds one of a gigantic layer cake, sliced with steep canyons (Zion, Bryce, the Grand Canyon), from which the layers have been peeled back by mass wasting and stream erosion, exposing the different levels of the plateau as a series of steep cliffs. These cliffs consist of hard sandstone or limestone strata capping weaker shales. The pace of cliff retreat is not uniform, however, because erosion is more active in zones of weakness caused by fractures and slight differences in rock composition.

Figure 14.5 A map of the western United States showing the Colorado Plateau and the Basin and Range Province.

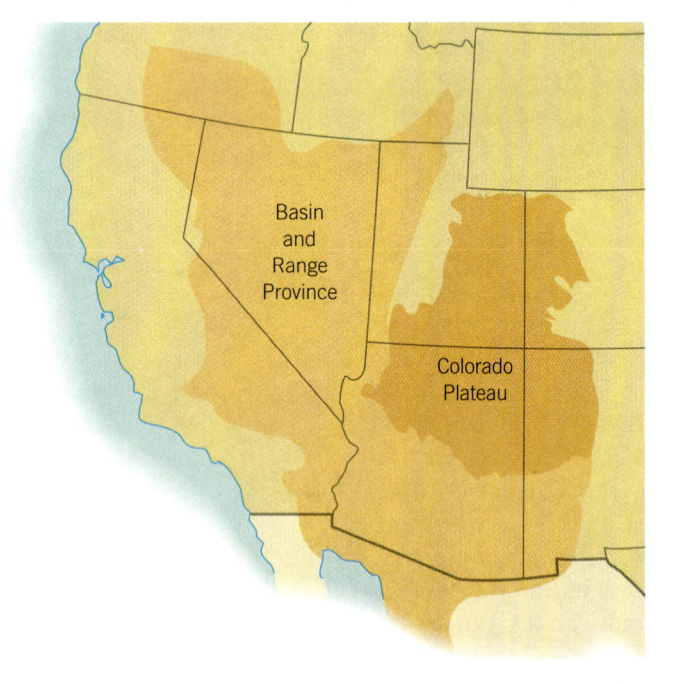

Left behind by the retreating cliffs are isolated erosional remnants. Resistant rocks that cap these features protect the softer strata beneath them from erosion and, at the same time, are undermined by the erosion of the soft strata. Depending on their size, the remnants are called *mesas, buttes,* or *chimneys* (Figure 14.6). Mesas (the Spanish word for "tables") encompass many square kilometers; as they are eroded, they evolve into the smaller buttes. Buttes, in turn, are reduced to smokestack-shaped chimneys. The empty spaces between these remnants are testament to the extensive erosion of the plateau.

The Colorado River and its tributaries—including every branch, gully, and rivulet that feeds into the main stream—are responsible for most of the plateau's erosion. The Colorado, whose headwaters are nourished by the snowy peaks of the Rocky Mountains, has incised its Grand Canyon by exploiting structural weaknesses in the plateau—several major fault zones that lie along the axis of a broad regional arch.

Some 30 million years ago, the plateau was at a lower altitude, and the ancestral Colorado River flowed sluggishly, east-west, across to destinations unknown. However, 5 million years ago, the Gulf of California opened to the south. Streams working northward from the gulf diverted the Colorado's flow to the south. Having a shorter, steeper route to the sea, the Colorado cut into the plateau with renewed vigor. At the same time,

Figure 14.6 Erosion of the flat-lying sediments of the Colorado Plateau. The mesas, buttes, and chimneys visible in the foreground are erosional remnants of the cliffs in the background.

the plateau continued to rise, so that the river sawed down through the layers of the plateau to create the canyon in about 4 million years. This is an extremely rapid rate, geologically speaking, if you compare it to the age of the 2-billion-year-old rocks at the bottom of the gorge.

Basin-and-Range Topography

Sheltered on the west by the high Sierra Nevada, the Basin and Range Province of the southwestern United States and northern Mexico is a classic rain shadow desert. As we learned in Chapter 9, the province is named for its rugged, bold mountains (horsts) and wide, intervening valleys (grabens), which extend north-northwest in hundreds of roughly parallel belts. The trend of the mountains and valleys follows major faults, formed as the crust has been stretched in an east-west direction. Unlike the crust beneath fold-mountain belts, the continental crust beneath the Basin and Range Province is thin, and it continues to stretch even thinner. The tectonic activity, which began about 30 million years ago, involves large-scale movements of the Pacific and North American plates and continues on an enormous scale. Geologists calculate that the mountains of the province are rising at the rate of 30 centimeters every 500 years.

The combination of tectonism and desert conditions has produced the distinctive structural, erosional, and depositional features of basin-and-range topography (Figure 14.7). In the early stages of basin-and-range topography, mountains uplifted by steep faults are separated by broad valleys. Occasional thunderstorms cause torrential rains that pour down the mountain slopes and gash narrow canyons in the bare cliffs. These

Figure 14.7 Racetrack Playa in the Cottonwood Mountains, west of Death Valley. Block faulting has created the prominent basin-and-range topography. A playa occupies the sediment-filled basin in the foreground.

bajada
Alluvial fans that coalesce to form a continuous apron along the base of mountains.

playa
A flat area at the bottom of a desert basin that becomes an intermittent (playa) lake in the wet season.

inselberg
An isolated knob or hill, remnant of a heavily eroded mountain in the desert regions; from the German meaning "island mountain."

pediment
A broad erosional surface formed by running water that slopes gently away from a receding mountain front; the bedrock may be exposed or thinly covered with debris.

short-lived streams—some more properly called mudflows—lose carrying power at the point where they emerge from the narrow canyon, and they spread, depositing their coarse sediment as gently sloping alluvial fans.

The fans coalesce to form a **bajada,** a continuous apron that runs along the base of the mountains. The bajadas grow outward and slope away from the mountains, leaving a shallow basin between them. After it rains, runoff from the intermittent streams accumulates in the basin, producing a temporary body of water called a playa lake. The lake dries up during the long interval between rains, leaving behind a dry lakebed, or **playa.** Thus unlike the Colorado Plateau, large areas of the Basin and Range Province exhibit internal drainage, because the intermittent streams have no outlet to the sea.

There are two major consequences of the erosion of the block mountains and the deposition of sediment in the adjacent basins. The first is that the basins fill with sediment. How much sediment? Death Valley, California, the lowest point in the United States at 85 meters below sea level, may provide a fair example. The present depth of the original valley floor is 2104 meters below the present floor. The difference between the two levels, 2019 meters, is the thickness of the sediment washed into the Death Valley basin, which clearly is not yet filled.

The second consequence of the erosion of the block mountains is that the formerly wide mountains are worn back. In time, all that are left of them are narrow divides capped by isolated remnants, **inselbergs** (Figure 14.8). The narrow divides form the crests of broad, gently sloping erosion surfaces called **pediments.** They extend from the former cliff lines and are created as the cliff retreats. In many places, a thin coating of sediment laid down on the pediments merges with the alluvial fans.

Deserts and Winds 313

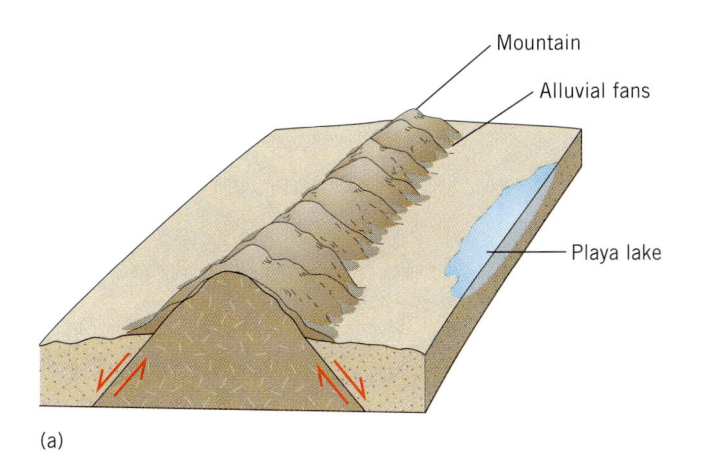

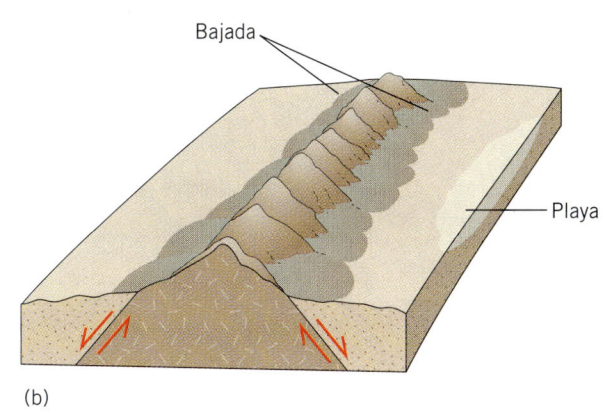

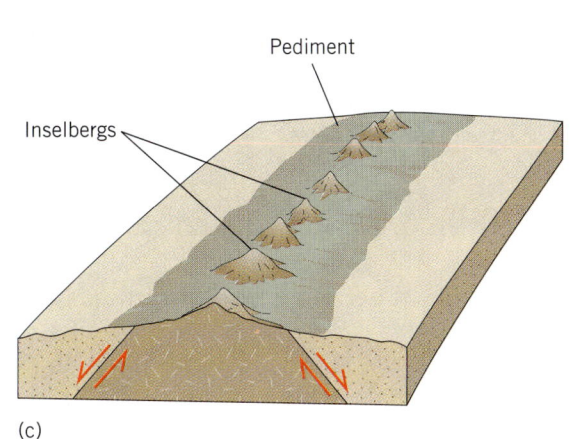

Figure 14.8 Stages in the erosion of a fault-block mountain in an arid climate. (a) Recent uplift produces bold mountains and alluvial fans at the mouths of steep canyons. (b) As the mountain is worn back, the fans coalesce to form a sediment ramp, or bajada. (c) Continued wearing back of the mountain creates an erosional ramp or rock pediment covered by a thin sediment veneer that is a continuation of the bajada. The divides of the mountains are breached, leaving isolated inselbergs.

The Geological Work of Wind in Arid Climates

Winds are moving fluids. They erode, transport, and deposit sediment in a manner similar to that other moving fluid in contact with the Earth's surface, running water. Like streams, winds transport large, heavy particles as *bed load* close to the ground; small, light particles are carried as *suspended load* within the main body of the fluid. Bed load particles are rolled, dragged, or pushed along the ground surface in *traction* or are skipped along in short leaps and bounces in *saltation*. Fine dust particles may be carried in suspension great distances before they settle out. Particles of African clay fall on Miami, and Gobi Desert dust has been detected in the soil of Hawaii.

Note: You may find it useful to read (or reread) the discussion of load transport on pages 244–245 of Chapter 11, Streams.

Features of Wind Erosion

Winds are most effective in the erosion of loose sediment. That is why wind erosion is a major factor in desert locations, which lack the vegetation and moisture to bind sediments to the ground. The removal of loose materials by wind is called **deflation**. Often left behind by the fleeing winds are broad fields of **desert pavement**—coarse fragments too heavy for wind transport, which protects the material below from further deflation. In regions of thick sand accumulations and strong winds, deflation produces *blowouts*, deep saucer-shaped depressions.

It is easy to underestimate the erosive work of wind if readily identifiable landforms like blowouts are missing. To identify former land surfaces, we must search for remnants, such as exposed tree roots that stand above the land surface, or mounds of soil protected by barriers to wind erosion, such as large boulders. Perhaps the most vivid

deflation
The removal of loose materials by wind, which often results in the lowering of the land surface.

desert pavement
The remaining pebble- to cobble-sized rock fragments covering a desert surface after removal of lighter fragments by wind.

Figure 14.9 A dust storm in Saskatchewan, Canada.

means of appreciating the erosive capability of winds is to view a dust storm in action (Figure 14.9). A sizable one can transport millions of tons of sediment.

Although winds play a minor role in the direct erosion of solid rock, their effects are nevertheless locally important in deserts. Torrents of saltating sand are capable of cutting down telephone poles and fence posts at their base. This natural sandblasting action abrades and pits rock outcrops close to the ground surface and converts rock fragments and boulders into angular, faceted **ventifacts** (Figure 14.10).

Features of Wind Deposition

Wind deposits are often so diffused that they are mixed with other sediments and soils and are thus not identifiable as massive features. Two exceptions are loess deposits and sand dunes.

Loess Deposits

Thick deposits of fine windblown dust are known as **loess.** Loess is a collection of angular, silt-sized rock and mineral fragments bound loosely by clay. When eroded, the loess

ventifact
A rock that has been polished and faceted by the action of windblown sand.

loess
A thick deposit of fine windblown dust consisting of unstratified silt-sized calcium carbonate and bits of clay, which has the ability to maintain nearly vertical walls despite its weak cohesion.

Figure 14.10 A large wind-sculpted ventifact. Notice the faceted planes.

forms vertical cliffs (Figure 14.11). The cohesiveness of loess also enables people to easily dig caves out of the cliffs. The insulated caves make comfortable habitats and were used for that purpose in central China and other regions. Loess, rich in nutrients, makes excellent agricultural soil. It underlies the midwestern U.S. corn belt, the wheat fields of eastern Washington State, and the vast farmlands of China.

The distribution and age (about 15,000 years) of loess deposits suggest that in many places, they are derived from glacial deposits. Typically, they are found in areas that border regions covered by ice during the last glacial advance, such as the Mississippi Valley and central Europe. They thin downwind from glaciated regions, which is strong evidence for their glacial origin. It is doubtful, however, that all loess is derived from glacial deposits. Some accumulates from windblown dust from deserts. The extensive,

Figure 14.11 Loess deposits, Shaanxi, China. The cohesiveness of these fine sediments allows them to be excavated for habitation.

Figure 14.12 Sand dunes, Colorado. This view displays the gentle windward slope and the steeper leeward slope, or slip face.

thick deposits of northern China, for example, have blown in from the Gobi Desert to the west. The great Yellow (Hwang-Ho) River of China erodes arid regions blanketed by loess. The color of the silt and clay loess particles carried away gives the river its name.

Sand Dunes

dune
A mound or hill of windblown sand.

slip face
The steeper downwind (or leeward) side of a dune.

Dunes are sandpiles formed where winds that carry abundant sand lose velocity because of an obstacle, depression, or friction with the land surface (Figure 14.12). In time, the sand itself acts as a barrier to the wind, trapping newly arrived sand. Although they come in a variety of sizes and shapes, most dunes are variations of a single plan, having a gentle windward slope and a steep downwind or leeward slope, also called the **slip face**.

It is not difficult to see how this basic architecture develops: wind gradually drives sand up the windward slope, but the slip face is protected, in a wind shadow. Sand simply cascades down the slip face under the influence of gravity and comes to rest at the natural angle of repose for sand, about 30°–35°. Additional sand supplied to the slip face triggers slumps and small slides, which serve to maintain that angle. If not pinned down by vegetation or moisture, the dune will slowly migrate downwind, with sand eroded from the windward face being tossed over to the leeward slip face. Because the winds are constantly shifting the position of the sand within the dune, sand dunes display a complex cross-bedding in which steeply dipping layers formed on the slip face alternate with gently dipping layers of the windward face.

So distinctive are dune cross-beds that their occurrence in ancient sediments is indicative of dry climates at the time and place that the sediment was deposited. The bold cliff of the Navajo sandstone that overlooks Zion National Park, Utah, is an exten-

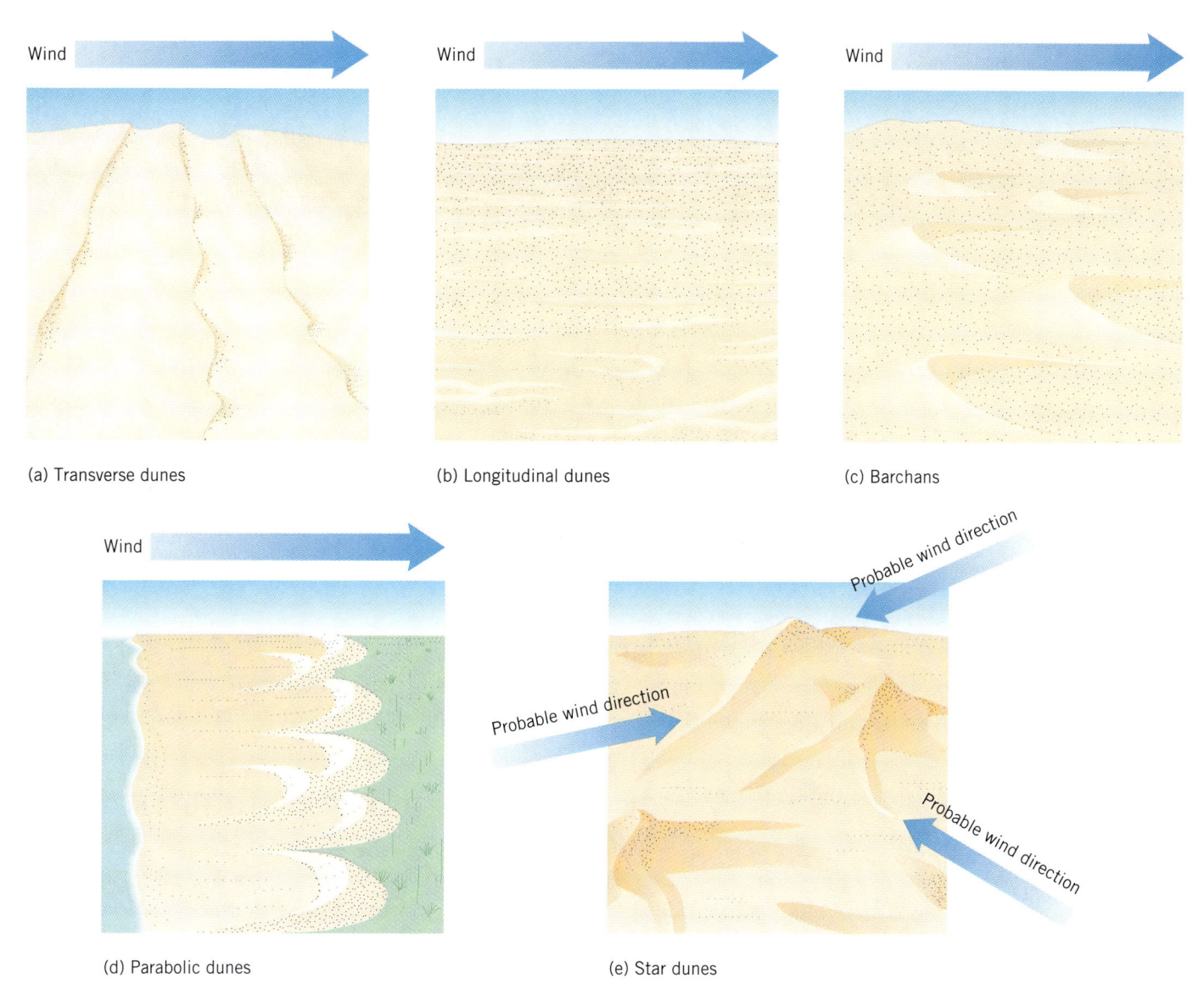

(a) Transverse dunes

(b) Longitudinal dunes

(c) Barchans

(d) Parabolic dunes

(e) Star dunes

Figure 14.13 The orientation of the major dune types relative to the prevailing wind direction.

sive rock formation of Jurassic age (about 150 million years old). Because the Navajo formation exhibits cross-bedding nearly identical to present-day windblown sands, we can assume it had a similar origin (see Chapter 7).

Sand dunes come in a variety of forms, depending upon the direction, strength, and constancy of the winds, sand supply, and vegetation. Wind direction is most important in determining the basic orientation of the dune and serves as a means of broadly classifying dunes. The long axis of a **transverse dune** is at right angles to the prevailing winds (Figure 14.13a). However, the long axis of a **longitudinal dune** lies parallel to the prevailing winds (Figure 14.13b).

Transverse dunes usually occur as a series of thick, nondescript ridges in regions of abundant sand. Some take the form of two of nature's more pleasing creations: the horned, crescent-shaped *barchan* and *parabolic dunes* (Figures 14.13c–d). Both develop in the face of steady winds that blow in an unchanging direction, but each tends to form in a different environment. Barchans generally develop on rocky or otherwise barren desert floors where sand supply is limited (Figure 14.14). Parabolic dunes tend to form in wetter climates than barchans, such as along beaches (see Figure 14.13d). The horns of the barchan crescent point downwind because the wind pushes their thin ends farther than their thicker centers. However, the ends of the parabolic dune tend to be anchored by surrounding vegetation, so their horns point upwind.

transverse dune
A long ridge of sand standing at right angles to the prevailing wind.

longitudinal dune
A long ridge of sand standing parallel to the prevailing wind.

Figure 14.14 Barchan dunes in the Namib Desert, western Africa. The gentler windward slope forms along the convex side.

Longitudinal dunes are long and narrow (see Figure 14.13b). They are particularly well developed in the deserts of North Africa and the Near East, where they are called *seifs* (pronounced "safes"). *Seif* is Arabic for "sword," which the dunes resemble.

Water Resources in Arid Climates

If deserts are so dry, where do the people who live in them find water to survive? The most obvious sources of water are the streams that flow through deserts. You may recall from Chapter 11 that the population of Egypt is mainly confined to the narrow fertile strip on either side of the Nile River and to the great delta of the Nile north of Cairo. Water is obtained directly from the river itself, from high-water-table wells along the banks, and from intricate canals and irrigation ditches.

In many arid regions, water comes from confined aquifers. However, as we saw in our discussion of the Ogallala sandstone, in Chapter 12, these aquifers are not an unlimited source of water, and overuse has led to serious depletion problems.

Other sources of water are the springs and shallow wells commonly found along the fringes of alluvial fans where sediments are capable of holding considerable quantities of water, like a huge sponge. Active faults may sever the water table of the fans and bring the upthrown side of the table to the surface. In these cases, springs may issue along the traces of the fault scarps. This type of water resource influenced the location of Salt Lake City, Utah, and Las Vegas, Nevada. Today, these growing cities face shortage problems as they continue to extract water at rates faster than it is replaced by rainfall. If wind deflation extends to the depth of the water table, it can also expose sources of groundwater. The water table then intersects the land surface and forms a spring or seep, producing a desert oasis.

What happens when the population of an arid region exerts demands for water that exceed the local supply? Since the days of the construction of the Roman aqueducts, the only feasible solution has been to import water from far away. But importation creates further problems of a technological, ecological, and political nature.

Consider, for example, what it takes to slake the seemingly insatiable thirst of southern California. Enormous quantities of water are diverted 1000 kilometers from rainy northern California and Owens Valley in the foothills of the Sierra Nevada. From the

It is not city dwellers who use most of southern California's water. Ninety percent of the water transfer is used to irrigate some of the world's richest cash-crop farmland, which supplies 25 percent of the United States' fruits and vegetables, as well as cotton, rice, soybeans, and sugar beets for export.

overused Colorado River, water is diverted 200 kilometers west to the Mojave Desert. These water transfers have transformed a virtual desert into one of the nation's most populated regions and most powerful economies. But this monumental accomplishment has come at a high price. The water transfers are bitterly opposed by many residents of the source regions and by people throughout the state who are sensitive to the complex ecological and economic issues raised by the water transfer. For example, many Californians are outraged that shipping the water south from northern California may deprive the rich farms and wetlands of the Sacramento Delta of sediment and water—leading to the drying and salting up of the state's most fertile region. In 1995, a compromise limiting the amount of water transferred to the south was reached. This agreement is expected to ensure the survival of the wetlands and delta.

On the Fringe of the Desert: The Sahel

Deserts grow and shrink over millennia, centuries, decades, or even years. Particularly sensitive to change are the broad semiarid regions bordering arid deserts. A striking example is the Sahel (Arabic for "fringe") of Africa, the transition zone between the Sahara Desert to the north and the savannah and rain forests to the south. For many years, portions of the broad Sahel and neighboring Somalia and Ethiopia have suffered drought, and the millions of people of these regions have suffered famine. Is this drought a long-term climate change? Is it a cyclical phenomenon? Or is it a result of human activity? Scientists of many disciplines are urgently investigating the problem.

The Contribution of Climate

Figure 14.15 graphs the fluctuating periods of rainfall and drought in the Sahel since 1950. The sparse rainfall of the Sahel is derived from the clash between the moist winds that blow over the land from the tropical Atlantic Ocean and the trade winds pushing southward. In the summer, the trade winds are warmed and rise off the land while relatively cooler, moist air creeps in from the Atlantic to take its place. The air masses interact over a broad, slowly shifting front, causing rain. The rainfall occurs almost exclusively from June through September, peaking in August. The dry season of the Sahel is initiated later in the year when the Sun passes well below the equator. This shifts the trade winds southward, and they block the moisture-laden Atlantic winds before they

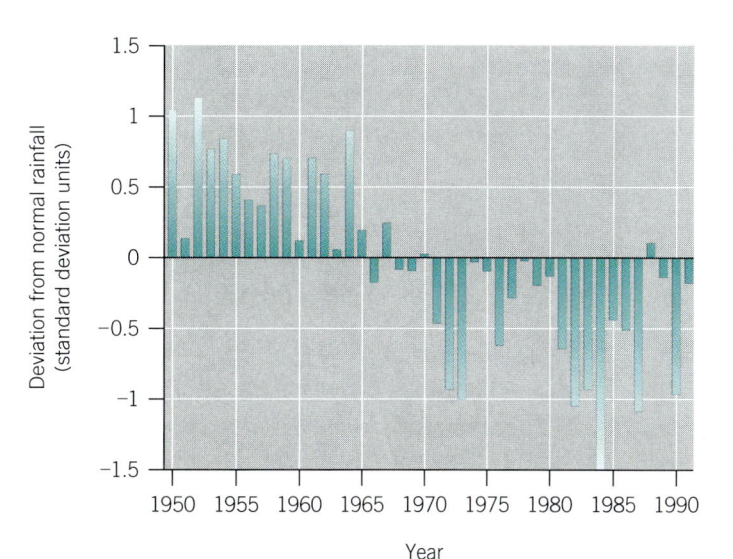

Figure 14.15 Fluctuating rainfall in the Sahel. Twenty years of higher-than-normal rainfall (1950 to 1970) were followed by 20 years of drought (1970 to 1990).

reach Africa. This seasonally reversing wind pattern is called the *monsoon effect.* There are indications that Sahel droughts occur when the trade winds blow farther south over the continent than is usual in summer, thus eliminating the Sahel rainy season.

The Human Contribution

Scientists agree that the severity and duration of Sahel droughts have been worsened through the destruction of the vegetative cover by overgrazing cattle, chopping down trees and brush for firewood, and overplanting in an effort to sustain failing crops. These practices have reduced the capacity of the soil to retain the moisture that is derived from the monsoon rains. As vegetation disappears, the soil becomes rock-hard; when the rains come, the water simply runs off the sunbaked surface.

Until recently, there was consensus among scientists that human activities in the Sahel create desert conditions. This destructive process, in which human activities accelerate the growth of deserts in marginal lands, is called **desertification.** According to this view, desertification of the Sahel, once begun, creates the very conditions that sustain it, and the Sahara Desert pushes ever farther south.

Is the desertification model correct? Like all hypotheses, it must be subjected to tests. From 1980 to 1990, scientists used satellite imagery to trace the southern boundary of the Sahara and see whether it was expanding or shrinking. They discovered that the Sahara had both expanded and contracted during this time and had exhibited no clear trend (Table 14.1). These findings do not support desertification. However, in human terms, the fact that permanent deserts have not been created in the Sahel by human activities is small comfort, because the cycle of land degradation has created what amounts to deserts—soils that are so dry and poor that crops cannot grow in them. No one can control the whims of the monsoon winds, but the human cycle of poverty must be broken if the cycle of land degradation is to be broken. Overgrazing, destruction of plant cover, and general misuse of the land are characteristic of poverty-stricken people trying to survive in a physical environment that cannot sustain their numbers (Figure 14.16).

In summary, drought is not the cause of the mass starvation in the Sahel. Volumes have been written on the misdirected government policies, warfare, and brutalities that have stripped many of the people of the Sahel of the capacity to feed themselves.

The human toll of land degradation is certainly not unknown in the United States. Dust bowl conditions milder than those of the Sahel, but not different in kind, affected the western American Plains states during the Depression years of the 1930s. Prolonged drought initiated the destruction of farmland, the uprooting of entire populations, and the mass migration of rural peoples from their homelands. The general origin of the dust bowl was much the same as in the Sahel of today—a combination of natural forces and human practices that destroyed vegetative cover and prevented rain. Consult John Steinbeck's *Grapes of Wrath* and the songs of Woody Guthrie, which describe the human impact of the dust bowl better than any scientific report.

desertification
Declining productivity of arable land due to mismanagement, leading to a state resembling a desert.

Table 14.1 Migration of the Boundary Between the Sahara (North) and the Sahel (South)

Migration Period	Distance (km)	Direction
From 1980 to 1984	240	Southward
From 1984 to 1985	110	Northward
From 1985 to 1986	30	Northward
From 1986 to 1987	55	Southward
From 1987 to 1988	100	Northward
From 1989 to 1990	77	Southward

Figure 14.16 The human effects of desertification in the Sahel.

STUDY OUTLINE

One-third of Earth's land surface is arid or semiarid. Running water, not wind, is the chief gradational agent of deserts. Not all deserts are hot, and only a small percentage of Earth's total desert area is covered by sand.

I. THE FORMATION OF DESERTS
A. A **desert** is a region with less than 25 centimeters of rainfall per year and whose land surface is too dry to support more than sparse vegetation.

B. The origin of deserts is connected to the planetary wind system.
1. As air at the equator converges, it is warmed, rises, and releases its moisture, creating the tropical rain belts. The dry air cools, sinks, and diverges just north and south of the equator, causing subtropical deserts.
2. As warm winds blowing from the ocean rise over a mountain range, they cool and release their moisture on the windward side, creating a **rain shadow** desert on the lee side.
3. Cold ocean currents stabilize air masses and cause coastal deserts.

II. WEATHERING AND EROSION IN ARID CLIMATES
A. In arid climates, mechanical and incomplete chemical weathering produce a blanket of loose material over the bedrock. Rapid water evaporation causes a highly alkaline soil. Mineral encrustations, called *desert varnish*, collect on rock surfaces.

B. **Intermittent** streams are dry between periods of rainfall. Most erosion is accomplished during floods. Different erosion rates of resistant and weak rock strata, combined with lack of vegetation, account for the rugged topography of deserts. Slopes in deserts retreat at constant angles, whereas slopes in humid regions become gentler.

C. The Colorado Plateau and the Basin and Range Province illustrate structural influences on desert topography.
1. The Colorado Plateau is an elevated region of relatively flat land composed of nearly horizontal layers of sedimentary rock formed into cliffs by mass

wasting and stream erosion. Isolated mesas, buttes, and chimneys form picturesque erosional remnants. The Grand Canyon was created by the simultaneous rising of the plateau and increased downcutting of the river.
2. The Basin and Range Province is a rain shadow desert. Its topography is caused by a combination of crustal uplift, erosion, and deposition.
   a. Alluvial fans coalesce to form **bajadas. Playas** are dry lakebeds. The Basin and Range exhibits internal drainage, because its intermittent streams have no outlet to the sea.
   b. Deposition causes the basins to fill with sediment, and erosion causes the mountains to be worn back, leaving narrow divides with sloping erosion surfaces called **pediments** and remnants called **inselbergs.**

III. THE GEOLOGICAL WORK OF WIND IN ARID CLIMATES

A. Like streams, winds transport heavier particles as bed load and lighter particles as suspended load. The amount of sediment and the size of the particles carried in suspension vary with wind velocity.

B. Removal of loose materials by wind is called **deflation. Desert pavement** is the remaining coarse fragments too heavy for wind transport. In regions of thick sand, deflation produces deep, saucer-shaped blowouts. Wind abrasion converts rocks into faceted **ventifacts.**

C. Features of wind deposition
   1. Thick deposits of windblown dust are called **loess.** When eroded, they form vertical cliffs.
   2. **Dunes** are formed where winds carrying sand lose velocity. Dunes, which migrate slowly downwind, have a gentle windward slope and a steep leeward slope, or **slip face.**
      a. Shifting winds cause cross-bedding, in which steeply dipping layers formed on the slip face alternate with gently dipping layers of the windward face.
      b. **Transverse dunes** lie at right angles to the wind. Some related types are crescent-shaped: barchan dunes point downwind, and parabolic dunes point upwind.
      c. **Longitudinal dunes** lie parallel to the wind.
      d. Star dunes are probably formed by regularly shifting winds.

IV. WATER RESOURCES IN ARID CLIMATES

A. Water is obtained in desert and semidesert environments from streams, confined aquifers, springs in alluvial fans, and oases.

B. When demand for water in an arid region exceeds supply, water may be imported from other regions. For example, water has been diverted from rainy northern California to dry southern California.

V. ON THE FRINGE OF THE DESERT: THE SAHEL.
The Sahel is a broad semiarid region bordering the Sahara Desert where drought and famine have been a severe problem for many years.

A. Drought probably occurs in the Sahel when the trade winds move farther south than usual, blocking the monsoon winds during the summer.

B. The severity of Sahel droughts has been worsened through the destruction of the vegetative cover, which reduces the capacity of the soil to retain moisture.
   1. Research findings do not support the theory of **desertification**—that destructive human activities accelerate the growth of deserts in marginal

lands. However, the cycle of land degradation has created what amounts to deserts: soils that are so dry and poor that crops cannot grow in them.
2. The same combination of natural forces and human practices caused the dust bowl conditions that affected the western American Plains states during the Depression years of the 1930s.

Study Terms

bajada (p. 312)
deflation (p. 313)
desert (p. 306)
desertification (p. 320)
desert pavement (p. 313)
dune (p. 316)
inselberg (p. 312)
intermittent stream (p. 309)

loess (p. 314)
longitudinal dune (p. 317)
pediment (p. 312)
playa (p. 312)
rain shadow (p. 307)
slip face (p. 316)
transverse dune (p. 317)
ventifact (p. 314)

Thought Questions

1. Explain why most of the world's deserts are found in the subtropics at approximately 30° north and south latitudes.
2. Explain the formation of a rain shadow desert.
3. Running water and not wind is the chief gradational agent of deserts. Explain.
4. Contrast the wind conditions that form transverse and longitudinal dunes; barchan and parabolic dunes.
5. The hills in humid climates wear down, whereas those of arid climates wear back. Explain.
6. Discuss the physical and social advantages and disadvantages of large-scale water transfer.
7. Describe the interaction of natural forces and human intervention that has led to starvation in the Sahel region of Africa.

Chapter 15

Wind-Driven Waves
  Headland Erosion and Bay Deposition

Beach Formation and Shoreline Processes
  The Nearshore Circulation Loop
  Beach Types
  The Sediment Budget
    Tampering with Beaches
  The Work of Tides

Coasts
  Tectonic Setting
  Sea-Level Changes

Coasts and Shoreline Processes

Consider the following facts.

- Many of the world's great cities are seaports with natural harbors. These cities are the destination of ships carrying the food, raw materials, and manufactured goods of world trade. (Unfortunately, the coastal ocean and nearshore ocean bottom have become the sewers and dumping ground for these cities and their factories. Just about everything nobody wants ends up there.)
- The life cycles of countless marine and land organisms, an immense food supply, are tied to tidal marshes, estuaries, and beaches. (Yet these are the very areas where human intervention is most prevalent.)
- An increasing number of Americans prefer to live near the shore, either permanently or when on vacation. (Yet many shorefront homeowners are losing their life savings to the sea, as the beach or cliffs erode out from under them.)

In this chapter, we describe the physical processes that affect shorelines and coasts: the work of waves, currents, tides, sea-level changes, and tectonic forces. We also consider the effects of human intervention, since in many parts of the world, the shore is shaped as much by the clash of human interests as by the crash of waves. We begin with a discussion of wind-driven waves because they are the principal agents that affect the shore on a daily basis.

Wind-Driven Waves

Wind-driven waves and associated currents are the chief means by which the sea erodes the land, builds beaches, and transports sediment along the coast. Most of this chapter is devoted to explaining this work. Figure 15.1 shows the anatomy of a wind-driven wave. The distance between wave crests is the *wavelength,* and the time it takes for successive crests to pass a given point (that is, the time for one complete vibration) is the

◂ Beach homes under attack during an Atlantic storm.

Figure 15.1 The components of a wind-driven wave.

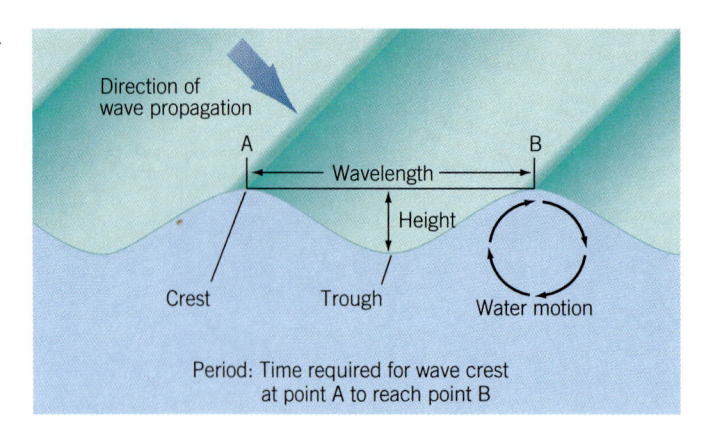

wave period. The vertical distance between the crest and the trough of the wave is the *wave height.*

Most wind-driven waves are generated during storms by friction between the moving air and the sea surface. Their height is dictated by the velocity and duration of the wind as well as the distance over which the wind is in contact with the sea surface. In the storm center, the sea surface is a chaotic mixture of wavelengths, periods, and heights. But if we observe carefully the pattern generated by the storm, we can see that the confused waves are transformed into a regular series of ripples as they move toward shore.

The greater the height of a wave in the open sea, the greater its energy. However, because only the sea surface makes contact with the wind, wave energy—and therefore wave motion—decreases with depth (Figure 15.2).

When a wave approaches a shoreline, another factor begins to influence its motion—the ocean bottom. As the water becomes shallower, contact between the wave and the ocean bottom increases. When the depth of the water is about half the wavelength, the wave begins to "feel bottom." Friction retards its motion, which leads to a series of transformations that intensify as the wave approaches shore. The incoming wave slows down drastically. Thus its length shortens and its height increases markedly, because the same amount of water is packed into a shrinking space. When the wave height reaches about one-seventh of the wavelength, the entire structure becomes gravitationally unstable, pitches forward, and collapses. Thus the energy contained in the wave is transferred to the shore.

The shallow sea bottom also changes the direction of incoming waves that approach the coast at an oblique angle. The segment of the wave that encounters shallow water first slows down, while the other segment continues to approach shore at deep-water speed. Thus the wave pivots so that its approach is nearly at right angles to the shoreline. This bending effect of waves upon entering shallow water is called *refraction* (Figure 15.3). You may recall that seismic waves display similar refraction effects as they speed up and slow down through the layers of the Earth's interior.

Figure 15.2 The changes in a wave as it approaches shore. As the wave begins to feel bottom, friction retards its motion and changes the shape of its orbit. The wave slows down and its height increases. Close to shore, the wave becomes gravitationally unstable and collapses.

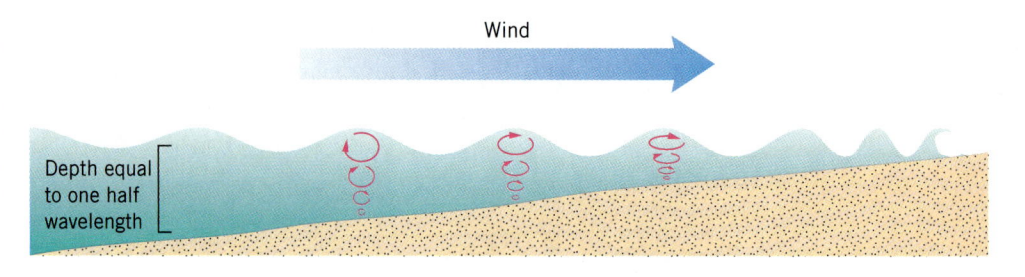

Coasts and Shoreline Processes 327

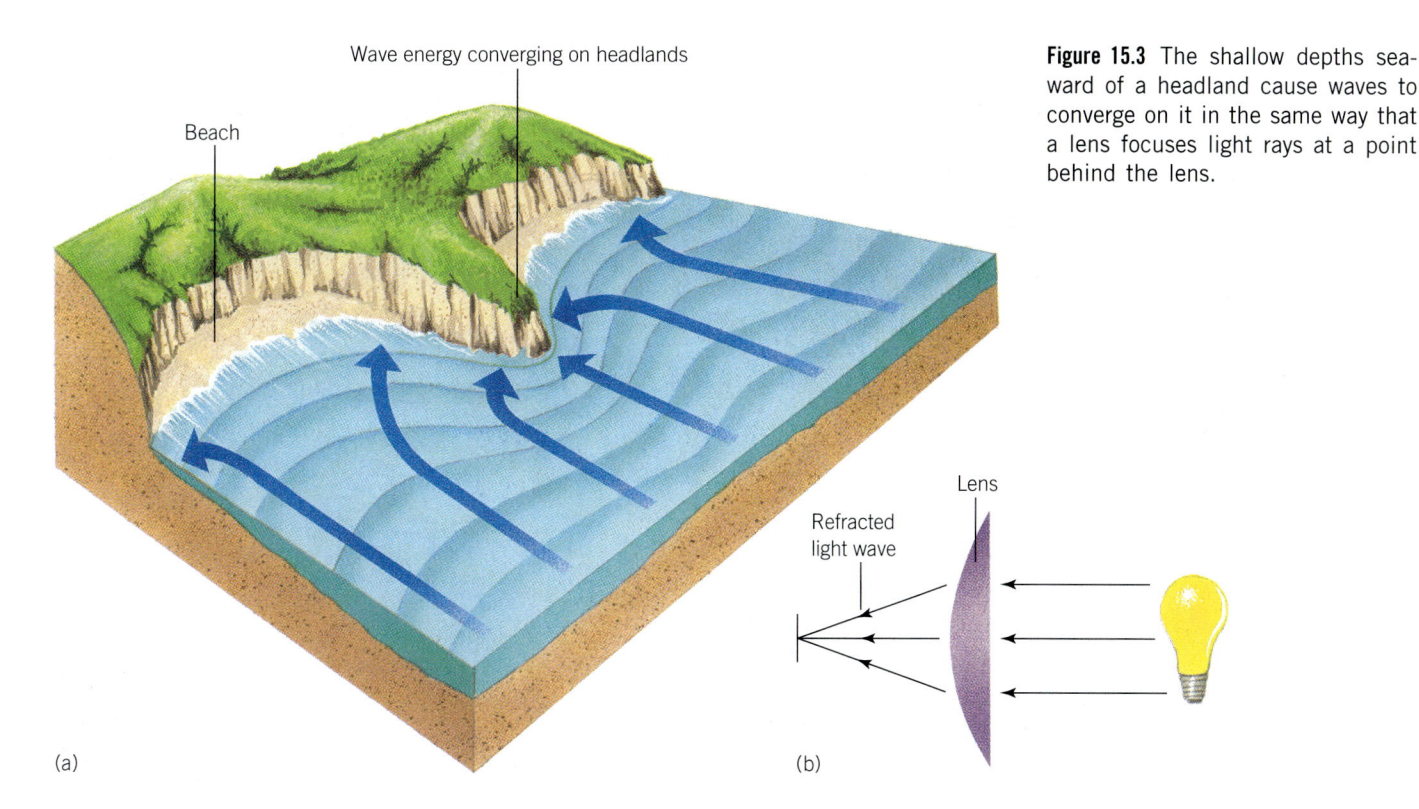

Figure 15.3 The shallow depths seaward of a headland cause waves to converge on it in the same way that a lens focuses light rays at a point behind the lens.

Headland Erosion and Bay Deposition

Figure 15.3 shows incoming waves attacking an irregular stretch of coast. The unprotected headlands are prime candidates for wave attack, because the shallow sea bottom that surrounds them slopes off into the deeper bays on either side. The depth change causes the waves to converge and concentrate energy at the headlands in a manner similar to the way a lens concentrates light at its focal point.

As waves break against the headlands, they erode rocks in a number of ways: by direct impact, by forcing compressed air into cracks, and by abrasion as rock fragments are hurled back against the cliffs. Headland erosion is also aided by chemical weathering and mass wasting in this moist environment. Gradually, a deep cut or *notch* will be carved in the headland at the high-tide watermark (Figure 15.4). The rock overlying the notch will eventually lose support and collapse, resulting in a near-vertical **wave-cut cliff** facing the ocean. Retreat of the cliff produces a gently sloping ramp planed by wave erosion—a *wave-cut platform*. Isolated remnants of the original headland may remain on the platform as *sea stacks* and *sea arches*. In eroding the notch, the waves may also hollow out soft portions of the wave-cut cliff to produce sea caves.

What happens to the fragments eroded from the cliff? Gradually, abrasion causes them to become rounded and smaller. Eventually, many of them are transported to calmer, deeper waters seaward of the wave-cut platform. Transport is accomplished by the back-and-forth motion of the waves and by water from spent breakers, which carries sediment off the platform. Gradually, a depositional *wave-built platform* forms as a seaward extension of the wave-cut platform.

At the same time that refraction concentrates energy at the headlands, wave action is diminished in the bays, because refraction has spread the energy over a wider area. Therefore, sediment is able to accumulate in this calmer, protected environment, forming a *beach*. The major sources of sediment are rock fragments eroded from the headlands and sand and silt swept into the bay by shallow currents close to shore. Given sufficient time, erosion of the headlands and deposition in the bays will lead to straightening of the coast—provided, of course, that the coast is of uniform composition and

wave-cut cliff
A coastal cliff formed by wave erosion.

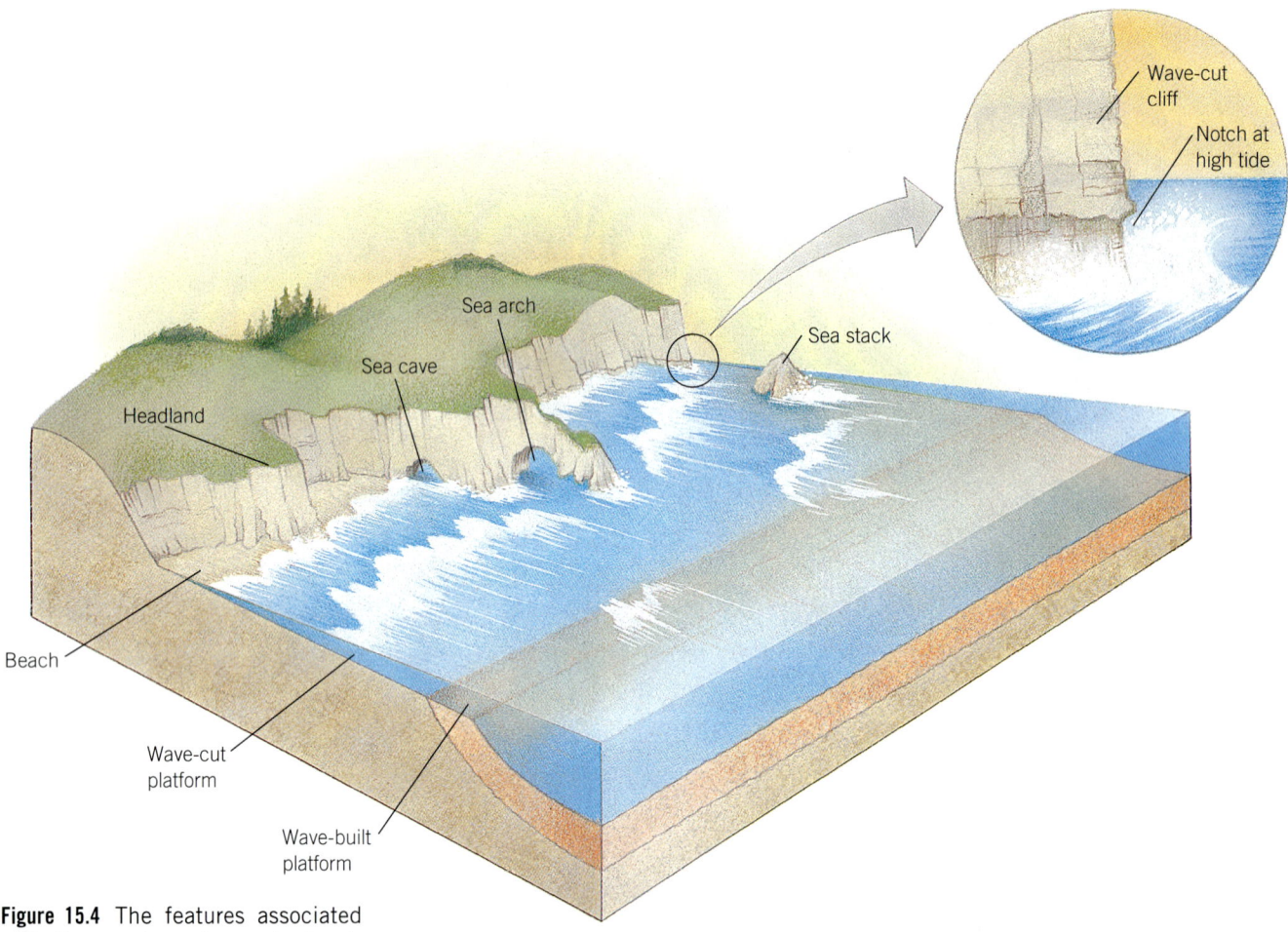

Figure 15.4 The features associated with the erosion of a rocky headland.

beach
A sloping portion of the shore composed of sediments deposited and moved by waves, tides, and longshore currents.

all the other factors that affect the coast remain constant. With straightening, wave energy is distributed evenly up and down the coast.

Beach Formation and Shoreline Processes

Oceanographer Douglas L. Inman succinctly describes most beaches as "essentially long rivers of sand that are moved by waves and currents, and are derived from the material eroded from the coast and brought to the sea by streams." More formally, a **beach** consists of fragmented material (it need not be sand) that is subject to wave action at the interface of land and sea. It extends from the land exposed at low tide to the highest point that storm waves can reach or the point where there is a distinct break in landform—for example, a sea cliff, a line of dunes, or permanent vegetation.

The profile of the beach is constantly adjusting to the incoming wave pattern, the controlling factor being the depths at which the waves break. Waves that arrive from nearby storms are short, high, and choppy; they break close to shore, wearing back the beach. During severe storms, they may even breach the dunes and beat against unprotected houses, highways, and sea walls. In extreme cases, the waves may strip away the entire beach. Most of the eroded sediment is not lost, but is carried offshore beyond the short breakers to form sand bars. This supply of sediment will be used to reconstruct the beach during calmer weather.

In bearing the brunt of breaking waves, giving ground during storms, and rebuilding during peaceful periods, beaches serve as energy sponges—flexible shock absorbers that protect the coast behind the beach from all but the fiercest attack.

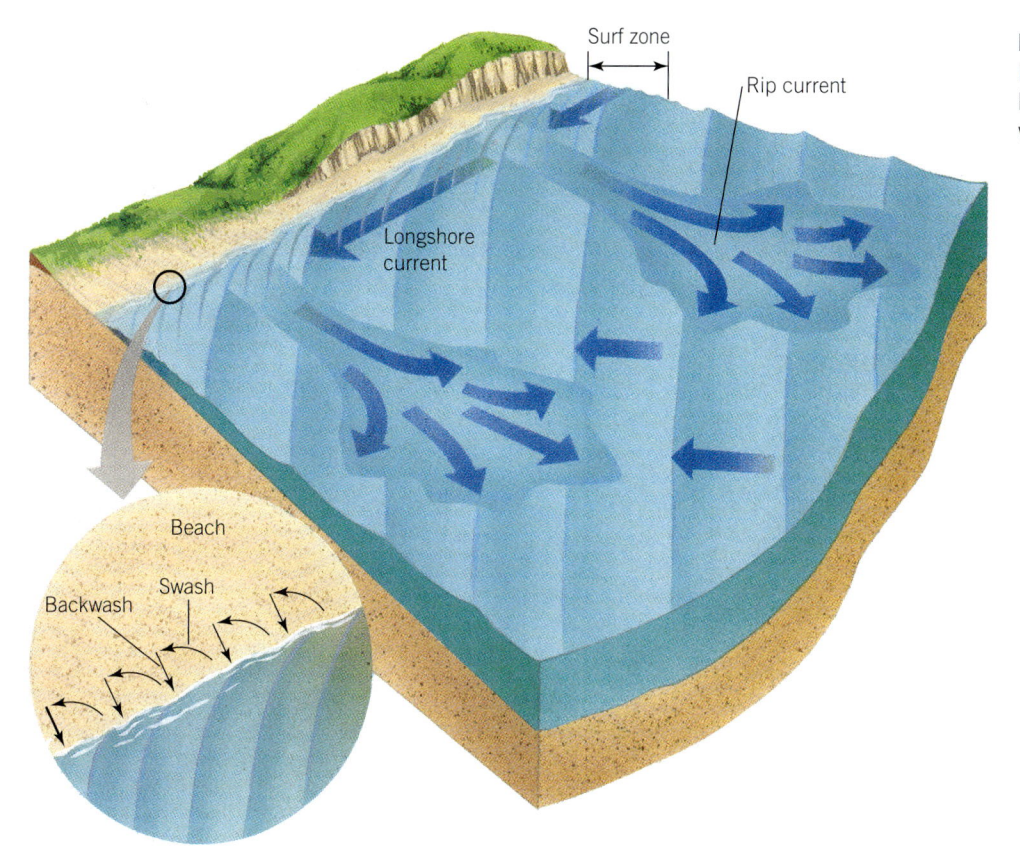

Figure 15.5 The nearshore circulation loop. Sediment moves parallel to the beach via swash and backwash, and via longshore currents.

The Nearshore Circulation Loop

A breaking wave drives water onto the beach as **swash,** which then is pulled back off the beach by gravity as **backwash** (Figure 15.5). If the wave strikes at an oblique angle, it also drives water down the beach as a **longshore current.** Backwash adds to and mixes with this current in the **surf zone.** At intervals along the beach, strong, narrow *rip currents* transfer the water pumped into the surf zone back out to sea. Beyond the surf zone, the rip current widens into a broad *rip head* that mixes with the calmer, deeper waters. The spacing of rip currents is determined by a complex set of circumstances. Basically, it depends upon beach and bottom contours, shore obstacles, and irregularities in the height and period of waves along the beach. In sum, the net effect of waves breaking against the shore at an oblique angle is to set up a nearshore circulation loop whose essential features include the breaking waves, longshore currents, rip currents, and rip heads.

Each portion of the circulation loop directly or indirectly contributes to the formation of the beach. The cycle of swash and backwash moves sediment landward, seaward, and also along the beach; it is primarily responsible for the construction of the beach profile. Longshore currents move sediments parallel to the shore and thus alter the beach in map view, creating a variety of features we will soon discuss. Triggered as they are by breaking waves, the two processes work in tandem. The swash and backwash of breaking waves combine with the longshore current into a system of *longshore sediment transport* parallel to the beach.

swash
The water from breaking waves that washes up onto the beach.

backwash
The return flow of breaking waves down the beach face.

longshore current
A shallow current parallel to the coast, caused by waves that approach the shore at an oblique angle.

surf zone
The zone of breaking waves.

Beach Types

In the process of longshore sediment transport, beaches and related features are molded into a variety of forms, dependent upon the configuration of the coast, the resistance of rocks to erosion, wave direction and intensity, tidal action, organic activity, sediment

Figure 15.6 Common beach forms. (The different types of beaches would not be found grouped this closely together in the natural world.)

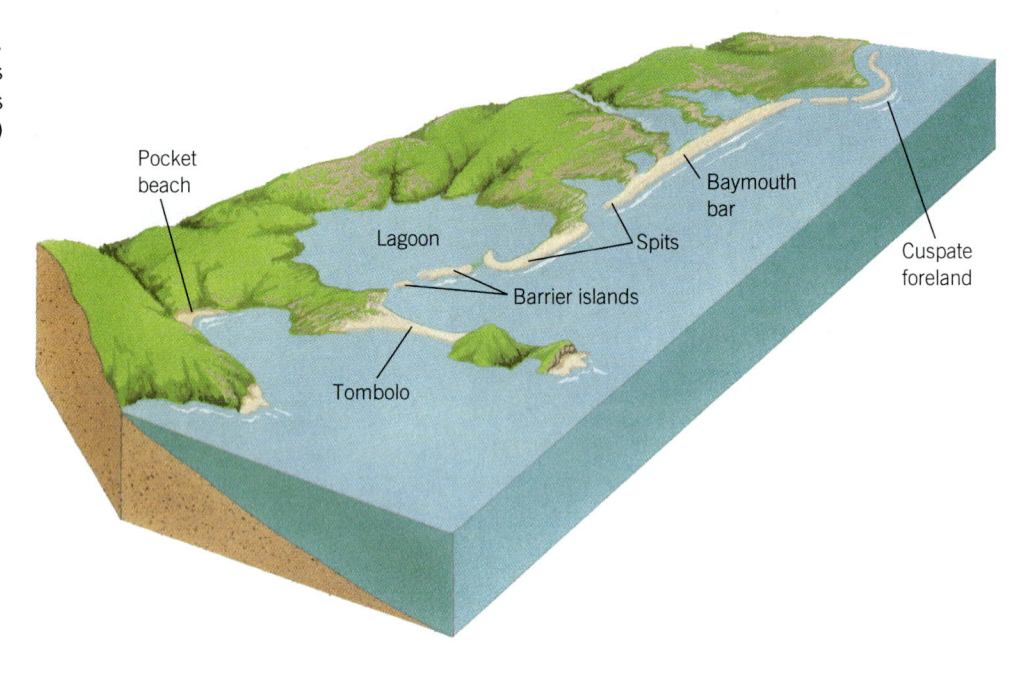

pocket beach
A small beach nestled between adjacent headlands.

tombolo
A strip of beach connecting the mainland to an island or islands to one another.

barrier island
A striplike island that parallels the coast.

spit
A curving, fingerlike projection of beach.

baymouth bar
A strip of beach extending from a headland into the mouth of a bay.

cuspate foreland
A coastal landform composed of seaward-projecting beaches that meet at a point.

supply, bottom topography, and sea-level changes. The following list describes some common beach forms (Figure 15.6).

- A **pocket beach** forms in the low-energy wave environment between bold headlands; much of its sediment comes from headlands eroded in the manner previously described.
- An island may be connected to the mainland by a strip of beach called a **tombolo**. The island, often a sea stack, acts as a barrier to breaking waves, and deposition occurs on the protected side between island and mainland.
- Between the mainland and the **barrier island** that extends more or less parallel to it is a protected lagoon. Many barrier islands are more than mere beaches, having dunes, thick vegetation, and swampy marshes on the side facing the lagoon. Where sediment supply is great, the lagoon shallow, and waves and current erosion low, the lagoon may be filled in to form a low, marshy wetland between mainland and barrier island (see Figure 15.8).
- A **spit** is a curved, fingerlike projection of beach that extends into the sea, elongating the shoreline. The glacial deposits that compose Cape Cod, Massachusetts, have been shaped by longshore currents into the classic example of a complexly curved spit. It is also the world's longest spit.
- A **baymouth bar** extends partially or completely across the mouth of a bay. In the former case, it is a type of spit; in the latter case, it connects headlands and straightens an irregular coast.
- When spits, coastal beaches, or barrier islands join, a seaward projection called a **cuspate foreland** is formed. Cape Hatteras and Cape Lookout, North Carolina, are spectacular examples, as is Cape Canaveral, Florida.

The Sediment Budget

A beach grows, shrinks, and migrates along the coast as part of a much larger sediment circulation system. Figure 15.7 illustrates a sand circulation system typical of the California coast. Sand enters the system via rivers and wave-cut cliffs and is temporarily stored on the beach. Longshore currents sweep the sediment to nearby submarine canyons, where bottom currents conduct it to deep-ocean basins. Note that the form of the beach is maintained *as the sediment passes through*. The beach grows when more

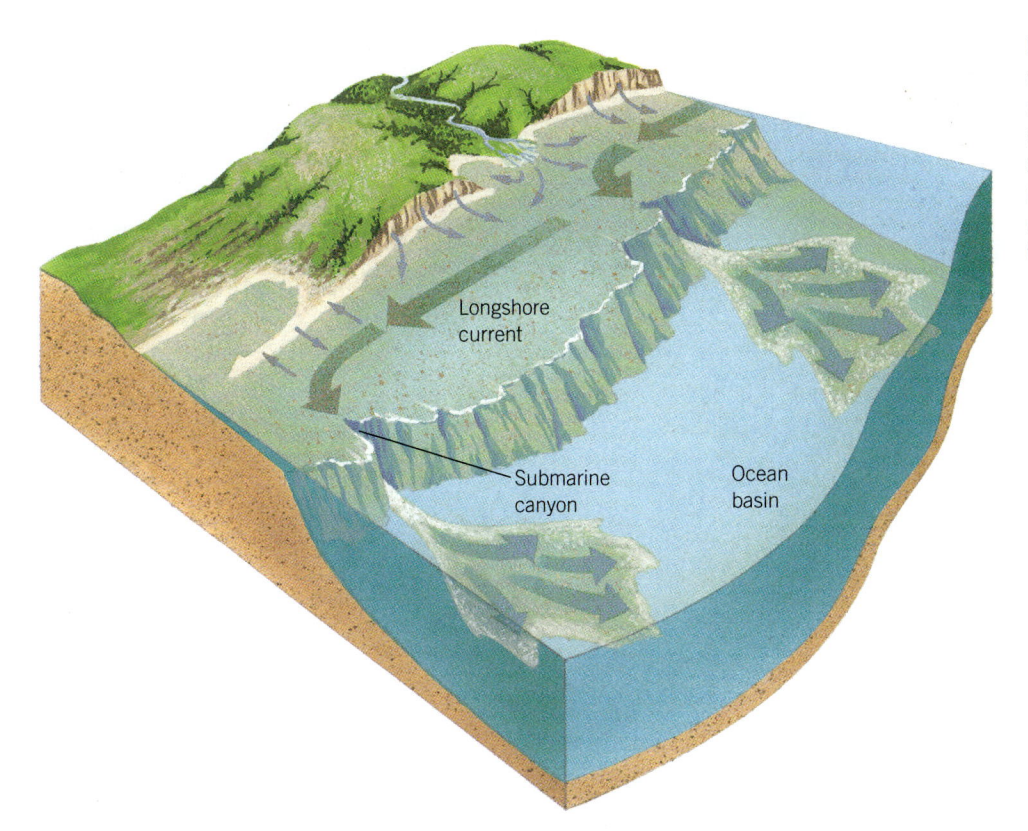

Figure 15.7 The sediment transport system off the California coast. Sediment is introduced to the system via streams and wave-cut cliffs. Longshore currents sweep the sediment into submarine canyons, where it is conducted via turbidity currents to the deep-sea floor.

sand arrives than waves, wind, and currents can carry away. The beach retreats when supply declines or when waves, winds, and currents intensify. In other words, a beach grows, shrinks, or remains static depending upon the balance between the amount of sand added to the system and the amount taken away. Its economy can be described in terms of a profit or loss budget.

Direct wave action tends to alternately build and erode the beach profile cyclically in accordance with the seasons. Over the long term, there is little net sediment loss or gain by this means. Longshore transport, more than any other factor, determines the long-term sediment balance of most beaches, although any of the factors on the beach may be dominant at a particular time or place.

This dynamic yet essentially stable arrangement is complicated by a rising sea level of about 15 centimeters per century on the east coast. The rise concentrates wave attack and tides at a higher level and drives the beaches inland. Barrier islands along the east coast illustrate the process (Figure 15.8). The rising sea forces waves to break higher on the beach and erode it. During great storms, the waves breach the dunes and spread sediments into the lagoons and bays as *overwash* fans behind the dunes. Storms cut through the entire island and link the lagoon to the open ocean via a *tidal inlet*. Eventually, the part of the barrier island that faced the lagoon becomes exposed to the open ocean as the rising sea forces the island to "roll over" toward the mainland. But there is no net sand loss.

Tampering with Beaches

The rather abstract concept of the beach budget has concrete consequences for those who own beachfront property. They are concerned with beach preservation and employ a variety of hard-stabilization techniques to achieve this end. Hard stabilization uses immovable objects in the attempt to hold static what is essentially a fluid system.

For example, suppose resort community A wishes to prevent the longshore current from eroding the beach. The typical solution is to catch sediment by building a barrier,

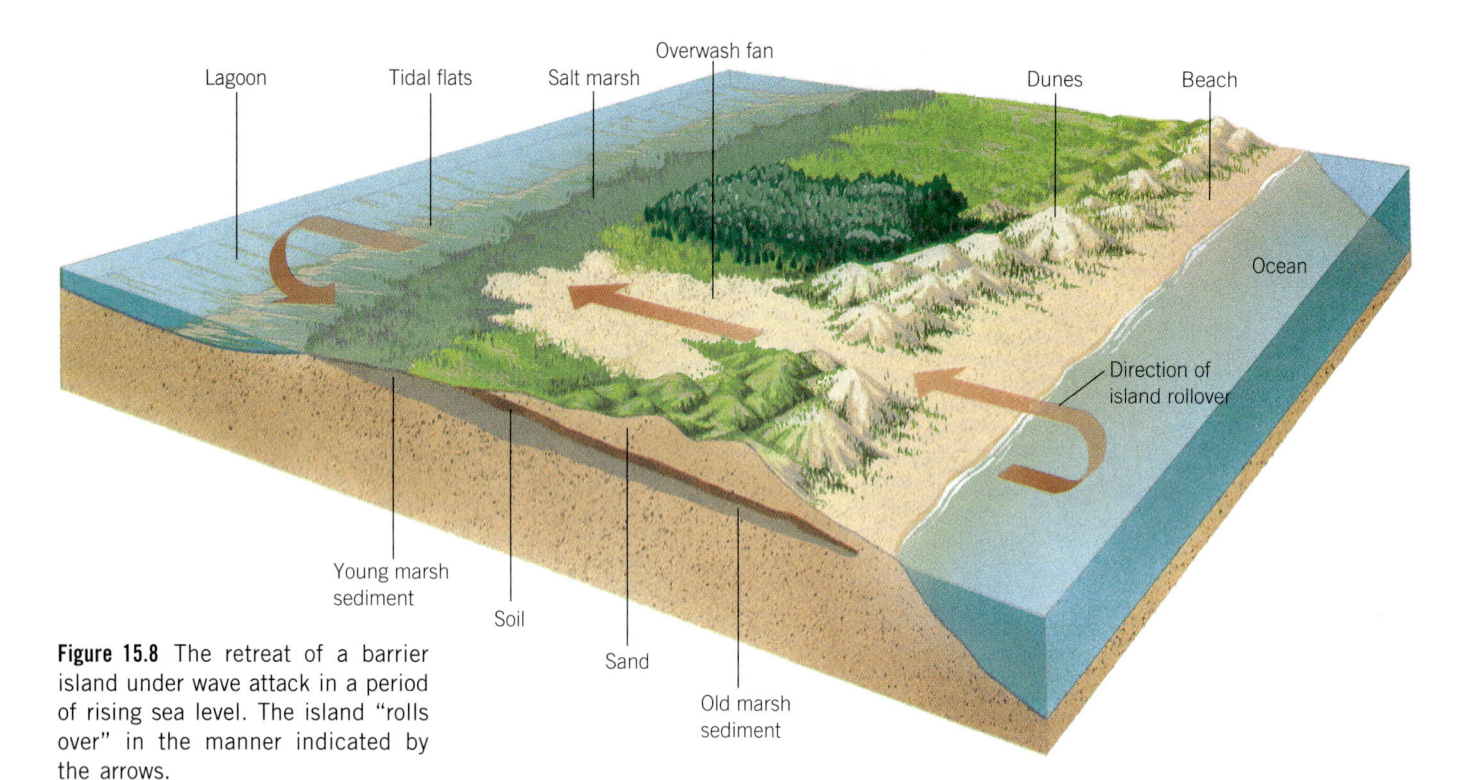

Figure 15.8 The retreat of a barrier island under wave attack in a period of rising sea level. The island "rolls over" in the manner indicated by the arrows.

groin
A structure projecting perpendicular to the beach, designed to trap sediments and prevent beach erosion.

called a **groin,** out to sea at right angles to the beach. The beach indeed grows on the upstream side of the groin, but at the expense of trapping the sediment that would ordinarily be deposited on the downstream side of the groin. It also intensifies the currents that sweep around the groin. These two results lead to accelerated erosion downstream of the groin and the destruction of once-stable beaches and human-made facilities (Figure 15.9). The downstream erosion sets up an unsightly domino effect as property owners construct groins to protect *their* stretch of beach. If the entire community's beach is eventually protected—at great expense—the problem is merely transferred, in

Figure 15.9 The installation of groins caused the condition that exists along Cape May, New Jersey. Deposition increases on the upcurrent side of the groin, whereas erosion is accelerated on the downcurrent side.

Figure 15.10 The results of hard stabilization. Forty years ago, Seabright, New Jersey, was known for its wide, beautiful beach. Then the Army Corps of Engineers built the retaining wall to protect the apartment houses from wave attack during the storms, but the wall precipitated the destruction of the beach. Today it is the only thing that stands between the buildings and the sea.

intensified form, to the neighboring beach community B, which is attacked by the same sediment-deprived longshore currents. It has been said that groins are like lawyers: nobody should get one, but once somebody does, we all need one.

However, environmental lawyer Katherine Stone has proposed another solution, the "sand rights" theory. According to traditional law, certain resources, like the air, belong to society at large. Stone argues that beach sand is such a resource, and that those who damage the beach by interrupting the natural flow of sand should be required to pay for the restoration of the beach. For example, community A should be required to pay the cost of restoring community B's beach; if that is not possible, the groin should be removed. The same principles would be applied to communities that construct dams that trap sediment intended for the beach.

The construction of retaining walls is another common hard-stabilization technique, designed to protect homes and shorefront facilities. They may indeed protect these structures for a time; but they do not prevent the erosion of sand from the beach, so that the beachfront property ends up minus its beach. The retaining wall itself is an artificial sea cliff subject to the energy of breaking waves. Eventually, it too will be undercut and eroded, leaving the shorefront structure exposed. The stretch of New Jersey coast shown in Figure 15.10 is an illustration of the sad results of relying on retaining walls to prevent beach erosion.

Mindful of the damages of hard stabilization, coastal engineers have developed soft-stabilization techniques that work in harmony with the natural processes that affect the beach. The simplest of them is to pump sand back onto the eroding beach. The source of the sand may be the far offshore, lagoons behind the beach, or sediment trapped by adjacent groins. Although this strategy is less harmful than hard stabilization, the pumped-in sand erodes more rapidly than the natural beach and in any case has little influence on the factors that cause the beach to erode. The high rate of replenishment and the huge sand volumes needed make this an expensive stopgap measure.

A number of geologists, environmentalists, and budget-conscious officials believe that hard and soft stabilizations are expensive and ultimately fail. The only rational solution, they believe, is to discourage settlement on beaches, and, where possible, relocate those who live there.

The Work of Tides

Wind-driven waves are the most active agents in building beaches and shaping coasts, as we have seen, and their work is greatly influenced by a number of factors. Chief among them are **tides**—the rhythmic rising and falling of the sea surface—which focus the point of wave attack and carry on a number of other important activities as well.

The energy of tides is supplied by the gravitational attraction between the Earth and the Moon and, to a lesser extent, between the Earth and the Sun. The attractions lead to the regular rise and fall of sea level observed along the shorelines of the oceans and large lakes of the world. The work of tides is cumulative and, with rare exceptions, underestimated, in that it is often masked by the effects of wind-driven waves.

We all know that the Moon revolves around the Earth. If you were to reason that the center of the Moon's orbit must be the center of the Earth, however, you would be incorrect. Mutual gravitational attraction causes the Earth and the Moon to revolve about a common point—that is, the center of mass of the Earth-Moon system. Because the Earth has a far greater mass than the Moon, the center point of the system is indeed located inside the Earth. On the other hand, the mass of the Moon exerts enough gravitational pull to draw the center of the system away from the center of the Earth. Thus each point on the Earth is subject to two forces: the attractive gravitational force that the Moon exerts on the Earth at that point and the outward-directed centrifugal force generated as the point revolves about the common center. At all points on the side of the Earth facing the Moon, gravitational force exceeds centrifugal force. The difference between the two forces generates a small *tidal force*. This force drives water toward point A, the point closest to the Moon; the water at point A produces a tidal bulge (or high tide).

The force relations are reversed on the side of the Earth facing away from the Moon. There, centrifugal force exceeds gravitational force. Thus the direction of the tidal forces is reversed, and a high tide is created at point B, opposite to point A. When A and B experience high tide, the region midway between the two is subject to low tides. Because the Earth completes one rotation about its axis every 24 hours, each point on Earth's surface *should* experience two high and two low tides per day.

Superimposed on these lunar tides is the gravitational effect of the Sun on the Earth, which exerts a tidal force of its own. **Neap tides,** having minimal tidal variation, occur twice each month when the Sun is at right angles to the Earth and Moon. **Spring tides,** having maximum tidal range, occur twice monthly when the Earth, Moon, and Sun are aligned; in this position, the tidal forces produced by the Sun add to those generated by the Moon.

Because tides feel bottom, they behave like shallow-water waves. They are affected by water depths and by the configuration of the coast as they approach the shore. When a high tide approaches a bay that is constricted by headland, barrier beach, or spit, the water level of the open ocean stands above the water in the bay. Thus water is forced through the constriction under great pressure. Swift **tidal currents** are generated that can scour the bottom to great depth: 112 meters at the mouth of San Francisco Bay; 390 meters in the straits between the islands of Kyusho and Shikoku, Japan. The currents reverse during low tide. Therefore, sediment is periodically washed in and out of the bay. Depending upon the strength of incoming and outgoing tidal currents, the bottom of the bay will either be filled in or flushed. For this reason, communities that dispose of their waste by carelessly dumping it in the bay during low tide may, to their surprise, find it returning to their shores at high tide.

Tidal currents help shape barrier beaches, spits, and deltas. In some situations, they are more potent agents for shaping coasts and beaches than longshore currents. Tidal currents are especially effective in conjunction with great storm waves. When a barrier island (such as Fire Island, New York, or Cape Hatteras, North Carolina) is breached by high waves during a storm, tidal currents pour through the gap and transport sediment from the open ocean and foreshore of the beach into the quiet waters behind the island. In time, longshore currents may seal the gap.

---

tide
The rhythmic rise and fall of the sea surface caused by the unequal gravitational attraction of the Moon—and, to a lesser degree, of the Sun.

neap tide
A low-amplitude, twice-monthly tide.

spring tide
A high-amplitude, twice-monthly tide.

tidal current
The current of a rising or falling tide forced through a narrow constriction of coast.

Figure 15.11 A satellite image of the drowned mid-Atlantic coast of the United States. Chesapeake Bay, an estuary, is the most prominent feature; as sea level rose following the last ice age, an intricate stream system of branching tributaries was invaded by the sea. Delaware Bay, to the north, is a similar, though less elaborate feature. Numerous beach forms, such as barrier islands and spits, are visible along the coast. The Susquehanna River, at the top left, cuts through resistant Appalachian ridges and enters into upper Chesapeake Bay.

Coasts

A *coast* is a strip of land bordering the ocean and directly subject to marine influences. The seaward border is the shoreline. The inland border is harder to define but is generally signaled by a marked change in topographic features. Nowhere else on Earth's surface do more geological processes and agents converge, which makes coasts difficult to classify. One way to study them is to divide them into two general types. *Primary coasts* owe their configuration mainly to nonmarine processes, such as land erosion or deposition, volcanic activity, or crustal movements. For example, Chesapeake Bay, on the eastern U.S. coast, was a river valley that was later submerged when the sea level rose following the retreat of the great North American ice sheet (Figure 15.11). *Secondary coasts* are molded primarily by marine influences, such as wave deposition or erosion, or the work of organisms, like coral reefs or mangrove trees. For example, just south of the Chesapeake Bay is Cape Hatteras, whose projecting shape and surrounding barrier beaches were formed by marine processes (Figure 15.12).

Tectonic Setting

Recall from Chapter 2 the placement of North and South America with respect to tectonic plate boundaries. The western edge of each continent is adjacent to a plate

Figure 15.12 Cape Hatteras, North Carolina, is at the apex of a cuspate foreland molded by waves and longshore currents.

submarine canyon
A steep canyon, resembling a V-shaped stream valley, that is cut into the continental shelf or slope.

boundary and has an *active margin*. The eastern edge of each continent is far from plate boundaries and, as a consequence, has a *passive margin*. These margins differ markedly from one another, and tectonic activity is the controlling factor.

Most stretches of the west coasts border on a subduction zone, where a deep-sea trench or similar structure is found almost directly offshore. Other stretches border rift zones or transform boundaries. These plate boundaries are the sites of intense tectonic activity and volcanism. As a consequence, the west coasts of North and South America are generally straight, mountainous, and extremely rugged. Few major streams enter directly into the ocean, and the west coast receives comparatively less sediment than the east coast. The continental margin is extremely narrow; and although there are broad beaches, they are of the wave-cut cliff and terrace variety. Some of these wave-cut cliffs and terraces have been uplifted; they ascend, steplike, high into the coast, which is striking evidence of recent uplift (Figure 15.13). This unstable tectonic environment does not favor the construction of a broad continental shelf. Instead, the sediment brought in by streams is funneled to deep trenches and abyssal fans via **submarine canyons** located close to shore.

The east coast runs along a passive continental margin located far from the Mid-Atlantic Ridge and other centers of tectonic activity. In this stable crustal environment, the margin was able to develop a broad continental shelf and slope. It is essentially a thick sediment wedge deposited over millions of years by streams emptying into the Atlantic Ocean. Except where ancient highlands intersect the shore, the coast is generally low and marked by a wide coastal plain, an extension of the continental shelf.

Sea-Level Changes

Because the Atlantic continental shelf is tectonically stable, it is a good place to trace the worldwide sea-level changes that have occurred since the last ice age. The emergence of the continental shelf happened recently, from a geological perspective, for the ice stood at its maximum a mere 18,000 years ago. During the ice ages, streams flowed across the shelf to the shelf break; like all streams, they cut channels and deposited sediment in accordance with this lower base level. Many of these effects of the vast sea-level retreat are still visible on the shelves (Figure 15.14).

Figure 15.13 The elevated marine terraces and wave-cut cliffs of San Clemente Island, California.

From analyses of sediments, fossils, and submerged shoreline features, we know that the seas have been advancing over the past 18,000 years, since the melting of the last ice sheet. The rate of advance was rapid until 6000 years ago, but has since leveled off. Sea level is currently rising about 4 millimeters per year off the Atlantic coast (about 16 inches per century). The effects of sea-level rise have been profound. For example, all of the low-lying coasts have been invaded by the sea. Many of today's islands were formerly inland hills. (In some cases, these hills were quite large—for example, the British Isles, which had been connected to mainland Europe.)

If you look at the East Coast of the United States on the map in Figure 15.15, you will notice that the coastline is highly irregular and indented. The landward fingers of the sea are **estuaries,** flooded river mouths where fresh- and saltwater mix and are controlled by tides. The Chesapeake Bay and the Hudson River are striking examples. In fact, the mouth of virtually every river in the world that emptied into the sea prior to this most recent sea-level rise is an estuary.

Sea-level rise is responsible for the numerous barrier islands that rim the East and Gulf coasts. Keep in mind that the slope of the shelf is so slight that a 1-meter sea-level rise causes the coastline to retreat by 1000 meters or more. The islands most probably originated as beaches on the emergent continental shelf during the ice age, when sea level was lower. The rising sea flooded the land behind the beaches, isolating them from the coast. The rising sea also elevated the point of wave attack on the beach faces, causing them to roll over in the manner discussed earlier. Thus the barrier islands have migrated toward the coast. Since migration and coastal retreat occur in tandem, the lagoons between them have been maintained. Occasionally, where a slightly higher coast has slowed retreat of the shoreline, the barrier islands have rolled up against the mainland and have closed out the lagoons and bays.

estuary
A mouth of a river invaded by the ocean, where fresh- and saltwater mix.

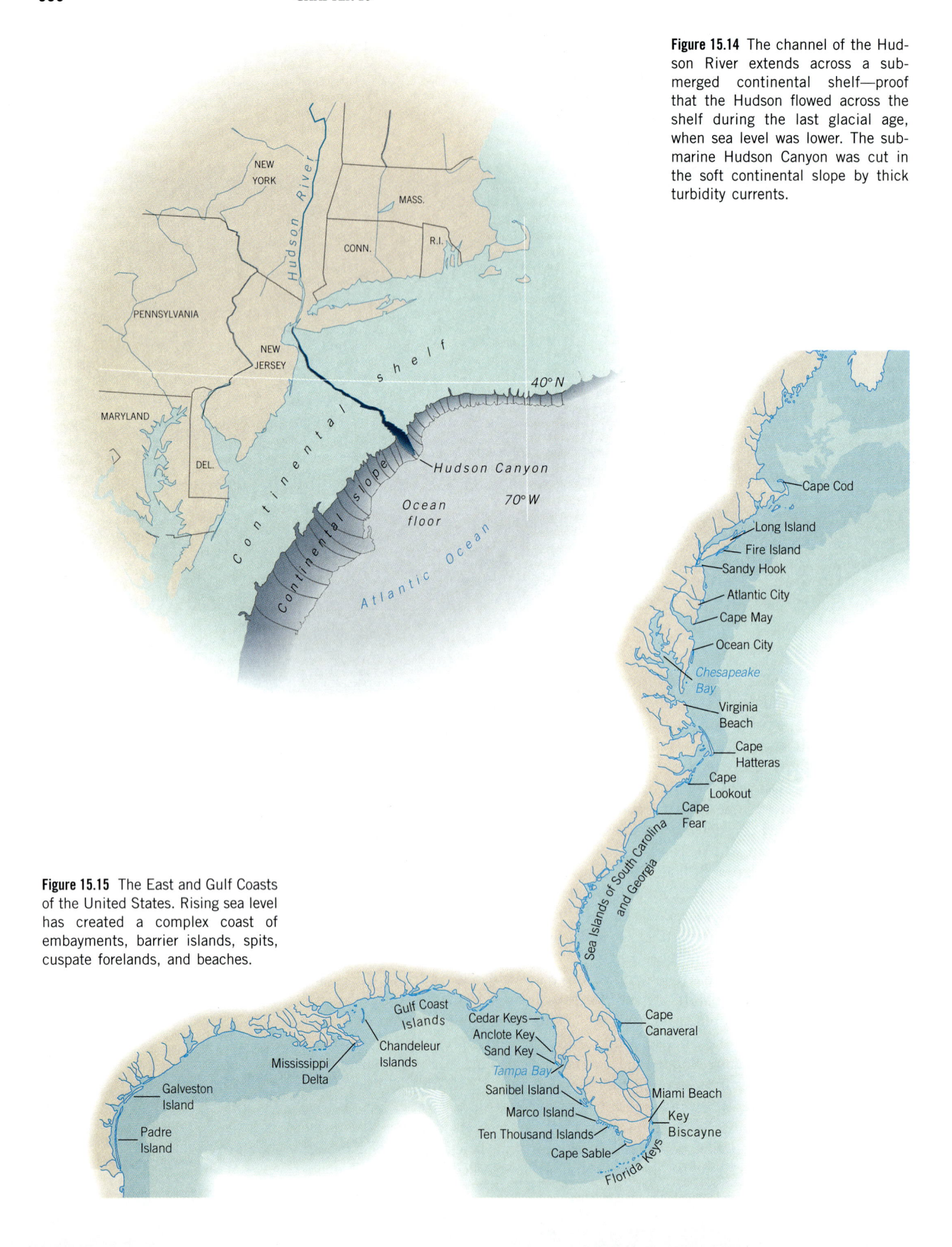

Figure 15.14 The channel of the Hudson River extends across a submerged continental shelf—proof that the Hudson flowed across the shelf during the last glacial age, when sea level was lower. The submarine Hudson Canyon was cut in the soft continental slope by thick turbidity currents.

Figure 15.15 The East and Gulf Coasts of the United States. Rising sea level has created a complex coast of embayments, barrier islands, spits, cuspate forelands, and beaches.

On the west coast, the mountains bordering Washington, Oregon, and Canada were covered by ice during the last glacial advance. Stream valleys were deepened by the ice, and streams flowed further west than at present. With the rise in sea level following melting of the ice, the valleys were flooded. Puget Sound is such a flooded valley. Thus the northwest coast is similar to the coasts of Maine and Nova Scotia, but far more rugged. As in the east coast, many small islands were cut off from the mainland by the sea-level rise.

Molded by so many factors—tectonic forces, waves, winds, currents, sediment supply, climate, tides, biological activity, sea-level changes, the works of human beings—coastal and shoreline processes are complex and vary with time. There is still much to learn about this subject. Perhaps the key concept to keep in mind is the dynamic nature of the coastal and shoreline environments. For this reason, short-term solutions to local problems can lead to disastrous mistakes on a wider scale. As with most choices involving conflicting interests, decisions are never clear-cut and often amount to a series of unsatisfactory trade-offs.

STUDY OUTLINE

I. **WIND-DRIVEN WAVES**
   A. Wind-driven water waves are vibrations by which energy is transferred from ocean to shore.
      1. Wave height is the vertical distance between crest (high point) and trough (low point); a wavelength is the distance between crests; wave period is the time it takes successive crests to pass a given point.
      2. At a storm center, waves of different lengths, periods, and heights interfere with one another and create a chaotic sea surface.
   B. As a wave approaches the shore, its movement is retarded by bottom friction. Close to shore, its length shortens and its height increases, but its period remains the same. Eventually, it collapses or breaks.
      1. A wave that approaches the shore at an oblique angle refracts; the segment closest to shore slows down first, and the entire wave pivots in a direction roughly parallel to the shoreline.
      2. Waves that approach an irregular coast erode the headlands and deposit sediment in the bays to form beaches. Erosional features of the headlands include **wave-cut cliffs**, *wave-cut platforms*, *sea stacks*, and *sea arches*. Erosion and deposition processes acting on an irregular coast lead to coastal straightening.

II. **BEACH FORMATION AND SHORELINE PROCESSES**
   A. Materials brought to the sea by streams and eroded from the coast are shaped by waves and currents into moving rivers of sand called **beaches.**
      1. In calm weather, long waves that break offshore and sweep sediment onto the foreshore flatten and broaden the beach. During storms, short and choppy waves attack the beach and carry sediments offshore.
      2. A wave striking the beach at an oblique angle sets in motion a nearshore circulation loop consisting of **swash** and **backwash, longshore current,** rip current, and rip head. The circulation loop, which encompasses the **surf zone,** is instrumental in *longshore sediment transport*. Swash and backwash move sediment in the direction of the longshore current.
      3. Longshore sediments mold beaches into a variety of forms that are also dependent on such factors as the resistance of rocks to erosion and wave direction and intensity. Some common beach forms include the **pocket beach,** the **tombolo,** the **barrier island,** the **spit,** the **baymouth bar,** and the **cuspate foreland.**

B. A typical beach is part of a larger sediment circulation system. Sand enters the system via wave-cut cliffs and rivers. Sand leaves the system via longshore and bottom currents.
   1. Rising sea level forces barrier islands to roll over and migrate landward.
   2. A **groin** is a barrier built seaward to prevent longshore drift. It leads to accelerated erosion downcurrent of the groin. Retaining walls protect structures but intensify beach erosion. Many geologists consider beaches to be inherently fluid systems from which construction should be banned under normal circumstances.

C. **Tides** are the rhythmic rising and falling of the sea surface caused by the interplay of the gravitational pull of the Moon and centrifugal forces generated as the Earth revolves about the center of mass of the Earth-Moon system.
   1. The Sun also produces a tidal effect. Very high **spring tides** and very low **neap tides** each occur twice monthly.
   2. A swift **tidal current** in conjunction with storm waves may breach a barrier island and deposit sediment in the protected waters behind the island.

III. **COASTS.** A coast is a strip of land bordering the ocean.

A. Primary coasts are molded mainly by nonmarine processes: land erosion and deposition, volcanism, and tectonism. Secondary coasts are formed primarily by marine influences: wave deposition or erosion, coral reefs, and the like.

B. Proximity to tectonic plate boundaries provides the broad structural setting within which coastal processes occur.
   1. The west coasts of North and South America are parts of tectonically active continental margins and are generally straight and narrow. Sediment brought to the narrow continental shelf is funneled by way of **submarine canyons** to the deep trenches or basins offshore.
   2. The east coasts are parts of passive continental margins. Far from activity associated with plate boundaries, the margin consists of a broad continental shelf, slope, and rise. The coasts are low, intricately configured, and marked by coastal plains that are extensions of the continental shelves.

C. The sea-level rise that accompanied the melting of the last ice sheet caused invasion of the older river valleys, forming **estuaries,** or drowned river coasts and barrier islands, which were subsequently beaten back toward the mainland.

STUDY TERMS

backwash (p. 329)
barrier island (p. 330)
baymouth bar (p. 330)
beach (p. 328)
cuspate foreland (p. 330)
estuary (p. 337)
groin (p. 332)
longshore current (p. 329)
neap tide (p. 334)
pocket beach (p. 330)

spit (p. 330)
spring tide (p. 334)
submarine canyon (p. 336)
surf zone (p. 329)
swash (p. 329)
tidal current (p. 334)
tide (p. 334)
tombolo (p. 330)
wave-cut cliff (p. 327)

Thought Questions

1. Describe the sediment budget, of which the beach-nearshore system is a part. What natural changes can affect it? What human-induced activities can change it?
2. What is your opinion of the sand rights proposition mentioned in this chapter? Explain.
3. Discuss the pros and cons of hard- and soft-stabilization techniques.
4. Describe some of the geological work of tides.
5. Describe how plate tectonics influences coastal development. What features are common to active margins? To passive margins?
6. Why are barrier islands more prevalent on the East and Gulf coasts than on the West Coast of the United States?

Chapter 16

Relative Time
- Deciphering Local Geologic History
  - Horizontality and Superposition
  - Unconformities
  - Cross-Cutting Relationships
- Correlation
  - Stratigraphic Correlation and Faunal Succession
  - The Relative Geologic Time Scale
  - Correlation Tools: Index Fossils and Overlapping Ranges
  - The Terminology of the Time Scale

Absolute Time
- Modes of Decay
- The Decay Principle
- Radioisotopes Useful in Dating
  - Uranium- (and Thorium-) to-Lead Dating
  - Potassium-to-Argon Dating
  - Rubidium-87–to–Strontium-87 Dating
  - Carbon-14 Dating
- Dating the Geologic Time Scale
- The Age of the Earth

Geologic Time

Recognition of the immense span of geologic time—and, by implication, our minute share of it—is probably geology's key contribution to human knowledge. Its impact on our view of reality is comparable to the earlier realization that the Earth is more a speck of dust than the center of the universe. In this chapter, we describe how geologists tell time. First, we discuss how they place rocks, and the geologic events recorded in them, in chronological order (**relative time**). Then we explain how they use radiometric methods to determine the ages of rocks and geologic events (**absolute time**).

relative time
The chronological ordering of events without reference to their age in years.

absolute time
The ages of objects or events as measured in time units such as years; determined by radiometric-dating techniques.

Relative Time

Relative dating concerns the chronological ordering of features and events without regard to their age in years. The general procedure is to work out the relative ages of the rocks in one locality or region and then to compare, or correlate, these rocks with those of other regions. The skills required depend ultimately on the few commonsense principles described in this chapter.

Deciphering Local Geologic History

How do geologists decipher the age relations and geologic history of a region? The first step is to identify the basic rock units, or **formations,** present. Formations are mappable bodies of rock that have distinctive mineralogical, structural, or textural characteristics

formation
A distinctive mappable rock unit.

◄ A geologist examines an instant, frozen in time some two hundred million years ago when a dinosaur left tracks across a mudflat, now preserved as an ancient rock layer in the Painted Desert, Arizona. Behind the geologist are younger layers, worn back to expose the tracks; beneath the tracks are older layers. What can this sequence of strata reveal about the region's past?

that set them apart from the rock bodies with which they are in contact. Often, a formation consists of a single igneous, metamorphic, or sedimentary rock type, such as granite, slate, or sandstone. But it may also consist of a number of closely related rocks examined as a single unit—for example, a sequence of alternating shale and sandstone strata. The point is that rocks of a specific formation have a common origin.

The second step in deciphering the geologic history of a region is to place these formations in chronological order. Geologists begin by applying a few simple principles that relate the arrangement of rock bodies as observed in the field to their relative ages: original horizontality, superposition, and cross-cutting relationships.

Horizontality and Superposition

The character of a sedimentary rock depends upon its depositional environment. You may recall from Chapter 7 some of the many environments where sediments are deposited on today's Earth: alluvial fans, flood plains, deltas, the shallow continental shelves, and the deep-ocean floor. Though markedly different from one another in texture and mineralogy, these sediments all share an important trait: They are deposited in layers parallel to the Earth's surface—which, on our spherical planet, is taken to be a horizontal plane. This simple observation is the basis of the **principle of original horizontality,** which states that sediments are deposited in approximately horizontal layers; those found to deviate from this orientation have been disturbed at a later date.

Horizontality is a consequence of gravitation, in which all particles are attracted toward the Earth's center (the vertical direction) but strike the Earth's surface en route. Therefore, if a blizzard of particles settles out of water (or air), it will blanket the Earth's surface and form a horizontal layer. It is self-evident that the layer deposited next must rest upon the previously deposited layer. Thus, in a sequence of sedimentary rocks or lavas, each layer is younger than the layer beneath it and older than the one above it. This concept forms the **principle of superposition.** It holds for all layers deposited on the Earth's surface, no matter how thick the sequence, and it is the principle upon which nearly all relative-age relationships involving sedimentary rock are based.

Unconformities

The fact that superposition determines the order in which sedimentary rocks are deposited in a sequence does not imply that all layers have been deposited continuously over time. Indeed, the geologic record in any relatively large region is typically replete with gaps, as if pages were either torn from the book or were never included. A gap in the rock record is called an **unconformity;** it is identified by an erosional surface between rocks of markedly different ages, and it signifies a major depositional break between the rocks above and below that surface. Figure 16.1 shows three common unconformities, each of which represents a significant period of erosion.

The strata below the **angular unconformity** in Figure 16.1(a) had to have been folded, uplifted, and eroded prior to deposition of the strata above them. Missing from the record (the gap) is the time it took to accomplish these changes. The metamorphic rock below the **nonconformity** in Figure 16.1(b) formed deep within the crust, beneath several kilometers of rocks, all of which had to have been stripped away prior to deposition of the overlying strata. In this case, too, an immense time interval is missing from the record. The **disconformity** in Figure 16.1(c) shows that the underlying strata had to have been uplifted and eroded in order to produce the uneven surface upon which the overlying strata were deposited. If they are of marine origin, then the underlying rocks must also have been depressed below sea level following the erosion period.

Although an unconformity is evidence of a major gap in the record, we have no way of knowing the magnitude of that gap—that is, how many years are missing—unless we can date the ages of the rocks above and below the unconformity. Determining numerical age requires radiometric-dating techniques, as we will see later in this chapter.

---

principle of original horizontality
The principle that sediments are deposited in horizontal layers, parallel to the Earth's surface.

principle of superposition
The principle that in a sequence of sedimentary strata, the oldest layer is located on the bottom and is followed in turn by successively younger layers, up to the top of the sequence.

unconformity
An erosion surface bounded by rocks of markedly different age and signifying a break in the geologic record.

angular unconformity
A surface formed by the deposition of sediments on the eroded, upturned edges of older, tilted strata.

nonconformity
An unconformity with stratified rocks above and igneous or metamorphic rocks below.

disconformity
An unconformity between parallel sedimentary layers.

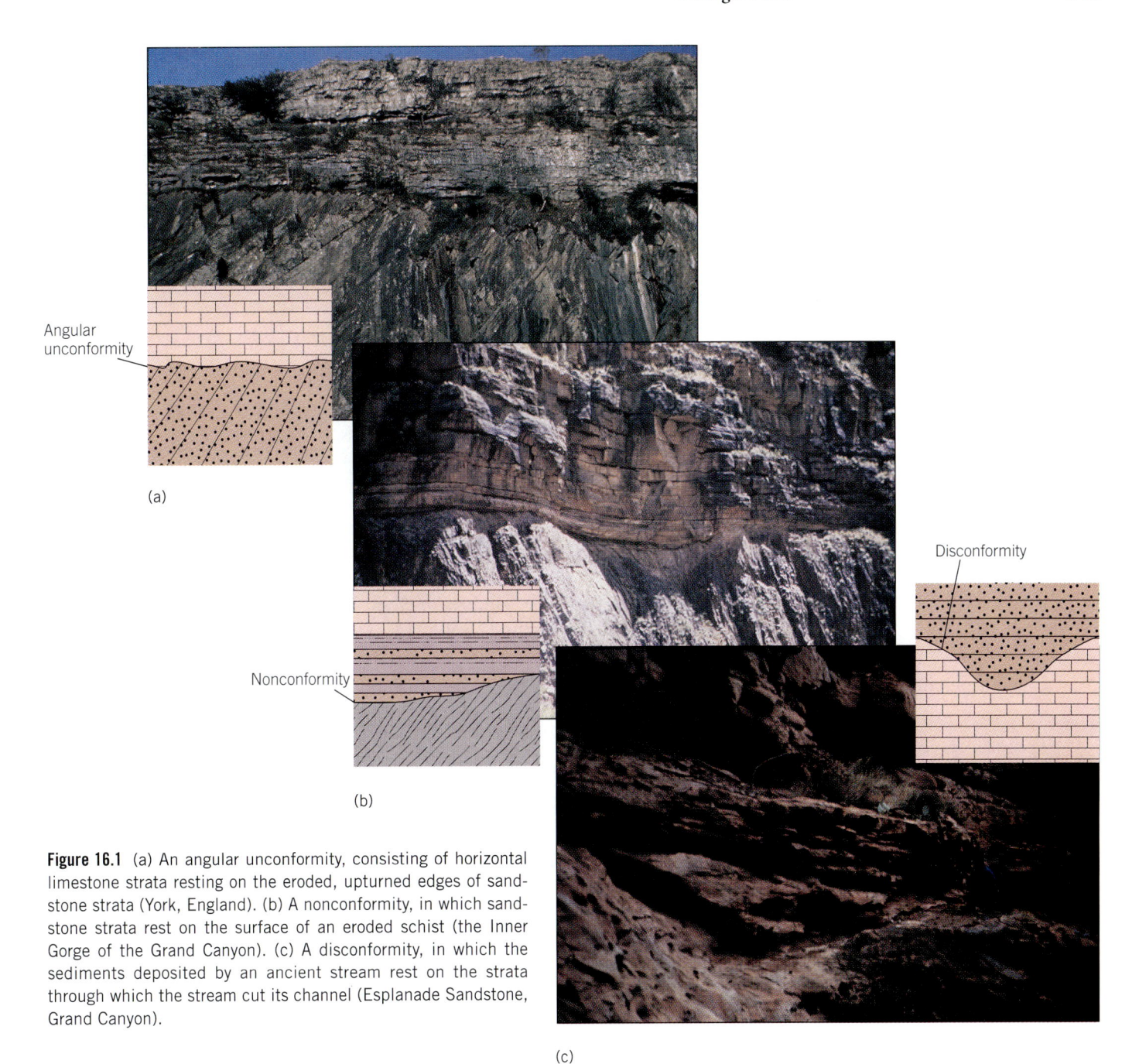

Figure 16.1 (a) An angular unconformity, consisting of horizontal limestone strata resting on the eroded, upturned edges of sandstone strata (York, England). (b) A nonconformity, in which sandstone strata rest on the surface of an eroded schist (the Inner Gorge of the Grand Canyon). (c) A disconformity, in which the sediments deposited by an ancient stream rest on the strata through which the stream cut its channel (Esplanade Sandstone, Grand Canyon).

Cross-Cutting Relationships

Not all rocks are sedimentary, of course. Suppose, during fieldwork, we come across an outcrop like the one shown in Figure 16.2, in which an igneous dike cuts across a sequence of sedimentary rocks. Close inspection reveals that a halo of baked sedimentary rock surrounds the dike, and inclusions of dislodged sedimentary rock (xenoliths) are found within the dike. This evidence tells us that the dike formed when magma intruded the formation and later cooled—in other words, it is younger than the strata. We notice also that both dike and strata have been displaced by a fault; so we conclude that the fault is younger than these features. In working out the sequence of events recorded at this outcrop, we have applied the **principle of cross-cutting relationships,** which states that a fault or intrusive body is younger than the rocks it cuts across or intrudes.

principle of cross-cutting relationships
The principle that an intrusion or fault is younger than the rock that it cuts.

Figure 16.2 The principle of cross-cutting relationships applied to a diagram of a road cut. The dike, an igneous intrusion, cuts across—and is therefore younger than—the sedimentary strata. The fault displaces the intrusion and the strata and is younger than both of them.

Correlation

Thus far we have seen that one of the first tasks for geologists studying a region is to identify the formations present and determine their chronological order. The next task is to correlate the formations with those of other regions. **Correlation** is the process of demonstrating equivalence, and the equivalence we are most concerned with here is age equivalence.

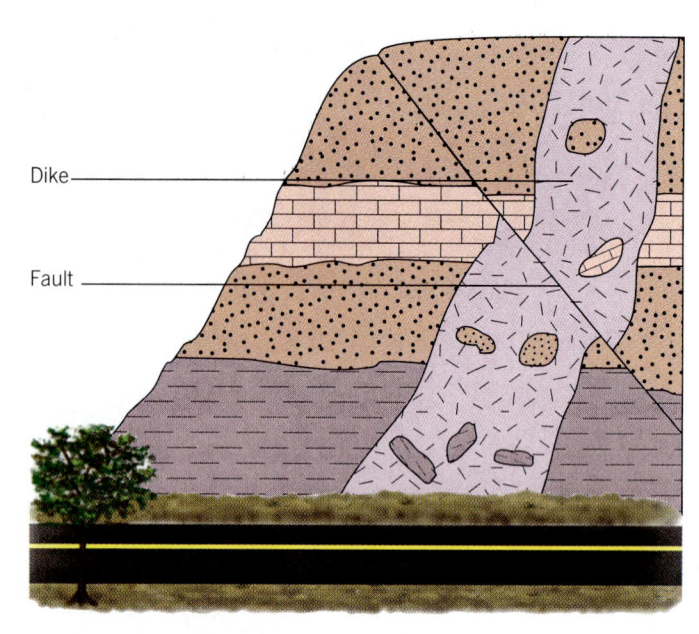

How could geologists prove that formation A of region 1 is equivalent to formation B of region 2 (Figure 16.3)? The simplest method is to physically trace the strata from A to B. Geologists call this method "walking it out," or more formally, proving *lateral continuity* (Figure 16.3a). This method is effective in regions where bedrock is continuously exposed—for example, in the bare cliffs of the Grand Canyon. However, most rocks are not continuously exposed, because they have been eroded or covered (Figure 16.3b). In those cases, geologists must rely on their *physical similarity,* provided that the outcrops are closely spaced, or their *similarity of sequence* (Figure 16.3c).

Correlation on the basis of physical characteristics has great value in oil, mineral, and groundwater exploration, as well as in civil engineering work—for example, in tunneling, securing foundations, or storing hazardous waste. It is also an essential step in organizing data to interpret geologic history. But it does not necessarily establish age equivalence between formations that are widely separated and lacking physical connection, because the same kinds of rocks may have formed repeatedly throughout geologic time. In these cases, physical similarity merely points to similarity in the processes that have formed these rocks.

Stratigraphic Correlation and Faunal Succession

Toward the close of the eighteenth century, the surveyor and canal engineer William Smith made careful notes of the shells and impressions he collected at various levels of strata and over widely separated regions of the English countryside. On the basis of his observations, Smith proposed the **principle of faunal succession:** that each formation contains a unique assemblage of fossils and that the fossil assemblages succeed one another in an orderly and predictable fashion.

What had Smith correlated by comparing the fossils in strata of widely separated regions? Recall from superposition that younger strata overlie older strata in a sedimentary sequence. So the same must be true of the fossils preserved in the strata. Thus, in matching fossil assemblages, Smith was correlating strata of the same relative age. He had discovered that fossils were a means of objectively dating the relative age of strata and, by extension, the events recorded in the strata.

Faunal succession constitutes material proof of organic evolution, for there is no other reason that the fossils in older strata should be different from, and yet related to, the fossils in younger strata and that fossils should succeed one another in an orderly fashion the world over. Furthermore, evolution is a one-way street—never have scientists observed mammals to devolve back to reptiles or multicellular forms to devolve

---

correlation
The establishing of equivalence, either in age or rock type, of separated rock units.

principle of faunal succession
The principle that fossil organisms succeed one another in definite and recognizable order, so that rocks containing identical fossils are identical in age.

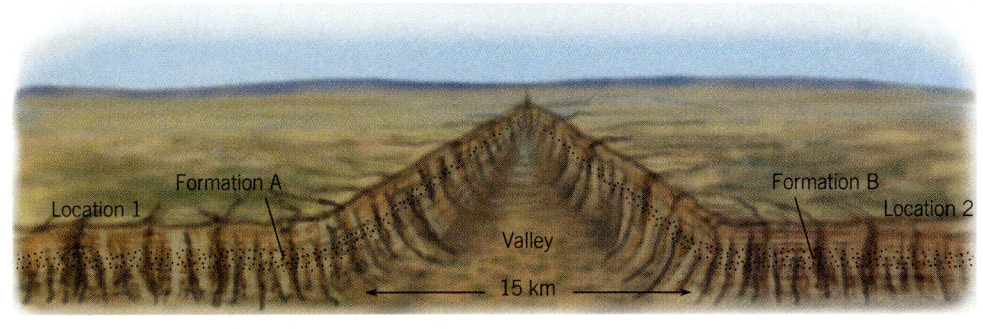

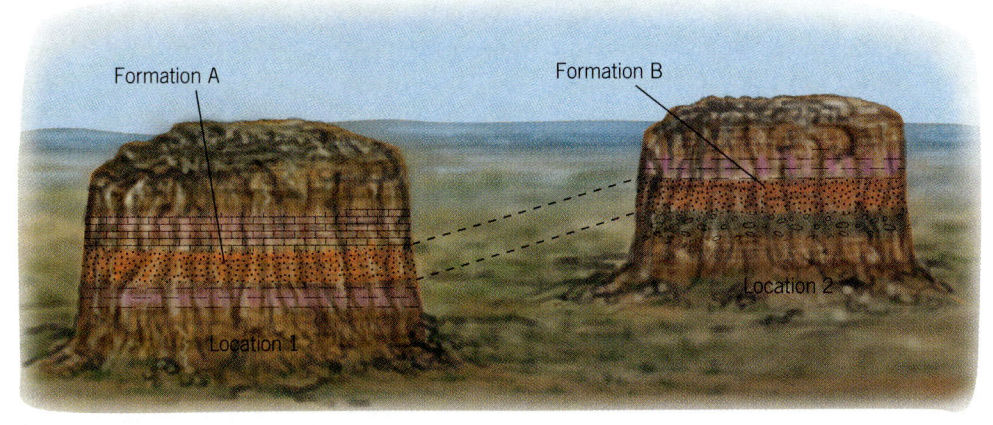

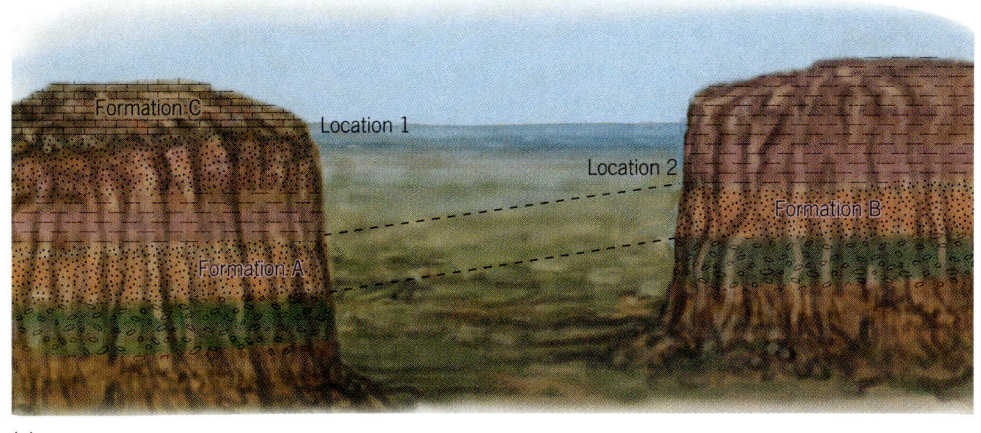

Figure 16.3 Three ways to correlate strata that are not distantly separated. (a) Lateral continuity. We can prove that formation A correlates with formation B by tracing A into B as we walk around the rim of the valley. (b) Physical similarity. These two isolated exposures are not widely separated, and their isolation was obviously caused by erosion. Thus the rocks of formations A and B, having the same distinctive characteristics, can be correlated by the similarity of their appearance. (c) Locations 1 and 2 are several kilometers apart. However, we can correlate the sandstone formations A and B on the basis of similarity of sequence—each is sandwiched between a conglomerate formation below and a shale formation above.

back to single-celled organisms. Likewise, geologists have never found bones of ancestral horses, tigers, and primates beneath dinosaur bones in undisturbed strata. That would be akin to an archeologist finding pocket calculators and VCRs in King Tut's tomb. For these and other reasons, biologists and paleontologists consider evolution a fact of Earth history; only the specific whys and hows are debated today.

The Relative Geologic Time Scale

Superposition and the fossil record form the basis of the *relative geologic time scale,* a chart that lists the events of Earth history in chronological order, oldest on bottom, youngest on top (Figure 16.4). The construction of the time scale would have been simple if a region existed where a sequence of strata contained an uninterrupted record of the most ancient to the most recent fossil assemblages. These assemblages could then have served as a standard to date the fossils and events of other regions. Unfortunately, no such region exists, so geologists were forced to do the next best thing and piece together the assemblages of many different regions, mentally stacking them so that the time scale or column contains no time gaps (Figure 16.5). Strata from the various localities that contain these assemblages constitute *standard sections* for given intervals of the time scale. Like the reference volumes in a library, their fossil assemblages serve as a means to correlate rocks of other regions.

Correlation Tools: Index Fossils and Overlapping Ranges

If fossils succeed one another in recognizable order, each species must thus have existed for a certain time interval and then become extinct. The first and last appearance of each fossil therefore represents a fixed range on the geologic time scale. An **index fossil** is particularly useful in correlation because it has a narrow range but a wide geographical distribution. A narrow time range ensures precision in dating, and wide distribution makes such fossils useful far from the standard sections—as far away as separated continents. Fossil A in Figure 16.6 is an index fossil because it meets these specifications. Fossil B, however, has a broader time range and does not qualify as an index fossil.

Another way of correlating strata is to look for fossils whose ranges overlap. For example, there is a narrow time interval that fossils C and D in Figure 16.6 have in common. Therefore, the age of the stratum that contains both of them can be bracketed.

The Terminology of the Time Scale

The relative time scale is divided into eons, eras, periods, and epochs, in order of decreasing magnitudes of time (see Figure 16.5). The term **Precambrian** refers to the time preceding the first appearance of abundant, easily recognized fossils.

Precambrian time is divided as follows: the Hadean eon (4600 to 3960 m.y.a.), which marks

index fossil
A fossil used to accurately establish the relative age of the stratum within which it is found.

Precambrian
All geologic time prior to the beginning of the Paleozoic era (about 570 million years ago).

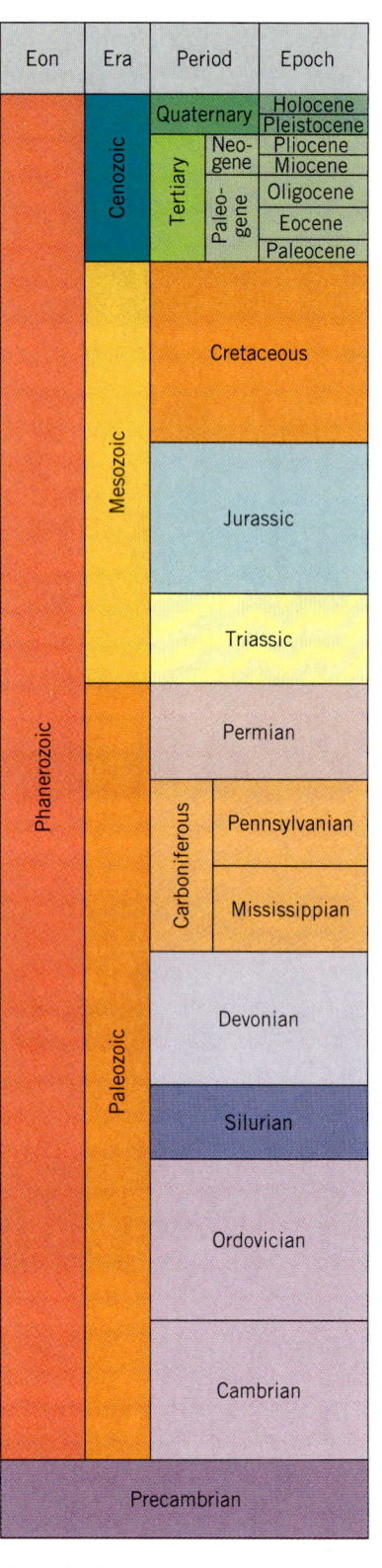

Figure 16.4 The relative geologic time scale. The scale is divided into eons, eras, periods, and epochs, in order of decreasing time magnitude.

Eon	Era	Period	Epoch
Phanerozoic	Cenozoic	Quaternary	Holocene
			Pleistocene
		Tertiary (Neogene)	Pliocene
			Miocene
		Tertiary (Paleogene)	Oligocene
			Eocene
			Paleocene
	Mesozoic	Cretaceous	
		Jurassic	
		Triassic	
	Paleozoic	Permian	
		Carboniferous (Pennsylvanian)	
		Carboniferous (Mississippian)	
		Devonian	
		Silurian	
		Ordovician	
		Cambrian	
	Precambrian		

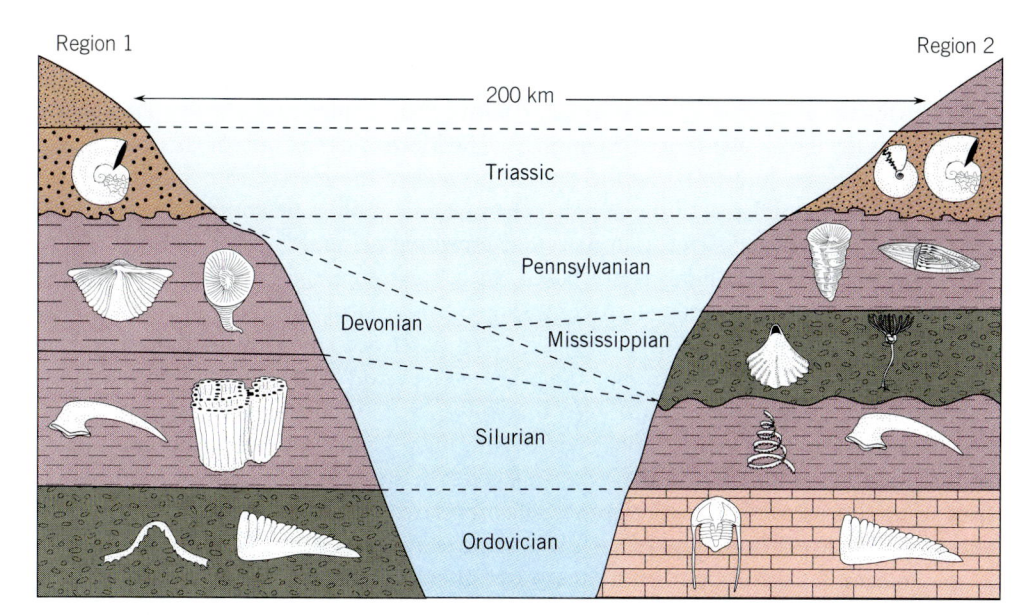

Figure 16.5 Similarity of fossil assemblages can be used to determine the relative ages of strata in widely separated regions. Between them, the two regions contain an unbroken record from the Ordovician through the Pennsylvanian periods.

the interval between the formation of the Earth and the oldest known Earth rocks; the Archean eon (3960 to 2500 m.y.a.), which is characterized by primitive bacteria and algae; and the Proterozoic eon (2500 to 570 m.y.a.), which is characterized by the development of the soft-bodied, multicelled organisms. Not until well into the twentieth century did radiometric studies reveal that the Precambrian comprises *about 85 percent of geologic time.*

Sedimentary rocks of the **Phanerozoic eon** contain abundant, easily recognizable fossils—virtually the entire record of advanced life. (The name *phanerozoic* is derived from the Greek words meaning "visible life.") This eon is divided into the Paleozoic, Mesozoic, and Cenozoic eras, on the basis of the fossil content of rocks deposited during those times.

Phanerozoic eon
That part of geologic time represented by rocks in which the evidence of life is abundant.

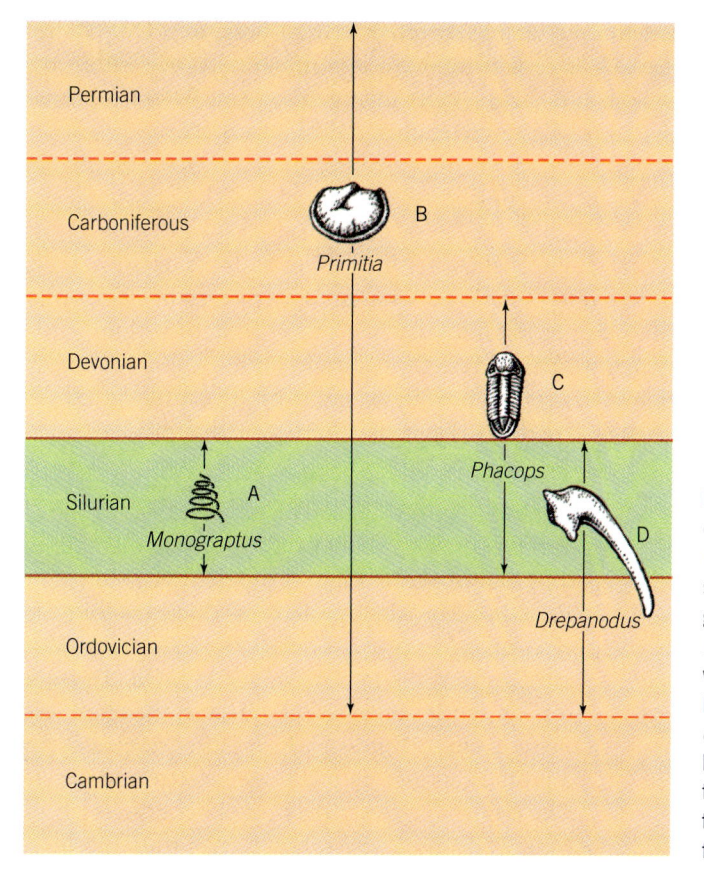

Figure 16.6 Correlation through the use of index fossils and overlapping ranges. *Monograptus* (fossil A) is an index fossil that inhabited a narrow interval of geologic time (the Silurian period). *Primitia* (fossil B), by contrast, has a wide range and is less useful for correlation purposes. *Phacops* and *Drepanodus* (fossils C and D, respectively) have different ranges, but because these ranges overlap, they can be used to bracket the age of rocks where both fossils are found.

Paleozoic era
The second of the geologic time eras, extending from the end of the Precambrian (about 570 million years ago) to the beginning of the Mesozoic (about 225 million years ago).

Mesozoic era
The era preceding the Cenozoic, extending from about 225 million years to 65 million years ago.

Cenozoic era
The most recent of the four eras of geologic time, beginning at the end of the Mesozoic era (65 million years ago) and extending to the present.

The boundaries of these eras are marked by major changes in the presence of life-forms, including both extinctions and the development of new lineages. The **Paleozoic era** is the age of extinct shelled marine organisms, primitive fish, amphibians, reptiles, and plants. (*Paleozoic* is the Greek word for "ancient life.") The **Mesozoic era** ("middle life") is the age of reptiles (the best known being the dinosaurs) and of the invertebrate ammonite (the convoluted, ancient relative of the modern chambered nautilus and squid). The **Cenozoic era** ("recent life") is the age of mammals and flowering plants.

Boundaries between periods are based on less extreme change. The names of the periods were assigned by nineteenth-century geologists. Most were named for the places where the rocks representing these ages were first described or for distinctive rock types within these regions. For example, the Devonian period was named for rock strata exposed in Devonshire in southwestern England.

Periods are in turn divided into epochs. Although there are many epochs, only those of the Cenozoic era are listed in Figure 16.4. The Pleistocene epoch, for example, is the time of the ice ages—the most recent ice advance having retreated only 10,000 to 15,000 years ago. The lower boundary of the Pleistocene is marked by abrupt changes in microscopic marine fossils brought on by the onset of cold climates.

Geologists find it useful to distinguish between the abstract units of geologic time—eras, periods, and epochs—and the rocks deposited during those time units. For this reason, the rocks deposited during the Silurian period are referred to as belonging to the Silurian *system*. The rocks deposited during a lesser time unit, the epoch, are referred to as belonging to a *series*.

Absolute Time

Neither superposition, faunal succession, nor cross-cutting relationships enable us to date the age of rocks and geologic events in years. They merely enable us to arrange them in chronological order. To determine absolute age, we need geological clocks that record time at a steady pace over long periods without being affected by heat, pressure, chemical reactions, or other crustal processes. Only radioactive atoms—which are preserved in certain minerals and, in some cases, within organic remains, such as bone, wood, and shells—meet this requirement. Before we consider radiometric dating, then, let us briefly review atoms and radioactivity.

Modes of Decay

Recall from Chapter 4, Minerals, that atoms of the same element that differ in mass are referred to as *isotopes*. Some isotopes of certain elements are unstable, which means that they do not last forever. Their nuclei emit, or in some instances capture, subatomic particles, which changes their atomic number, their atomic mass, or both. (See Chapter 4 for review.) The process of emitting or capturing those subatomic particles is called **radioactive decay**. There are three principal modes of decay:

radioactive decay
The disintegration of certain isotopes by the emission of subatomic particles.

1. *Alpha decay,* in which two protons and two neutrons are emitted as an alpha particle (Figure 16.7a). The atomic mass of the isotope is thus decreased by 4, and the atomic number by 2. Because the atomic number is changed, the isotope is transformed into a different element.
2. *Beta decay,* in which a neutron decays within the nucleus to a proton and an electron: $n^0 \to p^+ + e^-$ (Figure 16.7b). However, the electron, or beta particle, escapes from the nucleus and is emitted. The atomic mass is not changed, but one proton is added to the nucleus; therefore, the element is transformed.
3. *Beta (electron) capture,* which is the reverse of beta decay (Figure 16.7c). Here a proton captures an inner-shell electron to produce a neutron ($p^+ + e^- \to n^0$). The element is thus transformed because its atomic number has been decreased by 1. The atomic mass, however, is unchanged.

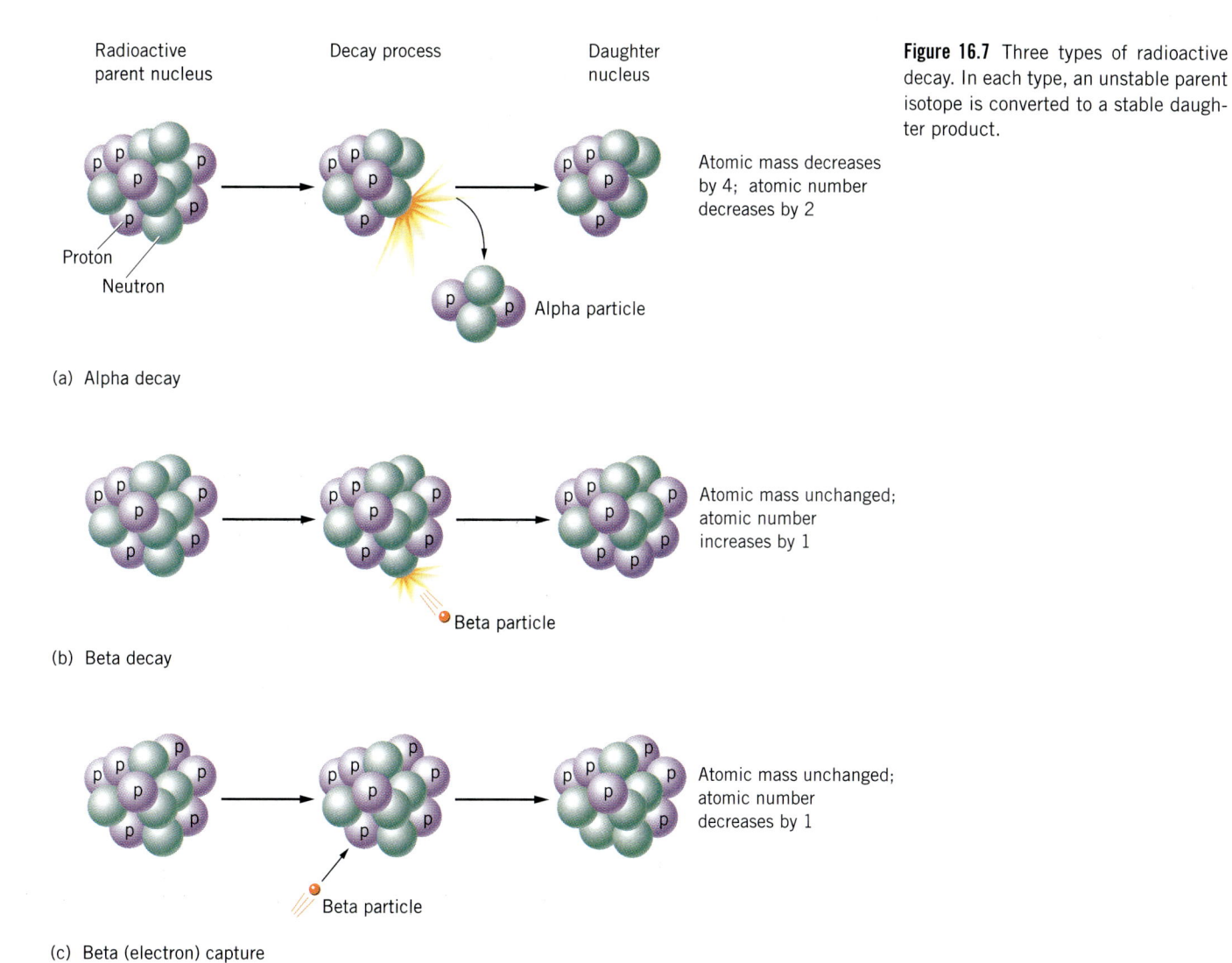

Figure 16.7 Three types of radioactive decay. In each type, an unstable parent isotope is converted to a stable daughter product.

(a) Alpha decay — Atomic mass decreases by 4; atomic number decreases by 2

(b) Beta decay — Atomic mass unchanged; atomic number increases by 1

(c) Beta (electron) capture — Atomic mass unchanged; atomic number decreases by 1

All radioactive decay, including electron capture, is accompanied by the release of energy in the form of heat. This heat is responsible for much of the Earth's high internal temperature.

As a result of emitting or capturing subatomic particles, a radioactive **parent isotope** is eventually transformed into a stable (nonradioactive) **daughter product.** The transformation may occur in a single step, or it may require a series of steps. In either case, parent and daughter form an identifiable pair within a sample. For example, the uranium isotope uranium-238 always decays to lead-206 after five alpha particle emissions, and rubidium-87 decays to strontium-87 after a single beta emission. For potassium-40, there are two daughter products: argon-40 and calcium-40; nevertheless, the percentage of each is well known and can be identified.

parent isotope
An isotope undergoing radioactive decay.

daughter product
An element formed from the decay of a parent isotope.

The Decay Principle

The fundamental principle of radiometric dating is that a given parent isotope will decay into a stable daughter product at a constant rate. The age of a sample, then, can be determined by knowing the decay rate of the parent and the amount of parent and daughter in the sample. The number of atoms that decay is proportional to the number of atoms that remain. That is, if more atoms remain, more atoms decay; if fewer

Figure 16.8 Hypothetical illustration of the half-life principle. A 16-gram sample of parent isotope A decays to daughter product B and has a half-life of 2 hours. If the sample is collected at 12 P.M., the end of the first half-life interval occurs at 2 P.M., and 50 percent of the parent remains; at the end of the second half-life interval (4 P.M.), 25 percent of the parent remains; and at the end of the next half-life interval (6 P.M.), 12.5 percent remains. Daughter atom B accumulates at the same rate that the parent decays. Therefore, if we examine a sample containing 2 grams of parent isotope A and 14 grams of daughter product B, we will know that the sample is 3 half-lives, or 6 hours, old.

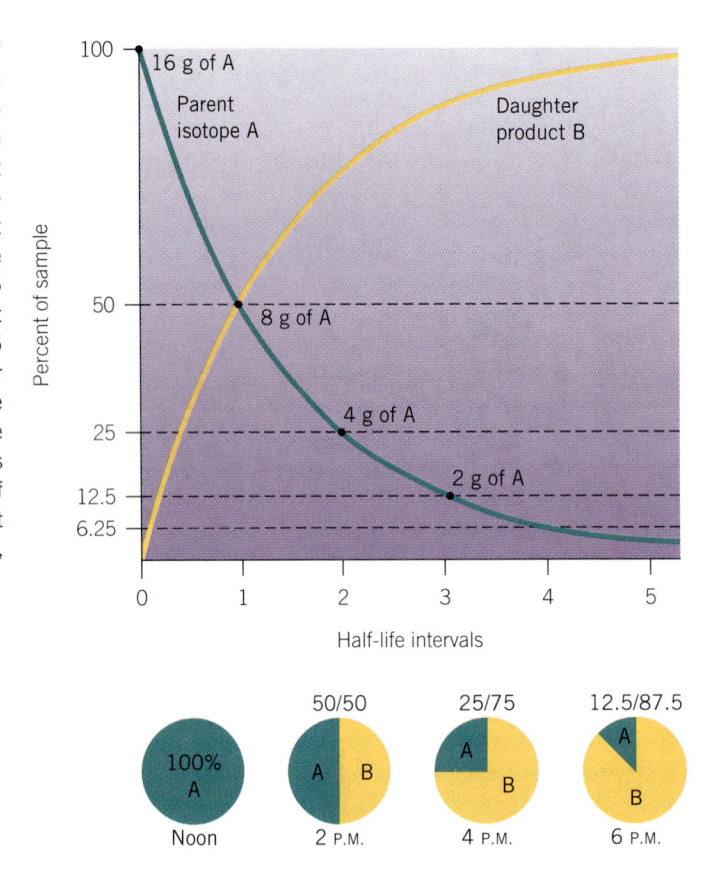

half-life
The time required for half of a given parent isotope to decay to its daughter product.

remain, fewer decay. The decay rate is expressed in terms of the **half-life** of the parent, which is the time it takes for half (or 50 percent) of the remaining parent to decay. Figure 16.8 illustrates how the half-life principle is used to determine the age of a sample.

Radioisotopes Useful in Dating

Table 16.1 lists the radioactive isotopes that are useful in dating ancient objects, together with their half-lives, daughter products, mineral occurrences, and effective time ranges. *Effective time range* is the interval over which a radioactive isotope yields useful dates. For example, the range of carbon-14, whose half-life is 5730 years, may stretch back 80,000 years; beyond that, the amount of carbon-14 remaining would be below the detection limits of our instruments. At the other extreme is uranium-238, which decays to lead-206 and has a half-life of 4.5 billion years. This isotope would not be useful in dating a 10,000-year-old lava flow because lead-206 would not have accumulated in measurable amounts. The effective range of uranium-238 begins with objects that are about 10 million years old and dips far back into the deep well of time. The effective time ranges of radioactive isotopes are not constant but expand with improved analytical techniques.

In general, the most reliable dates are derived from the minerals of igneous rocks, because radioactive isotopes are trapped within them at the time they crystallize from magma. Each crystallization is a clear-cut event. The parent-daughter ratio gives the time elapsed since the parent element began ticking away within the mineral. However, not every igneous mineral that contains radioactive isotopes is good for dating purposes; it must have remained impervious to weathering or physical damage throughout its history, so that no parent or daughter isotopes can have entered or escaped. Otherwise, chemical analysis will yield a parent-daughter ratio that does not reflect the correct age of the mineral.

Table 16.1 Some Parent and Daughter Isotopes Used for Radiometric Dating

Parent Isotope	Daughter Isotope	Half-Life (years)	Occurrence in Minerals and Other Materials	Effective Range (years)
Rubidium-87	Strontium-87	47 billion	Muscovite, biotite, lepidolite, microcline, glauconite, whole metamorphic rock	>10 million
Uranium-238	Lead-206	4.5 billion	Zircon, uraninite, pitchblende	>10 million
Uranium-235	Lead-207	700 million	Zircon, uraninite, pitchblende	>10 million
Thorium-232	Lead-208	14 billion	Zircon, uraninite, pitchblende	>20 million
Potassium-40	Argon-40, calcium-40	1.3 billion	Muscovite, biotite, hornblende, glauconite, sanidine, whole volcanic rock	>100,000
Carbon-14	Nitrogen-14	5730	Organic materials, glacial ice, groundwater, ocean water	0–80,000

Uranium- (and Thorium-) to-Lead Dating

Zircon is a tough, hard, chemically resistant mineral that is a minor constituent of granite and other igneous rocks. Usually, it contains small quantities of uranium and thorium—both of which decay to lead—and it is widely used in radiometric dating. As in all uranium-bearing minerals, zircon contains closely related parent-daughter pairs: uranium-235 to lead-207 and uranium-238 to lead-206. Each pair has a different half-life, and each can be used to date the mineral.

Because uranium-235 decays more rapidly than uranium-238, lead-207 accumulates in the mineral more rapidly than lead-206. Thus the changing ratio between these daughter isotopes also reveals the mineral's age (Figure 16.9).

Potassium-to-Argon Dating

Potassium is an abundant element, found in a wide variety of common minerals such as feldspar, mica, and amphibole and in the rarer sedimentary mineral glauconite. It consists of three isotopes: potassium-39, potassium-40, and potassium-41. Only potassium-40, which decays to argon-40, is radioactive. Because isotopes of the same element mix freely, potassium-40 is present in minerals containing potassium. Thus, the potassium-argon method has wide application in radiometric dating.

Rubidium-87–to–Strontium-87 Dating

Though relatively rare, rubidium and strontium are chemically similar to potassium and calcium, respectively. Both occur in a wide variety of common minerals, such as feldspar and mica. The 47-billion-year half-life of rubidium-87, which decays to strontium-87, makes it well suited for dating Precambrian rocks.

Carbon-14 Dating

The 5730-year half-life of carbon-14 makes this isotope extremely useful for measuring relatively recent events; in fact,

Figure 16.9 Because lead-207 and lead-206 are produced at different rates, the ratio of the two daughter products changes with time. Thus the ratio of the two lead isotopes extracted from a sample gives the sample's age.

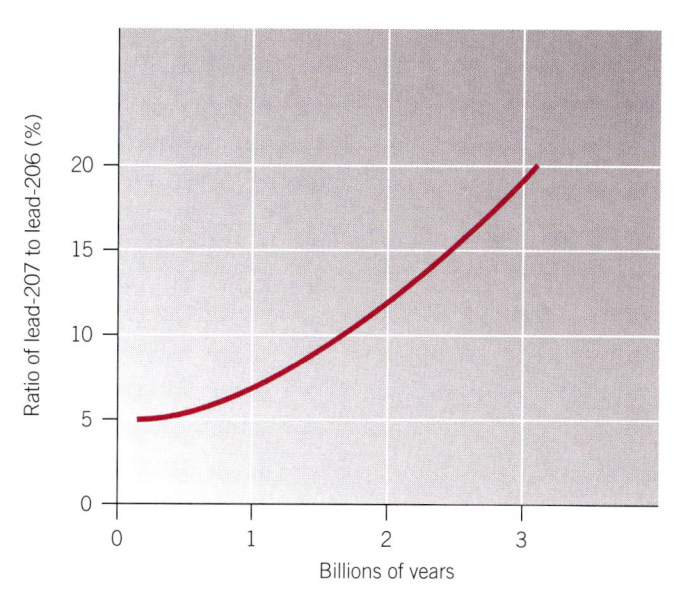

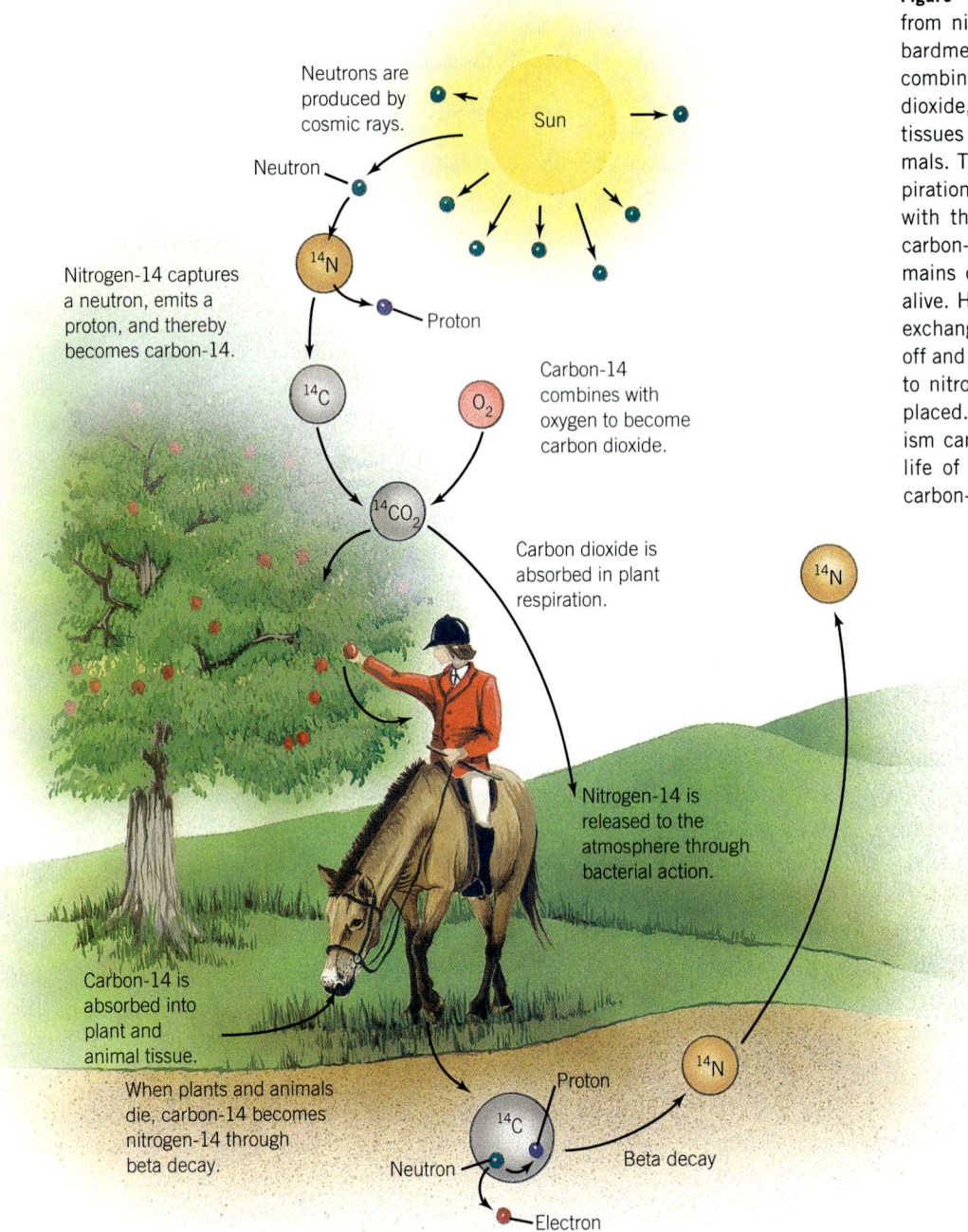

Figure 16.10 Carbon-14 is produced from nitrogen-14 by cosmic ray bombardment. The carbon isotope then combines with oxygen to form carbon dioxide, which is incorporated into the tissues and shells of plants and animals. Through photosynthesis and respiration, there is constant interchange with the atmosphere, so the ratio of carbon-14 to other carbon isotopes remains constant while the organism is alive. However, upon death and burial, exchange with the atmosphere is cut off and the carbon-14 that decays back to nitrogen in the organism is not replaced. Therefore, the age of the organism can be determined from the half-life of carbon-14 and the amount of carbon-14 remaining.

we can go back approximately 75,000 years before it decays into quantities too small to measure. This time is sufficient for dating such things as

- the last glacial advance and retreat
- campsites of early Homo sapiens that allow us to trace migration routes
- historical documents made of skin (such as the Dead Sea Scroll parchments)
- climatic and sea-level changes dating back well into the last ice age
- plant and animal migrations
- volcanic eruptions
- deep-ocean circulation
- earthquake recurrence intervals

Carbon-14 is derived from nitrogen-14 through bombardment by cosmic rays in the upper atmosphere (Figure 16.10). But carbon-14 atoms immediately begin to decay back to nitrogen-14 atoms; as with all unstable isotopes, the number of atoms that decay is proportional to the number of atoms present. Thus a balance is reached eventually between the carbon-14 that is manufactured and the carbon-14 that decays. At that point, carbon-14 in the atmosphere remains constant.

Carbon-14 is also in constant ratio to the dominant carbon isotope in the atmosphere, carbon-12. Because carbon-12 and carbon-14 are chemically identical, they combine with oxygen in the same ratio to form carbon dioxide (CO_2). Through photosynthesis and respiration, both are rapidly incorporated (or fixed) into the cells and tissues of plants and animals. And since interchange between organisms and atmosphere is open, the carbon-14/carbon-12 ratio within these living things equals that of the atmosphere.

Because photosynthesis and respiration end permanently with death, carbon-14 in a dead organism is cut off from the atmosphere. As it decays to nitrogen-14, the number of carbon-14 atoms in the organism declines. We can calculate the time since the organism died by comparing the percentage of carbon-14 relative to carbon-12 in the sample with the percentage of carbon-14 to carbon-12 in the atmosphere. For example, a drop of 50 percent means the sample is one half-life interval, or 5730 years, old.

Dating the Geologic Time Scale

Remember that the geologic time scale was worked out on the basis of superposition and fossil assemblages. How, then, do geologists date, in years, the boundaries between eras and periods of the scale?

One technique might be to analyze radioactive isotopes in the rock strata. Unfortunately, because extensive chemical weathering obscures their true age, sedimentary rocks do not lend themselves to accurate radiometric dating, so the next best thing is *bracketing*—combining the radiometric ages of igneous intrusions with the principles of cross-cutting relationships. Figure 16.11 shows that rock stratum B overlies intrusion

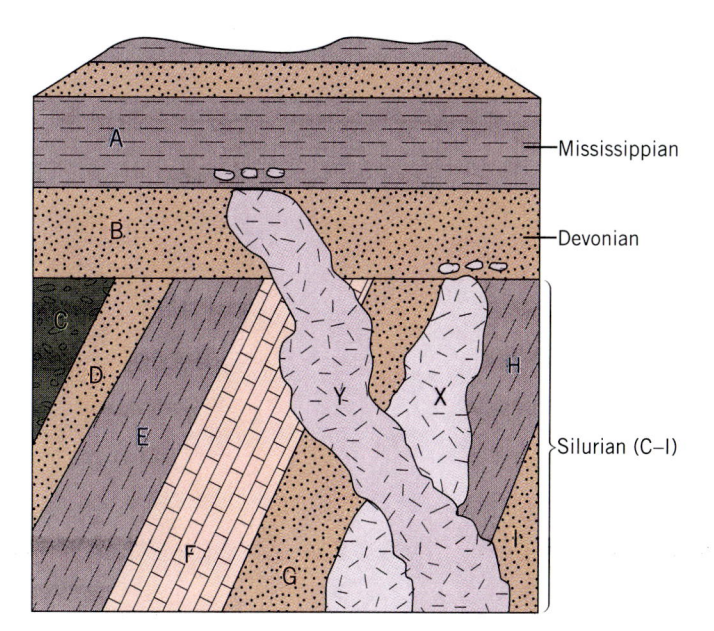

Figure 16.11 Cross-cutting relationships and radiometric dating are used to bracket strata whose relative ages were determined by fossil analysis. Through radiometric dating, we can determine that intrusion X is 400 million years old and intrusion Y is 350 million years old. Therefore, Layers C–I (of the Silurian period) are greater than 400 million years old. Layer B (of the Devonian period) is between 400 and 350 million years old, and Layer A (of the Mississippian period) is less than 350 million years old.

356 CHAPTER 16

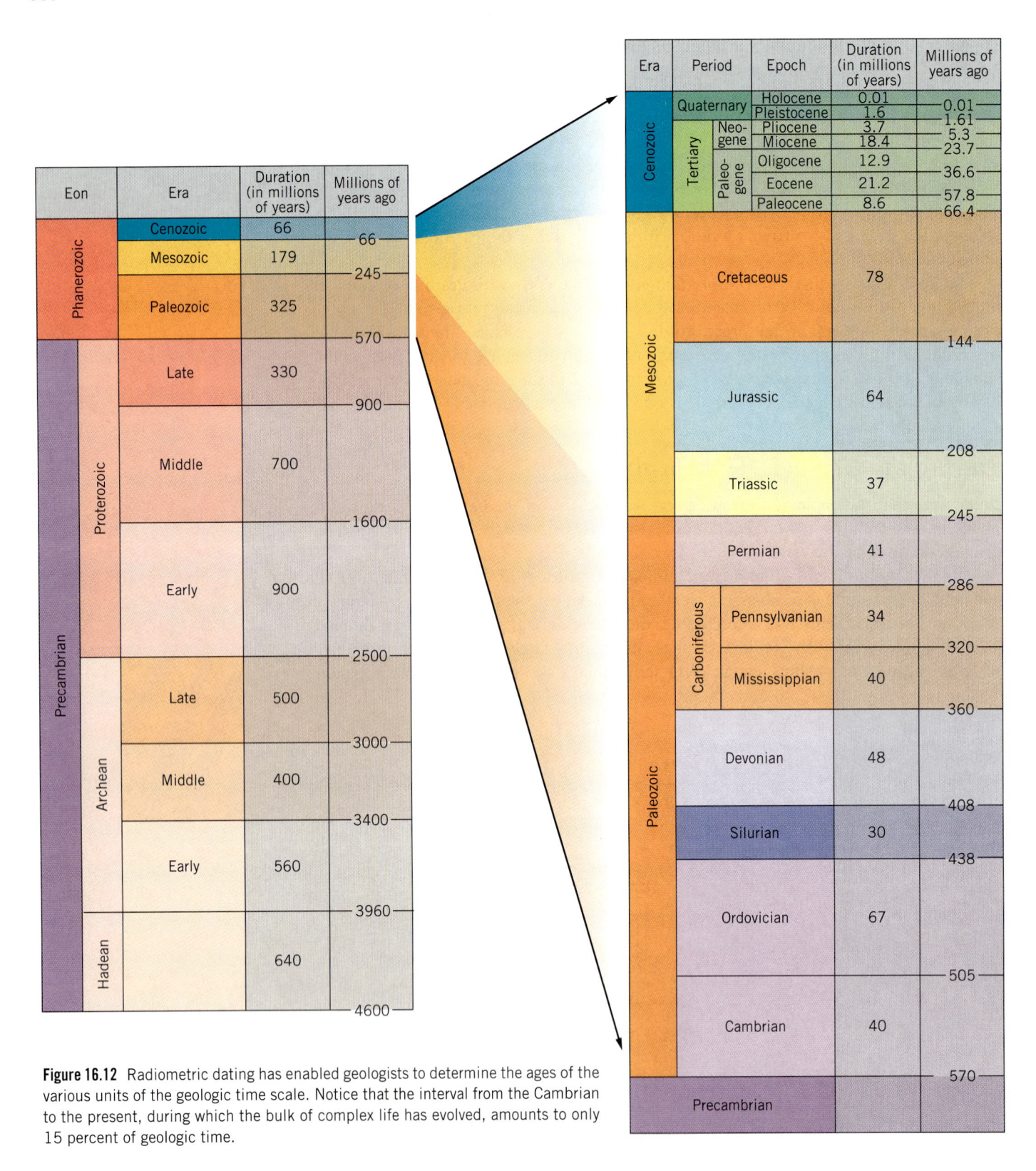

Eon	Era		Duration (in millions of years)	Millions of years ago
Phanerozoic		Cenozoic	66	—66—
		Mesozoic	179	—245—
		Paleozoic	325	—570—
Precambrian	Proterozoic	Late	330	—900—
		Middle	700	—1600—
		Early	900	—2500—
	Archean	Late	500	—3000—
		Middle	400	—3400—
		Early	560	—3960—
	Hadean		640	—4600—

Era	Period	Epoch	Duration (in millions of years)	Millions of years ago
Cenozoic	Quaternary	Holocene	0.01	—0.01—
		Pleistocene	1.6	—1.61—
	Tertiary (Neogene)	Pliocene	3.7	—5.3—
		Miocene	18.4	—23.7—
	Tertiary (Paleogene)	Oligocene	12.9	—36.6—
		Eocene	21.2	—57.8—
		Paleocene	8.6	—66.4—
Mesozoic	Cretaceous		78	—144—
	Jurassic		64	—208—
	Triassic		37	—245—
Paleozoic	Permian		41	—286—
	Carboniferous (Pennsylvanian)		34	—320—
	Carboniferous (Mississippian)		40	—360—
	Devonian		48	—408—
	Silurian		30	—438—
	Ordovician		67	—505—
	Cambrian		40	—570—
Precambrian				

Figure 16.12 Radiometric dating has enabled geologists to determine the ages of the various units of the geologic time scale. Notice that the interval from the Cambrian to the present, during which the bulk of complex life has evolved, amounts to only 15 percent of geologic time.

X but is cut by intrusion Y. Clearly, B is younger than X and older than Y. If intrusion X is dated radiometrically at 400 million years, and intrusion Y at 350 million years, then the age of B must fall somewhere between the two. If the fossils in B prove to be of the Devonian period, then we have an approximate date for that period. Obviously, the closer we can bracket the strata in future studies, the more closely we will be able to date the Devonian period. By applying these methods, geologists have dated the entire geologic time scale (Figure 16.12).

The Age of the Earth

Radiometric dating has enabled us to determine the age of the oldest rocks that we can find. A few years ago, geologists thought that the 4-billion-year-old metamorphic and igneous rocks of Canada's Great Slave Lake region contained the Earth's oldest known objects. At present, that distinction belongs to Australia's Big Stubby deposit, which contains 4.3-billion-year-old zircons. Certainly, it is possible that we will find vestiges of older rocks, but geologists estimate the Earth's age as approximately 4.6 billion years. How do they justify this estimate if the Earth's oldest known minerals fall more than 300 million years short of this date?

The answer lies in the numerous meteorites they have dated. Originating in the asteroid belt beyond Mars, these meteorites strike the Earth's surface when their paths intersect the Earth's gravitational field. They are the oldest objects that have been physically dated, and the oldest of them invariably dates at 4.6 billion years. This result checks out using both uranium-lead and rubidium-strontium methods. Lunar soils also yield dates of about 4.55 billion years; these are the oldest materials brought from the moon. So presumably, the largest planetary bodies of the Solar System all solidified at about that time.

Some people argue on religious grounds that the Earth is only 5000 to 10,000 years old, and geologists are often drawn into public debates as advocates of their scientific estimates. Sometimes, the position of geologists in this debate is misunderstood—for in the final analysis, geologists have no stake in how old the Earth is. They simply want to *know* how old it is! If the Earth is only 5000 years old, as some who take a literal interpretation of the Scriptures claim, so be it. However, the evidence points overwhelmingly to the contrary.

STUDY OUTLINE

Measurement of **relative time** is based on the fact that sedimentary rocks are deposited in layers, younger above older, and contain fossils that are of distinctive ages. Because a specific radioactive element decays into a specific stable element at a known rate, both can be used to measure **absolute time,** the age of objects in years.

I. RELATIVE TIME

A. By determining the relative ages of rocks in one locality or region and then comparing or correlating them with those of other regions, geologists established the chronological ordering of geologic events in the Earth's history.

B. Deciphering local geologic history
  1. **Formations** are bodies of rock with distinctive characteristics that set them apart from neighboring rock bodies.
  2. The **principle of original horizontality** states that sediments are deposited in layers parallel to the Earth's surface and, according to the **principle of superposition,** each layer is younger than the layer beneath it and older than the one above it.
  3. Geologists can learn a great deal about the history of rock formations from the way in which the rock layers make contact. An **unconformity** is an erosion surface between layers of markedly different ages.
     a. In an **angular unconformity,** strata rest upon the upturned edges of strata beneath.
     b. In a **nonconformity,** strata rest on eroded metamorphic or igneous rocks.
     c. A **disconformity** is an irregular contact between parallel strata.
  4. The **principle of cross-cutting relationships** states that a fault or intrusive body must be younger than the rocks it affects.

C. **Correlation** is the process of demonstrating age equivalence. There are three methods of local correlation: lateral continuity, physical similarity, and similarity of sequence. Fossils are used to correlate widely separated rocks.

D. The **principle of faunal succession** states that fossil assemblages succeed one another in an orderly fashion. Strata of the same relative age can be correlated by matching fossil assemblages.
   1. Superposition and the fossil record form the basis of the relative geologic time scale, the events of the Earth's history listed in chronological order. Strata from various localities constitute standard sections for given intervals of the relative geologic time scale.
   2. **Index fossils** are useful in correlation because they have narrow ranges but wide geographical distributions.
   3. The relative time scale is divided into eons, eras, periods, and epochs, in order of decreasing magnitude. The **Precambrian** comprises 85 percent of the geologic time scale. The **Phanerozoic eon** is divided into the **Paleozoic, Mesozoic,** and **Cenozoic eras.**

II. ABSOLUTE TIME

A. To determine absolute age, geologists use radioactive atoms that decay at a steady pace over long periods.

B. The process by which unstable atoms emit or capture subatomic particles is called **radioactive decay.** There are three forms: alpha decay, beta decay, and beta (electron) capture.
   1. As a result of radioactive decay, a **parent isotope** is eventually transformed into a stable (nonradioactive) **daughter product.** The age of a sample is determined by the ratio of parent to daughter and by knowledge of the decay rate.
   2. The decay rate is expressed in terms of the **half-life** of the parent, which is the time it takes for half (50 percent) of the remaining parent to decay.

C. The effective time range is the interval over which a radioactive isotope yields useful dates.
   1. Some radioactive isotopes used in dating are uranium-235 and uranium-238, which decay to lead-207 and lead-206, respectively; potassium-40, which decays to argon-40; and rubidium-87, which decays to strontium-87.
   2. The 5730-year half-life of carbon-14 makes this isotope extremely useful for dating relatively young materials of organic origin.

D. *Bracketing* combines the radiometric ages of igneous intrusions with the principle of cross-cutting relationships to date the geologic time scale.

E. The 4.6-billion-year age of the Earth has been determined by dating meteorites from the Solar System and from lunar soil samples.

STUDY TERMS

absolute time (p. 343)	half-life (p. 352)
angular unconformity (p. 344)	index fossil (p. 348)
Cenozoic era (p. 350)	Mesozoic era (p. 350)
correlation (p. 346)	nonconformity (p. 344)
daughter product (p. 351)	Paleozoic era (p. 350)
disconformity (p. 344)	parent isotope (p. 351)
formation (p. 343)	Phanerozoic eon (p. 349)

Precambrian (p. 348)
principle of cross-cutting relationships (p. 345)
principle of faunal succession (p. 346)
principle of original horizontality (p. 344)

principle of superposition (p. 344)
radioactive decay (p. 350)
relative time (p. 343)
unconformity (p. 344)

Thought Questions

1. In what sense does the principle of faunal succession derive from the principle of superposition?
2. In what sense does the principle of superposition derive from the principle of horizontality?
3. A zircon fragment is found within a granite pebble within a well-cemented conglomerate within a folded sequence of strata. The zircon is radiometrically dated at 1 billion years. What can you say about the age of the granite pebble? The relative age of the conglomerate? Of the folded strata? What sequence of events can you deduce from these dates?
4. Two rock sequences separated by thousands of kilometers contain no fossils in common. Does this discrepancy necessarily mean that their ages differ? Why or why not?
5. Give three reasons that radiometric dates, when properly determined, are generally reliable.
6. Suppose that instead of Europeans, the Incas of South America had discovered the principles of relative dating and had applied them to the Earth. Suppose as well that they developed their own geologic time scale using their own names and standard sections. Would their scale differ from the present one? Explain.

Chapter 17

Subdivisions of the Continental Crust
  Cratons
  Mountain Belts
  Continental Margins
    Passive Margins
    Active Margins

Mountain Building and Plate Tectonics
  Subduction and Mountain Building

Mountain Building by Accretion
  Accreted Terranes of Western North America
  The Himalayan Orogeny
  The Appalachian Orogeny

Growth and Evolution of the Continental Crust

Formation of the Continental Crust

As we have seen, the present ocean basins are relatively ephemeral features that encompass only the most recent 5 percent of geologic time, whereas the continents contain traces of rocks 4 billion years old. The oceanic crust passes through relatively rapid cycles of birth, growth, and decline; but the continental crust endures. Too light to sink, it floats on the mantle surface and accumulates the records of "deep time."*

The dominant processes of continental evolution are accretion and synthesis, as juxtaposed to the divergence and spreading that characterize the evolution of the ocean basins. We will begin our exploration of the structure and evolution of the continental crust by examining the anatomy of its parts.

Subdivisions of the Continental Crust

Though varied in detail, the continental crust consists of just three units, each with its own characteristic combination of age, rock types, and structure. The units are cratons, mountain belts, and continental margins (Figure 17.1).

---

*As coined by John McPhee in his book *Basin and Range*.

◀ The world's highest peak, Mount Everest of the Himalayas, is composed of sedimentary rocks deposited in the long-vanished Tethys Ocean.

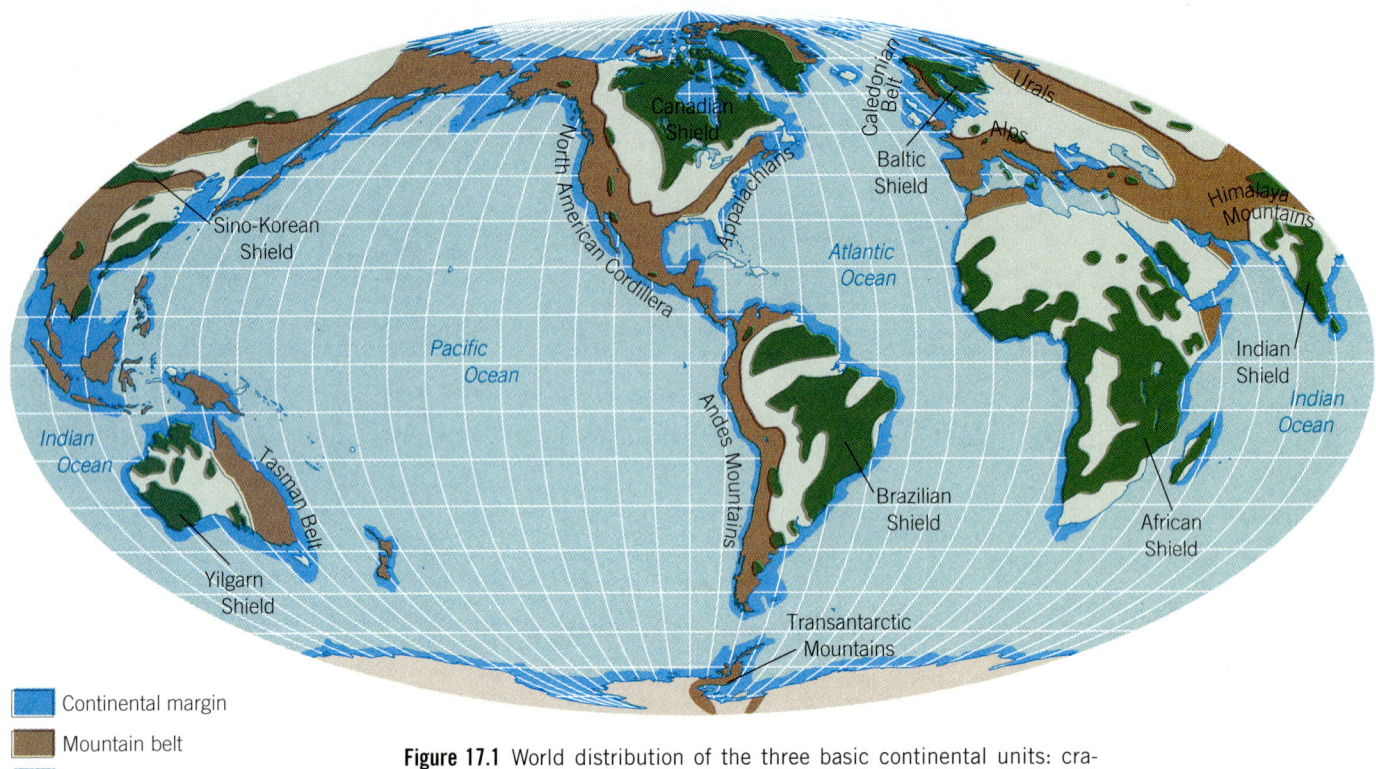

Figure 17.1 World distribution of the three basic continental units: cratons, mountain belts, and shields. The theory of plate tectonics explains the origins of these features and their role in continental evolution.

shield
A large region of exposed metamorphic and igneous basement rocks, generally having a gently convex surface and surrounded by a sediment-covered platform.

platform
That part of the craton consisting of essentially horizontal sedimentary strata overlying the older basement rocks.

craton
The stable core of the continental crust that includes basement rock, shield, and platform.

Cratons

All the continents consist of a more or less central foundation of ancient *basement rock.* The exposed regions of basement rock, the **shields,** are so called because they are broad, low, and very gently arched (Figure 17.2). Their relatively subdued topography belies their structural complexity, however, for shields are composed of the oldest and most complex rocks on Earth. Basically, they are mixtures of granites, gneisses, schists, ancient sedimentary rocks, and interbedded lavas.

Vast regions of basement rock covered by flat-lying or gently tilting sedimentary strata constitute the interior **platform** of the continent (Figure 17.3). The younger platform strata cover the ancient igneous and metamorphic rocks like a thin blanket. For the most part, they were deposited in shallow seas that washed over parts of the Precambrian shields at intervals between 600 million years ago and the more recent past. These strata are not products of today's oceans, although it is clear that many were formed in a manner similar to present continental shelf deposits. Platform and shield together form the **craton,** the core of the continent that has remained tectonically stable and suffered little or no deformation for a geologically long time-period—perhaps hundreds of millions of years. Figure 17.4 is a generalized cross section of the North American craton, which includes the Canadian Shield. The Cambrian strata of the platform that rest upon the shield are undisturbed, which indicates that the craton has been stable for at least 600 million years.

Mountain Belts

Continents also contain *mountain belts,* most of which weave graceful arcs along the outer edges of the cratons. Most people think of mountain belts as simply terrain of higher elevation than the surrounding region. Geologists, however, think about moun-

Formation of the Continental Crust 363

Figure 17.2 A LandSat image of the Canadian Shield, in the vicinity of northeast Manitoba. Pink granitic rocks intrude darker-colored metamorphic rocks. Lakes, their basins scoured in bedrock by glaciers, appear dark and accentuate the regional structure.

Figure 17.3 An aerial view of typical farmland in the midwestern United States. The flat topography is underlain by sedimentary rocks of the interior platform.

tains in terms of their structure—folds, faults, igneous intrusions, and metamorphism (Figure 17.5). Such features are often of greater significance than elevation, for they enable geologists to investigate the causes of mountain building. They also provide evidence of past mountain-building episodes, or **orogenies,** in regions that are no longer mountainous.

There are many kinds of mountains. Some are volcanic in origin, as we learned in Chapter 6. Others are primarily of tectonic origin and bear the distinctive marks of the kind of stress that created them. Fault-block mountains, for example, are the products of crustal extension and are associated with the rifting of divergent plate boundaries or back-arc basins. In this category are the East African Rift Valley, the Mid-Atlantic Ridge, and the mountains of the Basin and Range of Nevada (discussed in Chapters 2 and 15).

Still other mountain belts are the products of compression generated at convergent plate boundaries; they are characterized by folds, thrust faults, granitic batholiths, and

orogeny
The process of mountain building.

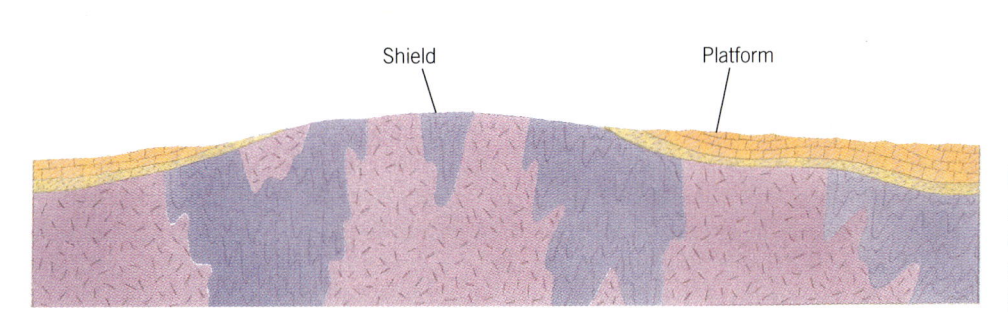

Figure 17.4 A cross section showing the relationship between complex shield and basement rocks and undeformed platform strata. (The vertical scale is greatly exaggerated.)

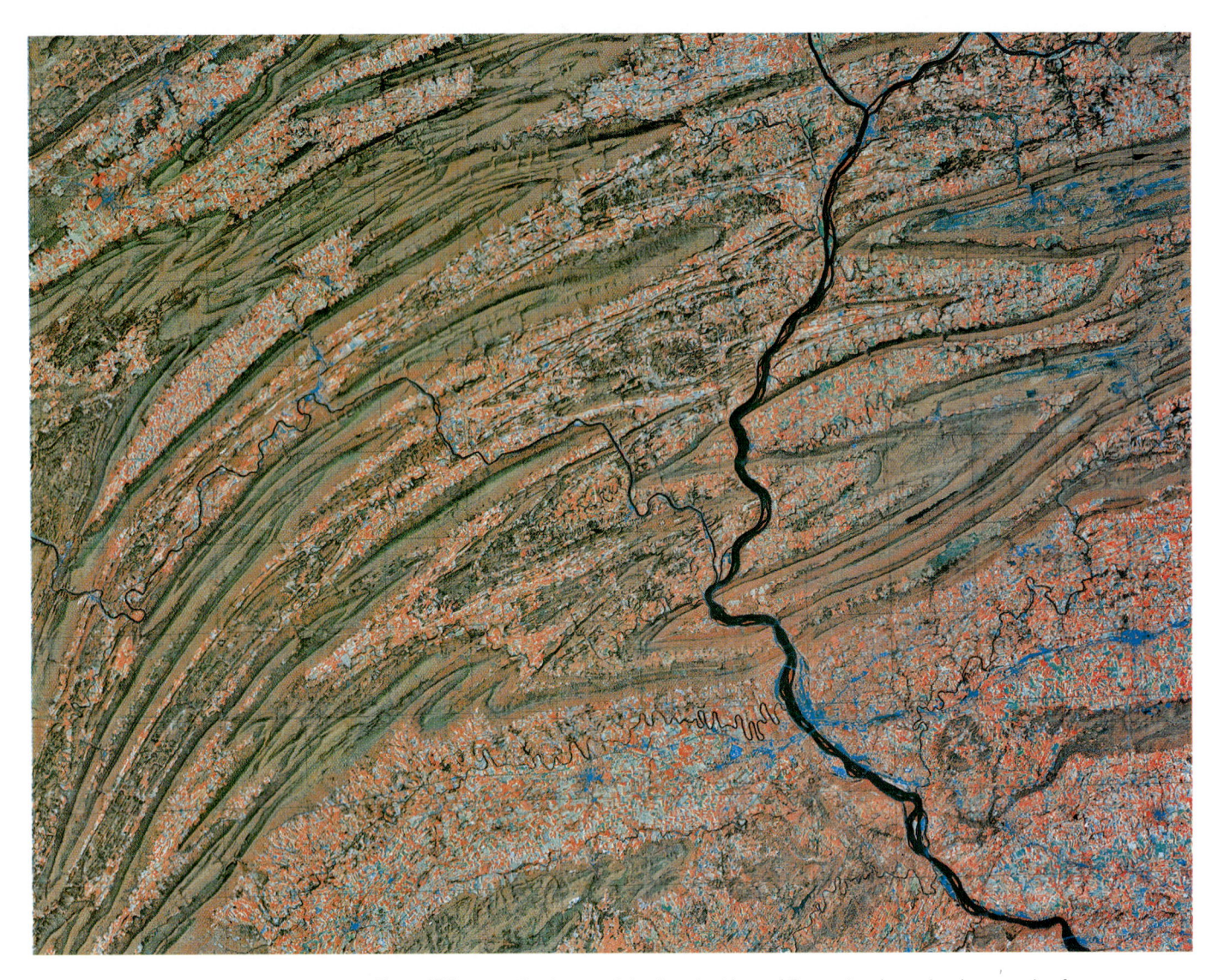

Figure 17.5 A satellite image of the Appalachians of Pennsylvania, a classic example of an extensively eroded fold-mountain belt. Resistant sandstone ridges delineate the zig-zag pattern of the plunging anticlines and synclines. The low-lying metamorphic rocks of the Appalachian-Piedmont are found to the southeast.

regional metamorphism. These are the mountains of the loftiest elevation and the most significant evidence of continent formation. When we examine these fold-mountain belts in detail, we find that they are commonly created over many millions of years and are the products of more than one episode of mountain building.

Continental Margins

Between the high-standing continental blocks and the deep-sea basins lie the continental margins, transition zones whose extent and shape are greatly influenced by their tectonic settings. As described in Chapter 2, active margins are commonly located close to plate boundaries, typically subduction zones, such as the one that runs directly offshore of western South America. Some active margins, such as the Gulf of California, border rift zones. Passive margins are located far from plate boundaries in regions of little tectonic activity. The eastern margins of North and South America, for instance, are thousands of kilometers from the actively spreading Mid-Atlantic Ridge.

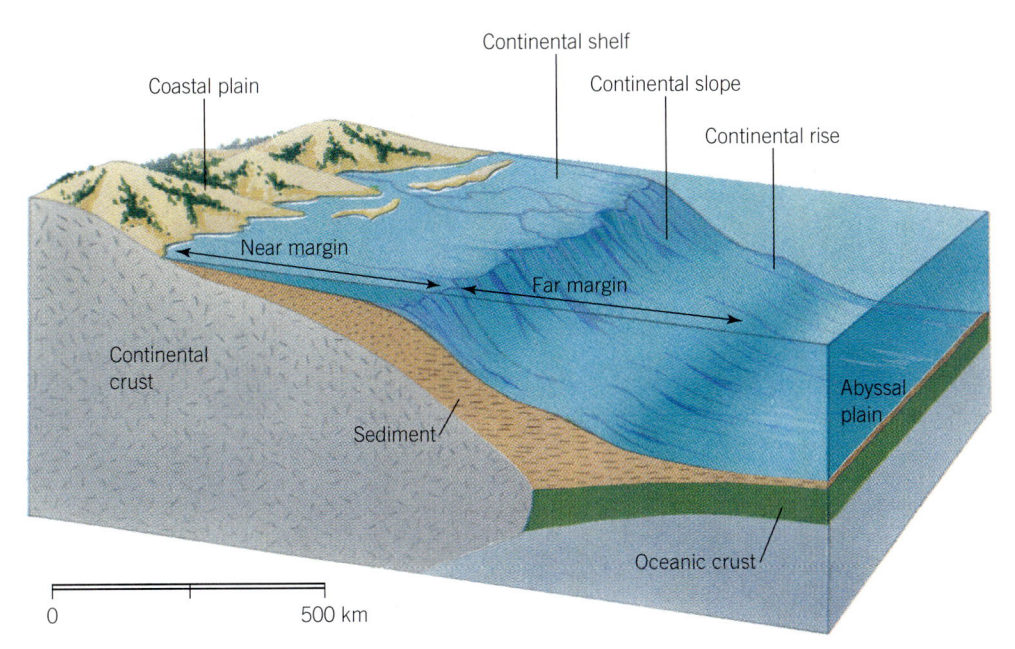

Figure 17.6 Anatomy of a passive margin. (The exaggeration of the vertical scale magnifies the ocean depth and makes the continental slope appear much steeper than it is in reality.)

Passive Margins

A passive margin consists of a thick wedge of land-derived sediment that slopes gently seaward so that its tapered edge rests on oceanic crust. The wedge is often thousands of meters thick, yet examination of the sediment reveals that much of it was deposited at (or close to) sea level. Therefore, the crust must have subsided as the load accumulated. Passive margins exhibit broad, flat continental shelves and gently seaward-tilting continental slopes that trail down to great depths. Their lower boundaries are marked by a continental rise at the base of the slope, and the rise tapers into the abyssal plains of the ocean floor.

A passive continental margin is divided into two broad environment zones: the *near margin*, which consists of the shallow continental shelf and upper slope, and the *far margin*, which consists of the deeper slope and rise (Figure 17.6). Near-margin sediments, brought to the coast by rivers and glacial ice, are generally clean, well-sorted gravels, sands, and interbedded shales that reflect the work of breaking waves, currents, storms, and tides. Where detrital sediments are lacking, and where the water is warm, extensive coral reef and other carbonate sediments may cover the near shelf. Far-margin deposits are a mixture of fine clay, silt, and microscopic shells that floated far from shore; poorly sorted muds and sands deposited by swift undersea turbidity currents that flowed down the continental slope; and deep-ocean assemblages of brown clay, lava, volcanic ash, and biochemical oozes.

The Atlantic and Gulf Coast margin of North America is a classic example of a passive continental margin. Because the sea has retreated from its maximum level of the past, wide areas of the shelf are exposed today. This exposed region of the shelf is called the Atlantic and Gulf Coastal Plain. There the marine sediments, which were originally set down in predominantly shallow waters, cover ancient igneous and metamorphic basement rocks.

Active Margins

An active continental margin has a steep gradient down to a deep trench and is located in a zone of tectonic instability. These factors combine to inhibit development of the elaborate features of passive margins. Instead, active margins are characterized by a narrow shelf and thick accumulations of sediment in the bordering trench. The sediments are brought to the trench by turbidity currents that are triggered by frequent earthquakes and rapid uplifts. It seems as though the vibrating crust were shaking the sediments down to the trench. Trenches that rim the Pacific Ocean act as traps that partially prevent the spread of sediment out onto the wider ocean floor.

Mountain Building and Plate Tectonics

In the course of surveying the subcontinent of India about one hundred fifty years ago, British engineer George Everest was surprised by the lateral pull of gravity on his instruments exerted by the nearby Himalaya Mountains. It was *less* than he and other scientists had expected, considering the great mass of matter that towered above the surrounding plain (Figure 17.7a). The cause of the discrepancy was the subject of spirited debate among geophysicists of the time. Gradually, a consensus emerged: the light crustal rock of the mountains extended down into the mantle, displacing the denser mantle rock and causing an overall deficiency of mass in the subsurface—hence, the weaker gravitational force (Figure 17.7b). Today, we have seismic evidence to support their conclusion that continental mountain belts like the Himalayas are composed of thickened crust. Viewed in cross section, their roots sink into the mantle upon which they float. Like any floating object, they rise where matter is eroded from them, just as the regions that receive the sediment are depressed in response. This state of balance due to buoyancy is called isostatic equilibrium, or *isostasy* (see Chapter 1).

In examining the Himalayas and similar mountains, we see that the reason for the crustal thickening is that the rock strata composing the mountains have been compressed perpendicular to the trend of the chain. This compression is evident in the structure of the chain, which is composed of folds and thrust faults. The thickened crust adjusts isostatically by sinking deeper while rising higher.

It took geologists over a century after George Everest's discovery to grasp the reason for the thickening and compression of the crust that creates mountain belts. When the theory of plate tectonics became accepted in the 1960s, it revolutionized the concept of mountain building, just as surely as it changed ideas about the origin of oceanic crust. In fact, oceanic crust and mountain building are united by plate tectonics.

Figure 17.7 (a) If the Himalayas simply sat on top of the mantle, a plumb bob would be deflected by the gravitational attraction of the mountains and the denser mantle rocks beneath them. (b) The actual angle of deflection is found to be much smaller, which suggests that the lighter rocks of the mountains sink deep beneath the visible mountain peaks and displace denser mantle rocks. The deficiency of mass in the subsurface leads to weaker gravitational attraction.

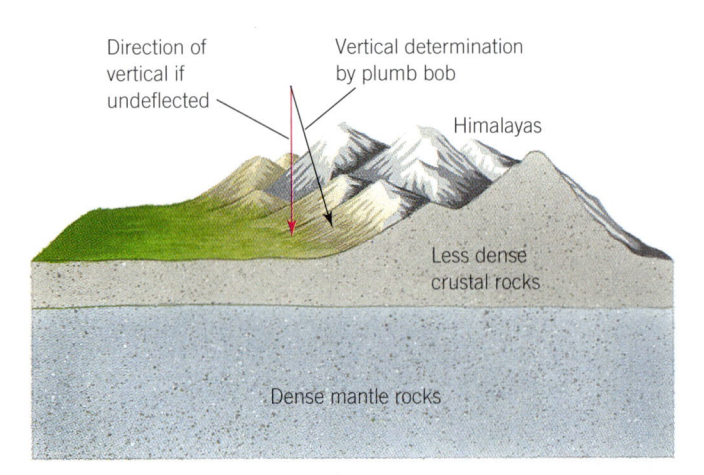

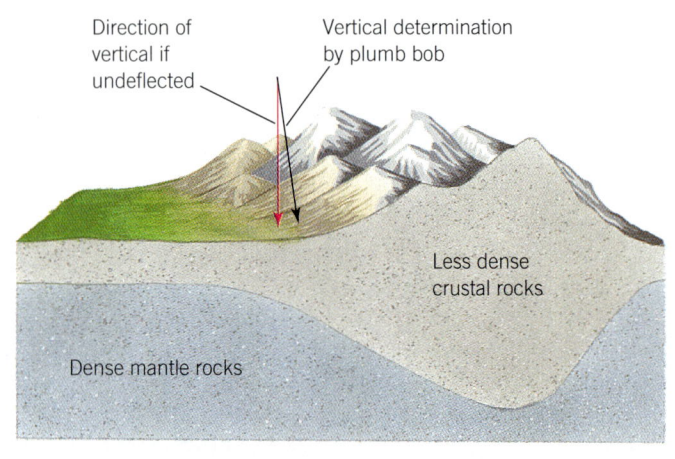

(a)

(b)

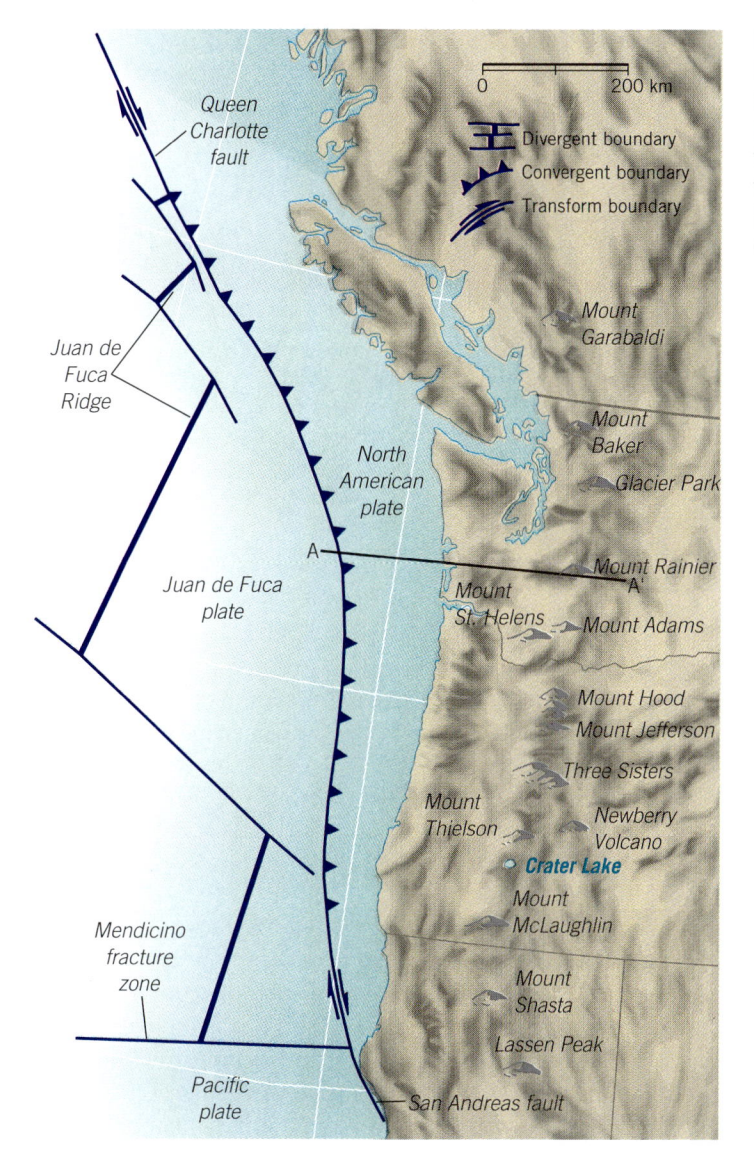

Figure 17.8 Subduction of the Juan de Fuca plate beneath the North American plate is responsible for three roughly parallel features: a sediment-filled trench, the coastal Olympic Range, and the Cascade Range farther inland. (See Figure 17.9 for an explanation of the cross section A to A'.)

Subduction and Mountain Building

Geologists learn about mountain belts by observing evidence of mountain-building processes at convergent boundaries and by reconstructing the events recorded in the deeply eroded rocks of ancient mountains. These studies reveal patterns consistent with the theory of plate tectonics.

The convergent boundary just off the coast of Oregon and Washington is a good place to study mountain-building processes (Figure 17.8). Here we see the spatial relationship among three parallel features: (1) an offshore subduction zone, (2) a coastal range made of sedimentary wedges, and (3) a higher inland mountain range having a granitic metamorphic core. As described in Chapter 2, this is a site of ocean-continent convergence. The dense Juan de Fuca plate is subducting beneath the light continental crust of the North American plate, setting a number of events in motion (Figure 17.9). The coastal Olympic Range was formed when the subducting plate plastered land-derived sediments, submarine lavas, and slivers of oceanic crust onto the edge of the neighboring North American plate as an **accretionary wedge** (also known as a **mélange** because of the chaotic structure of the rocks).

At depths of about 80 to 120 kilometers in the mantle, the descending Juan de Fuca plate has caused plumes of granitic magma to rise and intrude the continental crust

accretionary wedge
A large mass of sediment and lava scraped off a descending plate that accumulates on the margin of a continent or island arc bordering a subduction zone.

mélange
A chaotic assemblage of accretionary wedge sediments and oceanic crust thrown up onto land at a subduction zone.

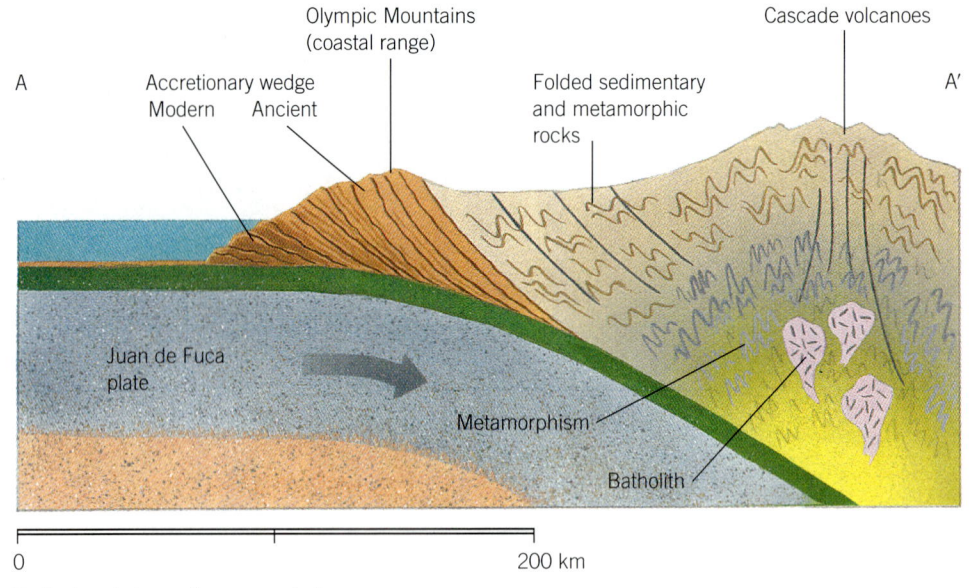

Figure 17.9 Cross section A to A′ in Figure 17.8. Subduction of the Juan de Fuca plate has set a number of events in motion: (1) the plastering of an accretionary wedge onto the continent to form the Olympic Range; (2) the creation of magma bodies that rise in the crust, forming batholiths and the Cascade volcanoes; (3) the metamorphism of crustal rocks surrounding the batholiths; and (4) uplift, folding, and faulting of the overlying sedimentary rocks.

above, forming batholiths or, where the magmas have reached the surface, the Cascade volcanoes. Deep in the continental crust, the heat released by the rising magmas has combined with the compression exerted by the converging plates to metamorphose the surrounding rocks. Shales and lavas have been converted to slates, schists, and gneisses. Higher in the crust, the same compression has thrown the rocks of the overriding continent into complex folds and thrust faults. The crustal thickening caused by the subduction, the emplacement of batholiths, and the compression has resulted in the rise of the complex mountain belt.

This generalized pattern of subduction, sedimentary accretion, and complex mountain building applies as well to the west coast of South America and to complex island arcs. The Japan Islands, for example, consist of several generations of volcanoes resting on platforms of granitic rock generated by the subduction of the Pacific plate. Sediments eroded from the islands and slivers of oceanic crust have been plastered back onto the islands, metamorphosed, uplifted, and eroded through several cycles. To geologists, Japan is a continent in miniature, a *microcontinent*.

Mountain Building by Accretion

The Cascades, Andes, and island arc microcontinents illustrate the importance of subduction, magma generation, and compression in orogenesis. But these processes are not the entire cause of mountain building. In Chapter 2, we compared a lithospheric plate to a vast, slow-motion conveyor belt that moves from a mid-ocean ridge to a deep-sea trench. Transported to the trench by this belt is its cargo: continents and microcontinents, island arcs, extinct seamount chains, submarine lava plateaus, deep-ocean sediments, and continental margin sediments.

What happens to all this material when it is brought to a trench adjacent to a continent? For a fragment of continental crust embedded in the descending plate, the

accreted terrane
A body of rock that is foreign to its surroundings and is bordered on all sides by faults.

Note: *The term* terrane *should not be confused with* terrain, *a general term for an area of land characterized by a particular feature, such as mountainous terrain or rocky terrain.*

answer is straightforward. Because it is not dense enough to sink into the mantle, it must collide with the continent on the overriding plate. As for the denser materials, such as basaltic volcanoes and marine lava plateaus, it was first believed that they were ground up in the deep-sea trench and forced down into the mantle. However, recent studies have shown that although some of the material is forced into the trench, most of it is sheared off the descending plate and driven onto the continent on the overriding plate. In other words, the continent grows by the addition of incoming fragments of continental and oceanic crust, large and small. These fragments are called **accreted terranes**. Having been brought in by the oceanic conveyor belt, an accreted terrane is foreign to the rocks that surround it and is separated from them by faults. These faults are the "skids" along which the terrane was transported to its present position. Still, how are mountains created from accreted terranes?

First, recall that huge quantities of sediment—about 70 percent of the world's total—are deposited on continental margins. Much of this sediment accumulates on passive margins that later become active margins as the continents are carried toward convergent plate boundaries. It is not surprising, therefore, that most fold-mountain belts are composed mainly of margin sediments uplifted and folded by collision with accreted terranes.

Second, remember that all these events occur within the relatively narrow zone of plate convergence. Margin sediments and accreted terranes are likely to be distorted further by the relentless stress within that zone. A terrane may be shortened in the direction of compression but stretched, or elongated, at right angles to it. Pieces of the terrane may be sheared off, rotated, and moved great distances from where it originally landed. New sets of folds may be imposed on uplifted sediments and older terranes by later-arriving terranes. Subduction zone magmas rising from the base of the crust may intrude and metamorphose these assemblages. These adjustments pack terranes and sediments of the mountain belt tightly; they also bind accreted materials to the continent.

Accreted Terranes of Western North America

Accreted terranes extending from Alaska to Mexico compose approximately 70 percent of western North America (Figure 17.10). The truly intriguing challenge is trying to discover where they came from. For this task, we rely on essentially the same techniques developed by the pioneering geologists who argued the case for continental drift: analysis of the ancient magnetism, fossils, and structural or depositional features of the terranes.

A typical terrane may consist of a sliver of basaltic oceanic crust—formerly a lava platform or seamount—buried beneath layers of oceanic sediment. Recall from Chapters 1 and 2 that as a lava cools, its magnetite crystals acquire the orientation of the Earth's magnetic field—an orientation that

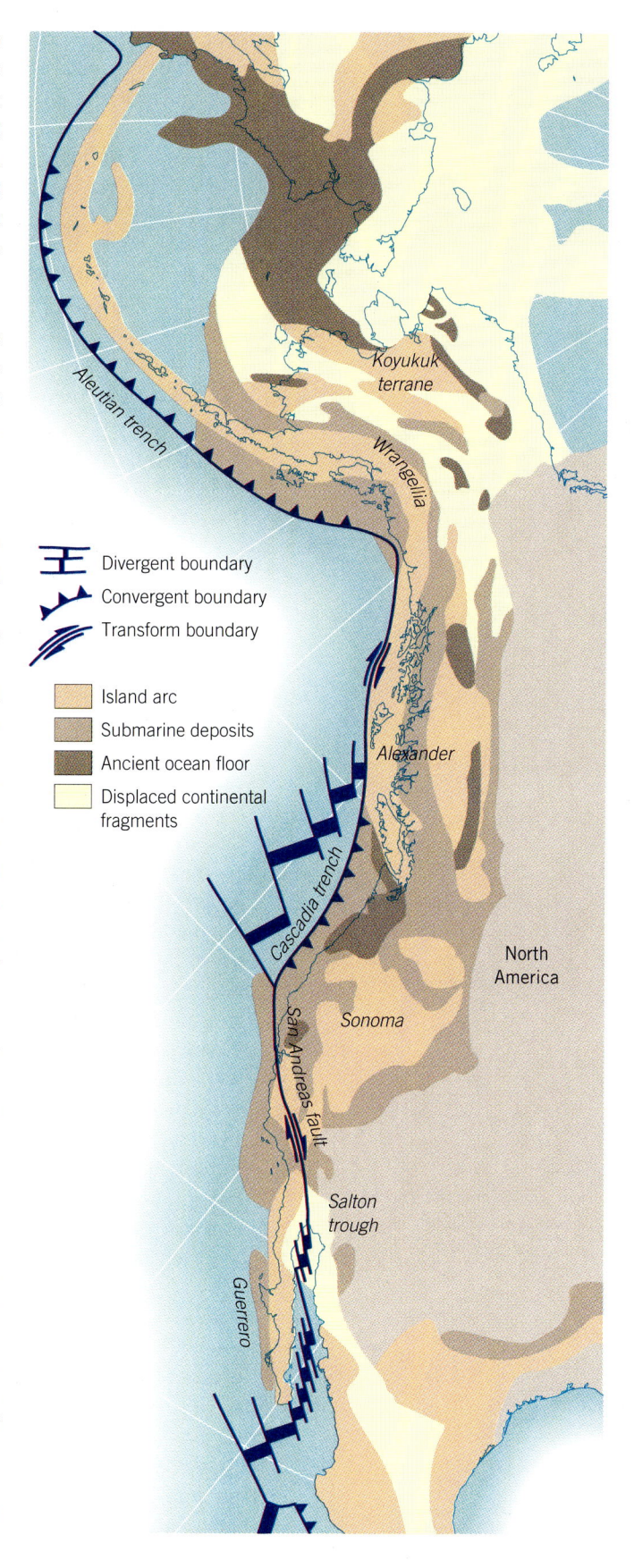

Figure 17.10 Most of western North America is an amalgam of accreted terranes.

depends on the location of the lava. For example, magnetite crystals that form at the magnetic equator acquire a horizontal magnetization, while those that cool at the poles acquire vertical magnetization.

Thus the magnetic properties of the lava enable us to determine the approximate latitude at which the oceanic platform was formed. Furthermore, by dating the fossils in the sediments directly overlying the basalts, we can also determine the approximate age of the basalts. Therefore, by analyzing the terrane from top to bottom, we can trace its journey from point of origin to its present location.

One especially interesting terrane is Wrangellia, named for the Wrangell Mountains of Alaska. Geologists believe it originated near the equator some 200 million years ago as a large chunk of oceanic crust that traveled 9000 kilometers parallel to the west coasts of South and North America before docking in Alaska. Along the way, pieces of Wrangellia were sheared off, rotated, and embedded in other regions of North America; patches of it have been found in Oregon and British Columbia, among other places. This, by the way, is a conservative interpretation; some geologists, relying on similarity of fossils, believe Wrangellia originated in the South Pacific around the latitude of present-day Indonesia.

The Himalayan Orogeny

The term *accreted terrane* is generally applied to rock bodies that are small in relation to the host landmass. However, a subducting lithospheric plate may drag an entire continent along with it. Large as it is, the continent is still less dense than the mantle and cannot sink. If the trench borders a continent on the overriding plate, the two continents have no choice but to collide. Obviously, the actual motion of convergence and collision is excruciatingly slow by human standards. As a result of the convergence, the sediments that were deposited on the intervening ocean basin and on the margins of the advancing continents are buckled and uplifted into huge fold-mountain chains.

Figure 17.11 shows India on its collision course with Asia that began during the Jurassic period some 150 million years ago. The journey was initiated in the wake of the fragmentation of Pangaea, which led to the modern configuration of the continents and ocean basins. A vast region of oceanic crust that was between Asia and India, having been subducted beneath Asia, no longer exists. Working backward from the fact of this enormous mountain range, we note several pieces of evidence as we reconstruct the events that led to the Himalayan orogeny. The first is the presence of a belt of *ophiolites,* which are metamorphosed basalts, gabbros, and peridotites. Found in the Indus-Tsangpo valley just to the north of the Himalayas, they are slices of oceanic crust that were sheared off the descending Indian plate that subducted beneath Asia. Together with mélange sediments, they mark the boundary, or **suture zone,** between the colliding Indian and Eurasian landmasses. The granitic batholiths that intrude the Eurasian plate north of this zone were generated by the subduction of the Indian plate in advance of the collision (see Figure 17.11c).

suture zone
The narrow region that marks the juncture of two colliding blocks of continental crust.

The next piece of evidence is the tremendous thickness of the crust beneath the Himalayas, approximately 60 kilometers, almost double the average of the continents as a whole. The great thickness was caused by the underthrusting of the Indian plate beneath the Eurasian plate and by shortening due to compression. Too light to sink, the doubled-up mass of granitic crust has risen to great heights, in part because of isostatic adjustment and in part because the weight of the Himalayas is supported by the strength of the underthrust Indian plate. The thickened crust forms the platform that supports the Tibetan plateau north of the Himalayas. The plateau is itself higher than any mountain range of the United States. Finally, notice the large fold and thrust-faulted mountains south of the suture zone. The upper zone of the range is composed mainly of sediments that accumulated on the passive continental margin of India and on the deep-ocean floor during that continent's long journey. At greater depth are metamorphic gneisses and schists of the deep crust, also thrown into huge folds. The Indian

Formation of the Continental Crust 371

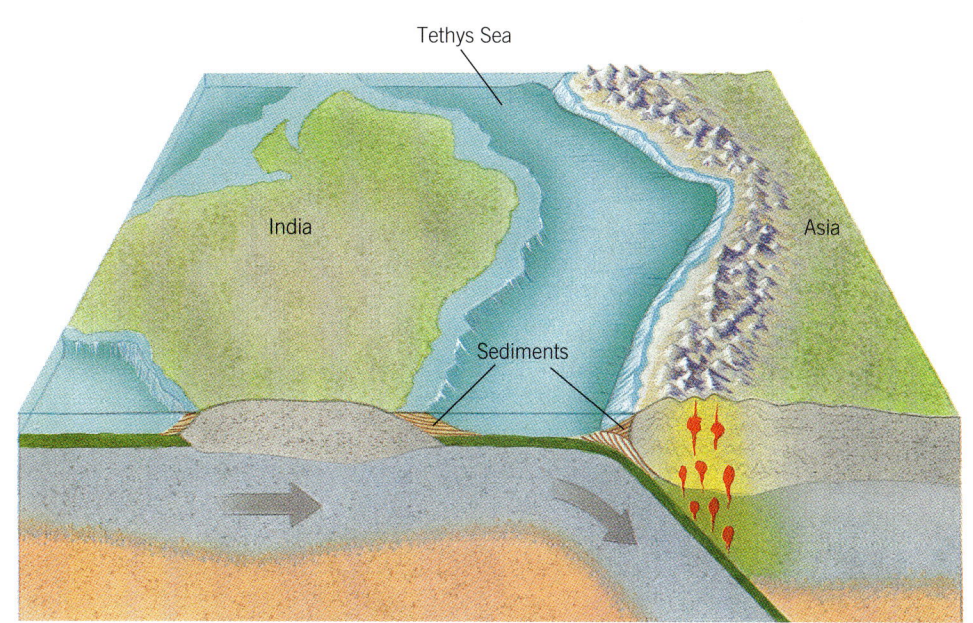

Figure 17.11 The Himalayan orogeny. (a) India and Asia prior to collision. The lithosphere that subducted beneath Asia generated granitic batholiths that intruded the continental crust. At the same time, sediment was deposited in the Tethys Sea. (b) India and Asia collided, causing crustal uplift. (c) Today, the Himalayas are constructed of deformed Tethys sediment and granitic batholiths. The thickened continental crust is responsible for the great heights of the Himalayas and the Tibetan Plateau. Subduction of the Indian lithosphere has ceased.

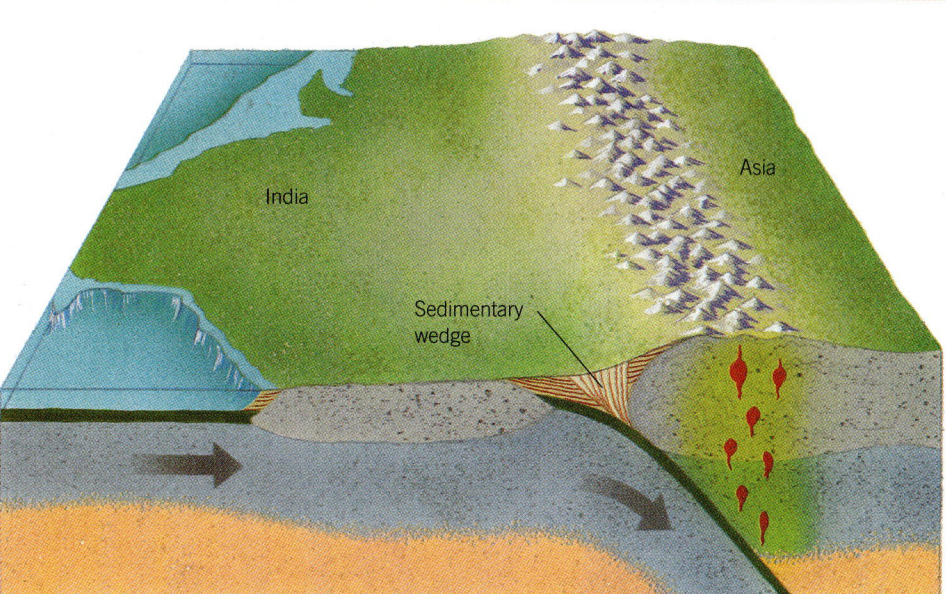

The Appalachian Orogeny

The Appalachian mountain range extends from Newfoundland to Alabama and then continues a short distance below the Gulf Coastal Plain, where it is covered by younger sediments. Another mountain system, the Ouachitas, similar in age and structure to the Appalachians, extends across Arkansas, Oklahoma, and Texas down into northern Mexico (Figure 17.12). The two mountain systems are divided along their lengths and widths into a number of provinces that differ in rock type, structure, and age. Most geologists believe that these provinces were formed by a series of terranes that collided with the North American craton. The collisions, which occurred at irregular intervals, spanned the Paleozoic era between 570 and 250 million years ago. Each of the terranes that was added to the continent brought with it the rocks, structures, and fossils reflective of its history. The problem for the geologist is deciphering the complicated chain of events that created the mountains we see today.

Seismic reflection profiling methods have determined that the metamorphic Blue Ridge and Piedmont regions of the eastern Appalachians are each bounded by major thrust faults. The extent of displacement along the Brevard fault—which separates the metamorphic Blue Ridge Mountains from the sediments of the Valley and Ridge Province—is enormous, at least 250 kilometers. Moreover, seismic surveys reveal that sedimentary rocks lie beneath the shallow, nearly flat fault plane. These rocks are continuous with those exposed in the Valley and Ridge Province to the west, and the metamorphic and igneous rocks of the Blue Ridge and Piedmont rest on top of them. The only interpretation that makes sense of these data is that the older Blue Ridge and Piedmont rocks were driven westward over the preexisting sedimentary rocks.

Figure 17.13 employs the modern theory of plate tectonics to explain the origin of these huge faults. During the late Precambrian (800–700 million years ago), the ancestor of the Atlantic Ocean, the *proto-Atlantic,* opened and spread so that what are now the African and North American cratons were separated. However, several continental fragments and island arcs were left close to North America. The whole region probably resembled the configuration of islands, microcontinents, and true continents akin to those surrounding the Southeast Asian mainland today.

Throughout the Paleozoic era, the proto-Atlantic slowly closed, and the continental fragments collided with North America. What the collisions ultimately led to was the reassemblage of the continental crust into the landmass we know as Pangaea. The first great series of collisions, some 450 million years ago, involved relatively small microcontinents and island arcs. The second series involved large segments of what is now northern Europe. Finally, by the end of the Paleozoic era, the proto-Atlantic closed out, and Africa collided directly against North America, completing the enormous jigsaw puzzle. The total of these orogenies left behind an Appalachian mountain range that rivaled the Alpine-Himalayan belt of today.

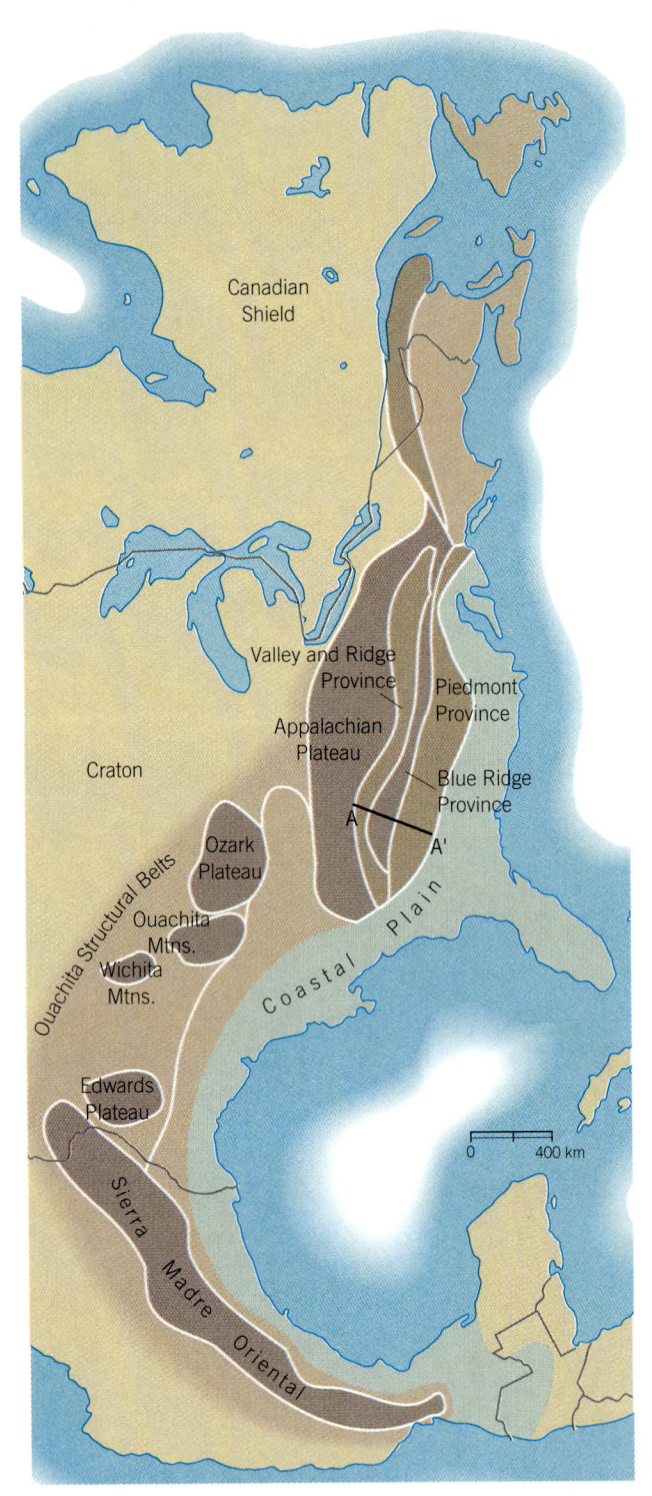

Figure 17.12 The Appalachian mountain belt (and associated structures) is a complex feature that trends from Labrador to Mexico.

Formation of the Continental Crust

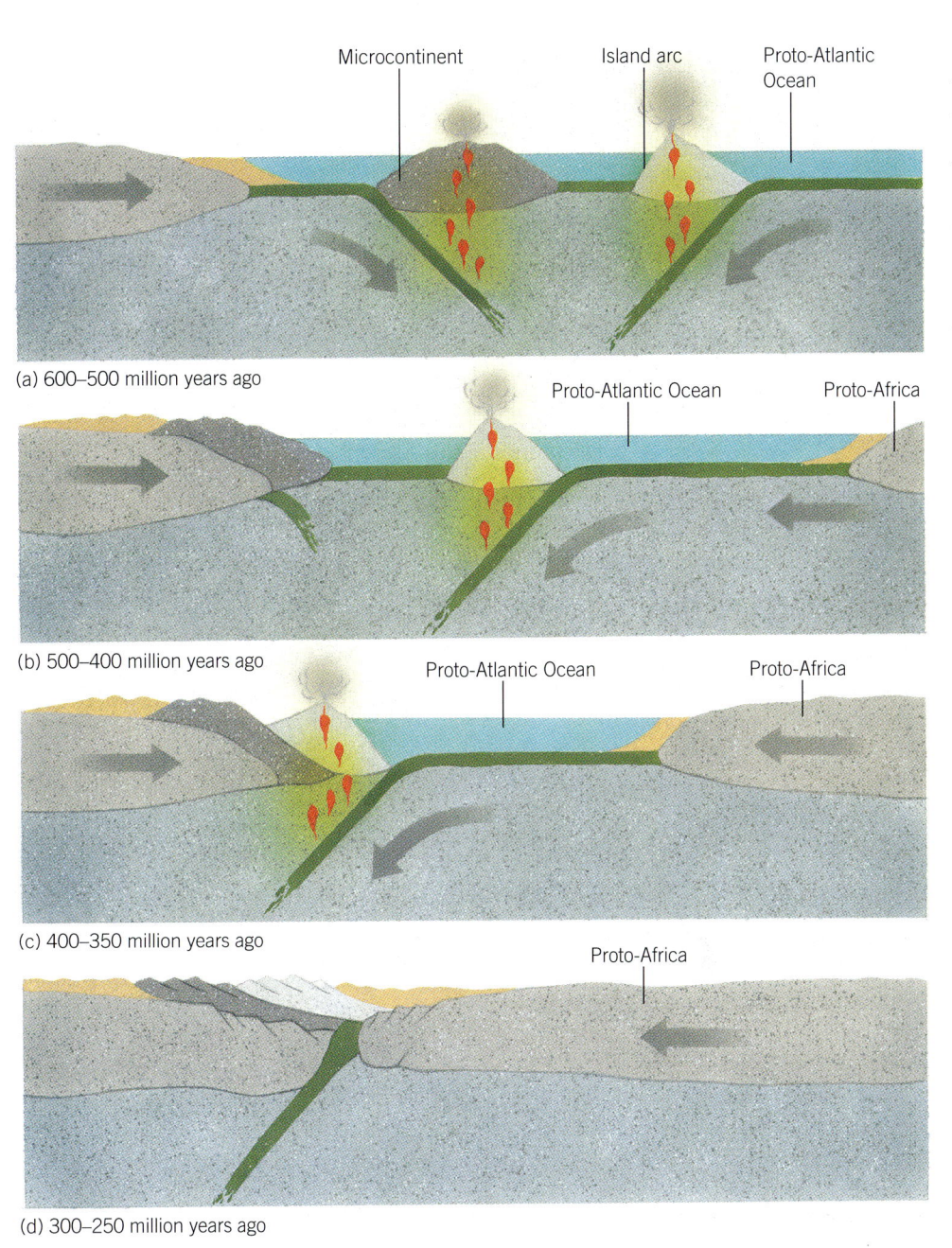

Figure 17.13 A schematic depiction of the origin of the southern Appalachians. The range was constructed over hundreds of millions of years during the Paleozoic era. A series of collisions involving microcontinents, and finally Africa, drove terranes that had arrived earlier westward along thrust faults. Ancient shelf sediments, decoupled and deformed by the collisions, became the fold-and-thrust belt of the Valley and Ridge Province. Sediments eroded from the high mountains to the east created the deltaic rocks exposed in the Appalachian Plateau.

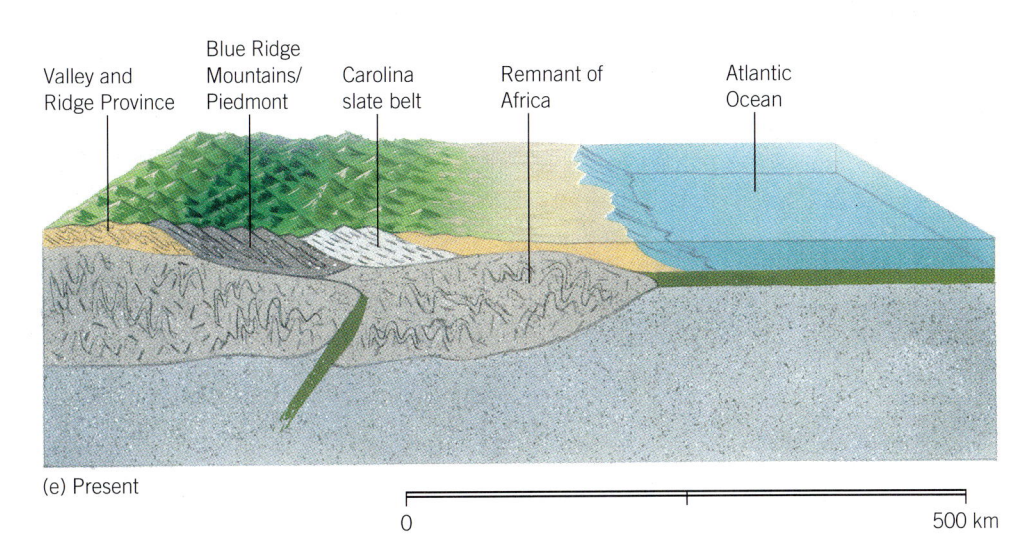

Throughout the long history of mountain building, each newly arrived terrane bulldozed the one ahead of it farther inland. As a result, the ancient shelf sedimentary rocks of the continent were decoupled from the hard basement rock upon which they had been deposited and wrinkled into huge folds in the process. Sedimentary rocks to the west of the fold belt were gently uplifted with a slightly westward tilt; this thick wedge now constitutes the Appalachian Plateau (see Figure 17.12). Formerly a vast deltaic plain, it is constructed of the sediment eroded from the mountains to the east that were created by the collisions of the various incoming terranes.

Following the period of Appalachian mountain building, the supercontinent Pangaea fragmented; the modern Atlantic opened up; and North America, Europe, and Africa drifted toward their present positions. As described in Chapter 2, the breakup of the supercontinent was initiated by continental rifting. Sediment-filled grabens and lava flows caused by the rifting are preserved along the east coast, from Nova Scotia to North Carolina. Indeed, each of the modern continents contains within it fragments that had previously belonged to other continents. Fragments of ancient Africa that were left behind in North America during the breakup include the Yucatan Peninsula, Florida, patches of the Appalachian-Piedmont, and parts of the southeastern New England states.

The Appalachians have been tectonically quiet for the past several hundred million years—which should be expected, because eastern North America has moved far from plate boundaries. As North America drifted westward from the Mid-Atlantic Ridge, new river systems were established, with their outlets in the present-day Atlantic Ocean. These rivers have stripped thousands of meters of sediment from the Appalachian Mountains. The present topography of the Appalachians reflects the effects of prolonged erosion.

Growth and Evolution of the Continental Crust

The hard-rock cores of the continents, the cratons, are composed of deeply eroded microcontinents and larger crustal fragments. Preserved within them are the records of plate collisions and orogenies that occurred billions of years ago (Figure 17.14). Suture zones mark the ancient plate boundaries. Granitic batholiths and regionally metamorphosed sedimentary rocks and lavas mark ancient episodes of subduction. Rift zones filled with basaltic lava mark plate divergences. Each event left its imprint on those that occurred earlier. Thus the geologist seeking to interpret the history of a collision must first subtract from the rock record the impact of any later episodes of collision.

Figure 17.15 shows the age pattern of the ancient shields and mountain belts of North America. Proceeding from the margins inward, we encounter progressively older mountain belts and terranes: the Cenozoic Pacific coast ranges, the Mesozoic Cordillera, the Paleozoic Appalachians, the terranes of the Precambrian craton. We are led to the inescapable conclusion that our continent consists of a conglomeration of terranes, the younger ones welded to the older ones by subduction, compression, and later collisions. For this reason, geologist Paul Hoffman refers to our nation as "the united *plates* of America."*

So we see that plate tectonics elegantly explains the evolution of continents, just as it does the evolution of ocean basins. Consistent with the uniformitarian axiom that the present is the key to the past, the processes responsible for this evolution are readily observable on today's Earth. We can see how plate subduction produces simple island arcs (such as the Aleutians). We also see how these arcs grow into microcontinents (such

---

*As quoted in Robert H. Dott and Donald R. Prothero, *Evolution of the Earth,* 5th ed., New York: McGraw-Hill, 1994, p. 172.

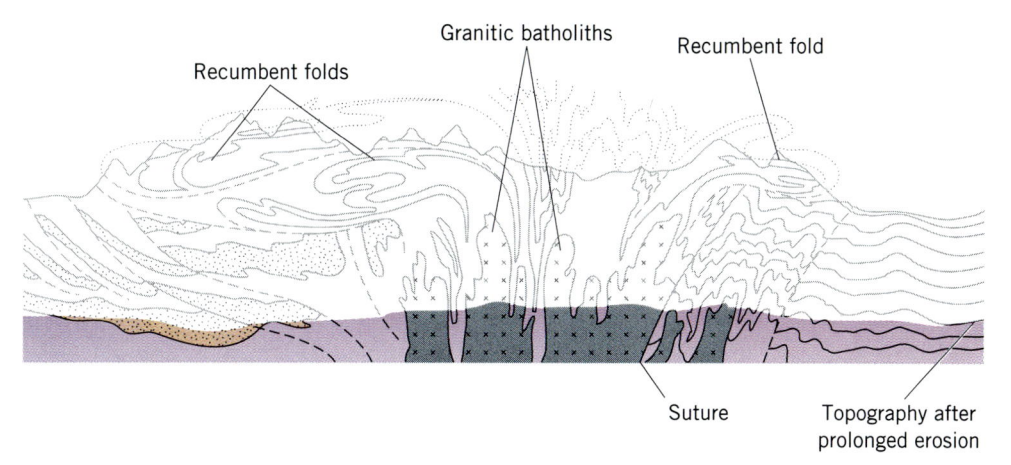

Figure 17.14 Shield-basement complexes are basically the remnants of former great mountain chains. Deep erosion has stripped most of the thick, elaborately folded sedimentary cover. Only the metamorphic rocks and batholiths of the deep crust remain. (Adapted from Robert H. Dott and Donald R. Prothero, *Evolution of the Earth,* Fifth Edition, 1994, p.476, fig.15.30. Reproduced with permission of The McGraw-Hill Companies.)

as Japan), and then continents (such as North America), through complex volcanism, erosion, metamorphism, the addition of sediments, and the accretion of other terranes. Finally, we see how huge continental blocks riding on lithospheric plates may collide and form supercontinents (such as Asia/India).

In this chapter, we have envisioned the continents as collages of accreted terranes, swept together at irregular intervals by descending plates. Remember, however, that continental collages are not permanent works of art. The continents are continually being torn apart and reassembled in different forms. Crustal fragments that were once combined may later be separated by rifting or may be moved sideways over great distances by transform faulting; all are subject to the whims of the convection currents that drive the plates. The latest reshuffling of the crust, involving the assembly of the modern continents following the breakup of Pangaea, has yet to be completed. Crustal fragments are still fleeing toward Asia and North America around the rim of the Pacific and Indian oceans, as well as around the Mediterranean Sea.

The total volume of continental crust continues to grow. The subducting plates cause the production of huge volumes of magma that are added directly to the crust as granitic batholiths. Magma added to the oceanic crust through igneous activity at mid-ocean ridges and hot spots are accreted to the continents. Even sediments recently eroded from the continents will eventually find their way back. Through subduction, plate collisions, metamorphism, and igneous intrusion, these sediments will be pushed back and welded to the continents. As long as its supply of internal energy holds up, the Earth will go through many cycles of plate tectonics. The history of these cycles will be preserved in the hard-rock continental crust.

Figure 17.15 The "united plates of America." The continent is mainly an amalgam of terranes welded together through subduction, compression, and later collisions. In general, the age of the terranes is younger toward the margins of the continent, but the pattern is disrupted by rifting and subsequent rearrangements.

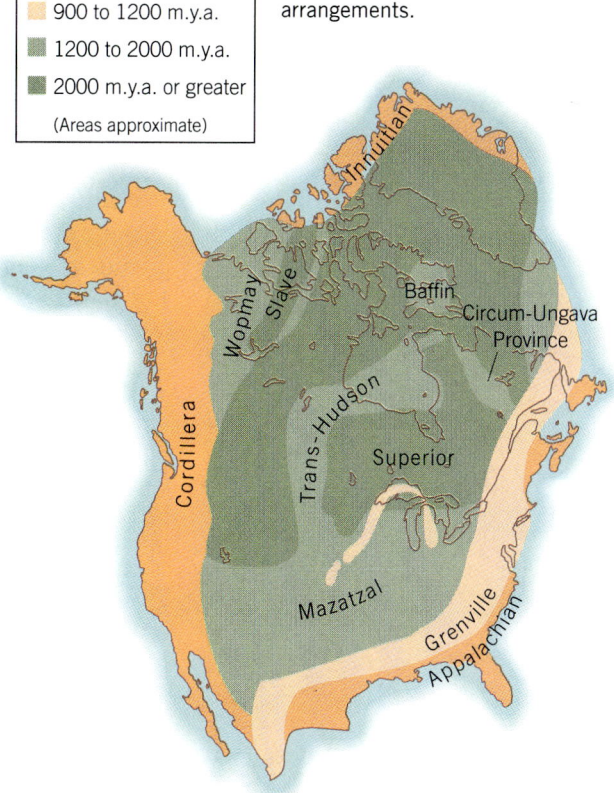

Study Outline

I. **SUBDIVISIONS OF THE CONTINENTAL CRUST**
   A. The continental crust consists of three units: cratons, mountain belts, and continental margins. **Cratons** are the central, tectonically stable cores of the continents. They consist of foundations of ancient igneous and metamorphic basement rock, blanketed by **platforms** of sedimentary strata. The exposed regions of basement rock are called **shields.**

   B. In mountain belts, geologists see folds, faults, igneous intrusions, and metamorphism as evidence of intense deformation that enables them to investigate the causes of past mountain-building episodes, or **orogenies.**
      1. Volcanoes are the products of the igneous activity associated with subduction zones.
      2. Fault-block mountains are the products of the crustal extension associated with rifting at divergent plate boundaries.
      3. Fold mountains are the products of compression generated at convergent plate boundaries.

   C. The extent and shape of the continental margins are influenced by their tectonic settings.
      1. Passive margins, located far from plate boundaries, consist of thick, broad, sloping sediment wedges that are divided into continental shelf, slope, and rise.
      2. Active margins are characterized by narrow, steep continental shelves. Land-derived sediments are carried to the nearby trenches by turbidity currents.

II. **MOUNTAIN BUILDING AND PLATE TECTONICS**
   A. Continental mountain belts are composed of thickened crust; their roots sink into the mantle upon which they float in a state of isostatic equilibrium. The crustal thickening is due to compression perpendicular to the trend of the chain.

   B. Western North America, a typical site of ocean-continent convergence, is characterized by three parallel features: a deep-sea trench, an **accretionary wedge** forming a coastal range, and a higher inland fold-mountain range having a granitic/metamorphic core.
      1. Land-derived sediments, submarine lavas, and slivers of oceanic crust from the subducting plate are plastered onto the edge of the overriding plate as an accretionary wedge, or **mélange.**
      2. Plumes of granitic magma intrude the continental crust, forming batholiths or volcanoes.

   C. Continents grow by the addition of fragments sheared off the descending plate, or **accreted terranes**. Margin sediments and accreted terranes are distorted and uplifted by later-arriving terranes.

   D. The Himalayan range resulted from the collision of continents. Slices of oceanic crust found in a valley north of the Himalayas and mélange sediments mark the **suture zone** between the colliding Indian and Eurasian landmasses.

   E. Seismic reflection profiling provides strong evidence that the Appalachians consist of a series of terranes that collided with the North American craton between 570 and 250 million years ago.
      1. Each newly arrived terrane bulldozed the one ahead of it farther inland. Ancient shelf sediments were wrinkled into huge folds and uplifted in the process.

2. The present topography of the Appalachians reflects the effects of prolonged erosion.

III. GROWTH AND EVOLUTION OF THE CONTINENTAL CRUST

A. Plate tectonics explains the evolution of continents, just as it does the evolution of ocean basins. Plate subduction produces simple island arcs that grow into microcontinents and then into continents through complex volcanism and metamorphism, the accretion of sediments, the arrival of other terranes, and continental collision.

B. As long as the Earth has sufficient internal heat, the continents will continue to be torn apart and reassembled in different forms. Crustal fragments once combined may later be separated, and the total volume of continental crust will continue to grow.

STUDY TERMS

accreted terrane (p. 368)
accretionary wedge (p. 367)
craton (p. 362)
mélange (p. 367)
orogeny (p. 363)
platform (p. 362)
shield (p. 362)
suture zone (p. 370)

THOUGHT QUESTIONS

1. What conditions would be necessary to change the passive Atlantic continental margin into an active one?
2. Looking at a world map, locate those regions other than the Himalayas that are the probable site of continent-continent convergences.
3. What sections of the North American continent show signs of becoming accreted terranes in the future?
4. How would you identify the line of convergence, or suture zone, between two continental blocks?
5. What kind of evidence might prove where an accreted terrane came from? How old it was? Its environment of formation?
6. What is meant by the description of the continent as the "united plates of America"? What evidence supports the phrase?
7. What evidence suggests that the Earth's crust was much warmer 4 billion years ago than it is today?

Chapter 18

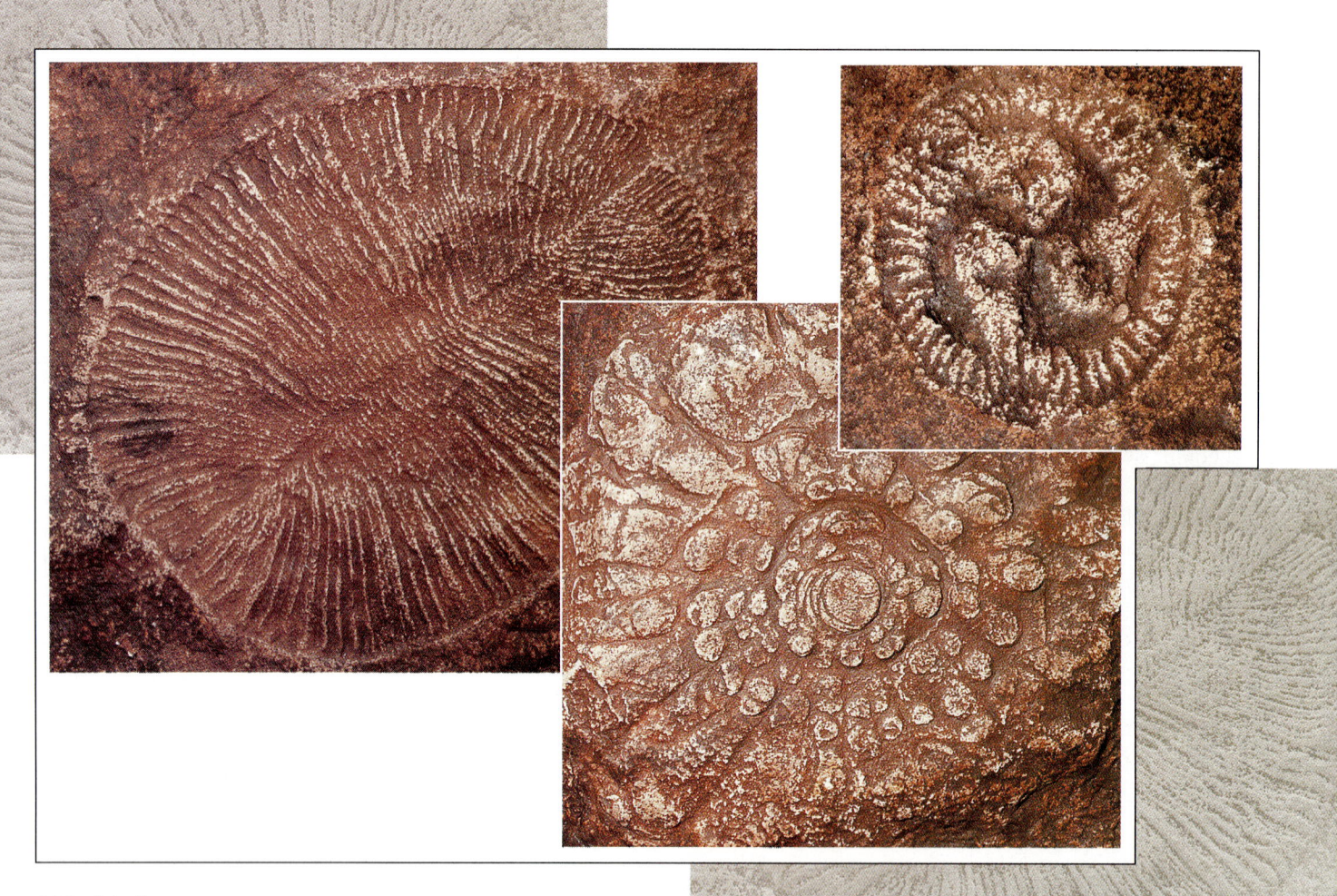

- **Origin of the Planets**
- **Primitive Planet Earth: The Hadean Eon**
- **The Archean Eon**
- **The Proterozoic Eon**
  - The Precambrian Atmosphere and Oceans
  - The Origin of Life
  - The Late-Proterozoic World
  - Proterozoic Life
- **The Evolution of Life on Earth**
  - Evidence of Evolution in Fossils and Comparative Anatomy
  - Transitional Fossils
  - Vestigial Organs
  - Evidence of Evolution in Embryology
  - Evidence of Evolution in Genetics
  - The Mechanisms of Evolution
  - Geology and Natural Selection
  - The Status of Evolution

Earth History: From the Origin of the Planet Through the Proterozoic Eon

The primary focus of this book so far has been physical geology—that is, the study of the processes that affect the materials, internal structure, and surface features of the Earth. You can see this focus reflected in the chapter titles: Minerals, Plate Tectonics and the Ocean Floor, Volcanism, Rock Deformation, Streams, Glaciers and Climate, and so on. In Chapters 18 and 19, we employ our knowledge of the Earth's materials and processes to summarize briefly its history, from the time of its formation to the present. Our approach follows the guiding principle of uniformitarianism, formulated by James Hutton in 1788, which states that the present is the key to the past.

Using Hutton's principle, we begin our summary of Earth history by asking: What observations of the universe today shed light on the origin of the Earth and the other planets of the Solar System?

Origin of the Planets

In certain regions of space today, astronomers can see what appear to be new stars forming (Figure 18.1). Surrounding the young stars are glowing clouds of gases and dust particles, flattened into wide orbiting disks, that are maintained by the stars' strong gravitational attraction. Mixed in with the overwhelming abundance of hydrogen and helium of the cloud are minute percentages of the same elements that compose the planets of our Solar System. There is every reason to believe a similar disk of gaseous matter surrounded our embryo Sun in the past. Nearly all of the planets and their satellites orbit the Sun in what is very nearly a common plane—a feature we would expect if they condensed from the same orbiting disk.

◀ Late-Proterozoic Ediacaran fossils. These soft-bodied organisms became extinct at the close of the Proterozoic and do not seem related to the hard-shelled Cambrian organisms that succeeded them. Why this is so remains an unsolved mystery.

CHAPTER 18

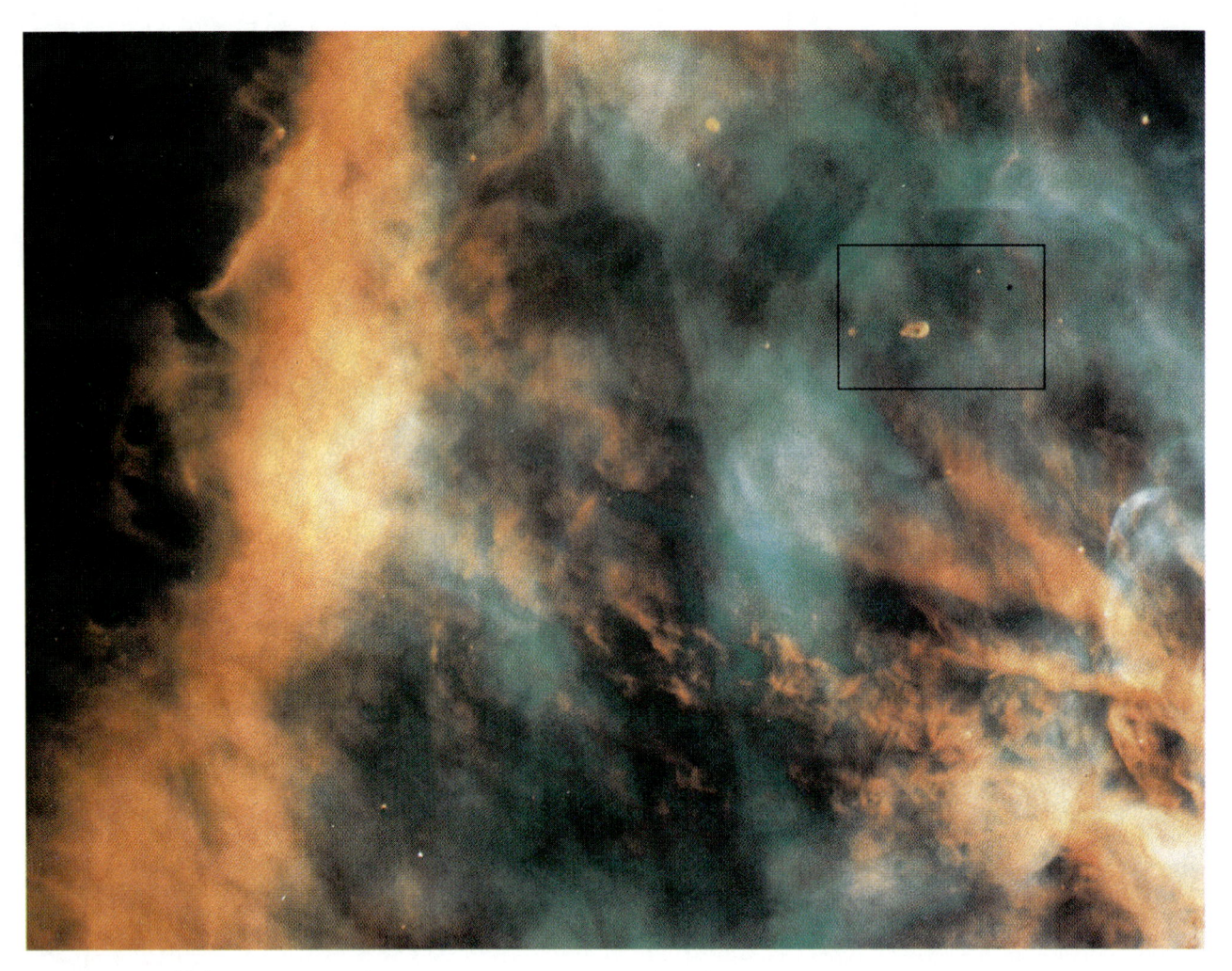

Figure 18.1 A stellar nebula from which primitive solar systems condense. The disklike structure in the rectangle at the upper right is a gaseous ring encircling a sun. Spectral analysis helps determine composition: Blue codes the presence of oxygen; red, the presence of nitrogen.

Planetary geologists believe that solid particles condensed out of the gaseous disk surrounding the primitive Sun in a temperature-controlled sequence. Closest to the Sun came the light metals, such as aluminum, titanium, calcium, and their oxides. Somewhat farther from the Sun followed iron, nickel, and silicate minerals; included among the latter were the hydrous silicates, which hold chemically bound water within them. These metals and silicates constituted a tiny percentage of the dust cloud, but the close-in terrestrial planets and asteroids are composed of them. At still greater distances, with temperatures approaching 200 kelvin (-73 °C), substances we normally think of as gaseous were converted to ice and formed the distant Jovian planets. These planets consist mainly of layered hydrogen, ammonia, and methane ice wrapped around tiny rock-iron cores.

Support for the condensation hypothesis comes from samples of the Solar System in primitive condition in the form of meteorites, hunks of rock and iron that have fallen from space. Mostly derived from the asteroid belt, they are relics left over from the birth of the Sun and the planets. An analysis of the decay of radioactive elements found within them shows that they were formed 4.6 billion years ago—the probable age of the Solar System. The structure of the most common type, the *chondrites,* is an aggregate of condensed, pea-sized mineral matter—the original hot rain of the cooling dust cloud. In addition, the chemical composition of an important subclass, the carbonaceous chondrites, closely matches the composition of the Sun minus its gaseous hydrogen and helium. This analysis strongly suggests that the *chondrites* and the Sun have a common origin; by extension, so do the Sun and the other rocky bodies of the Solar System.

The largest asteroids are but a few kilometers in diameter, which leaves open the question of how the planets evolved from the droplets of the original dust cloud into the large bodies they are today (Figure 18.2). One popular hypothesis favors the collection through gravitational attraction of condensed matter into small asteroid-sized bodies called *planetesimals* that collided, fragmented, and re-formed over a period of a few million years. Some gradually acquired enough mass to sweep up the smaller bodies in the surrounding space like gravitational vacuum cleaners. They became the primitive planets.

We do not know the precise steps whereby the primitive planet Earth was transformed into the body we see today. We are fairly certain, however, that it went through an early process of remelting and core formation. The necessary heat was supplied by meteoroid impacts, which, in this early period of the Solar System, were far more numerous than they are today; by the compaction of matter in the planet; and by the decay of abundant radioisotopes. The heat derived from these diverse sources melted the iron scattered within the interior of the planet. Being denser than the silicates of the planet, the molten iron sank, forming a dense planetary core. The sinking process liberated still more heat, which led to the total or partial melting of the silicate minerals. Being lighter than the core, they rose to form the mantle and lithosphere.

The challenge that confronts geologists is to trace the Earth's evolution from this primitive condition to its present state. Unfortunately, they face a formidable obstacle: Radiometric dating of meteorites has established that the Earth formed approximately 4.6 billion years ago, but the oldest Earth rocks discovered so far are only about 3.96 billion years old. This gap of 640 million years in the geologic record is greater than the combined time span of the Paleozoic, Mesozoic, and Cenozoic eras!

Primitive Planet Earth: The Hadean Eon

Geologists call the 640-million-year interval between the Earth's origin and the oldest preserved rocks the **Hadean eon.** This eon is the first large-scale interval of Precambrian time (Figure 18.3). With little or no tangible record to guide them, geologists can only speculate about the appearance and structure of the early Earth. The prevailing hypothesis is that the Hadean Earth was a still-accreting planet, pock-marked with randomly distributed impact craters. Basaltic lava flows and volcanoes spewed water vapor and other gases, some of which were retained to form a primitive atmosphere and ocean. The dead surfaces of the Moon and Mercury, and portions of the surfaces of Mars and Venus, preserve their primitive features. On the Earth, later recycling and erosion destroyed all remnants of the planet's early basaltic crust.

The term *Hadean* was derived from *hadeas*, the Latin name for hell. The term is appropriate, for the Earth was probably a very hot place during this early stage. We make this assumption because the sources of the Earth's internal heat—meteoroid impact, compaction, core formation, and radioactive decay—were far more active than at present. Today, it is primarily the decay of long-lived radioactive isotopes that powers the Earth's internal engine.

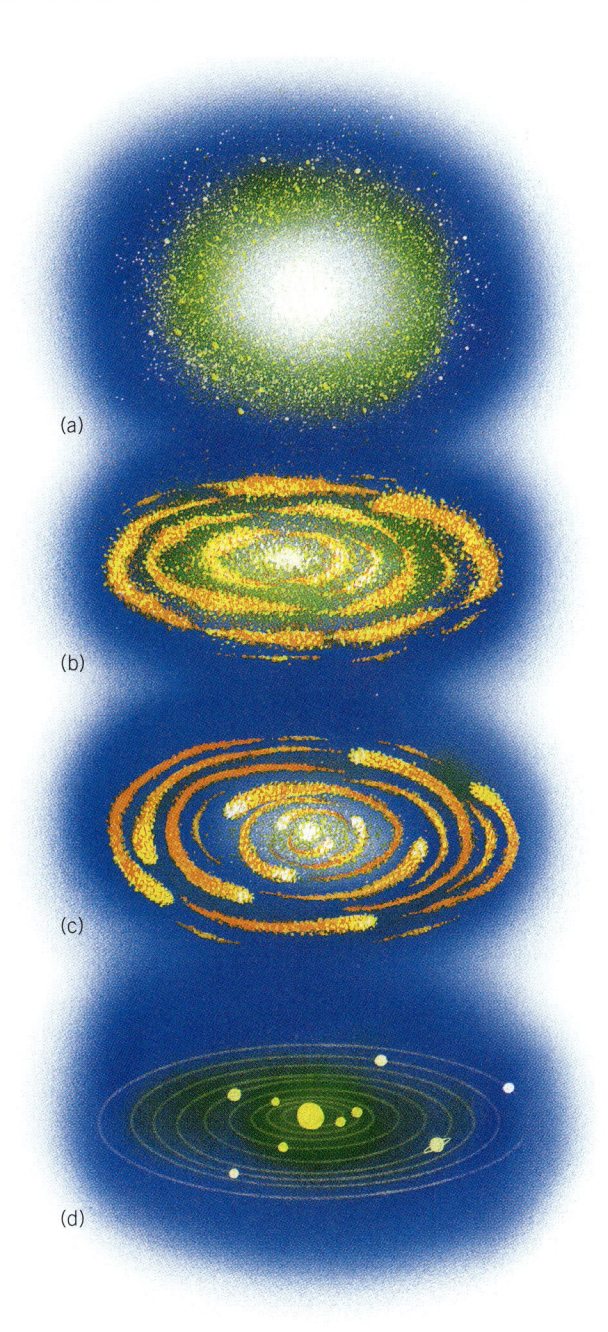

Figure 18.2 The condensation hypothesis. (a) The primitive solar nebula. (b) The nebula flattens into a rotating disk. Most matter streams in toward the center to form the primitive Sun; some matter remains outside. (c) The disk cools, forming tiny particles that grow into planetesimals. (d) A few of the planetesimals grow large enough to attract others, as well as debris from the surrounding space. Eventually they grow into the primitive planets.

Hadean eon
The 640-million-year interval between the time of the Earth's origin (4.6 b.y.a.) and the age of the oldest preserved Earth rocks (about 3.96 billion years).

Eon	Era	Period	Epoch	From — To (m.y.a.) (m.y.a.)	Geological Events	Biological Events
Phanerozoic	Cenozoic	Quaternary	Holocene	0.01 — Pres.	End of last Ice Age	
			Pleistocene	1.61 — 0.01		
		Tertiary	Pliocene	5.3 — 1.61	Ice Ages begin	Earliest humans
			Miocene	23.7 — 5.3		
			Oligocene	36.6 — 23.7	Formation of Alpine-Himalyan Mt. Belt	
			Eocene	57.8 — 36.6		
			Paleocene	66.4 — 57.8		First primates/Extinction of dinosaurs and other species
	Mesozoic	Cretaceous		144 — 66.4		First flowering plants
		Jurassic		208 — 144	Onset of Cordilleran orogeny	First birds
						First mammals
		Triassic		245 — 208	Pangaea begins to break up	First dinosaurs
	Paleozoic	Permian		286 — 245	Formation of Pangaea / Culmination of Appalachian mountain-building	Extinction of trilobites and many other marine and land animals
		Carboniferous / Pennsylvanian		320 — 286	Coniferous forests buried and converted to coal	
						First reptiles
		Carboniferous / Mississippian		360 — 320		
		Devonian		408 — 360		First amphibians
		Silurian		438 — 408		First land plants
		Ordovician		505 — 438	Proterozoic supercontinent begins to break up	First fishes
		Cambrian		570 — 505	Formation of Proterozoic supercontinent	First marine invertibrates, trilobites
Proterozoic				2500 — 570	Formation of the North American craton	First multicellular organisms
Archean				3960 — 2500		First single-celled organisms
					Age of the oldest known rocks	
Hadean				4600 — 3960	Formation of the Earth	

Figure 18.3 The geologic time scale with major geological and biological events displayed in the right-hand columns. The various mountain-building episodes, as well as the formation and disassembly of the continents, actually spanned many millions of years. The Appalachian Mountains, for example, were in the process of forming throughout the Paleozoic era.

The Archean Eon

The age of the oldest known rock (3.96 billion years) is considered by geologists to mark the onset of the **Archean eon,** which lasted about 1.5 billion years, ending 2.5 billion years before the present. Widely exposed in the shields and cratons of the continents, Archean rocks are the most primitive rocks of the continental crust (Figure 18.4; see also Figure 17.2). What can we learn from them? Essentially, they tell us that the Archean eon was a time of transition during which the hot Earth cooled considerably, and, in so doing, began to evolve toward its present state. During the Archean, the enduring continental crust formed, the atmosphere and oceans began to develop, and the earliest life forms emerged.

The Archean regions of the present-day continents, such as the Lake Superior district, display a broadly similar sequence of rock types and structural patterns. The oldest rocks in the sequence are typically banded gneisses, which form the structural foundation of the Archean crust. These metamorphic rocks were derived from igneous rocks (quartz diorite) of fairly high silica content. The modern counterparts of these igneous rocks crystallize from magmas that are generated in subjection zones and form island arcs. Consequently, Archean gneisses provide good evidence that subduction was an important Archean process.

Preserved within deeply eroded, down-faulted troughs and synclines in the gneisses are *greenstone belts,* solidified submarine lava flows that erupted from the mantle through fissures in the gneissic crust. Many of the minerals in the lavas have been converted to green chlorite, which gives these metamorphosed igneous rocks their distinctive color. The greenstone lavas were richer in iron and magnesium than today's basaltic lavas. In fact, they were similar in composition to today's upper mantle. This fact supports the hypothesis that enough heat remained available in some subduction zones and rift valleys during the Archean to cause the total melting of upper mantle rocks—a condition that has rarely recurred since that time.

With a few exceptions, the only surviving Archean sedimentary rocks are poorly sorted *turbidites* that are found interbedded with—and overlying—the greenstone lavas. Their mode of occurrence and their texture and composition closely resemble those of the present-day sediments that fill the deep trenches adjacent to the island arcs of the Pacific and Indian oceans. These sediments are transported downward to the deep ocean floor in the form of dense slurries of mud that hug the submarine slopes of the volcanic arcs. The spatial and age relationships of the Archean greenstones, turbidites, and banded gneisses offer good evidence that the eon was characterized by unstable arch-trench systems bordering subduction zones.

Because of the high internal temperatures that prevailed, most geologists do not envision the Archean world as supporting large-scale lithospheric plates like the present-day Earth. Instead, they envision a more chaotic arrangement featuring thinner and shorter-lived *platelets.* The platelets were created by closely spaced convection currents, whose rising limbs brought fresh magma to the surface. The platelets then spread laterally and were soon recycled as they descended back into the mantle. In the Hadean and early Archean, mantle temperatures were high enough to remelt the platelets almost completely. As the interior cooled in the late Archean, however, the platelets melted only partially when they descended. This process generated the silica-rich magmas that rose to form the Archean crust and are today preserved as the metamorphic banded gneisses. Thus the Archean crust, like its modern counterpart, was the distillate of a churning process that remained on the surface because it was too light to be subducted.

The surviving Archean crust initially formed as a series of island arcs. Through the accretion of sediment, the addition of magma derived from the descending platelets, and collision with other island arcs, the island arcs rapidly evolved into microcontinents—perhaps similar in size and complexity to present-day Japan. Some 2.5 billion years ago, a number of these microcontinents combined as terranes to form the core of

Archean eon
The unit of geologic time dating from the end of the Hadean eon (3.96 b.y.a.) to the beginning of the Proterozoic eon (2.5 b.y.a.).

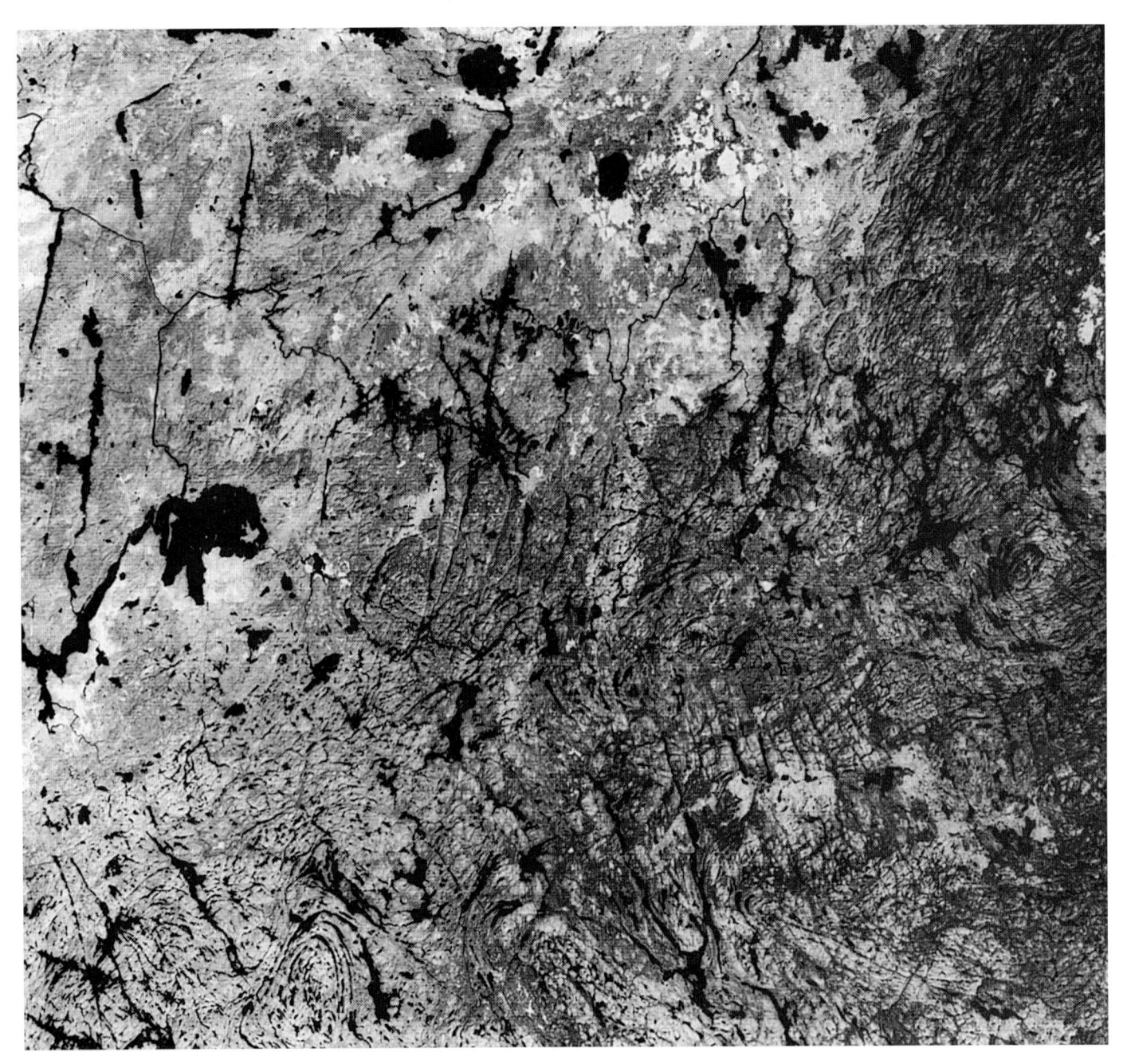

(a) (b)

Figure 18.4 (a) Landsat image of the Grenville front, boundary between the Precambrian Superior and Grenville provinces of the Canadian Shield, western Quebec. (b) Simplified geologic map of the region. The smooth Superior terrain to the northwest is underlain mainly by Archean gneisses and granites, and by an Archean greenstone belt. The rougher, more structurally complex Grenville terrain to the southeast is underlain by metamorphosed sedimentary and igneous rocks of the Proterozoic age. The front itself marks a zone of late Proterozoic plate convergence. Prior to the Paleozoic, the entire region was eroded to form a low, gently undulating surface. The numerous lakes were scoured out of zones of weak rock during the Pleistocene ice ages.

Earth History: From the Origin of the Planet Through the Proterozoic Eon

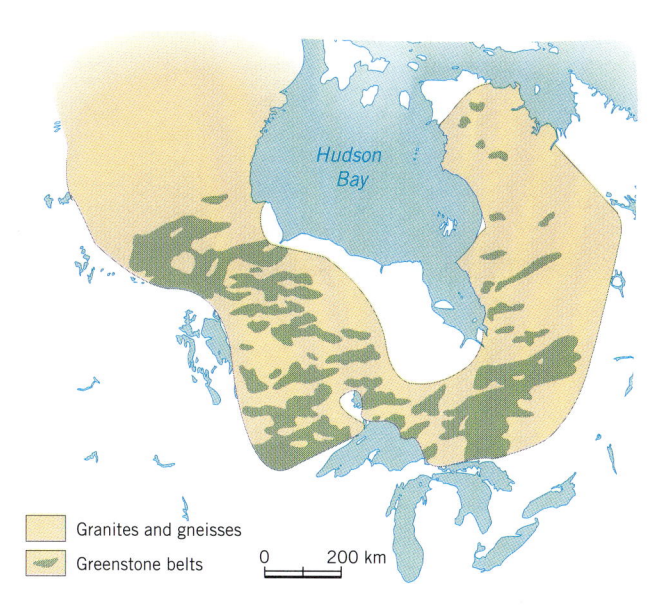

Figure 18.5 The Archean core of North America, the Superior Province, consists of a series of deeply eroded island arcs (greenstones) and microcontinents (granite-gneiss complexes) that were sutured together by the collisions of small plates by about 2.5 billion years ago—in an event known as the Kenoran orogeny.

the North American craton (Figure 18.5). This event, called the *Kenoran orogeny*, brought together the Superior, Churchill, and Slave provinces of the Canadian Shield. During the orogeny, vast quantities of granite intruded into this assemblage, increasing the craton's volume, buoyancy, and rigidity. Similar episodes of assembly and granitic intrusion formed the cratons of other continents at roughly the same time. For this reason, the Kenoran orogeny has been chosen to mark the end of the Archean and the beginning of the Proterozoic eon.

The Proterozoic Eon

The **Proterozoic eon** is difficult to describe briefly because it extended from 2.5 billion to 570 million years ago, an interval encompassing more than 40 percent of geologic time. Perhaps the safest generalization we can make is that, during the Proterozoic, the Earth evolved into a planet very much like the one we now inhabit. By the close of the eon, it supported full-fledged oceans, large continents, and fully developed tectonic and hydrologic cycles. Several periods of widespread glaciation occurred as well.

The fundamental internal process that led to this modernization of the Proterozoic world involved the continued growth of continental crust that had begun toward the close of the Archean. The growth was accomplished by essentially the same processes as seen during earlier eons, but acting at a slower pace. The Proterozoic continents became thick, rigid, and stable, much like the present continents. At the same time, the small, rapidly churning platelets that characterized the Archean gave way to large, modern-style lithospheric plates. These changes reflected the considerable cooling of the interior of the Earth in the billion years since the Hadean eon, and mantle convection that proceeded at a reduced rate (Figure 18.6).

Geologists reconstruct the movement of the Proterozoic continents by combining the techniques used to prove the much later construction and breakup of Pangaea. One technique involves deciphering the magnetization of Proterozoic sediments and lavas to determine where the rocks were located at the time of their original formation. A second involves identifying suture zones and accretionary wedges that mark Proterozoic plate collisions. A third technique involves matching up distinctive Proterozoic rocks and structural trends that were once continuous but now appear on separated continents (Figure 18.7). Geologists agree that the Proterozoic continents combined—and separated—a number of times, and that they were probably joined as a single land mass toward the close of the Proterozoic eon, 570 million years ago.

Proterozoic eon
The unit of geologic time dating from the close of the Archean eon (2.5 b.y.a.) to the onset of the Paleozoic eon (570 m.y.a.).

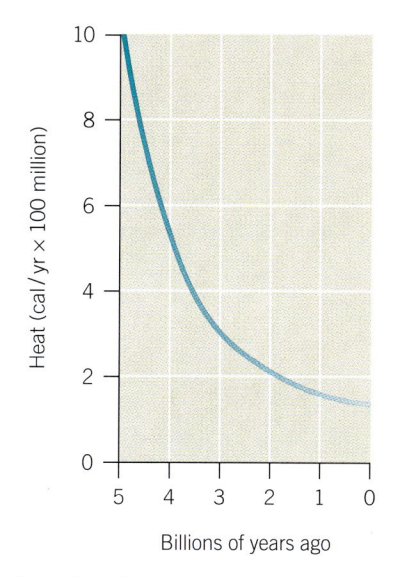

Figure 18.6 The decline of internal heat production over geologic time. The decline was steep during the first two billion years, but has long since leveled off.

Figure 18.7 Reconstruction of a former landmass from continental blocks that are at present widely separated. Evidence includes the matching of identical formations, ancient rifts, and failed arms.

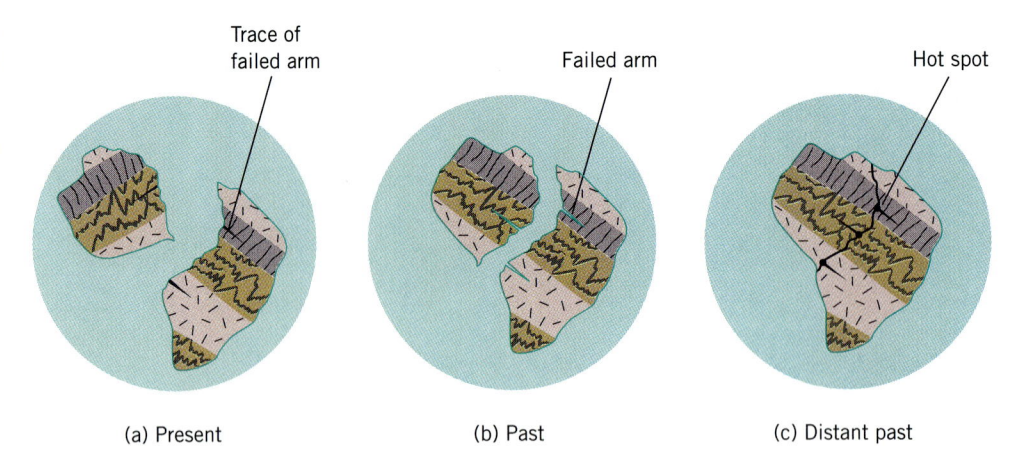

(a) Present (b) Past (c) Distant past

The Precambrian Atmosphere and Oceans

The Earth's primitive atmosphere most likely formed early in the Hadean by **outgassing**—that is, through the release of gases by volcanoes, when the Earth passed through its early molten or semimolten stage (Figure 18.8). Based on the composition of present-day volcanic gases and our knowledge of mantle chemistry, geologists believe that these escaping gases consisted mainly of methane (CH_4), water vapor (H_2O), ammonia (NH_3), and hydrogen sulfide (H_2S). The Sun's ultraviolet radiation dissociated (split apart) some of the water vapor and ammonia molecules in the upper atmosphere. These reactions released light, fast-moving hydrogen, which drifted into space, but left the heavier gases behind. The remaining gases included nitrogen, which constitutes approximately 79 percent of the modern atmosphere, and a small quantity of oxygen, which recombined with some of the methane to form carbon dioxide. Thus no free oxygen existed.

Outgassing during the Hadean also released more water than the atmosphere could hold. The excess fell to the Earth as rain and flowed downhill to the low places on the Earth's surface—the basaltic regions of the crust—to form the world's oceans. The running water contained dissolved ions weathered from bedrock. These ions remained in the oceans for time spans that depended upon their solubilities—with sodium, chlorine, and sulfate ions persisting for the longest periods. The same process applies today, which is why the oceans are salty.

Although outgassing represents the traditional explanation for the origin of the Earth's atmosphere and oceans, other possibilities exist as well. In 1985, Halley's Comet passed close enough to the Earth to allow planetary geologists to get a good look at it. It proved to be a huge melting snowball—that is, a disintegrating mass of ice, roughly the same size as Manhattan, wrapped around a rocky core. It is plausible that impacting comets seeded the Hadean Earth with water and possibly with primitive organic compounds.

outgassing
The release of gases and water vapor from molten rocks that led to the formation of the atmosphere and oceans.

Figure 18.8 Outgassing of mantle-derived methane, water vapor, ammonia, and hydrogen sulfide by volcanic eruptions formed the primitive atmosphere. Ultraviolet radiation from the Sun triggered a series of reactions that converted these gases to an atmosphere rich in nitrogen and carbon dioxide. Condensation of water vapor formed the primitive oceans.

The Origin of Life

All living things are composed of the same basic elements—carbon, oxygen, nitrogen, and hydrogen. These elements were present in the primitive atmosphere and ocean in the form of water (H_2O), carbon dioxide (CO_2), methane (CH_4), and ammonia (NH_3). In organisms, these elements are combined in the form of *amino acids*. Amino acids, in turn, combine to form the complex molecules of life: carbohydrates, fats, proteins, and nucleic acids (DNA and RNA).

Experiments have shown that amino acids can be synthesized from the basic elements with the aid of a variety of energy sources, including heat from the interior of the Earth that escapes through undersea hydrothermal vents, solar radiation, and lightning discharges. So far, however, the combination of free amino acids into the large molecules of life has neither been observed in nature nor accomplished in the laboratory. As far as we know, only living things are capable of synthesizing amino acids. How the *first* amino acids formed the *first* large molecules of life remains an unsolved mystery.

All living things share three characteristics: They take in materials for energy, growth, and repair, and expel what they do not need; they reproduce; and they respond to stimuli in their environment. The evolution from complex molecules to advanced life, however, involved the development of the *cell*.

The cell's permeable wall protected its interior from the external environment while allowing nutrients to enter and wastes to leave. The interior contained separate structures that could store energy and release it as necessary, which enabled the cell to reproduce. In modern cells, the energy storage function is accomplished by adenosine triphosphate (ATP). Reproduction is accomplished via deoxyribonucleic acid (DNA). The DNA is structured to pass on a precise set of instructions, thereby ensuring the genetic continuity of future generations. Changes in the genetic code, or *mutations*, led to *evolution*, the change in organisms through time (to be discussed later in this chapter).

The Late-Proterozoic World

Although we would still need an oxygen mask to breathe, we would certainly find the late-Proterozoic world more comfortable than the Hadean and early Archean worlds. The difference can most likely be attributed to the gradual conversion from an atmosphere lacking free oxygen—that is a *reducing atmosphere*—to one in which oxygen began to accumulate. Evidence of this conversion is preserved in Precambrian sedimentary rocks and is intricately connected to the evolution of life, as we will soon see.

Tangible evidence that a reducing atmosphere existed from Archean to early Proterozoic time is found in the **banded iron formations** of the Lake Superior district and other continents. The iron layers deposited exclusively during this time span consist of extremely thin layers of iron minerals (magnetite) that were precipitated in freshwater lakes or in the ocean. These minerals alternate with thicker layers of chert—a rock composed of precipitated silica.

banded iron formation
A rock that consists of alternating bands of iron-rich minerals and chert or quartz.

The significance of the banded irons lies in the chemical environment necessary for iron to be carried in solution and later deposited. Iron dissolves readily in the absence of free oxygen–that is, under reducing conditions. In contrast, it precipitates readily in the presence of free oxygen—that is, under oxidizing conditions. So we may assume that during the Archean, a reducing atmosphere dissolved the iron that was exposed on the surface of rocks, and the iron was subsequently transported to the ocean or lakes, where it remained in solution. Early in the Proterozoic, free oxygen was generated in the oceans and lakes allowing the iron to precipitate.

What generated the oxygen? According to the fossil record, the growth in the atmosphere's free oxygen supply was caused by the evolution of organisms capable of *photosynthesis*, the process by which the energy of sunlight is used to extract carbon dioxide from the air or water. As part of this process, oxygen is released as a by-product. Precambrian photosynthesis was accomplished by single-celled cyanobacteria (also called blue-green algae) that proved capable of surviving in a reducing environment (Figure 18.9). Like the reefs built from the shells of corals, structures called **stromatolites** are formed by colonies of these bacteria. Stromatolites have been extracted from the 3.5 billion-year-old Archean sedimentary rocks of western Australia. They are among the oldest fossils and appear similar to modern stromatolites.

stromatolite
A reef-like sedimentary structure produced by photosynthesizing bacteria.

Cyanobacteria were most likely responsible for the release of the oxygen in the water that precipitated the banded iron formations. After the dissolved iron in the water was used up, the excess oxygen, in the common molecular form, O_2, escaped into the atmosphere.

Figure 18.9 This present-day coastal desert in Baja California displays stromatolites remarkably similar to Proterozoic forms.

Some of it drifted to the upper atmosphere, absorbed the Sun's incoming ultraviolet radiation, and was converted to ozone, O_3. This *ozone layer* shielded the Earth's land surface from deadly ultraviolet radiation and made possible the eventual colonization by plants and animals. By supplying the molecular oxygen necessary for supporting animal life and for shielding terrestrial life from harmful radiation, cyanobacteria transformed the Earth to a degree achieved by no other organisms since then, including humans.

Proterozoic Life

Like other forms of bacteria and viruses, the cyanobacteria of the Archean eon are called **prokaryotes,** because their cells lacked a nucleus. The steadily increasing free oxygen content of the atmosphere throughout the Proterozoic eon spurred the evolution of a more versatile group of organisms, the **eukaryotes,** whose cells have a well-defined nucleus that encloses DNA and other structures. The highly regarded symbiotic theory envisions these organisms emerging from combinations of mutually dependent prokaryotic forms. In any event, the eukaryotes process energy far more efficiently than the prokaryotes and, unlike the prokaryotes, many are capable of sexual reproduction. In contrast to simple cell division, which produces identical organisms, sexual reproduction mingles the DNA of separate organisms, thereby permitting varied genetic and evolutionary possibilities. All the more advanced single-celled organisms, such as green algae and Protistas (including amoebas, ciliates, foraminifera, and diatoms) are eukaryotes. Likewise, all multicellular organisms, including plants and animals, are eukaryotes. Clearly the evolution of the eukaryotes marks a critical juncture in the history of life.

The scant fossil record of the Proterozoic consists primarily of single-celled and multicelled algae, with the late Proterozoic **Ediacaran fauna** constituting a major exception (see the photo at the beginning of this chapter). Named for the region where the fossils were first discovered, in the Ediacaran hills of South Australia, this fauna comprises a great diversity of more complex, soft-bodied marine animals, none of which appears to have survived the Proterozoic eon. Rocks containing Ediacaran fauna are relatively rare; in most localities, Proterozoic strata are barren of fossils except for occasional stromatolites and bacteria. In some localities, barren Proterozoic strata grade upward, without a physical break, into strata that differ by only one feature: The upper strata contain numerous, easily recognizable fossils—fossils of animals that had hard shells or what biologists call external skeletons—that are markedly distinct from the Ediacaran fauna. The first occurrence of these hard-shelled fossils in the rock strata marks the base of the Cambrian system, the oldest rocks of the Paleozoic era.

prokaryote
A single-celled organism without a nuclear membrane.

eukaryote
A multicelled organism with the nucleus of each cell enclosed in a protective membrane.

Ediacaran fauna
A variety of soft-bodied animals that died out before the end of the Proterozoic eon.

Why the soft-bodied Ediacaran fauna died out and why hard-shelled Cambrian fauna appeared so abruptly in the fossil record are among the major unsolved mysteries in the history of life. It is certain that the Cambrian fauna, as eukaryotic forms, evolved from Precambrian organisms, but the connections remain obscure.

The Evolution of Life on Earth

The theory of **evolution** is the unifying explanation of life on Earth. In its simplest form, it is the proposition that all organisms living today have descended by gradual changes from ancient ancestors quite unlike themselves. With 150 years of accumulated evidence to support this theory, modern geologists and biologists not only accept evolution, but also employ it every day as the foundation for their research on the frontiers of the life and Earth sciences.

Nevertheless, every 20 years or so, objections to the theory heat up and hit the front pages of newspapers as a "controversy" over the validity of the theory and a "debate" about the teaching of evolution in U.S. schools. The objections come mainly from those individuals who want either limits placed on the teaching of evolution in science courses or creationism taught alongside evolution as a legitimate scientific theory.

Creationism is a system of beliefs based on a literal interpretation of the Bible—meaning that everything written in the Bible is the exact truth, as told by God to the prophets. The creationist system of belief leaves no room for the concept of geologic time, because the Bible says that God created the heavens and the Earth (in six days) approximately 6000 years ago. Creationists also believe in the doctrine of fixed or immutable species, which states that God created each animal, plant, bacterium, and virus in the same form in which it is found today. Thus the creationist system of belief cannot accept the theory of biological evolution.

The idea that evolution has occurred and is the organizing principle of life is accepted by all scientists familiar with the subject—not only by biologists and geologists, but also by paleontologists, ecologists, embryologists, biochemists, and geneticists. They are convinced the theory is correct, not because they are anti-religion, but because mountains of evidence have shown that evolution is the only logical, all-encompassing explanation of the variety, complexity, and history of life on this planet.

You may recall from Chapter 1 that the term *theory*, as used by scientists, differs from the everyday sense of the word, which a scientist would call an educated guess or hypothesis. In science, a hypothesis is not elevated to the level of a theory until it is supported by a substantial body of evidence, has demonstrated its wide predictive and explanatory value, and has withstood rigorous testing. Note, however, that creationists do not apply the same criteria to their "theory," because they consider its basic assumptions, which come directly from the Bible, to be unassailable. For this reason, creationism is not a science. In science, no such thing as an unassailable assumption exists; nothing is off limits, in the sense that it must be accepted as an article of faith and not subjected to scientific scrutiny.

The predictive-explanatory value of a theory and its ability to stand up under rigorous testing are crucial criteria, because they provide the means to discover *falsification*, or proof that the theory is wrong. Falsification requires only the discovery of evidence that the theory is incapable of explaining, or evidence that runs contrary to predictions based on the theory. Let us see how the theory of evolution withstands a number of falsification tests.

evolution
The process by which populations of organisms change in form and function over time.

Evidence of Evolution in Fossils and Comparative Anatomy

Fossils offer a good place to begin the search for falsification. If the theory is incorrect and present-day life has not evolved from past life, then we should expect to find no change in the fossil record with time, and all rock strata should contain fossils exactly like the animals, plants, and bacteria of the present. The fossil record proves, however, that living things have changed dramatically through time. We find that fossils replace one another in definite and recognizable order as we proceed upward in time from older through younger strata (the law of faunal succession).

A distinct pattern emerges out of this vast fossil record: the sequential appearance of the major plant and animal groups. Relatively simple bacteria and blue-green algae appear first, in Archean rocks, followed by marine invertebrate fossils in late-Proterozoic–early-Cambrian rocks. Fossils of the first backboned animals and jawless fish appear in late-Cambrian strata, jawed fish with bony skeletons in Silurian strata, amphibians in Devonian strata, reptiles in Carboniferous strata, mammals in Triassic strata, and birds in Jurassic strata. (The mammalian order to which humans belong, the primates, has been traced to the Paleocene.)

The case for evolution grows even stronger when the anatomical relationships between fossils and living organisms are examined in detail. Figure 18.10 illustrates the structural similarities in the forelimbs of various living mammals, birds, reptiles, and amphibians. Clearly these forms are descended from the ancestor of all land vertebrates, the Devonian crossopterygian fish.

Transitional Fossils

The primitive amphibian forelimb in Figure 18.10 is but one example of the clear linkages that have been discovered between extinct ancestral forms and their modern descendants. The Jurassic *Archaeopteryx* (see Figure 7.19) is clearly the bridge between reptiles and primitive birds. The size of a crow, it had feathers, a wishbone, and a bird's pelvis; however, it also had such reptilian features as teeth, claws, an abdominal rib cage, and a long bony tail. In fact, *Archaeopteryx* bears a close anatomical resemblance to a small species of dinosaur; in a sense, then, the dinosaurs survive today as birds.

The abundant Cenozoic terrestrial deposits of the high plains of North America contain one of the best records of evolutionary descent, one leading to *Equus,* the modern horse (Figure 18.11). The fossil record begins 54 million years ago in Eocene strata with the bones of *Hyracotherium*. About the size of a small dog, this animal had four toes and short legs. Fossils in successively younger strata show a clear trend toward an increase in body size, longer legs, reduction of the toes to a single functioning digit on each foot (the hoof), and the development of teeth with intricate enameled crowns. Like this year's new car, *Equus* is clearly a modification of previous designs.

The same sediments containing the bones of the horse family also provide evidence of the environmental pressures that spurred the evolution of this species. As we proceed upward in time from Eocene into Pleistocene sediments, we find that the Cenozoic climate was growing drier. Plant remains, pollen, ancient soil profiles, and other features attest to a changeover from a tropical forest to a tall grassland habitat. In response, the ancestral horse evolved from a deer-like browser, feeding on tender foliage, to a grazing plains animal, feeding on tough grasses. On the open plain, speed, strength, and the kind of teeth that could grind down the grasses were highly desirable characteristics, and they favored—through mechanisms discussed below—the evolutionary changes illustrated in Figure 18.11.

Vestigial Organs

The blend of fossil and anatomical evidence derived from the analysis of *Archaeopteryx, Equus,* and crossopterygian bones—along with numerous similar examples—is generally sufficient to convince most objective observers of the reality of evolution. The creationist might argue that each species was created separately and that the similarities among

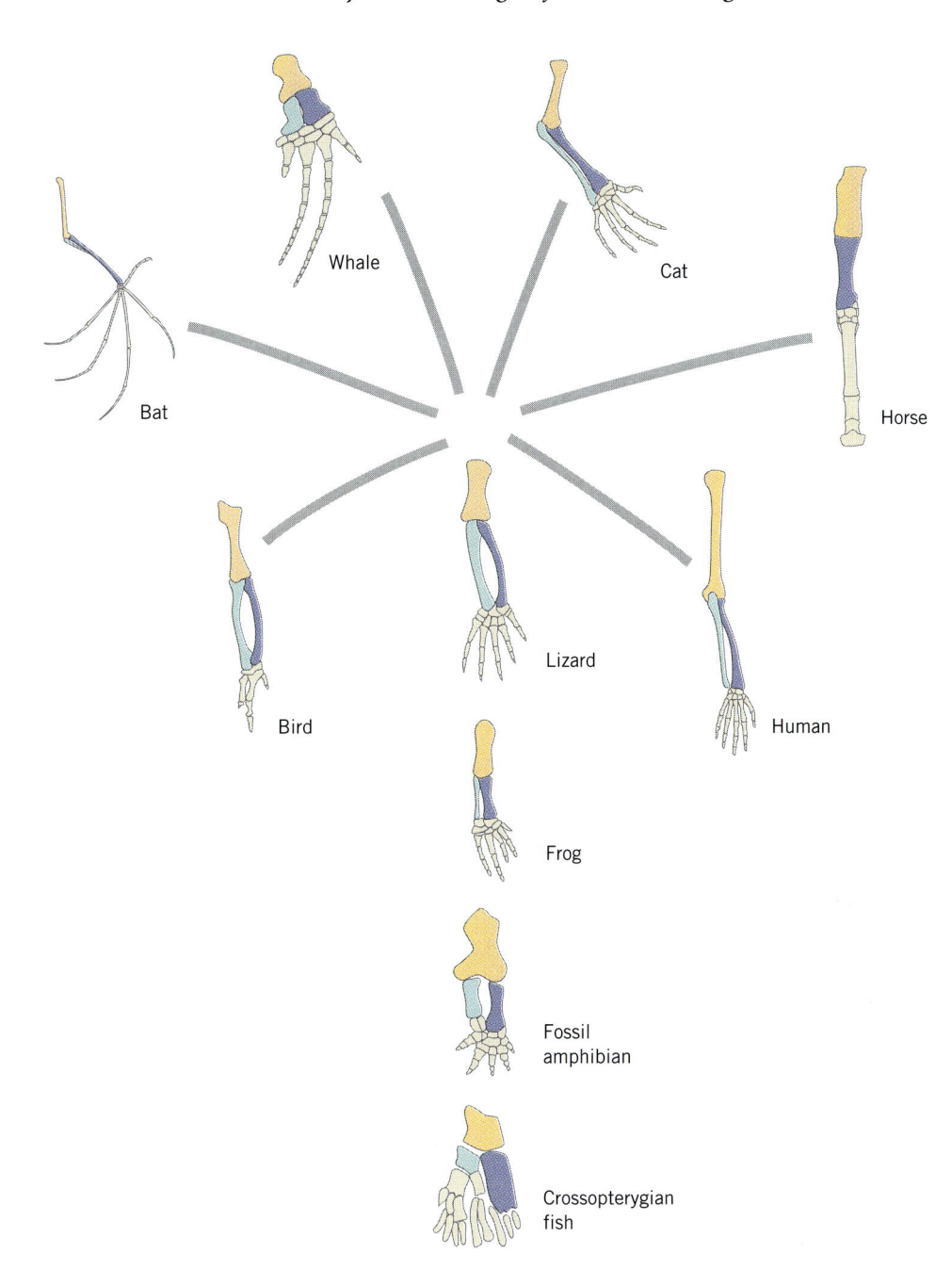

Figure 18.10 Structural similarities in the forelimbs of the vertebrates. (For the purpose of comparing their forms, the forelimbs have been drawn at different scales.) The best explanation for the consistency of structure is that the various organisms were all derived from a common (crossopterygian) ancestor. The differences among the forelimbs are evolutionary adaptations over long time periods. (Reprinted from *Evolution and the Myth of Creationism* by Tim M. Berra with the permission of the publishers, Stanford University Press. © 1990 by the Board of Trustees of the Leland Stanford Junior University.)

them are merely reflective of God's overall design. This hypothesis, however, cannot account for the presence of *vestigial organs*—diminished organs that serve no discernible function in a living species, but were a vital part of the anatomy in an earlier, extinct ancestor. What need, for example, has a whale or a snake for pelvic and thigh bones (Figure 18.12)? The plausible explanation is that whales and snakes are descended from four-legged ancestors. The human body itself contains numerous vestigial organs, including wisdom teeth, ear-wiggling muscles, and a tail bone at the base of the vertebral column—hardly a confirmation that we were created separately and distinct from other organisms.

The evidence of evolution we have advanced so far has come from geological research (this is, after all, a geology textbook), but the evidence drawn from numerous other branches of science is equally compelling. Furthermore, the evidence of evolution is internally consistent from discipline to discipline. Let us examine two of the many examples: one from embryology and the other from genetics.

18.11 Evolutionary changes leading to the modern horse, *Equus*. The major trends included an increase in body size, as shown in the family tree (a), and a reduction in the number of toes and the development of teeth with intricate, enameled crowns as shown in (b). ("Evolution of the horse" from *Biology* by Neal Campbell. Copyright © 1987 by The Benjamin/Cummings Publishing Company. Reprinted by permission.)

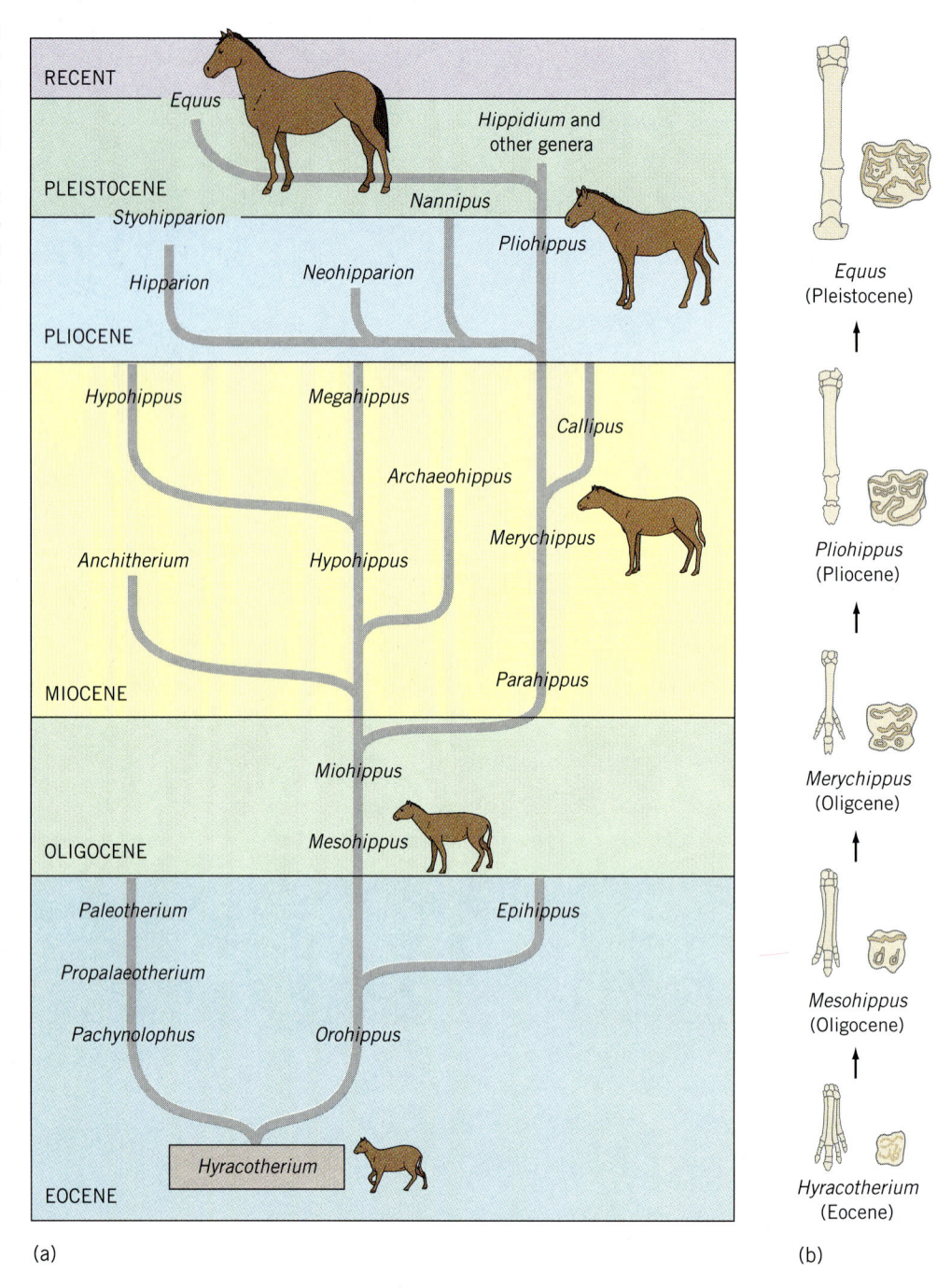

(a)

(b)

Evidence of Evolution in Embryology

Embryology is the study of organisms in their early stages of development, generally while they remain enclosed within a seed, egg, or uterus. What can this discipline tell us about the validity of the theory of evolution? Figure 18.13 compares the five classes of vertebrates—fish, amphibians, reptiles, birds, and mammals—at three parallel stages of embryonic development. Note the remarkable *consistency of pattern,* and a sequence of development consistent with the appearance of these vertebrates in the fossil record. "The early embryos of all vertebrate classes resemble one another markedly (stage 1). The embryos of vertebrates that do not respire by means of gills (reptiles, birds, and mammals) nevertheless pass through a gill-slit stage complete with aortic arches and a two-chambered heart, like those of a fish (stage 2). The passage through a fish-like stage

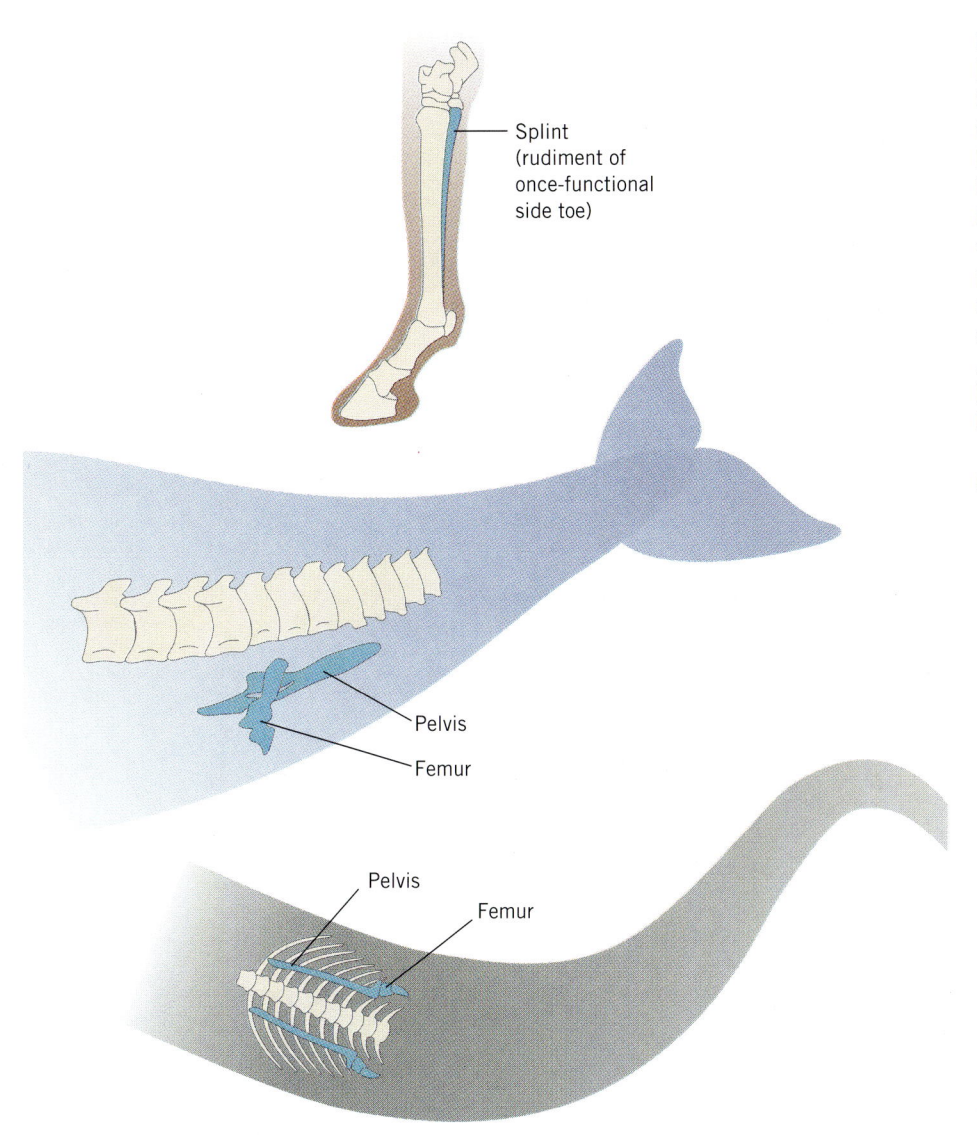

Figure 18.12 Vestigial organs of the horse (top), the whale (middle), and the python (bottom). In the horse's leg, the splints on either side of the leg bone are rudiments of the once-functional side toes of the horse's three-toed ancestor. The rudimentary pelvises and femurs in the whale and the python reflect the fact that both animals are descended from four-legged ancestors. (Reprinted from *Evolution and the Myth of Creationism* by Tim M. Berra with the permission of the publishers, Stanford University Press. © 1990 by the Board of Trustees of the Leland Stanford Junior University.)

by the embryos of the higher vertebrates is not explained by creation, but is readily accounted for as an evolutionary relic."*

With the exception of the fish and salamander, only in stage 3—relatively late in the developmental sequence—does each embryo display some of the features characteristic of its own species. For example, the forelimb structures of the various vertebrates depicted in Figure 18.10 are derived from the same embryonic cell layers at the same late stage of development.

Evidence of Evolution in Genetics

Genetics is the study of heredity, or how the characteristics of organisms are transmitted from one generation to the next. The genetic code of an organism, and hence all its heritable traits, are found in deoxyribonucleic acid (DNA). The DNA molecule consists of two strands of sugar and phosphate chains, twisted into a helical structure and linked together by four *nucleotide* bases, like the rungs of a ladder. Each rung consists of two

---

*Tim M. Berra, *Evolution and the Myth of Creationism* (Stanford, CA: Stanford University Press, 1990), 22.

Figure 18.13 Similarities of features in the embryos of six vertebrates during three stages of development. The sequence of changes that occur as the embryo develops reflects the sequence of changes that occurred as the species evolved.

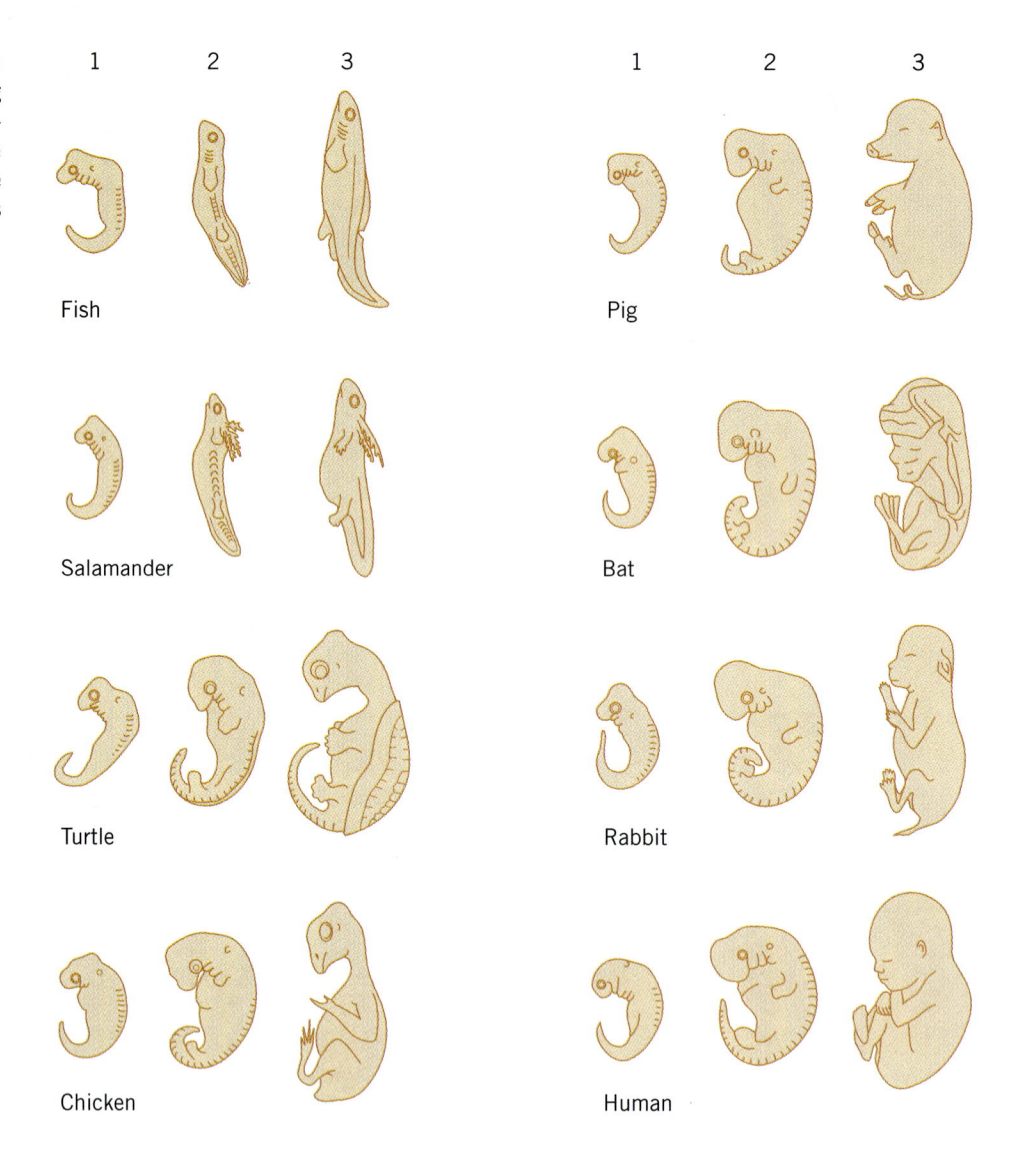

1 2 3   1 2 3

Fish   Pig

Salamander   Bat

Turtle   Rabbit

Chicken   Human

complementary nucleotides. Traits are transmitted by *genes,* which are specific sequences of nucleotides (or "rungs") located along the strands. They control the synthesis of the organism's protein.

A species is a plant or animal population that reproduces only with other members of the population, thus maintaining the group's unique set of characteristics. In the process of sexual reproduction, a male's sperm cell fertilizes a female's egg cell. The nucleotide of each DNA strand of the male cell links up perfectly with its complement on each strand of the female cell to form a complete, new DNA molecule. Because the strands form very tight bonds, a high degree of energy (heat) is required to separate them. When the DNA strands of two distinct species are combined experimentally, they form loose bonds because not all of their nucleotides hook up. As a result, the strands separate easily when a little heat is applied. Hence, the temperature of separation of the strands provides a measure of DNA similarity and, presumably, how closely the species are related.

The technique we have described, *DNA–DNA hybridization,* has been applied to primate DNA, and the results match the fossil record (Figure 18.14). For example, the branch of great apes whose DNA is least related to us, the orangutans, is shown by the fossil evidence to have diverged from our common ancestor 17 million years ago. Fossils of the Old World monkeys, whose DNA is even more distant, show an earlier divergence, occurring some 24 million years ago. It also seems that humans are more closely

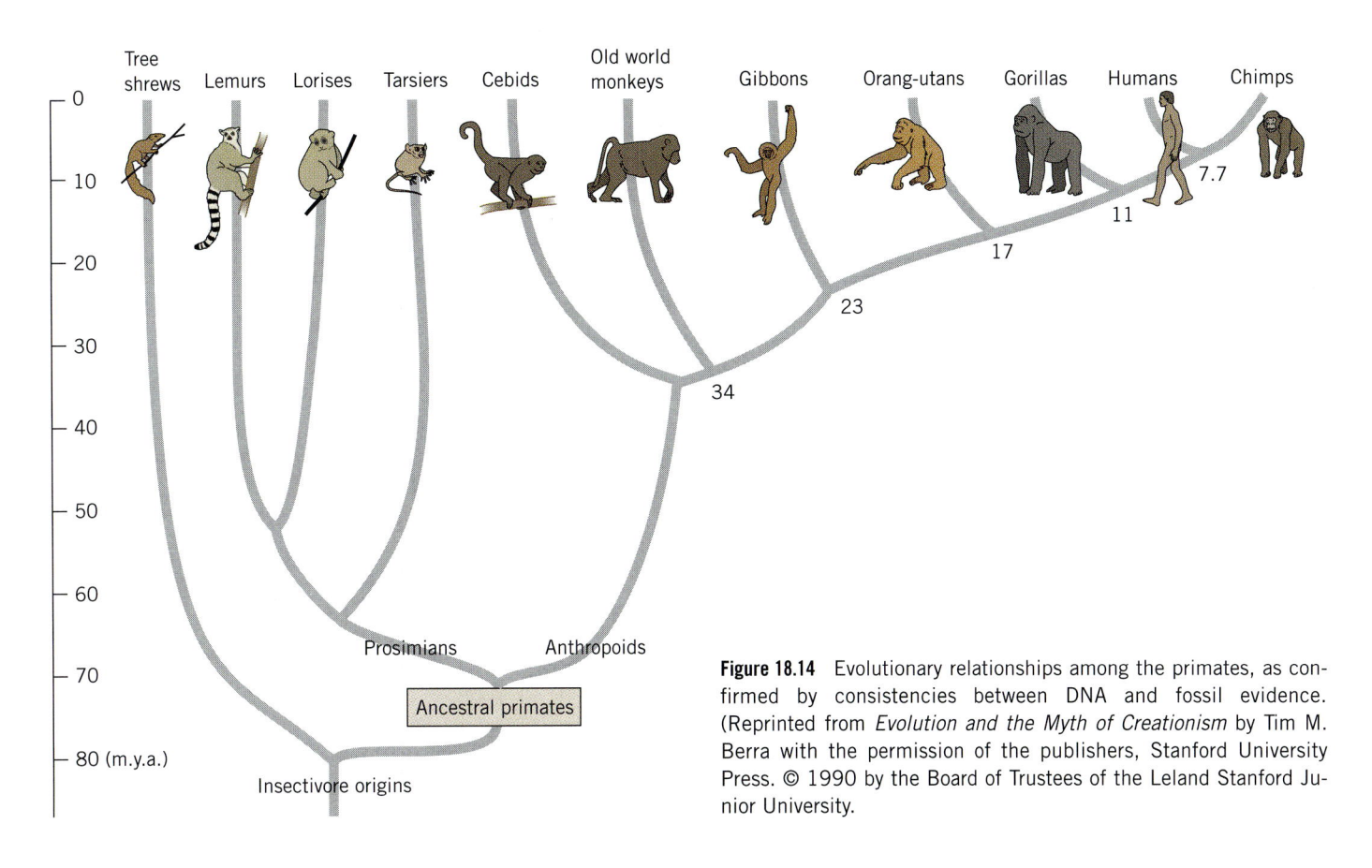

Earth History: From the Origin of the Planet Through the Proterozoic Eon 395

Figure 18.14 Evolutionary relationships among the primates, as confirmed by consistencies between DNA and fossil evidence. (Reprinted from *Evolution and the Myth of Creationism* by Tim M. Berra with the permission of the publishers, Stanford University Press. © 1990 by the Board of Trustees of the Leland Stanford Junior University.

related to chimpanzees than either species is to the gorillas. We share 97 percent of our DNA with chimps; chimps and humans diverged from one another about 7.7 million years ago.

The Mechanisms of Evolution

A glance around your classroom will confirm that no single individual contains all the heritable traits of the species (its gene pool); genetic variation is readily apparent among individuals. It is also true that, in each generation, more individuals are born to a given population in the wild than can survive. This scenario leads to competition (either direct or indirect) among individuals of both the same and different species for available resources—water, food, mates, shelter, and sunlight. Those individuals having heritable traits that enable them to survive long enough to reproduce will pass on these traits to the next generation; those having unfavorable traits are likely to die before reproducing. Thus unfavorable traits tend not to be handed down to subsequent generations.

Charles Darwin and Alfred Russell Wallace independently identified this process in the mid-nineteenth century. They named it *natural selection*, and in 1858 they jointly proposed that it was the basic mechanism of biological evolution. They emphasized that, given sufficient time and an ever-changing environment, the small changes from generation to generation would add up to major evolutionary changes.

For a while, skeptics claimed that the Darwin-Wallace theory did not explain the evolution of *new* species, merely the selection of traits *within* species. For example, humans have selectively bred all manner of dogs from the original parent wolf stock, but they are still dogs! Let them revert to the wild and mate at will for a few generations, however, and a blending of traits will become evident. What the Darwin-Wallace theory was unable to explain was how small changes in traits arise in nature—until the advent of modern genetics, that is, and the discovery of *mutations*. A mutation is basically a "mistake" during DNA replication that alters the basic unit of inheritance, the gene. Geneticists have calculated that it occurs spontaneously, about once every 10^6 to 10^8 replications. Exposure to natural nuclear radiation, X rays, ultraviolet light, and certain chemicals can increase the frequency of mutations.

Most mutations are detrimental or neutral, but a few are advantageous. These changes are incorporated into the gene pool and spread among the population at speeds that depend on the reproduction rate of the organism and on the *selective pressure* to which the population is subjected. For example, bacteria and insects mutate rapidly in response to selective pressure from the antibiotics and pesticides we devise to kill them, forcing biochemists to constantly change the formulas of our weapons.

Geneticist J. B. S. Haldane has calculated that if a mutation produces individuals whose ability to survive has been increased by 1/1000, and the mutation is present in 1/10,000 of the population, it can spread to 99 percent of the population in 23,400 generations. For bacteria that reproduce every few hours, this process may only take a few years. Changes in more complex organisms take a long time from the human perspective, but a short amount of time from a geological point of view.

Geology and Natural Selection

Geological processes play a major role in natural selection, because they exert a subtle control over terrestrial and marine environments. Virtually every form of life is affected by these processes, and any geological change brings about corresponding genetic adjustments or *adaptations*. These adaptations are what we ultimately mean by evolution. A few examples out of many will suffice to convey the interplay between geological change and natural selection.

Consider once again the evolution of the horse, this time in relationship to the Cenozoic orogeny that produced the Rocky Mountains. As the massive peaks increased in height, they grew ever more efficient at blocking the moist Pacific winds that had kept the North American Great Plains as a lush forest—a process leading to the dry conditions that persist to this day. As the peaks rose inch by inch, the plains slowly dried up, and *Equus* evolved in response to its gradually changing habitat.

Geographical isolation as a result of geological processes can also spur evolution. Relatively recent volcanism produced the Galapagos Islands, which are located 1000 kilometers west of Ecuador. A species of small South American ground finch that somehow made its way to these islands found little or no competition from other birds (Figure 18.15). Although the originals were seed eaters, their descendants diversified into new species with a variety of feeding modes and habitats. Different beak size and shape, body size, and other features enabled the new finch species to better exploit the food-gather-

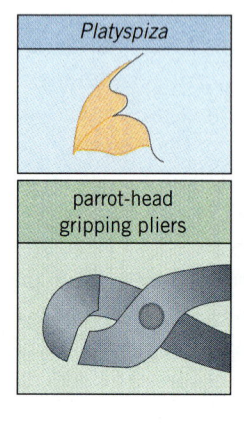

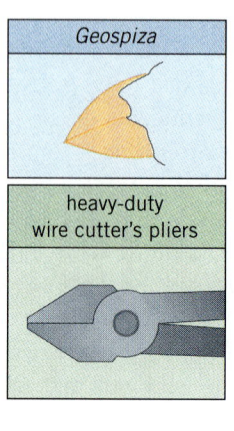

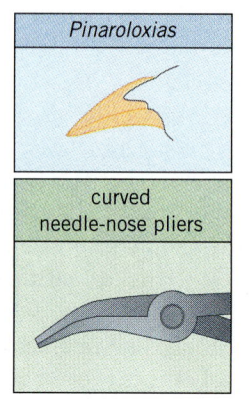

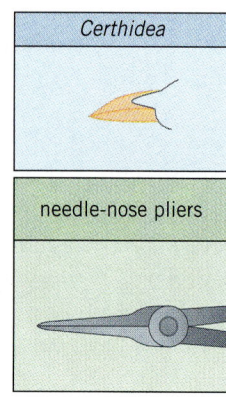

Figure 18.15 Charles Darwin observed that, just as humans devise different tools for different jobs, the beaks of various Galapagos finch species evolved into different shapes, each specialized for gathering that species' particular type of food.

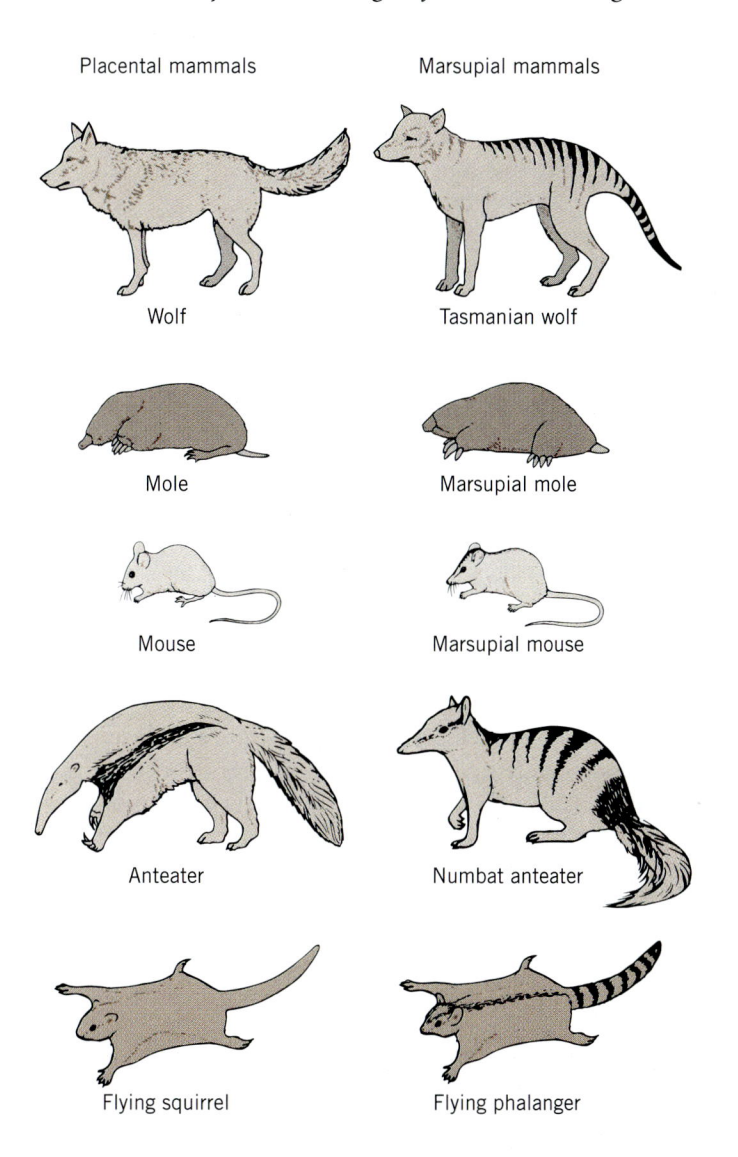

Figure 18.16 Evolutionary convergence between the placental mammals of North America and Eurasia and the marsupial (pouched) mammals of Australia. The similarity of form reflects the adaptations of distantly related organisms to similar habitats and environmental pressures. ("Similarities of marsupials and placentals" from P. H. Raven and G. B. Johnson, *Biology*, 2nd ed., 1989. Reproduced with permission of The McGraw-Hill Companies.)

ing opportunities (what biologists call *ecological niches*) of the island. For example, one species grew smaller and developed a thin, tubular, hummingbird-like beak that enabled its members to hover over flowers and siphon nectar. Another species of tree-dwelling finch developed a hard, pointed, woodpecker-like beak useful for hammering tree trunks and digging out insects.

"Seeing this gradation and diversity in one, small, intimately related group of birds, one might readily fancy that from an original paucity of birds in the archipelago, one species has been taken and modified to different needs," Darwin wrote in *Voyage of the Beagle* after visiting the islands in 1835. Darwin's finches, as they are called, became the model of *adaptive radiation,* or evolutionary divergence: one species evolving into many in response to a freshly opened habitat.

When Europeans first explored Australia and New Zealand in the eighteenth century, they discovered not only kangaroos, but also animals remarkably similar to those found on other continents: cats, moles, wolves, mice, anteaters, groundhogs, and squirrels (Figure 18.16). These Australian mammals turned out to be very different from the mammals on all the other continents, however. They were, like the kangaroos, marsupials—pouched mammals that diverged 100 million years ago from a common ancestor of their placental look-alikes.

The marsupials evolved in isolation from placental mammals after Australia separated from Gondwanaland early in the Cenozoic era. The similarity in appearance to their placental relatives reflects evolutionary adaptations to similar habitats and environmental conditions. The Australian marsupials are a classic example of *convergent evolution.*

The mechanism of evolution need not be isolation, however. Meteoroid impact, volcanism, or both are probably responsible for the extinction of the dinosaurs and reptiles at the close of the Cretaceous period. As we will learn in the next chapter, the sudden death of their competition gave small mammals the opportunity to radiate explosively during the early Cenozoic.

The Status of Evolution

In the more than 150 years since Darwin and Wallace proposed evolution through natural selection, the theory has withstood every falsification test thrown at it. No alternative theory with even an iota of its explanatory and predictive power has been proffered. Over these many years the theory has, of course, been modified to keep pace with our improving knowledge of geology, ecology, genetics, and biochemistry; nevertheless, the foundations of the theory are firmer than ever.

As with all great theories, evolution encourages, rather than inhibits, research. Paleontologists continue to seek greater insights into the causes of the extinctions of species. Is the dying out of species purely random, or are underlying mechanisms at work? They are also interested in the *pace* of evolution. Is it primarily gradual, as Darwin envisioned, or does it proceed in fits and starts, with explosive change followed by long periods of little change (what the backers of this hypothesis call *punctuated equilibrium*)?

Geoscientists are also interested in the origin of life. How did it evolve from nonliving compounds? Researchers have a fair grasp of most of the chemical steps involved, but much remains to be learned. Darwin hypothesized that life probably began in a warm pond somewhere. His view is shared by many of today's scientists who believe that the prerequisites for the origin of life are present on the Earth's surface under ordinary conditions. Recent discoveries have led some geoscientists to other conclusions, however. They are taking a closer look at the communities surrounding the black smokers venting from mid-ocean rift valleys and analyzing organisms associated with hot springs and geysers. These exotic habitats contain primitive bacteria and algae identical to Archean forms. Other geoscientists are deeply engaged in exploring the possibility of primitive life on other planetary bodies inside and outside the Solar System. It is clear that these geoscientists will have a full plate well into the next century, but it is doubtful that Darwin would be surprised at any of it.

STUDY OUTLINE

I. ORIGIN OF THE PLANETS. The planets began as a cloud of hot gases and dust particles that condensed with cooling.

  A. Light metals, iron, and silicate minerals condensed closer to the Sun to form the terrestrial planets; hydrogen, ammonia, water vapor, and methane gases condensed farther out to form the Jovian planets.

  B. Gravitational attraction further condensed matter into planetesimals, some of which acquired enough mass to continue growing to the size of a planet.

  C. Through heat supplied by meteoroid impact and the decay of abundant radioisotopes, all of the planets went through core formation and remelting early in their histories.

II. **PRIMITIVE PLANET EARTH: THE HADEAN EON (4.6 TO 3.96 B.Y.A.).** As no rock evidence remains from the **Hadean eon,** geologists can only speculate that the surface of the still-hot Earth was dotted with impact craters and with basaltic lava flows and volcanoes that spewed water vapor and other gases.

III. **THE ARCHEAN EON (3.96 TO 2.5 B.Y.A.).** The age of the oldest known rock (3.96 b.y.) marks the beginning of the **Archean eon.**
   A. Archean rocks, which are today found in the shields and cratons of the continents, provide evidence that during this eon, the Earth cooled considerably, continental crust formed, the atmosphere and oceans began to develop, and the earliest life forms emerged.
      1. Archean gneisses, metamorphosed from high-silica igneous rocks, are evidence of early subduction. Preserved within the gneisses are *greenstone belts*—solidified submarine lava flows.
      2. Internal temperatures were still high and the crust consisted of small, rapidly recycling platelets that later formed island arcs and microcontinents.
   B. At the close of the eon, in the Kenoran orogeny, a number of microcontinents combined to form the core of the North American craton.

IV. **THE PROTEROZOIC EON (2.5 B.Y.A. TO 570 M.Y.A.)**
   A. The **Proterozoic eon** encompassed the development of modern-style ocean basins and continental blocks, and full-scale plate tectonics.
      1. A cooling interior permitted the development of thicker, more rigid lithospheric plates.
      2. During the Proterozoic, the continents combined, broke apart, and recombined several times.
   B. The Precambrian Atmosphere and Oceans. In the Hadean and Archean eons, **outgassing** (the release of gases by volcanoes) formed a reducing atmosphere—one devoid of free oxygen.
      1. Excess water vapor in the atmosphere fell to Earth to form primitive oceans.
      2. Life began in the primitive oceans. All living things:
         a. are composed of the same elements (carbon, oxygen, nitrogen, and hydrogen).
         b. take in materials for energy and growth, and expel what they do not need.
         c. reproduce.
         d. respond to stimuli in their environment.
      3. Evolution from complex molecules to advanced life began with the development of the cell.
      4. Early in the Proterozoic, cyanobacteria, which were capable of living in a reducing environment, produced free oxygen as a by-product of photosynthesis.
      5. As free oxygen proliferated in the oceans, iron that had been held in solution in the reducing environment was able to precipitate to form sedimentary layers called **banded iron formations.**
   C. Proterozoic Life
      1. The cyanobacteria were **prokaryotes.** Their cells lacked a nucleus and they reproduced by simple cell division. The increase of free oxygen spurred the evolution of the **eukaryotes,** the ancestors of all multicellular organisms. Eukaryotes' cells have a well-defined nucleus and DNA, and are able to reproduce sexually—characteristics that allowed for varied genetic and evolutionary possibilities.

2. The late Proterozoic saw the proliferation of **Ediacaran fauna,** complex soft-bodied marine animals.

V. **EVOLUTION OF LIFE ON EARTH.** The theory of **evolution,** one of science's greatest accomplishments, is the unifying explanation of life on Earth.
   A. Evolution is supported by a wide variety of evidence, such as:
      1. Faunal succession—the unfailing appearance of fossils in strata of decreasing age in the same sequential order, from simple bacteria and blue-green algae to marine invertebrates, to the first backboned animals and jawless fish, to jawed fish with bony skeletons, to amphibians, to reptiles, and finally to birds and mammals.
      2. Comparative Anatomy
         a. Anatomical similarities between fossils and living organisms are clear evidence that all land vertebrates are related and are descended from the Devonian crossopterygian fish.
         b. The evolution of individual species—such as the evolution of the modern horse from its ancestor *Hyracotherium*—can be demonstrated through comparative anatomy.
      3. Vestigial organs—diminished organs in modern animals that serve no purpose suggest that the animals evolved from extinct forms in which the organs were important.
      4. Comparative embryology—the fact that fish, amphibians, reptiles, birds, and mammals all proceed through the same three initial stages of embryonic development suggests that all evolved from a common ancestor.
      5. Genetics—measures of DNA similarities among species match closely the evolutionary relationships of species established by the fossil record.
   B. There are two basic mechanisms of evolution:
      1. Natural selection—the process by which successive generations within a species change to adapt to changing environments.
      2. Mutation—the process of accidental genetic change that may result in the proliferation of a new species if a change provides a competitive advantage.
   C. By exerting control over terrestrial and marine environments, geological processes play a major role in biological evolution.

STUDY TERMS

Archean eon (p. 383)
banded iron formation (p. 387)
Ediacaran fauna (p. 388)
eukaryote (p. 388)
evolution (p. 389)

Hadean eon (p. 381)
outgassing (p. 386)
prokaryote (p. 388)
Proterozoic eon (p. 385)
stromatolite (p. 387)

THOUGHT QUESTIONS

1. What surface features did the terrestrial planets share in their early history? Why do the surfaces of the Moon, Mars, and Venus retain a better record of their early history than does the Earth's surface?
2. Describe the relationship between temperature changes in the interior and changes in the Earth's lithosphere and atmosphere during the Archean eon.
3. What were the Archean cyanobacteria and how did they make the world livable for advanced forms of life?

4. From cyanobacteria to mammals, the biological history of the Earth can be viewed as a succession of evolutionary innovations that gave organisms a survival advantage. List in order of appearance three of the biological structures discussed in the chapter and explain why each was an evolutionary innovation. Then choose one of the structures and describe how biological evolution might have proceeded without it.
5. Employing your knowledge of relative and absolute dating techniques (Chapter 16), explain how paleontologists have determined that the human and chimpanzee species diverged about 7.7 million years ago.
6. Whereas viruses, bacteria, and other simple organisms can evolve to meet environmental changes in a matter of months, mammals take hundreds of thousands—even millions—of years for evolutionary changes to take hold. Considering the critical time factor involved in biological adaptation, name some of the environmental changes brought about by human technological advances and the adaptational challenges faced by our species.

Chapter 19

The Paleozoic Era
- Plate Tectonics of the Paleozoic Era
- Paleozoic Life
  - The Vertebrates
  - Early Amphibians and Reptiles
  - Causes of Paleozoic Radiations and Extinctions

The Mesozoic and Cenozoic Eras
- Plate Tectonics of the Mesozoic and Cenozoic Eras
  - The North American Cordillera
- Mesozoic and Cenozoic Climate
- Mesozoic and Cenozoic Life

Earth History: From the Paleozoic Era to the Present

In this chapter, we pick up our summary of Earth history at the beginning of the Paleozoic era and continue through the Mesozoic era to the present, the Cenozoic era. During this time span, the supercontinent Pangaea formed, then broke apart to create the present ocean basins, continents, and mountain chains. Also during this time span, the complex life that presently inhabits the Earth evolved from clearly recognizable ancestral forms.

The Paleozoic Era

Snails. Clams. Squids. Crabs. Corals. Sponges. Worms. Starfish. Fishes (including sharks). Amphibians. Reptiles. Land plants. Trees. Insects. The ancestors of these life forms common to our world—and of other life forms that died out along the way—evolved during the Paleozoic era, the era of ancient life. The variety of life on land and in the oceans made the Paleozoic world seem very much like our present one.

The Paleozoic era encompassed the interval between 570 and 245 million years ago. It began with the abrupt appearance of hard-shelled organisms and ended with the extinction of 95 percent of all marine species and 50 percent of all land species in existence at the time. This "time of the great dying," which was spread out over several million years, was the worst setback for life on Earth in geological history.

A number of hypotheses seek to explain the great explosion of life during the Paleozoic and the great extinction that brought it to a close. Although they vary in detail, most explanations point to environmental changes controlled by plate tectonics and the drift of the continents.

◀ This could have been the plains of Nebraska during the early Cenozoic era about 40 million years ago. The hyracotherium (middleground) is the ancestor of the modern horse. However, some of the mammal species shown (such as the huge titanotheres in the background) represent evolutionary dead ends.

Figure 19.1 Approximate configuration of the late-Proterozoic supercontinent.

Plate Tectonics of the Paleozoic Era

Recall from Chapter 18 that a supercontinent, clustered astride the equator, existed toward the close of the Proterozoic. From this starting point, the tectonic history of the Paleozoic may be broadly divided into three stages: (1) rifting of the Proterozoic supercontinent; (2) wide separation of the smaller continental blocks; and (3) reassembly, culminating in the supercontinent, Pangaea (see Figure 1.12). Each tectonic stage is recorded in Paleozoic rocks and structures.

During the first stage, *rifting,* the Proterozoic supercontinent divided into six continental blocks: Gondwana, Laurentia, Baltica, Siberia, Kazakhstania, and China (Figure 19.1). The margins of these ancient blocks display features associated with the early phases of plate divergence, such as steep faults and sediment- and lava-filled failed arms similar in appearance to the modern East African Rift Valley (see Figure 2.18).

Paleomagnetic data reveal that during the second stage, *separation,* Gondwana drifted toward the South Pole and the other continents migrated northward. This long interval of continental separation and rearrangement led to an important feature of Paleozoic history—the widespread advance and retreat of shallow seas over the low-lying interiors of the continents. Indeed, Laurentia and Baltica were largely underwater for most of this era. We can only speculate about why these submergences occurred, but plate tectonics theory does tell us that rapidly spreading plates are warmer and more buoyant than slower-spreading plates. They also possess wider mid-ocean ridges. These differences resulted in shallower ocean basins of lesser volume. The excess ocean water was displaced and crept inland over the continental interiors. A second reason for the periodic submergence of the continents was related to Gondwana's drift toward the South Pole. Glacial deposits on Gondwana prove that extensive ice sheets covered that continent intermittently throughout the era. Periodic advances and retreats of these ice sheets raised and lowered worldwide sea level, much like modern episodes of glacial advance and retreat.

The long periods of sea-level advance and retreat left a detailed sedimentary record on the continental interiors. A typical early Paleozoic *platform sequence* consists of a thin blanket of sandstone that was deposited as the sea advanced over a subsiding Precambrian igneous and metamorphic basement (Figure 19.2). The sandstone then grades upward into shale and/or thick limestone strata, which were deposited as the water slowly deepened. These sequences are stacked one on top of another throughout the midcontinental regions of North America and other continents. The scarcity of land-derived sediment is good evidence that the continents had been worn low prior to the marine invasion and could not supply much sediment to the encroaching seas. Virtually all that we know of Paleozoic life comes from fossils dug out of platform rocks.

During the third major stage of Paleozoic history, *reassembly,* the oceans that divided the six continental blocks gradually closed, and the continents converged, creating Pan-

Figure 19.2 This Cambrian sandstone at the Wisconsin Dells is a typical Paleozoic platform deposit. Cross-bedding, ripple marks, and other features indicate that it was deposited in a shallow, transgressing sea.

gaea. Most of this action occurred during the latter part of the era. Laurentia and Baltica collided, welding together a large continent consisting of what is today the greater part of North America and northern Europe. At roughly the same time, the convergence of China, Siberia, and Kazakhstania produced Asia. Asia then smashed into Baltica, forming the huge Northern Hemisphere landmass, Laurasia. Pangaea became complete at the close of the Paleozoic era, when Gondwana rotated and drifted northward, far enough for Africa (and South America) to collide with Laurasia.

Mountain building accompanied each reassembly event, as the subducting plates brought the continents together along convergent boundaries. The branching North Atlantic mountain belt, which includes the Appalachians, was formed primarily by the two major collisions that completed Pangaea (Figure 19.3). The first collision, which merged Baltica and Laurentia, took place in the Devonian period. It can be traced from the northern Appalachians into northern Europe. The second collision, which joined Africa with these continents, took place in the Permian period. It can be traced from the southern Appalachians into southern Europe and northern Africa.

The history of the North Atlantic mountain belt is even more complex and intricate than the above description implies, because collisions of smaller terranes with Laurasia and Gondwana occurred both prior to, and between, the two main events. Folding and shallow thrust faulting were the principal styles of deformation, as each terrane that docked onto a continent pushed earlier arrivals inland. For example, an island arc crashed into eastern North America during the late Ordovician period. This collision initiated the *Taconic orogeny*, which formed the Taconic Mountains of New York, the Berkshire Mountains of New England, and various other structures extending all the way to Newfoundland. The details of Paleozoic assembly as they affected eastern North America are described in Chapter 17.

The orogenies that accompanied Pangaea's creation uplifted the margins of the colliding continents. The shallow seas that had covered the continental interiors for most of the Paleozoic gradually retreated. By the close of the era, the continents stood emergent. Streams issuing from the rising North Atlantic mountain belt spread sediment onto the adjoining lowlands of North America, Europe, and Africa, forming swampy deltaic plains that supported huge forests (Figure 19.4). This environment favored the creation of coal and the evolution of amphibians and reptiles. In North America, these thick deltaic deposits formed the great Appalachian Plateau (known locally as the Catskills, Poconos, and Cumberland Mountains). Beneath the plateau sediments lie the early Paleozoic carbonate platform deposits of the continental interior; beneath those deposits is the Precambrian basement.

Figure 19.3 Two major collisions formed the complex, branching North Atlantic Mountain Belt. The first welded Eurasia to North America and created Laurasia. The second welded Africa and South America to Laurasia, creating Pangaea. The Appalachians of North America bare the evidence of these and other smaller-scale collisions. (Adapted from E. C. Bullard, J. E. Everett, and A. G. Smith, "The Fit of Continents Around the Atlantic," in P. M. S. Blacket, E. C. Bullard, and S. K. Runcorn, eds., *A Symposium on Continental Drift,* London: Royal Society of London, 1965, vol. 1088, p. 41.)

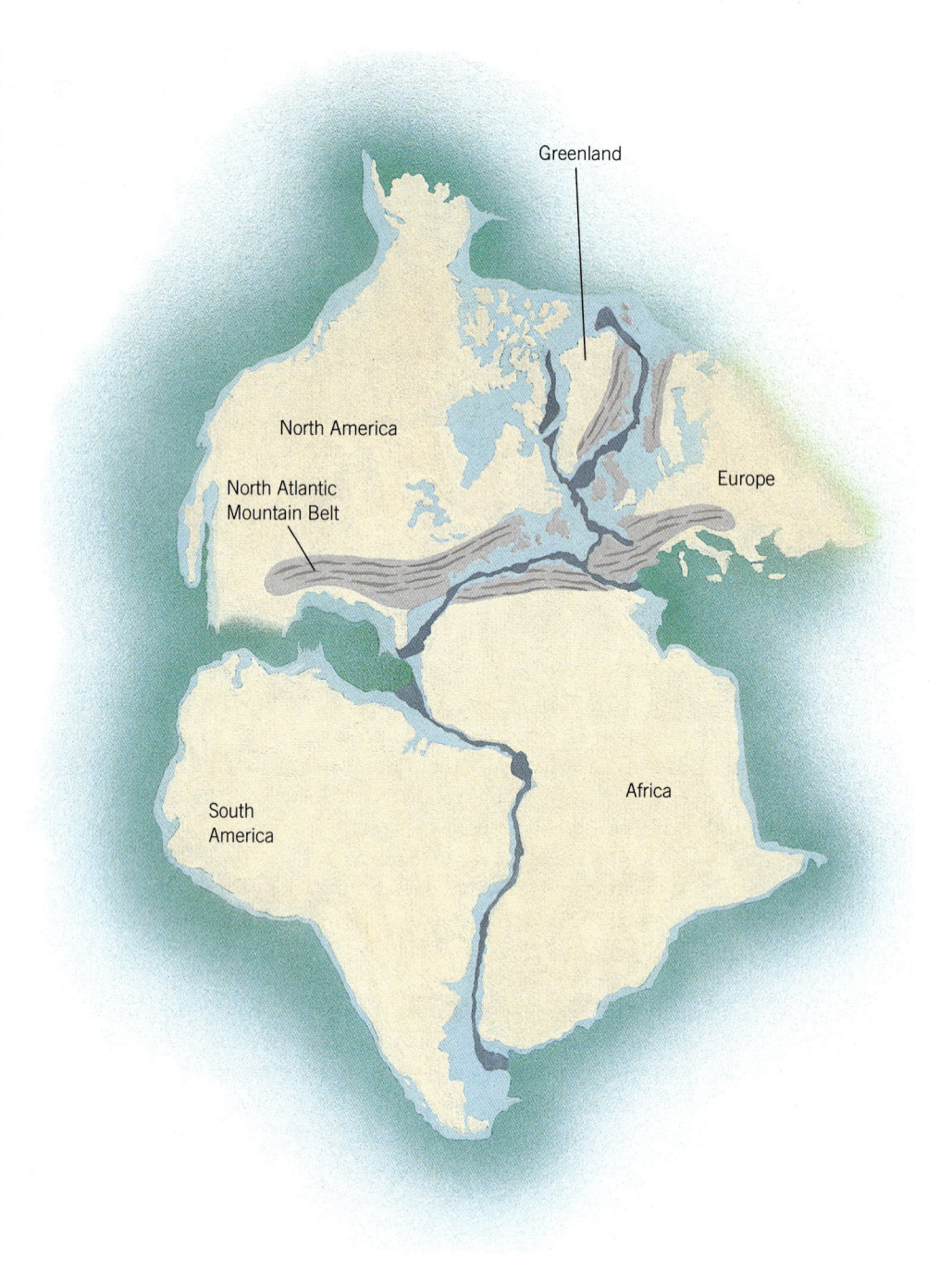

Paleozoic Life

During the Cambrian, the first period of the Paleozoic era, the giant Proterozoic supercontinent began to fragment, creating many shallow seas and thousands of square kilometers of continental shelf. The *Burgess shale,* which is preserved today in the Rocky Mountains of British Columbia, was one of those shelf or platform deposits. It contains the world's richest and most diverse collection of mid-Cambrian fossils.

What do we find as we split and sift through its thinly bedded shale layers? Mainly multicellular organisms with body plans of great structural complexity and diversity. Remnants and imprints of nerve cells, gills, muscles, and circulatory and respiratory systems are all present in the Burgess shale fossils. Some Cambrian body plans were failures and rapidly vanished; they do not appear in younger deposits. Other basic designs proved more adaptable than their rivals, and can be found in organisms that have survived to this day.

Figure 19.4 Reconstruction of a forest during the Pennsylvania period some 300 million years ago. The remains of ancient conifers, ferns, and cycads were buried and converted to coal in the oxygen-depleted swamps.

Several major lineages of **invertebrates,** animals without backbones, had evolved by the beginning of the Cambrian: the first sponges, worms, ancestral starfish, sea urchins, snails, clams, and squid. The best-known animals of this period are the *trilobites,* the first representatives of the phylum Arthropoda, or "jointed-leg animals," whose modern members include crustaceans, spiders, and insects (Figure 19.5).

Whereas the opening of the Cambrian had ushered in an explosion of life, a series of extinction events toward the latter half of the period caused nearly half of the existing animal families, including many trilobites, to die out. The Ordovician period that followed saw dramatic evolutionary spurts within each major surviving family. Each spurt can be viewed as an episode of *adaptive radiation,* in which surviving species diversified and replaced the extinct forms.

The invertebrate radiation that began in the Ordovician continued throughout the Paleozoic era (Figure 19.6). In turn, the mass extinctions that closed out the Paleozoic during the Permian period set the stage for the more modern invertebrate fauna that characterized the Mesozoic and Cenozoic eras.

invertebrate
An animal that does not have a backbone.

Figure 19.5 Early Paleozoic sea-bottom community. Trilobites (center foreground) coexisted with a variety of other animals and plants. The straight-shelled predators with branching arms are nautiloid cephalopods, ancestors of modern squid. Also prominent are solitary corals (upright tentacles), branching bryozoa colonies (colored orange), and coiled gastropods. The organisms that look like plants with long, thin stems are actually distant echinoderm relatives of modern sea urchins and starfish.

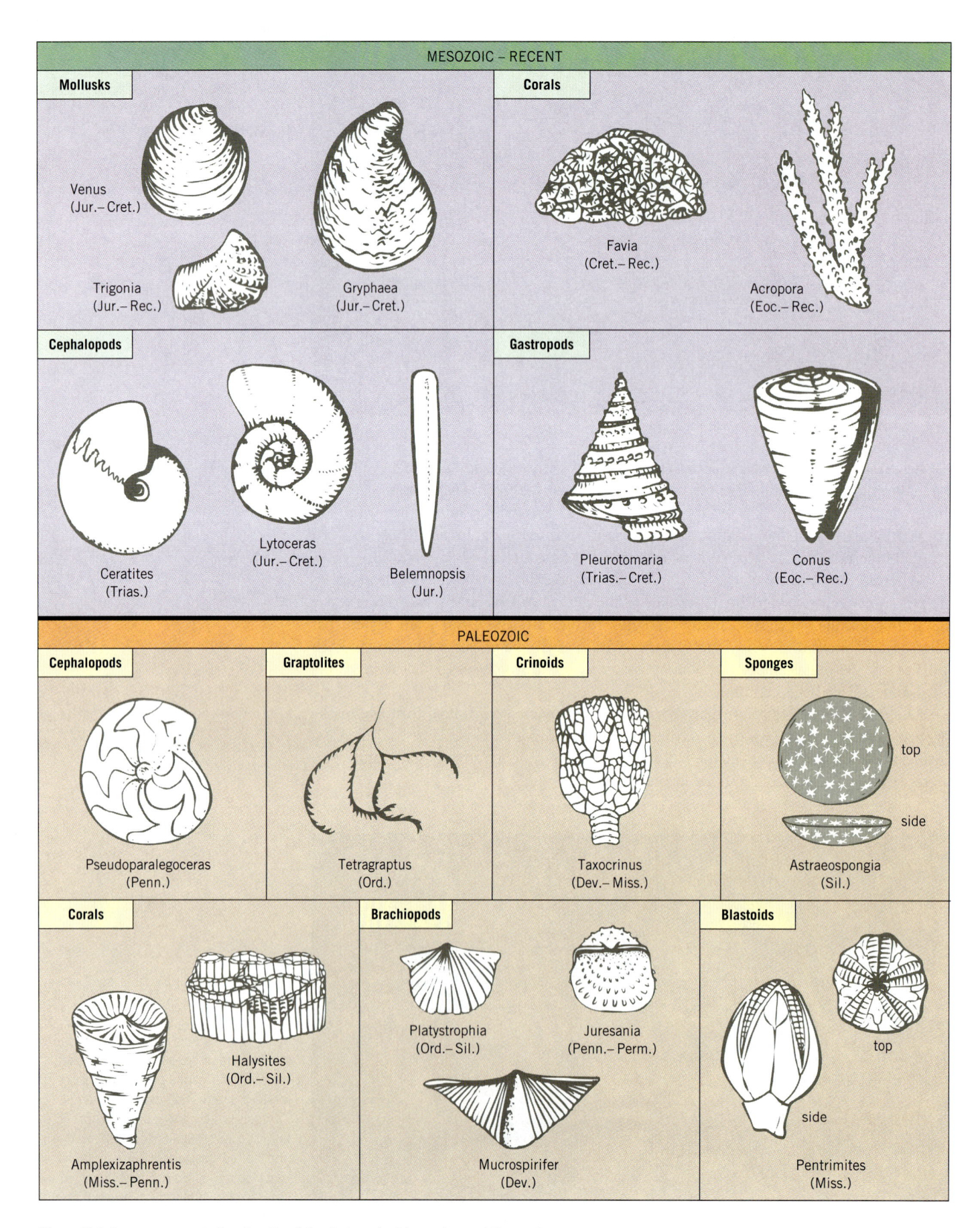

Figure 19.6 Some representative fossils of the Paleozoic, Mesozoic, and Cenozoic eras.

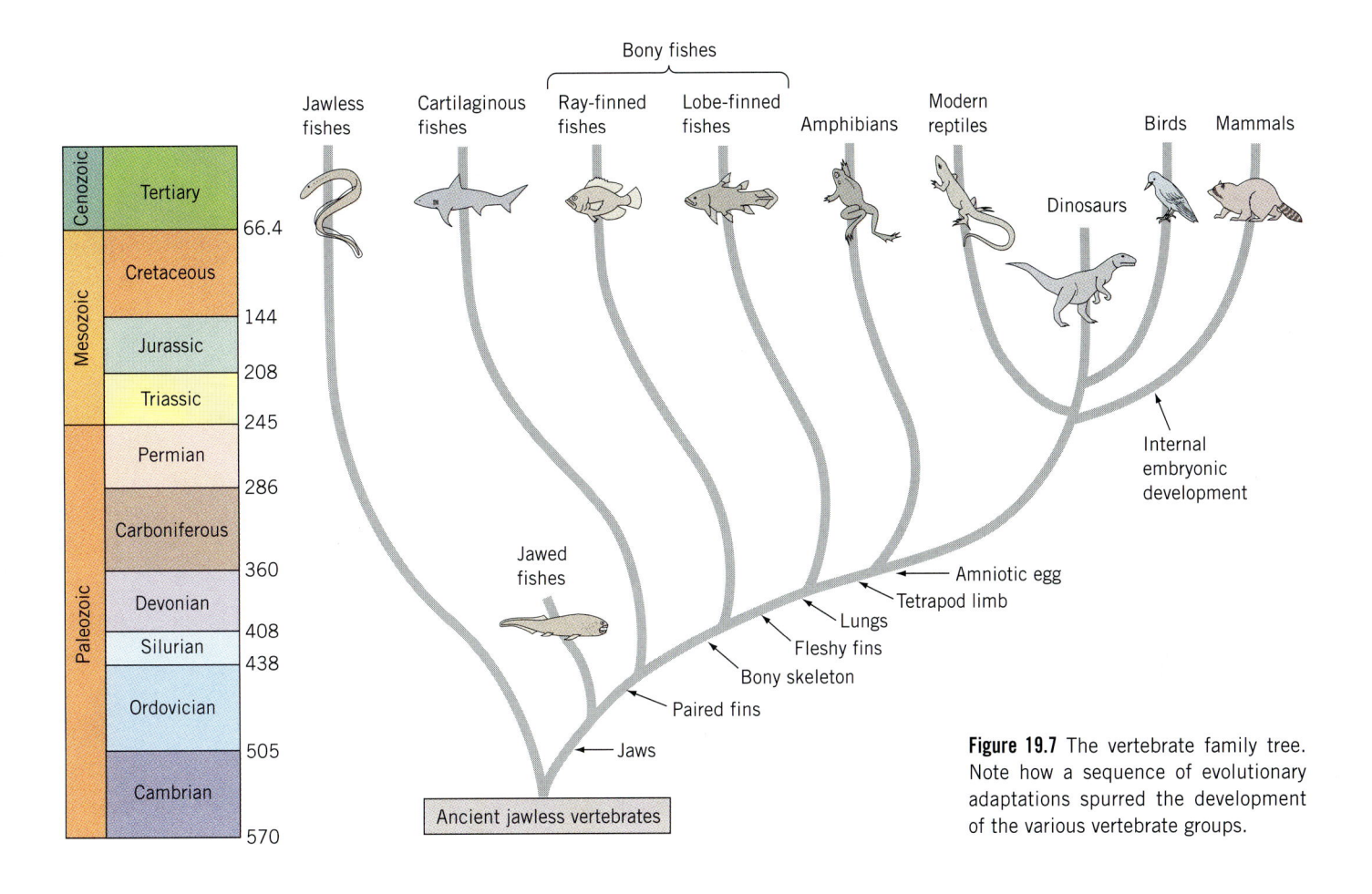

Earth History: From the Paleozoic Era to the Present 409

Figure 19.7 The vertebrate family tree. Note how a sequence of evolutionary adaptations spurred the development of the various vertebrate groups.

The Vertebrates

Three major classes of **vertebrates**—animals having a spinal column—evolved during the Paleozoic: fishes, amphibians, and reptiles. The origins of the vertebrates remain obscure. Early vertebrate fossils are rare, although recent finds prove that fishes existed as far back as the early Cambrian! From that period, we can trace, albeit with gaps, the lineage of all the modern vertebrates (Figure 19.7). As shown in the figure, each evolutionary innovation, which distinguishes one vertebrate class from another, was a trait that gave organisms a survival advantage—one that enabled them either to adapt better than other organisms in the same environmental niche, or to diversify into new environments.

The first Ordovician fish, the *ostracoderms,* were jawless and had isolated bony armor plates or scales embedded in their skin to ward off predatory invertebrates, such as the giant sea scorpions, the *eurypterids.* Ostracoderms flourished within freshwater habitats for millions of years, and many branches evolved from the limited Ordovician stock. An unknown ancestor within that stock gave rise to primitive, jawed fishes with superior armor plating, the *placoderms* (Figure 19.8). For a time the jawless varieties held their own. By the mid-Devonian, however, the placoderms had gained adaptive superiority and the ostracoderms became extinct. The placoderms, in turn, suffered an identical fate. After flourishing in the Devonian—some species reached three meters in length—they lost out to more efficient jawed fishes. The descendants of the latter group populate our modern oceans, lakes, and streams.

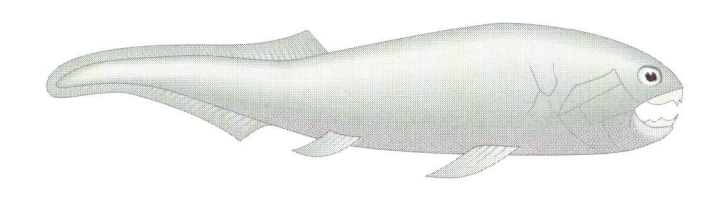

vertebrate
An animal with a backbone, which is an internal skeleton of bone or cartilage.

Figure 19.8 The Devonian armor-plated placoderm was at least 3 meters long and had massive, powerful jaws. It must have been a feared predator.

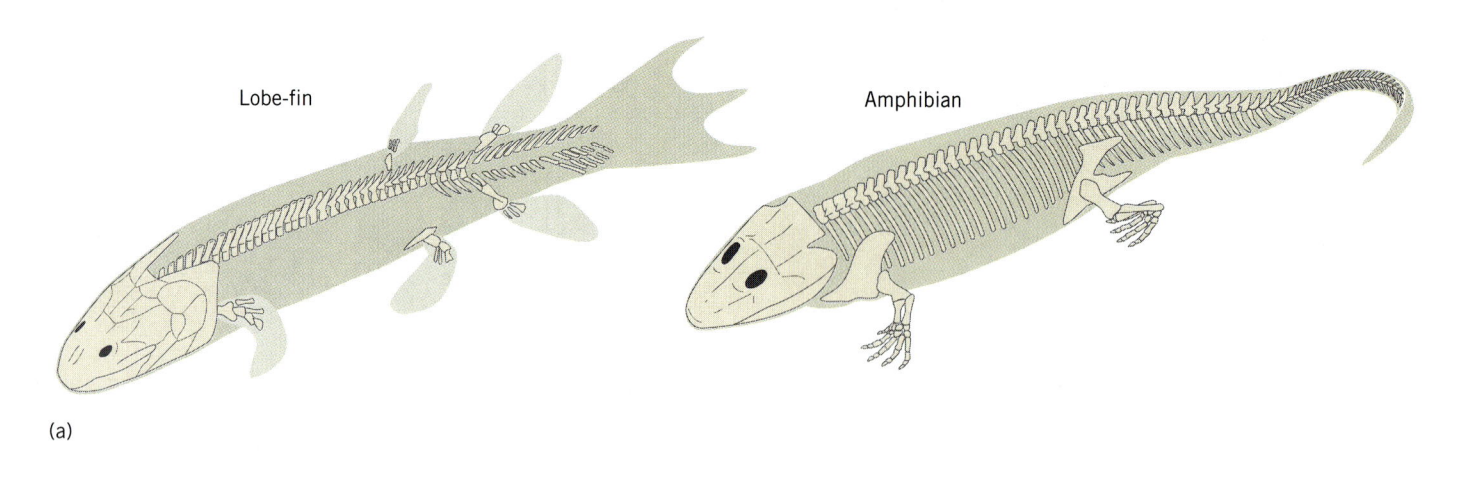

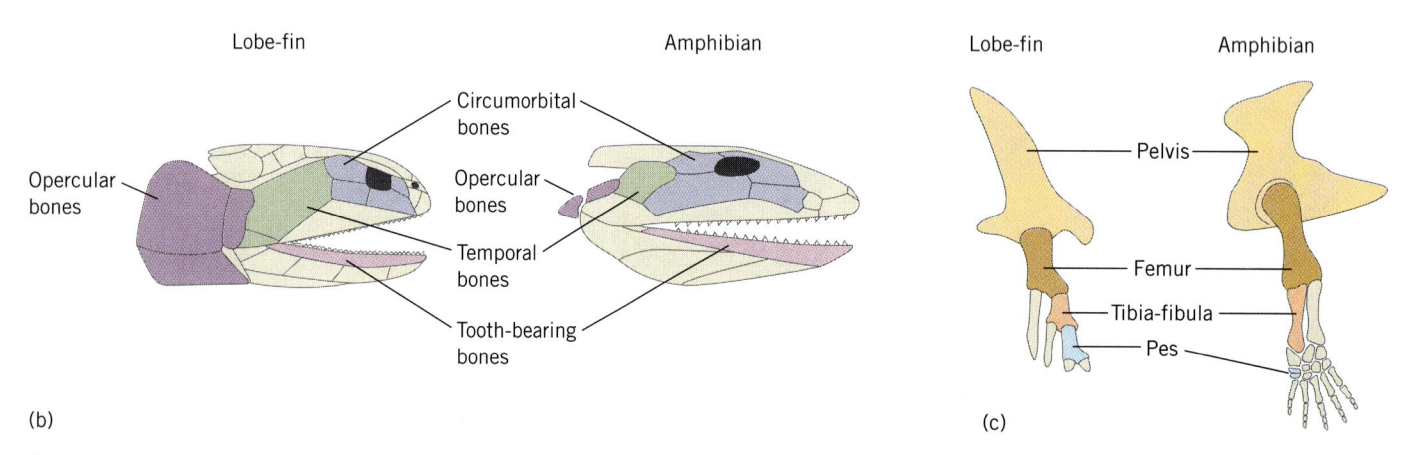

Figure 19.9 (a) Anatomical details showing the striking similarities between a Devonian lobe-finned fish (crossopterygian) and a Permian amphibian (labyrinthodont). (b) Comparison of skulls and lower jaws. (c) Comparison of pelvis and hindlimb. (Comparison of Crossopterygian and early amphibian from E. H. Colbert, *Evolution of the Vertebrates - A History of the Backboned Animals Through Time,* 3rd ed., fig. 26, p. 71, 1980. Copyright © 1980. Reprinted by permission of Wiley-Liss, Inc., a division of John Wiley & Sons, Inc.)

Some of the innovations shared by the more advanced jawed fishes include paired fins that allow for greater maneuverability, a jaw firmly attached to the brain case, and, for the vast majority, a flexible bony skeleton that provides better support. An evolutionary success story, the biological class of bony fishes includes more species than any other vertebrate group.

Most of today's bony fishes are ray-finned varieties, but a subclass has lungs and stubby, fleshy fins—the lobe-finned fishes. These *crossopterygians* have been traced back to the late Silurian and early Devonian periods, and it is highly probable that an extinct ancestor of the crossopterygian gave rise to the amphibians and thus to all land vertebrates. Convincing evidence appears when we compare the skeleton of an extinct crossopterygian with that of an early amphibian (Figure 19.9). A nearly bone-for-bone match is observed in the shoulder, limbs, digits, skull, and jaws. The basic tetrapod, or four-limbed, body plan continued in all higher vertebrates, including humans.

Early Amphibians and Reptiles

The evolutionary adaptations of the amphibians included legs and lungs that develop in adulthood. Unlike fishes, amphibians could crawl overland, from marsh to marsh or lake to lake, during a dry season. The amphibians evolved in response to the seasonal or swampy conditions that prevailed on the lowlands and flood plains of the late-Paleozoic landmasses. They flourished in the late Paleozoic and Triassic periods, and many such as frogs, toads, newts, and salamanders have, of course, survived to this day. The amphibians never became fully emancipated from their water habitat, however. As with fishes, their eggs depend upon the watery environment in which they were laid. If the water dries up, the eggs desiccate and die.

As the amphibians diversified, reptiles evolved from them and began their long-lived and enormously successful reign (Figure 19.10). This great class radiated explosively

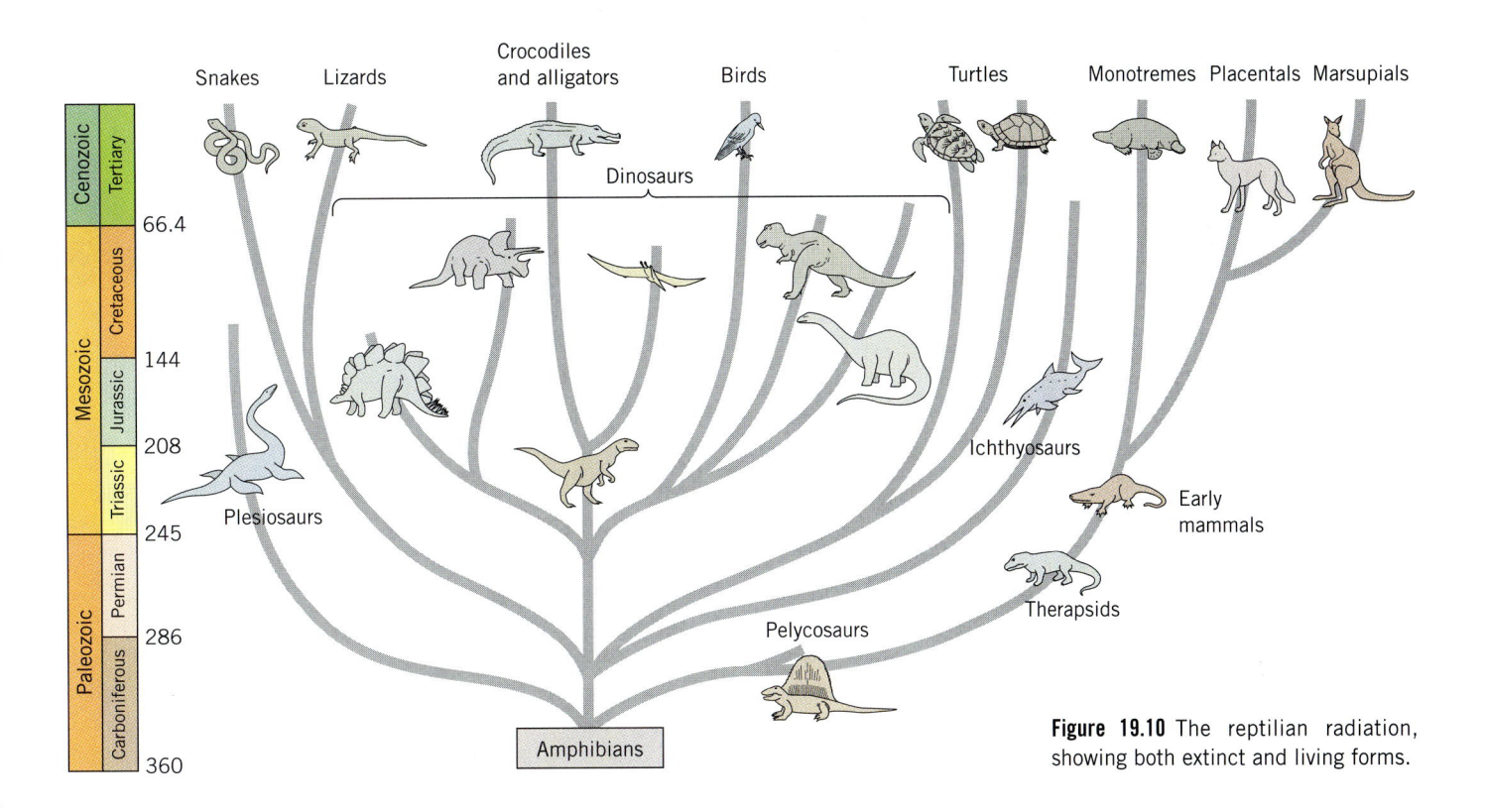

Earth History: From the Paleozoic Era to the Present 411

Figure 19.10 The reptilian radiation, showing both extinct and living forms.

from the second half of the Pennsylvanian into the Permian period, and then continued expanding throughout the Mesozoic era. The key adaptation that spurred their radiation was the *amniotic egg*, an air-tight sack, wrapped in three protective membranes, within which the embryo floats in a nutrient-rich, water environment (Figure 19.11). Amniotic eggs serve land vertebrates in much the same way that seeds serve trees and flowering plants—by preventing fluid loss and thereby making it possible to reproduce on dry land.

The next great reproductive adaptation afforded far better protection of both the embryo and the newborn, but did not occur until well into the Mesozoic. It involved *internal embryonic development*, a key characteristic of mammals.

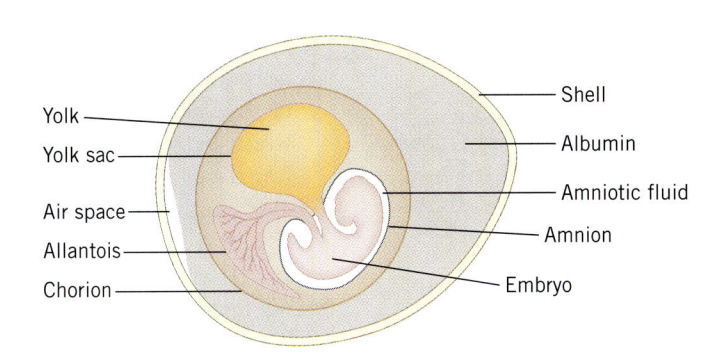

Figure 19.11 Anatomy of the amniotic egg.

Causes of Paleozoic Radiations and Extinctions

Although the cause (or causes) of the great Paleozoic marine radiations and extinctions remains elusive, one explanation relates them to the rhythm of the drifting continents, which separated and collided several times during the era. For example, during the separation phase of the early and mid-Paleozoic, numerous continents were distributed around the globe, each standing alone, surrounded by water. Thus, on each continental margin a *provincial fauna* evolved, specially adapted to the prevailing environmental conditions. In the separation phase, marine species proliferated. In the late Paleozoic, however, these continents collided, reducing the continental margin space (Figure 19.12). Now many species that had never before encountered one another were forced into fierce competition over shrinking habitats. Relatively few survived.

Another well-regarded hypothesis suggests that the huge lava eruptions that created the Siberian traps contributed to the mass extinctions. Ash and dust from the eruptions may have cooled the atmosphere substantially, adding a sudden change in climate to the environmental stresses that were already wreaking havoc on weakened species. A recent

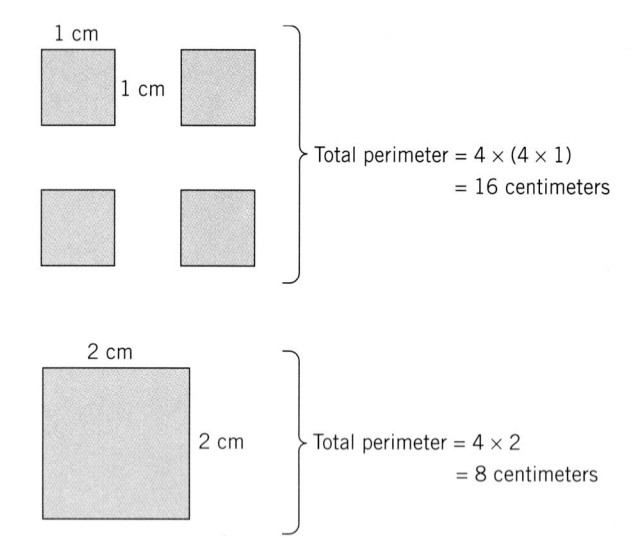

Figure 19.12 Explanation of how living space along the continental margins was reduced as separate continental blocks collided to form a supercontinent. The total perimeter of four one-centimeter squares is 16 centimeters. When the four squares are pushed together to form one large square, the perimeter is reduced to 8 centimeters.

proposal suggests that similar mantle-derived plumes erupted onto the sea floor, belching carbon dioxide and methane into the ocean and atmosphere. These excess gases caused a rapid reduction in available oxygen and the extinction of many marine species. The same basic hypotheses—close competition for available habitat, intensive climatic change, and introduction of excess carbon dioxide—apply to the terrestrial extinctions that also occurred at the close of the Paleozoic.

The Mesozoic and Cenozoic Eras

Taken together, the Mesozoic era (from 245 to 66 million years ago) and the Cenozoic era (from 66 million years ago to the present) constitute about 5 percent of geologic time. Nevertheless, within this relatively short interval, the world we know has taken shape. During the Mesozoic and Cenozoic eras, the modern ocean basins were created, the continents drifted away from Pangaea to their present alignments, the great mountain systems of the world arose, the dinosaurs came and went, and mammals, birds, flowers, fruits, grasses, and grains evolved. Mesozoic and Cenozoic rocks cover most older rocks and are the least eroded. As a result, they generally yield a more complete record of the Earth's environment and processes than do Paleozoic and Precambrian rocks.

Plate Tectonics of the Mesozoic and Cenozoic

The tectonic history of the Mesozoic and Cenozoic eras is a relatively uncomplicated story if we stick to the major events and skip over the details. We begin with Pangaea at the close of the Paleozoic. This supercontinent lasted less than 80 million years. Its disintegration, which began late in the Triassic period, started with a series of rifts that separated Gondwanaland from Laurasia and fragmented Gondwanaland itself into a number of independent continental blocks. All the continents remained close together throughout the Mesozoic, and some were even partially connected, but the rifting continued into the Cenozoic era. It produced a widening Atlantic Ocean and divided Pangaea into seven separate continental blocks—the five continents of today, plus India and New Zealand.

The theory of plate tectonics tells us that it is impossible to have continental drift without the formation and destruction of oceanic crust. Each episode of rifting and separation is accompanied by the production of oceanic crust; each episode of continental collision is preceded by the subduction and destruction of oceanic crust. For this reason, geologic history is as much the history of the ocean basins as it is the history of the continents.

Figure 19.13 is the by-now-familiar map of the early Mesozoic world. Notice Pangaea was surrounded by a world ocean, *Panthalassa,* and a triangular extension of Panthalassa, the Tethys Sea, was wedged between Gondwana and Eurasia. This configuration changed as Pangaea broke apart during the Mesozoic and early Cenozoic. As Africa and India collided with Eurasia, the old Tethys crust was destroyed in the subduction zone and the Alpine-Himalayan mountain belt was produced. The eastern Mediterranean Sea basin and the Black Sea are the sole remnants of this once great seaway. The destruction of the Tethys crust, however, was accompanied by the creation of younger oceanic crust in the rift zones *behind* the drifting continents. This younger crust underlies the modern Indian Ocean.

A similar, though more complicated, process is associated with the Atlantic and Pacific oceans. The Pacific is essentially a shrinking remnant of Panthalassa. Since the early Mesozoic, its width has been reduced by the expansion of the Atlantic Ocean,

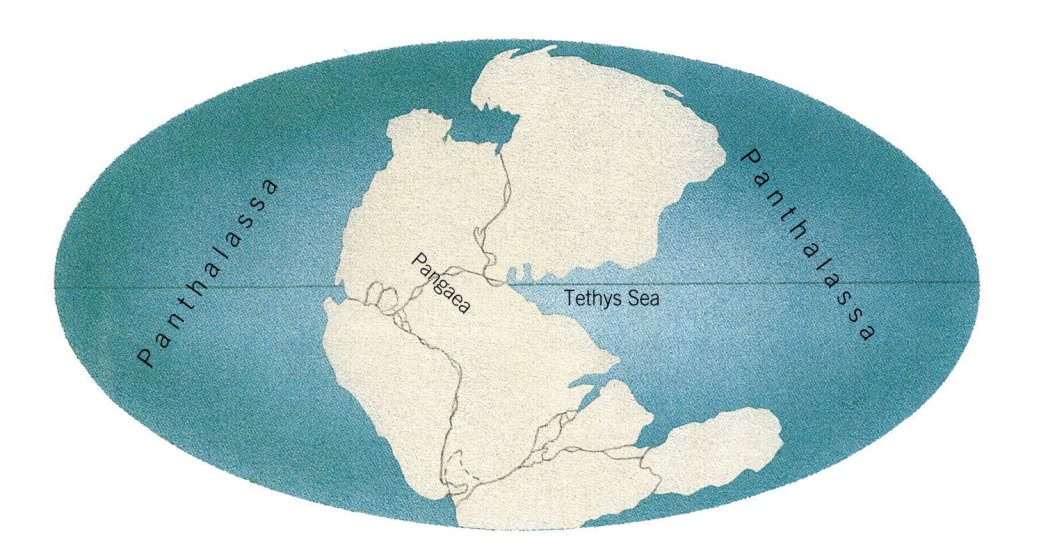

Figure 19.13 The Pangaea supercontinent.

which has caused North and South America to drift westward and Eurasia to drift eastward. During this time, new oceanic crust has been produced continuously along the East Pacific Rise. As this crust spreads from the rise, it is subducted beneath the margins of the Pacific basin. The Mesozoic and Cenozoic volcanism associated with this subduction has formed the island arcs of the western Pacific and the Cascade and Andes ranges of North and South America.

An eastward projection of this Pacific oceanic crust underlies the Atlantic-Caribbean basin, and the volcanism associated with its subduction created the Lesser Antilles. A later episode of subduction that occurred west of the basin built the volcanic backbone of Central America, which both isolated the Caribbean Sea and simultaneously connected North and South America. This episode occurred just a few million years ago, during the Pliocene and Pleistocene epochs.

The opening of the Central American land bridge led to the southward migration of North American mammals. These animals eventually supplanted many native South American species—a classic example of how geological events influence biological evolution.

The North American Cordillera

The western mountain system that extends from Alaska to Mexico, the North American Cordillera, has had an extraordinarily complex history (Figure 19.14). This series of orogenies, which spans the Mesozoic and Cenozoic eras, has varied in style with time and place. We can classify this style into three broad categories: subduction, crustal extension and transform faulting, and terrane accretion.

Subduction During the Mesozoic and early Cenozoic eras, as the Atlantic opened, the North American plate spread westward until it impinged upon the eastward-spreading Farallon plate of the Pacific. The Farallon plate responded by slowly subducting beneath the formerly passive margin of western North America, which was then located far to the east of the present margin (Figure 19.15). This subduction produced a sequence of mountain-building events similar to those observed in the modern Andes. The late Jurassic *Nevadan disturbance* produced the volcanoes and granite batholiths that form the core of the present-day Sierras, and the Cretaceous *Sevier orogeny* of Nevada produced the belt of fold- and thrust-fault mountains east of the Sierras. Although only the eroded stubs of these features remain today, they offer enough evidence for us to piece together the general picture. A final compressional pulse, the **Laramide orogeny,** lasted from the close of the Mesozoic until well into early Cenozoic time and produced the Rocky Mountains farther to the east (Figure 19.16). The Nevadan, Sevier, Laramide, and similar mountain-building episodes together constitute the **Cordilleran orogeny.**

Broad, shallow seas advanced and retreated across western North America both before and during these mountain-building episodes. The thick strata deposited within them and along their margins record the environmental history of the region—wind-blown deserts,

Laramide orogeny
The late-Cretaceous mountain-building episode that formed the Rocky Mountains.

Cordilleran orogeny
The complex series of mountain-building episodes, from the Jurassic period through the mid-Cenozoic era, that produced the mountains of western North America.

coal-bearing swamps, huge stagnant lakes, flood plains, and deltas. Many of these deposits originated in the rising mountain belts. As the mountain building intensified and broadened, the seas retreated and the deposits were uplifted. Today they are found in the high plains that border the Rockies and within the mountain chain itself. The *Morrison formation*, a thick coastal deposit, contains one of the world's finest collections of Jurassic dinosaur skeletons.

Crustal Extension and Transform Faulting The orogenic style of the western Cordillera has changed as the Cenozoic era has progressed. Although subduction continues to form the Cascade Range of Washington and Oregon, the tectonics of California is controlled by transform faulting, and the Basin and Range Province to the east is primarily being shaped by block faulting (Figure 19.17). The major reason for these changes is that the southern half of western North America has overridden the East Pacific Rise and the ancient Farallon plate is no longer subducting beneath that region. Consequently, the stress pattern has changed from compression to extension and horizontal shear. Westernmost California has been incorporated into the Pacific plate, which is moving northwestward relative to the North American plate along the San Andreas transform fault.

The block faulting and thin crust that characterize the Basin and Range Province provide strong evidence that the crust is being stretched in that region. Furthermore, seismic waves slow at shallow depths in the crust, indicating that semimolten rocks are present not many kilometers beneath the surface. What is the cause of the stretching? Could it be magma rising off remnants of the old Farallon plate? A mantle plume? The transform motion along the San Andreas fault? Geologists have yet to reach a definitive conclusion. Whatever the cause, the rift fractures of the Basin and Range Province extend into Idaho and eastern Washington, and they may have provided escape conduits for the lavas that formed the Columbia-Snake River Plateau during the Miocene period.

Accreted Terranes Another major complexity of Cordilleran mountain building, the accreted terranes of western North America, has increased the size of that region by at least 25 percent. These features are found throughout the entire western Cordillera from Alaska to Mexico and were transported during episodes of plate subduction and transform faulting. As in the Appalachians, these terranes collided with earlier terranes as they docked and bulldozed continental margin sediments into mountains. Deciphering their origins and the paths taken to their present locations will add greatly to our knowledge of Mesozoic and Cenozoic history.

Figure 19.14 Provinces of the North American Cordillera.

Mesozoic and Cenozoic Climate

Climate, which is essentially the long-term rainfall and temperature pattern of a region, is controlled mainly by the distribution of solar energy over the Earth's surface. The Earth's spherical shape causes the Sun's rays to strike it most directly at the equator and

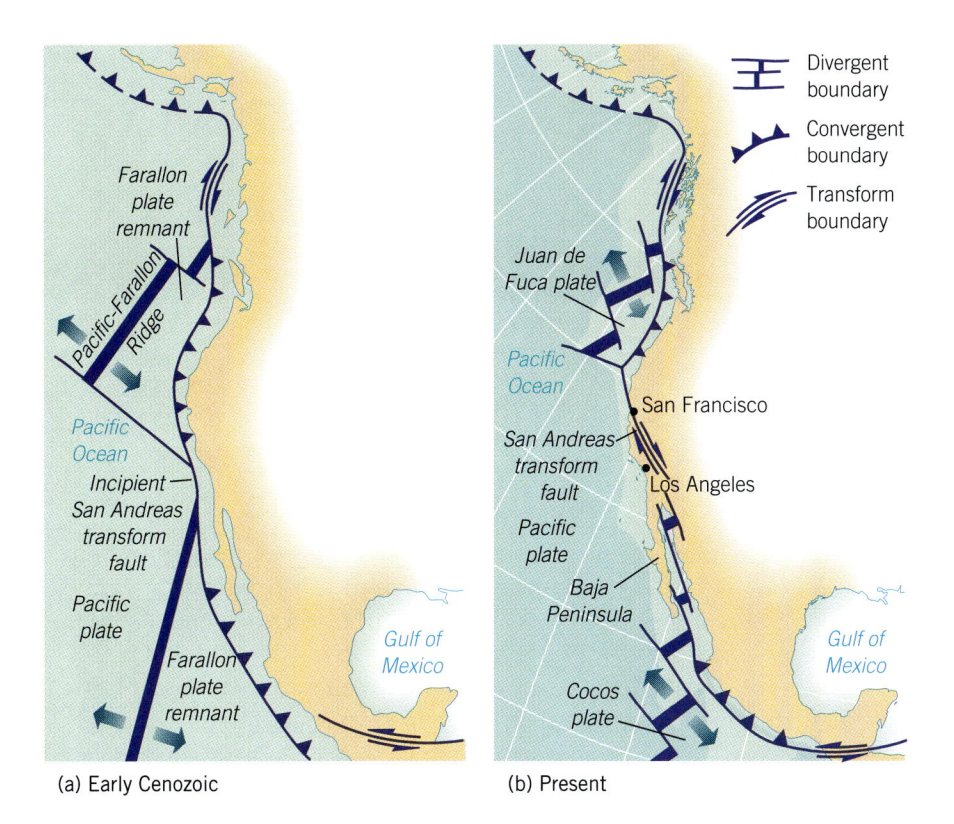

Figure 19.15 (a) Subduction of the Farallon plate beneath the western margin of North America resulted in the Cordilleran orogeny. (b) Later changes in plate motion produced the San Andreas fault.

(a) Early Cenozoic

(b) Present

least directly at the poles. For this reason, equatorial regions are, on average, warmer than polar regions. We can assume that this relationship has prevailed throughout geologic time. A number of astronomical, geological, oceanographical, and meteorological factors combine to modify climate: slight periodic shifts in the Earth's orbit, the positions and altitudes of the land masses, ocean currents, and prevailing wind direction are four important ones. Thus, even if we assume a constant output of solar energy, it is not surprising that the Earth's climate has fluctuated significantly at various times in the past.

Geologists have developed a variety of reliable methods to investigate past climates, the most obvious of which is the study of sediments. Redbeds and windblown sands are generally indicative of deserts. Coral reefs and other limestones are indicative of warm

Figure 19.16 Uplifted and deformed peaks of the northern Rocky Mountains.

Figure 19.17 Magmas and semimolten rock rising from the mantle may be the cause of crustal stretching beneath the Basin and Range Province of the western United States.

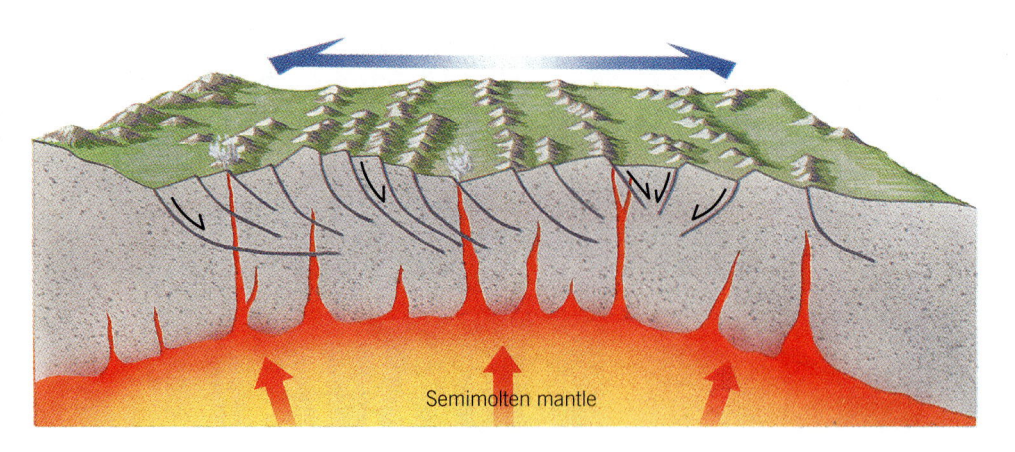

seas teeming with life. Marine sediments mixed with glacial till suggest a cold climate. Coal deposits indicate the existence of swamps, which in turn suggests a warm-to-temperate climate. Layers of red sediments alternating with layers of other, duller colors may have formed under seasonal wet and dry conditions.

Fossils also serve as climatic indicators. For example, finding a Cretaceous plant fossil similar to modern tropical species in rock strata provides a good clue that a mild climate prevailed in the part of the world where the strata were deposited. The most accurate climatic data have come from the chemical analysis of the fossilized shells of marine animals called foraminifera, because the varying oxygen isotope content of their shells reflects cooling and warming climates. The warmer the water, the more ^{18}O isotopes relative to ^{16}O isotopes these organisms incorporate into their shells. This information can be translated into precise readings of surface ocean temperatures.

The synthesis of these methods has revealed interesting results. The evidence suggests that the Permian climate was one of extremes, much like our present climate: hot and humid at the equator, cold and dry at the glacial South Pole, seasonal and variable in the intervening regions. Within the vast interior of mountainous Pangaea, far from the moderating influence of the oceans, temperatures were probably similar to those of the present interior of Siberia or the northern Plains of the United States, fluctuating between blistering heat in the summer and freezing cold in the winter.

This climate of harsh extremes was moderated during the Triassic and Jurassic. By the Cretaceous, the climate had become quite mild; the world was subtropical to temperate and apparently lacked polar ice caps. Geologists cite several reasons for the change. First, during the Cretaceous, shallow seas covered the now-separating continents for the first time since the Paleozoic. The seas absorbed and redistributed solar heat. Second, the Cretaceous was a time of intensive and widespread volcanism, during which huge quantities of carbon dioxide were released into the atmosphere, intensifying the greenhouse effect.

The slow poleward drift of the continents and their uplift caused worldwide climates to cool markedly during the early Cenozoic, however, and temperatures have followed a downward course since then. This environmental change favored organisms that were able to insulate themselves with fat and hair or fur, and that could maintain a constant body temperature—in other words, mammals. The global cooling late in the mid-Cenozoic may have also resulted from the isolation of Antarctica over the South Pole.

Whatever the causes, ice sheets began to grow on Antarctica toward the mid-Cenozoic and on northern Canada late in the Cenozoic. The latest epoch of the Cenozoic, the Quaternary, has been characterized by the extensive growth and retreat of Northern Hemisphere glaciers. At present we are past the midpoint of an interglacial age. We are also experiencing short-term, human-induced global warming that most experts attribute to the burning of fossil fuels. (The ice ages and glaciers are discussed more fully in Chapter 13, Glaciers and Climate.)

Mesozoic and Cenozoic Life

The Mesozoic is commonly referred to as the "Age of the Dinosaurs" or, less commonly, the "Age of the Reptiles." These terms do not fully convey the variety of life that existed during the era, however. Many paleontologists find the marine life that proliferated throughout the era equally interesting: the corals, for example, and the ammonites—ancient relatives of the chambered nautilus—with their incredibly intricate shell patterns (Figure 19.18).

The Mesozoic also included the life forms whose more modern descendants we associate with the Cenozoic era: mammals, including the dominant placental and marsupial orders; angiosperms (plants having flowers, seeds, and vascular systems); and birds. Their fossils lie deep in Jurassic and Cretaceous strata, indicating that they were successful competitors for roughly 100 million years before the start of the Cenozoic.

Just as the dinosaurs were not the only interesting organisms to have evolved during the Mesozoic, neither were they the only such life forms to have died out suddenly at the Cretaceous–Tertiary boundary (that is, the close of the Mesozoic era). Most groups—including microscopic plankton, plants, marine invertebrates, and vertebrates—were drastically reduced.

The extinctions had a negligible long-term effect on the physical geology of the Earth, but were of immense biological importance, clearing the way for the rise of mammals, flowering plants, and birds. Indeed, geologic change need not be drastic or catastrophic to have a significant effect on organic evolution. The East African Rift Valley began to form during the Miocene epoch. As the mountains and high ridges began to rise around the rift valley, they blocked the moisture-laden winds coming from the Indian Ocean. Gradually the climate to the west of the valley grew markedly drier. As the lush rain forests were replaced by high grassland savanna, forest animals were thrown into fierce competition within their shrinking habitats. One species in particular competed successfully by abandoning the forests and adapting to the opportunities offered by the savanna (Figure 19.19). This animal was a large primate whose pelvic structure allowed it to walk upright and whose brain was somewhat larger and more complex than its close relatives. The rest is history.

Figure 19.18 Typical ammonite fossils of the late-Mesozoic. This great subclass of the cephalopod mollusks died out abruptly at the close of the Mesozoic era.

Figure 19.19 An artist's conception of early human ancestors, *A. Aferensis*, walking upright in the East African Rift Valley. The active volcano in the background is symbolic of the fact that skeletons of this genus were preserved in ash deposits. Notice the savanna vegetation.

STUDY OUTLINE

I. THE PALEOZOIC ERA (570 to 245 M.Y.A.). This era is marked by the abrupt appearance of hard-shelled fossils in the rock strata.
   A. Three stages of the tectonic history of the Paleozoic are:
      1. *Rifting* of the Proterozoic supercontinent into six continental blocks.
      2. *Separation* of the smaller continental blocks and resulting advance and retreat of shallow seas over the low-lying interiors of the continents.
      3. *Reassembly* into the supercontinent Pangaea.

   B. Mountain building and draining of inland seas resulted from the collisions of continental blocks. Sediments transported by streams down from the North Atlantic mountain belt were deposited as swampy deltaic plains, an environment that supported the evolution of amphibians and reptiles.

   C. The Paleozoic era began with an explosion of life and ended with mass extinctions. Each stage was marked by an evolutionary innovation that spurred the radiation of a new class of animals.
      1. First to appear were the **invertebrates**—animals without backbones—such as sponges, snails, and **trilobites.**
      2. Next came the **vertebrates**—animals having a spinal column—beginning with fishes, amphibians, and reptiles.
         a. Amphibians evolved from the lobe-finned *crossopterygian* fishes. Amphibian adults could survive on land, but had to lay their eggs in water.
         b. Reptiles were the first true land animals because their amniotic eggs could be laid on dry land.
      3. Paleozoic marine life flourished as the Proterozoic supercontinent rifted and shallow seas covered the margins of the separating land masses.
      4. Two possible causes of the extinctions at the close of the period are:
         a. A decrease in margin space and the draining of the seas as the continental blocks recombined to form Pangaea.
         b. Massive lava and mantle plume eruptions that cooled the climate and increased the amount of carbon dioxide in the oceans and atmosphere.

II. THE MESOZOIC (245 TO 66 M.Y.A.) AND CENOZOIC (66 M.Y.A. TO PRESENT) ERAS.
   A. During these eras, the modern ocean basins were created, Pangaea broke up, and the continents moved to their present positions. Plate collisions created the Rocky, Himalayan, and Andean mountain belts.
      1. Subduction beneath the western margin of the North American plate resulted in the **Laramide orogeny.** This and other mountain-building events contributed to the **Cordilleran orogeny.**
      2. Today western North America is being shaped by crustal extension and transform faulting.

   B. Mesozoic and Cenozoic Climate
      1. At the close of the Paleozoic, the climate of Pangaea was one of extremes: hot and humid at the equator, cold and dry at the glacial South Pole, variable in the intervening regions.
      2. As the Mesozoic progressed, shallow seas covered the continents as they drifted away from Pangaea, and the global climate became uniformly mild.
      3. As plate collisions caused continental uplift, Antarctica became isolated over the South Pole and North America and Eurasia drifted northward. Ice sheets then developed on these continents and worldwide climates once again became extreme.

   C. The Mesozoic and Cenozoic eras encompass the rise and fall of the dinosaurs and the radiation of mammals, birds, and flowering plants.

STUDY TERMS

Cordilleran orogeny (p. 413)
invertebrate (p. 407)
Laramide orogeny (p. 413)
vertebrate (p. 409)

THOUGHT QUESTIONS

1. Compare the tectonic processes that formed three of the great mountain belts of the world: the North American Cordillera, the Appalachians, and the Himalayas.
2. Throughout the midcontinental regions of North America and other continents, you will find early Paleozoic platform sequences that consist of alternating layers of thin sandstone and thick shale or limestone. What does this sedimentary record tell us about the early Paleozoic environment? How did reassembly of the continents into Pangaea change that environment?
3. Compare and contrast the extent and possible causes of the late Paleozoic and late Cretaceous extinctions.
4. North and South America were isolated during most of the Cenozoic, and distinctive mammal species evolved on each continent. When the continents were joined by the emergence of Central America, many North American mammal species migrated south and soon supplanted the native South American species. From your knowledge of superposition, faunal succession, and facies changes, describe how geologists were able to piece together this scenario.

Appendix A

Conversion Table for Metric and English Units

Length

1 kilometer (km)	= 0.6214 miles
	= 328 yards
	= 1000 meters
1 meter (m)	= 1.0936 yards
	= 3.2808 feet
	= 100 centimeters
1 centimeter (cm)	= 0.3937 inches
1 mile (mi)	= 1.6093 kilometers
	= 1760 yards
1 yard (yd)	= 0.9144 meters
	= 3 feet
1 foot (ft)	= 0.3048 meters
	= 12 inches
1 inch (in.)	= 2.54 centimeters

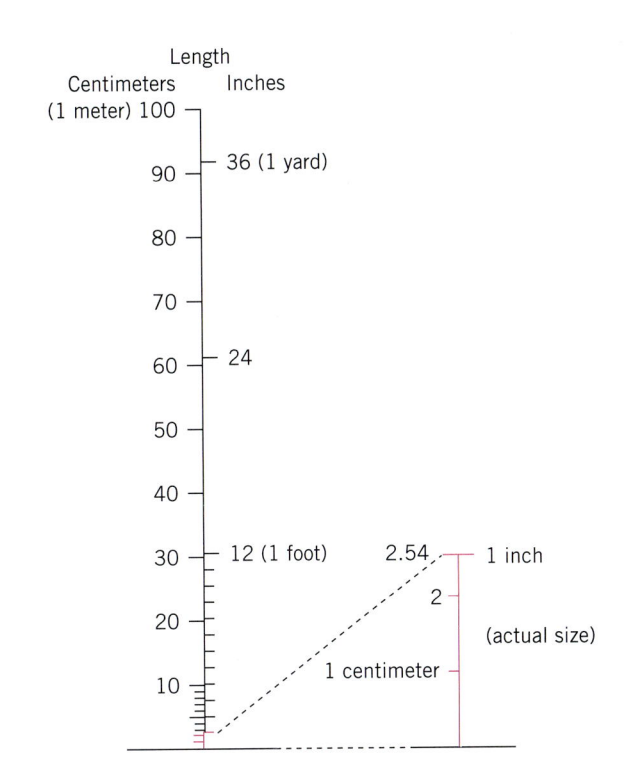

Area

1 square kilometer (km^2)	= 0.386 square miles
	= 100 hectares
1 hectare	= 2.471 acres
	= 10,000 square meters
1 square meter (m^2)	= 1.20 square yards
	= 10.76 square feet
	= 10,000 square centimeters
1 square centimeter (cm^2)	= 0.15 square inches
1 square mile (mi^2)	= 2.59 square kilometers
	= 640 acres
1 acre	= 4033 square meters
	= 4840 square yards
1 square yard (yd^2)	= 0.833 square meters
	= 9 square feet
1 square foot (ft^2)	= 0.093 square meters
	= 144 square inches
1 square inch ($in.^2$)	= 6.45 square centimeters

Volume

1 cubic kilometer (km^3)	= 0.24 cubic miles
	= 1,000,000,000 cubic meters
1 cubic meter (m^3)	= 1.31 cubic yards
	= 35.31 cubic feet
	= 1000 liters
1 liter (L)	= 0.26 gallons
	= 1000 cubic centimeters
1 cubic centimeter (cm^3)	= 0.06 cubic inches
1 cubic mile (mi^3)	= 4.17 cubic kilometers
	= 5,460,000,000 cubic yards
1 cubic yard (yd^3)	= 0.76 cubic meters
	= 27 cubic feet
1 cubic foot (ft^3)	= 0.028 cubic meters
	= 1728 cubic inches
1 cubic inch ($in.^3$)	= 16.67 cubic centimeters
1 gallon (gal)	= 3.785 liters

APPENDIX A

Mass

1 metric ton (t)	= 1.1 tons
	= 1000 kilograms
1 kilogram (kg)	= 2.205 pounds
	= 1000 grams
1 gram (g)	= 0.035 ounces
1 ton	= 0.91 metric tons
	= 2000 pounds
1 pound (lb)	= 0.45 kilograms
	= 16 ounces
1 ounce (oz)	= 28.35 grams

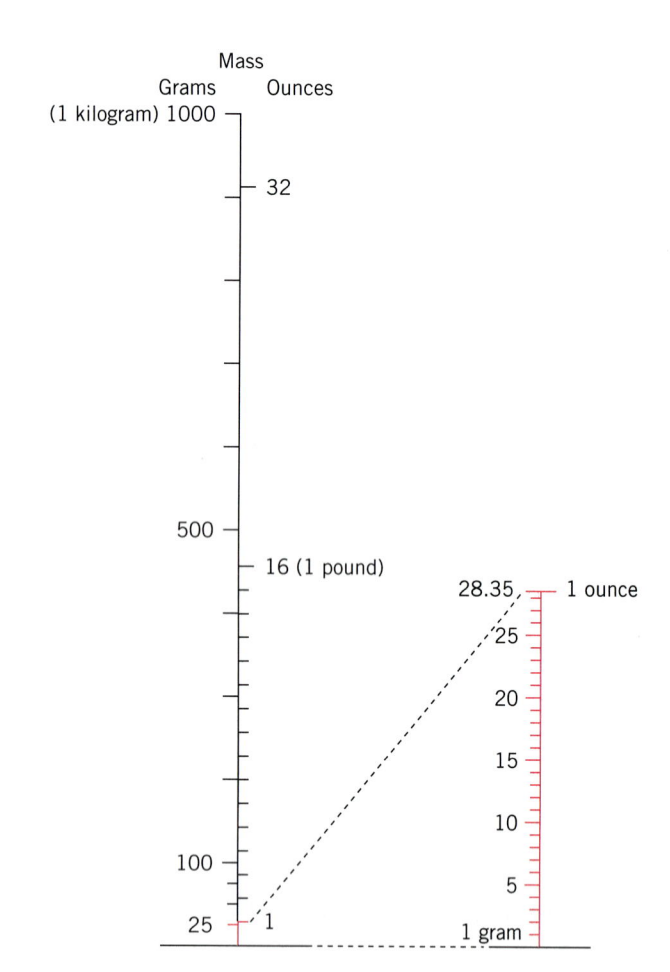

Temperature

Kelvin (K)	= °C + 273
°C	= $\frac{5}{9}$ (°F − 32)
°F	= $\frac{5}{9}$ °C + 32

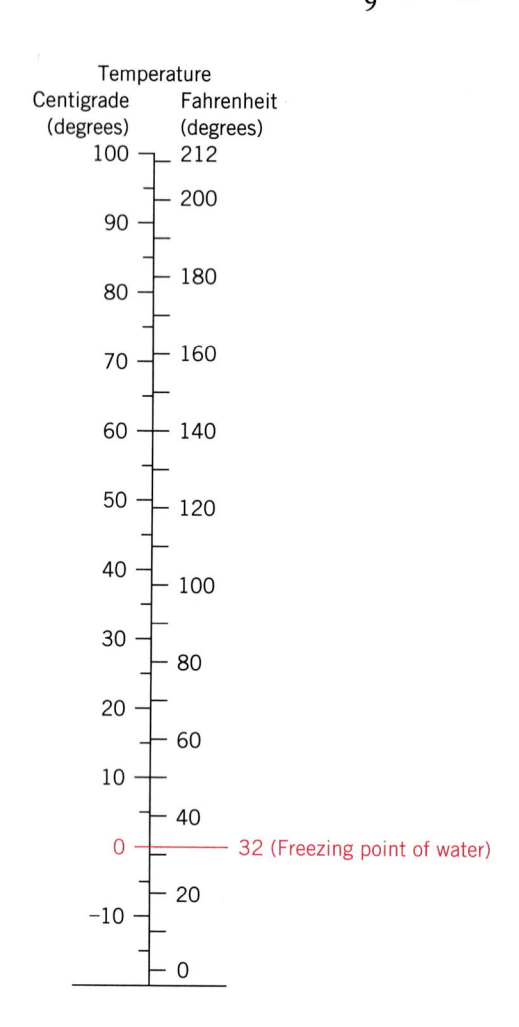

APPENDIX B

Periodic Table of the Elements

Period	IA	IIA	IIIB	IVB	VB	VIB	VIIB	VIIIB			IB	IIB	IIIA	IVA	VA	VIA	VIIA	VIIIA (Inert elements)
1	1 **H** 1.008																	2 **He** 4.003
2	3 **Li** 6.941	4 **Be** 9.012											5 **B** 10.81	6 **C** 12.01	7 **N** 14.01	8 **O** 16.00	9 **F** 19.00	10 **Ne** 20.18
3	11 **Na** 22.99	12 **Mg** 24.31											13 **Al** 26.98	14 **Si** 28.09	15 **P** 30.97	16 **S** 32.06	17 **Cl** 35.45	18 **Ar** 39.95
4	19 **K** 39.10	20 **Ca** 40.08	21 **Sc** 44.96	22 **Ti** 47.88	23 **V** 50.94	24 **Cr** 52.00	25 **Mn** 54.94	26 **Fe** 55.85	27 **Co** 58.93	28 **Ni** 58.69	29 **Cu** 63.55	30 **Zn** 65.38	31 **Ga** 69.72	32 **Ge** 72.59	33 **As** 74.92	34 **Se** 78.96	35 **Br** 79.90	36 **Kr** 83.80
5	37 **Rb** 85.47	38 **Sr** 87.62	39 **Y** 88.91	40 **Zr** 91.22	41 **Nb** 92.91	42 **Mo** 95.94	43 **Tc** (98)	44 **Ru** 101.1	45 **Rh** 102.9	46 **Pd** 106.4	47 **Ag** 107.9	48 **Cd** 112.4	49 **In** 114.8	50 **Sn** 118.7	51 **Sb** 121.8	52 **Te** 127.6	53 **I** 126.9	54 **Xe** 131.3
6	55 **Cs** 132.9	56 **Ba** 137.3	57 **La*** 138.9	72 **Hf** 178.5	73 **Ta** 180.9	74 **W** 183.9	75 **Re** 186.2	76 **Os** 190.2	77 **Ir** 192.2	78 **Pt** 195.1	79 **Au** 197.0	80 **Hg** 200.6	81 **Tl** 204.4	82 **Pb** 207.2	83 **Bi** 209.0	84 **Po** (209)	85 **At** (210)	86 **Rn** (222)
7	87 **Fr** (223)	88 **Ra** 226	89 **Ac†** (227)	104 **Unq**	105 **Unp**	106 **Unh**	107 **Uns**	108 **Uno**	109 **Une**									

Transition metals

Metals ← → Nonmetals

*Lanthanides

58 **Ce** 140.1	59 **Pr** 140.9	60 **Nd** 144.2	61 **Pm** (145)	62 **Sm** 150.4	63 **Eu** 152.0	64 **Gd** 157.3	65 **Tb** 158.9	66 **Dy** 162.5	67 **Ho** 164.9	68 **Er** 167.3	69 **Tm** 168.9	70 **Yb** 173.0	71 **Lu** 175.0

†Actinides

90 **Th** 232.0	91 **Pa** (231)	92 **U** 238.0	93 **Np** (237)	94 **Pu** (244)	95 **Am** (243)	96 **Cm** (247)	97 **Bk** (247)	98 **Cf** (251)	99 **Es** (252)	100 **Fm** (257)	101 **Md** (258)	102 **No** (259)	103 **Lr** (260)

Manufactured

Atomic number —— 1
Element symbol —— **H**
Atomic mass —— 1.008

Table of Atomic Masses*

Element	Symbol	Atomic Number	Atomic Mass	Element	Symbol	Atomic Number	Atomic Mass	Element	Symbol	Atomic Number	Atomic Mass
Actinium	Ac	89	(227)†	Hafnium	Hf	72	178.5	Promethium	Pm	61	(145)
Aluminum	Al	13	26.98	Helium	He	2	4.003	Protactinium	Pa	91	(231)
Americium	Am	95	(243)	Holmium	Ho	67	164.9	Radium	Ra	88	226
Antimony	Sb	51	121.8	Hydrogen	H	1	1.008	Radon	Rn	86	(222)
Argon	Ar	18	39.95	Indium	In	49	114.8	Rhenium	Re	75	186.2
Arsenic	As	33	74.92	Iodine	I	53	126.9	Rhodium	Rh	45	102.9
Astatine	At	85	(210)	Iridium	Ir	77	192.2	Rubidium	Rb	37	85.47
Barium	Ba	56	137.3	Iron	Fe	26	55.85	Ruthenium	Ru	44	101.1
Berkelium	Bk	97	(247)	Krypton	Kr	36	83.80	Samarium	Sm	62	150.4
Beryllium	Be	4	9.012	Lanthanum	La	57	138.9	Scandium	Sc	21	44.96
Bismuth	Bi	83	209.0	Lawrencium	Lr	103	(260)	Selenium	Se	34	78.96
Boron	B	5	10.81	Lead	Pb	82	207.2	Silicon	Si	14	28.09
Bromine	Br	35	79.90	Lithium	Li	3	6.941	Silver	Ag	47	107.9
Cadmium	Cd	48	112.4	Lutetium	Lu	71	175.0	Sodium	Na	11	22.99
Calcium	Ca	20	40.08	Magnesium	Mg	12	24.31	Strontium	Sr	38	87.62
Californium	Cf	98	(251)	Manganese	Mn	25	54.94	Sulfur	S	16	32.06
Carbon	C	6	12.01	Mendelevium	Md	101	(258)	Tantalum	Ta	73	180.9
Cerium	Ce	58	140.1	Mercury	Hg	80	200.6	Technetium	Tc	43	(98)
Cesium	Cs	55	132.9	Molybdenum	Mo	42	95.94	Tellurium	Te	52	127.6
Chlorine	Cl	17	35.45	Neodymium	Nd	60	144.2	Terbium	Tb	65	158.9
Chromium	Cr	24	52.00	Neon	Ne	10	20.18	Thallium	Tl	81	204.4
Cobalt	Co	27	58.93	Neptunium	Np	93	(237)	Thorium	Th	90	232.0
Copper	Cu	29	63.55	Nickel	Ni	28	58.69	Thulium	Tm	69	168.9
Curium	Cm	96	(247)	Niobium	Nb	41	92.91	Tin	Sn	50	118.7
Dysprosium	Dy	66	162.5	Nitrogen	N	7	14.01	Titanium	Ti	22	47.88
Einsteinium	Es	99	(252)	Nobelium	No	102	(259)	Tungsten	W	74	183.9
Erbium	Er	68	167.3	Osmium	Os	76	190.2	Uranium	U	92	238.0
Europium	Eu	63	152.0	Oxygen	O	8	16.00	Vanadium	V	23	50.94
Fermium	Fm	100	(257)	Palladium	Pd	46	106.4	Xenon	Xe	54	131.3
Fluorine	F	9	19.00	Phosphorus	P	15	30.97	Ytterbium	Yb	70	173.0
Francium	Fr	87	(223)	Platinum	Pt	78	195.1	Yttrium	Y	39	88.91
Gadolinium	Gd	64	157.3	Plutonium	Pu	94	(244)	Zinc	Zn	30	65.38
Gallium	Ga	31	69.72	Polonium	Po	84	(209)	Zirconium	Zr	40	91.22
Germanium	Ge	32	72.59	Potassium	K	19	39.10				
Gold	Au	79	197.0	Praseodymium	Pr	59	140.9				

*The values given here are to four significant figures.
†A value given in parentheses denotes the mass of the longest-lived isotope.

APPENDIX C
Mineral Identification Table

Mineral	Composition	Cleavage/Fracture	Hardness	Color/Streak	Miscellaneous Properties
Actinolite	Complex Ca, Mg, Fe hydrous silicate	Prismatic (at 56° and 124°)/subconchoidal, uneven	5–6	White to light green/colorless	Slender glassy crystals; vitreous to pearly luster
Amphiboles (see actinolite, tremolite)					
Andalusite	$AlO(SiO_4)$	One direction, distinct	7.5	Flesh, reddish brown, olive green/colorless	Blunt, nearly square prisms; vitreous luster
Anhydrite	$CaSO_4$	Three directions (at 90°)/uneven, sometimes splintery	3–3.5	Colorless to bluish or violet/grayish white	Brittle, somewhat greasy; vitreous to pearly luster
Apatite	$Ca_3(F \cdot Cl \cdot OH)(PO_4)_3$	One direction, imperfect/conchoidal and uneven	5	Green or brown/white	Prominent pyramidal crystals; vitreous to subresinous luster
Augite	$Ca(Mg,Fe,Al)(Si,Al)_2O_6$	Prismatic (at 87° and 93°)/uneven to conchoidal	5–6	Dark green to black/greenish gray	Squarish cross section; vitreous luster
Azurite	$Cu_3(CO_3)_2(OH)_2$	One direction, perfect/conchoidal	3.5–4	Intense azure/blue	Effervesces in HCl; vitreous luster
Barite	$BaSO_4$	Two directions, perfect/uneven	3–3.5	Colorless, white, blue, yellow, red/white	Tabular crystals, also globular; vitreous luster; 4.5 specific gravity
Bauxite	$Al_2O_3 \cdot 2H_2O$	Uneven	1–3	White, gray, yellow, red/colorless	Pisolitic, concretionary grains; dull to earthy luster
Beryl	$Be_2Al_2(Si_6O_{18})$	Poor/conchoidal to uneven	7.5–8	Bluish green or yellow/white	Hexagonal crystal form; vitreous luster
Biotite	$K(Mg,Fe)_3(AlSi_3O_{10})(OH)_2$	One direction parallel to base/none	2.5–3	Dark green, brown, or black/colorless	Thin, flexible sheets; splendent luster
Calcite	$CaCO_3$	Rhombohedral, perfect/conchoidal	3 on cleavage; 2.5 on base	White, colorless/white or grayish	Effervesces in HCl; vitreous to earthy luster
Chalcopyrite	$CuFeS_2$	Sometimes distinct/uneven	3.5–4	Brass yellow, often tarnished/greenish black	Brittle; metallic luster; 4.1–4.3 specific gravity
Chlorite	$(Mg,Fe)_5(Al,Fe)_2Si_3O_{10}(OH)_8$	One direction	2–2.5	Light to dark green, yellowish/white or colorless	Flaky

APPENDIX C

Mineral	Composition	Cleavage/Fracture	Hardness	Color/Streak	Miscellaneous Properties
Cinnabar	HgS	One direction, perfect	2.5	Vermilion to brownish red/scarlet	Slightly sectile; adamantine to dull earthy luster; 8.1 specific gravity
Copper	Cu	None/hackly	2.5–3	Copper red on fresh surface/red	Highly ductile and malleable; metallic luster
Corundum	Al_2O_3	None	9	Brown, pink, blue, or yellow/colorless	Rhombohedral parting; adamantine to vitreous luster
Diamond	C	Octahedral, perfect/conchoidal	10	White or colorless to pale yellow/white	Very high refractive index; adamantine to greasy luster
Diopside	$CaMg(Si_2O_6)$	Prismatic (at 87° and 93°), imperfect/uneven to conchoidal	5–6	White to light green/grayish green	Squarish cross section; vitreous luster
Dolomite	$CaMg(CO_3)_2$	Rhombohedral, perfect/subconchoidal	3.5–4	Pink, flesh, colorless, or white/colorless	Does not effervesce in cold HCl; vitreous luster
Epidote	Complex Ca, Al, Fe hydrous silicate	Perfect, parallel to base/uneven	6–7	Yellowish green to blackish green/colorless	Peculiar yellowish green color; vitreous luster
Feldspar group (see plagioclase feldspars, potassium feldspars)					
Fluorite	CaF_2	Octahedral, perfect/flat conchoidal	4	Green, yellow, or purple/white	Cubic crystals; vitreous luster
Galena	PbS	Cubic/flat subconchoidal	2.5	Lead gray/lead gray	Brittle; metallic luster; 7.4–7.6 specific gravity
Garnet group	General formula $A_3B_2(SiO_4)_3$	None/conchoidal to uneven	6.5–7	Red or brown/white	12-sided crystals; vitreous to resinous luster; 3.5–4.3 specific gravity
Goethite (limonite)	$FeO(OH)mH_2O$	None	5.5–6	Dark brown to black/yellowish brown	Globular form; silky, often submetallic, earthy luster
Gold	Au	None/hackly	2.5–3	Various shades of yellow/golden yellow	Very malleable and ductile; metallic luster; 19.3 specific gravity when pure
Graphite	C	Basal, perfect	1–2	Black to steel gray/black	Greasy feel, foliated; metallic or dull earthy luster
Gypsum	$CaSO_4·2H_2O$	Four directions; three unequal/fibrous in one direction; conchoidal	2	Colorless, white, gray/white	Sometimes in fibrous masses; vitreous, pearly, silky luster

APPENDIX C

Mineral	Composition	Cleavage/Fracture	Hardness	Color/Streak	Miscellaneous Properties
Halite	NaCl	Cubic, perfect/conchoidal	2.5	Colorless or white/colorless	Salty taste; transparent to translucent
Hematite	Fe_2O_3	None	5.5–6	Reddish brown to black/light to dark red	Breaks with almost cubic angles; metallic to dull earthy luster; 5.3 specific gravity
Hornblende	Complex Ca, Mg, Fe, Al hydrous silicate	Prismatic (at 56° and 124°) perfect/subconchoidal, uneven	5–6	Dark green to black/colorless	Long prisms; vitreous or silky luster
Kaolinite	$Al_2(Si_2O_5)OH)_4$	Coarse masses, direction seldom distinct/none	2–2.5	White with colored impurities/colorless	Earthy, soapy feeling; dull earthy luster
Magnetite	Fe_3O_4	Not distinct/subconchoidal to uneven	6	Iron black/black	Strongly magnetic, octahedral parting; metallic and splendent luster; 5.2 specific gravity
Malachite	$Cu_2CO_3(OH)_2$	One direction, perfect/subconchoidal to uneven	3.5–4	Bright green/pale green	Effervesces in HCl; adamantine to vitreous luster
Micas (see muscovite, biotite)					
Muscovite	$KAl_2(AlSi_3O_{10})(OH)_2$	One direction parallel to base/none	2–2.5	Colorless, yellow, brown, or green/colorless	Thin, flexible sheets; vitreous, silky, or pearly luster
Olivine group	$(Mg:Fe)_2(SiO_4)$	Poor/conchoidal	6.5–7	Pale yellow green to brownish black/white or gray	Granular texture; vitreous luster; 3.8–4.4 specific gravity
Plagioclase feldspars					
Albite	$(NaAlSi_3O_8)$ Varies between above 100% sodium aluminum silicate and below 100% calcium silicate	Two directions (at 93°), good/uneven to conchoidal	6	Colorless, white, gray, greenish, or yellow/colorless	Striations on cleavage; vitreous to pearly luster
Anorthite	$(CaAl_2Si_2O_8)$	Two directions (at 94°), good/conchoidal to uneven	6	Colorless, white, gray, or reddish/colorless	Striations on cleavage; vitreous to pearly luster
Potassium feldspars (aluminosilicates)					
Microcline	$K(AlSi_3O_8)$	Three directions (at 89°), good/uneven	6	White, yellow, pink, or green/colorless	Twinned crystals (reversed pairs); vitreous luster
Orthoclase	$K(AlSi_3O_8)$	Two directions (at 90°), good/conchoidal to uneven	6	Colorless, white, gray, flesh, or red/colorless	Twinned crystals (reversed pairs); vitreous luster

Mineral	Composition	Cleavage/Fracture	Hardness	Color/Streak	Miscellaneous Properties
Pyrite	FeS$_2$	Cubic, poor/conchoidal to uneven	6–6.5	Pale brass yellow/greenish or brownish black	Brittle, cubic crystals; metallic, splendent luster; 5.0 specific gravity
Pyroxene (see augite, wollastonite)					
Quartz	SiO$_2$	None/conchoidal	7	Colorless or white, with colored impurities/colorless	Crystals end in prisms; vitreous, greasy, splendent luster
Serpentine	Mg$_3$(Si$_2$O$_5$)(OH)$_4$	Sometimes distinct/conchoidal or splintery	2–5	Mixed shades of green/white	Fibrous or platy variety; greasy, silky, and waxy luster
Sillimanite	Al$_2$SiO$_5$	One direction	6–7	White to gray/white	Long, slender crystals or fibrous; vitreous luster
Sphalerite	ZnS	Six directions, perfect/conchoidal	3.5–4	Yellow brown, dark brown, or black/white or yellow to brown	Brittle; resinous to adamantine luster; 3.9–4.1 specific gravity
Sulfur	S	None/irregular	1.5–2.5	Bright yellow/colorless	Insoluble; resinous to dull luster
Talc	Mg$_3$(Si$_4$O$_{10}$)(OH)$_2$	One direction parallel to base/none	1	Light green, silvery white/white	Soapy feeling; pearly, greasy luster
Topaz	Al$_2$(SiO$_4$)(F·OH)$_2$	Perfect, parallel to base/subconchoidal to uneven	8	Colorless to yellow/colorless	Prismatic crystals; vitreous luster
Tourmaline	Complex silicate of B and Al	Poor/subconchoidal to uneven	7–7.5	Commonly black/colorless	Crystals triangular in cross section; vitreous to resinous luster
Tremolite	Complex Ca, Mg hydrous silicate	Prismatic (at 56° and 124°) perfect/splintery surface	5–6	White to light green/colorless	Radiating, bladed columns; vitreous or silky luster
Wollastonite	Ca(SiO$_3$)	Two directions (at 84° and 96°), perfect/uneven	5–5.5	Colorless, white, or gray/white	Sometimes fibrous; vitreous or silky luster
Zircon	Zr(SiO$_4$)	Poor/conchoidal	7.5	Brown, colorless, or gray/colorless	Terminated prisms; adamantine luster; 4.7 specific gravity

Appendix D

Basic Guide to Geologic Maps*

A geologic map uses a combination of colors, lines, and symbols to depict the composition, structure, and age relationships of the rock units within a given portion of the Earth's crust. It shows the Earth as it would appear from the air with all of the loose materials stripped away to reveal the bedrock below. In some regions—such as the American West—this bedrock is already extensively exposed and we are able to trace the various rock units for many kilometers. In most other regions, however, rock exposures are widely scattered, and the map must be constructed from observing cliff faces, ledges, stream banks, road cuts, and other excavations.

Map 1 (Figure D.1) illustrates the first phase in the construction of a geologic map. Each of the symbols (circles, dots, dashed lines, wavy lines, etc.) represents an outcrop that a team of geologists has examined. Each symbol, plotted in its correct relationship to the stream, house, roads, and other outcrops, represents a different kind of rock. Dots, for example, are often used to represent sandstone; circles, conglomerate; straight lines, shale; and wavy lines, schist or gneiss. The symbols depict the basic rock units, or *formations*, of the region.

The flat T-shaped symbols with numbers like 3°, 10°, 25°, and 40° are called *strike-and-dip* symbols and give the amount and direction of slope of the layers in the various outcrops. If such a symbol were used to describe a sloping roof, the long line (the *strike* line) would parallel the direction of the ridge pole, and the short line (the *dip* line) would point directly down the slope of the roof in the direction that water would drain. The number gives the inclination (or dip) of the roof in degrees measured below a horizontal plane. For a flat roof, the dip would be 0°, and for a vertical wall, 90°.

The strike-and-dip symbols tell whether the layers at each place go down into the Earth at a steep or gentle angle and in what direction. The geologists can predict the subsurface geology by the strike-and-dip symbols and the distribution pattern of the rock formations, even though no cliffs or other steep cuts show the vertical dimensions and depths directly. The other numbers (82 to 96) near some of the outcrops refer to entries in the geologists' field notebooks in which they record numerous additional observations on the rocks that cannot be conveniently represented by map symbols. For example, the geologists might notice that the rocks along a line running north-south, just west of the house shown in Map 1 appear more fractured than normal, suggesting that these outcrops lie near an important break or *fault* in the rocks. They may also note interruptions in the characteristic fold patterns of these formations; formations of different ages abut against one another.

Their next step is to construct Map 2 (Figure D.2) by sketching in the inferred boundaries or contacts between the various formations. By drawing cross sections in different directions across the area of Map 2, we can see how it shows the subsurface geology (Figure D.3). We can predict what might be seen in the walls of imaginary trenches 120 or 150 meters deep if they were dug along the lines A–A' and B–B'. The top line of each cross section shown here represents the surface of the ground. Proceeding from A' toward A' along this top line, the map shows schist, symbolized by wavy lines. About 75 meters from A, the schist gives way to sandstone, symbolized by a stipple pattern. The strike-and-dip symbol just north of the stream, Map 2, indicates that this sandstone layer is inclined toward the east at an angle of 45°. It is known by scaling from the map (see bar scale, Map 2) how far it is from A to the top and bottom of

---

*Adapted from Walter S. White, *Geologic Maps: Portrait of the Earth,* U.S. Geological Survey pamphlet, U.S. Government Printing Office, 0-333-138: QL 3, 1992.

430 APPENDIX D

Figure D.1

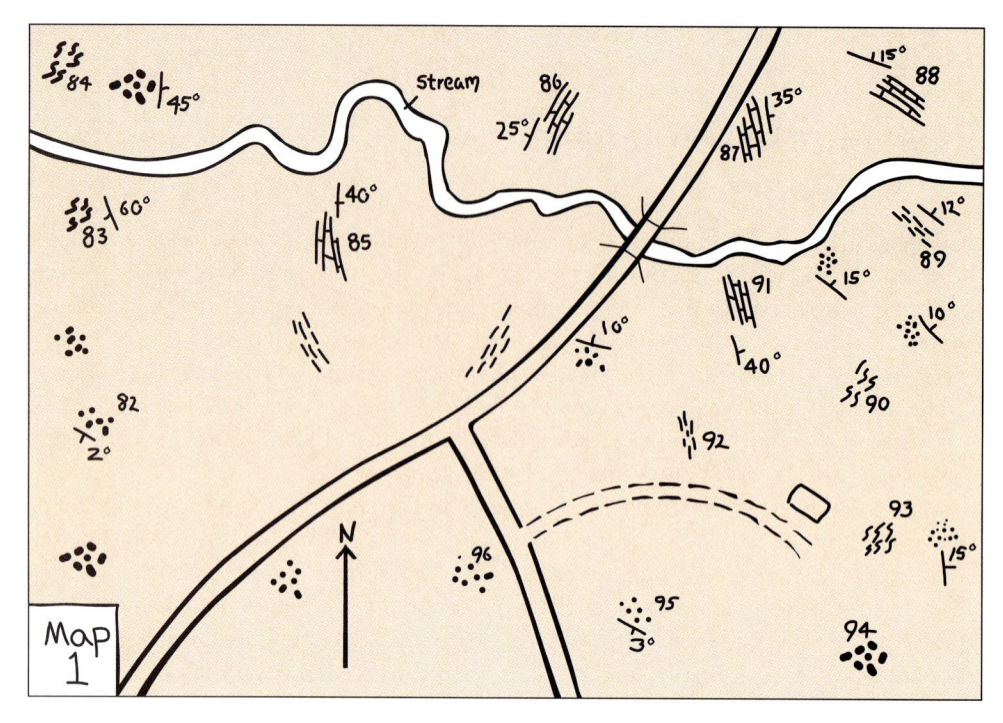

Figure D.2

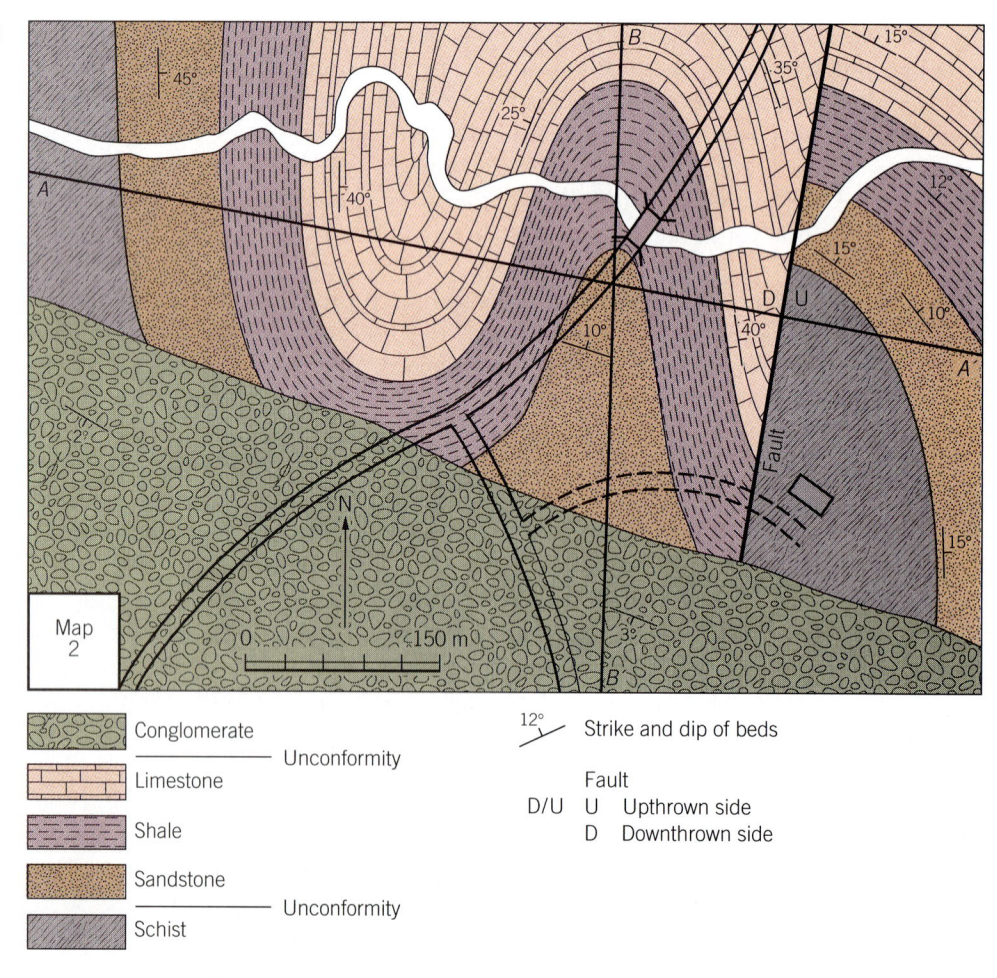

Conglomerate
―――― Unconformity
Limestone
Shale
Sandstone
―――― Unconformity
Schist

⌐12° Strike and dip of beds

Fault
D/U   U Upthrown side
      D Downthrown side

Figure D.3

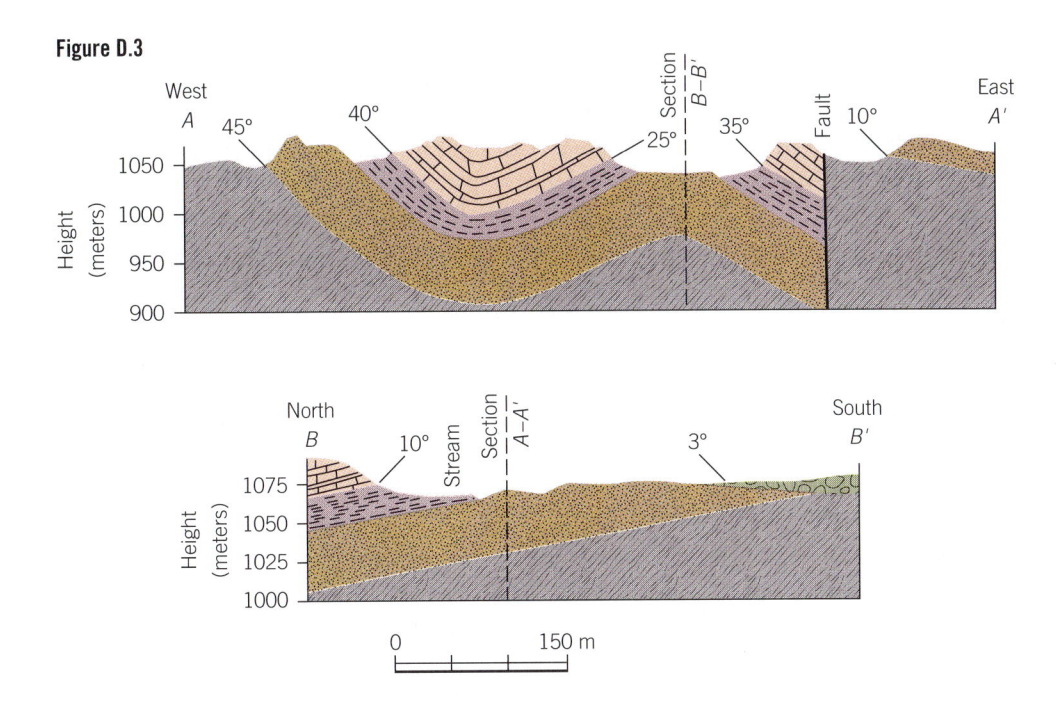

the sandstone layer, and it is known that the sandstone extends downward into the Earth toward the east at an angle of 45°; therefore we can tentatively draw it in with this inclination for at least a few hundred feet below the surface in the cross section. Proceeding eastward in the same way, we can fill out the near-surface part of the section all the way from A to A'. (The heavy line labeled "fault" on the map represents a break in the Earth's crust. The strike-and-dip symbols do not show the fault's attitude, but let us assume that it is vertical.) If the boundary lines of the various rock units are extended downward parallel to their dips, and the patterns are filled in, we can see how rock formations in the western two-thirds of the section probably join below the surface in a trough-shaped structure, or a syncline. Familiarity with similar folded structures elsewhere leads the geologists to round off the bottom as is shown in the cross-sectional drawing.

Eastward, the east side of the *syncline* is the west side of an upwarp, or *anticline*, and everything down to the top of the sandstone can be drawn without any serious stretch of imagination. From the cross section alone, the thickness of the sandstone is not known. However, if the geologists know the thickness of the sandstone in other places where it comes to the surface within and outside the map area, they may be able to predict its thickness in the anticline and complete this part of the cross section with some confidence. Otherwise they might use the thickness of the sandstone layer found in the western part of the section and dash the base of the section to show uncertainty. The upper surface of the schist is the floor on which the sandstone was deposited. The attitude of the schist layers suggested by wavy lines need not conform to the dip of the layers in the sandstone and overlying rocks.

The fault is a break or dislocation in the rock layers, and each rock layer ends against it on the west side and starts anew between the fault and point A'. If the layers almost join across the fault, the displacement is small; if the fit is poor, the displacement is large. Estimates of the actual vertical displacement can be made by measuring how far the rocks on one side of the fault must be shifted to bring the base of the sandstone to the same level. Cross sections can be drawn in other directions and places to determine what lies below the surface in all parts of the area. In section B–B', drawn nearly north–south, the dip of 10° measured in sandstone gives the same thickness for the shale that was obtained from the construction of section A–A'. Such a check for internal consistency between geologic maps and sections makes both the map and the cross section more credible.

The conglomerate formation (open circle pattern) in the southern part of the area is shown as a thin sheet inclined gently south, as the 3° dip of the bedding suggests. Map 2 shows a discontinuity between the conglomerate and the rocks to the north—the northward limit of the conglomerate is a straight line that cuts across other formation boundaries. This suggests a fault, but the map does not indicate that a fault is necessarily present. If the conglomerate included fossils younger than the other rocks, or if it contained pebbles of rock units to the north, it would be possible to interpret the conglomerate as a broad gravel bank that lies on top of the other rocks; under these circumstances, no fault need be present. This latter interpretation is shown in cross section *B–B'*. The existence of these two possible interpretations shows how a geologic map may correctly represent the distribution of rocks at the surface and still be ambiguous.

We can reconstruct the geologic history of the area encompassed by the map by applying some basic geological concepts to the cross sections. For example, from the principle of superposition, we know that the sandstone in section *A–A'* was originally laid down as a sheet on a relatively flat surface that cuts across the layering in the schist, marking a great time-break, or unconformity. Still earlier, the schist must have been deformed, and erosion then carved a level surface on this rock before the sandstone was deposited. Layers of shale were deposited on the sandstone and this shale was overlain, in turn, by limestone. Detailed study of these formations would probably reveal whether they were laid down on the sea floor, in lakes, or on land, because each kind of rock has physical or chemical characteristics that reflect its environment of deposition. Fossils in the rock then provide information about both the environment and the time of deposition.

After the limestone was deposited, the area was crumpled into folds and later faulted. Again the area was subjected to erosion, and a relatively flat surface was cut across all the rocks. The conglomerate was deposited on this surface, which marks the second unconformity of the region. The area was then tilted slightly toward the south to give the conglomerate its present low dip in that direction. The latest event that produced the land surface was the erosion that is still going on.

In this brief account of how geologists make and use maps to organize field observations for study and analysis, we have described what they do as a series of steps taken in a regular sequence, with each completed before the next is begun. Actually, because a geologic map represents an interpretation, it is used as a diagram to test hypotheses. It may be drawn and redrawn many times before a version satisfies all pertinent facts. If more than one version survives these tests, additional field or laboratory observations may be needed to select the best depiction.

Glossary

aa lava A lava flow with a rough, jagged surface. (p. 134)

ablation The loss, or wastage, of ice from a glacier through melting, evaporation, and calving. (p. 286)

abrasion The mechanical wearing or grinding of rock surfaces by friction and impact of rock particles transported by wind, ice, waves, running water, or gravity. (p. 289)

absolute age The age of object or event in years as determined by radiometric dating. (p. 8)

absolute time The ages of objects or events as measured in time units such as years; determined by radiometric-dating techniques. (p. 343)

abyssal plain A flat, level area of the ocean floor that begins at the foot of the continental rise. (p. 39)

accreted terrane A body of rock that is foreign to its surroundings and is bordered on all sides by faults. (p. 368)

accretionary wedge A large mass of sediment and lava scraped off a descending plate that accumulates on the margin of a continent or island arc bordering a subduction zone. (pp. 49, 367)

accumulation The mass of ice gained by a glacier because of snowfall. (p. 286)

active continental margin A boundary between an ocean and a continent marked by plate interaction and thus by frequent volcanic or seismic activity. (p. 50)

aftershocks A series of lower-magnitude earthquakes that may directly follow a higher-magnitude earthquake. (p. 68)

A horizon The uppermost zone of the soil horizon. (p. 221)

alluvial fan A gently sloping, fan-shaped mass of sediment deposited by a stream where it issues from a narrow canyon onto a plain or valley floor. (p. 246)

alpine glacier A glacier in mountainous terrain that flows downslope in a valley previously made by a stream. (p. 285)

amphibole A group of double-chain ferromagnesian silicates whose cleavage planes intersect at approximately 120° and 60°. (p. 101)

andesite A dark, fine-grained, extrusive igneous rock composed of intermediate plagioclase and a ferromagnesian mineral. (p. 114)

angle of repose The maximum slope, or angle, at which a pile of loose material remains stable. (p. 225)

angular unconformity A surface formed by the deposition of sediments on the eroded, upturned edges of older, tilted strata. (p. 344)

anticline A fold with a core of stratigraphically older rocks; usually its convex side is upward. (p. 196)

aphanitic An igneous rock texture in which the mineral components are too small to be identified by the unaided eye. (p. 112)

aquifer A permeable body of subsurface sediment or fractured rock that conducts water. (p. 264)

Archean eon The unit of geologic time dating from the end of the Hadean eon (3.96 b.y.a.) to the beginning of the Proterozoic eon (2.5 b.y.a.). (p. 383)

arête A divide between adjacent valleys, narrowed to a knife-edged ridge by the backward erosion of the valley walls. (p. 290)

artesian well A well under sufficient hydrostatic pressure to force the water to rise above the top of the aquifer. (p. 264)

ash Pyroclastic material less than 2 millimeters in diameter. (p. 135)

asthenosphere A semimolten zone of the mantle just below the lithosphere whose depth extends from about 100 to 200 kilometers and is characterized by diminished seismic wave velocities. (pp. 43, 81)

atom The smallest unit of an element that still retains the properties of the element; a dense, positively charged nucleus surrounded by negatively charged electrons. (p. 89)

atomic mass The average mass of the atoms of an element; to a close approximation, the number of protons plus the number of neutrons in an atom of the element. (p. 90)

atomic number The number of protons in the nucleus of an atom. (p. 90)

aureole A halolike region of contact metamorphism surrounding an igneous intrusion. (p. 184)

axial plane An imaginary planar surface that divides a fold symmetrically. (p. 196)

backwash The return flow of breaking waves down the beach face. (p. 329)

bajada Alluvial fans that coalesce to form a continuous apron along the base of mountains. (p. 312)

banded iron formation A rock that consists of alternating bands of iron-rich minerals and chert or quartz. (p. 387)

barrier island A striplike island that parallels the coast. (p. 330)

basal slip Movement in which a glacier decouples from the valley floor and slides downslope. (p. 286)

basalt A dark, fine-grained, mafic, extrusive igneous rock composed largely of plagioclase feldspar, pyroxene, and olivine. (p. 114)

base level The limiting level below which a stream cannot erode its bed. (p. 243)

basin A synclinal circular structure with rocks dipping gently toward the center. (p. 199)

batholith An igneous intrusion with a large mass, a surface area greater than 100 km^2, and no known floor. (p. 110)

baymouth bar A strip of beach extending from a headland into the mouth of a bay. (p. 330)

beach A sloping portion of the shore composed of sediments deposited and moved by waves, tides, and longshore currents. (p. 328)

bedding The stratification or layering of sedimentary rock. (p. 161)

bedding plane The plane that marks the boundary of each layer of stratified rock. (p. 161)

bed load Large or dense particles that are transported by streams on or immediately above the stream bed or by winds on or immediately above the ground. (p. 244)

beds (strata) Visually distinguishable layers of sedimentary rock. (p. 161)

belt of soil moisture The thin layer of moisture just beneath the land surface; the uppermost subdivision of the zone of aeration. (p. 263)

B horizon The zone of the soil profile consisting of enriched clays and precipitates leached from the A horizon. (p. 221)

biochemical sediment A sediment precipitated directly or indirectly by the activities of organisms. (p. 150)

block A pyroclastic fragment larger than 64 millimeters in diameter that is ejected in a solid state. (p. 135)

body wave A seismic wave that travels through the Earth's interior. (p. 61)

bomb A clump of partially molten lava, ejected from a volcano, whose shape is streamlined in flight; larger than 64 millimeters in diameter. (p. 135)

Bowen's reaction series A proposed sequence of mineral crystallization from basaltic magma, based on experimental evidence. (p. 117)

braided stream A stream that divides into branching and intertwining subchannels separated by islands or sandbars. (p. 246)

breccia A detrital rock consisting of angular pebble-size or larger rock fragments commonly set in a matrix of silt or sand. (p. 152)

brittle response The fracturing of a rock in response to stress with little or no permanent deformation prior to its rupture. (p.194)

burial metamorphism Metamorphism that results in response to the pressure exerted by the weight of the overlying rock. (p. 178)

calcification The encrusting of a soil with alkaline compounds, particularly calcium oxides. (p. 223)

caldera A volcanic crater larger than 1 kilometer in diameter, usually formed by explosion or collapse. (p. 131)

calving The process through which blocks of ice break off from ice shelves and float out to sea as icebergs. (p. 288)

capacity The transporting ability of a stream as measured by the quantity of sediment that the stream carries in a given time interval. (p. 244)

capillary fringe The thin belt just above the water table within which moisture is drawn up by surface tension and partly fills the voids in sediments and rocks. (p. 263)

cap rock An impermeable rock layer overlying oil-bearing reservoir rock. (p. 169)

catastrophism The doctrine that a series of sudden, violent, and short-lived worldwide events are responsible for the state of the Earth's crust and for the variety of life-forms that live on it. (p. 4)

cave A natural cavity beneath the surface of the Earth, usually formed by the dissolution of limestone by groundwater; large caves are also called caverns. (p. 270)

cementation The process by which precipitates bind together the grains of a sediment, converting it into rock. (p. 151)

Cenozoic era The most recent of the four eras of geologic time, beginning at the end of the Mesozoic era (66 million years ago) and extending to the present. (p. 350)

chemical bond The forces exerted between atoms that hold the atoms together. (p. 92)

chemical sediment A sediment composed of ions precipitated directly from water. (p. 150)

chemical weathering The surface process that decomposes rocks and minerals through chemical reactions. (p. 213)

chert A hard sedimentary rock formed from the lithification of biochemical silica. (p. 160)

C horizon The layer of weathered bedrock at the base of the soil horizon. (p. 221)

cinder Glassy, vesicular airborne fragment from 4 to 32 millimeters in diameter. (p. 135)

cinder cone A steep-sided volcano formed by the accumulation of ash, cinders, and other debris close to the vent. (p. 139)

cirque A deep, curving, steep-walled depression scooped out of the bedrock at the head of a valley glacier. (p. 289)

clastic texture Texture of a rock composed mainly of fragments of other rocks and minerals; most commonly used to describe detrital rocks. (p. 152)

cleavage The tendency of a mineral to break along parallel planes of weak bonding. (p. 95)

coal A combustible carbonaceous rock formed from the compaction of altered plant remains. (p. 167)

columnar joints Cracks that form as the result of contraction during the cooling of lava flows, dividing the lava into columns. (p. 134)

compaction Reduction in volume of sediments resulting from the weight of newly deposited sediments above. (p. 151)

competence The transporting ability of a stream, as measured by the largest particles that it can carry. (p. 245)

composite cone See *stratovolcano*.

compound A substance formed by the chemical combination of two or more elements in definite proportions and commonly having properties different from those of its constituent elements. (p. 92)

compressive stress The stress generated by forces directed toward one another on opposite sides of a real or imaginary plane. (pp. 177, 193)

concordant Pertaining to igneous rock bodies that intruded and solidified parallel to the layers of country rock. (p. 110)

cone of depression A concentric indentation in the water table that develops around a pumping well. (p. 265)

confining bed An impermeable layer adjacent to an aquifer that prevents escape of groundwater from the aquifer. (p. 264)

conglomerate A detrital rock consisting of rounded pebble-size or larger rock fragments commonly set in a matrix of silt or sand. (p. 152)

contact metamorphism The transformation of rocks caused by heat escaping from an igneous intrusion. (p. 178)

continental drift A hypothesis suggesting that the continents move over the Earth's surface. (p. 14)

continental glacier See *ice sheet*.

continental rise That part of the continental margin that extends gently downward from the continental slope to the abyssal plain. (p. 48)

continental shelf A very gently sloping surface that extends from the shoreline to the continental slope. (p. 48)

continental slope A more steeply sloping surface that extends from the continental shelf to the continental rise or oceanic trench. (p. 48)

convection A circular heat transfer mechanism in which material heated from below rises because it is less dense than the cooler material above it. The warm material then surrenders heat, becomes denser, and sinks down to the bottom, where it is reheated. (p. 82)

convection current The flow resulting from the rise of warmer, less dense material and the descent of cooler, denser material. (p. 52)

convergent boundary The border between two plates that are colliding with one another. Crustal material is subducted back into the mantle at these boundaries. (p. 43)

Cordilleran orogeny The complex series of mountain-building episodes, from the Jurassic period through the mid-Cenozoic era, that produced the mountains of western North America. (p. 413)

correlation The establishing of equivalence, either in age or rock type, of separated rock units. (p. 346)

covalent bond A type of chemical bond in which electrons are shared by atoms. (p. 93)

crater A circular depression; a volcanic crater contains the vent or vents of the volcano. (p. 131)

craton The stable core of the continental crust that includes basement rock, shield, and platform. (p. 362)

creep The imperceptibly slow downslope movement of soil and rock particles, mainly occurring in cold climates where water alternately freezes and thaws. (p. 232)

crevasse A fracture in the rigid outer layer of a glacier caused by glacier movement. (p. 285)

cross-bedding Thin strata laid down by currents of wind or water at an oblique angle to the main bed. (p. 161)

cross-cutting relationships The principle that an intrusion or fault is younger than the rock that it cuts. (p. 345)

crust The outermost layer of the Earth, about 2 to 50 kilometers thick, consisting of continental and oceanic components of distinctly different density and composition. The crust represents less than 0.1 percent of the Earth's total volume. (pp. 13, 79)

crystal A solid element or compound whose atoms display a definite, orderly atomic arrangement repeated throughout the solid. (p. 94)

crystal face The smooth surface of an unbroken crystal. (p. 94)

cuspate foreland A coastal landform composed of seaward-projecting beaches that meet at a point. (p. 330)

cutbank A steep slope caused by erosion of the outer bank of a meander. (p. 247)

data Items of factual or statistical information. (p. 23)

daughter product An element formed from the decay of a parent isotope. (p. 351)

debris flow Downslope movement of a viscous mass of rock and soil particles, more than half of which are larger than a sand grain. (p. 231)

deep-sea trench A deep, narrow, elongated depression in the sea floor. (p. 34)

deflation The removal of loose materials by wind, which often results in the lowering of the land surface. (p. 313)

delta A triangle-shaped landform built up over time where a stream enters a calmer body of water and deposits its sediment load. (p. 252)

dendritic drainage A stream drainage pattern resembling a branching tree that is common to regions underlain by materials of uniform composition or horizontal layers. (p. 255)

deposition The gravitational settling of rock-forming materials out of such natural agents as water, wind, or ice. (pp. 9, 150)

desert A region with an average annual precipitation of less than 25 centimeters, too dry to support more than sparse vegetation. (Climatologists define deserts more complexly on the basis of temperature-precipitation-evaporation ratios.) (p. 306)

desertification Declining productivity of arable land due to mismanagement, leading to a state resembling a desert. (p. 320)

desert pavement The remaining pebble- to cobble-size rock fragments covering a desert surface after removal of lighter fragments by wind. (p. 313)

desert varnish A dark, shiny, enamel-like film of iron oxide and manganese that coats the surface of rocks in desert regions. (p. 308)
detrital sediment Fragments derived from the weathering of rocks, transported by water, wind, or ice, and deposited in loose layers on the Earth's surface. (p. 150)
dike A tabular, discordant igneous intrusion. (p. 110)
dip The angle formed by the intersection of a bedding or fault plane and the horizontal plane; measured in a vertical plane perpendicular to strike. (p. 195)
dip-slip fault A fault in which the movement is parallel to the dip of the fault plane. (p. 200)
direct solution The dissolving of rock or mineral materials. (p. 217)
discharge The volume of water moving past a particular point in a stream over a given time interval; the product of the cross-sectional area of the channel and the stream velocity. (p. 265)
discharge area The area—primarily stream channels—that receives groundwater from the zone of saturation and conducts it away. (p. 243)
disconformity An unconformity between parallel sedimentary layers. (p. 344)
discordant Pertaining to igneous rock bodies that cut across the layers of country rock. (p. 110)
disseminated deposit An ore finely scattered within a host rock body. (p. 125)
dissolved load The portion of a stream load carried in solution. (p. 244)
distributary One of a system of small channels that carry water and sediment from the mainstream channel and spread them over the delta surface. (p. 252)
divergent boundary The border between two plates that are moving away from one another as new crust is formed. (p. 43)
dome An anticlinal circular structure with rocks dipping gently away from the center. (p. 199)
drainage basin The total area that contributes water to a stream system. (p. 241)
drift Glacial deposit of rock fragments. (p. 290)
drumlin A rounded, low, and elongated hill of compacted till, with its blunter end pointing upstream and its tapering end pointing in the direction of ice flow. (p. 293)
ductile response The permanent deformation of a body, without fracture, in the shape of a solid. (p. 194)
dune A mound or hill of windblown sand. (p. 316)

earthflow Downslope movement of a loose mass consisting mainly of rock fragments and soil in a semifluid state. (p. 230)
earthquake Vibrations within the Earth set in motion by the sudden release of accumulated strain energy. (p. 59)
Ediacaran fauna A varietey of soft-bodied animals that died out before the end of the Proterozoic eon. (p. 388)
effluent stream A stream that intersects the water table and gains water from the zone of saturation. (p. 265)
elastic limit The maximum amount of stress that a material can withstand before it deforms permanently. (p. 194)
elastic-rebound theory The concept that earthquakes are generated by the sudden slippage of rocks on either side of a fault plane. In the process, the rocks release gradually accumulated strain energy and are returned to an unstrained condition. (p. 61)
elastic response The deformation of a body in proportion to the applied stress and its recovery once the stress is removed. (p. 194)
electron A tiny particle with a negative charge—equal to a proton's positive charge—that orbits the nucleus of the atom. (p. 89)
electron shell A region surrounding the nucleus occupied by electrons having approximately the same energy. (p. 90)
element A substance made up entirely of atoms of the same atomic number that cannot be decomposed into a simpler substance by ordinary chemical or physical means. (pp. 13, 89)
epicenter The point on the Earth's surface that is directly above the focus of an earthquake. (p. 61)
erosion The wearing down of rocks or soil, by weathering, mass wasting, running water, wind, and ice. (p. 9)

esker A snakelike ridge of roughly stratified gravel and sand, left behind when glacial ice melted. (p. 294)
estuary A mouth of a river invaded by the ocean, where fresh- and saltwater mix. (p. 337)
eukaryote A mulitcelled organism with the nucleus of each cell enclosed in a protective membrane. (p. 388)
evaporite Deposit from the evaporation of aqueous solutions. The most common example is rock salt. (p. 159)
evolution The process by which populations of organisms change in form and function over time. (p. 389)
exfoliation The flaking or stripping of a rock body in concentric layers. (p. 216)
extrusive Igneous rock formed from lava flows or pyroclastic materials that were spewed to the Earth's surface. (p. 112)

fall The free downward movement of detached rock fragments through the air from a cliff or steep slope. (p. 224)
fault A fracture in bedrock along which rocks on one side have moved relative to the other side. (pp. 30, 194)
fault breccia A metamorphic rock consisting of angular fragments that are the result of the grinding and shattering action that occurs along active fault zones. (p. 181)
fault scarp A cliff created by the movement along a fault. (p. 59)
faunal succession The principle that fossil organisms succeed one another in definite and recognizable order, so that rocks containing identical fossils are identical in age. (pp. 6, 346)
feldspar A common group of aluminum silicate rock-forming minerals that contain potassium, sodium, or calcium and display right-angle cleavage. (p. 99)
ferromagnesian Category of silicate minerals rich in iron and magnesium. (p. 99)
firn The pelletlike form assumed by snow in its transition to glacial ice. (p. 284)
fissure eruption A volcanic eruption through a long fracture rather than a central vent. (p. 139)
fjord A deep, narrow arm of the sea, formed when a glacial trough that extended to the sea became submerged after melting of the ice. (p. 290)
flood basalt Highly fluid basaltic lava produced during a fissure eruption. (p. 139)
flood plain Layers of sediment beds deposited across a valley as a stream periodically overflows its banks and meanders laterally. (p. 249)
flow Downslope movement of loose rock and soil as a viscous fluid mass. (p. 230)
focus The initial point of rupture within the Earth from which seismic waves are generated in an earthquake. (p. 61)
fold Permanent wavelike deformation in layered rock or sediment. (p. 194)
fold axis A line formed by the intersection of the axial plane and a layer or bed surface. (p. 196)
foliation The arrangement of a rock in parallel planes or layers; in metamorphic rocks, caused by parallel alignment of the minerals. (p. 177)
footwall The rock mass beneath an inclined vein or fault. (p. 200)
foreshocks A series of lower-magnitude earthquakes that may directly precede a higher-magnitude earthquake. (p. 68)
formation A distinctive mappable rock unit. (p. 343)
fossil The remains, trace, or imprint of a plant or animal preserved in rock. (pp. 6, 164)
fossil fuel Any hydrocarbon that can be used for fuel. (p. 167)
fracture The manner in which minerals break other than along planes of cleavage. (p. 95)
fracture zone A region of closely spaced cracks or faults in rocks. Also, a prominent crack or fault that runs perpendicular to, and offsets, the mid-ocean ridge. (p. 31)
frost wedging A process by which water seeps into rock joints, freezes, and wedges the rock apart. (p. 214)
fumarole A vent or ground opening that spews volcanic fumes or vapors. (pp. 136, 274)

geologic time scale A chronological ordering of geologic events and time units listed in sequence from the oldest on the bottom to the youngest on the top. (p. 7)

geologist One who investigates the materials, processes, products, and history of the Earth. Geologists conduct basic scientific research in order to increase our understanding of the Earth, and they apply their knowledge to improve our lives in many ways. (p. 3)

geology The study of the materials, processes, products, and history of the Earth. (p. 3)

geothermal gradient The rate of increase of the temperature in the Earth with depth; it varies from region to region. The geothermal gradient averages 25 °C per kilometer of depth in the crust. (p. 81)

geyser A periodic eruption of hot water and steam, caused by the heating of pressurized groundwater in a network of underground channels in contact with hot rock or magma. (p. 273)

glacial age An interval of geologic time in which glacial ice sheets advance. (p. 296)

glacial erratic A large rock fragment or boulder, carried by glacial ice away from its place of origin and usually deposited on bedrock of a different type. (p. 293)

glacial striations Parallel grooves and scratches that were gouged in the bedrock by rock fragments attached to the bottom of the ice as the glacier slid over it. (p. 289)

glacial terminus The extremity or outer margin of a glacier; the glacial front. (p. 286)

glacial trough A U-shaped valley carved by a mountain glacier in what was previously a V-shaped stream valley. (p. 290)

glacier A large ice mass formed on land by the compaction and recrystallization of snow that survives from year to year and shows evidence of present or past flow. (p. 284)

glassy The texture of an igneous rock having a high content of glass. (p. 112)

gneiss A foliated metamorphic rock in which bands of granular minerals alternate with bands of flaky minerals. (p. 180)

graben An elongated, depressed block bounded by normal faults on its long sides. (p. 202)

gradation The balance between erosion and deposition that maintains a general slope of equilibrium trending toward sea level. (p. 211)

graded bedding Stratification in which particle size changes from coarse to fine from the bottom to the top of each bed. (p. 161)

graded stream A stream whose slope and channel are so adjusted that it has just enough energy to transport its load; no excess energy is present to erode the channel, nor is there a deficiency of energy that leads to deposition of sediment in the channel. (p. 245)

gradient The steepness or slope over a specific length of a stream channel. (p. 243)

granite A coarse-grained, intrusive igneous rock containing quartz and feldspar (primarily potassium feldspar). (p. 115)

greenhouse effect The warming of the Earth's atmosphere through the presence of atmospheric gases that absorb and re-radiate the heat rising from the surface. (p. 143)

groin A structure projecting perpendicular to the beach, designed to trap sediments and prevent beach erosion. (p. 332)

ground moraine A rough blanket of till that accumulated under a glacier and was exposed as the terminus receded toward the head of the glacier. (p. 292)

groundwater Underground water within the zone of saturation, below the water table. (p. 262)

guyot A sea-floor volcano that at one time rose above sea level, where its top was planed by wave erosion. (p. 40)

Hadean eon The 640-million-year interval beween the time of the Earth's origin (4.6 b.y.a.) and the age of the oldest preserved Earth rocks (about 3.96 billion years). (p. 381)

half-life The time required for half of a given parent isotope to decay to its daughter product. (p. 352)

hanging valley A glacial valley, formed in what was previously a tributary stream valley that was cut off and left stranded at a higher elevation from the main glacial valley. (p. 290)

hanging wall The rock mass overlying an inclined vein or fault. (p. 200)

hardness The resistance of the surface of a mineral to scratching. (p. 96)

headward erosion The lengthening of a stream or gully by erosion at the head of the stream valley. (p. 255)

horn A steeply carved rock peak, formed by the walls of three or more cirques. (p. 290)

hornfels Fine-grained rock formed by contact metamorphism. (p. 182)

horst An uplifted block bounded by normal faults on its long sides. (p. 202)

hot spot An area of volcanic activity produced by a plume of magma rising from the mantle. (p. 41)

hot spring A spring whose waters have been heated above body temperature (36.7 °C) by hot rock or magma. (p. 274)

hydrologic cycle The continuous circulation pattern of the world's water, from ocean to atmosphere to Earth's land surface to ocean again. (p. 8)

hydrolysis The reaction between minerals (especially silicates) and water. (p. 218)

hydrothermal metamorphism The transformation of rock through the action of high-temperature solutions. (p. 179)

hypothesis A tentative explanation of a phenomenon or process that is tested for validity by repeated observation or experimentation. (p. 22)

iceberg A block of glacial ice floating on a body of water, with 80 percent or more of its volume below the surface. (p. 288)

ice sheet A glacier more than 50,000 square kilometers in area that spreads in all directions unconfined by underlying topography. (p. 285)

ice shelf A sheet of thick ice terminating in steep cliffs. One end remains attached to the land; the other end extends out into the sea, where it floats. (p. 288)

igneous rock Rock that has cooled from the molten, or magmatic, state. (p. 10)

impact cratering The crustal deformation caused by a meteoroid collision with the Earth's surface. (p. 204)

incised meander A deep, narrow winding stream valley caused by the combination of regional uplift and stream downcutting. (p. 254)

index fossil A fossil used to accurately establish the relative age of the stratum within which it is found. (p. 348)

index mineral A mineral that characterizes a given intensity of metamorphism, having developed under a specific range of temperature and pressure conditions. (p. 187)

influent stream A stream that lies above the water table and loses water to the zone of saturation. (p. 265)

inner core The solid spherical center of the Earth, extending from the outer core to the Earth's crust. (pp. 14, 79)

inselberg An isolated knob or hill, remnant of a heavily eroded mountain in a desert region; from the German meaning "island mountain." (p. 312)

interglacial age An interval of geologic time in which glacial ice sheets retreat. (p. 296)

intermittent stream A stream whose channel conducts water during periods of high rainfall and is dry during intervening periods. (p. 309)

intraplate earthquake An earthquake whose epicenter is far from a lithospheric plate boundary. (p. 70)

intrusive An igneous rock derived from magma that solidified within the mantle or crust. (p. 112)

invertebrate An animal that does not have a backbone. (p. 407)

ion An atom with either a positive or a negative charge caused by the loss or gain of electrons. (p. 93)

ionic bond A chemical bond that holds two oppositely charged ions together through electron transfer. (p. 93)

island arc A curved chain of islands with its convex side to the ocean, emerging from the deep-sea floor. Generally located close to a continent. (p. 34)

isograd A line on a map connecting points of equal metamorphic intensity; usually indicated by the first appearance of a given index mineral. (p. 188)

isostasy The condition of balance, or equilibrium, in which the crust floats on the mantle. (p.13)

isotope A variety of an element that differs from other varieties of the same element in the number of neutrons it contains and thus in its atomic mass. (p. 90)

joint A fracture in a rock, without noticeable movement along the plane of fracture. (p. 194)

joint set A group of joints, generally parallel. (p. 200)

kame A mound or short ridge of stratified sand and gravel deposited by water streaming under or trapped within glacial ice; from the Scottish word meaning "steep-sided monument." (p. 295)

karst topography Topography formed by the intensive dissolution of underlying limestone bedrock, featuring sinkholes, intricate cave networks, and diversion of surface drainage underground. (p. 272)

kettle A depression in a moraine or outwash plain formed when a large block of ice was isolated from the retreating glacier and buried in the drift, melting afterward. (p. 295)

laccolith A mushroom-shaped, concordant igneous intrusion that has domed the overlying crustal rocks. (p. 110)

lahar A mudflow of volcanic material. (p. 136)

landslide The rapid downslope movement along shear planes of a mass of rock fragments and soil. (p. 225)

lapilli Pyroclastic particles that range in size from 2 to 64 millimeters in diameter. (p. 135)

Laramide orogeny The late-Cretaceous mountain building episode that formed the Rocky Mountains. (p. 413)

lateral moraine A long, narrow mound of till that lies perpendicular to the glacier front, formed by plucking and rockfalls along the valley walls. (p. 293)

laterite A highly weathered red soil typical of tropical and semitropical climates, enriched in iron and aluminum oxides. (p. 222)

lava Magma that flows out onto the surface of the Earth; also refers to the rock body formed after the magma cools. (p. 131)

lava dome A convex structure of solidified lava, extruded from the vent of a volcanic crater. (p. 134)

lava plateau An elevated, flat-topped region composed of a thick succession of horizontal lava flows. (p. 139)

leaching The transfer in solution of organic matter and other elements from upper to lower soil levels. (p. 221)

limestone The consolidated product of calcareous sand, limy mud, and/or crushed shells. (p. 157)

liquefaction The transformation of saturated sediment or soil to liquid when ground shaking causes the particles to lose contact. (p. 74)

lithification The process by which sediment is converted to sedimentary rock through compaction, cementation, or crystallization. (pp. 10, 151)

lithosphere A rigid zone of the Earth that includes the crust and a sliver of upper mantle and that rests directly on the asthenosphere. (pp. 42, 79)

lithostatic stress The uniform stress in the Earth's crust, caused by the weight of the overlying rocks. (p. 177)

local base level A transitory, local feature, such as a lake, dam, or resistant rock layer, below which a stream cannot erode its channel. (p. 243)

loess A thick deposit of fine windblown dust consisting of unstratified silt-size calcium carbonate and bits of clay, which has the ability to maintain nearly vertical walls despite its weak cohesion. (p. 314)

longitudinal dune A long ridge of sand standing parallel to the prevailing wind. (p. 317)

long profile A cross section of a stream channel showing the gradient from the source to the mouth of the stream. (p. 243)

longshore current A shallow current parallel to the coast, caused by waves that approach the shore at an oblique angle. (p. 329)

longshore sediment transport The movement of sediment parallel to the shore by longshore currents, swash, and backwash. (p. 329)

lower mantle The part of the mantle that extends from 670 to 2900 kilometers in depth. (p. 81)

luster The quality and intensity of light reflected from the surface of a mineral. (p. 97)

magma Molten (hot-liquid) rock material, generated within the Earth, that forms igneous rocks when solidified. (pp. 10, 109)

magnetic reversal A switch in the direction of the Earth's magnetic field such that a compass needle that today points north would, at the time of reversal, have pointed south. (p. 37)

magnitude A measure of earthquake strength as interpreted from the maximum wave amplitude recorded by a seismograph. (p. 65)

mantle The layer of the Earth located between the crust and the outer core whose depth extends from about 20 to 2900 kilometers. (pp. 13, 78)

mantle plume A pipe-shaped mass of heat-softened light rock that rises from the mantle toward the crust. (p. 82)

marble A metamorphic rock consisting mainly of recrystallized calcite and/or dolomite. (p. 182)

mass wasting The downslope movement of rocks and soil caused by gravity alone, without the aid of a transport medium such as a stream, glacier, or lava flow. (p. 9)

matrix The fine-grained material surrounding larger grains in a rock. (p. 151)

meander *As a verb:* to flow in a winding, sinuous course. *As a noun:* a loop created as a stream winds back and forth across a valley. (p. 247)

meander cutoff A new channel created when a stream takes the shorter route across the neck of a meander. (p. 248)

mechanical weathering The combination of physical processes that disintegrates a rock without chemical change. (p. 213)

medial moraine The joining of lateral moraines where two valley glaciers merge. (p. 293)

mélange A chaotic assemblage of accretionary wedge sediments and oceanic crust thrown up onto land at a subduction zone. (p. 367)

Mercalli intensity scale A 12-point scale that measures earthquake severity in terms of the damage inflicted. (p. 66)

Mesozoic era The era preceding the Cenozoic, extending from about 245 million years to 66 million years ago. (p. 350)

metallic bond A chemical bond created in electron-donating elements through the merging of electron shells. (p. 94)

metamorphic rock A rock that has been altered from its original state through metamorphism. (p. 11)

metamorphic zone An area of equal metamorphic intensity between isograds in which the mineral content of rocks remains constant. (p. 187)

metamorphism The structural and mineralogical changes that occur in solid rock through the action of heat, pressure, and chemically active fluids. (p. 11)

mid-ocean ridge A seismically and volcanically active mountain range, marked by a central rift valley, that extends continuously through the major ocean basins. (p. 30)

migmatite From the Greek *migma*, meaning "mixture"; a complex mixture of metamorphic and igneous rock components (usually granite). (p. 180)

mineral A naturally occurring and inorganic solid with a definite chemical composition and orderly internal atomic arrangement. (p. 89)

model A hypothesis expressed as a visual or statistical simulation, or as a description by analogy of phenomena or processes that are difficult to observe and describe directly. (p. 22)

Mohorovičić discontinuity (Moho) The surface that defines the boundary between the Earth's crust and mantle. (p. 79)

monocline A sudden steepening in an otherwise gently dipping strata. (p. 199)

moraine A landform composed of till left behind by a retreating glacier. (p. 292)

mud crack Polygonal crack formed by the drying and shrinking of mud, silt, or clay. (p. 162)

mudflow Downslope movement of a fluidized mass of clay and other fine-grained materials. (p. 230)

multiple working hypotheses An approach to geologic research in which several possible explanations of a phenomenon are developed and evaluated simultaneously and impartially. (p. 23)

natural levee A ridge of sand and silt that parallels the stream channel and that is deposited over time, when the stream overflows its banks during floods. (p. 250)

neap tide A low-amplitude, twice-monthly ocean tide. (p. 334)

neutron A particle in an atomic nucleus with a mass virtually equal to the proton's but with no electric charge. (p. 89)

nonconformity An unconformity with stratified rocks above and igneous or metamorphic rocks below. (p. 344)

normal fault A fault in which the hanging wall has been moved downward relative to the footwall. (p. 200)

nucleus The dense center of an atom composed of protons and neutrons. Nearly all of the mass of an atom is concentrated in the nucleus. (p. 90)

nuée ardente A turbulent, ground-hugging, gaseous cloud erupted from a volcano. (p. 136)

obsidian A volcanic glass, either black or dark-colored, usually of rhyolite composition, and characterized by conchoidal fracture. (p. 134)

O horizon The top layer of soil consisting of decayed plant matter. (p. 221)

oil field A geologically and spatially related feature containing two or more oil accumulations. (p. 169)

oil sand Broadly defined as a porous sand deposit from which petroleum can be extracted; also refers to certain sedimentary deposits impregnated with tarlike petroleum residue. (p. 169)

oil shale A fine-grained sedimentary rock from which oil or gas can be distilled. (p. 169)

oil trap A structural or stratigraphic arrangement within a sedimentary basin, consisting of permeable reservoir rock overlain by impermeable cap rock, which allows oil to accumulate and prevents its escape. (p. 169)

olivine A ferromagnesian silicate common in mafic and ultramafic rocks having an internal structure of isolated tetrahedra. (p. 101)

original horizontality The principle that sediments are deposited in horizontal layers, parallel to the Earth's surface. (p. 344)

orogeny The process of mountain building. (p. 363)

outer core The upper layer of the Earth's core whose depth extends from about 2900 to 5140 kilometers. It is presumed to be molten. (pp. 14, 78)

outgassing The release of gases and water vapor from molten rocks that led to the formation of the atmosphere and oceans. (p. 386)

outwash plain A broad, gently sloping sheet of stratified sand and gravel sediment deposited by meltwater streaming out of the front of a glacier. (p. 294)

overturned fold A fold in which one limb has rotated past the perpendicular, so that both limbs dip in the same direction. (p. 196)

oxbow lake A crescent-shaped abandoned meander channel isolated from the stream channel by a meander cutoff and sedimentation. (p. 248)

oxidation The process by which an element combines with oxygen. (p. 219)

pahoehoe lava A lava flow with a smooth, ropy surface. (p. 134)

paleomagnetism The study of the Earth's past magnetism as recorded in rocks at the time of their formation. (p. 20)

Paleozoic era The second of the geologic time eras, extending from the end of the Precambrian (about 570 million years ago) to the beginning of the Mesozoic (about 245 million years ago). (p. 350)

Pangaea The name of the supercontinent that existed between 200 and 300 million years ago, which has since fragmented into the present continents. (p. 14)

parent isotope An isotope undergoing radioactive decay. (p. 351)

passive continental margin A margin between an ocean and a continent that does not include a plate boundary. It is also free of volcanic or seismic activity. (p. 47)

paternoster lake A small lake in a depression carved in a glacial valley floor. (p. 290)

pediment A broad erosional surface formed by running water that slopes gently away from a receding mountain front; the bedrock may be exposed or thinly covered with debris. (p. 312)

perched water table The upper limit of a local groundwater body stranded above the regional water table by an impermeable layer of rock or sediment. (p. 264)

peridotite A dark, coarse-grained, intrusive igneous rock composed mainly of olivine, small amounts of pyroxene, and little or no plagioclase. (p. 114)

periodic table of the elements A chart of the elements arranged in order of increasing atomic number and according to similarities of electron structure. As a consequence, the elements are divided into groups having similar chemical properties. (p. 91)

permafrost A condition of permanently frozen soil or subsoil occurring in cold climates. (p. 232)

permeability The capacity of a material to conduct a fluid; measured by the volume of fluid that will move through a cross-sectional area in a given period of time, subject to a given pressure difference. (p. 264)

petroleum A naturally occurring liquid of complex hydrocarbon composition. Depending on context, the term may include natural gas. (p. 168)

phaneritic An igneous rock texture in which the mineral components are visible to the unaided eye. (p. 112)

Phanerozoic eon That part of geologic time represented by rocks in which the evidence of life is abundant. (p. 349)

phreatic eruption A volcanic eruption of steam, mud, or ash initiated by the contact of water with magma or hot rock. (p. 132)

pillow lava A term applied to lavas of ovoid or pillow shape. (p. 132)

placer deposit An accumulation of mineral particles, derived from the mechanical weathering of a mother lode or source rock, that is transported and deposited by running water. (p. 248)

plagioclase A group of feldspars in which calcium and sodium substitute for one another in solid solution. (p. 103)

plate A rigid segment of the lithosphere that moves as a unit over the asthenosphere. (p. 42)

plate tectonics The theory that proposes that the lithosphere is divided into plates that interact with one another at their boundaries, producing tectonic activity. (p. 41)

platform That part of the craton consisting of essentially horizontal sedimentary strata overlying the older basement rocks. (p. 362)

playa A flat area at the bottom of a desert basin that becomes an intermittent (playa) lake in the wet season. (p. 312)

plucking An erosion process in which meltwater trickles into rock fractures, refreezes, and expands, breaking off rock fragments. (p. 289)

plunging fold A fold whose axis is inclined rather than horizontal. (p. 196)

pluton An intrusive rock body. (p. 110)

pluvial lake A body of water in a nonglaciated region formed by abundant rainfall due to the cooler climate caused by the growth of ice sheets. (p. 295)

pocket beach A small beach nestled between adjacent headlands. (p. 330)

podzolization The process of nutrient leaching of the A horizon that produces the ashy gray podzol soils of coniferous forests in cool, moist climates. (p. 224)

point bar A low, crescent-shaped deposit of sand and gravel developed on the inner bank of a meander where the water velocity is low. (p. 247)

porosity The ratio of open space to total volume of a subsurface material. (p. 263)

porphyritic An aphanitic igneous-rock texture in which large crystals are embedded within an aphanitic or glassy matrix. (p. 112)

pothole A smooth, bowl-shaped depression in the bed of a stream caused by abrasion when turbulent currents circulate stones or coarse sediment. (p. 244)

Precambrian All geologic time prior to the beginning of the Paleozoic era (about 570 million years ago). (p. 348)

primary (P) wave A seismic wave that propagates through the Earth as a series of compressions and expansions. (p. 61)

principle of cross-cutting relationships See *cross-cutting relationships*.

principle of faunal succession See *faunal succession*.

principle of original horizontality See *original horizontality*.

principle of superposition See *superposition*.

prokaryote A single-celled organism without a nuclear membrane. (p. 388)

Proterozoic eon The unit of geologic time dating from the close of the Archean eon (2.5 b.y.a.) to the onset of the Paleozoic eon (570 m.y.a.). (p. 385)

proton A positively charged particle in an atomic nucleus. (p. 89)

pumice A very light, cellular, glassy rock; often floats in water and is commonly used as an abrasive. (p. 134)

pyroclastic Pertaining to rock material formed by an explosive ejection from a volcanic vent. (p. 135)

pyroclastic rock A rock of any size formed from the cementation or welding of volcanic fragments. (p. 136)

pyroxene A ferromagnesian silicate common in mafic and ultramafic rocks having single-chain structure and right-angle cleavage. (p. 101)

quartz A mineral composed exclusively of silicon dioxide, whose tetrahedra are linked in a framework structure. (p. 99)

quartzite A metamorphic rock composed mainly of quartz, formed by the recrystallization of sandstone. (p. 182)

radial drainage A stream pattern that radiates like the spokes of a wheel from the summit of a volcano. (p. 256)

radioactive decay The disintegration of certain isotopes by the emission of subatomic particles. (pp. 90, 350)

rain shadow A dry region on the leeward side of a mountain range where rainfall is significantly less than on the windward side; the region is said to be "in the rain shadow" of the mountain range. (p. 307)

recessional moraine A lateral or end moraine accumulated during a pause in a glacier's retreat. (p. 292)

recharge area The area—mainly the broad upland between stream valleys—that receives precipitation and adds water to the zone of saturation. (p. 265)

recrystallization The formation of new crystalline mineral grains in a rock. (p. 151)

rectangular drainage A drainage pattern commonly found in homogeneous igneous and metamorphic rocks in which tributary streams display right-angled bends that follow joints and faults. (p. 256)

recumbent fold An overturned fold whose axial plane is horizontal. (p. 196)

reflection The return of a wave to its original medium upon striking the boundary of another medium. (p. 77)

refraction The change of direction that occurs when a wave passes from one medium to another; also, the bending of water waves upon entering shallow water. (pp. 77, 326)

regional metamorphism Metamorphism of an extensive area of the crust, generally associated with intensive compression and mountain building. (p. 178)

regolith All the fragmented and unconsolidated material overlying bedrock. (p. 221)

relative age The age of an object or event expressed relative to the age of other objects or events but not in relation to time units such as years. (p. 6)

relative time The chronological ordering of events without reference to their age in years. (p. 343)

reservoir rock Any porous and permeable rock, such as sandstone, that might contain oil or natural gas deposits. (p. 169)

reverse fault A fault in which the hanging wall has been raised relative to the footwall. (p. 200)

Richter magnitude scale A logarithmic scale of earthquake magnitudes based on the maximum-amplitude wave recorded by a standard seismograph and corrected for distance of the seismograph from the epicenter. (p. 65)

ripple mark A corrugated form displayed in sedimentary rocks caused by currents of air or water that moved over the sediments prior to lithification. (p. 162)

rock cycle A model that describes the formation, breakdown, and reformation of a rock as a result of sedimentary, igneous, and metamorphic processes. (p. 11)

rockslide The rapid downslope movement along shear planes of a mass consisting mostly of large chunks of bedrock. (p. 225)

saltation Transport by streams in which bed load particles are moved in short skips and bounces. (p. 245)

saltwater encroachment The displacement in the zone of saturation of freshwater by saltwater. (p. 268)

sandstone A detrital rock consisting primarily of sand held together by a cementing agent. (p. 152)

schist A foliated, coarse-textured metamorphic rock; most of the minerals display a pronounced parallelism. (p. 180)

scientific method A process of investigation in which a problem is identified, data are collected and analyzed, and a hypothesis is formulated and tested. (p. 22)

sea arch A bridgelike erosional remnant of coastal rocks. (p. 327)

sea-floor spreading The theory that new oceanic crust is created at the mid-ocean ridges, spreads laterally, and descends back into the mantle at the deep-sea trenches. (p. 35)

seamount A sea-floor volcano that has never risen above sea level. (p. 40)

sea stack An erosional remnant of a former headland. (p. 327)

secondary (S) wave A seismic wave that causes the components of a rock to vibrate perpendicularly to the direction of wave propagation. Also called a shear wave. (p. 61)

sediment Particles that are mechanically transported by water, wind, or ice, or chemically precipitated from solution, or secreted by organisms, and deposited in loose layers on the Earth's surface. (pp. 4, 149)

sedimentary rock A layered rock formed from the consolidation of sediment. (p. 4)

seismic belt A long, narrow zone of earthquake activity associated with lithospheric plate boundaries. (p. 42)

seismic gap A seismically inactive segment of a fault within which strain is accumulating; these gaps are bracketed by the epicenters of relatively recent great earthquakes. (p. 71)

seismic wave A wave generated within the Earth by sudden fault slippage or an explosion. (p. 61)

seismogram A record of seismic waves detected by a seismograph. (p. 62)

seismograph A device that detects and records seismic waves. (p. 62)

settling velocity The velocity required for a suspended particle of a given size to sink to the bottom of the stream channel. If the stream velocity is greater than this value, the particle will remain in suspension; if less, it will sink. (p. 245)

shadow zone A region between 105° and 142° from the epicenter of an earthquake where no direct P and S waves reach the Earth's surface, owing to the properties of the outer core. (p. 78)

shale A fine-textured detrital rock with a layered structure resulting from compaction of clay, mud, or silt. (p. 152)

shear metamorphism The transformation of rocks within the shear zone associated with active fault movement; mainly involves grinding, pulverizing, and recrystallization of the rocks. (p. 178)

shear stress Stress (force per unit area) that acts parallel to a (fault) plane and tends to cause the rocks on either side of the plane to slide by one another. (p. 193)

sheeting A type of jointing parallel to the rock surface caused by pressure release. Similar to exfoliation. (p. 216)

shield A large region of exposed metamorphic and igneous basement rocks, generally having a gently convex surface and surrounded by a sediment-covered platform. (p. 362)

shield volcano A broad, low-profile volcanic cone, commonly composed of basaltic flows. (p. 137)

shock metamorphism Changes in rock and minerals caused by shock waves from high-velocity impacts, mainly from meteoroids. (p. 179)

silicate A compound containing silicon and oxygen ions arranged as negatively charged ions. (p. 99)

silicon-oxygen tetrahedron An arrangement in which four oxygen ions surround one silicon ion, forming a four-sided structure of negative charge. (p. 101)

sill A tabular, concordant igneous intrusion. (p. 110)

sinkhole A circular surface depression that occurs when the roof of an underground cave collapses or dissolves. (p. 271)

slate A fine-grained, foliated metamorphic rock; mostly formed from the transformation of shale. (p. 180)

slide The rapid downslope movement along shear planes of more or less consolidated rock or fragmented materials. (p. 225)

slip face The steeper downwind (or leeward) side of a dune. (p. 316)

slump Downslope movement of rock or loose debris along a concave plane. (p. 228)

snowline Altitude above which the snow is permanent. (p. 284)

soil Unconsolidated material capable of supporting vegetation. (p. 221)

soil horizon A soil layer with physical and chemical properties that differ from those of adjacent soil layers. (p. 221)

soil profile A vertical cross section of soil that displays all soil horizons. (p. 221)

solifluction The slow downslope movement of waterlogged soil caused by repeated freezing and thawing in cold climates. (p. 233)

sorting The process by which the agents of transportation (principally, running water) separate sediments according to shape, size, or density. (p. 153)

specific gravity The density of a substance compared with the density of water. (p. 97)

spheroidal weathering The peeling of small rock bodies into onionlike layers. (p. 216)

spit A curving, fingerlike projection of beach. (p. 330)

spreading center The segment of the mid-ocean ridge that is the site of active rifting and sea-floor spreading. (p. 46)

spring A place where the land surface intersects the water table and groundwater seeps or flows naturally out of the ground. (p. 265)

spring tide A high-amplitude, twice-monthly ocean tide. (p. 334)

stalactite A calcite deposit that projects from a cave roof, precipitated from groundwater supersaturated with calcium carbonate. (p. 271)

stalagmite A cone-shaped calcite deposit, growing up from a cave floor, precipitated from groundwater supersaturated with calcium carbonate. (p. 271)

strain The result of stress applied to a body, causing the deformation of its shape and/or a change of volume. (p. 194)

strata (beds) Visually distinguishable layers of sedimentary rock. (pp. 4, 161)

stratified drift Deposits left by streams or pools of glacial meltwater that are sorted and layered according to size. (p. 292)

stratovolcano A volcanic cone consisting of alternating layers of pyroclastic deposits and lava. (p. 140)

streak The color of the powder of a mineral when scratched on a porcelain plate. (p. 96)

stream Flowing water within a channel of any size. (p. 241)

stream divide The boundary that separates adjacent stream valleys. (p. 255)

stream piracy The diversion of a stream into another stream that has a steeper gradient. (p. 256)

stream terrace An elevated shelflike surface upon which the stream formerly flowed; typically, an ancient flood-plain remnant abandoned as the stream cut downward and established a course at a lower level. (p. 254)

stress The force applied to a plane divided by the area of the plane. (p. 193)

strike The direction of the line formed by the intersection of a horizontal plane with a bedding or fault plane. (p. 195)

strike-slip fault A fault in which the movement is parallel to the strike of the fault plane. (p. 200)

stromatolite A reef-like sedimentary structure produced by photosynthesizing bacteria. (p.387)

subduction The downward plunging of one lithospheric plate under another. (p. 48)

subduction zone The region where one lithospheric plate thrusts downward under another. (p. 48)

submarine canyon A steep canyon, resembling a V-shaped stream valley, that is cut into the continental shelf or slope. (p. 336)

subsidence The sinking of an area of the Earth's surface upon compaction or dissolution of subsurface materials, often caused by groundwater extraction. (p. 267)

superposition The principle that in a sequence of sedimentary strata, the oldest layer is located on the bottom and followed in turn by successively younger layers, on up to the top of the sequence. (pp. 5, 344)

surface wave A seismic wave that travels along the Earth's surface and affects the interior to depths that are dependent upon its wavelength. (p. 61)

surf zone The zone of breaking waves. (p. 329)

surge A brief period of rapid glacial flow. (p. 286)

suspended load The fine particles carried in the mass of a stream and kept aloft by turbulence. (p. 244)

suture zone The narrow region that marks the juncture of two colliding blocks of continental crust. (p. 370)

swash The water from breaking waves that washes up onto the beach. (p. 329)

syncline A fold with a core of stratigraphically younger rocks; usually its concave side is upward. (p. 196)

talus Coarse angular rock fragments that collect at the base of the cliffs from which they were dislodged. (p. 225)

tarn A lake in a cirque. (p. 290)

tensile stress The stress generated by forces directed away from one another on opposite sides of a real or imaginary plane. (p. 193)

tephra A general term that refers to all airborne pyroclastic debris. (p. 136)

terminal moraine A thick pile of till deposited at the line of maximum glacial advance. (p. 292)

theory A widely accepted explanation for a group of known facts. A theory is a hypothesis that has been elevated to a high level of confidence by repeated confirmation through testing and experimentation. (p. 24)

thrust fault A low-angle reverse fault that generally dips at about 15° or less. (p. 200)

tidal current The current of a rising or falling tide forced through a narrow constriction of coast. (p. 334)

tide The rhythmic rise and fall of the sea surface caused by the unequal gravitational attraction of the Moon—and, to a lesser degree, of the Sun. (p. 334)

till Unsorted, unlayered drift deposited directly by glacial ice. (p. 292)

tombolo A strip of beach connecting the mainland to an island or islands to one another. (p. 330)

transform boundary The border between two plates that are sliding by one another horizontally without either creating or destroying oceanic crust. It connects offset ridge segments, offset trenches, or ridge segments to trenches. (p. 43)

transform fault Type of fault where the plates slide by one another horizontally without either creating or destroying oceanic crust. (p. 52)

transition zone The upper mantle region, between 400 and 700 kilometers in depth, characterized by a series of steplike increases in seismic wave velocities and the conversion of minerals to denser forms. (p. 81)

transverse dune A long ridge of sand standing at right angles to the prevailing wind. (p. 317)

trellis drainage A drainage pattern in which the main stream cuts across the regional structure (typically folded or tilted strata), and tributaries follow parallel belts of weak strata that lie perpendicular to the main stream. (p. 255)

tributary stream A stream that flows into a larger stream. (p. 241)

tsunami A large, often deadly sea wave set in motion by an undersea earthquake. (p. 73)

tuff A rock of consolidated volcanic ash. (p. 136)

turbidite The deposited sediment of a turbidity current; characterized by graded bedding and light sorting. (p. 156)

ultimate base level The lowest possible level to which a stream can erode its channel; with rare exceptions, sea level. (p. 243)

unconformity An erosion surface bounded by rocks of markedly different age and signifying a break in the geologic record. (p. 344)

uniformitarianism The principle that is based on the concept that past geological events can be explained by forces occurring today. "The present is the key to the past." (p. 4)

upper mantle The outermost part of the mantle whose depth extends from about 20 to 670 kilometers. (p. 79)

valley glacier See *alpine glacier*.

vent A conduit through which magma rises to the surface. (p. 130)

ventifact A rock that has been polished and faceted by the action of wind-blown sand. (p. 314)

vertebrate An animal with a backbone, which is an internal skeleton of bone or cartilage. (p. 409)

volcanic breccia A pyroclastic rock composed of angular fragments that are larger than 64 millimeters in diameter. (p. 136)

volcanic cone An accumulation of lava and/or pyroclastics around a volcanic vent. (p. 129)

volcano A vent in the surface of the Earth through which magma, gases, and rock fragments erupt; also the term for the landform that develops around the vent. (p. 130)

water table The surface marking the upper limit of the zone of saturation. (p. 262)

wave-cut cliff A coastal cliff formed by wave erosion. (p. 327)

wave-cut platform A more or less smooth sloping surface planed by wave erosion. (p. 327)

weathering The physical and chemical alteration of rocks exposed to the atmospheric influences on the Earth's surface. (pp. 9, 150)

whole-mantle convection The theory that the entire mantle circulates and mixes in the course of bringing heat from the outer core to the Earth's surface. (p. 82)

zone of aeration The subsurface zone between the water table and the surface; retains little moisture except within the belt of soil moisture and the capillary fringe. (p. 262)

zone of fracture The rigid outer layer of a glacier where the ice fractures as a result of stress caused by glacial movement. (p. 285)

zone of plastic flow The inner mass of a glacier where the ice moves without fracturing. (p. 285)

zone of saturation A subsurface zone where groundwater accumulates and completely fills the voids in sediments and rocks. (p. 262)

Selected Readings

Chapter 1

Adams, F. D. 1954. *The birth and development of the geological sciences.* New York: Dover Press (originally published in 1938).

Badash, L. 1989. The age-of-the-Earth debate. *Scientific American* 261 (August): 90–96.

Berggren, W. A., and V. A. Couvering, eds. 1984. *Catastrophes and Earth history: The new uniformitarianism.* Princeton, N.J.: Princeton University Press.

Brice, W. R. 1982. Bishop Usher, John Lightfoot and the age of creation. *Journal of Geological Education* 30:18–24.

Carpi, J. 1993. Research vacations. *Earth* 2, no. 3 (May): 50–53.

Dietz, R. S., and J. C. Holden. 1970. The break-up of Pangaea. *Scientific American* 222 (April): 30–41.

Gould, S. J. 1965. Is uniformitarianism necessary? *American Journal of Science* 263:223–228.

Hurley, P. M. 1968. The confirmation of continental drift. *Scientific American* 218 (April): 52–64.

Marvin, U. 1973. *Continental drift: The evolution of a concept.* Washington, D.C.: Smithsonian Institution Press.

Morrison, P., and P. Morrison. 1987. *The ring of truth: An inquiry into how we know what we know.* New York: Random House.

Sullivan, W. 1974. *Continents in motion.* New York: McGraw-Hill.

Wegener, A. 1966. *The origin of the continents and oceans.* New York: Dover Press (originally published in German, 1915).

Wilson, J. T. 1963. Continental drift. *Scientific American* 208 (April): 86–100.

Chapter 2

Ballard, R. D. 1983. *Exploring our living planet.* Washington, D.C.: National Geographic Society.

Bonatti, E. 1987. The rifting of continents. *Scientific American* 256 (March): 74–81.

Burke, K. C., and J. T. Wilson. 1976. Hotspots on the Earth's surface. *Scientific American* 235 (August): 46–57.

Cox, A., and R. B. Hart. 1986. *Plate tectonics: How it works.* Palo Alto: Blackwell.

Hallan, A. 1973. *A revolution in the earth sciences: From continental drift to plate tectonics.* New York: Oxford.

Moores, E., ed. 1990. *Plate tectonics: Readings from Scientific American.* New York: W. H. Freeman.

Vine, F. J. 1966. Spreading of the ocean floor: New evidence. *Science* 154:1405–1415.

Vink, G. E., et al. 1985. The Earth's hot spots. *Scientific American* 252 (April): 50–57.

Chapter 3

Allen, C. R. (chairman). 1976. *Predicting earthquakes: A scientific and technical evaluation with implications for society.* Washington, D.C.: National Academy of Sciences.

Bolt, B. A. 1993. *Earthquakes,* 2d ed. New York: W. H. Freeman.

Clark, S. P. 1970. *Structure of the Earth.* Englewood Cliffs, N.J.: Prentice Hall.

Davidson, K. 1994. Predicting earthquakes. *Earth* 3, no. 3: 56–63.

Dawson, J. 1993. Cat scanning the Earth. *Earth* 2 (May): 37–41.

Elsasser, W. M. 1958. The Earth as a dynamo. *Scientific American* 199 (May): 50–62.

Fischman, J. 1992. Falling into the gap. *Discover* (October): 57–63.

Jeanloz, R., and T. Lay. 1993. The core-mantle boundary. *Scientific American* 268 (May): 48.

Johnston, A. 1992. New Madrid: The rift, the river and the earthquake. *Earth* 1, no. 1:34–43.

McKenzie, D. P. 1983. The Earth's mantle. *Scientific American* offprint 249 (3):50–62.

Monastersky, R. 1993. Lessons from Landers. *Earth* 2, no. 3: 41–47.

———. 1994. Los Angeles quake: A taste of the future? *Science News* 145:53.

Ringwood, A. E. 1983. The Earth's core: Its composition, formation, and bearing on the origin of the Earth. *Proceedings of the Royal Society.* London. 395:1–46.

Soren, D. 1988. The day the world ended for Kourion: Reconstructing an ancient earthquake. *National Geographic* (July): 30–53.

USGS Staff. 1964. *Alaska's Good Friday earthquake, March 27, 1964.* U.S. Geological Survey Circular 491.

———. 1990. The Loma Prieta California earthquake: An anticipated event. *Science* 247:286–293.

Vogel, S. 1994. The big flush. *Earth* 3, no. 2:39–43.

Wyllie, P. 1976. *The way the Earth works.* New York: John Wiley and Sons.

Chapter 4

Berry, L. G., B. Mason, and R. V. Dietrich. 1983. *Mineralogy,* 2d ed. New York: W. H. Freeman.

Chesterman, C. W. 1979. *The Audubon Society field guide to North American rocks and minerals.* New York: Alfred A. Knopf.

Dana, J. D. 1985. *Manual of mineralogy,* 20th ed. (revised by C. S. Hurbut, Jr., and C. Klein). New York: John Wiley and Sons.

Ernst, W. G. 1969. *Earth materials.* Englewood Cliffs, N.J.: Prentice Hall.

Sorrell, C. A., and G. F. Sandstrom. 1973. *Rocks and minerals: A guide to field identification.* New York: Golden Press.

Chapter 5

Barker, D. S. 1983. *Igneous rocks.* Englewood Cliffs, N.J.: Prentice Hall.

Bonatti, E. 1994. The Earth's mantle below the oceans. *Scientific American* 270, no. 3:44–52.

Bowen, N. L. 1928. *The evolution of igneous rocks.* Princeton, N.J.: Princeton University Press (reprinted by Dover Press).

Holden, A., and P. Singer. 1960. *Crystals and crystal growing.* New York: Doubleday.

MacKenzie, W. S., C. H. Donaldson, and C. Guilford. 1982. *Atlas of igneous rocks and their textures.* New York: Halstead Press.

Chapter 6

Blong, R. J. 1984. *Volcanic hazards.* Sydney: Academic Press.

Coffin, M. E., and O. Eldholm. 1993. Large igneous provinces. *Scientific American* 269, no. 4 (October): 42–49.

Decker, R., and B. Decker. 1989. *Volcanoes.* New York: W. H. Freeman.

Earthquakes and volcanoes (a bimonthly publication). U.S. Geological Survey.

MacDonald, G. A. 1972. *Volcanoes.* Englewood Cliffs, N.J.: Prentice Hall.

Simkin, T., et. al. 1981. *Volcanoes of the world.* Stroudsburg, Pa.: Hutchinson Ross.

Tilling, R. I. 1985. *Eruptions of Mount Saint Helens: Past, present, and future.* Washington, D.C.: U.S. Geological Survey.

Tilling, R. I., C. Heliker, and T. L. Wright. 1987. *Eruptions of Hawaiian volcanoes.* Washington, D.C.: U.S. Geological Survey.

Williams, H., and A. R. McBirney. 1979. *Vulcanology.* San Francisco: Freeman Cooper.

Chapter 7

Blatt, H., G. V. Middleton, and R. C. Murray. 1992. *Origin of sedimentary rocks,* 3d ed. Englewood Cliffs, N.J.: Prentice Hall.

Howell, D. G., K. T. Bird, and D. L. Gautier. 1993. Oil: Are we running out? *Earth* 2, no. 2:26–33.

Pettijohn, F. J. 1975. *Sedimentary rocks,* 3d ed. New York: Harper and Row.

Prothero, D. R. 1990. *Interpreting the stratigraphic record.* New York: W. H. Freeman.

Weiner, J. 1990. *The next one hundred years: Shaping the future of our living earth.* New York: Bantam.

Chapter 8

Barth, T. F. 1978. Metamorphic rocks, metamorphism, and metasomatism. In *McGraw-Hill encyclopedia of the Earth sciences,* 5th ed., 462–475. New York: McGraw-Hill.

Best, M. G. 1982. *Igneous and metamorphic petrology.* San Francisco: W. H. Freeman.

Miyashiro, A. 1973. *Metamorphism and metamorphic belts.* New York: John Wiley and Sons.

Philpotts, A. R. 1990. *Principles of igneous and metamorphic petrology.* Englewood Cliffs, N.J.: Prentice Hall.

Chapter 9

Billings, M. P. 1973. *Structural geology,* 3d ed. Englewood Cliffs, N.J.: Prentice Hall.

Chew, B. 1993. Anatomy of a mountain range. *Earth* 2, no. 1 (January): 36–43.

Davis, G. H. 1984. *Structural geology of rocks and regions.* New York: John Wiley and Sons.

Oberlander, T. M. 1985. Origin of drainage transverse to structures in orogens. In *Tectonic geomorphology,* M. Morisawa and J. T. Hack, eds., 55–182. London: Allyn and Irwin.

Chapter 10

Birkeland, P. W. 1984. *Soils and geomorphology.* New York: Oxford University Press.

Brown, L. R., and E. C. Wolf. 1984. Soil erosion: Quiet crisis in the world economy. In *Worldwatch Paper* 60: Worldwatch Institute.

Goldich, S. S. 1938. A study of rock weathering. *Journal of Geology* 46:17–58.

Hunt, C. B. 1972. *Geology of soils.* San Francisco: W. H. Freeman.

Kiersch, G. A. 1965. The Vaiont Reservoir disaster. In *Mineral Information Service* 18, no. 7:129–138. California Division of Mines and Geology.

Repetto, R. 1990. Deforestation of the tropics. *Scientific American* 262, no. 4:36–42.

Schuster, R. L., ed. Landslides: Analysis and control. In *Transportation Board Special Report* 176:11–33. Washington D.C.: National Academy of Sciences.

Small, R. J., and M. J. Clark. 1982. *Slopes and weathering.* Cambridge, U.K.: Cambridge University Press.

Waters, T. 1993. Roof of the world: How the Himalayas change our climate. *Earth* 2, no.4:26–35.

Voight, B., ed. 1978. *Rockslides and avalanches: Part I, natural phenomena.* New York: Elsevier.

Chapter 11

Holloway, M. 1994. Nurturing nature. *Scientific American* 270, no. 4:98–108.

Leopold, L. B., M. G. Wolman, and J. P. Miller. 1964. *Fluvial processes in geomorphology.* San Francisco: W. H. Freeman.

Mackin, J. H. 1948. Concept of the graded river. *Bulletin of the Geological Society of America* 59:463–512.

McPhee, J. 1989. *The control of nature.* New York: Farrar Straus Giroux.

Morgan, J. P. 1970. Deltas: A resource. *Journal of Geological Education* 18, no. 3:107–117.

Morisawa, M. 1986. *Rivers: Form and process.* New York: Longman.

Chapter 12

Armstead, C. H. 1978. *Geothermal energy.* New York: John Wiley and Sons.

Environmental Protection Agency. 1990. *Citizen's guide to groundwater protection.* Washington, D.C.: Environmental Protection Agency.

Fetter, C. W. 1988. *Applied hydrology,* 2d. ed. Columbus, Ohio: Merrill.

Grossman, D., and S. Shulman. 1994. Verdict at Yucca Mountain. *Earth* 3, no. 2:54–63.

Heath, R. C. 1983. Basic groundwater hydrology. In *Water Supply Paper* 2220. Washington, D.C.: U.S. Geological Survey.

Sweeting, M. 1972. *Karst landforms.* New York: Columbia University Press.

Chapter 13

Flint, R. F. 1971. *Glacial and Pleistocene geology.* New York: John Wiley and Sons.

Hays, J. D., J. Imbrie, and N. J. Shacketon. 1976. Variations in the Earth's orbit: Pacemaker of the ice ages. *Science* 194:1121–1132.

Imbrie, J., and K. P. Imbrie. 1986. *Ice ages: Solving the mystery.* Cambridge, Mass.: Harvard University Press.

National Academy of Sciences. 1989. *Ozone depletion, greenhouse gases, and climate change.* Washington, D.C.: National Academy Press.

———. 1992. *Global environmental change.* Washington, D.C.: National Academy Press.

Chapter 14

Bagnold, R. A. 1941. *The physics of windblown sand and desert dunes.* London: Methune and Company.

Byers, H. R. 1974. *General meteorology.* New York: McGraw-Hill.

El-Ashry, M., and D. C. Gibbons, eds. 1988. *Water and arid lands of the western United States.* Washington, D.C.: World Resources Institute.

Ellis, W. 1987. Africa's stricken Sahel. *National Geographic* 172:140–179.

Kotyakou, V. M. 1991. The Aral Sea: A critical environmental zone. *Environment* 33, no. 1:4–9, 36–39.

Mabbutt, J. A. 1977. *Desert landforms.* Cambridge, Mass.: MIT University Press.

Walker, A. S. 1992. *Deserts: Geology and resources.* Washington, D.C.: U.S. Geological Survey.

Chapter 15

Bird, E. F. 1984. *Coasts: An introduction to coastal geomorphology.* Oxford, U.K.: Blackwell.

Flanagan, R. 1993. Beaches on the brink. *Earth* 2, no. 6:24–33.

Inman, D. L. 1988. Nearshore sedimentary processes. In *McGraw-Hill encyclopedia of the geological sciences,* 2d. ed., 407–414. New York: McGraw-Hill.

Shepard, F. P. 1971. *Our changing shorelines.* New York: McGraw-Hill.

Chapter 16

Brush, S. G. 1982. Finding the age of the Earth by physics or faith?! *Journal of Geologic Education* 30:34–58.

Dalrymple, G. B. 1991. *The age of the Earth.* Palo Alto: Stanford University Press.

Eicher, D. C. 1976. *Geologic time,* 2d ed. Englewood Cliffs, N.J.: Prentice Hall.

Gould, S. J. 1987. *Time's arrow, time's cycle: Myth and metaphor in the discovery of geological time.* Cambridge, Mass.: Harvard University Press.

Schoch, R. M. 1989. *Stratigraphy: Principles and methods.* New York: Van Nostrand Reinhold.

Chapter 17

Cook, F. A., L. D. Brown, and J. C. Oliver. 1980. The southern Appalachians and the growth of continents. *Scientific American* 243, no. 4:156–168.

Davidow, B. 1993. The high one (Denali National Park, Alaska, how it was stitched together). *Earth* 2, no. 3:44–51.

Dewey, J. F., and J. M. Bind. 1970. Mountain belts and the new global tectonics. *Journal of Geophysical Research* 75: 2625–2647.

Hoffman, P. 1988. United plates of America, the birth of a craton: Early Proterozoic assembly and growth of Laurentia. *Annual Review of Earth and Planetary Sciences* 16:543–604.

Howell, D. C. Terranes. 1985. *Scientific American* 253, no. 5:116–126.

Jones, D. L., et al. 1982. The growth of western North America. *Scientific American* 247, no. 5:70–128.

Molnar, P., and P. Jupponier. 1977. The collision between India and Eurasia. *Scientific American* 236, no. 4:30–41.

Chapter 18

Allegre, C. J., and S. H. Schneider. 1995. The evolution of the Earth. *Scientific American* 271, no. 4:66–75.

Berra, T. M. 1990. *Evolution and the myth of creationism.* Stanford, CA: Stanford University Press.

Eicher, D. L., A. L. McAlester, and M. L. Rottman. 1984. *The history of the Earth's crust.* Englewood Cliffs, N.J.: Prentice Hall.

Gould, S. J. 1994. The evolution of life on the Earth. *Scientific American* 271, no. 4:84–91.

Kasting, J. F. 1993. Earth's early atmosphere. *Science* 259:920–926.

Kattermole, P. 1995. *Earth and other planets.* Oxford: Oxford University Press.

Milner, R. 1996. Charles Darwin and associates, ghostbusters. *Scientific American* 275, no. 4:96–101.

Whitfield, P. 1993. *From so simple a beginning, the book of evolution.* New York: Macmillan.

Chapter 19

Alvarez, W., and F. Asaro. 1990. What caused the mass extinction? An asteroid impact. *Scientific American* 262, no. 10:78–84.

Courtillot, V. E. 1990. What caused the mass extinction? A volcanic eruption. *Scientific American* 262, no. 10:85–92.

Erwin, D. H. 1996. The mother of mass extinctions. *Scientific American* 275, no. 1:72–78.

Larson, R. L. 1995. The mid-Cretaceous superplume episode. *Scientific American* 272, no. 2:82–86.

Pendick, D. 1997. Rocky mountain why? *Earth* 6, no. 5 (June):50–56.

Windley, B. F. 1995. *The evolving continents,* 3d ed. New York: John Wiley and Sons.

Acknowledgments

pp. ii–iii, Elizabeth Morales

Chapter 1

p. 2, Stephen Trimble; **p. 5** (Fig. 1.1), Landform Slides; (Fig. 1.2), Elizabeth Morales; **p. 6** (Fig. 1.3), Sinclair Stammers/Science Photo Library/Photo Researchers Inc.; **p. 9** (Fig. 1.5), Elizabeth Morales; **p. 10** (Fig. 1.6a), David R. Frazier/Photo Researchers Inc.; (Fig. 1.6b), Richard Gross; **p. 11** (Fig. 1.7a), William Ferguson; (Fig. 1.7b), Richard Gross; (Fig. 1.8a), John Shelton; (Fig. 1.8b), E. R. Degginger; **p. 12** (Fig. 1.9), Elizabeth Morales; **p. 13** (Fig. 1.10), Elizabeth Morales; **p. 15** (Fig. 1.12), Elizabeth Morales; **p. 17** (Fig. 1.13), Elizabeth Morales; **p. 19** (Fig. 1.14), Elizabeth Morales; (Fig. 1.15), Illustration by Juan Barbera, *New York Times,* July 19, 1988; **p. 20** (Fig. 1.16), Elizabeth Morales; **p. 21** (Fig. 1.18), Elizabeth Morales

Chapter 2

p. 28, NOAA map; **p. 30** (Fig. 2.1), Courtesy of Marie Tharp; **pp. 32–33** (Fig. 2.3), Courtesy of Marie Tharp; **p. 34** (Fig. 2.4), Peter Ryan/Science Photo Library/Photo Researchers Inc.; **p. 35** (Fig. 2.6), Elizabeth Morales; **p. 36** (Fig. 2.7), Elizabeth Morales; **p. 40** (Fig. 2.11), Douglas Faulkner/Photo Researchers Inc.; **p. 41** (Fig. 2.12), Elizabeth Morales; **p. 42** (Fig. 2.13), Elizabeth Morales; **p. 46** (Fig. 2.17), Elizabeth Morales; **p. 47** (Fig. 2.18a–b), Elizabeth Morales; (Fig. 2.18c), Courtesy of Marie Tharp; **p. 48** (Fig. 2.19), Elizabeth Morales; **p. 49** (Fig. 2.20), Elizabeth Morales; **p. 50** (Fig. 2.21), Elizabeth Morales; **p. 51** (Fig. 2.22), Elizabeth Morales

Chapter 3

p. 58, Haruyoshi Yamaguchi/SYGMA; **p. 60** (Fig. 3.1), USGS Photo Library; **p. 61** (Fig. 3.2), G. K. Gilbert/USGS; **p. 62** (Fig. 3.3), Elizabeth Morales; **p. 63** (Fig. 3.4), Elizabeth Morales; (Fig. 3.6b), Cindy Charles/Gamma-Liaison; **p. 69** (Fig. 3.12), Georg Gerster/COMSTOCK; **p. 72** (Fig. 3.15), Phil Green/Science Photo Library/Photo Researchers Inc.; **p. 73** (Fig. 3.16), Elizabeth Morales; **p. 74** (Fig. 3.17), Wide World Photos; **p. 75** (Fig. 3.18), Figure from *Earthquakes: A Primer* by Bruce Bolt, p. 103. Copyright © 1978 by W. H. Freeman and Company. Used with permission; **p. 76** (Fig. 3.19), David Kennerly/Gamma-Liaison; **p. 77** (Fig. 3.21), E. R. Degginger; **p. 80** (Fig. 3.24), Elizabeth Morales

Chapter 4

p. 88, Erica and Harold van Pelt; **p. 95** (Fig. 4.8), E. R. Degginger; **p. 97** (Fig. 4.9), Paul Silverman/Photo Researchers Inc.; (Fig. 4.10), Breck Kent; (Fig. 4.11a), Breck Kent; (Fig. 4.11b), H. Rudolf Becker/Phototake; **p. 98** (Fig. 4.12), Breck Kent

Chapter 5

p. 108, Alfred Pasieka/Science Photo Library/Photo Researchers Inc.; **p. 110** (Fig. 5.1), Elizabeth Morales; **p. 111** (Fig. 5.2), Landform Slides; (Fig. 5.3), Tom Bean; **p. 112** (Fig. 5.5), E. R. Degginger; (Fig. 5.6), E. R. Degginger; **p. 113** (Fig. 5.7a), John Shelton; (Fig. 5.7b), Bruce Iverson; (Fig. 5.8), E. R. Degginger; (Fig. 5.9), Landform Slides; **p. 114** (Fig. 5.10), E. R. Degginger; (Fig. 5.11), E. R. Degginger; **p. 115** (Fig. 5.12), E. R. Degginger; (Fig. 5.13), E. R. Degginger; **p. 116** (Fig. 5.15a–b), Landform Slides; **p. 117** (Fig. 5.16), Landform Slides; (Fig. 5.17), Alfred Pasieka/Science Photo Library/Photo Researchers Inc.; **p. 119** (Fig. 5.19), BPS; **pp. 120–121** (Fig. 5.20), Elizabeth Morales; (Fig. 5.21), Elizabeth Morales; **p. 122** (Fig. 5.22), Peter Ryan/Scripps/Science Photo Library/Photo Researchers Inc.; **p. 123** (Fig. 5.24), William Ferguson; (Fig. 5.25), Landform Slides; **p. 124** (Fig. 5.26), Courtesy of the Woods Hole Oceanographic Institute; **p. 125** (Fig. 5.27), Institute of Oceanographic Sciences/NERC/Photo Researchers Inc.

Chapter 6

p. 128, Soames Summerhays/Photo Researchers Inc.; **p. 130** (Fig. 6.1), F. Gohier/Photo Researchers Inc.; **p. 131** (Fig. 6.2), Elizabeth Morales; **p. 133** (Fig. 6.5), William Ferguson; (Fig. 6.6), Mats Wibe Lund; **p. 134** (Fig. 6.7a–c), E. R. Degginger; **p. 135** (Fig. 6.8), Tom Till; (Fig. 6.9), John Shelton; **p. 136** (Fig. 6.10), David Weintraub/Photo Researchers Inc.; **p. 136** (Fig. 6.11), Breck Kent; **p. 137** (Fig. 6.12), USGS; **p. 138** (Fig. 6.13), Elizabeth Morales; (Fig. 6.14), John Shelton; **p. 139** (Fig. 6.15), Michael Freeman/Bruce Coleman Ltd.; (Fig. 6.16), USGS; **p. 141** (Fig. 6.18), USGS; (Fig. 6.19), USGS; **p. 142** (Fig. 6.20), Elizabeth Morales; **p. 144** (Fig. 6.21), Courtesy of the National Oceanic and Atmospheric Administration; **p. 145** (Fig. 6.22), Pat and Tom Leeson/Photo Researchers Inc.

Chapter 7

p. 148, Tom Bean; **p. 150** (Fig. 7.1), Elizabeth Morales; **p. 153** (Fig. 7.3a), K. H. Switak/Photo Researchers Inc.; (Fig. 7.3b), Landform Slides; **p. 155** (Fig. 7.5), William Ferguson; (Fig. 7.6), Bruno Maso/Photo Researchers Inc.; (Fig. 7.7), Landform Slides; **p. 156** (Fig. 7.9), William Ferguson; **p. 157** (Fig. 7.10), Tom Bean; **p. 158** (Fig. 7.11), William Ferguson; **p. 159** (Fig. 7.12), John Shelton; **p. 160** (Fig. 7.13a), Landform Slides; (Fig. 7.13b), Ray Simons/Photo Researchers Inc.; **p. 161** (Fig. 7.14), William Ferguson; **p. 162** (Fig. 7.16), Tom Bean; (Fig. 7.17), Tom Bean; (Fig. 7.18), Stephen Trimble; **p. 163** (Fig. 7.19a), Paul Zohl/Photo Researchers Inc.; (Fig. 7.19b), Tom Bean; (Fig. 7.19c), E. R. Degginger; (Fig. 7.19d), William Ferguson; (Fig. 7.19e), Tom Bean; (Fig. 7.19f), Landform Slides; **p. 164** (Fig. 7.20), Thomas Taylor/Photo Researchers Inc.; **p. 165** (Fig. 7.21a–b), E. R. Degginger; **p. 166** (Fig. 7.22), Tom Pantages; **p. 168** (Fig. 7.23), Elizabeth Morales

Chapter 8

p. 174, Massimo Borchi/Bruce Coleman Ltd.; (inset), Vandystadt/Photo Researchers Inc.; **p. 178** (Fig. 8.3), John Shelton; **p. 180** (Fig. 8.4), E. R. Degginger; (Fig. 8.5), E. R. Degginger; **p. 181** (Fig. 8.6a), Breck Kent; (Fig. 8.6b), Landform Slides; (Fig. 8.7), William Ferguson; **p. 182** (Fig. 8.8), Tom Bean; (Fig. 8.9a–d), John Shelton; **p. 183** (Fig. 8.10), A. W. Ambler/Photo Researchers Inc.; (Fig. 8.11a), E. R. Degginger; (Fig. 8.11b), Bruce Iverson; **p. 185** (Fig. 8.12), Three photomicrographs adapted from Myron G. Best, *Igneous and Metamorphic Petrology*, p. 416. Copyright © 1982 by W. H. Freeman and Company. Used with permission; **p. 186** (Fig. 8.15), John Shelton; (Fig. 8.16), Landform Slides; **p. 187** (Fig. 8.17a), E. R. Degginger; (Fig. 8.17b), Landform Slides; (Fig. 8.18), Stephen Trimble; **p. 188** (Fig. 8.19) John Shelton; **p. 188** (Fig. 8.20), Elizabeth Morales; **p. 189** (Fig. 8.21), Elizabeth Morales

Chapter 9

p. 192, John Shelton; **p. 197** (Fig. 9.6), USGS; (Fig. 9.7), Landform Slides; **p. 198** (Fig. 9.8a), Aerofilms; (Fig. 9.9), Elizabeth Morales; **p. 199** (Fig. 9.10a), Landform Slides; (Fig. 9.10b), Elizabeth Morales; (Fig. 9.11a), H. R. Joesting/USGS; (Fig. 9.11b), Elizabeth Morales; **p. 202** (Fig. 9.13a), Jules Cowan/Bruce Coleman Ltd.; (Fig. 9.13b), Elizabeth Morales; **p. 204** (Fig. 9.16), Elizabeth Morales; (Fig. 9.17), Breck Kent; **p. 206** (Fig. 9.19), John Shelton

Chapter 10

p. 210, Owen Franken/Stock, Boston; **p. 211** (quote), *Love Is Here To Stay*, words & music by George Gershwin and Ira Gershwin © 1938 Chappell & Co. (ASCAP) Copyright Renewed. Rights for Extended Renewal Term in United States controlled by George Gershwin Music and Ira Gershwin Music. All Rights administered by WB Music Corp. All Rights outside United States controlled by Chappell & Co. All Rights Reserved. Used by Permission WARNER BROS. PUBLICATIONS U.S. INC., Miami, FL 33014; **p. 212** (Figure 10.1), Elizabeth Morales; **p. 213** (Fig. 10.2), Jules Cowan/Bruce Coleman, Ltd.; **p. 214** (Fig. 10.4), Landform Slides; **p. 215** (Fig. 10.5), Tom Bean; **p. 216** (Fig. 10.6), Landform Slides; **p. 217** (Fig. 10.7), Gregory Dimigian/Photo Researchers Inc.; **p. 220** (Fig. 10.10a–b), Courtesy of Charles Adler; **p. 223** (Fig. 10.12), George Whiteley/Photo Researchers Inc.; **p. 224** (Fig. 10.13), Bill Bachman/Photo Researchers Inc.; **p. 225** (Fig. 10.14), J. Serreo/Photo Researchers, Inc.; **p. 226** (Fig. 10.15), John Shelton; **p. 227** (Fig. 10.17a), USGS; **p. 229** (Fig. 10.18d), USGS; **p. 230** (Fig. 10.19), Bouvet/Hires/Duclos/Gamma-Liaison; **p. 231** (Fig. 10.20), USGS; **p. 232** (Fig. 10.21), Elizabeth Morales; **p. 233** (Fig. 10.22), John Shelton; **p. 234** (Fig. 10.23), Fletcher & Baylis/Photo Researchers Inc.; **p. 235** (Fig. 10.24), Elizabeth Morales; **p. 236** (Fig. 10.25), Gary Williams/Gamma-Liaison

Chapter 11

p. 240, Jim Steinberg/Photo Researchers Inc.; **p. 244** (Fig. 11.5), Landform Slides; **p. 246** (Fig. 11.7a), Norman Tomalin/Bruce Coleman Ltd.; (Fig. 11.7b), Bruce Coleman Ltd.; **p. 247** (Fig. 11.8), Tom Bean; **p. 248** (Fig. 11.9), John Shelton; **p. 249** (Fig. 11.10), Elizabeth Morales; **p. 251** (Fig. 11.12), Jeff Christensen/Gamma-Liaison; **p. 255** (Fig. 11.16), Tom Bean

Chapter 12

p. 260, Tom Till; **p. 262** (Fig. 12.1), Elizabeth Morales; **p. 264** (Fig. 12.3), Elizabeth Morales; **p. 268** (Fig. 12.7), Elizabeth Morales; **p. 270** (Fig. 12.9), Elizabeth Morales; **p. 271** (Fig. 12.10), Kevin Downey; **p. 272** (Fig. 12.11), AP/Wide World Photos; **p. 273** (Fig. 12.12), Landform Slides; **p. 274** (Fig. 12.13), Biological Photo Service; **pp. 276–277** (Fig. 12.15), Elizabeth Morales

Chapter 13

p. 282, Tom Bean; **p. 285** (Fig. 13.2), Tom Bean; **p. 287** (Fig. 13.3), Elizabeth Morales; **p. 288** (Fig. 13.4), Elizabeth Morales; **p. 289** (Fig. 13.5), Paul Hanny/Gamma-Liaison; **p. 290** (Fig. 13.6), Tom Bean; **p. 291** (Fig. 13.7), Elizabeth Morales; **p. 292** (Fig. 13.8), Tom Till; **p. 293** (Fig. 13.9), Tom Bean; **p. 294** (Fig. 13.10), Elizabeth Morales; **p. 295** (Fig. 13.11), Terraphotographics/BPS

Chapter 14

p. 304, Jim Steinberg/Photo Researchers Inc.; **p. 307** (Fig. 14.2), Elizabeth Morales; **p. 308** (Fig. 14.3), Jim Brandenburg/Minden Pictures; **p. 309** (Fig. 14.4), Stephen Trimble; **p. 311** (Fig. 14.6), Phil Degginger; **p. 312** (Fig. 14.7), John Shelton; **p. 314** (Fig. 14.9), John Eastcott & Yva Momatiuk/The Image Works; (Fig 14.10), John Shelton; **p. 315** (Fig. 14.11), Christopher Liu/ChinaStock; **p. 316** (Fig. 14.12), E. R. Degginger; **p. 318** (Fig. 14.14), Landform Slides; **p. 321** (Fig. 14.16), Reza/Gamma-Liaison

Chapter 15

p. 324, Randy Taylor/Gamma-Liaison; **pp. 327–332** (Figs. 15.3–15.8), Elizabeth Morales; **p. 332** (Fig. 15.9), John Shelton; **p. 333** (Fig. 15.10), Phil Degginger; **p. 335** (Fig. 15.11), Earth Satellite Corporation; **p. 336** (Fig. 15.12), Earth Satellite Corporation; **p. 337** (Fig. 15.13), John Shelton

Chapter 16

p. 342, Tom Bean; **p. 345** (Fig. 16.1a), Landform Slides; (Fig. 16.1b), John Shelton; (Fig. 16.1c), USGS; **p. 354** (Fig. 16.10), Elizabeth Morales

Chapter 17

p. 360, William Thompson; **p. 363** (Fig. 17.2), USGS; (Fig. 17.3), Terraphotographics/BPS; (Fig. 17.4), Elizabeth Morales; **p. 364** (Fig. 17.5), USGS; **p. 365** (Fig. 17.6), Elizabeth Morales; **p. 366** (Fig. 17.7), Elizabeth Morales; **p. 368** (Fig. 17.9), Elizabeth Morales; **p. 371** (Fig. 17.11), Elizabeth Morales; **p. 373** (Fig. 17.13), Elizabeth Morales

Chapter 18

p. 378 (3 photos), William E. Ferguson; **p. 380** (Fig. 18.1), Courtesy of NASA; **p. 381** (Fig. 18.2), Elizabeth Morales; **p. 382** (Fig. 18.3), Patrice Rossi Calkin/Rossi Illustration; **p. 384** (Fig. 18.4), Courtesy of NASA; **p. 385** (Fig. 18.5), GeoSystems Global Corp.; **p. 386** (Figs. 18.7, 18.8), Patrice Rossi Calkin/Rossi Illustration; **p. 388** (Fig. 18.9), BPS; **pp. 391–397** (Figs. 18.10–18.16), Patrice Rossi Calkin/Rossi Illustration

Chapter 19

p. 402, Breck P. Kent; **p. 404** (Fig. 19.1), Patrice Rossi Calkin/Rossi Illustration; **p. 405** (Fig. 19.2), E. R. Degginger/ Bruce Coleman, Inc.; **p. 406** (Fig. 19.3), Elizabeth Morales; **p. 407** (Fig. 19.4), Ludek Pesek/Science Photo Library/Photo Researchers, Inc.; (Fig. 19.5), Breck P. Kent; **pp. 408–412** (Figs. 19.6–19.12), Patrice Rossi Calkin/Rossi Illustration; **p. 413** (Fig. 19.13), Elizabeth Morales; **p. 414** (Fig. 19.14), GeoSystems Global Corp.; **p. 415** (Fig. 19.16), Pat & Tom Leeson/Photo Researchers, Inc.; **p. 416** (Fig. 19.17), Elizabeth Morales; **p. 417** (Fig. 19.18), Stephen J. Krasemann/Photo Researchers, Inc.; (Fig. 19.19), "Lucy" diorama, D. Finnin/C. Chesek, © American Museum of Natural History

Inside back cover

Adapted from David G. Howell, "Terranes," from *Scientific American*, November 1985, pp. 122–125. Copyright © 1985 by Scientific American, Inc. All rights reserved.

Index

Note to the Reader: Throughout this index *italicized* page numbers refer to illustrations.

A horizon, *221,* 221–222
Aa lava, 134, *135*
Ablation of glaciers, 286, *288*
Abrasion by glaciers, 289
Absolute age, 8, 350. *See also* Absolute time
    radioisotopes for determining, 352–355, *353–354*
Absolute time, 8, 343, 350–357. *See also* Absolute age
    age of Earth, 357
    and geologic time scale, 355–356, *356*
    radioactive decay and, 350–352, *351–352*
Abyssal plains, 39, 165
Accreted terranes, 166, 368–369, *369, 371–373*
    of Appalachians, 372–373, *372–374*
    of Himalayas, 370–372, *371*
    in North American Cordillera, 414
    of western North America, *369,* 369–370
Accretion in mountain building. *See* Mountain building, by accretion
Accretionary wedges, 49, 367
Accumulation
    of glaciers, 286, *288*
    of groundwater, 261–264, *262–264*
Active plate margins, 50, 165, 365–366
    and coasts, 336
    sedimentary environments in, 165
Adaptations, 396
Adaptive radiation, 397, 407
Afar Triangle, 46, *47*
Aftershocks, earthquake, 68
Agassiz, Lake, 295
Age of Earth, 357
Air-fall deposits, 136
Alaska, accreted terranes in, 370
Algae, 160
Alluvial fans, 154, *154*
    from braided streams, 246–247, *247*
    in deserts, 318
Alpha decay, 350, *351*
Alpine glaciers, 285, *285*
Altyn Tagh fault, 202
Amazon basin, placer deposits at, 249
American Geophysical Union, 23
Amino acids, 386–387
Amniotic egg, 411, *411*
Amphibians, early, 410–411
Amphibole group, structure of, 101, *102*
Amplitude, 64–65
Anatomy, evidence of evolution in fossils and comparative, 389–391, *391–392*
Anchorage, Alaska, slump at, 228, *229*
Andesite, 114
Andesite-diorite rock family, 114, *115*

Andesitic volcanoes, 140–142
Angle of repose on slopes, 225
Angles in mineral crystals, 94
Angular unconformities, 344, *345*
Anhydrous minerals, 179
Animals in mechanical weathering, 217, *217*
Animas River, 247, *248*
Annular drainage patterns, 256, *256*
Antarctic ice sheet, 283, *284,* 288
Anthracite coal, 167
Anticlines, 196, *196,* 197–198, *198*
Aphanitic texture, 112, *112*
Aplite, 115
Appalachian Mountains, *364*
    drainage pattern of, 256
    orogeny of, 372–373, *372–374*
Appalachian Plateau, *372,* 374
Apparent polar-wandering curve, 21, *21*
Aquifers, 264, *264,* 265–270, *267, 269,* 318
Archean eon, 383–385
Arches, sea, 327
Arêtes, 290, *291*
Argon, 353
Arid climates. *See* Deserts
Arkose sandstones, 155, *155*
Armenia, earthquake in, 74
Arsenic in drinking water standards, 278
Artesian wells, 264, 266–267, *267*
    hydraulic pressure loss in, *269,* 269–270
    saltwater encroachment in, *268,* 268–269
Ash, 135
Ash flows, 142
Asperity, 68
Asthenosphere, 14, 22, 43, *46,* 81
Asymmetrical folds, 196, *197*
Atchafalaya River, 253, *254*
Atlantic and Gulf Coast margin, 365
Atlantic Ocean, ancestor of, 372
Atmosphere
    and climate, 298–299
    in ice sheets, 283
    reducing, 387
    and volcanic activity, *143,* 143–144
    water quantity in, 262
Atolls, 41, *41*
Atomic mass, 90
Atomic number, 90
Atomic structure, 90–92, *91*
Atoms, 89–92, *91*
Aureoles, 184
Australia in glacial age, *284*
Axial planes, 196, *196*
Axial tilt, 297, *297*
Azurite, 96, *97*

B horizon, *221,* 222
Backwash, 329, *329*
Bacon, Francis, 16
Bajadas, 312, *313*

Banded iron formations, 387
Barchan dunes, 317, *317–318*
Barrier islands, 330, *330,* 331, *332,* 337, *338*
Barringer (Meteor) Crater, *206,* 207
Basal slip in glaciers, 286, *287*
Basalt, 114
    flood, 139
    texture of, *112*
Basalt dikes in lithosphere, 120, *121*
Basalt-gabbro rock family, 114
Basaltic lavas
    flows of, 134
    porosity of, 263
Basaltic magmas
    in lithosphere, 120, *121*
    at mid-ocean ridges, 120, *120*
Basaltic rocks, 13
Basaltic volcanoes, 120, 137–139, *138–139*
Base level of streams, 243
Basement rock, 362
Basin and Range Province, 202, *203,* 310, 311–312, *312,* 363, *416*
Basin-and-range topography, 202, *310,* 311–312, *312*
Basins, 199–200, *201*
Batholiths, 110–111, *111*
Bauxite, formation of, 222, *223*
Baymouth bars, 330, *330*
Bays, deposition in, 327–328
Beaches, 327, 328. *See also* Coasts
    sediment budget at, 330–333, *331–333*
    storm damage at, 333, *333*
    tampering with, 331–333, *331–333*
    tides and, 334
    types of, 329–330, *330*
Bed load
    stream, 244
    wind, 313
Bedding, 161–162, *162*
Bedding planes, 161
Beds
    confining, 264
    delta, 252
    sediment, 161
Belt of soil moisture, *262,* 263
Benioff, Hugo, 34
Benioff zone. *See* Wadati-Benioff zone
Beta decay, 350, *351*
Beta (electron) capture, 350, *351*
Bicarbonate ions, *212*
Big Stubby deposit, 357
Big Thompson River, *240*
Bighorn Mountains, *199*
Biochemical limestones, 158, *158*
Biochemical sedimentary rocks, 151, 157–160, *158–160*
Biochemical sediments, 150
Bituminous coal, 167
Black Hills, 200

449

Black Mountains, *304*
Blocks, 135
Blowouts, 313
Blue Ridge region, 372, *372–373*
Blueschist, 181
    in subduction zones, 190
Body waves, 61, *63*
Bombs, 135–136
Bonding, 92–94, *93–94*
Bonneville Salt Flats, 160, 295
Bottomset beds, 252
Boulder fields, 214, *214*
Bowen, Norman Levi, 117
Bowen's reaction series, 117–118, *118*
Bowling Green, Kentucky, water pollution at, 272–273
Bracketing, in geologic dating, 355–356
Braided streams, 246–247, *247*
Breccia, *113*, 152, *155*
    fault, 181, *182*
    volcanic, 136, *137*
Brevard fault, 372
Brittle response to rock stress, 194, *195*
Brown clay in deep-ocean sediments, 165–166
Bryce Canyon, joint sets at, 200, *202*
Budget, glacial, 286–288, *288*
Bunker Hill, 293
Burgess shale, 406
Burial metamorphism, 178
Burrows in fossilization, *163*
Bushveld Complex, 125
Buttes, 310, *311*

C horizon, *221*, 222
Cadmium in drinking water standards, 278
Calcareous oozes, *165*, 166
Calcification, 223
Calcite, double refraction in, 98, *98*
Calcium in groundwater, 275
Caldera, 131
Caldera eruptions, 142, *142*
California
    coastal sediment transport in, 330–331, *331*
    water transfer in, 318, *319*
Calving, ice, 288, *289*
Canadian Rockies, 203, *204*
Canadian Shield, *363*, 385
Canyons, submarine, 336
Capacity of streams, 244–245
Cape Canaveral, Florida, 330
Cape Cod, Massachusetts, 330
Cape Hatteras, North Carolina, 330, 335, *336*
Cape Lookout, North Carolina, 330
Cape May, New Jersey, 332
Capillary fringe, *262*, 263
Carbon
    in coal, 167
    isotopes of, 90, *91*
Carbon cycle, 212, *212*

Carbon dioxide
    in carbon cycle, 212, *212*
    in cave formation, 270
    and climate, *298*, 298–299
    and global warming, 299
    in hydrolysis, 218
    from volcanic activities, 143–144
Carbon-14 isotopes, 353–355, *354*
Carbonates, 100, 104, *104*
Carbonic acid
    in cave formation, 270
    in hydrolysis, 218
Carbonization fossilization, *163*
Cascade Range, building of, 367–368, *367–368*
Casts in fossilization, *163*
Catastrophism, 4
Caves
    formation of, 270–272, *270–272*
    in loess deposits, 315, *315*
    from salt crystallization, 215, *215*
    sea, 327
Cell, 387
Cementation, 151, *152*
Cenozoic era, 350, 412
    climate in, 414–416
    fossils of, *408*
    life in, 417, *417*
    North American Cordillera of, 413–414, *414*
    plate tectonics of, 412–413, *413*
Central mound, 205
Central rift valley, 30
Centrifugal force, tides from, 334
Chalcedony, 164
Chalk, 158
Channel erosion in streams, 243–244, *244*
Channels, distributary, 252
Chemical bonds, 92
Chemical limestones, 159, *160*
Chemical sedimentary rocks, 151, 157–160, *158–160*
Chemical sediments, 150
Chemical weathering, 150, 213–214, 217–219, *218*
    and dissolved loads, 244
    and headland erosion, 327
Chemically active fluids, metamorphism from, 177–178
Chert, 160, *161*
Chesapeake Bay, 335, *335*, 337
Chief Mountain, 203, *204*
Chimneys, 310, *311*
Chlorofluorocarbons (CFCs), 299
Chondrites, 380–381
Cinder cones, 139, *139*
Cinders, 135
Circulation loop, nearshore, 329, *329*
Cirques in glaciers, *285*, 287, 289, 290
Classification
    of igneous rocks, 113–116, *116*
    of metamorphic rocks, 180–183, *180–183*

    of minerals, 100
    of sedimentary rocks, 151–160, *152–153*
Clastic texture, 152
Clays
    in deep-ocean sediments, 165–166
    in detrital rocks, 152–153, 154
    in hydrolysis, 218–219
    porosity and permeability of, 264
Cleavage
    of minerals, 95
    slaty, 180, 184, *186*
Cliff-dwelling Indians, 215, *215*
Cliffs, wave-cut, 327
Climate, 296, 319–320
    atmospheric variation in, *298*, 298–299
    Cenozoic, 414–416
    Earth orbit effects on, 296–298, *297*
    future changes in, 299–301
    Mesozoic, 414–416
    plate tectonic effects on, 296
    and soil formation, 222–224
    volcanic activity effects on, *143*, 143–144
Coal, 160, 167, *168*
Coarse detrital rocks, 154–156, *154–156*
Coarse sediments, 152
Coasts, 335, *335*. *See also* Beaches
    nearshore circulation loop of, 329, *329*
    and sea-level changes, 336–339, *338*
    tectonic settings for, 335–336, *337*
    tides at, 334
    wind-driven waves at, 325–328, *326–328*
Cocos plate, 49, *68*
Color
    of metamorphic rocks, 181
    of minerals, 96, *97*
Colorado Plateau, 199, 310–311, *310–311*
Colorado River
    suspended load of, 244
    water diverted from, 319
Columnar joints, 134, *135*
Compaction
    in glacier formation, 284
    in lithification, 151
Competence of streams, 244, *245*
Composite cones, 140
Compounds, 92
Compressive stresses, 177, *177*, 193, *194*
Concave stream profiles, 243, *243*
Conchoidal fracture, 95
Concordant plutons, 110
Concretions, 164
Cones of depression in wells, 265, *266*
Confined aquifers, 264, *264*, 265–270, *267*, *269*, 318
Confining beds, 264
Conglomerates, 152, *155*
Consistency of pattern in fossil records, 392
Constancy of interfacial angles in minerals, 94
Contact metamorphic rocks, 184, *185*
Contact metamorphism, 178
Continent-continent convergence, 50–52, *51*

INDEX

Continental crust, 13, 361, *362*
   cratons in, 362, *362*
   growth and evolution of, 374–375
   minerals in, 99
   mountain belts in, *362,* 362–364, *364*
Continental drift, 14–16, *15*
   paleomagnetic evidence of, 20–21, *20–21*
   physical evidence of, 16–20, *17, 19–20*
   and plate tectonics, 21–22. *See also* Plate tectonics
Continental glaciers, 285
Continental margins, 50, 364–366, *365*
   and coasts, 336
   sedimentary environments in, 165
Continental rise, 48
Continental shelf, 48, *284*
Continental slope, 48
Continuous reaction series, 117
Convection currents, 52, *53*
Convection models
   of core, *83,* 83–84
   of mantle, 82, *83*
Convergent boundaries, 43. *See also* Subduction zones
   earthquakes at, 67
   processes of, *46,* 48–52, *49–51,* 54
Convergent evolution, 398
Coquinas, 158, *158*
Coral reefs, 158–159, *159*
Cordilleran orogeny, 413
Core, 14, 79
   convection models of, *83,* 83–84
   of craters, 205
   detection of, *79, 80*
   temperatures in, 81–82
Correlation of rock units, 346–350, *347–349*
   and faunal succession, 346–347
   index fossils and overlapping ranges in, 348, *349*
Covalent bonds, 93–94, *94*
Cox, Alan, 38
Crandell, Dwight R., 143
Crater Lake, 129–130, *130*
Crater rim, 205
Craters
   meteoroid, 204–207, *205–206*
   volcanic, 131
Cratons, 362, *362*
Creationism, 389
Creep, 232–233, *233*
Crests of waves, 325–326, *326*
Crevasses in glaciers, 285, *287*
Cross-bedding, 161–162, *162*
Cross-cutting relationships, 345, *346,* 355, *355*
Crossopterygians, 410, *410*
Crust, 13. *See also* Continental crust; Oceanic crust
   detection of, *78,* 79
   lithosphere in, 42–43
   minerals in, 98–100

Crustal extension in North American Cordillera, 414, *416*
Crustal uplift and climate, 296
Crystal faces, 94
Crystal form of minerals, 94–95, *95*
Crystallization and crystals, 94–95, *95*
   in igneous rocks, *116–117,* 116–119, *119*
   in magma, 111–113
   in metamorphism, 179, 184
Currents, ocean
   longshore, 329, *329*
   rip, 329, *329*
   tidal, 334
Cuspate forelands, 330, *330*
Cutbanks, 247
Cycle
   carbon, 212, *212*
   hydrologic, 8, *9*
   Milankovitch, 296–298, *297*
   nutrient, 221, *222*
   rock, 10–12, *10–12*

Dacite, 115
Dakota sandstone, 266–267
Dana, James, 41
Darwin, Charles, 40, 395, *396,* 397, 398
Data collection, 23
Dating. *See* Absolute time; Relative time
Daughter products, 351–353
Death Valley, California, sediment in, 312
Debris flows, 231, *231*
Decay, radioactive, 90, 350–352, *351–352*
Deep-sea oozes, 165, 165–166, *166*
Deep-sea trenches, 31–34, *32–33*
Deep time, 361
Deflation by wind, 313
Deformation of rocks, 193
   folds, *196–199,* 196–200, *201*
   fractures, 200–203, *202–204*
   impact craters, 204–207, *205–206*
   map depiction of, 195, *195*
   from stress, 193–194, *194*
Delaware Bay, 335
Deltaic crossbeds, 252
Deltas, formation of, 252–253, *253–254*
Dendritic drainage patterns, 255, *256*
Density
   and specific gravity, 97–98
   and subduction zone magmas, 122, *122*
Deposition of sediments, 9, 150, *151*
   in bays, 327–328
   by glaciers, 290–295, *292–295*
   and gradation, 211–213
   by streams, 244–245
   by wind, 314–318, *315–318*
Depositional environments, *154,* 154–156
Deposits
   loess, 314–316, *315*
   placer, 248–249, *250*
   turbidite, *154,* 156, 161
Desert pavement, 313
Desert varnish, 308, *309*

Desertification, 320
Deserts, *304,* 305–306
   Basin-and-Range Province, *310,* 311–312, *312*
   borders of, 319–320, *320–321*
   Colorado Plateau, 310–311, *310–311*
   distribution of, *306*
   rain shadow, 307, *307*
   soil formation in, 223, *224*
   water resources in, 309, 318–319, *319*
   weathering and erosion in, 308–313, *309–313*
   and winds, 306–308, *307–308,* 313–318
Detrital sedimentary rocks, 150, 151, 152–157, *153–157*
Diabase, 114
Diapirs, 122–123
Diatoms, 160
Diatremes, 110
Dietz, Robert, 35
Differential weathering, *220,* 220–221
Dikes, 110, *111*
   in cross-cutting relationships, 345, *346*
   oceanic, in lithosphere, 120, *121*
Diorite, 114, *115*
Dip, 195, *195*
Dip-slip faults, 200
Direct preservation fossilization, *163*
Direct solution weathering, 217–218, *218*
Discharge areas for groundwater, 265
Discharge of streams, 243, *244*
Disconformities, 344, *345*
Discontinuous reaction series, 116
Discordant plutons, 110
Disseminated deposits, 125
Dissolved ions in hydrolysis, 219
Dissolved load of streams, 244
Distributary channels, 252
Divergent boundaries, 43
   metamorphism at, 188
   processes, 46–48, *46–48,* 54
DNA-DNA hybridization, 394, *395*
Domes, *199,* 199–200, *201*
   lava, 134, *136*
Double refraction in minerals, 98, *98*
Drainage basins, 241, *242*
Drainage patterns of streams, 254–256, *256–257*
Drift
   continental. *See* Continental drift
   glacial, 290–292, 293–295, *294–295*
   landforms of, 293–295, *294–295*
Drinking water standards, 278
Drumlins, 293, *293*
du Toit, Alexander, 14
Ductile response to rock stress, 194, *195*
Dunes, 316–318, *316–318*
Dust
   and climate, 299
   volcanic, 143–144, *144*
Dust storms, 314, *314*

Earth
 age of, 357
 as dynamo, 83, *83*
 evolution of life on, 389–398
 and Hadean eon, 381
 interior of, 12–14, *13*, 77–78, *77–84*, *80–83*
 as machine, 8–12
Earth's orbit, climate changes from, 296–298, *297*
Earthflows, 230
Earthquakes, 59
 associated with volcanoes, 34, 132, 138, 142, 143
 causes of, 59–61, *60*
 construction factors in, 74–76, *76*
 forecasting, 70–73, *72*
 intraplate, 69–70, *70*
 locating, 64, *65*
 and Los Angeles faults, 68–69, *69*
 magnitude and energy measurements of, 64–66, *65–66*
 minimizing damage from, 74–76
 from ocean-continent convergence, 50
 and plate tectonics, 66–70, *67*
 precursor studies of, 72–73, *73*
 recurrence studies of, 71–72, *72*
 and San Andreas fault, 68, *68–69*
 seismic gaps in, 71
 seismic waves from, 61–66, *63–65*
 slides from, 228
 and trenches, *32–33*, 34–36
East African Rift Valley, 30, *31*, 363
East Pacific Rise, 52
Eccentricity of Earth's orbit, 297, *297*
Eclogite in subduction zones, 190
Ecological niches, 397
Ediacaran fauna, *378*, 388–389
Effective time ranges of radioisotopes, 352, *353*
Effluent streams, 265
El Capitan reef, 159, *159*
Elastic limit, 194
Elastic-rebound earthquake theory, 61, *62*
Elastic response to rock stress, 194
Electric charges, 90
Electron capture. *See* Beta (electron) capture
Electron decay. *See* Beta decay
Electron shells, 90, *91*
Electrons, 89
Elements, 13, 89–92
Elevation in stream energy, 242
Embryology, evidence of evolution in, 392–393, *394*
Emperor Seamount chain, age of, 41, *42*
End moraines, 292
Energy
 in earthquakes, measurements of, 64–66, *65–66*
 in streams, 242–245, *243–244*
 in volcanic activity, 144
 in waves, 327–328
Energy resources
 coal, 160, 167, *168*
 petroleum, 168–170, *169*
 in sedimentary rocks, 167–170
Eons, 348, *348–350*
Epicenters of earthquakes, 61, *63*, 64, *65*
Epochs, 348, *348*, 349, 350
Equigranular texture, 180, 183
Equus, 390, *392*, 396
Eras, 348, *348*, 350
Erosion, 9
 from beach control, 331–333, *331–333*
 in cave formation, 271
 in deserts, 308–313, *309–313*
 by glaciers, 289–290, *291*
 and gradation, 211
 of headlands, 327–328, *327–328*
 and sedimentary rocks, 151
 in streams, 243–244, *244*, 247, 255
 at talus slopes, 225
 by wind, 313–314, *314*
Eruptions, volcanic
 fissure, *138*, 139
 forecasting, 142–143
 materials in, 133–136, *134–137*
 mechanics of, 132–133, *133*
 Mount Saint Helens, 140–142, *141*
Eskers, 294, *294*
Estuaries, 337
Eukaryotes, 388
Eurypterids, 409
Evaluation by scientific community, 23
Evaporites, 159–160
Everest, George, 366
Everest, Mount, *360*, 361
Evolution, 387
 evidence of, in embryology, 392–393, *394*
 evidence of, in fossils and comparative anatomy, 389–391, *391–392*
 evidence of, in genetics, 393–395, *395*
 and faunal succession, 346–347
 and geology and natural selection, 396–398
 of life on Earth, 389–398
 mechanisms of, 395–396
 status of, 398
 theory of, 389, 398
Exfoliation, 216
External processes, 8–10
Extrusive rocks, 112, 133–134

Failed arms, 47, *47–48*
Falls, 224–225, *225*
Falsification, 389
Far margins, 365
Fault breccia, 181, *182*
Faults, 59, *60*, 193, 194, 200–202, *203*
 of accreted terranes, 369
 in central rift valley, 30
 at Los Angeles, 68–69, *69*
 San Andreas fault, 68, *68–69*
 transform, 52, *52*, 414, *416*
Fauna, provincial, 411
Faunal succession, 6, 346–347
Feldspars
 in crust, 99, 103
 hydrolysis of, 151, 218–219
Felsenmeer, 214, *214*
Ferromagnesian minerals, 99
Field occurrence of metamorphic rocks, 183–187
Fine detrital rocks, 156–157, *157*
Fine sediments, 152
Finger Lakes, formation of, 295
Firn, 284
Fissure eruptions, *138*, 139
Fjords, 290
Flanking cones, 131, *131*
Flood basalt, 139
Flood plains, 249, 249–252
Floods, controlling, 250–252, *251–252*
Florida, sinkholes in, 271–272, *272*
Flow(s), 230–231, *230–231*
 in glaciers, 285–286, *287*
 in streams, 243–244, *244*
Flow segregation in fractionation, 118
Focus of an earthquake, 61, *63*, 64
Fold axes, 196
Fold-mountain belts, 184, *185–186*
Folds, 193, 194, 196, *196*
 anticlines and synclines, 196, *196*
 monoclines, domes, and basins, 199, 199–200, *201*
 plunging, 196–198, *197–198*
Foliated rocks, 180–181, *180–182*
Foliation, 177, *177–178*
Footwalls, 200, *203*
Forecasting
 earthquakes, 70–73, *72*
 mass wasting, 233–235, *235*
 volcanic eruptions, 142–143
Foreset beds, 252
Foreshocks, earthquake, 68
Formations, 343–344
Fossil fuels, 167
Fossil records, 149
 consistency of pattern in, 392
Fossils, 6, *6*, *163*, 164
 and comparative anatomy, evidence of evolution in, 389–391, *391–392*
 as continental drift evidence, 18–20, *19–20*
 in faunal succession, 346–347
 index, 348, *349*
 of Paleozoic, Mesozoic, and Cenozoic eras, *408*
 from Phanerozoic eon, 349
 in Precambrian time, 348
 transitional, 390, *391*
Fractionation, 118
Fracture in minerals, 95, *95*

Fracture zones, 31
  in glaciers, 285, *287*
Fractures, 200–203, *202–204*
Framework structures of silicates, 101, *102*
Freezing
  and creep, 232–233
  and frost wedging, 214
Friction
  in artesian wells, 267, *267*
  in streams, 243–244
  in wind-driven waves, 326, *326*
Fringing reefs, 40–41
Fronts, 300
Frost wedging, 214, *214*
Fuels. *See* Energy resources
Fumaroles, 136, 274–275

Gabbro, *108*, 114
  in lithosphere, 120, *121*
  texture of, *112*
Galena, 97
Galeras volcano, 125
Gaps, seismic, 71
Gases, volcanic, 136
Genes, 394
Genetics, evidence of evolution in, 393–395, *395*
Geodes, 164, *164*
Geologic time, 6–7, *6–8*, 343, 355–356, *356*
Geologic time scale, 7, *7–8*, *382*
  relative, *348*, 348–350
Geological Society of America, 23
Geological surveys for mass wasting predictions, 234
Geologists, 3
Geology, 3
  and natural selection, 396–398
Geothermal gradient, 81–82
Geyserite, 274
Geysers, 261, 273–274, *274–275*
Giant's Causeway, *135*
Gilbert, G. K., 59–61
Glacial ages, causes of, 296–298, *297*
Glacial budget, 286–288, *288*
Glacial erratics, *292*, 293
Glacial lakes, 290, 295
Glacial striations, 289, *290*
Glacial terminus, 286, *287*
Glacial till, 153, *153*
Glacial troughs, 290, *291*
Glaciers, 283–285, *284–285*
  budget for, 286–288, *288*
  as continental drift evidence, 18, *19*
  depositional features of, *287*, 290–295, *292–295*
  erosion by, 289–290, *291*
  moraines from, *287*, 292–293, *294*
  and mountain topography, 289–290, *291*
  movement of ice in, 285–286, *287*
  water quantity in, 262
Glass, volcanic, 134, *134*
Glassy texture, 112, *113*

Global warming
  human-induced, 299–301
  from volcanic activities, 144
Glomar Challenger (ship), 39
Glossopteris as continental drift evidence, 19, *20*
Gneiss, *11*, 180, *180*, 186, *188*
Gobi Desert, formation of, 307
Gondwana succession, 20
Gondwanaland, 14, *15*
Grabens, 202, *203*
Gradation from weathering, 211–212
Graded bedding, 161, *161*
Graded streams, 245
Gradient
  geothermal, 81–82
  hydraulic, 265, *266*
  stream, 243
Grand Canyon
  formation of, 310
  groundwater springs at, *260*
Granite, 115, *115*, *117*, 186, *186*–187, *188*
  porosity and permeability of, 264
Granite-rhyolite rock family, 115, *115*
Granitic magmas, 119
Granitic rocks, 13
Granodiorite, 115, 122, *123*
Gravel, porosity and permeability of, 264
Gravestones, differential weathering effects in, *220*, 220–221
Gravitational force of mountains, 366, *366*
Gravitational settling in fractionation, 118
Gravity
  in creep process, 232
  and principle of original horizontality, 344
  in stream energy, 242
  and tides, 334
Gravity sliding model of plate movement, 52, *53*
Great Barrier Reef, 159
Great Dismal Swamp, 167
Great earthquakes, 65
Great Lakes, formation of, 295
Great Salt Lake, 295
Great Slave Lake region, 357
Greenhouse effect, 299
  human-induced, 299–301
  from volcanic activities, 143
Greenland ice sheet, 283
Greenschist, 181, *181*
  in subduction zones, 190
Greenstone, 183
Greenstone belts, 383, *384*
Grenville front, *384*
Groins, beach erosion from, 332, *332*, 333
Ground moraines, 292, *294*
Ground tilting, 143
Groundwater, *260*, 261
  accumulation of, 261–264, *262–264*
  caves from, 270–272, *270–272*

  in geysers, hot springs, and fumaroles, 273–275, *274–275*
  karst topography from, 272–273, *273*
  movement of, 265, *265*
  pollution of, 276–278, *276–278*
  quality of, 275–278, *276–278*
  water quantity in, 262
  for wells, 265–270, *266–269*
Gulf of Aden, 46, 47, *47*
Gulf of California, margins at, 364
Gulf of Mexico, oil-forming at, 168
Guyots, 40
Gypsum in groundwater, 275

Hadean eon, 381
Haicheng, China, earthquake at, 70–71
Haldane, J. B. S., 396
Half-life principle, 90, 352, *352–353*
Halides, 100, 104, *104*
Hanging valleys, 290, *291*
Hanging walls, 200, *203*
Hard-stabilization beach control techniques, 333, *333*
Hardness
  of groundwater, 275
  of minerals, 96
Hawaiian Islands
  age of, 41, *42*
  formation of, 119, 138, *138*
Hazardous-waste disposal, 276–277, *276–278*
Headlands erosion, 327–328, *327–328*
Headward stream erosion, 255
Heat. *See also* Temperature
  and continental crust formation, 375
  internal. *See* Internal heat
  metamorphism from, 176
  from radioactive decay, 351
Heating, winds from, 306
Heezen, Bruce, 29, 30, 120
Height, wave, 326, *326*
Hess, Harry, 35
Himalayas
  accreted terranes of, 370–372, *371*
  from continent-continent convergence, 50–51, *51*
  gravitational force of, 366, *366*
Hoffman, Paul, 374
Holocene interglacial age, 296
Horizontality, principle of original, 344
Hornfels, 183
Horns, 290, *291*
Horsts, 202, *203*
Hot spots, 41, *42*, 119, *120*
  and lava plateaus, 139, *139*
  volcanoes from, 137–138, *138*
Hot springs, 261, 274, *275*
Hudson River, 243, *243*, 337, *338*
Hutton, James, 4, 8, 10, 379
Huttonian revolution, 4–6
  Earth as machine in, 8–12

Huttonian revolution *(continued)*
  geologic time in, 6–8, *7*
  and interior of Earth, 12–14, *13*
Hwang-Ho River, 316
Hydraulic gradient, 265, *266*
Hydraulic pressure loss in artesian wells, *269*, 269–270
Hydrologic cycle, 8, *9*
  streams in, 241
Hydrolysis, 150–151, 218–219
Hydrothermal metamorphism, 179
Hydrous minerals, 179
Hydroxides, 100
Hypotheses, 22–23

Ice
  in frost wedging, 214
  in glacial deposition, 290, 292
  in glaciers, movement of, 285–286, *287*
Ice ages, causes of, 296–298, *297*
Ice calving, 288, *289*
Ice-sculpted mountains, 289–290, *291*
Ice sheets, 283–285, *284–285*, 286–288
Ice shelves, 288
Icebergs, 288, *288–289*
Igneous activity. See also Volcanoes and volcanism
  geologic settings of, 119–123, *120*
  at mid-ocean ridges, 120, *121–122*
  mineral concentration and, 124–125, *124–125*
  in subduction zones, 122–123, *122–123*
Igneous rocks, 10, *10*, 109–111, *110*
  classification of, 113–116, *116*
  crystallization in, *116–117*, 116–119, *119*
  for dating, 352–353
  formation of, 111–119
  fractionation in, 118
  texture of, 111–113, *112–113*
  weathering rates of, *219*, 219–220
Ignimbrite, 136, *137*
Impact crater(s), 204–205, *205–206*
  frequency of, 206–207
  structure of, 205–206
Incised meanders, 254, *255*
Index fossils, 348, *349*
Index minerals, 187
India, path of, 370–372, *371*
Indonesia in glacial age, *284*
Industrial waste in groundwater, 276–278, *276–278*
Inert elements, 91, 92
Influent streams, 265
Inman, Douglas L., 328
Inner core, 14, *78*, 79
Inner ring, 205
Inselbergs, 312, *313*
Insoluble residues, 219
Interglacial ages, 296–298, *297*
Interior
  in Huttonian revolution, 12–14, *13*
  seismic waves in, *77–78*, 77–79
  structural and compositional divisions of, 79–81
Intermittent streams, 309
Internal embryonic development, 411
Internal heat
  core convection models of, *83*, 83–84
  geothermal gradient, 81–82
  mantle convection models of, *82*, 82–83
  sources of, 83
Intraplate earthquakes, 69–70, *70*
Intrusive rocks, 112, 133–134
Invertebrates, 407
Ionic bonds, 93, *93*
Ions, 93
  bicarbonate, *212*
  in dissolved loads, 244
  in hydrolysis, 219
Iron
  formation of, 222
  in groundwater, 275
Irrigated soils, salinization in, 223, *224*
Isacks, Don, 42
Island arc, 34
Islands, barrier, 330, *330*, 331, *332*, 337, *338*
Isograds, 187
Isostasy, 13, *14*, 366
Isotopes, 90, *91*, 350
  in dating, 352–353

Japan
  in glacial age, *284*
  as microcontinent, 368
Joint sets, 200, *202*
Joints, 193, 194, 200, *202*
Juan de Fuca plate, 49, *68*
  subduction of, 367–368, *367–368*
Juan de Fuca Ridge
  magnetism in, 36, *37*
  transform fault boundary at, 52
Jupiter, comet collision on, 207

Kagangel Atoll, *40*
Kame terraces, 295
Kames, *294–295*, 295
Karst topography, 272–273, *273*
Kenoran orogeny, 385, *385*
Kerogen, 169
Kettles, *294–295*, 295
Kilauea volcano, 138, *138*
Knickpoints, 245, *245–246*
Kobe, Japan, earthquake at, *58*, 76
Krakatoa, eruption of, 142
Kurile trench, 34, *35*
Kwangsi Province, 272, *273*

Laccoliths, 110
Lagoons, 330, *330*, 332
Lahars, 136, 230, *230*
Lakes. See also Kettles; Paternoster lakes; Pluvial lakes; Tarns
  glacial, 290, 295
  oxbow, 248, *248–249*
  playa, 312, *312–313*
  in sinkholes, 271
  water quantity in, 262
Land subsidence, 267–268
Landers, California, earthquake at, 74
Landfills, liquefaction of, 74, *75*
Landforms of stratified drift, 293–295, *294–295*
Landslides, 225–228, *226–227*, *232*
  on volcanoes, 141
Lapilli, 135
Laramide orogeny, 413
Las Vegas, Nevada, water for, 318
Lateral continuity in correlation, 346, *347*
Lateral moraines, 293
Laterite soils, 222–223, *223*
Laurasia, 14, *15*
Lava domes, 134, *136*
Lava fountains, 137
Lava plateaus, 110, 139, *139*
Lavas, 110, 131, 133–134, *134*
  basaltic, 134, 263
  crystals in, 113
  magnetism in, 37
  pillow, 120, *122*, 132, *133*
  porosity of, 263
  silicic, 134, *136*
Leaching in soil formation, 221, 222
Lead
  in dating, 353
  in drinking water standards, 278
LePichon, Xavier, 42
Levees, *249*, 250–252, *252*
Lewis thrust fault, *204*
Life
  in Cenozoic era, 417, *417*
  on Earth, evolution of, 389–398
  in Mesozoic era, 417, *417*
  origin of, 386–387
  in Paleozoic era, 406–412, *407–412*
  Proterozoic, 388–389
Life-form distributions as continental drift evidence, 18–20, *19–20*
Lignite coal, 167
Limbs of folds, 196
Limestone caves, 270–272, *270–272*
Limestones, 157–159, *158–160*
  in groundwater, 275
  porosity and permeability of, 264
Limonite, *97*
Liquefaction, 74
Lithic sandstone, 155
Lithification, 10, 151, *152*
Lithosphere, 14, 42–43, *46*, 79–80, *80*
Lithospheric plates, 22. See also Plate tectonics
Lithostatic stress, 177, *177*, 194, *194*
Little Ice Age, 299
Loads
  stream, 244
  wind, 313

Local base levels of streams, 243
Local geologic history, 343–345
Lockport dolomite, 246
Loess deposits, 314–316, *315*
Loma Prieta, California, earthquake at, 71, 72, 74
Long profiles of streams, 243, *243*
Longitudinal dunes, 317, *317*
Longshore currents, 329, *329*
Longshore sediment transport, 329
Los Angeles
    earthquake recurrence studies of, 71–72
    faults at, 68–69, *69*
Lower mantle, 81
Luster of minerals, 97, *97*
Lystrosaurus as continental drift evidence, 18, 19, *19*, 20

Mafic magma, 80
Mafic rocks, 114
Magma chambers, 132
Magmas, 10, *10*, 109. *See also* Igneous rocks; Volcanoes and volcanism
    crystals in, 111–113
    formation of rocks from, 111–119
    and hot springs, 274
    at mid-ocean ridges, 120, *121–122*
    at subduction zones, 122–123, 140
    in upper mantle, 80–81
    viscosity of, 132
Magnesium in groundwater, 275
Magnetic fields
    as continental drift evidence, 20–21, *20–21*
    and convection models, *83*, 83–84
    as sea-floor spreading evidence, 36–39, *37–39*
Magnetic inclination, *20*, 20–21
Magnetic reversals
    and core convection, 83
    in polarity reversal time scale, 37–38, *37–39*
Magnetites, *20*, 21
Magnitude of earthquakes, 64–66, *65–66*
Malaysia in glacial age, *284*
Malleability of metals, 94
Manganese nodules, 124, *125*, 166, *166*
Mantle, 13–14
    asthenosphere in, 43, 81
    convection models of, 82, *83*
    detection of, *78*, 79
    lower, 81
    temperatures in, 82
    transition zone, 81
    upper, 79, 80–81, *82*
Mantle plumes, 82, *83*
    and hot spots, 119, *120*
Maps of planar features, 195, *195*
Marble, 175, 183, *183*, 220–221
Margins, continental, 50, 364–366, *365*
    and coasts, 336
    sedimentary environments in, 165

Marianas trench, 34
Marsupials, Australian, *397*, 397–398
Mason, Ronald, 36
Mass
    of atoms, 90
    in stream energy, 242
Mass wasting, 9, 224
    creep, 232–233, *233*
    falls, 224–225, *225*
    flows, 230–231, *230–231*
    in headland erosion, 327
    predicting and preventing, 233–235, *235*
    slides, 225–229, *226–227*, *229*
Matrix in sedimentary rocks, 151
Matthews, Drummond, 38
Mauna Loa volcano, *128*, 138, *140*
Mazama, Mount, 129–130, *130*, 131
McKenzie, Dan, 42
Meander cutoffs, 248, *249*
Meander loops, placer deposits at, *250*
Meander scars, 248, *249*
Meandering streams, 247–248, *248–249*, 254, *255*
Mechanical weathering, 213, 214, *214*
    frost wedging, 214, *214*
    from plants and animals, 217, *217*
    salt crystallization, 215, *215*
    sheeting, exfoliation, and spheroidal, 216, *216*
Medial moraines, *287*, 293
Medieval Warm Period, 299
Mélanges, 166, 189–190, 367
Meltwater in glacial deposition, 290–292, 293–294
Mercalli earthquake intensity scale, 66
Mercury (element) in drinking water standards, 278
Mesa Verde National Park, *215*
Mesas, 310, *311*
Mesopotamia, salinization in, 223
Mesozoic era, 350, 412
    climate in, 414–416
    fossils of, *408*
    life in, 417, *417*
    North American Cordillera of, 413–414, *414*
    plate tectonics of, 412–413, *413*
Metallic bonds, 94
Metals, 100
    and periodic table, 91
    from volcanic activity, 144
Metamorphic grade, 187
Metamorphic rocks, 11, *11*, 175–176, 179
    classification of, 180–183, *180–183*
    contact, 184, *185*
    field occurrence of, 183–187
    foliated, 180–181, *180–182*
    minerals of, 179
    nonfoliated, 183, *183*
    regional, 184–187, *186–188*
    textures of, 180
    zones, 187

Metamorphic zones, 187
Metamorphism, 11, 175–176
    from chemically active fluids, 177–178
    from heat, 176
    from meteoroid impact, 179
    and plate tectonics, 188–190
    from pressure, 177, *177*
    pressure-temperature relationships in, *176*, 187
    of sea floor, 189, *189*
    subduction zone, 189–190
    types of, 178–179
Meteor, 205
Meteorites, 8, 204–205
Meteoroid impacts, 204–207, *205–206*
    metamorphism from, 179
Meteoroids, 204, *205*
Methane, 299
Mexico City, Mexico, earthquake at, 65, 75
Michigan basin, 200, *201*
Micrite, 158
Microcontinents, 368
Microseisms, 64
Mid-Atlantic Ridge, *28*, 29–30, *30–31*, 363
Mid-ocean ridges, 30, 120, *121–122*
Migmatite, 181, 186, *188*
Milankovitch, Milutin, 297
Milankovitch cycle, 296–298, *297*
Mineral concentration
    and igneous activity, 124–125, *124–125*
    at subduction zones, 124–125
Mineral deposits in streams, 248–249, *250*
Minerals, 88, 89
    atoms and elements in, 89–92, *91*
    bonding in, 92–94, *93–94*
    classification of, 100
    cleavage of, 95
    color of, 96, *97*
    of crust, 98–100
    crystal form of, 94–95, *95*
    fracture in, 95, *95*
    hardness of, 96
    index, 187
    luster of, 97, *97*
    of metamorphic rocks, 179
    origin of, 104
    other groups of, 103–104
    physical properties of, 94–98
    specific gravity of, 97–98
    streak of, 96
    structure of, 100–103, *101–102*
Mississippi River
    delta of, 253, *254*
    drainage by, 241, *242*
    flooding of, 250–251, *251*
    sediment from, 241
Models, developing, 22
Moenkopi Formation, *157*
Mohorovičić, Andrija, 79
Mohorovičić discontinuity, 79, *80*
Mohs scale of mineral hardness, 96

Molds in fossilization, 163
Monoclines, *199*, 199–200
Monograptus fossils, *349*
Monsoon effect, 320
Moon
    age of, 357
    and tides, 334
Moraines, *287*, 292–293, *294*
Morgan, Jason, 42
Morrison formation, 414
Mountain belts, *362*, 362–364, *364*
    metamorphic rocks in, 184, *186*
    plunging fold systems in, 197–198
Mountain building, 366
    by accretion, 368–374, *369*, *371–373*
    in Appalachians, *372–373*, 372–374
    in Himalayas, 370–372, *371*
    from subduction, 367–368, *367–368*
    in western North America, *369*, 369–370
Mountain glaciers, 285, *285*
Mountains, ice-sculpted, 289–290, *291*
Movement
    of glacial ice, 285–286, *287*
    of groundwater, 265, *265*
    of rocks. *See* Mass wasting
Mud cracks in sediments, 162, *162*
Mudflow deposits, 136
Mudflows, 230, *230*, 236
Mudstone, 152–153
Mullineaux, Donald R., 143
Multiple working hypotheses, 23
Mutations, 387, 396
Mylonite, 181, *182*

Namib desert, 308, *308*, 318
Natural levees, 249, *250*
Natural selection, 395
    geology and, 396–398
Nazca plate, 42, 49
Neap tides, 334
Near margins, 365
Nearshore circulation loop of beaches, 329, *329*
Nebulas, *380*, *381*
Negative electric charges, 90
Neutrons, 89, *91*
    in radioactive decay, 350, *351*
Nevadan disturbance, 413
Nevado del Ruiz, lahar at, 230, *230*
New Madrid, Missouri, earthquake at, 70, *70*
Niagara Falls, 246, *246*
Nile River
    delta of, 252
    profile of, 243, *243*
Nitrate in drinking water standards, 278
Nitrous oxide, 299
Nodules, 164
    manganese, 124, *125*, 166, *166*
Nonconformities, 344, *345*
Nonfoliated rocks, 183, *183*
Nonmetals, 91, 100
Normal faults, 200, 202, *203*

North America
    accreted terranes of, *369*, 369–370
    tectonic setting along, 335–336
North American Cordillera, 413–414, *414*
Northern Rockies, 192
Northridge, California, earthquake at, 69, 76, *76*
Notches in headland erosion, 327
Nuclear clocks, 90
Nucleotides, 393–394
Nucleus of atoms, 90
Nuée ardente, 136, *137*
Nutrient cycling, 221, *222*

O horizon, 221, *221*
Oblique angles with wind-driven waves, 326
Observations, 22
Obsidian, 134, *134*
Ocean-continent convergence, 49–50, *50*
Ocean floor, 31–34, *32–33*
    metamorphism of, 189, *189*
    and plate tectonics, 41–54
    sea-floor spreading hypothesis for, 35–41, *36–41*
Ocean-ocean convergence, 48–49, *49*
Oceanic crust, 13
Oceans
    sedimentary environments in, 164–167, *165–166*
    water quantity in, 262
    wind-driven currents in, 307–308
    wind-driven waves in, 325–328
Ogallala aquifer, 269, *269–270*
Oil. *See* Petroleum
Oil fields, 169
Oil sands, 169
Oil shales, 169
Oil traps, 169, *169*
Oliver, Jack, 42
Olivine
    crystallization of, 116, *116*
    hydrolysis of, 219
    in lithosphere, *121*
    in peridotite rock family, 114
    structure of, 101, *102*
Olympic Range
    building of, 367–368, *367–368*
    rain shadow deserts near, 307
Ooids, 159, *160*
Oolitic limestones, 159, *160*
Opaque minerals, 98
Ophiolite, 124, 189, 370
Orbits of Earth, climate changes from, 296–298, *297*
Organic evolution, 6
Origin of the Continents and Oceans, The (Wegener), 14
Orinoco River, oil at, 168
Orogenies, 363. *See also* Mountain building
Ostracoderms, 409
Ouachitas mountain system, 372, *372*
Outer core, 14, *78*, 79

Outer ring, 205
Outgassing, 386, *386*
Outwash plains, 294, *294*
Outwashes, 287
Overlapping ranges in relative time, 348, *349*
Overturned folds, 196, *197*
Overwash fans, 331, *332*
Oxbow lakes, 248, *248–249*
Oxidation, 219
Oxides, 100, 104, *104*
Oxygen in minerals, 98–99
Ozone layer, 388

P waves
    interior detection by, 77, *77*, 79
    and precursor studies, 72
Pacific plate, 42, 49
Packing as porosity factor, 263, *263*
Pahoehoe lava, 134, *135*
Painted Canyon, *178*
Paleomagnetic evidence of continental drift, 20–21, *20–21*
Paleoseismology, 71
Paleozoic era, 350, 403
    causes of radiations and extinctions in, 411–412, *412*
    fossils of, *408*
    life in, 406–412, *407–412*
    plate tectonics of, 404–405, *404–407*
Palisades Sill, 225, *225*
Pallet Creek, California, earthquake effects on, 71–72
Pangaea, 14, *15*, 413
    creation of, 404–405
Panthalassa, 412
Papers, scientific, 23
Parabolic dunes, 317, *317*
Parent isotopes, 351–353
Parker, Robert, 42
Particle shape, 153–154
Particle size, 152, *153*
Passive margins, 47–48, 165, 365, *365*
    and coasts, 336
    sedimentary environments in, 165
Paternoster lakes, 290
Peat, 167
Pediments, 312, *313*
Pegmatite, 115, 118–119, *119*
Pelee, Mount, eruption of, 140
Perched water tables, 264, *264*
Peridotite, 114, *114*
    from gravitational settling, 118
    in lithosphere, 120, *121*
Periodic table of the elements, 90, *92*
Periods (age), 348, *348*, 350
Periods (wave)
    of ocean wave, 325–326
    of seismic waves, 65
Permafrost, 232–233
Permeability in groundwater accumulation, 264, *264*

Peru-Chile trench, 34
Petrification, *163*
Petroleum, *170*
    generation of, 168–169
    in oil shales and oil sands, 169
    and oil traps, 169, *169*
    and plate tectonics, 170
    supply of, 170, *170*
Phaneritic texture, 112, *112*, 115
Phanerozoic eon, 349
Phosphates, 100
Photosynthesis, 387
Phreatic volcanic eruptions, 132–133, *133*
Phyllite, 186, *187*
Physical similarity in correlation, 346, *347*
Piedmont region, 372, *373*, 374
Pillow lavas, 120, *122*, 132, *133*
Pinatubo, Mount
    eruption of, 143, *144*
    lahar at, 230
Pine forests, soil for, 223–224
Pipes, 110
Piracy, stream, 256, *257*
Placer deposits, 248–249, *250*
Placoderms, 409, *409*
Plagioclase feldspar, 117, *117*
Plagioclase group, 103
Planar features, map depiction of, 195, *195*
Planetesimals, 381, *381*
Planets, origin of, 379–381, *380–381*
Plants, mechanical weathering from, 217, *217*
Plastic flow zones in glaciers, 285–286, *287*
Plate, 42
Plate boundaries, 43, 54, 335–336
Plate motion
    causes of, 52–53, *53*
    convection for, 82, *83*
Plate tectonics, 41–43, *44–45*
    of Cenozoic era, 412–413, *413*
    and climate, 296
    and coasts, 335–336, *337*
    and continental drift, 21–22
    convergent boundaries process in, 43, *46*, 48–52, *49–51*, 54
    divergent boundaries process in, 43, 46–48, *46–48*, 54
    and earthquakes, 66–70, *67*
    of Mesozoic era, 412–413, *413*
    and metamorphism, 188–190
    and mountain building. *See* Mountain building
    and ocean floor, 41–54
    of Paleozoic era, 404–405, *404–407*
    in petroleum process, 170
    plate motion causes in, 52–53, *53*
    and sedimentation, 166–167
    theory of, 24
    transform boundaries process in, 43, *46*, 52, *52*, 54
    and volcanism, 131, *132*
Platelets, 383

Platforms, 362, *363*
Platform sequence, 404
Playas, 312, *312–313*
Pleistocene epoch, 296, 350
Plucking by glaciers, 289
Plunging folds, 196–198, *197–198*
Plutons, 110
Pluvial lakes, 295
Pocket beaches, 330, *330*
Podzolization, 224
Point bars, 247, *248–249*
Polarity reversals. *See* Magnetic reversals
Pollution of groundwater, 276–278, *276–278*
Pools in petroleum process, 169
Porosity in groundwater accumulation, 263, *263*
Porphyritic textures, 112, *113*
Positive electric charges, 90
Potassium feldspar, 218–219
Potassium-40 isotopes, 353
Potential energy of streams, 242–243
Potholes, 244
    formation of, *244*
    placer deposits at, *250*
Precambrian atmosphere and oceans, 386, *386*
Precambrian time, 348–349
Precession of Earth axis, *297*, 297–298
Precipitated sedimentary rocks, 157–160, *158–160*
Precipitation and deserts, 305
Precursor studies of earthquakes, 72–73, *73*
Predicting
    earthquakes, 70–73, *72*
    mass wasting, 233–235, *235*
    volcanic eruptions, 142–143
Pressure
    in artesian wells, *269*, 269–270
    in glacier formation, 284
    in glacier plucking, 289
    metamorphism from, 177, *177*
Pressure-temperature relationships in metamorphism, *176*, 187
Primary coasts, 335
Primary (P) waves, 61–62, *63–64*, 64
Principle of cross-cutting relationships, 345, *346*
Principle of faunal succession, 346–347
Principle of original horizontality, 344
Principle of superposition, 344
Profiles
    of soil, *221*, 221–222
    of streams, 243, *243*
Prokaryotes, 388
Proterozoic eon, 385, *385*
    late-world of, 387–388
    life in, 388–389
    origin of life in, 386–387
    Precambrian atmosphere and oceans in, 386, *386*
Proto-Atlantic Ocean, 372
Protons, 89, *91*

Provincial fauna, 411
Puerto Rico, mudslide in, *236*
Pumice
    porosity of, 264
    texture of, *113*, 134, *134*
Pumping period with wells, 265–266, *266*
Punctuated equilibrium, 398
Pyrite
    luster of, 97, *97*
    oxidation of, 219
Pyroclastic materials, 135–136, *137*
Pyroclastic rocks, 136, *137*
Pyroclastic texture, 113, *113*
Pyroxene, 116, *116*
    in lithosphere, *121*
    structure of, 101, *102*

Quartz, 186
    color of, 96, *97*
    in crust, 99
    crystals, angles between, 94, *95*
    in detrital rocks, 154
    dissolution of, 177
    hydrolysis of, 219
    weathering rate of, 220
Quartz conglomerates, 155
Quartz sandstone, 155, *156*
Quartzite, 183, *183*

Racetrack Playa, *312*
Radial drainage patterns, 256, *256*
Radiating dikes, 110
Radiation, adaptive, 397, 407
Radicals, 99
Radioactive decay, 90, 350–352, *351–352*
Radioactive isotopes, 83, 352–355, *353–354*
Radioactive waste disposal, 276
Radioisotopes. *See* Radioactive isotopes
Radiolaria, 160
Radiometric dating, *7*, *8*, 343, 352–356, *353–355*
Rain forests
    soil formation in, 223
    and winds, 306
Rain shadow deserts, 307, *307*, 311
Rainfall
    and deserts, 305
    groundwater accumulation from, 261–262
    in the Sahel, 319, *319*
Rainier, Mount, size of, *140*
Reaction rims, 116
Reaction series, Bowen's, 117–118, *118*
Reassembly, 404–405
Recessional moraines, 292, *294*
Recharge areas for groundwater, 265
Recrystallization in rocks, 151
Rectangular drainage patterns, 256, *256*
Recumbent folds, 196, *197*
Recurrence studies of earthquakes, 71–72, *72*
Red Sea, 46, 47, *47*

INDEX

Reducing atmosphere, 387
Reefs, 158–159, *159*
Reflection of seismic waves, 77–78, *77–78*
Refraction
    in minerals, 98, *98*
    of seismic waves, 77–78, *77–78*
    of wind-driven waves, 326, *327*
Regional metamorphic rocks, 184–187, *186–188*
Regional metamorphism, 178, *178*
Regolith, 221
Reid, H. Fielding, 61
Rejuvenation of streams, 254, *255*
Relative age, 6. *See also* Relative time
Relative geologic time scale, 348, *348*
    terminology of, 348–350
Relative time, 343
    correlation in, 346–350, *347–349*
    cross-cutting relationships in, 345, *346*, 355, *355*
    and faunal succession, 346–347
    horizontality and superposition in, 344
    local geologic history, 343–345
    scale for, *348–349*, 348–350
    unconformities in, 344, *345*
Reptiles, early, 410–411, *411*
Research and theory, 22–24
Reservoir rocks, 169
Resonance effects of earthquakes, 76
Retaining walls, beach erosion from, 333, *333*
Reverse faults, 200, *203*
Rhyolite, 115
    texture of, *113*
Richter, Charles, 65
Richter magnitude scale, 65
Ridge-push plate model of plate movement, 52, *53*
Rifting, 404
Rills, 255
Ring dikes, 110
Ring of Fire, 34, 131, *132*
Rio Grande River, profile of, *243*
Rip currents, 329, *329*
Rip heads, 329
Ripple marks in sediments, 162, *162*
Rivers. *See* Streams
Rock cycle, 10–12, *10–12*
Rock families, 113
Rock flour, 289
Rocks
    as continental drift evidence, 17–18
    deformation of. *See* Deformation of rocks
    igneous. *See* Igneous rocks
    metamorphic. *See* Metamorphic rocks
    minerals in. *See* Minerals
    movement of. *See* Mass wasting
    sedimentary. *See* Sedimentary rocks
    weathering of. *See* Weathering
Rockslides, 225–226, *226*
Rocky Mountains, *415*
Rollover of barrier islands, 331, *332*

Roots, mechanical weathering from, 217, *217*
Roundness of sediment particles, 153, *154*
Rubidium-87 isotopes, 353

S waves, interior detection by, 77, *77*, *79*
Sahel region, 319–320, *320–321*
Saint Helens, Mount, *136*
    eruption at, 140–142, *141*
    lahar, at, 230
    life on, after eruption, *145*
    nuée ardente from, *137*
Salinization, 223
Salt, crystallization of, 215, *215*
Salt Lake City, Utah, water for, 318
Saltation
    in streams, 245
    in wind, 313
Saltwater
    encroachment, *268*, 268–269
    in irrigated soils, 223, *224*
    in wells, *268*, 268–269, 276
San Andreas fault, 68, *68–69*, 415
    earthquake recurrence studies of, 71
    as fracture zone, 31
    and San Francisco earthquake, 61
    as strike-slip fault, 202
    as transform fault, 52
San Clemente Island, *337*
San Francisco, California
    earthquake at, 61, *61*, 65
    earthquake recurrence studies of, 72
San Joaquin Valley, land subsidence at, 268
San Juan River, 254, *255*
Sand dunes, 316–318, *316–318*
Sand rights theory, 333
Sands, porosity and permeability of, 263
Sandstone, 152, 155, *156*
    in artesian well systems, 266–267
    and salt crystallization, 215
Scarps, 59
Schist, 180, *181*
    folding of, 186, *187*
Schistosity, 180
Scientific method, 22–23, *23*
Scientific papers, 23
Scoria, texture of, 134, *134*
Sea arches, 327
Sea caves, 327
Sea-floor spreading hypothesis, 35–36, *36*. *See also* Ocean floor
    magnetic evidence of, 36–39, *37–39*
    sediment evidence of, 39
    topographical evidence of, 39–41, *40–41*
Sea level
    coasts affected by, 336–339, *338*
    and ice sheet melting, 283, *284*
    rate of change, *338*
    and sediment supply, 330–331
Sea stacks, 327
Seabright, New Jersey, beach erosion at, 333, *333*

Seamounts, 40
Seas. *See* Oceans
Secondary coasts, 335
Secondary (S) waves, 61–62, *63–64*, 64
Sedimentary rocks, 4, *148*, 149
    classification of, 151–160, *152–153*
    detrital, 150, 152–157, *153–157*
    energy resources in, 167–170
    from lithification, 10, *11*, 151, *152*
    precipitated, 157–160, *158–160*
    from weathering, transport, and deposition, 150–151
Sediments, 4, 149
    at beaches, budget for, 330–333, *331–333*
    as continental drift evidence, 20
    from Mississippi River, 241
    in nearshore circulation loop, 329
    oceanic, 164–167, *165–166*
    and plate tectonics, 166–167
    in sea-floor spreading hypothesis, 39
    structure of, 160–164, *161–164*
    in wind, 313–314
Seifs, 318
Seismic activity. *See* Earthquakes
Seismic belts, 42
Seismic gaps, 71
Seismic waves, 13, *13*, 61–66, *63–65*
    in interior, 77–78, *77–79*
    refraction and reflection of, 77–78, *77–78*
    seismographs for, 62–64, *63*
Seismograms, 62–64, *64*
Seismographs, 62–64, *63*
Selective pressure, 396
Selenium in drinking water standards, 278
Self-sustaining dynamo, Earth as, 83, *83*
Semiarid regions
    Sahel, 319–320, *320–321*
    soil formation in, 223, *224*
Separation of continental blocks, 404
Series, 350
Serpentinite, 114, *114*
Settling velocity of streams, 245
Sevier orogeny, 413
Shaanxi, China, loess deposits at, *315*
Shadow zones, *78*, 79
Shales, 152–153
    oil, 169
    permeability of, 264
Shape as porosity factor, 263, *263*
Shear metamorphism, 178
Shear stress in rock deformation, 193, *194*
Shear waves, 61
Sheet silicates, structure of, 101, *102*
Sheeting, 216
Shells, rocks from, 158
Shield volcanoes, 137–138, *138*
Shields, 362, *362–363*
Shock metamorphism, 179
Shoemaker-Levy 9 comet, 207
Shoreline. *See* Coasts
Siccar Point, Scotland, 4–6, *5*
Sieh, Kerry, 71–72

Sierra Nevada mountains, rain shadow deserts near, 307
Silica in subduction zone magmas, 123
Silicates, 99, 100
　hydrolysis of, 219
　structure of, 100–103, *101–102*
Siliceous oozes, *165*, 166
Silicic lava flows, 134, *136*
Silicic rocks, 115
Silicic volcanoes, 140–142
Silicon in minerals, 99
Silicon-oxygen tetrahedrons, 101, *101*
Sills, 110, *111*
Siltstone, 152
Silurian system, 350
Similarity of sequence in correlation, 346, *347*
Sinkholes, 271–272, *272*
Size, classification of rocks by, 152, *153*
Slab-pull model of plate movement, 52, *53*
Slate, 180, *180*
　and regional metamorphism, 184, *186*
Slaty cleavage, 180, 184, *186*
Slides, 225–228, *226–227*
　conditions favoring, 228
　slumps, 228–229, *229*
Slip faces, 316
Slopes, talus, 154, *154*, 225, *226*
Slumps, 228–229, *229*
Snowline, 284–285, 286, *288*
Soft-stabilization beach control techniques, 333
Soil, porosity and permeability of, 263
Soil formation, 221
　in arid and semiarid climates, 223, *224*
　climate in, 222–224
　in cold to temperate climates, 223–224
　nutrient cycling in, 221
　soil profile, *221*, 221–222
　in tropical, humid climates, 222–223, *223*
Soil horizon, 221–222
Solar system, 379–381
Solenhofen limestone, 158, *163*
Solid-state reactions, 179
Solifluction, 233, *234*
Solifluction lobes, 233, *234*
Sorting
　as porosity factor, 263, *263*
　in sediments, 153, *153*
South America, tectonic setting along, 335–336
Specific gravity of minerals, 97–98
Sphericity of sediment particles, 153–154
Spheroidal weathering, 216, *216*
Spits, 330, *330*
Spreading centers, 46. *See also* Divergent boundaries
Spring tides, 334
Springs
　in deserts, 318
　formation of, 265, *266*
　hot, 261, 274, *275*

Stalactites, 271, *271*
Stalagmites, 271, *271*
Standard sections, 348
Star dunes, 317, *317*
Stocks, 111
Stone, Katherine, 333
Stoping, 123
Storms, wind-driven waves from, 326
Strain, 61, 194
Strata, 4, *5*, 161
Stratified drift
　from glaciers, 292
　landforms of, 293–295, *294–295*
Stratigraphic correlation and faunal succession, 346–347
Stratosphere, 306
Stratovolcanoes, 140
Streak of minerals, 96
Streak plates, 96
Stream divides, 255
Stream piracy, 256, *257*
Stream terraces, 254
Streams, 241–242
　channel erosion in, 243–244, *244*
　competence of, 244, 245
　deltas from, 252–253, *253–254*
　in deserts, 318
　discharge of, 243, *244*
　drainage patterns of, 254–256, *256–257*
　energy of, 242–245, *243–244*
　flood plains of, 249–252, *251*
　flow in, 243–244, *244*
　graded, 245
　intermittent, 309
　meandering, 247–248
　mineral deposits in, 248–249, *250*
　profiles of, 243, *243*
　rejuvenation of, 254, *255*
　systems of, 255
　transport and deposition in, 244–245
　types of, 246–248, *247–248*
　water quantity in, 262
　at water table, 265
Stress
　earthquakes from, 61, *62*
　in glaciers, 285
　rock deformation from, 193–194, *194*
　rock response to, 194
Strike, 195, *195*
Strike-slip faults, 200, 202, *203*
Stromatolites, 387, *388*
Strontium-87, 353
Structure
　of atoms, 90–92, *91*
　of minerals, 100–103, *101–102*
　of sediments, 160–164, *161–164*
　of volcanoes, 137–142, *138–142*
Subduction, 48
　in North American Cordillera, 413–414, *415*
Subduction zones, 48
　and coasts, 336

　earthquakes near, 67
　igneous activity at, 122–123, *122–123*
　magmas at, 122–123, 140
　metamorphism at, 189–190
　mineral concentration at, 124–125
　mountain building at, 367–368, *367–368*
　volcanoes at, 119, 131
Sublimation in glaciers, 286, *288*
Submarine canyons, 336
Submarine fans, 156
Subsidence, land, 267–268
Sugarloaf Mountain, *210*
Sulfates, 100
Sulfides, 100, 104, *104*
Sulfur in coal, 167
Sun and tides, 334
Superposition, 5, 344
Supersaturated solutions, 111
Surf zones, 329, *329*
Surface waves, 61, 62, 64, *64*
Surge deposits, 136
Surges by glaciers, 286
Surtsey Island, creation of, 133, *133*
Suspended loads
　in streams, 244
　in wind, 313
Susquehanna River, *335*
Sutter's Mill, California, 248
Suture zones, 370
Swamps in coal process, 167
Swash, 329, *329*
Sykes, Lynn, 42
Symmetrical folds, *192*, 196
Synclines, 196, *196*, 197–198, *198*
Systematic observations, 22

Tablemounts, 40
Taconic orogeny, 405
Talus slopes, 154, *154*, 225, *226*
Tambora, Mount, eruption of, 143
Tangshan, China, earthquake at, 70, 71
Tarns, 290
Tectonics. *See* Plate tectonics
Temperate climates, soil formation in, 223–224
Temperature. *See also* Climate; Heat
　Earth. *See* Internal heat
　and magma viscosity, 132
　and rock deformation, 194
　and volcanic activity, 143–144, *144*
Tensile stresses in rock deformation, 193, *194*
Tephra, 136
Terminal moraines, *287*, 292, *294*
Terraces, stream, 254
Terranes, accreted, 166, 368–369, *369, 371–373*
　of Appalachians, *372–373*, 372–374
　of Himalayas, 370–372, *371*
　in North American Cordillera, 414
　of western North America, *369*, 369–370
Tethys Sea, 14, *15*, 50–51
Tetrahedral linkages, 101, *102*

INDEX

Textures
　of igneous rocks, 111–113, *112–113*
　of metamorphic rocks, 180
Tharp, Marie, 29–30, 31, 120
Thawing in creep process, 232–233
Theories
　development of, 24
　in science, 389
Thorium-232 isotopes, 353
Thrust faults, 69, 200, 203, *203–204*
Tibetan plateau, 50, *51*, 370, *371*
Tidal currents, 334
Tidal force, 334
Tidal inlets, 331, *332*
Tides, 334
Till, 292
Tilt of Earth's axis, 297, *297*
Tiltmeters, 143
Time
　absolute. *See* Absolute time
　geologic, *6–7*, 6–8, 343, 355–356, *356*
　relative. *See* Relative time
　and rock stress, 194
Tombolos, 330, *330*
Topography, 199–200
　as sea-floor spreading evidence, 39–41, *40–41*
Topset beds, 252
Tracks in fossilization, *163*
Traction, 313
Transform boundaries, 43
　earthquakes at, 67
　process of, *46*, 52, *52*, 54
Transform faults, 52, *52*
　in North American Cordillera, 414, *416*
Transition zone, 81
Transitional fossils, 390, *391*
Translucent minerals, 98
Transparent minerals, 98
Transport
　in headland erosion, 327
　in nearshore circulation loop, 329
　and sedimentary rocks, 150, 151
　by streams, 244–245
Transverse dunes, 317, *317*
Traps, 139
Travel time of seismic waves, 77, *77*
Trees, mechanical weathering from, 217, *217*
Trellis drainage patterns, 255–256, *256*
Trenches, deep sea, 31–34, *32–33*
Tributary glaciers, *287*, 290, *291*
Tributary streams, 241
Trilobites, 407, *407*
Triple junction, 46–47, *47*, *48*
Tropical climates, soil formation in, 222–223, *223*
Troughs
　glacial, 290, *291*
　wave, *326*
Tsunamis, 73, *73*
Tuff, 136, *137*
Tundra, permafrost damage to, 233

Turbidites, *154*, 156, 161, 383
Turnagain Heights, Alaska, slump at, 228–229, *229*

Ultimate base levels of streams, 243
Ultramafic rock, 80–81, 99, 114
Unaltered hard parts in fossilization, *163*
Unconfined aquifers, 264, *264*
Unconformities, 4, 344, *345*
Underground water. *See* Groundwater
Uniformitarianism, 4
United States, coal in, 167
Upheaval Dome, *199*
Upper mantle, 79, 80–81, *82*
Uranium isotopes, 353

Vaiont Dam, slide at, 226–228, *227*
Valley glaciers, 285, *285*
Valley trains, 294
Valleys
　hanging, 290, *291*
　plunging fold systems in, 197–198
Variable composition of minerals, 103
Ventifacts, 314, *314*
Vents, volcanic, 130–131, *131*
Vertebrates, 409–410, *409–410*
Vestigial organs, 390–391, *393*
Vine, Fred, 38
Viscosity of magma, 132
Volatiles, 136
Volcanic breccia, 136, *137*
Volcanic cones, 129
Volcanic gases, 136
Volcanoes and volcanism, 110, 129–130
　anatomy of, 130–131, *131*
　andesitic, 140–142
　basaltic, 120, 137–139, *138–139*
　and climate, *143*, 143–144
　constructive aspects of, 144
　and earthquakes, 34, 132, 138, 142, 143
　eruption forecasting, 142–143
　eruption materials in, 133–136, *134–137*
　eruption mechanics in, 132–133, *133*
　eruption styles in, 137–142, *138–142*
　gases from, 136
　guyots, 40
　mudflows from, 230, *230*
　silicic, 140–142
　structure of, 137–142, *138–142*
　at subduction zones, 119, 131
　tectonic settings of, 131, *132*

Wadati, Kiyoo, 34
Wadati-Benioff zone, 34, *35*
Walden Pond, 295
Wallace, Alfred Russell, 395, 398
Waste in groundwater, 276, *276–277*
Water
　in creep process, 232
　in deserts, 309, 318–319, *319*
　in flows, 230–231

　in frost wedging, 214
　in glacial deposition, 290–292
　in hydrologic cycle, 8, *9*
　metamorphism from, 177
　in minerals, 103
　in rock deformation, 194
　in soil formation, 222–223
　and subduction zone magmas, 122–123
　transferring, 318–319
　underground. *See* Groundwater
Water pressure, slides from, 228
Water tables, 262, *262*, 264, *264*, 265
　and caves, 270–271
　influences on, 262
　and wells, 265–270, *266–269*
Waterfalls, 245–246, *245–246*, *250*
Wave-built platforms, 327
Wave-cut cliffs, 327
Wave-cut platforms, 327
Wave height, 326, *326*
Wave period, 325–326
Wavelength, 325, *326*
Waves
　seismic. *See* Seismic waves
　wind-driven, 325–328, *326–328*
Weathering, 9, 150, 211–213
　chemical, 213–214, 217–219, *218*
　in deserts, 308–313, *309–313*
　and dissolved loads, 244
　and gradation, 211–212
　and headland erosion, 327
　mechanical, 213, 214–217, *214–217*
　in mineral origin, 104
　rates of, *219*, 219–220
　sedimentary rocks from, 150–151
　and soil formation. *See* Soil formation
Wegener, Alfred, 14, 16, 19, 22, 41
Welded tuff, 136, *137*
Wells, 264, 265–267, *266–267*
　in deserts, 318
　extraction problems with, 267–270, *268–269*
　hydraulic pressure loss in, *269*, 269–270
　saltwater encroachment in, *268*, 268–269
Wentworth scale for rock classification, 152, *153*
Western North America, accreted terranes of, *369*, 369–370
Wetlands and flood control, 251–252
Whole-mantle convection, 82, *83*
Wilson, J. Tuzo, 41, 52
Wind gaps, 256
Winds, 313
　deposition by, 314–318, *315–318*
　deserts from, 306–308, *307–308*, 313–318
　erosion by, 313–314, *314*
Wizard Island, 130, *130*
Wrangell Mountains, *285*
Wrangellia terrane, 370

Xenoliths, 123, *123*

Yazoo tributaries, 250
Yellow River, 316
Yellowstone National Park, rhyolite tuff deposits at, 142, *142*
Yellowstone River, *213*
Yosemite Falls, 246, *246*
Yosemite Valley, *123*

Yucatan, meteor crater at, 207
Yungay, Peru, debris flow at, 231, *231*

Zagros Mountains, 197–198, *198*
Zeolite, metamorphic facies of, 189
Zion National Park, 316–317
Zircon for dating, 353

Zones of aeration, 262, *262*
Zones in crystals, 117, *117*
Zones of fracture in glaciers, 285, *287*
Zones of metamorphic rocks, 187
Zones of plastic flow in glaciers, 285–286, *287*
Zones of saturation, 262, *262*